AF711266

REVISION RECORD
FOR THE STATE OF CALIFORNIA
ERRATA

January 1, 2020

2019 Title 24, Part 4, California Mechanical Code

General Information:

1. The date of this erratum is for identification purposes only. See the History Note Appendix on the backside or accompanying page.

2. This erratum is issued by the California Building Standards Commission in order to correct non-substantive printing errors or omissions in California Code of Regulations, Title 24, Part 4, of the 2019 California Mechanical Code. Instructions are provided below.

3. Health and Safety Code Section 18938.5, establishes that only building standards in effect at the time of the application for a building permit may be applied to the project plans and construction. This rule applies to both adoptions of building standards for Title 24 by the California Building Standards Commission, and local adoptions and ordinances imposing building standards. An erratum to Title 24 is a non-regulatory correction because of a printing error or omission that does not differ substantively from the official adoption by the California Building Standards Commission. Accordingly, the corrected code text provided by this erratum may be applied on and after the stated effective date.

4. You may wish to retain the superseded material with this revision record so that the prior wording of any section can be easily ascertained.

Title 24, Part 4

Remove Existing Pages	Insert Buff-Colored Pages
7 – 8	7 – 8
45 – 46	45 – 46
319 – 320	319 – 320
507 – 508	507 – 508

1.4.0 Department of Consumer Affairs.

1.4.1 *Specific scope of application of the agency responsible for enforcement, the enforcement agency, and the specific authority to adopt and enforce such provisions of this code, unless otherwise stated.*

Board of Pharmacy.

Application *– Pharmacies.*

Enforcing Agency *– State or local agency specified by applicable provisions of law.*

Authority Cited *– Business and Professions Code Section 4005.*

References *– Business and Professions Code Sections 4005, 4127.7, and 4201.*

1.4.2 Adopting Agency Identification. *The provisions of this code applicable to buildings identified in this section will be identified in the Matrix Adoption Tables under the acronym CA.*

1.5.0 Reserved for California Energy Commission.

1.6.0 Reserved for Department of Food and Agriculture.

1.7.0 California Department of Public Health.

1.7.1 *Specific scope of application of the agency responsible for enforcement, the enforcement agency, and the specific authority to adopt and enforce such provisions of this code, unless otherwise stated.*

Department of Public Health.

Application *– Food establishments.*

Enforcing Agency *– Department of Health Services and the local health agency.*

Authority Cited *– Health and Safety Code Sections 110065, 113710, and 131200.*

References *– Health and Safety Code Sections 110065, 113705, 114140, and 131200.*

1.7.2 Adopting Agency Identification. *The provisions of this code applicable to buildings identified in this section will be identified in the Matrix Adoption Tables under the acronym DPH.*

1.8.0 Department of Housing and Community Development [HCD].

1.8.1 Purpose. *The purpose of this code is to establish the minimum requirements necessary to protect the health, safety, and general welfare of the occupants and the public by governing the erection, construction, reconstruction, enlargement, conversion, alteration, repair, moving, removal, demolition, ventilation, and maintenance or use of heating, ventilating, cooling, refrigeration systems, incinerators, and other miscellaneous heat-producing appliances.*

1.8.2 Authority and Abbreviations.

1.8.2.1 General. *The Department of Housing and Community Development is authorized by law to promulgate and adopt building standards and regulations for several types of building applications. The applications under the authority of the Department of Housing and Community Development are listed in Sections 1.8.2.1.1 through 1.8.2.1.3.*

1.8.2.1.1 Housing Construction.

Application *– Hotels, motels, lodging houses, apartments, dwellings, dormitories, condominiums, shelters for homeless persons, congregate residences, employee housing, factory-built housing, and other types of dwellings containing sleeping accommodations with or without common toilet or cooking facilities including accessory buildings, facilities, and uses thereto. Sections of this code which pertain to applications listed in this section are identified using the abbreviation "HCD 1".*

Enforcing Agency *– Local building department or the Department of Housing and Community Development.*

Authority Cited *– Health and Safety Code Sections 17040, 17050, 17920.9, 17921, 17921.5, 17921.6, 17921.10, 17922, 17922.6, 17922.12, 17922.14, 17926, 17927, 17928, 18300, 18552, 18554, 18620, 18630, 18640, 18670, 18690, 18691, 18865, 18871.3, 18871.4, 18873, 18873.1 through 18873.5, 18938.3, 18944.11, and 19990; and Government Code Section 12955.1.*

References *– Health and Safety Code Sections 17000 through 17062.5, 17910 through 17995.5, 18200 through 18700, 18860 through 18874, and 19960 through 19997; Civil Code Sections 1101.4, 1101.5 and 1954.201; and Government Code Sections 12955.1 and 12955.1.1.*

1.8.2.1.2 Housing Accessibility.

Application *– Covered multifamily dwellings as defined in Chapter 2 of the California Code of Regulations, Title 24, Part 2, also known as the California Building Code including, but not limited to, lodging houses, dormitories, timeshares, condominiums, shelters for homeless persons, congregate residences, apartments, dwellings, employee housing, factory-built housing, and other types of dwellings containing sleeping accommodations with or without common toilet or cooking facilities.*

Sections of this code identified by the abbreviation "HCD 1-AC" require specific accommodations for persons with disabilities as defined in Chapter 2 of the California Building Code. The application of such provisions shall be in conjunction with other requirements of this code and apply only to newly-constructed covered multifamily dwellings as defined in Chapter 2 of the California Building Code. "HCD 1-AC" applications include, but are not limited to, the following:

(1) All newly-constructed covered multifamily dwellings as defined in Chapter 2 of the California Building Code.

(2) New common use areas as defined in Chapter 2 of the California Building Code serving existing covered multifamily dwellings.

(3) Additions to existing buildings, where the addition alone meets the definition of covered multifamily dwellings as defined in Chapter 2 of the California Building Code.

(4) New common use areas serving new covered multifamily dwellings.

(5) Where any portion of a building's exterior is preserved, but the interior of the building is removed, including all structural portions of floors and ceilings, the building is considered a new building for determining the application of California Building Code, Chapter 11A.

"HCD 1-AC" building standards generally do not apply to public use areas or public accommodations such as hotels and motels, and public housing. Public use areas, public accommodations, and public housing, as defined in Chapter 2 of the California Building Code, are subject to the Division of the State Architect (DSA-AC) in Chapter 11B, and are referenced in Section 1.9.1.

Newly constructed covered multifamily dwellings, which can also be defined as public housing, shall be subject to the requirements of Chapter 11A and Chapter 11B.

***Enforcing Agency** – Local building department or the Department of Housing and Community Development.*

***Authority Cited** – Health and Safety Code Sections 17040, 17050, 17920.9, 17921, 17921.5, 17921.6, 17921.10, 17922, 17922.6, 17922.12, 17922.14, 17926, 17927, 17928, 18300, 18552, 18554, 18620, 18630, 18640, 18670, 18690, 18691, 18865, 18871.3, 18871.4, 18873, 18873.1 through 18873.5, 18938.3, 18944.11, and 19990; and Government Code Section 12955.1.*

***References** – Health and Safety Code Sections 17000 through 17062.5, 17910 through 17995.5, 18200 through 18700, 18860 through 18874, and 19960 through 19997; Civil Code Sections 1101.4, 1101.5 and 1954.201; and Government Code Sections 12955.1 and 12955.1.1.*

1.8.2.1.3 Permanent Buildings in Mobilehome Parks and Special Occupancy Parks.

***Application** – Permanent buildings, and permanent accessory buildings or structures, constructed within mobilehome parks and special occupancy parks that are under the control and ownership of the park operator. Sections of this code which pertain to applications listed in this section are identified using the abbreviation "HCD 2".*

***Enforcing Agency** – The Department of Housing and Community Development, local building department or other local agency that has assumed responsibility for the enforcement of Health and Safety Code, Division 13, Part 2.1, commencing with Section 18200 for mobilehome parks and Health and Safety Code, Division 13, Part 2.3, commencing with Section 18860 for special occupancy parks.*

***Authority Cited** – Health and Safety Code Sections 17040, 17050, 17920.9, 17921, 17921.5, 17921.6, 17921.10, 17922, 17922.6, 17922.12, 17922.14, 17926, 17927, 17928, 18300, 18552, 18554, 18620, 18630, 18640, 18670, 18690, 18691, 18865, 18871.3, 18871.4, 18873, 18873.1 through 18873.5, 18938.3, 18944.11, and 19990; and Government Code Section 12955.1.*

***References** – Health and Safety Code Sections 17000 through 17062.5, 17910 through 17995.5, 18200 through 18700, 18860 through 18874, and 19960 through 19997; Civil Code Sections 1101.4, 1101.5 and 1954.201; and Government Code Sections 12955.1 and 12955.1.1.*

1.8.3 Local Enforcing Agency.

***1.8.3.1 Duties and Powers.** The building department of every city, county, or city and county shall enforce all the provisions of law, this code, and the other rules and regulations promulgated by the Department of Housing and Community Development pertaining to the installation, erection, construction, reconstruction, movement, enlargement, conversion, alteration, repair, removal, demolition, or arrangement of apartments, condominiums, hotels, motels, lodging houses, and dwellings, including accessory buildings, facilities, and uses thereto.*

The provisions regulating the erection and construction of dwellings and appurtenant structures shall not apply to existing structures as to which construction is commenced or approved prior to the effective date of these regulations. Requirements relating to use, maintenance, and occupancy shall apply to all dwellings and appurtenant structures approved for construction or constructed before or after the effective date of this code.

For additional information regarding the use and occupancy of existing buildings and appurtenant structures, see California Code of Regulations, Title 25, Division 1, Chapter 1, Subchapter 1, commencing with Section 1.

***1.8.3.2 Laws, Rules, and Regulations.** Other than the building standards contained in this code, and notwithstanding other provisions of law, the statutory authority and location of the laws, rules, and regulations to be enforced by local enforcing agencies are listed by statute in Sections 1.8.3.2.1 through 1.8.3.2.5.* ||

***1.8.3.2.1 State Housing Law.** Refer to the State Housing Law, California Health and Safety Code, Division 13, Part 1.5, commencing with Section 17910, and California Code of Regulations, Title 25, Division 1, Chapter 1, Subchapter 1, commencing*

CALIFORNIA MECHANICAL CODE – MATRIX ADOPTION TABLE
CHAPTER 3 – GENERAL REGULATIONS

(Matrix Adoption Tables are non-regulatory, intended only as an aid to the code user. See Chapter 1 for state agency authority and building applications.)

Adopting Agency	BSC	BSC-CG	SFM	HCD			DSA			OSHPD						BSCC	DPH	AGR	DWR	CEC	CA	SL	SLC
				1	2	1-AC	AC	SS	SS/CC	1	1R	2	3	4	5								
Adopt Entire Chapter	X																						
Adopt Entire Chapter as amended (amended sections listed below)			X	X	X			X	X	X	X	X	X	X	X								
Adopt only those sections that are listed below																							
Chapter/Section																							
303.2, Exception										X	X	X	X	X	X								
303.7			X					X	X														
303.7.1				X	X																		
303.13 *Exception*				X	X																		
311.2				X	X																		
312.1				X	X			X	X														
316.2				X	X																		
316.5				X	X																		
316.9				X	X																		
318.0										X	X	X	X	X	X								
319.1										X	X			X	X								
319.2												X		X									
320.1										X				X									
320.2												X		X									
320.3													X										
320.4										X				X									
320.5										X	X	X		X	X								
321.0										X	X	X	X	X	X								
322.0										X	X	X	X	X	X								
322.1										X		X	X	X									

This state agency does not adopt sections identified with the following symbol: †

The Office of the State Fire Marshal's adoption of this chapter or individual sections is applicable to structures regulated by other state agencies pursuant to Section 1.11.0.

CALIFORNIA MECHANICAL CODE – MATRIX ADOPTION TABLE
CHAPTER 17 – REFERENCED STANDARDS

(Matrix Adoption Tables are non-regulatory, intended only as an aid to the code user. See Chapter 1 for state agency authority and building applications.)

Adopting Agency	BSC	BSC-CG	SFM	HCD			DSA			OSHPD						BSCC	DPH	AGR	DWR	CEC	CA	SL	SLC
				1	2	1-AC	AC	SS	SS/CC	1	1R	2	3	4	5								
Adopt Entire Chapter	X			X	X																		
Adopt Entire Chapter as amended (amended sections listed below)			X					X	X	X	X	X	X	X	X								
Adopt only those sections that are listed below																							
Chapter/Section																							
Table 1701.1	X		X				X	X	X	X	X	X	X	X									

This state agency does not adopt sections identified with the following symbol: †

The Office of the State Fire Marshal's adoption of this chapter or individual sections is applicable to structures regulated by other state agencies pursuant to Section 1.11.0.

HISTORY NOTE APPENDIX

CALIFORNIA MECHANICAL CODE
(CALIFORNIA CODE OF REGULATIONS, TITLE 24, PART 4)

For prior code history, see the History Note Appendix to the California Mechanical Code, 2016 Triennial Edition, effective January 1, 2017.

1. *(BSC 03/18, HCD 01/18, DSA-SS 03/18, OSHPD 04/18, SFM 04/18) Adoption by reference of the 2018 Uniform Mechanical Code with necessary amendments to become the 2019 California Mechanical Code, and repeal of the 2015 edition of the Uniform Mechanical Code; effective on January 1, 2020.*
2. *Erratum to correct editorial errors in Matrix Adoption Tables and miscellaneous corrections throughout chapters 1, 3, and 17, effective January 1, 2020.*

The Best Tools in Training for the 2019 California Plumbing and Mechanical Codes!

2018 Uniform Plumbing Code
Illustrated Training Manual
Item: 30-18TCT
Member: $155.20
Non-Member: $194.00

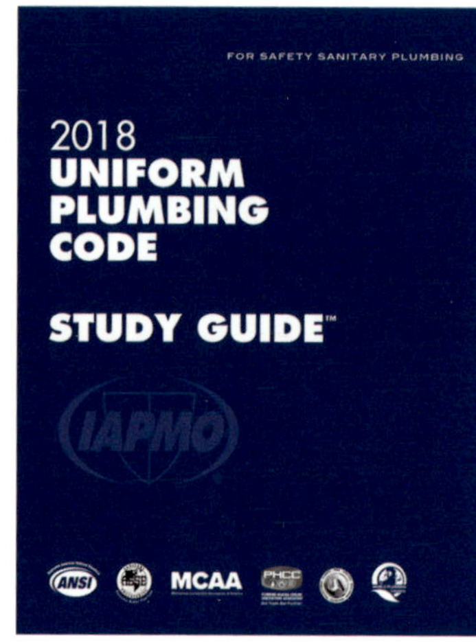

2018 Uniform Plumbing Code
Study Guide
Item: 30-18IT
Member: $64.80
Non-Member: $81.00

2018 Uniform Mechanical Code
Illustrated Training Manual
Item: 30-18TMT
Member: $136.00
Non-Member: $170.00

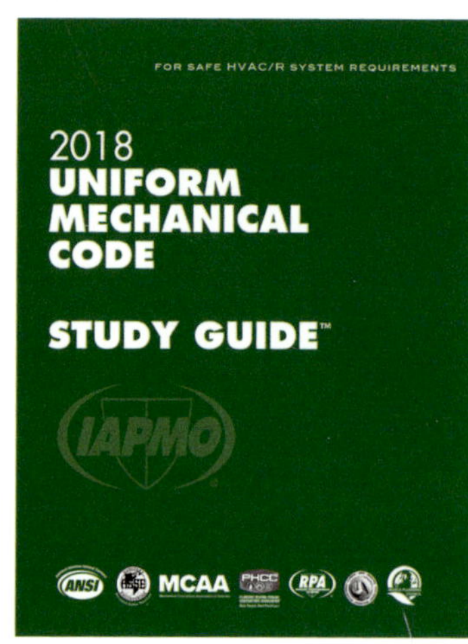

2018 Uniform Mechanical Code
Study Guide
Item: 30-18MT
Member: $64.80
Non-Member: $81.00

These Supporting Books Will Help in All Your Training and Education Needs. Never Stop learning, Visit www.iapmolearn.org for Upcoming Training in Your Area.

IMPORTANT NOTICE

Your purchase of this code includes a free subscription for all State issued supplements and errata.

You must register online at www.iapmomembership.org/reg to keep your code up-to-date!

2019 CALIFORNIA MECHANICAL CODE

CALIFORNIA CODE OF REGULATIONS
TITLE 24, PART 4

Based on the 2018 Uniform Mechanical Code®

California Building Standards Commision

Effective January 1, 2020

For Errata and Supplement effective dates see the History Note Appendix

International Association of Plumbing and Mechanical Officials
4755 E. Philadelphia Street • Ontario, CA 91761-2816 – USA

First Printing, July 2019
Printed in The United States

PREFACE

This document is Part 4 of thirteen parts of the official triennial compilation and publication of the adoptions, amendments and repeal of administrative regulations to *California Code of Regulations, Title 24*, also referred to as the *California Building Standards Code*. Part 4 is known as the California Mechanical Code and incorporates, by adoption, the 2018 edition of the *Uniform Mechanical Code* of the International Association of Plumbing and Mechanical Officials with the California amendments.

The *California Building Standards Code* is published in its entirety every three years by order of the California legislature, with supplements published in intervening years. The California legislature delegated authority to various state agencies, boards, commissions and departments to create building regulations to implement the State's statutes. These building regulations or standards, have the same force of law, and take effect 180 days after their publication unless otherwise stipulated. The *California Building Standards Code* applies to occupancies in the State of California as annotated.

A city, county, or city and county may establish more restrictive building standards reasonably necessary because of local climatic, geological or topographical conditions. Findings of the local condition(s) and the adopted local building standard(s) must be filed with the California Building Standards Commission to become effective and may not be effective sooner than the effective date of this edition of the *California Building Standards Code*. Local building standards that were adopted and applicable to previous editions of the *California Building Standards Code* do not apply to this edition without appropriate adoption and the required filing.

Should you find publication (e.g., typographical) errors or inconsistencies in this code or wish to offer comments toward improving its format, please address your comments to:

California Building Standards Commission
2525 Natomas Park Drive, Suite 130
Sacramento, CA 95833-2936
Phone: (916) 263-0916
Web Page: www.dgs.ca.gov/bsc
Email: cbsc@dgs.ca.gov

ACKNOWLEDGEMENTS

The 2019 *California Mechanical Code* (Code) was developed through the outstanding collaborative efforts of the Department of Housing and Community Development, Division of State Architect, Office of the State Fire Marshal, Office of Statewide Health Planning and Development, California Energy Commission, California Department of Public Health, California State Lands Commission, Board of State and Community Corrections, and the California Building Standards Commission (Commission).

This collaborative effort included the assistance of the Commission's Code Advisory Committees and many other volunteers who worked tirelessly to assist the Commission in the production of this Code.

Governor Edmund G. Brown Jr.

Members of the Building Standards Commission
Secretary Marybel Batjer – Chair
Steven Winkel – Vice-Chair

Rajesh Patel | *Juvilyn Alegre*
Elley Klausbruckner | *Cheryl Roberts*
Larry Booth | *Erick Mikiten*
James Barthman | *Kent Sasaki*
Peter Santillan

Mia Marvelli – Executive Director
Michael L. Nearman – Deputy Executive Director

For questions on California state agency amendments, please refer to the contact list on page iv.

California Code of Regulations Title 24

California Agency Information Contact List

The following state agencies may propose building standards for publication in Title 24. Request notice of such activity with each agency of interest. See Sections 1.2 through 1.14 of the California Building Code (Part 2 of Title 24) for more detailed information on the regulatory jurisdiction of each state agency.

Board of State and Community Corrections
www.bscc.ca.gov ....(916) 445-5073
Local Adult and Juvenile Detention Facility Standards

California Building Standards Commission
www.dgs.ca.gov/bsc ....(916) 263-0916
State Buildings including UC and CSU Buildings, Parking Lot and Walkway Lighting, Green Building Standards for Non-residential Buildings

California Energy Commission
*www.energy.ca.gov ....**Energy Hotline** (800) 772-3300*
Building Efficiency Standards
Appliance Efficiency Standards
Compliance Manual/Forms

California State Lands Commission
www.slc.ca.gov...(562) 499-6312
Marine Oil Terminals Standards

California State Library
www.library.ca.gov ....(916) 323-9843

Department of Consumer Affairs:

Acupuncture Board
www.acupuncture.ca.gov ....(916) 515-5200
Office Standards

Board of Pharmacy
www.pharmacy.ca.gov ....(916) 574-7900
Pharmacy Standards

Bureau of Barbering and Cosmetology
www.barbercosmo.ca.gov ....(800) 952-5210
Barber and Beauty Shop, and College Standards

Bureau of Household Goods and Services
www.bhgs.dca.ca.gov.............................(916) 999-2041
Insulation Testing Standards

Structural Pest Control Board
www.pestboard.ca.gov ....(800) 737-8188
Structural Standards

Veterinary Medical Board
www.vmb.ca.gov....................................(916) 515-5220
Veterinary Hospital Standards

Department of Food and Agriculture
www.cdfa.ca.gov

(916) 900-5004
Meat & Poultry Packing Plant Standards

(916) 900-5008
Rendering & Collection Center Standards, Dairy Standards

Department of Housing and Community Development
www.hcd.ca.gov..(916) 445-9471
Residential - Hotels, Motels, Apartments, Single-Family Dwellings; and Permanent Structures in Mobilehome & Special Occupancy Parks

(916) 445-3338
Factory-Built Housing, Manufactured Housing & Commercial Modular

Mobilehome - Permits & Inspections
Northern Region - (916) 255-2501
Southern Region - (951) 782-4420

(916) 445-9471
Employee Housing Standards

Department of Public Health
www.dph.ca.gov ....(916) 449-5661
Organized Camps Standards
Public Swimming Pools Standards

Division of the State Architect
www.dgs.ca.gov/dsa ....(916) 445-8100

Access Compliance
Fire and Life Safety
Structural Safety

Public Schools Standards
Essential Services Building Standards
Community College Standards

State Historical Building Safety Board
www.dgs.ca.gov.....................................(916) 445-8100
Historical Rehabilitation, Preservation, Restoration or Relocation Standards

Office of Statewide Health Planning and Development
www.oshpd.ca.gov..(916) 440-8356
Hospital Standards
Skilled Nursing Facility Standards & Clinic Standards

Office of the State Fire Marshal
www.osfm.fire.ca.gov ....(916) 568-3800
Code Development and Analysis
Fire Safety Standards

How to Distinguish Between Model Code Language and California Amendments

To distinguish between model code language and the incorporated California amendments, including exclusive California standards, California amendments will appear in italic font print.

[BSC] *This is an example of a state agency acronym used to identify an adoption or amendment by the agency. The acronyms will appear at California Amendments and in the Matrix Adoption Tables. Sections 1.2.0 through 1.14.0 in Chapter 1, Division 1 of this code, explain the used acronyms, the application of state agency adoptions to building occupancies or building features, the enforcement agency as designated by state law (may be the state adopting agency or local building or fire official), the authority in state law for the state agency to make the adoption, and the specific state law being implemented by the agency's adoption. The following acronyms are used in Title 24 to identify the state adopting agency making an adoption.*

Legend of Acronyms of Adopting State Agencies

BSC	*California Building Standards Commission (see Section 1.2.0)*
BSC-CG	*California Building Standards Commission-CALGreen (see Section 1.2.3)*
BSCC	*Board of State and Community Corrections (see Section 1.3.0)*
SFM	*Office of the State Fire Marshal (see Section 1.11.0)*
HCD-1	*Department of Housing and Community Development (see Section 1.8.2.1.1)*
HCD-2	*Department of Housing and Community Development (see Section 1.8.2.1.3)*
HCD-1/AC	*Department of Housing and Community Development (see Section 1.8.2.1.2)*
DSA-AC	*Division of the State Architect-Access Compliance (see Section 1.9.1)*
DSA-SS	*Division of the State Architect-Structural Safety (see Section 1.9.2)*
DSA-SS/CC	*Division of the State Architect-Structural Safety/Community Colleges (see Section 1.9.2.2)*
OSHPD 1	*Office of Statewide Health Planning and Development (see Section 1.10.1)*
OSHPD 1R	*Office of Statewide Health Planning and Development (see Section 1.10.1)*
OSHPD 2	*Office of Statewide Health Planning and Development (see Section 1.10.2)*
OSHPD 3	*Office of Statewide Health Planning and Development (see Section 1.10.3)*
OSHPD 4	*Office of Statewide Health Planning and Development (see Section 1.10.4)*
OSHPD 5	*Office of Statewide Health Planning and Development (see Section 1.10.5)*
DPH	*Department of Public Health (see Section 1.7.0)*
AGR	*Department of Food and Agriculture (see Section 1.6.0)*
CEC	*California Energy Commission (see Section 100 in Part 6, the California Energy Code)*
CA	*Department of Consumer Affairs (see Section 1.6.0):*
	Board of Barbering and Cosmetology
	Board of Examiners in Veterinary Medicine
	Board of Pharmacy
	Acupuncture Board
	Bureau of Home Furnishings
	Structural Pest Control Board SPCB
SL	*State Library (see Section 1.12.0)*
SLC	*State Lands Commission (see Section 1.14.0)*
DWR	*Department of Water Resources (see Section 1.13.0 of Chapter 1 of the California Plumbing Code in Part 2 of Title 24)*

The state agencies are available to answer questions about their adoptions. Contact information is provided on page iv of this code.

To learn more about the use of this code refer to the following pages. Training materials on the application and use of this code are available at the website of the California Building Standards Commission www.dgs.ca.gov/bsc.

CALIFORNIA MATRIX ADOPTION TABLES

Format of the Matrix Adoption Tables

The matrix adoption tables, examples of which follow, are non-regulatory aids intended to show the user which state agencies have adopted and/or amended given sections of model code. An agency's statutory authority for certain occupancies or building applications determines which chapter or section may be adopted, repealed, amended or added. See Chapter 1, Division I, Sections 1.2.0 through 1.14.0, for agency authority, building application and enforcement responsibilities.

The side headings identify the scope of the state agencies' adoption as follows:

Adopt the entire UMC chapter without state amendments.

If there is an "X" under a particular state agency's acronym in this row, it means that particular state agency has adopted the entire model code chapter without any state amendments.

Example:

CALIFORNIA MECHANICAL CODE – MATRIX ADOPTION TABLE
CHAPTER 1 - GENERAL CODE PROVISIONS

(Matrix Adoption Tables are non-regulatory, intended only as an aid to the user. See Chapter 1 for state agency authority and building application.)

Adopting Agency	BSC	BSC-CG	SFM	HCD			DSA			OSHPD						BSCC	DPH	AGR	DWR	CEC	CA	SL	SLC
				1	2	1-AC	AC	SS	SS/CC	1	1R	2	3	4	5								
Adopt Entire Chapter			X																				
Adopt Entire Chapter as amended (amended sections listed below)							S	A	M	P	L	E											
Adopt only those sections that are listed below																							
Chapter/Section																							
202																							

Adopt the entire UMC chapter as amended, state-amended sections are listed below:

If there is an "X" under a particular state agency's acronym in this row, it means that a particular state agency has adopted the entire model code chapter, with state amendments.

Each state-amended section that the agency has added to that particular chapter is listed. There will be an "X" in the column, by that particular section, under the agency's acronym, as well as an "X" by each section that the agency has adopted.

Example:

CALIFORNIA MECHANICAL CODE – MATRIX ADOPTION TABLE
CHAPTER 1 - GENERAL CODE PROVISIONS

(Matrix Adoption Tables are non-regulatory, intended only as an aid to the user. See Chapter 1 for state agency authority and building application.)

Adopting Agency	BSC	BSC-CG	SFM	HCD			DSA			OSHPD						BSCC	DPH	AGR	DWR	CEC	CA	SL	SLC
				1	2	1-AC	AC	SS	SS/CC	1	1R	2	3	4	5								
Adopt Entire Chapter																							
Adopt Entire Chapter as amended (amended sections listed below)	X		X				S	A	M	P	L	E											
Adopt only those sections that are listed below																							
Chapter/Section																							
202	X		X																				

Adopt only those sections which are listed below:

If there is an "X" under a particular state agency's acronym in this row, it means that particular state agency is adopting only specific model code or state-amended sections within this chapter. There will be an "X" in the column under the agency's acronym, as well as an "X" by each section that the agency has adopted.

Example:

CALIFORNIA MECHANICAL CODE – MATRIX ADOPTION TABLE
CHAPTER 1 - GENERAL CODE PROVISIONS

(Matrix Adoption Tables are non-regulatory, intended only as an aid to the user. See Chapter 1 for state agency authority and building application.)

Adopting Agency	BSC	BSC-CG	SFM	HCD			DSA			OSHPD						BSCC	DPH	AGR	DWR	CEC	CA	SL	SLC
				1	2	1-AC	AC	SS	SS/CC	1	1R	2	3	4	5								
Adopt Entire Chapter																							
Adopt Entire Chapter as amended (amended sections listed below)							S	A	M	P	L	E											
Adopt only those sections that are listed below	X																						
Chapter/Section																							
101	X																						
102	X																						

HOW TO DISTINGUISH MODEL CODE LANGUAGE FROM CALIFORNIA AMENDMENTS

To distinguish between model code language and incorporated California Amendments, including exclusive California standards, California amendments will appear in italics font print.

Symbols in the margins indicate the status of code changes as follows:

[BSC] This symbol identifies which State agency(s) by its "acronym" that has amended a section of the model code. For a complete listing of State agency acronyms, see the Application Section within Chapter 1, Division I.

‖ This symbol indicates that a change has been made to a California amendment.

| This symbol indicates that a section, paragraph, or table has been revised or relocated within IAPMO model code language.

> This symbol indicates deletion of California language.

→ This symbol indicates where an entire section, paragraph, or table has been deleted from IAPMO model code language.

Revision Markings
A double right angle («) in the margin indicates that the text or a table has been relocated within the code. The table found on page xiii points out the relocations in the 2018 edition of the Uniform Mechanical Code.

TIA TIA indicates that the revision is the result of a Tentative Interim Amendment.
TIA For further information on tentative interim amendments see Section 5 of the IAPMO
TIA Regulations Governing Committee Projects available at http://codes.iapmo.org/
TIA

A reference in brackets [ ] following a section or paragraph indicates material that has been extracted from another document. This reprinted material is not the complete and official position of the source document on the referenced subject that is represented by the standard in its entirety. Material contained in this document that is taken or extracted from NFPA standards is used with permission of the National Fire Protection Association. This material is not the complete and official position of the NFPA on the reference subject, which is represented solely by the relevant standard in its entirety. NFPA standards can be accessed at www.nfpa.org. In addition, this extracted material may include revisions or modifications developed through IAPMO's standards development process. Therefore, NFPA disclaims responsibility for the content of this Code.

Text that is extracted pursuant to IAPMO's Extract Guidelines, but outside of the regular revision process is denoted with the use of the source document in the margin. This text is not fully processed by IAPMO in accordance with ANSI's public announcement consensus requirements for an American National Standard (ANS) nor approved by ANSI's Board of Standards Review. The next revision cycle processes such text in accordance with those requirements.

Important Notices and Disclaimers

The 2018 edition of the *Uniform Mechanical Code* is developed through a consensus standards development process approved by the American National Standards Institute. This process brings together volunteers representing varied viewpoints and interests to achieve consensus on mechanical issues. While the International Association of Plumbing and Mechanical Officials (IAPMO) administers the process and establishes rules to promote fairness in the development of consensus, it does not independently test, evaluate, or verify the accuracy of any information or the soundness of any judgments contained in its codes and standards.

IAPMO disclaims liability for any personal injury, property, or other damages of any nature whatsoever, whether special, indirect, consequential, or compensatory, directly or indirectly resulting from the publication, use of, or reliance on this document. IAPMO also makes no guarantee or warranty as to the accuracy or completeness of any information published herein.

In issuing and making this document available, IAPMO is not undertaking to render professional or other services for or on behalf of any person or entity. Nor is IAPMO undertaking to perform any duty owed by any person or entity to someone else. Anyone using this document should rely on his or her own independent judgment or, as appropriate, seek the advice of a competent professional in determining the exercise of reasonable care in any given circumstances.

Updating IAPMO Codes

Users of IAPMO codes should be aware that IAPMO codes may be amended from time to time through the issuance of Tentative Interim Amendments or corrected by Errata. IAPMO codes consist of the current edition of the document together with any Tentative Interim Amendment and any Errata in effect.

In order to determine whether an IAPMO code has been amended through the issuance of Tentative Interim Amendments or corrected by Errata, please visit the IAPMO Group codes information pages on IAPMO's website (www.iapmo.org). The codes information pages provide a list of IAPMO codes with up-to-date specific information including any issued Tentative Interim Amendments and Errata.

To access the code information pages for a specific code, go to http://codes.iapmo.org to select from the list of IAPMO codes. For Tentative Interim Amendments, go to the standard council decisions. For Errata, select the archived revision information.

Origin and Development

The industry has long recognized the advantages of a statewide adopted mechanical code. The first edition of the *Uniform Mechanical Code*® (UMC®) was adopted by IAPMO in 1967. The widespread use of this code over the past five decades by jurisdictions throughout the United States and internationally is testament to its merit.

Publishing the 2003 *Uniform Mechanical Code*, is a significant milestone because it is the first time in the history of the United States, a mechanical code was developed through a true consensus process. The 2018 edition represents the most current approaches in the mechanical field and is the sixth edition developed under the ANSI consensus process. Contributions to the content of this code consists of diverse interests as consumers, enforcing authorities, installers/maintainers, labor, manufacturers, research/standards/testing laboratories, special experts, and users.

The *Uniform Mechanical Code* provides consumers with complete requirements for the installation and maintenance of heating, ventilating, cooling, and refrigeration systems, while, at the same time, allowing latitude for innovation and new technologies. The public at large is invited and encouraged to take part in IAPMO's open consensus code development process. This code is updated every three years. The *Uniform Mechanical Code* is dedicated to all those who in working to achieve "the ultimate mechanical code" have unselfishly devoted their time, effort, and personal funds to create and maintain this, the finest mechanical code in existence today.

The *Uniform Mechanical Code* updates every three years in revision cycles that begin twice each year that takes two years to complete.

Each revision cycle advances according to a published schedule that includes final dates for all major events and contains four basic steps as follows:
1. Public and Committee Proposal Stage;
2. Comment Stage;
3. Association Technical Meeting;
4. Council Appeals and Issuance of Code.

IAPMO develops "full consensus" codes built on a foundation of maximum participation and agreement by a broad range of interests. This philosophy has led to producing technically sound codes that promote health and safety, yet do not stifle design or development.

It is important to stress that, the process remains committed to the principles of consensus code development where consensus Technical Committees and Correlating Committees revise codes. The public and membership is offered multiple opportunities to debate, provide input and raise concerns through Amending Motions at the annual Assembly Consideration Session. Anyone may submit an appeal related to the issuance of a document through the IAPMO Standards Council.

The 2018 *Uniform Mechanical Code* is supported by the Mechanical Contractors Association of America (MCAA), the Plumbing-Heating-Cooling Contractors National Association (PHCC-NA), the United Association (UA), and the World Plumbing Council (WPC). The presence of these logos, while reflecting support, does not imply any ownership of the copyright to the UMC, which is held exclusively by IAPMO. Further, the logos of these associations indicate the support of IAPMO's open consensus process being used to develop IAPMO's codes and standards.

FORMAT OF THE UNIFORM MECHANICAL CODE

The format of the *Uniform Mechanical Code* (UMC) arranges each chapter in accordance with a specific subject matter. However, Chapter 3 is dedicated to general requirements that are applicable to every chapter. The subject matters are divided as follows:

CHAPTERS	SUBJECTS
1	Administration
2	Definitions
3	General Regulations
4	Ventilation Air
5	Exhaust Systems
6	Duct Systems
7	Combustion Air
8	Chimneys and Vents
9	Installation of Specific Appliances
10	Boilers and Pressure Vessels
11	Refrigeration
12	Hydronics
13	Fuel Gas Piping
14	Process Piping
15	Solar Energy Systems
16	Stationary Power Plants
17	Referenced Standards
Appendix A	Residential Plan Examiner Review Form for HVAC System Design
Appendix B	Procedures to be Followed to Place Gas Equipment in Operation
Appendix C	Installation and Testing of Oil (Liquid) Fuel-Fired Equipment
Appendix D	Fuel Supply: Manufactured/Mobile Home Parks and Recreational Vehicle Parks
Appendix E	Sustainable Practices
Appendix F	Sizing of Venting Systems and Outdoor Combustion and Ventilation Opening Design
Appendix G	Example Calculation of Outdoor Air Rate

The following is a summary of the scope and intent of the provisions addressed within the chapters and appendices of the *Uniform Mechanical Code*:

Chapter 1 Administration.
Chapter 1 regulates the application, enforcement, and administration of subsequent requirements of the code. As well as establishing the scope of the code, this chapter is concerned with enforcing the requirements contained in the body of the code. A mechanical code, as with any other code, is intended to be adopted as a legally enforceable document to safeguard health, safety, property and public welfare. The code cannot be effective without satisfactory provisions for its administration and enforcement. The Authority Having Jurisdiction is to review the proposed and completed work and to decide whether a mechanical system conforms to the code requirements. As a public servant, the Authority Having Jurisdiction enforces the code in an unbiased, proper manner. The design professional is responsible for the design of a safe mechanical system. The contractor is responsible for installing the system in accordance with the plans.

Chapter 2 Definitions.
To maintain consistency and encourage the use of common terminology, Chapter 2 establishes definitions to provide clarity of terms and promote the use of a common language throughout the code. Understanding definitions within the context of their application enables greater collaboration, efficiency, standardization and interpretation in applying and enforcing terms used throughout the code. Codes are technical documents, and every term can impact the meaning of the code text. Terms not defined have a normally accepted meaning.

Chapter 3 General Regulations.
Chapter 3 regulates the general requirements, not specific to other chapters, for installing mechanical systems. Many regulations are not specific mechanical requirements, but relate to the overall mechanical system. This chapter contains safety requirements for appliance location and installation, appliance and system access, condensate disposal, and clearances to combustibles, and return or outside air used in mechanical systems. Listing method of approval, based on applicable nationally recognized standards, for the safe and proper installation of mechanical systems is essential to ensure protection of public health, safety, and welfare. The safety requirements provide protection for piping, material, and structures, with provisions for installation practices, removing stress and strain of the pipe, sleeving, and hanger support. The building's structural stability is protected by the regulations for cutting and notching of structural members.

Chapter 4 Ventilation Air.
Chapter 4 regulates the minimum requirements for ventilation air supply, exhaust, and makeup air for spaces within a building. Building ventilation is one important factor affecting the relationship between airborne transmission of respiratory infections and the health and productivity of workers. Ventilation air may be composed of mechanical or natural ventilation, infiltration, recirculated air, transfer air, or a suitable combination of that. Providing a comfortable and healthy indoor environment for building occupants is of primary concern. When considering how much ventilation should be supplied, typical and unusual significant sources of indoor pollution need to be controlled. Areas such as kitchens, bathrooms, and laundries are all built to allow specific functions. These spaces produce pollutants such as moisture, odors, volatile organic compounds, particles, or combustion byproducts. The purpose of local exhaust is to control concentrates of these pollutants in the room into which they were emitted in and to reduce the spread of the pollutants into other parts of the occupancy. Local exhaust ventilation is the source control for pollution that is expected in certain rooms. Using local exhaust to extract contaminants before they can mix with the indoor environment is essential.

Chapter 5 Exhaust Systems.
Chapter 5 regulates the minimum requirements for exhaust systems. Chapter 5 contains two parts: part I provides exhaust requirements for environmental air ducts and product conveying ducts; part II provides exhaust requirements for commercial kitchens. Environmental air ducts include exhaust ducts used for transporting the air from domestic kitchens, bathrooms, and clothes dryers. Systems that carry nonabrasive exhaust, such as smoke, moderately abrasives such as sawdust, and high abrasives such as manganese or acid vapors use product-conveying ducts. Part II provides the minimum fire safety requirements related to the design, installation, inspection, and maintenance of grease-type operations, such as cooking, for both fuel-gas and solid fuel. Cooking produces a significant amount of smoke, fumes, vapors, heat, and other pollutants. Therefore, acceptable kitchen ventilation is necessary to provide the occupants protection from smoke, unpleasant odors, pollutants, dangerous gases, and to prevent fires from the build-up of grease. There are two types of exhaust hoods (Type I and Type II) used in commercial kitchen applications. Type I hoods are intended to be installed above equipment or

appliances that generate grease or smoke. Type II hoods are intended to be installed above equipment or dishwashers that generate steam, heat, or products of combustion, or where grease or smoke is not present. Type II hood exhaust system requirements are addressed in Section 519.0.

Chapter 6 Duct Systems.

Chapter 6 regulates requirements for ducts and plenums that are portions of a heating, cooling, absorption or evaporative cooling, or exhaust system. This chapter contains material and installation requirements for metal, gypsum, factory-made, flexible, and plastic ducts. It also contains fire protection requirements, smoke dampers, and automatic shutoff for the building's air distribution system.

Chapter 7 Combustion Air.

Chapter 7 regulates combustion air requirements for the ventilation and dilution of flue gases for appliances installed in buildings. It applies to fuel-gas appliances except for direct vent appliances and clothes dryers. Chapter 5 provides makeup air for clothes dryers. Chapter 7 provides acceptable methods for supplying satisfactory combustion air to ensure proper combustion. Combustion air can be supplied by using indoor combustion air or by introducing the air from the outdoors.

Combustion is the rapid oxidation of fuel to release energy. The oxygen required to release the energy from the fuel normally comes from the air. Incomplete combustion of fuel occurs when inadequate oxygen is provided to the appliance. Combustion is needed to provide ventilation cooling for the casing and internal controls. When a lack of oxygen occurs, some of the carbon is not oxidized, and carbon monoxide forms.

Chapter 8 Chimneys and Vents.

Chapter 8 regulates the installation, design, and construction of venting systems for fuel-burning appliances. The provisions addressed within this chapter follow procedures an installer would use to design or evaluate a venting system. Many requirements apply to the design and construction of venting systems, chimneys, installation of gas vents, and the sizing of venting system for a Category I appliance. Sizing venting systems require rigorous engineering calculations. However, the venting sizing requirements and sizing tables in this chapter already perform the calculations for the benefit of the end user.

Combustion appliances produce products of incomplete combustion, including potentially harmful carbon monoxide (CO). It is desirable to vent these products to the outdoors. Although the gas is clean-burning fuel, the products of combustion must not be allowed to collect within a building.

Chapter 9 Installation of Specific Appliances.

Chapter 9 regulates the minimum requirements for the design, construction and installation of specific appliances. The provisions address the minimum requirements for gas-fired appliances, oil-fired appliances, wood-fired appliances, and electric-type appliances. In addition to the requirements of this chapter, appliances are also required to comply with the general requirements of Chapter 3.

Chapter 10 Boilers and Pressure Vessels.

Chapter 10 regulates the construction, installation, operation, repair, and alteration of boilers and pressure vessels. A low-pressure boiler provides steam at a pressure that does not exceed 15 psig, a gauge pressure more than 160 psi (1103 kPa) or heats water to a temperature more than 250°F (121°C). Potable water heaters are free from the requirements of Chapter 10 as they are within the scope of the Uniform Plumbing Code (UPC).

Pressure vessels store large amounts of energy and must comply with ASME Boiler and Pressure Vessel Code (BPVC) Section VIII. The stored energy must be contained to prevent disastrous failures. Boilers must comply with ASME BPVC Section I, ASME BPVC Section IV, or NFPA 85. Installing a safety relief valve and expansion tank prevents pressures in the tank from exceeding the design threshold.

Chapter 11 Refrigeration.

Chapter 11 regulates the design, installation, and construction requirements of refrigeration systems and the installation and construction of cooling towers. Refrigeration is a method used for achieving heat transfer to cool spaces. Refrigerants are the most common medium used to transfer the heat energy from the low-temperature level to the high-temperature level. Table 1102.3 lists the most commonly used refrigerants and is labeled by a number. The concentration limits provided in Table 1102.3 are useful for the quantity of refrigerant required to cool a volume of space safely.

A major milestone reached in the industry was the introduction of lower flammability refrigerants such as A2L and B2L. The industry's concerns to address lower Global Warming Potential (GWP) refrigerants led to the development of these lower flammability refrigerants. Chapter 11 does not only include these lower flammability refrigerants in Table 1102.3, but it also contains specific A2L and B2L provisions to address the precautions necessary for the proper use of these products.

In addition, Chapter 11 addresses other minimum requirements for refrigeration systems such as refrigeration machinery rooms (including ventilation), relief valves, and pressure vessels. Apart from refrigerants, the chapter also addresses minimum requirements for systems that use other mediums such as ammonia and brine systems. For ammonia systems, such systems are required to comply with IIAR 2, IIAR 3, IIAR4, and IIAR 5.

Chapter 12 Hydronics.

Chapter 12 regulates hydronic systems that are part of heating, cooling, ventilation, and conditioning systems. Such piping systems include steam, hot water, chilled water, steam condensate, and the ground source heat pumps systems. The ground source heat pumps provisions in this chapter apply to the hydronics portions of the system. It is worth noting the Uniform Solar Energy & Hydronics Code (USEHC) address added provisions for ground source heat pumps in its geothermal energy system chapter.

Materials for piping and tubing must meet the working temperature and pressure of the system. In addition, materials must be compatible with the transfer medium to prevent deformation, bursting, or any chemical action between the material and the transfer medium. The allowable joining methods for piping or tubing are provided for application and enforcement purposes.

Chapter 13 Fuel Gas Piping.

Chapter 13 regulates gas piping systems in a building, structure or within the property lines of buildings up to 5 psi. Gas piping systems must supply the minimum volume of gas required by each gas appliance to perform their proper operation under working conditions without exceeding the maximum pressure specified by each manufacturer. Because of the hazards associated with fuel gas, it is important to ensure the gas system has been inspected and tested, and that it is safe to turn on the gas supply to the building.

Chapter 14 Process Piping.

Chapter 14 regulates process piping that typically is found in refineries. Process piping is considered the piping or tubing portion that transports liquid or gas, which is used directly in research, laboratory, or production process. This chapter may be used with another chapter of the code. For example, refrigeration piping, fuel gas piping, or fuel oil piping may need to comply with this chapter for process piping and the applicable chapter for materials, design, and installation.

Chapter 15 Solar Energy Systems.

Chapter 15 refers the user to the Uniform Solar Energy and Hydronics Code (USEHC) for installations about solar energy systems. Reference is also made to Section 1203.0 as the heat source provisions are applicable to solar energy systems.

Chapter 16 Stationary Power Plants.

This chapter provides minimum requirements for stationary fuel cell power plants. A fuel cell is a device that produces electricity by a chemical reaction. Fuel cells have been used mostly for satellites. However, the costs of fuel cells have decreased to a point where they are now economically attainable. With the increase in electricity costs, fuel cells are being considered as an alternative for producing electricity.

Chapter 17 Referenced Standards.

Chapter 17 provides two comprehensive tables with referenced standards. The standards listed in Table 1701.1 are applied as indicated in the applicable reference section(s). A list of additional standards, publications, practices, and guides that are not referenced in specific sections appear in Table 1701.2 and shall be submitted to the Authority Having Jurisdiction for approval.

Referenced standards set forth specific details of accepted practices, materials specifications, or test methods in many specialized applications. Standards provide an efficient method of conveying complex information and specifications on the performance requirements for materials, products, systems, application, and installation. The manner and purpose for a standard's use and, in turn, code compliance, must be definitive in all references to the standard. If the standard is intended to be a requirement for judging code compliance, the code must state its intent for use. The standard should adequately address a defined need and at the same time specify the minimum performance requirements, technical characteristics and methods of testing, and required test results.

The referenced standards table is organized in a manner that makes it easy to find specific standards in alphabetical order, and by acronym of the publishing agency of the standard. The table lists the title of the standard, the edition, any addenda, and the section or sections of the code that reference the standard. Contact information for each publishing agency is provided at the end of the chapter.

Appendix A Residential Plan Examiner Review Form for HVAC System Design.
A residential plan examiner review form is located in this appendix as an example to assist users in identifying whether the HVAC system has followed the approved procedures for system design (loads, equipment, and ducts).

Appendix B Procedures to be Followed to Place Gas Equipment in Operation.
Appendix B provides requirements for the procedures that apply after an appliance is installed in place, piped, and connected to its venting system. The requirements include adjusting the burner input, air adjustments, verifying operation of safety shutoffs, automatic ignition, and protective devices, checking draft for vent-connected appliances, and operating instructions.

Appendix C Installation and Testing of Oil (Liquid) Fuel-Fired Equipment.
Appendix C governs the installation, testing, or repair of oil or liquid fuel-burning equipment used in buildings or structures and equipment.

Appendix D Fuel Supply: Manufactured/Mobile Home Parks and Recreational Vehicle Parks.
The provisions of this appendix apply to the fuel gas piping systems of mobile home and recreational vehicle parks. These provisions also apply to the use, maintenance, and installation for supplying fuel gas for accessory buildings or structures, and building components.

Appendix E Sustainable Practices.
This appendix provides a comprehensive set of technically sound provisions that encourage sustainable practices and works toward improving the design and construction of mechanical systems that result in a positive long-term environmental impact. Environmental sustainability is important because it involves natural resources that human beings need for economic or manufactured capital. Their sustainability is defined by their reliance on infinitely available resources that are naturally occurring, constant, and free to access.

Appendix F Sizing of Venting Systems and Outdoor Combustion and Ventilation Opening Design.
Appendix F provides added information on the sizing of gas vents. This appendix is useful to the end user for the proper sizing of venting systems. A series of examples are given that show how to use the tables and other requirements of Chapter 8. In addition, Appendix F shows an example of how to determine the required combination of indoor and outdoor combustion air opening sizes for appliances under Chapter 7. The combustion air example also provides a table that contains the required volume of space per the appliance BTU/h input based on the standard method.

Appendix G Example Calculation of Outdoor Air Rate.
Appendix G gives an example of how to calculate the required outdoor air rate under Chapter 4.

These lists represent the membership at the time the Committee was balloted on the final text of this edition. Since that time, changes in the membership may have occurred.

IAPMO Standards Council
Linden Raimer, Chairman
Raimer Consulting Services, LLC, [U]

Carl Crimmins, MN State Pipe Trades [SE]
Bill Erickson, CJ Erickson Plumbing Company [U]
James Majerowicz, UA Instructor [L]
Rich Prospal, ASSE [C]
Ron Rice, City of St. Paul [C]
Robert "Bud" Riestenberg, Piping Systems, Inc. [U]
Bob Siemsen, City of Lincoln [E]
Kevin Tindall, Tindall & Ranson Plumbing, Heating & A/C, Inc. [I/M]

Nonvoting

Gabriella M. Davis, Secretary
International Association of Plumbing and Mechanical Officials

Lynne Simnick, Recording Secretary
International Association of Plumbing and Mechanical Officials

IAPMO Uniform Mechanical Code Technical Committee
April Trafton, Chairperson
American Society of Plumbing Engineers [SE]

Bob Adler, City of San Jose, California [E]
Mike Afonso, UA Local 342 [L]
DJ Berger, New Orleans Pipe Trades [L]
Maggie Carroll, Underwriters Laboratories [R/S/T]
Ian Chang, Intertek Testing Services [R/S/T]
Richard Church, Plastic Pipe & Fittings Association [M]
David Delaquila, National Propane Gas Association [U]
David Dias, Sheet Metal Workers Local 104 [L]
Pennie Feehan, Copper Development Association [M]
Roel Garcia, City of Houston [E]
Eli Howard, SMACNA [R/S/T]
Ralph Koerber, Air Duct Council [M]
Harvey Kreitenberg, Harvey Kreitenberg & Associates [I/M]
David Mann, CA State Pipe Trades [L]
John Nielsen, State of Idaho [E]
James Pavesic, United Association [L]
Phil Ribbs, PHR Consultants [SE]
Anthony Scarano, Plastic Piping Consultant [SE]
Robert Sewell, Plumbers and Steamfitters, Local 159 [L]
Chris Smith, Lubrizol Corporation [M]
Donald Surrena, National Association of Home Builders [U]
Don Taylor, Dittman Plumbing [SE]
Chris Van Rite, Air Distribution Institute [M]
Bob Wiseman, Air Conditioning Contractors of America [I/M]
Osama Younan, City of Los Angeles - Department of Building and Safety [E]
Randy Young, Northern CA JATC [L]

Alternates

David Bixby, Air Conditioning Contractors of America [I/M]
Dan Buuck, National Association of Home Builders [U]
Phil Campbell, United Association [L]
Michael Cudahy, Plastic Pipe & Fittings Association [M]
TJ Dodd, Austin Area Plumbing & Pipe Fitters JATC [L]
John Hamilton, NEMIC [L]
Shawn Hargis, City of Los Angeles - Department of Building and Safety [E]
Vickie Lovell, Air Movement and Control Association International, Inc. [M]
Leonard A. Ramociotti, LAR Consulting [SE]
Jason Shelton, Air Duct Council [M]
John Taecker, Underwriters Laboratories [R/S/T]
Che Timmons, Plumbers and Steamfitters Local 342 [L]

Nonvoting

Hugo Aguilar, IAPMO Staff Liaison
Eric Nette, NFPA [R/S/T]
DJ Nunez, Ex -Officio IAPMO [E]
Laura Moreno, NFPA [R/S/T]

COMMITTEE MEMBERSHIP CLASSIFICATION ABBREVIATIONS

These classifications apply to Technical Committee members and represent their principal interest in the activity of a committee.

M *Manufacturer*: A representative of a maker or marketer of a product, assembly or system, or portion thereof, that is affected by the standard.

U *User*: A representative of an entity that is subject to the provisions of the standard or that voluntarily uses the standard.

I/M *Installer/Maintainer*: A representative of an entity that is in the business of installing or maintaining a product, assembly or system affected by the standard.

L *Labor*: A labor representative or employee concerned with safety in the workplace.

R/S/T *Research/Standards/Testing Laboratory*: A representative of an independent research organization; an organization that develops codes, standards or other similar documents; or an independent testing laboratory.

E *Enforcing Authority*: A representative or an agency or an organization that promulgates and/or enforces standards.

C *Consumer*: A person who is, or represents, the ultimate purchaser of a product, system, or service affected by the standard, but who is not included in the User classification.

SE *Special Expert*: A person not representing any of the previous classifications, but who has special expertise in the scope of the standard or portion thereof.

2018 Location	2015 Location
505.2, 505.3	505.1.1, 505.1.2
505.1.1	505.1.3
505.6.1	505.2
505.8 – 505.11	505.3 – 505.6
Table 505.9	Table 505.4
506.5 – 506.7	506.3 – 506.5
506.8.1 – 506.8.4	506.5.1 – 506.5.4
506.8	506.6
506.10.2	506.7
506.11.1	506.7.1
506.11.2, 506.11.3	506.7.3, 506.7.4
506.11.6 – 506.11.6.3	506.7.7 – 506.7.7.3
506.9	506.8
507.2 – 507.6	507.1 – 507.5
519.1(1), 5191(2)	508.1(3), 508.1(4)
519.2	508.4
508.4	508.5
508.3.1 – 508.3.4	508.6 – 508.9
Figure 508.3.3	Figure 508.8
508.5 – 508.5.1.5	508.10 – 508.10.1.5
Table 508.5.1.2	Table 508.10.1.2
Table 508.5.1.3	Table 508.10.1.3
Table 508.5.1.4	Table 508.10.1.4
Table 508.5.1.5	Table 508.10.1.5
519.3	508.10.1.6
508.5.2 – 508.5.3	508.10.2 – 508.10.3
508.3.5 – 508.3.5.3.3	508.11 – 508.11.3.3
508.6, 508.7	508.12, 508.13
519.4	510.1.7
510.1.7	510.1.8
519.5	510.10
510.10	510.11
508.1(3), 508.1(4)	519.1(1), 519.1(2)
602.2.4	602.2.5
902.13	702.0, 702.1
902.16	704.0, 704.1
802.2.4 – 802.2.7	802.2.2 – 802.2.5
802.6.1 – 802.6.5	802.6.2 – 802.6.6
Figure 802.6.1, Table 802.6.1	Figure 802.6.2, Table 802.6.2
Figure 802.6.3.1, Figure 802.6.3.2	Figure 802.6.4.1, Figure 802.6.4.2
802.7.3.3 – 802.7.3.5	802.7.3.4 - 802.7.3.6

2018 Location	2015 Location
Table 802.7.3.3	Table 802.7.3.4
803.2.7 – 803.2.27	803.2.6 – 803.2.26
902.6	902.5
902.8 – 902.12	902.6 – 902.10
904.2.1 – 904.2.9	904.2(1) – 904.2(9)
904.7 – 904.7.3	904.7(1) – 904.7(4)
905.6 – 905.6.3	905.6(1) – 905.6(4)
303.11 – 303.12	905.8 – 905.9
905.8 – 905.8.2	905.10 – 905.10.2
907.1 – 907.1.4	907.1(1) – 907.1(5)
911.2.1 – 911.2.3	911.2(1) – 911.2(3)
914.3 – 9.14.3.3	914.3(1) – 914.3(4)
914.7.1, 914.7.2	914.7(1), 914.7(2)
915.7.1, 915.7.2	915.7(1), 915.7(2)
918.3.1	918.3(2)
920.0 – 926.1	921.0 – 927.1
1312.2	927.2
926.2 – 938.1	927.3 – 939.1
926.3.1	927.4(2)
927.3.1	928.3(2)
1102.3	1102.2
1106.10	1106.2
1106.12	1106.3
1106.5, 1106.6	1106.6, 1106.7
1106.2.5.1	1107.2
1106.4	1107.3
1106.7 – 1106.9	1107.9 – 1107.11
1109.1.3, 1109.1.4	1109.1.2, 1109.1.3
1112.6 – 1112.14	1112.5 – 1112.13
Table 1112.12.3	Table 1112.11.3
1217.1.1	1201.4
1210.4	1205.4
1206.2	1206.3
1211.4 – 1211.14.2	1211.3 – 1211.13.2
1213.2 – 1213.3	1213.3 – 1213.4
1201.5	1213.5
1217.4 – 1217.6.2	1217.5 – 1217.7.2
Table 1217.4	Table 1217.5
1220.1 – 1220.3	1220.2 – 1220.4
Table 1220.2.1	Table 1220.3.1
1308.5.10.5, 1308.5.10.6	1308.5.10.1, 1308.5.10.2
1308.7.1 – 1308.7.8	1308.7.2 – 1308.7.9
1308.10 – 1308.13.1	1308.8 – 1308.11.1

2018 Location	2015 Location
902.15	1310.0, 1310.1
1310.0 – 1310.1.3	1311.0 – 1311.1.3
1310.1.4 – 1310.2.4.2	1311.1.4 – 1311.2.4.2
Table 1310.2.4.1	Table 1311.2.4.1
1310.2.5 – 1310.5	1311.2.5 – 1311.5
1310.6 – 1312.1.2	1311.6 – 1313.1.2
1312.3 – 1312.3.2	1313.2 – 1313.2.2
1312.5 – 1312.11	1313.3 – 1313.9
Figure 1312.9	Figure 1313.7
902.5	1313.9.1
1313.0 – 1315.6	1314.0 – 1316.6
Table 1313.6.1	Table 1314.6.1
Figure 1315.1.1	Figure 1316.1.1
Table 1315.3	Table 1316.3
Equation 1315.3(1) – Equation 1315.3(2)	Equation 1316.3(1) – Equation 1316.3(2)
E 201.3 – E 201.7	E 201.2 – E 201.6
E 201.9 – E 201.12	E 201.7 – E 201.10
E 201.14	E 201.11
F 101.2 – F 101.5	F 101.1 – F 101.4
Figure F 101.2(1) – Figure F 101.2(14)	Figure F 101.1(1) – Figure F 101.1(14)

RELATED PUBLICATIONS AND SERVICES

The Documents Listed are Not by this Reference Adopted by The State of California.

IAPMO provides a variety of other products which are useful for inspectors, building officials, architects, engineers, manufacturers, contractors, plumbers, and apprentices.

IAPMO Headquarters Directory
4755 East Philadelphia Street, Ontario, California 91761-2816
Publication Order Desk Phone: 800-85-IAPMO
Publication Toll Free Fax: 877-85-CODES
E-mail: iapmo@iapmo.org
Website: www.iapmo.org
Free Code Question: 800-201-0335

Uniform Mechanical Code – 2018 Edition:
The Uniform Mechanical Code contains complete requirements for the installation and maintenance of heating, ventilating, cooking and refrigeration systems.

Uniform Mechanical Code Illustrated Training Manual:
Contains technical diagrams and illustrations that demonstrate the intent and use of the UMC. A great reference for everyone involved in Mechanical HVACR design and installation.

Uniform Mechanical Code Study Guide:
The Study Guide is a complete self-study course for learning the UMC. A big help in getting ready for a certification exam! This book is the perfect complement to the UMC Illustrated Training Manual.

Guide to Important Code Questions to the Uniform Mechanical Code:
An excellent reference for learning and understanding mechanical code changes and identifies code changes between editions. It is a useful tool for preparing code change proposals.

Uniform Plumbing Code – 2018 Edition:
The Uniform Plumbing Code is the most widely adopted plumbing code in the world. The 2018 edition contains complete "turnkey" requirements for the installation and maintenance of plumbing systems, all in one easy to use book.

Uniform Plumbing Code Illustrated Training Manual:
The UPC Illustrated Training Manual is an excellent reference for anyone involved in the plumbing industry. It contains an extensive definitions section and several hundred comprehensive technical diagrams and illustrations. It serves as a textbook, and it also is useful as a valuable tool for explaining the intent and use of the Code.

Uniform Plumbing Code Study Guide:
This book is the perfect complement to the UPC Illustrated Training Manual. Alone, it constitutes a complete self-study course for learning the UPC. It has hundreds of questions, general practice exams, and plumbing math, pipe sizing exercises and fitting identification. A big help in getting you ready for a certification exam!

Guide to Important Code Questions to the Uniform Plumbing Code:
An excellent reference for learning and understanding plumbing code changes and identifies code changes between editions. It is a useful tool for preparing code change proposals.

RELATED PUBLICATIONS AND SERVICES

The Documents Listed are Not by this Reference Adopted by The State of California.

Uniform Swimming Pool, Spa, and Hot Tub Code – 2018 Edition:
The Uniform Swimming Pool, Spa and Hot Tub Code contains complete and current requirements for the erection, installation, alteration, repair, relocation, replacement, addition to and use or maintenance of these systems.

Uniform Solar, Hydronics and Geothermal Code – 2018 Edition:
The Uniform Solar, Hydronics and Geothermal Code contains complete requirements for the erection, installation, alteration, addition, repair, relocation, replacement, addition to, use or maintenance of these systems.

Drain Waste and Vent Calculator:
A slide-rule style calculator provides quick and simple access to the fixture unit and sizing tables of Chapter 7 of the UPC.

Water Sizing Calculator:
Sizing water systems becomes a much simpler task with the use of this handy slide-rule style calculator.

Natural Gas Pipe Sizing Calculator:
This Natural Gas Pipe Sizing Calculator is presented in an easy to use and read slide-rule format. Designed for systems with a supply pressure of six to eight inches of water column, all pipe capacities are given in cubic feet per hour.

Trailer Standards:
IAPMO has developed a number of standards for specialty products used in manufactured homes and recreational vehicles. These standards are available individually, or they may be purchased as a set.

Education and Training Seminars:
IAPMO specializes in conducting training and education, including fulfilling continuing education requirements, on the UPC and UMC as well as all of the most commonly utilized codes in the Construction Trades. Special seminars on industry or code subjects are expeditiously created and delivered upon request. Please call to obtain the seminar schedule or to arrange for a custom special seminar.

Backflow Prevention Institute – IAPMO:
The IAPMO Backflow Prevention Institute provides professional education and training for the control of hazards to our water supply and is dedicated to water-system safety worldwide. The Institute provides training in backflow prevention and cross-connection control. This program includes backflow preventers, testers, repairers, cross-connection control surveyors, program administrators and fire-sprinkler backflow preventer testers. Each of these courses provides compliance certification to American Society of Sanitary Engineering (ASSE) Series 5000 Professional Qualification Standards.

Backflow Prevention Reference Manual:
This unique 300-page manual contains full-color illustrations and sections relating to real-world installations, backflow prevention, testing cross-connection control, repair and inspections. The manual is an excellent textbook and a reference the inspectors in every jurisdiction should have access too. It is also an important resource for system designers, urban planners and utility companies – professionals on both sides of the water meter.

Backflow Prevention Tech-Wheel:
Find quick answers to your backflow preventer questions! The Backflow Tech-Wheel contains a wealth of information in a graphic, quick reference format. One side delivers solutions to device application questions; the other a troubleshooting guides offering solutions to backflow problems.

ANSI Z124 Standards:
These standards are written to specifically address a variety of plastic plumbing fixtures and components. They are available individually, or may be grouped for quantity discounts.

RELATED PUBLICATIONS AND SERVICES

The Documents Listed are Not by this Reference Adopted by The State of California.

IAPMO Installation Standards:
IAPMO standards committees have formulated installation standards for a wide variety of commonly used plumbing materials and systems. The IAPMO installation standards are included after the text of the Uniform Plumbing Code, or can be purchased separately.

Material and Property Standards:
IAPMO does not generally develop material and property standards, but when a need exists the Association will take a leadership role by filling the void. They are available, are subject to amendments and are withdrawn when recognized consensus standards are formulated.

OFFICIAL Magazine:
IAPMO's bimonthly publication features informative articles related to Plumbing and Mechanical HVACR installations, award winning general interest features, technical columns and industry updates. Subscriptions are available and all IAPMO members receive a free copy of every issue!

Membership in IAPMO:
IAPMO membership is open to all interested persons. Membership categories include Student, Individual, Senior, Organizational and Governmental. Member benefits include discounted prices on IAPMO publications, a subscription to Official magazine, and numerous educational and training opportunities.

HOW TO CONTACT IAPMO

Mailing Address: 4755 East Philadelphia Street, Ontario, California 91761-2816
Main Number: 909-472-4100 • **Publication Orders:** 800-85-IAPMO • **Publication Fax:** 877-85-CODES
Website: www.iapmo.org • **Office Hours:** Monday – Friday, 8am – 5pm Pacific Time

TABLE OF CONTENTS

CALIFORNIA MECHANICAL CODE – MATRIX ADOPTION TABLE
CHAPTER 1 – ADMINISTRATION

(Matrix Adoption Tables are non-regulatory, intended only as an aid to the code user. See Chapter 1 for state agency authority and building applications.)

Adopting Agency	BSC	BSC-CG	SFM	HCD			DSA			OSHPD						BSCC	DPH	AGR	DWR	CEC	CA	SL	SLC
				1	2	1-AC	AC	SS	SS/CC	1	1R	2	3	4	5								
Adopt Entire Chapter																							
Adopt Entire Chapter as amended (amended sections listed below)																							
Adopt only those sections that are listed below	X		X	X	X			X	X	X		X	X	X									
Chapter/Section																							
1.1.0	X		X	X	X			X	X	X		X	X	X									
1.1.1	X		X	X	X			X	X	X		X	X	X									
1.1.2	X		X	X	X			X	X	X		X	X	X									
1.1.3	X		X	X	X			X	X	X		X	X	X									
1.1.3.1	X		X	X	X			X	X	X		X	X	X									
1.1.3.2	X		X	X	X			X	X	X		X	X	X									
1.1.4	X		X	X	X			X	X	X		X	X	X									
1.1.5	X		X	X	X			X	X	X		X	X	X									
1.1.6	X		X	X	X			X	X	X		X	X	X									
1.1.7	X		X	X	X			X	X	X		X	X	X									
1.1.7.1	X		X	X	X			X	X														
1.1.7.2	X		X	X	X			X	X														
1.1.7.3	X		X	X	X			X	X														
1.1.8	X		X	X	X			X	X	X		X	X	X									
1.1.9	X			X	X			X	X	X		X	X	X									
1.1.9 Exception				X	X																		
1.1.10	X		X	X	X			X	X	X		X	X	X									
1.1.11	X		X	X	X			X	X	X		X	X	X									
1.1.12	X		X	X	X			X	X	X		X	X	X									
1.2.0 and 1.2.1.4	X																						
1.2.2		X																					
1.2.3	X																						
1.8.0 – 1.8.10 and subsections				X	X																		
1.9.0								X	X														
1.9.1							X																
1.9.2								X	X														
1.9.2.1 & 1.9.2.1.1								X															
1.9.2.2 & 1.9.2.2.1									X														
1.10.1 and subsections										X	X												
1.10.2 and subsections												X											
1.10.3 and subsections													X										
1.10.4 and subsections														X									
1.10.5 and subsections															X								
1.11.0			X																				

CALIFORNIA MECHANICAL CODE – MATRIX ADOPTION TABLE
CHAPTER 1 – ADMINISTRATION (continued)

(Matrix Adoption Tables are non-regulatory, intended only as an aid to the code user. See Chapter 1 for state agency authority and building applications.)

Adopting Agency	BSC	BSC-CG	SFM	HCD			DSA			OSHPD						BSCC	DPH	AGR	DWR	CEC	CA	SL	SLC
				1	2	1-AC	AC	SS	SS/CC	1	1R	2	3	4	5								
Adopt Entire Chapter																							
Adopt Entire Chapter as amended (amended sections listed below)																							
Adopt only those sections that are listed below	X		X	X	X			X	X	X		X	X	X									
Chapter/Section																							
Division II – Administration																							
101.0 – 104.1										X	X	X	X	X	X								
104.2 (Items 1 – 5)				X	X																		
104.3 – 107.0										X		X	X	X									

This state agency does not adopt sections identified with the following symbol: †

The Office of the State Fire Marshal's adoption of this chapter or individual sections is applicable to structures regulated by other state agencies pursuant to Section 1.11.0.

CHAPTER 1

ADMINISTRATION

DIVISION I
CALIFORNIA ADMINISTRATION

1.1.0 General.

1.1.1 Title. *These regulations shall be known as the California Mechanical Code, may be cited as such and will be referred to herein as "this code." The California Mechanical Code is Part 4 of thirteen parts of the official compilation and publication of the adoption, amendment, and repeal of mechanical regulations to the California Code of Regulations, Title 24, also referred to as the California Building Standards Code. This part incorporates by adoption the 2018 Uniform Mechanical Code of the International Association of Plumbing and Mechanical Officials with necessary California amendments.*

1.1.2 Purpose. *The purpose of this code is to establish the minimum requirements to safeguard the public health, safety and general welfare through structural strength, means of egress facilities, stability, access to persons with disabilities, sanitation, adequate lighting and ventilation, and energy conservation; safety to life and property from fire and other hazards attributed to the built environment; and to provide safety to fire fighters and emergency responders during emergency operations.*

1.1.3 Scope. *The provisions of this code shall apply to the construction, alteration, movement, enlargement, replacement, repair, equipment, use and occupancy, location, maintenance, removal, and demolition of every building or structure or any appurtenances connected or attached to such buildings or structures throughout the State of California.*

1.1.3.1 Nonstate-Regulated Buildings, Structures, and Applications. *Except as modified by local ordinance pursuant to Section 1.1.8, the following standards in the California Code of Regulations, Title 24, Parts 2, 2.5, 3, 4, 5, 6, 9, 10 and 11 shall apply to all occupancies and applications not regulated by a state agency.*

1.1.3.2 State-Regulated Buildings, Structures, and Applications. *The model code, state amendments to the model code, and/or state amendments where there are no relevant model code provisions shall apply to the following buildings, structures, and applications regulated by state agencies as specified in Sections 1.2.0 through 1.14.0, except where modified by local ordinance pursuant to Section 1.1.8. When adopted by a state agency, the provisions of this code shall be enforced by the appropriate enforcing agency, but only to the extent of authority granted to such agency by the state legislature.*

Note: *See Preface to distinguish the model code provisions from the California provisions.*

(1) State-owned buildings, including buildings constructed by the Trustees of the California State University, and to the extent permitted by California laws, buildings designed and constructed by the Regents of the University of California, and regulated by the Building Standards Commission. See Section 1.2.0 for additional scope provisions.

(2) Section 1.3.0 is reserved for the Board of State and Community Corrections.

(3) Section 1.4.0 is reserved for the Department of Consumer Affairs.

(4) Section 1.5.0 is reserved for the California Energy Commission.

(5) Section 1.6.0 is reserved for the Department of Food and Agriculture.

(6) Organized camps, laboratory animal quarters, public swimming pools, radiation protection, commissaries serving mobile food preparation vehicles, and wild animal quarantine facilities regulated by the Department of Public Health. See Section 1.7.0 for additional scope provisions.

(7) Hotels, motels, lodging houses, apartments, dwellings, dormitories, condominiums, shelters for homeless persons, congregate residences, employee housing, factory-built housing and other types of dwellings containing sleeping accommodations with or without common toilets or cooking facilities. See Section 1.8.2.1.1 for additional scope provisions.

(8) Accommodations for persons with disabilities in buildings containing newly constructed covered multifamily dwellings, new common use areas serving existing covered multifamily dwellings, additions to existing buildings where the addition alone meets the definition of covered multifamily dwellings and new common-use areas serving new covered multifamily dwellings which are regulated by the Department of Housing and Community Development. See Section 1.8.2.1.2 for additional scope provisions.

(9) Permanent buildings and permanent accessory buildings or structures constructed within mobilehome parks and special occupancy parks regulated by the Department of Housing and Community Development. See Section 1.8.2.1.3 for additional scope provisions.

(10) Accommodations for persons with disabilities regulated by the Division of the State Architect. See Section 1.9.1 for additional scope provisions.

(11) Public elementary and secondary schools, community college buildings and state-owned or state-leased essential service buildings regulated by the Division of the State Architect. See Section 1.9.2 for additional scope provisions.

(12) Reserved for the State Historical Building Safety Board with the Division of the State Architect.

(13) General acute care hospitals, acute psychiatric hospitals, skilled nursing and/or intermediate care facilities, clinics licensed by the Department of Public Health and correctional treatment centers regulated by the Office of Statewide Health Planning and Development. See Section 1.10.0 for additional scope provisions.

(14) Applications regulated by the Office of State Fire Marshal include but are not limited to the following in accordance with Section 1.11.0:

(1) Buildings or structures used or intended for use as an:

1.1. Asylum, jail, prison.

1.2. Mental hospital, hospital, home for the elderly, children's nursery, children's home or institution, school or any similar occupancy of any capacity.

1.3. Theater, dancehall, skating rink, auditorium, assembly hall, meeting hall, nightclub, fair building or similar place of assemblage where 50 or more persons may gather together in a building, room, or structure for the purpose of amusement, entertainment, instruction, deliberation, worship, drinking or dining, awaiting transportation, or education.

1.4. Small family day care homes, large family day-care homes, residential facilities and residential facilities for the elderly, residential care facilities.

1.5. State institutions or other state-owned or state-occupied buildings.

1.6. High rise structures.

1.7. Motion picture production studios.

1.8. Organized camps.

1.9. Residential structures.

(2) Tents, awnings, or other fabric enclosures used in connection with any occupancy.

(3) Fire alarm devices, equipment, and systems in connection with any occupancy.

(4) Hazardous materials, flammable, and combustible liquids.

(5) Public school automatic fire detection, alarm, and sprinkler systems.

(6) Wildland-urban interface fire areas.

(15) Section 1.12.0 is reserved for public libraries constructed and renovated using funds from the California Library Construction and Renovation Bond Act of 1988 and regulated by the State Librarian.

(16) Section 1.13.0 is reserved for the Department of Water Resources.

(17) For applications listed in Section 1.9.1 regulated by the Division of the State Architect – Access Compliance, outdoor environments and uses shall be classified according to accessibility uses described in Chapters 11A and 11B.

(18) Section 1.14.0 is reserved for Marine Oil Terminals regulated by the California State Lands Commission.

1.1.4 Appendices. *Provisions contained in the appendices of this code shall not apply unless specifically adopted by a state agency or adopted by a local enforcing agency in compliance with Health and Safety Code Section 18901 et. seq. for Building Standards Law, Health and Safety Code Section 17950 for State Housing Law and Health and Safety Code Section 13869.7 for Fire Protection Districts. See Section 1.1.8 of this code.*

1.1.5 Referenced Codes. *The codes, standards and publications adopted and set forth in this code, including other codes, standards, and publications referred to therein are, by title and date of publication, hereby adopted as standard reference documents of this code. When this code does not specifically cover any subject related to building design and construction, recognized architectural, or engineering practices shall be employed. The National Fire Codes, standards, and the Fire Protection Handbook of the National Fire Protection Association are permitted to be used as authoritative guides in determining recognized fire prevention engineering practices.*

1.1.6 NonBuilding Standards, Orders and Regulations. *Requirements contained in the Uniform Mechanical Code, or in any other referenced standard, code or document, which are not building standards as defined in Health and Safety Code Section 18909, shall not be construed as part of the provisions of this code. For nonbuilding standards, orders, and regulations, see other titles of the California Code of Regulations.*

1.1.7 Order of Precedence and Use.

1.1.7.1 Differences. *In the event of any differences between these building standards and the standard reference documents, the text of these building standards shall govern.*

1.1.7.2 Specific Provisions. *Where a specific provision varies from a general provision, the specific provision shall apply.*

1.1.7.3 Conflicts. *When the requirements of this code conflict with the requirements of any other part of the California Building Standards Code, Title 24, the most restrictive requirements shall prevail.*

1.1.7.3.1 Detached One- and Two-Family Dwellings. *Detached one-and two-family dwellings, efficiency dwelling units, lodging houses, live/work units, townhouses not more than three stories above grade plane in height with a separate means of egress, and their accessory structures, shall not be required to comply with the California Residential Code if constructed in accordance with the California Building Code.*

1.1.8 City, County, or City and County Amendments, Additions or Deletions. *The provisions of this code do not limit the authority of city, county, or city and county governments to establish more restrictive and reasonably necessary differences to the provisions contained in this code pursuant to complying with Section 1.1.8.1. The effective date of amendments, additions, or deletions to this code by a city, county, or city and county filed pursuant to Section 1.1.8.1 shall be the date filed. However, in no case shall the amendments, additions, or deletions to this code be effective any sooner than the effective date of this code.*

Local modifications shall comply with Health and Safety Code Section 18941.5 for Building Standards Law, Health and Safety Code Section 17958 for State Housing Law or Health and Safety Code Section 13869.7 for Fire Protection Districts.

1.1.8.1 Findings and Filings.

(1) The city, county, or city and county shall make express findings for each amendment, addition, or deletion based upon climatic, topographical, or geological conditions.

Exception: *Hazardous building ordinances and programs mitigating unreinforced masonry buildings.*

(2) The city, county, or city and county shall file the amendments, additions or deletions expressly marked and identified as to the applicable findings. Cities, counties, cities and counties, and fire departments shall file the amendments, additions, or deletions, and the findings with the California Building Standards Commission at 2525 Natomas Park Drive, Suite 130, Sacramento, CA 95833.

(3) Findings prepared by fire protection districts shall be ratified by the local city, county, or city and county and filed with the California Department of Housing and Community Development, Division of Codes and Standards, P.O. Box 278180, Sacramento, CA 95827-8180 or 9342 Tech Center Drive, Suite 500, Sacramento, CA, 95826.

1.1.8.2 Locally Adopted Energy Standards – California Energy Code, Part 6. *In addition to the provisions of Section 1.1.8.1 of this part, the provisions of this section apply to cities, counties, and cities and counties amending adopted energy standards affecting buildings and structures subject to the California Energy Code, Part 6.*

Applicable provisions of Public Resources Code Section 25402.1 and applicable provisions of Chapter 10 of the California Administrative Code, Part 1 apply to local amendment of energy standards adopted by the California Energy Commission.

1.1.9 Effective Date of This Code. *Only those standards approved by the California Building Standards Commission that are effective at the time an application for building permit is submitted shall apply to the plans and specifications for, and to the construction performed under, that permit. For the effective dates of the provisions contained in this code, see the History Note page of this code.*

Exception: *Plans approved by the Department of Housing and Community Development or a Department-approved design approval agency for factory built housing as defined by Health and Safety Code Section 19971. Approved plans, pursuant to the California Code of Regulations, Title 25, Division 1, Chapter 3, Subchapter 1, Article 3, Section 3048 remain valid for a period of 36 months from the date of plan approval.*

1.1.10 Availability of Codes. *At least one complete copy each of Titles 8, 19, 20, 24, and 25 with all revisions shall be maintained in the office of the building official responsible for the administration and enforcement of this code. Each state department concerned and each city, county, or city and county shall have an up-to-date copy of the code available for public inspection, see Health and Safety Code Section 18942(e)(1) and (2).*

1.1.11 Format. *This part fundamentally adopts the Uniform Mechanical Code by reference on a chapter-by-chapter basis. When a specific chapter of the Uniform Mechanical Code is not printed in the code and is marked "Reserved", such chapter of the Uniform Mechanical Code is not adopted as a portion of this code. When a specific chapter of the Uniform Mechanical Code is marked "Not adopted by the State of California" but appears in the code, it may be available for adoption by local ordinance.*

Note: *Matrix Adoption Tables at the front of each chapter may aid the code user in determining which chapter or sections within a chapter are applicable to buildings under the authority of a specific state agency, but they are not to be considered regulatory.*

1.1.12 Validity. *If any chapter, section, subsection, sentence, clause, or phrase of this code is for any reason held to be unconstitutional, contrary to statute, exceeding the authority of the state as stipulated by statutes or otherwise inoperative, such decision shall not affect the validity of the remaining portion of this code.*

1.2.0 Building Standards Commission.

1.2.1 BSC. *Specific scope of application of the agency responsible for enforcement, the enforcement agency, and the specific authority to adopt and enforce such provisions of this code, unless otherwise stated.*

(1) State Buildings for All Occupancies.

Application *– State buildings (all occupancies), including buildings constructed by the Trustees of the California State University (CSU) and the Regents of the University of California (UC) where no state agency has the authority to adopt building standards applicable to such buildings.*

Enforcing Agency *– State or local agency specified by the applicable provisions of law.*

Authority Cited *– Health and Safety Code Section 18934.5.*

Reference *– Health and Safety Code, Division 13, Part 2.5, commencing with Section 18901.*

(2) ***University of California, California State Universities, and California Community Colleges.***

Application *– Standards for lighting for parking lots and primary campus walkways at the University of California, California State Universities, and California Community Colleges.*

Enforcing Agency *– State or local agency specified by the applicable provisions of law.*

Authority Cited *– Government Code Section 14617.*

Reference *– Government Code Section 14617.*

(3) ***Existing State-Owned Buildings, including those owned by the University of California and by the California State University*** *– Building seismic retrofit standards including abating falling hazards of structural and nonstructural components and strengthening of building structures. See also Division of the State Architect.*

Enforcing Agency *– State or local agency specified by the applicable provisions of law.*

Authority Cited *– Government Code Section 16600.*

References *– Government Code Sections 16600 through 16604.*

(4) ***Unreinforced Masonry Bearing Wall Buildings.***

Application *– Minimum seismic strengthening standards for buildings specified in the California Existing Building Code, except for buildings subject to building standards adopted pursuant to Part 1.5 (commencing with Section 17910).*

Enforcing Agency *– State or local agency specified by the applicable provisions of law.*

Authority Cited *– Health and Safety Code Section 18934.7.*

References *– Health and Safety Code Division 13, Part 2.5 commencing with Sections 18901.*

1.2.1.1 State Building. *For purposes of this code, a "state building" is a structure for which a state agency or state entity has authority to construct, alter, enlarge, replace, repair or demolish.*

1.2.1.2 Enforcement. *[CSU, UC, Judicial Council and California Department of Corrections Rehabilitation] state agencies or state entities authorized to construct state buildings may appoint a building official who is responsible to the agency for enforcement of the provisions of the California Building Standards Code.*

Exception: *State buildings regulated by other sections of this code remain the enforcement responsibility of the designated entities.*

1.2.1.3 Enforcement, Reserved for DGS.

1.2.1.4 Adopting Agency Identification. *The provisions of this code applicable to buildings identified in this section will be identified in the Matrix Adoption Tables under the acronym BSC.*

1.2.2 BSC-CG. *Specific scope of application of the agency responsible for enforcement, the enforcement agency, and the specific authority to adopt and enforce such provisions of this code, unless otherwise stated.*

1. ***Green Building Standards for Nonresidential Occupancies.***

Application *– All occupancies where no other state agency has the authority to adopt green building standards applicable to those occupancies.*

Enforcing Agency *– State or local agency specified by the applicable provisions of law.*

Authority Cited *– Health and Safety Code Sections 18930.5, 18938 and 18940.5.*

Reference *– Health and Safety Code, Division 13, Part 2.5, commencing with Section 18901.*

2. ***Graywater Systems for Nonresidential Occupancies.***

Application *– The construction, installation, and alteration of graywater systems for indoor and outdoor uses in nonresidential occupancies.*

Enforcing Agency *– State or local agency specified by the applicable provisions of law.*

Authority Cited *– Health & Safety Code Section 18941.8.*

Reference *– Health & Safety Code Section 18941.8.*

1.2.3 Alternative Materials, Design, and Methods of Construction and Equipment. *The provisions of this code are not intended to prevent the installation of any material or to prohibit any design or method of construction not specifically prescribed by this code, provided that any such alternative has been approved. An alternative material, design, or method of construction shall be approved where the building official finds that the proposed design is satisfactory and complies with the intent of the provisions of this code, and that the material, method or work offered is, for the purpose intended, at least the equivalent of that prescribed in this code in quality, strength, effectiveness, fire resistance, durability and safety.*

1.2.3.1 Research Reports. *Supporting data, where necessary to assist in the approval of materials or assemblies not specifically provided for in this code, shall consist of valid research reports from approved sources.*

1.2.3.2 Tests. *Whenever there is insufficient evidence of compliance with the provisions of this code, or evidence that a material or method does not conform to the requirements of this code, or in order to substantiate claims for alternative materials or methods, the building official shall have the authority to require tests as evidence of compliance to be made at no expense to the jurisdiction. Test methods shall be as specified in this code or by other recognized test standards. In the absence of recognized and accepted test methods, the building official shall approve the testing procedures. Tests shall be performed by an approved agency. Reports of such tests shall be retained by the building official for the period required for retention of public records.*

1.3.0 Reserved for Corrections Standards Authority.

1.4.0 Department of Consumer Affairs.

1.4.1 *Specific scope of application of the agency responsible for enforcement, the enforcement agency, and the specific authority to adopt and enforce such provisions of this code, unless otherwise stated.*

Board of Pharmacy.

Application *– Pharmacies.*

Enforcing Agency *– State or local agency specified by applicable provisions of law.*

Authority Cited *– Business and Professions Code Section 4005.*

References *– Business and Professions Code Sections 4005, 4127.7, and 4201.*

1.4.2 Adopting Agency Identification. *The provisions of this code applicable to buildings identified in this section will be identified in the Matrix Adoption Tables under the acronym CA.*

1.5.0 Reserved for California Energy Commission.

1.6.0 Reserved for Department of Food and Agriculture.

1.7.0 California Department of Public Health.

1.7.1 *Specific scope of application of the agency responsible for enforcement, the enforcement agency, and the specific authority to adopt and enforce such provisions of this code, unless otherwise stated.*

Department of Public Health.

Application *– Food establishments.*

Enforcing Agency *– Department of Health Services and the local health agency.*

Authority Cited *– Health and Safety Code Sections 110065, 113710, and 131200.*

References *– Health and Safety Code Sections 110065, 113705, 114140, and 131200.*

1.7.2 Adopting Agency Identification. *The provisions of this code applicable to buildings identified in this section will be identified in the Matrix Adoption Tables under the acronym DPH.*

1.8.0 Department of Housing and Community Development [HCD].

1.8.1 Purpose. *The purpose of this code is to establish the minimum requirements necessary to protect the health, safety, and general welfare of the occupants and the public by governing the erection, construction, reconstruction, enlargement, conversion, alteration, repair, moving, removal, demolition, ventilation, and maintenance or use of heating, ventilating, cooling, refrigeration systems, incinerators, and other miscellaneous heat-producing appliances.*

1.8.2 Authority and Abbreviations.

1.8.2.1 General. *The Department of Housing and Community Development is authorized by law to promulgate and adopt building standards and regulations for several types of building applications. The applications under the authority of the Department of Housing and Community Development are listed in Sections 1.8.2.1.1 through 1.8.2.1.3.*

1.8.2.1.1 Housing Construction.

Application *– Hotels, motels, lodging houses, apartments, dwellings, dormitories, condominiums, shelters for homeless persons, congregate residences, employee housing, factory-built housing, and other types of dwellings containing sleeping accommodations with or without common toilet or cooking facilities including accessory buildings, facilities, and uses thereto. Sections of this code which pertain to applications listed in this section are identified using the abbreviation "HCD 1".*

Enforcing Agency *– Local building department or the Department of Housing and Community Development.*

Authority Cited *– Health and Safety Code Sections 17040, 17050, 17920.9, 17921, 17921.5, 17921.6, 17921.10, 17922, 17922.6, 17922.12, 17922.14, 17926, 17927, 17928, 18300, 18552, 18554, 18620, 18630, 18640, 18670, 18690, 18691, 18865, 18871.3, 18871.4, 18873, 18873.1 through 18873.5, 18938.3, 18944.11, and 19990; and Government Code Section 12955.1.*

References *– Health and Safety Code Sections 17000 through 17062.5, 17910 through 17995.5, 18200 through 18700, 18860 through 18874, and 19960 through 19997; Civil Code Sections 1101.4, 1101.5 and 1954.201; and Government Code Sections 12955.1 and 12955.1.1.*

1.8.2.1.2 Housing Accessibility.

Application *– Covered multifamily dwellings as defined in Chapter 2 of the California Code of Regulations, Title 24, Part 2, also known as the California Building Code including, but not limited to, lodging houses, dormitories, timeshares, condominiums, shelters for homeless persons, congregate residences, apartments, dwellings, employee housing, factory-built housing, and other types of dwellings containing sleeping accommodations with or without common toilet or cooking facilities.*

Sections of this code identified by the abbreviation "HCD 1-AC" require specific accommodations for persons with disabilities as defined in Chapter 2 of the California Building Code. The application of such provisions shall be in conjunction with other requirements of this code and apply only to newly-constructed covered multifamily dwellings as defined in Chapter 2 of the California Building Code. "HCD 1-AC" applications include, but are not limited to, the following:

(1) All newly-constructed covered multifamily dwellings as defined in Chapter 2 of the California Building Code.

(2) New common use areas as defined in Chapter 2 of the California Building Code serving existing covered multifamily dwellings.

(3) Additions to existing buildings, where the addition alone meets the definition of covered multifamily dwellings as defined in Chapter 2 of the California Building Code.

(4) New common use areas serving new covered multifamily dwellings.

(5) Where any portion of a building's exterior is preserved, but the interior of the building is removed, including all structural portions of floors and ceilings, the building is considered a new building for determining the application of California Building Code, Chapter 11A.

"HCD 1-AC" building standards generally do not apply to public use areas or public accommodations such as hotels and motels, and public housing. Public use areas, public accommodations, and public housing, as defined in Chapter 2 of the California Building Code, are subject to the Division of the State Architect (DSA-AC) in Chapter 11B, and are referenced in Section 1.9.1.

Newly constructed covered multifamily dwellings, which can also be defined as public housing, shall be subject to the requirements of Chapter 11A and Chapter 11B.

***Enforcing Agency** – Local building department or the Department of Housing and Community Development.*

***Authority Cited** – Health and Safety Code Sections 17040, 17050, 17920.9, 17921, 17921.5, 17921.6, 17921.10, 17922, 17922.6, 17922.12, 17922.14, 17926, 17927, 17928, 18300, 18552, 18554, 18620, 18630, 18640, 18670, 18690, 18691, 18865, 18871.3, 18871.4, 18873, 18873.1 through 18873.5, 18938.3, 18944.11, and 19990; and Government Code Section 12955.1.*

***References** – Health and Safety Code Sections 17000 through 17062.5, 17910 through 17995.5, 18200 through 18700, 18860 through 18874, and 19960 through 19997; Civil Code Sections 1101.4, 1101.5 and 1954.201; and Government Code Sections 12955.1 and 12955.1.1.*

1.8.2.1.3 Permanent Buildings in Mobilehome Parks and Special Occupancy Parks.

***Application** – Permanent buildings, and permanent accessory buildings or structures, constructed within mobilehome parks and special occupancy parks that are under the control and ownership of the park operator. Sections of this code which pertain to applications listed in this section are identified using the abbreviation "HCD 2".*

***Enforcing Agency** – The Department of Housing and Community Development, local building department or other local agency that has assumed responsibility for the enforcement of Health and Safety Code, Division 13, Part 2.1, commencing with Section 18200 for mobilehome parks and Health and Safety Code, Division 13, Part 2.3, commencing with Section 18860 for special occupancy parks.*

***Authority Cited** – Health and Safety Code Sections 17040, 17050, 17920.9, 17921, 17921.5, 17921.6, 17921.10, 17922, 17922.6, 17922.12, 17922.14, 17926, 17927, 17928, 18300, 18552, 18554, 18620, 18630, 18640, 18670, 18690, 18691, 18865, 18871.3, 18871.4, 18873, 18873.1 through 18873.5, 18938.3, 18944.11, and 19990; and Government Code Section 12955.1.*

***References** – Health and Safety Code Sections 17000 through 17062.5, 17910 through 17995.5, 18200 through 18700, 18860 through 18874, and 19960 through 19997; Civil Code Sections 1101.4, 1101.5 and 1954.201; and Government Code Sections 12955.1 and 12955.1.1.*

1.8.3 Local Enforcing Agency.

***1.8.3.1 Duties and Powers.** The building department of every city, county, or city and county shall enforce all the provisions of law, this code, and the other rules and regulations promulgated by the Department of Housing and Community Development pertaining to the installation, erection, construction, reconstruction, movement, enlargement, conversion, alteration, repair, removal, demolition, or arrangement of apartments, condominiums, hotels, motels, lodging houses, and dwellings, including accessory buildings, facilities, and uses thereto.*

The provisions regulating the erection and construction of dwellings and appurtenant structures shall not apply to existing structures as to which construction is commenced or approved prior to the effective date of these regulations. Requirements relating to use, maintenance, and occupancy shall apply to all dwellings and appurtenant structures approved for construction or constructed before or after the effective date of this code.

For additional information regarding the use and occupancy of existing buildings and appurtenant structures, see California Code of Regulations, Title 25, Division 1, Chapter 1, Subchapter 1, commencing with Section 1.

***1.8.3.2 Laws, Rules, and Regulations.** Other than the building standards contained in this code, and notwithstanding other provisions of law, the statutory authority and location of the laws, rules, and regulations to be enforced by local enforcing agencies are listed by statute in Sections 1.8.3.2.1 through 1.8.3.2.5 below:*

***1.8.3.2.1 State Housing Law.** Refer to the State Housing Law, California Health and Safety Code, Division 13, Part 1.5, commencing with Section 17910, and California Code of Regulations, Title 25, Division 1, Chapter 1, Subchapter 1, commencing* *with Section 1, for the erection, construction, reconstruction, movement, enlargement, conversion, alteration, repair, removal, demolition, or arrangement of apartments, condominiums, hotels, motels, lodging houses, and dwellings, including accessory buildings, facilities, and uses thereto.*

1.8.3.2.2 Mobilehome Parks Act. *Refer to the Mobilehome Parks Act, California Health and Safety Code, Division 13, Part 2.1, commencing with Section 18200 and California Code of Regulations, Title 25, Division 1, Chapter 2, commencing with Section 1000 for mobilehome park administrative and enforcement authority, permits, plans, fees, violations, inspections, and penalties both within and outside mobilehome parks.*

Exception: *Mobilehome parks where the Department of Housing and Community Development is the enforcing agency.*

1.8.3.2.3 Special Occupancy Parks Act. *Refer to the Special Occupancy Parks Act, California Health and Safety Code, Division 13, Part 2.3, commencing with Section 18860 and California Code of Regulations, Title 25, Division 1, Chapter 2.2, commencing with Section 2000 for special occupancy park administrative and enforcement authority, permits, fees, violations, inspections, and penalties both within and outside of special occupancy parks.*

Exception: *Special occupancy parks where the Department of Housing and Community Development is the enforcing agency.*

1.8.3.2.4 Employee Housing Act. *Refer to the Employee Housing Act, California Health and Safety Code, Division 13, Part 1, commencing with Section 17000 and California Code of Regulations, Title 25, Division 1, Chapter 1, Subchapter 3, commencing with Section 600 for employee housing administrative and enforcement authority, permits, fees, violations, inspections, and penalties.*

1.8.3.2.5 Factory-Built Housing Law. *Refer to the Factory-Built Housing Law, California Health and Safety Code, Division 13, Part 6, commencing with Section 19960 and California Code of Regulations, Title 25, Division 1, Chapter 3, Subchapter 1, commencing with Section 3000 for factory-built housing administrative and enforcement authority, permits, fees, violations, inspections, and penalties.*

1.8.4 Permits, Fees, Applications, and Inspections.

1.8.4.1 Permits. *A written construction permit shall be obtained from the enforcing agency prior to the erection, construction, reconstruction, installation, relocation, or alteration of any mechanical system.*

Exceptions:

(1) Work exempt from permits as specified in Chapter 1, Administration, Division II, Section 104.2 Items (1) through (5) of this code.

(2) Changes, alterations, or repairs of a minor nature not affecting structural features, egress, sanitation, safety, or accessibility as determined by the enforcing agency.

Exemptions from permit requirements shall not be deemed to grant authorization for any work to be done in any manner in violation of other provisions of law or this code.

1.8.4.2 Fees. *Subject to other provisions of law, the governing body of any city, county, or city and county may prescribe fees to defray the cost of enforcement of rules and regulations promulgated by the Department of Housing and Community Development. The amount of the fees shall not exceed the amount reasonably necessary to administer or process permits, certificates, forms, or other documents, or to defray the costs of enforcement. For additional information, see State Housing Law, Health and Safety Code, Division 13, Part 1.5, Section 17951 and California Code of Regulations, Title 25, Division 1, Chapter 1, Subchapter 1, Article 3, commencing with Section 6.*

1.8.4.3 Plan Review and Time Limitations. *Subject to other provisions of law, provisions related to plan checking, prohibition of excessive delays, and contracting with or employment of private parties to perform plan checking are set forth in the State Housing Law, Health and Safety Code Section 17960.1, and for employee housing, in Health and Safety Code Section 17021.*

1.8.4.3.1 Retention of Plans. *The building department of every city, county, or city and county shall maintain an official copy, microfilm, or electronic or other type of photographic copy of the plans of every building, during the life of the building, for which the department issued a building permit.*

Exceptions:

(1) Single or multiple dwellings not more than two stories and basement in height.

(2) Garages and other structures appurtenant to buildings listed in Exception 1.

(3) Farm or ranch buildings appurtenant to buildings listed in Exception 1.

(4) Any one-story building where the span between bearing walls does not exceed 25 feet (7620 mm), except a steel frame or concrete building.

All plans for common interest developments as defined in Section 4100 of the California Civil Code shall be retained. For additional information regarding plan retention and reproduction of plans by an enforcing agency, see Health and Safety Code Sections 19850 through 19852.

1.8.4.4 Inspections. *Construction or work for which a permit is required shall be subject to inspection by the building official and such construction or work shall remain accessible and exposed for inspection purposes until approved. Approval as a result of an inspection shall not be construed to be an approval of a violation of the provisions of this code or other regulations of the Department of Housing and Community Development.*

1.8.5 Right of Entry for Enforcement.

1.8.5.1 General. *Subject to other provisions of law, officers and agents of the enforcing agency may enter and inspect public and private properties to secure compliance with the rules and regulations promulgated by the Department of Housing and Community Development. For limitations and additional information regarding enforcement, see the following:*

(1) For applications subject to State Housing Law as referenced in Section 1.8.3.2.1 of this code, refer to Health and Safety Code, Division 13, Part 1.5, commencing with Section 17910 and California Code of Regulations, Title 25, Division 1, Chapter 1, Subchapter 1, commencing with Section 1.

(2) For applications subject to the Mobilehome Parks Act as referenced in Section 1.8.3.2.2 of this code, refer to Health and Safety Code, Division 13, Part 2.1, commencing with Section 18200, and California Code of Regulations, Title 25, Division 1, Chapter 2, commencing with Section 1000.

(3) For applications subject to the Special Occupancy Parks Act as referenced in Section 1.8.3.2.3 of this code, refer to Health and Safety Code, Division 13, Part 2.3, commencing with Section 18860, and California Code of Regulations, Title 25, Division 1, Chapter 2.2, commencing with Section 2000.

(4) For applications subject to the Employee Housing Act as referenced in Section 1.8.3.2.4 of this code, refer to Health and Safety Code, Division 13, Part 1, commencing with Section 17000 and California Code of Regulations, Title 25, Division 1, Chapter 1, Subchapter 3, commencing with Section 600.

(5) For applications subject to the Factory-Built Housing Law as referenced in Section 1.8.3.2.5 of this code, refer to Health and Safety Code, Division 13, Part 6, commencing with Section 19960 and California Code of Regulations, Title 25, Division 1, Chapter 3, Subchapter 1, commencing with Section 3000.

1.8.6 Local Modification by Ordinance or Regulation.

1.8.6.1 General. *Subject to other provisions of law, a city, county, or city and county may make changes to the provisions adopted by the Department of Housing and Community Development. If any city, county, or city and county does not amend, add, or repeal by local ordinances or regulations the provisions published in this code or other regulations promulgated by the Department of Housing and Community Development, those provisions shall be applicable and shall become effective 180 days after publication by the California Building Standards Commission. Amendments, additions, and deletions to this code adopted by a city, county, or city and county pursuant to California Health and Safety Code Sections 17958.5, 17958.7 and 18941.5, together with all applicable portions of this code, shall also become effective 180 days after publication of the California Building Standards Code by the California Building Standards Commission.*

1.8.6.2 Findings, Filings, and Rejections of Local Modifications. *Prior to making any modifications or establishing more restrictive building standards, the governing body shall make express findings and filings, as required by California Health and Safety Code Section 17958.7, showing that such modifications are reasonably necessary due to local climatic, geological, or topographical conditions. No modification shall become effective or operative unless the following requirements are met:*

(1) The express findings shall be made available as a public record.

(2) A copy of the modification and express finding, each document marked to cross-reference the other, shall be filed with the California Building Standards Commission for a city, county, or city and county, and with the Department of Housing and Community Development for fire protection districts.

(3) The California Building Standards Commission has not rejected the modification or change.

Nothing in this section shall limit the authority of fire protection districts pursuant to California Health and Safety Code Section 13869.7(a).

1.8.7 Alternate Materials, Designs, Tests, and Methods of Construction.

1.8.7.1 General. *The provisions of this code as adopted by the Department of Housing and Community Development are not intended to prevent the use of any alternate material, appliance, installation, device, arrangement, design, or method of construction not specifically prescribed by this code. Consideration and approval of alternates shall comply with Section 1.8.7.2 for local building departments and Section 1.8.7.3 for the Department of Housing and Community Development.*

1.8.7.2 Local Building Departments. *The building department of any city, county, or city and county may approve alternates for use in the erection, construction, reconstruction, movement, enlargement, conversion, alteration, repair, removal, demolition, or arrangement of apartments, condominiums, hotels, motels, lodging houses, dwellings, or accessory structures, except for the following:*

(1) Structures located in mobilehome parks as defined in California Health and Safety Code Section 18214.

(2) Structures located in special occupancy parks as defined in California Health and Safety Code Section 18862.43.

(3) Factory-built housing as defined in California Health and Safety Code Section 19971.

1.8.7.2.1 Approval of Alternates. *The consideration and approval of alternates by a local building department shall comply with the following procedures and limitations:*

(1) The approval shall be granted on a case-by-case basis.

(2) *Evidence shall be submitted to substantiate claims that the proposed alternate, in performance, safety, and protection of life and health, conforms to, or is at least equivalent to, the standards contained in this code and other rules and regulations promulgated by the Department of Housing and Community Development.*

(3) *The local building department may require tests performed by an approved testing agency at the expense of the owner or owner's agent as proof of compliance.*

(4) *If the proposed alternate is related to accessibility in covered multifamily dwellings or facilities serving covered multifamily dwellings as defined in Chapter 2 of the California Building Code, the proposed alternate must also meet the threshold set for equivalent facilitation as defined in Chapter 2 of the California Building Code.*

For additional information regarding approval of alternates by a local building department pursuant to the State Housing Law, see California Health and Safety Code Section 17951(e) and California Code of Regulations, Title 25, Division 1, Chapter 1, Subchapter 1.

1.8.7.3 Department of Housing and Community Development. *The Department of Housing and Community Development may approve alternates for use in the erection, construction, reconstruction, movement, enlargement, conversion, alteration, repair, removal, or demolition of apartments, condominiums, hotels, motels, lodging houses, dwellings, or accessory structures thereto, and permanent buildings in mobilehome parks and special occupancy parks. The consideration and approval of alternates shall comply with the following:*

(1) *The department may require tests at the expense of the owner or owner's agent to substantiate compliance with the California Building Standards Code.*

(2) *The approved alternate shall, for its intended purpose, be at least equivalent in performance and safety to the materials, designs, tests, or methods of construction prescribed by this code.*

1.8.8 Appeals Board.

1.8.8.1 General. *Every city, county, or city and county shall establish a process to hear and decide appeals of orders, decisions, and determinations made by the enforcing agency relative to the application and interpretation of this code and other regulations governing construction, use, maintenance and change of occupancy. The governing body of any city, county, or city and county may establish a local appeals board and a housing appeals board to serve this purpose. Members of the appeals board(s) shall not be employees of the enforcing agency and shall be knowledgeable in the applicable building codes, regulations and ordinances as determined by the governing body of the city, county, or city and county.*

Where no such appeals boards or agencies have been established, the governing body of the city, county, or city and county shall serve as the local appeals board or housing appeals board as specified in California Health and Safety Code Sections 17920.5 and 17920.6.

1.8.8.2 Definitions. *The following terms shall for the purposes of this section have the meaning shown.*

Housing Appeals Board. *The board or agency of a city, county, or city and county which is authorized by the governing body of the city, county, or city and county to hear appeals regarding the requirements of the city, county, or city and county relating to the use, maintenance, and change of occupancy of buildings and structures, including requirements governing alteration, additions, repair, demolition, and moving. In any area in which there is no such board or agency, "Housing Appeals Board" means the local appeals board having jurisdiction over the area.*

Local Appeals Board. *The board or agency of a city, county, or city and county which is authorized by the governing body of the city, county, or city and county to hear appeals regarding the building requirements of the city, county, or city and county. In any area in which there is no such board or agency, "Local Appeals Board" means the governing body of the city, county, or city and county having jurisdiction over the area.*

1.8.8.3 Appeals. *Except as otherwise provided by law, any person, firm, or corporation adversely affected by a decision, order, or determination by a city, county, or city and county relating to the application of building standards published in the California Building Standards Code, or any other applicable rule or regulation adopted by the Department of Housing and Community Development, or any lawfully enacted ordinance by a city, county, or city and county, may appeal the issue for resolution to the local appeals board or housing appeals board as appropriate.*

The local appeals board shall hear appeals relating to new building construction and the housing appeals board shall hear appeals relating to existing buildings.

1.8.9 Unsafe Buildings or Structures.

1.8.9.1 Authority to Enforce. *Subject to other provisions of law, the administration, enforcement, actions, proceedings, abatement, violations, and penalties for unsafe buildings and structures are contained in the following statutes and regulations:*

(1) *For applications subject to the State Housing Law as referenced in Section 1.8.3.2.1 of this code, refer to Health and Safety Code, Division 13, Part 1.5, commencing with Section 17910 and California Code of Regulations, Title 25, Division 1, Chapter 1, Subchapter 1, commencing with Section 1.*

(2) *For applications subject to the Mobilehome Parks Act as referenced in Section 1.8.3.2.2 of this code, refer to Health and Safety Code, Division 13, Part 2.1, commencing with Section 18200 and California Code of Regulations, Title 25, Division 1, Chapter 2, commencing with Section 1000.*

(3) *For applications subject to the Special Occupancy Parks Act as referenced in Section 1.8.3.2.3 of this*

code, refer to Health and Safety Code, Division 13, Part 2.3, commencing with Section 18860 and California Code of Regulations, Title 25, Division 1, Chapter 2.2, commencing with Section 2000.

(4) For applications subject to the Employee Housing Act as referenced in Section 1.8.3.2.4 of this code, refer to Health and Safety Code, Division 13, Part 1, commencing with Section 17000 and California Code of Regulations, Title 25, Division 1, Chapter 1, Subchapter 3, commencing with Section 600.

(5) For applications subject to the Factory-Built Housing Law as referenced in Section 1.8.3.2.5 of this code, refer to Health and Safety Code, Division 13, Part 6, commencing with Section 19960 and California Code of Regulations, Title 25, Division 1, Chapter 3, Subchapter 1, commencing with Section 3000.

***1.8.9.2 Actions and Proceedings.** Subject to other provisions of law, punishments, penalties, and fines for violations of building standards are contained in the following statutes and regulations:*

(1) For applications subject to the State Housing Law as referenced in Section 1.8.3.2.1 of this code, refer to Health and Safety Code, Division 13, Part 1.5, commencing with Section 17910 and California Code of Regulations, Title 25, Division 1, Chapter 1, Subchapter 1, commencing with Section 1.

(2) For applications subject to the Mobilehome Parks Act as referenced in Section 1.8.3.2.2 of this code, refer to Health and Safety Code, Division 13, Part 2.1, commencing with Section 18200 and California Code of Regulations, Title 25, Division 1, Chapter 2, commencing with Section 1000.

(3) For applications subject to the Special Occupancy Parks Act as referenced in Section 1.8.3.2.3 of this code, refer to Health and Safety Code, Division 13, Part 2.3, commencing with Section 18860 and California Code of Regulations, Title 25, Division 1, Chapter 2.2, commencing with Section 2000.

(4) For applications subject to the Employee Housing Act as referenced in Section 1.8.3.2.4 of this code, refer to Health and Safety Code, Division 13, Part 1, commencing with Section 17000 and California Code of Regulations, Title 25, Division 1, Chapter 1, Subchapter 3, commencing with Section 600.

(5) For applications subject to the Factory-Built Housing Law as referenced in Section 1.8.3.2.5 of this code, refer to Health and Safety Code, Division 13, Part 6, commencing with Section 19960 and California Code of Regulations, Title 25, Division 1, Chapter 3, Subchapter 1, commencing with Section 3000.

1.8.10 Other Building Regulations.

***1.8.10.1 Existing Structures.** Notwithstanding other provisions of law, the replacement, retention, and extension of original materials and the use of original methods of construction for any existing building or accessory structure, or portions thereof, shall be permitted in accordance with the provisions of this code and the California Existing Building Code, as adopted by the Department of Housing and Community Development. For additional information, see California Health and Safety Code, Sections 17912, 17920.3, 17922 and 17958.8.*

***1.8.10.2 Moved Structures.** Subject to the requirements of California Health and Safety Code Sections 17922, 17922.3 and 17958.9, local ordinances or regulations relating to a moved residential building or accessory structure thereto, shall permit the replacement, retention, and extension of original materials and the use of original methods of construction so long as the structure does not become or continue to be a substandard building.*

1.9.0 Division of the State Architect.

1.9.1 Reserved for the Division of the State Architect – Access Compliance.

1.9.2 Division of the State Architect – Structural Safety.

1.9.2.1 DSA-SS [Division of the State Architect – Structural Safety].

***Application** – Public elementary and secondary schools, community college buildings, and state-owned or state-leased essential services buildings.*

***Enforcing Agency** – Division of the State Architect – Structural Safety (DSA-SS).*

The Division of the State Architect has been delegated the responsibility and authority by the Department of General Services to review and approve the design and oversee and observe the construction of public elementary and secondary schools, community colleges, and state-owned or state-leased essential services buildings.

***Authority Cited** – Education Code Sections 17310 and 81142, and Health & Safety Code Section 16022.*

***References** – Education Code Sections 17280 through 17317 and 81130 through 81147, and Health & Safety Code Sections 16000 through 16023.*

***1.9.2.1.1 Adopting Agency Identification.** The provisions of this code applicable to buildings identified in this Subsection 1.9.2.1 will be identified in the Matrix Adoption Tables under the acronym DSA-SS.*

1.9.2.2 DSA-SS/CC [Division of the State Architect – Structural Safety/Community Colleges].

***Application** – Community Colleges.*

The Division of the State Architect has been delegated the authority by the Department of General Services to promulgate alternate building standards for application to community colleges, which a community college may elect to use in lieu of standards promulgated by DSA-SS, in accordance with Section 1.9.2.1. Refer to Title 24, Part 2, Section 1.9.2.2.

***Enforcing Agency** – Division of the State Architect – Structural Safety/Community Colleges (DSA-SS/CC).*

The Division of the State Architect has been delegated the authority by the Department of General Services to review and approve the design and oversee and observe the construction of community colleges electing to use the alternative building standards as provided in this section.

Authority Cited *– Education Code Section 81053.*

References *– Education Code Sections 81052, 81053, and 81130 through 81147.*

1.9.2.2.1 Adopting Agency Identification. *The provisions of this code applicable to buildings identified in this Subsection 1.9.2.2 will be identified in the Matrix Adoption Tables under the acronym DSA-SS/CC.*

1.10.0 Office of Statewide Health Planning and Development.

1.10.1 OSHPD 1 and OSHPD 1R. *Specific scope of application of the agency responsible for enforcement, enforcement agency, specific authority to adopt and enforce such provisions of this code, unless otherwise stated.*

OSHPD 1 and OSHPD 1R

Application *– **[OSHPD 1]** General acute-care hospital buildings. **[OSHPD 1R]** Non-conforming hospital buildings that have been removed from acute care service.*

Enforcing Agency *– Office of Statewide Health Planning and Development (OSHPD). The office shall enforce the Division of the State Architect access compliance regulations and the regulations of the Office of the State Fire Marshal for the above stated facility types.*

1.10.1.1 Applicable Administrative Standards:

(1) Title 24, Part 1, California Code of Regulations: Chapters 6 and 7.

(2) Title 24, Part 2, California Code of Regulations: Sections 1.1.0 and 1.10.0, Chapter 1, Division I, and as indicated in the adoption matrix for Chapter 1, Division II.

1.10.1.2 Applicable Building Standards. *California Building Standards Code, Title 24, Parts 2, 3, 4, 5, 6, 9, 10 and 11.*

Authority Cited *– Health and Safety Code Sections 127010, 127015, 1275, and 129850.*

References *– Health and Safety Code Sections 19958, 127010, 127015, 129680, 1275, and 129675 through 130070.*

1.10.1.3 Adopting Agency Identification. *The provisions of this code applicable to buildings identified in this Subsection 1.10.1 will be identified in the Matrix Adoption Tables under the acronym OSHPD 1, and OSHPD 1R.*

1.10.2 OSHPD 2. *Specific scope of application of the agency responsible for enforcement, enforcement agency, specific authority to adopt and enforce such provisions of this code, unless otherwise stated.*

OSHPD 2

Application *– Skilled nursing facilities and intermediate – care facility buildings.*

Enforcing Agency *– Office of Statewide Health Planning and Development (OSHPD). The office shall also enforce the Division of the State Architect access compliance regulations and the regulations of the Office of the State Fire Marshal for the above stated facility type.*

1.10.2.1 Applicable Administrative Standards:

(1) Title 24, Part 1, California Code of Regulations: Chapter 7.

(2) Title 24, Part 2, California Code of Regulations: Sections 1.1.0 and 1.10.0, Chapter 1, Division I, and as indicated in the adoption matrix for Chapter 1, Division II.

1.10.2.2 Applicable Building Standards. *California Building Standards Code, Title 24, Parts 2, 3, 4, 5, 6, 9, 10, and 11.*

Authority Cited *– Health and Safety Code Sections 127010, 127015, 1275, and 129850.*

References *– Health and Safety Code Sections 127010, 127015, 1275, and 129680.*

1.10.2.3 Adopting Agency Identification. *The provisions of this code applicable to buildings identified in this Subsection 1.10.2 will be identified in the Matrix Adoption Tables under the acronym OSHPD 2.*

1.10.3 OSHPD 3. *Specific scope of application of the agency responsible for enforcement, enforcement agency, specific authority to adopt and enforce such provisions of this code, unless otherwise stated.*

OSHPD 3

Application *– Licensed clinics and any freestanding building under a hospital license where outpatient clinical services are provided.*

Enforcing Agency *– Local building department.*

1.10.3.1 Applicable Administrative Standards.

(1) Title 24, Part 1, California Code of Regulations: Chapter 7.

(2) Title 24, Part 2, California Code of Regulations: Sections 1.1.0 and 1.10.0, Chapter 1, Division I, and as indicated in the adoption matrix for Chapter 1, Division II.

1.10.3.2 Applicable Building Standards. *California Building Standards Code, Title 24, Parts 2, 3, 4, 5, 6, 9, 10, and 11.*

Authority Cited *– Health and Safety Code Sections 127010, 127015, and 1226.*

References *– Health and Safety Code Sections 127010, 127015, 129885, and 1226, Government Code Section 54350, and State Constitution Article 11 Section 7.*

1.10.3.3 Adopting Agency Identification. *The provisions of this code applicable to buildings identified in this Subsection 1.10.3 will be identified in the Matrix Adoption Tables under the acronym OSHPD 3.*

1.10.4 OSHPD 4. *Specific scope of application of the agency responsible for enforcement, enforcement agency, specific authority to adopt and enforce such provisions of this code, unless otherwise stated.*

OSHPD 4

Application *– Correctional Treatment Centers.*

Enforcing Agency *– Office of Statewide Health Planning and Development (OSHPD). The Office shall also enforce the Division of the State Architect access compliance regulations and the regulations of the Office of the State Fire Marshal for the above stated facility types.*

1.10.4.1 Applicable Administrative Standards:

(1) Title 24, Part 1, California Code of Regulations: Chapter 7.

(2) Title 24, Part 2, California Code of Regulations: Sections 1.1.0 and 1.10.0, Chapter 1, Division I, and as indicated in the adoption matrix for Chapter 1, Division II.

1.10.4.2 Applicable Building Standards. *California Building Standards Code, Title 24, Parts 2, 3, 4, 5, 6, 9, 10, and 11.*

Authority Cited *– Health and Safety Code Sections 127010, 127015, and 129790.*

References *– Health and Safety Code Sections 127010, 127015, 1275, and 129675 through 130070.*

1.10.4.3 Adopting Agency Identification. *The provisions of this code applicable to buildings identified in this Subsection 1.10.4 will be identified in the Matrix Adoption Tables under the acronym OSHPD 4.*

1.10.5 OSHPD 5. *Specific scope of application of the agency responsible for enforcement, enforcement agency and the specific authority to adopt and enforce such provisions of this code, unless otherwise stated.*

OSHPD 5

Application *– Acute psychiatric hospital buildings.*

Enforcing Agency *– Office of Statewide Health Planning and Development (OSHPD). The office shall also enforce the Division of the State Architect – Access Compliance regulations and the regulations of the Office of the State Fire Marshal for the above-stated facility types.*

1.10.5.1 Applicable Administrative Standards.

(1) Title 24, Part 1, California Code of Regulations: Chapter 7.

(2) Title 24, Part 2, California Code of Regulations: Sections 1.1 and 1.10, Chapter 1, Division I, and as indicated in the adoption matrix for Chapter 1, Division II.

1.10.5.2 Applicable Building Standards. *California Building Standards Code, Title 24, Parts 2, 3, 4, 5, 6, 9, 10 and 11.*

Authority Cited *– Health and Safety Code Sections 1275 and 129850.*

References – Health and Safety Code Sections 129680, 1275 and 129675 through 130070.

1.10.5.3 Adopting Agency Identification. *The provisions of this code applicable to buildings identified in this Subsection 1.10.5 will be identified in the Matrix Adoption Tables under the Acronym OSHPD 5.*

1.11.0 Office of the State Fire Marshal.

1.11.1 SFM – Office of the State Fire Marshal. *Specific scope of application of the agency responsible for enforcement, the enforcement agency and the specific authority to adopt and enforce such provisions of this code, unless otherwise stated.*

Application:

Institutional, Educational, or any Similar Occupancy. *Any building or structure used or intended for use as an asylum, jail, mental hospital, hospital, sanitarium, home for the aged, children's nursery, children's home, school, or any similar occupancy of any capacity.*

Authority Cited *– Health and Safety Code Section 13143.*

Reference *– Health and Safety Code Section 13143.*

Assembly or Similar Place of Assemblage. *Any theater, dancehall, skating rink, auditorium, assembly hall, meeting hall, nightclub, fair building or similar place of assemblage where 50 or more persons may gather together in a building, room or structure for the purpose of amusement, entertainment, instruction, deliberation, worship, drinking or dining, awaiting transportation, or education.*

Authority Cited *– Health and Safety Code Section 13143.*

Reference *– Health and Safety Code Section 13143.*

Small Family Day Care Homes.

Authority Cited *– Health and Safety Code Sections 1597.45, 1597.54, 13143, and 17921.*

Reference *– Health and Safety Code Section 13143.*

Large Family Day Care Homes.

Authority Cited *– Health and Safety Code Sections 1597.46, 1597.54, and 17921.*

Reference *– Health and Safety Code Section 13143.*

Residential Facilities and Residential Facilities for the Elderly.

Authority Cited *– Health and Safety Code Section 13133.*

Reference *– Health and Safety Code Section 13143.*

Any State Institution, Other State-Owned or State-Occupied Building.

Authority Cited *– Health and Safety Code Section 13108.*

Reference *– Health and Safety Code Section 13143.*

High-Rise Structures.

Authority Cited *– Health and Safety Code Section 13211.*

Reference *– Health and Safety Code Section 13143.*

Motion Picture Production Studios.

Authority Cited *– Health and Safety Code Section 13143.1.*

Reference *– Health and Safety Code Section 13143.*

Organized Camps.

Authority Cited *– Health and Safety Code Section 18897.3.*

Reference *– Health and Safety Code Section 13143.*

Residential. *All hotels, motels, lodging houses, apartment houses, and dwellings, including congregate residences and buildings and structures accessory thereto. Multiple-story structures existing on January 1, 1975, let for human habitation, including and limited to, hotels, motels and apartment houses, less than 75 feet (22 860 mm) above the lowest floor level having building access, wherein rooms used for sleeping are let above the ground floor.*

Authority Cited *– Health and Safety Code Sections 13143.2 and 17921.*

Reference *– Health and Safety Code Section 13143.*

Residential Care Facilities. *Certified family care homes, out-of-home placement facilities, halfway houses, drug and/or alcohol rehabilitation facilities and any building or structure used or intended for use as a home or institution for the housing of any person of any age when such person is referred to or placed within such home or institution for protective social care and supervision services by any governmental agency.*

Authority Cited *– Health and Safety Code Section 13143.6.*

Reference *– Health and Safety Code Section 13143.*

Tents, Awnings, or other Fabric Enclosures Used in Connection with any Occupancy.

Authority Cited *– Health and Safety Code Section 13116.*

Reference *– Health and Safety Code Section 13143.*

Fire Alarm Devices, Equipment, and Systems in Connection with any Occupancy.

Authority Cited *– Health and Safety Code Section 13114.*

Reference *– Health and Safety Code Section 13143.*

Hazardous Materials.

Authority Cited *– Health and Safety Code Section 13143.9.*

Reference *– Health and Safety Code Section 13143.*

Flammable and Combustible Liquids.

Authority Cited *– Health and Safety Code Section 13143.6.*

Reference *– Health and Safety Code Section 13143.*

Public School Automatic Fire Detection, Alarm, and Sprinkler Systems.

Authority Cited *– Health and Safety Code Section 13143 and California Education Code Article 7.5, Sections 17074.50, 17074.52, and 17074.54.*

References *– Government Code Section 11152.5, Health and Safety Code Section 13143 and California Education Code Chapter 12.5, Leroy F. Greene School Facilities Act of 1998, Article 1.w.*

Wildland-Urban Interface Fire Area.

Authority Cited *– Health and Safety Code Sections 13143, 13108.5(a), 18949.2(b), (c) and Government Code Section 51189.*

References *– Health and Safety Code Sections 13143, Government Code Sections 51176, 51177, 51178, and 51179 and Public Resources Code Sections 4201 through 4204.*

1.11.2 Duties and Powers of the Enforcing Agency.

1.11.2.1 Enforcement.

1.11.2.1.1 *The responsibility for enforcement of building standards adopted by the State Fire Marshal and published in the California Building Standards Code relating to fire and panic safety and other regulations of the State Fire Marshal shall except as provided in Section 1.11.2.1.2 be as follows:*

(1) The city, county, or city and county with jurisdiction in the area affected by the standard or regulation shall delegate the enforcement of the building standards relating to fire and panic safety and other regulations of the State Fire Marshal as they relate to Group R-3 occupancies, as described in Section 310.1 of Part 2 of the California Building Standards Code, to either of the following:

1.1. The chief of the fire authority of the city, county, or city and county, or an authorized representative.

1.2. The chief building official of the city, county, or city and county, or an authorized representative.

(2) The chief of any city or county fire department or of any fire protection district, and authorized representatives, shall enforce within the jurisdiction the building standards and other regulations of the State Fire Marshal, except those described in Item 1 or 4.

(3) The State Fire Marshal shall have authority to enforce the building standards and other regulations of the State Fire Marshal in areas outside of corporate cities and districts providing fire protection services.

(4) The State Fire Marshal shall have authority to enforce the building standards and other regulations of the State Fire Marshal in corporate cities and districts providing fire protection services on request of the chief fire official or the governing body.

(5) Any fee charged pursuant to the enforcement authority of this section shall not exceed the estimated reasonable cost of providing the service for which the fee is charged pursuant to Section 66014 of the Government Code.

1.11.2.1.2 Pursuant to Health and Safety Code Section 13108, and except as otherwise provided in this section, building standards adopted by the State Fire Marshal published in the California Building Standards Code relating to fire and panic safety shall be enforced by the State Fire Marshal in all state-owned buildings, state-occupied buildings, and state institutions throughout the state. Upon the written request of the chief fire official of any city, county or fire protection district, the State Fire Marshal may authorize such chief fire official and his or her authorized representatives, in their geographical area of responsibility, to make fire prevention inspections of state-owned or state-occupied buildings, other than state institutions, for the purpose of enforcing the regulations relating to fire and panic safety adopted by the State Fire Marshal pursuant to this section and building standards relating to fire and panic safety published in the California Building Standards Code. Authorization from the State Fire Marshal shall be limited to those fire departments or fire districts which maintain a fire prevention bureau staffed by paid personnel.

Pursuant to Health and Safety Code Section 13108, any requirement or order made by any chief fire official who is authorized by the State Fire Marshal to make fire prevention inspections of state-owned or state-occupied buildings, other than state institutions, may be appealed to the State Fire Marshal. The State Fire Marshal shall, upon receiving an appeal and subject to the provisions of Chapter 5 (commencing with Section 18945) of Part 2.5 of Division 13 of the Health and Safety Code, determine if the requirement or order made is reasonably consistent with the fire and panic safety regulations adopted by the State Fire Marshal and building standards relating to fire and panic safety published in the California Building Code.

Any person may request a code interpretation from the State Fire Marshal relative to the intent of any regulation or provision adopted by the State Fire Marshal. When the request relates to a specific project, occupancy or building, the State Fire Marshal shall review the issue with the appropriate local enforcing agency prior to rendering such code interpretation.

1.11.2.1.3 Pursuant to Health and Safety Code Section 13112, any person who violates any order, rule or regulation of the State Fire Marshal is guilty of a misdemeanor punishable by a fine of not less than $100.00 or more than $500.00, or by imprisonment for not less than six months, or by both. A person is guilty of a separate offense each day during which he or she commits, continues or permits a violation of any provision of, or any order, rule or regulation of, the State Fire Marshal as contained in this code.

Any inspection authority who, in the exercise of his or her authority as a deputy State Fire Marshal, causes any legal complaints to be filed or any arrest to be made shall notify the State Fire Marshal immediately following such action.

***1.11.2.2 Right of Entry.** The fire chief of any city, county, or fire protection district, or such person's authorized representative, may enter any state institution or any other state-owned or state-occupied building for the purpose of preparing a fire suppression preplanning program or for the purpose of investigating any fire in a state-occupied building.*

The State Fire Marshal, his or her deputies or salaried assistants, the chief of any city or county fire department or fire protection district and his or her authorized representatives may enter any building or premises not used for dwelling purposes at any reasonable hour for the purpose of enforcing this chapter. The owner, lessee, manager or operator of any such building or premises shall permit the State Fire Marshal, his or her deputies or salaried assistants and the chief of any city or county fire department or fire protection district and his or her authorized representatives to enter and inspect them at the time and for the purpose stated in this section.

1.11.2.3 More Restrictive Fire and Panic Safety Building Standards.

1.11.2.3.1 Any fire protection district organized pursuant to Health and Safety Code Part 2.7 (commencing with Section 13800) of Division 12 may adopt building standards relating to fire and panic safety that are more stringent than those building standards adopted by the State Fire Marshal and contained in the California Building Standards Code. For these purposes, the district board shall be deemed a legislative body and the district shall be deemed a local agency. Any changes or modifications that are more stringent than the requirements published in the California Building Standards Code relating to fire and panic safety shall be subject to Section 1.1.8.1.

1.11.2.3.2 Any fire protection district that proposes to adopt an ordinance pursuant to this section shall, not less than 30 days prior to noticing a proposed ordinance for public hearing, provide a copy of that ordinance, together with the adopted findings made pursuant to Section 1.11.2.3.1, to the city, county, or city and county where the ordinance will apply. The city, county, or city and county may provide the district with written comments, which shall become part of the fire protection district's public hearing record.

1.11.2.3.3 The fire protection district shall transmit the adopted ordinance to the city, county, or city and county where the ordinance will apply. The legislative body of the city, county, or city and county may ratify, modify or deny an adopted ordinance and transmit its determination to the district within 15 days of the determination. Any modification or denial of an adopted ordinance shall include a written statement describing the reasons for any modifications or denial. No ordinance adopted by the district shall be effective until ratification by the city, county, or city and county where the ordinance will apply. Upon ratification of an adopted ordinance, the city, county, or city and county shall file a copy of the findings of the district, and any findings of the city, county, or city and county, together with the adopted ordinance expressly marked and identified to which each finding refers, in accordance with Section 1.1.8.1, Item 3.

1.11.2.4 Request for Alternate Means of Protection. *Requests for approval to use an alternative material, assembly or materials, equipment, method of construction, method of installation of equipment or means of protection shall be made in writing to the enforcing agency by the owner or the owner's authorized representative and shall be accompanied by a full statement of the conditions. Sufficient evidence or proof shall be submitted to substantiate any claim that may be made regarding its conformance. The enforcing agency may require tests and the submission of a test report from an approved testing organization as set forth in Title 19, California Code of Regulation, to substantiate the equivalency of the proposed alternative means of protection.*

When a request for alternate means of protection involves hazardous materials, the authority having jurisdiction may consider implementation of the findings and recommendations identified in a Risk Management Plan (RMP) developed in accordance with Title 19, Division 2, Chapter 4.5, Article 3.

Approval of a request for use of an alternative material, assembly of materials, equipment, method of construction, method of installation of equipment, or means of protection made pursuant to these provisions shall be limited to the particular case covered by request and shall not be construed as establishing any precedent for any future request.

1.11.2.5 Appeals. *When a request for an alternate means of protection has been denied by the enforcing agency, the applicant may file a written appeal to the State Fire Marshal for consideration of the applicant's proposal. In considering such appeal, the State Fire Marshal may seek the advice of the State Board of Fire Services. The State Fire Marshal shall, after considering all of the facts presented, including any recommendations of the State Board of Fire Services, determine if the proposal is for the purposes intended, at least equivalent to that specified in these regulations in quality, strength, effectiveness, fire resistance, durability and safety, and shall transmit such findings and any recommendations to the applicant and to the enforcing agency.*

1.11.3 Construction Documents.

1.11.3.1 Public Schools. *Plans and specifications for the construction, alteration, or addition to any building owned, leased, or rented by any public school district shall be submitted to the Division of the State Architect.*

1.11.3.2 Movable Walls and Partitions. *Plans or diagrams shall be submitted to the enforcing agency for approval before the installation of, or rearrangement of, any movable wall or partition in any occupancy. Approval shall be granted only if there is no increase in the fire hazard.*

1.11.3.3 New Construction High-Rise Buildings:

(1) Complete plans or specifications, or both, shall be prepared covering all work required to comply with new construction high-rise buildings. Such plans and specifications shall be submitted to the enforcing agency having jurisdiction.

(2) All plans and specifications shall be prepared under the responsible charge of an architect or a civil or structural engineer authorized by law to develop construction plans and specifications, or by both such architect and engineer. Plans and specifications shall be prepared by an engineer duly qualified in that branch of engineering necessary to perform such services. Administration of the work of construction shall be under the charge of the responsible architect or engineer except that where plans and specifications involve alterations or repairs, such work of construction may be administered by an engineer duly qualified to perform such services and holding a valid certificate under Chapter 7 (commencing with Section 65700) of Division 3 of the Business and Professions Code for performance of services in that branch of engineering in which said plans, specifications and estimates and work of construction are applicable.

This section shall not be construed as preventing the design of fire-extinguishing systems by persons holding a C-16 license issued pursuant to Division 3, Chapter 9, Business and Professions Code. In such instances, however, the responsibility charge of this section shall prevail.

1.11.3.4 Existing High-Rise Buildings:

(1) Complete plans or specifications, or both, shall be prepared covering all work required by Section 3412 for existing high-rise buildings. Such plans or specifications shall be submitted to the enforcing agency having jurisdiction.

(2) When new construction is required to conform with the provisions of these regulations, complete plans or specifications, or both, shall be prepared in accordance with the provisions of this subsection. As used in this section, "new construction" is not intended to include repairs, replacements or minor alterations which do not disrupt or appreciably add to or affect the structural aspects of the building.

1.11.3.5 Retention of Plans. *Refer to Building Standards Law, Health and Safety Code Sections 19850 and 19851 for permanent retention of plans.*

1.11.4 Fees.

1.11.4.1 Other Fees. *Pursuant to Health and Safety Code Section 13146.2, a city, county, or district which inspects a hotel, motel, lodging house, or apartment house may charge and collect a fee for the inspection from the owner of the structure in an amount, as determined by the city, county or district, sufficient to pay its costs of that inspection.*

1.11.4.2 Large Family Day Care. *Pursuant to Health and Safety Code Section 1597.46, Large Family Day Care Homes, the local government shall process any required permit as economically as possible, and fees charged for review shall not exceed the costs of the review and permit process.*

1.11.4.3 High-Rise. *Pursuant to Health and Safety Code Section 13217, High-rise Structure Inspection: Fees and costs, a local agency which inspects a high-rise structure pursuant to Health and Safety Code Section 13217 may charge and collect a fee for the inspection from the owner of the high-rise structure in an amount, as determined by the local agency, sufficient to pay its costs of that inspection.*

1.11.4.4 Fire Clearance Preinspection. *Pursuant to Health and Safety Code Section 13235, Fire Clearance Preinspection fee, upon receipt of a request from a prospective licensee of a community care facility, as defined in Section 1502, of a residential care facility for the elderly, as defined in Section 1569.2, or of a child day care facility, as defined in Section 1596.750, the local fire enforcing agency, as defined in Section 13244, or State Fire Marshal, whichever has primary jurisdiction, shall conduct a preinspection of the facility prior to the final fire clearance approval. At the time of the preinspection, the primary fire enforcing agency shall price consultation and interpretation of the fire safety regulations and shall notify the prospective licensee of the facility in writing of the specific fire safety regulations which shall be enforced in order to obtain fire clearance approval. A fee of not more than $50.00 may be charged for the preinspection of a facility with a capacity to serve 25 or fewer persons. A fee of not more than $100.00 may be charged for a preinspection of a facility with a capacity to serve 26 or more persons.*

1.11.4.5 Care Facilities. *The primary fire enforcing agency shall complete the final fire clearance inspection for a community care facility, residential care facility for the elderly, or child day care facility within 30 days of receipt of the request for the final inspection, or as of the date the prospective facility requests the final prelicensure inspection by the State Department of Social Services, whichever is later.*

Pursuant to Health and Safety Code Section 13235, a preinspection fee of not more than $50.00 may be charged for a facility with a capacity to serve 25 or less clients. A fee of not more than $100.00 may be charged for a preinspection of a facility with a capacity to serve 26 or more clients.

Pursuant to Health and Safety Code Section 13131.5, a reasonable final inspection fee, not to exceed the actual cost of inspection services necessary to complete a final inspection may be charged for occupancies classified as Residential Care Facilities for the Elderly (RCFE).

Pursuant to Health and Safety Code Section 1569.84, neither the State Fire Marshal nor any local public entity shall charge any fee for enforcing fire inspection regulations pursuant to state law or regulation or local ordinance, with respect to Residential Care Facilities for the Elderly (RCFE) which service six or fewer persons.

1.11.4.6 Requests of the Office of the State Fire Marshal. *Whenever a local authority having jurisdiction requests that the State Fire Marshal perform plan review and/or inspection services related to a building permit, the applicable fees for such shall be payable to the Office of the State Fire Marshal.*

1.11.5 Inspections. *Work performed subject to the provisions of this code shall comply with the inspection requirements of Title 24, Part 2, California Building Standards Code, Sections 109.1, 109.3, 109.3.4, 109.3.5, 109.3.6, 109.3.8, 109.3.9, 109.3.10, 109.5, and 109.6 as adopted by the Office of the State Fire Marshal.*

1.11.5.1 Existing Group I-1 or R Occupancies. *Licensed 24-hour care in a Group I-1 or R occupancy in existence and originally classified under previously adopted state codes shall be reinspected under the appropriate previous code, provided there is no change in the use or character which would place the facility in a different occupancy group.*

1.11.6 Certificate of Occupancy. *A Certificate of Occupancy shall be issued as specified in Title 24, Part 2, California Building Code, Section 111.*

Exception: *Group R, Division 3 and Group U occupancies.*

1.11.7 Temporary Structures and Uses. *See Title 24, Part 2, California Building Code, Section 107.*

1.11.8 Service Utilities. *See Title 24, Part 2, California Building Code, Section 112.*

1.11.9 Stop Work Order. *See Title 24, Part 2, California Building Code, Section 115.*

1.11.10 Unsafe Buildings, Structures, and Equipment. *See Title 24, Part 2, California Building Code, Section 116.*

1.11.11 Adopting Agency Identification. *The provisions of this code applicable to buildings identified in this Section 1.11.0 will be identified in the Matrix Adoption Tables under the acronym SFM.*

1.12.0 Reserved for the State Librarian.

1.13.0 Reserved for the Department of Water Resources.

1.14.0 Reserved for the State Lands Commission.

DIVISION II

ADMINISTRATION

101.0 General.

101.1 Title. This document shall be known as the "Uniform Mechanical Code," may be cited as such, and will be referred to herein as "this code."

101.2 Scope. The provisions of this code shall apply to the erection, installation, alteration, repair, relocation, replacement, addition to, use, or maintenance of mechanical systems within this jurisdiction.

101.3 Purpose. This code is an ordinance providing minimum requirements and standards for the protection of the public health, safety, and welfare.

101.4 Unconstitutional. Where a section, subsection, sentence, clause, or phrase of this code is, for a reason, held to be unconstitutional, such decision shall not affect the validity of the remaining portions of this code. The legislative body hereby declares that it would have passed this code, and each section, subsection, sentence, clause, or phrase thereof, irrespective of the fact that one or more sections, subsections, sentences, clauses, and phrases are declared unconstitutional.

101.5 Validity. Where a provision of this code, or the application thereof to a person or circumstance, is held invalid, the remainder of the code, or the application of such provision to other persons or circumstances, shall not be affected thereby.

102.0 Applicability.

102.1 Conflicts Between Codes. Where the requirements within the jurisdiction of this mechanical code conflict with the requirements of the plumbing code, the plumbing code shall prevail. In instances where this code, applicable standards, or the manufacturer's installation instructions conflict, the more stringent provisions shall prevail. Where there is a conflict between a general requirement and a specific requirement, the specific requirement shall prevail ***[OSHPD 1, 1R, 2, 3, 4 & 5]*** *See Chapter 1, Division I, Section 1.1.7.*

102.2 Existing Installations. Mechanical systems lawfully in existence at the time of the adoption of this code shall be permitted to have their use, maintenance, or repair continued where the use, maintenance, or repair is in accordance with the original design and location and no hazard to life, health, or property has been created by such mechanical system.

102.3 Maintenance. Mechanical systems, materials, and appurtenances, both existing and new, of a premise under the Authority Having Jurisdiction shall be maintained in operating condition. Devices or safeguards required by this code shall be maintained in accordance with the code edition under which installed.

The owner or the owner's designated agent shall be responsible for maintenance of mechanical systems. To determine compliance with this subsection, the Authority Having Jurisdiction shall be permitted to cause a mechanical system to be reinspected.

102.3.1 Commercial HVAC Systems. Commercial HVAC systems both existing and new, and parts thereof shall be inspected and maintained in operating condition in accordance with ASHRAE/ACCA 180. The owner or the owner's designated agent shall be responsible for maintenance of mechanical systems and equipment. To determine compliance with this subsection, the Authority Having Jurisdiction shall be permitted to cause a HVAC system to be reinspected.

102.3.2 Residential HVAC Systems. Residential HVAC systems both existing and new, and parts thereof shall be inspected in accordance with ACCA 4 QM. The owner or the owner's designated agent shall be responsible for maintenance of mechanical systems and equipment. To determine compliance with this subsection, the Authority Having Jurisdiction shall be permitted to cause a HVAC system to be reinspected.

102.4 Additions, Alterations, Renovations, or Repairs. Additions, alterations, renovations, or repairs shall conform to that required for a new system without requiring the existing mechanical system to be in accordance with the requirements of this code. Additions, alterations, renovations, or repairs shall not cause an existing system to become unsafe, insanitary or overloaded.

Additions, alterations, renovations, or repairs to existing mechanical system installations shall comply with the provisions for new construction, unless such deviations are found to be necessary and are first approved by the Authority Having Jurisdiction.

102.5 Health and Safety. Where compliance with the provisions of this code fails to eliminate or alleviate a nuisance, or other dangerous or insanitary condition that involves health or safety hazards, the owner or the owner's agent shall install such additional mechanical system facilities or shall make such repairs or alterations as ordered by the Authority Having Jurisdiction.

102.6 Changes in Building Occupancy. Mechanical systems that are a part of a building or structure undergoing a change in use or occupancy, as defined in the building code, shall be in accordance with the requirements of this code that are applicable to the new use or occupancy.

102.7 Moved Structures. Parts of the mechanical system of a building and part thereof that is moved from one foundation to another, or from one location to another, shall be in accordance with the provisions of this code for new installations and completely tested as prescribed elsewhere in this section for new work, except that walls or floors need not be removed during such test where equivalent means of inspection acceptable to the Authority Having Jurisdiction are provided.

102.8 Appendices. The provisions in the appendices are intended to supplement the requirements of this code and shall not be considered part of this code unless formally adopted as such.

103.0 Duties and Powers of the Authority Having Jurisdiction.

103.1 General. The Authority Having Jurisdiction shall be the Authority duly appointed to enforce this code. For such purposes, the Authority Having Jurisdiction shall have the powers of a law enforcement officer. The Authority Having Jurisdiction shall have the power to render interpretations of this code and to adopt and enforce rules and regulations supplemental to this code as deemed necessary in order to clarify the application of the provisions of this code. Such interpretations, rules, and regulations shall comply with the intent and purpose of this code.

In accordance with the prescribed procedures and with the approval of the appointing authority, the Authority Having Jurisdiction shall be permitted to appoint such number of technical officers, inspectors, and other employees as shall be authorized from time to time. The Authority Having Jurisdiction shall be permitted to deputize such inspectors or employees as necessary to carry out the functions of the code enforcement agency.

The Authority Having Jurisdiction shall be permitted to request the assistance and cooperation of other officials of this jurisdiction so far as required in the discharge of the duties required by this code or other pertinent law or ordinance.

103.2 Liability. The Authority Having Jurisdiction charged with the enforcement of this code, acting in good faith and without malice in the discharge of the Authority Having Jurisdiction's duties, shall not thereby be rendered personally liable for damage that accrues to persons or property as a result of an act or by reason of an act or omission in the discharge of such duties. A suit brought against the Authority Having Jurisdiction or employee because of such act or omission performed in the enforcement of provisions of this code shall be defended by legal counsel provided by this jurisdiction until final termination of such proceedings.

103.3 Applications and Permits. The Authority Having Jurisdiction shall be permitted to require the submission of plans, specifications, drawings, and such other information in accordance with the Authority Having Jurisdiction, prior to the commencement of, and at a time during the progress of, work regulated by this code.

The issuance of a permit upon construction documents shall not prevent the Authority Having Jurisdiction from thereafter requiring the correction of errors in said construction documents or from preventing construction operations being carried on thereunder where in violation of this code or of other pertinent ordinance or from revoking a certificate of approval where issued in error.

103.3.1 Licensing. Provision for licensing shall be determined by the Authority Having Jurisdiction.

103.4 Right of Entry. Where it is necessary to make an inspection to enforce the provisions of this code, or where the Authority Having Jurisdiction has reasonable cause to believe that there exists in a building or upon a premises a condition or violation of this code that makes the building or premises unsafe, insanitary, dangerous, or hazardous, the Authority Having Jurisdiction shall be permitted to enter the building or premises at reasonable times to inspect or to perform the duties imposed upon the Authority Having Jurisdiction by this code, provided that where such building or premises is occupied, the Authority Having Jurisdiction shall present credentials to the occupant and request entry. Where such building or premises is unoccupied, the Authority Having Jurisdiction shall first make a reasonable effort to locate the owner or other person having charge or control of the building or premises and request entry. Where entry is refused, the Authority Having Jurisdiction has recourse to every remedy provided by law to secure entry.

Where the Authority Having Jurisdiction shall have first obtained an inspection warrant or other remedy provided by law to secure entry, no owner, occupant, or person having charge, care or control of a building or premises shall fail or neglect, after a request is made as herein provided, to promptly permit entry herein by the Authority Having Jurisdiction for the purpose of inspection and examination pursuant to this code.

104.0 Permits.

104.1 Permits Required. It shall be unlawful for a person, firm, or corporation to make an installation, alteration, repair, replacement, or remodel a mechanical system regulated by this code except as permitted in Section 104.2, or to cause the same to be done without first obtaining a separate mechanical permit for each separate building or structure.

104.2 Exempt Work. A permit shall not be required for the following:

(1) A portable heating appliance, portable ventilating equipment, a portable cooling unit, or a portable evaporative cooler.

(2) A closed system of steam, hot, or chilled water piping within heating or cooling equipment regulated by this code.

(3) Replacement of a component part that does not alter its original approval and is in accordance with other applicable requirements of this code.

(4) Refrigerating equipment that is part of the equipment for which a permit has been issued pursuant to the requirements of this code.

(5) A unit refrigerating system.

Exemption from the permit requirements of this code shall not be deemed to grant authorization for work to be done in violation of the provisions of the code or other laws or ordinances of this jurisdiction.

104.3 Application for Permit. To obtain a permit, the applicant shall first file an application therefore in writing on a form furnished by the Authority Having Jurisdiction for that purpose. Such application shall:

(1) Identify and describe the work to be covered by the permit for which application is made.

(2) Describe the land upon which the proposed work is to be done by legal description, street address, or similar description that will readily identify and definitely locate the proposed building or work.

(3) Indicate the use or occupancy for which the proposed work is intended.

(4) Be accompanied by construction documents and other data in accordance with Section 104.3.1.

(5) Be signed by the permittee or the permittee's authorized agent. The Authority Having Jurisdiction shall be permitted to require evidence to indicate such authority.

(6) Give such other data and information in accordance with the Authority Having Jurisdiction.

104.3.1 Construction Documents. Construction documents, engineering calculations, diagrams, and other data shall be submitted in two or more sets with each application for a permit. The construction documents, computations, and specifications shall be prepared by, and the mechanical system designed by, a registered design professional. Construction documents shall be drawn to scale with clarity to identify that the intended work to be performed is in accordance with the code.

Exception: The Authority Having Jurisdiction shall be permitted to waive the submission of construction documents, calculations, or other data where the Authority Having Jurisdiction finds that the nature of the work applied for is such that reviewing of construction documents is not necessary to obtain compliance with the code.

104.3.2 Plan Review Fees. Where a plan or other data is required to be submitted in accordance with Section 104.3.1, a plan review fee shall be paid at the time of submitting construction documents for review.

The plan review fees for mechanical system work shall be determined and adopted by this jurisdiction.

The plan review fees specified in this subsection are separate fees from the permit fees specified in Section 104.5.

Where plans are incomplete or changed so as to require additional review, a fee shall be charged at the rate shown in Table 104.5.

104.3.3 Time Limitation of Application. Applications for which no permit is issued within 180 days following the date of application shall expire by limitation, plans and other data submitted for review thereafter, shall be returned to the applicant or destroyed by the Authority Having Jurisdiction. The Authority Having Jurisdiction shall be permitted to extend the time for action by the applicant for a period not to exceed 180 days upon request by the applicant showing that circumstances beyond the control of the applicant have prevented action from being taken. No application shall be extended more than once. In order to renew action on an application after expiration, the applicant shall resubmit plans and pay a new plan review fee.

104.4 Permit Issuance. The application, construction documents, and other data filed by an applicant for a permit shall be reviewed by the Authority Having Jurisdiction. Such plans shall be permitted to be reviewed by other departments of this jurisdiction to verify compliance with applicable laws under their jurisdiction. Where the Authority Having Jurisdiction finds that the work described in an application for permit and the plans, specifications, and other data filed therewith are in accordance with the requirements of the code and other pertinent laws and ordinances and that the fees specified in Section 104.5 have been paid, the Authority Having Jurisdiction shall issue a permit therefore to the applicant.

104.4.1 Approved Plans or Construction Documents. Where the Authority Having Jurisdiction issues the permit where plans are required, the Authority Having Jurisdiction shall endorse in writing or stamp the construction documents "APPROVED." Such approved construction documents shall not be changed, modified, or altered without authorization from the Authority Having Jurisdiction, and the work shall be done in accordance with approved plans.

The Authority Having Jurisdiction shall be permitted to issue a permit for the construction of a part of a mechanical system before the entire construction documents for the whole system have been submitted or approved, provided adequate information and detailed statements have been filed in accordance with pertinent requirements of this code. The holder of such permit shall be permitted to proceed at the holder's risk without assurance that the permit for the entire building, structure, or mechanical system will be granted.

104.4.2 Validity of Permit. The issuance of a permit or approval of construction documents shall not be construed to be a permit for, or an approval of, a violation of the provisions of this code or other ordinance of the jurisdiction. No permit presuming to give authority to violate or cancel the provisions of this code shall be valid.

The issuance of a permit based upon plans, specifications, or other data shall not prevent the Authority Having Jurisdiction from thereafter requiring the correction of errors in said plans, specifications, and other data or from preventing building operations being carried on thereunder where in violation of this code or of other ordinances of this jurisdiction.

104.4.3 Expiration. A permit issued by the Authority Having Jurisdiction under the provisions of this code shall expire by limitation and become null and void where the work authorized by such permit is not commenced within 180 days from the date of such permit, or where the work authorized by such permit is suspended or abandoned at a time after the work is commenced for a period of 180 days. Before such work is recommenced, a new permit shall first be obtained to do so, and the fee, therefore, shall be one-half the amount required for a new permit for such work, provided no changes have been made or will be made in the original construction documents for such work, and provided further that such suspension or abandonment has not exceeded 1 year.

104.4.4 Extension. A permittee holding an unexpired permit shall be permitted to apply for an extension of the time within which work shall be permitted to commence under that permit where the permittee is unable to commence work within the time required by this section. The Authority Having Jurisdiction shall be permitted to extend the time for action by the permittee for a period not exceeding 180 days upon written request by the permittee showing that circumstances beyond the control of the permittee have prevented action from being taken. No permit shall be extended more than once. In order to renew action on a permit after expiration, the permittee shall pay a new full permit fee.

104.4.5 Suspension or Revocation. The Authority Having Jurisdiction shall be permitted to, in writing, suspend or revoke a permit issued under the provisions of this code where the permit is issued in error or on the basis of incorrect information supplied or in violation of other ordinance or regulation of the jurisdiction.

104.4.6 Retention of Plans. One set of approved construction documents and computations shall be retained by the Authority Having Jurisdiction until final approval of the work is covered therein.

One set of approved construction documents, computations, and manufacturer's installation instructions shall be returned to the applicant, and said set shall be kept on the site of the building or work at times during which the work authorized thereby is in progress.

104.5 Fees. Fees shall be assessed in accordance with the provisions of this section and as set forth in the fee schedule, Table 104.5. The fees are to be determined and adopted by this jurisdiction.

104.5.1 Work Commencing Before Permit Issuance. Where work for which a permit is required by this code has been commenced without first obtaining said permit, a special investigation shall be made before a permit is issued for such work.

104.5.2 Investigation Fees. An investigation fee, in addition to the permit fee, shall be collected whether or not a permit is then or subsequently issued. The investigation fee shall be equal to the amount of the permit fee that is required by this code if a permit were to be issued. The payment of such investigation fee shall not exempt a person from compliance with other provisions of this code, nor from a penalty prescribed by law.

104.5.3 Fee Refunds. The Authority Having Jurisdiction shall be permitted to authorize the refunding of a fee as follows:

(1) The amount paid hereunder that was erroneously paid or collected.

(2) Refunding of not more than a percentage, as determined by this jurisdiction where no work has been done under a permit issued in accordance with this code.

The Authority Having Jurisdiction shall not authorize refunding of a fee paid except upon written application filed by the original permittee not to exceed 180 days after the date of fee payment.

105.0 Inspections and Testing.

105.1 General. Mechanical systems for which a permit is required by this code shall be inspected by the Authority Having Jurisdiction.

No mechanical system or portion thereof shall be covered, concealed, or put into use until inspected and approved as prescribed in this code. Neither the Authority Having Jurisdiction nor the jurisdiction shall be liable for expense entailed in the removal or replacement of material required to permit inspection. Mechanical systems regulated by this code shall not be connected to the energy fuel supply lines until authorized by the Authority Having Jurisdiction.

105.2 Required Inspections. New mechanical system work and such portions of existing systems as affected by new work, or changes, shall be inspected by the Authority Having Jurisdiction to ensure compliance with the requirements of this code and to ensure that the installation and construction of the mechanical system are in accordance with approved plans. The Authority Having Jurisdiction shall make the following inspections and other such inspections as necessary. The permittee or the permittee's authorized agent shall be responsible for the scheduling of such inspections as follows:

(1) Underground inspection shall be made after trenches or ditches are excavated and bedded, piping installed, and before backfill is put in place.

(2) Rough-in inspection shall be made prior to the installation of wall or ceiling membranes.

(3) Final inspection shall be made upon completion of the installation.

105.2.1 Uncovering. Where a mechanical system, or part thereof, which is installed, altered, or repaired, is covered or concealed before being inspected, tested, and approved as prescribed in this code, it shall be uncovered for inspection after notice to uncover the work has been issued to the responsible person by the Authority Having Jurisdiction. The requirements of this section shall not be considered to prohibit the operation of mechanical systems installed to replace existing equipment serving an occupied portion of the building in the event a request for inspection of such equipment has been filed with the Authority Having Jurisdiction not more than 72 hours after such replacement work is completed, and before a portion of such mechanical system is concealed by a permanent portion of the building.

105.2.2 Other Inspections. In addition to the inspections required by this code, the Authority Having Jurisdiction shall be permitted to require other inspections to ascertain compliance with the provisions of this code and other laws that are enforced by the Authority Having Jurisdiction.

105.2.3 Inspection Requests. It shall be the duty of the person doing the work authorized by a permit to notify the Authority Having Jurisdiction that such work is ready for inspection. The Authority Having Jurisdiction shall be permitted to require that a request for inspection be filed not less than 1 working day before such inspection is desired. Such request shall be permitted to be made in writing or by telephone, at the option of the Authority Having Jurisdiction.

It shall be the duty of the person requesting inspections in accordance with this code to provide access to and means for inspection of such work.

105.2.4 Advance Notice. It shall be the duty of the person doing the work authorized by the permit to notify the Authority Having Jurisdiction, orally or in writing that said work is ready for inspection. Such notification shall be given not less than 24 hours before the work is to be inspected.

105.2.5 Responsibility. It shall be the duty of the holder of a permit to make sure that the work will stand the test prescribed before giving the notification.

The equipment, material, and labor necessary for inspection or tests shall be furnished by the person to whom the permit is issued or by whom inspection is requested.

105.2.6 Reinspections. A reinspection fee shall be permitted to be assessed for each inspection or reinspection where such portion of work for which inspection is called is not complete or where required corrections have not been made.

This provision shall not be interpreted as requiring reinspection fees the first time a job is rejected for failure to be in accordance with the requirements of this code, but as controlling the practice of calling for inspections before the job is ready for inspection or reinspection.

Reinspection fees shall be permitted to be assessed where the approved plans are not readily available to the inspector, for failure to provide access on the date for which the inspection is requested, or for deviating from plans requiring the approval of the Authority Having Jurisdiction.

To obtain reinspection, the applicant shall file an application therefore in writing upon a form furnished for that purpose and pay the reinspection fee in accordance with Table 104.5.

In instances where reinspection fees have been assessed, no additional inspection of the work will be performed until the required fees have been paid.

105.3 Testing of Systems. Mechanical systems shall be tested and approved in accordance with this code or the Authority Having Jurisdiction. Tests shall be conducted in the presence of the Authority Having Jurisdiction or the Authority Having Jurisdiction's duly appointed representative.

No test or inspection shall be required where a mechanical system, or part thereof, is set up for exhibition purposes and has no connection with water or an energy fuel supply. In cases where it would be impractical to provide the required water or air tests, or for minor installations and repairs, the Authority Having Jurisdiction shall be permitted to make such inspection as deemed advisable in order to be assured that the work has been performed in accordance with the intent of this code. Joints and connections in the mechanical system shall be airtight, gastight, or watertight for the pressures required by the test.

105.3.1 Defective Systems. In buildings or premises condemned by the Authority Having Jurisdiction because of an insanitary condition of the mechanical system, or part thereof, the alterations in such system shall be in accordance with the requirements of this code.

105.3.2 Retesting. Where the Authority Having Jurisdiction finds that the work will not pass the test, necessary corrections shall be made, and the work shall be resubmitted for test or inspection.

105.3.3 Approval. Where prescribed tests and inspections indicate that the work is in accordance with this code, a certificate of approval shall be issued by the Authority Having Jurisdiction to the permittee on demand.

105.4 Connection to Service Utilities. No person shall make connections from a source of energy or fuel to a mechanical system or equipment regulated by this code and for which a permit is required until approved by the Authority Having Jurisdiction. The Authority Having Jurisdiction shall be permitted to authorize temporary connection of the mechanical system equipment to the source of energy or fuel for the purpose of testing the equipment.

106.0 Violations and Penalties.

106.1 General. It shall be unlawful for a person, firm, or corporation to erect, construct, enlarge, alter, repair, move, improve, remove, convert, demolish, equip, use, or maintain a mechanical system or permit the same to be done in violation of this code.

106.2 Notices of Correction or Violation. Notices of correction or violation shall be written by the Authority Having Jurisdiction and shall be permitted to be posted at the site of the work or mailed or delivered to the permittee or their authorized representative.

Refusal, failure, or neglect to comply with such notice or order within 10 days of receipt thereof, shall be considered a violation of this code and shall be subject to the penalties set forth by the governing laws of the jurisdiction.

106.3 Penalties. A person, firm, or corporation violating a provision of this code shall be deemed guilty of a misdemeanor, and upon conviction thereof, shall be punishable by a fine, imprisonment, or both set forth by the governing laws of the jurisdiction. Each separate day or a portion thereof, during which a violation of this code occurs or continues, shall be deemed to constitute a separate offense.

106.4 Stop Orders. Where work is being done contrary to the provisions of this code, the Authority Having Jurisdiction shall be permitted to order the work stopped by notice in writing served on persons engaged in the doing or causing such work to be done, and such persons shall forthwith stop work until authorized by the Authority Having Jurisdiction to proceed with the work.

106.5 Authority to Disconnect Utilities in Emergencies. The Authority Having Jurisdiction shall have the authority to disconnect a mechanical system to a building, structure, or equipment regulated by this code in case of emergency where necessary to eliminate an immediate hazard to life or property.

106.6 Authority to Condemn. Where the Authority Having Jurisdiction ascertains that a mechanical system or portion thereof, regulated by this code, has become hazardous to life, health, or property, or has become insanitary, the Authority Having Jurisdiction shall order in writing that such mechanical system either be removed or placed in a safe or sanitary condition. The order shall fix a reasonable time limit for compliance. No person shall use or maintain a defective mechanical system after receiving such notice.

Where such mechanical system is to be disconnected, written notice shall be given. In cases of immediate danger to life or property, such disconnection shall be permitted to be made immediately without such notice.

107.0 Board of Appeals.

107.1 General. In order to hear and decide appeals of orders, decisions, or determinations made by the Authority Having Jurisdiction relative to the application and interpretations of this code, there shall be and is hereby created a Board of Appeals consisting of members who are qualified by experience and training to pass upon matters pertaining to mechanical system design, construction, and maintenance and the public health aspects of mechanical systems and who are not employees of the jurisdiction. The Authority Having Jurisdiction shall be an ex-officio member and shall act as secretary to said board but shall have no vote upon a matter before the board. The Board of Appeals shall be appointed by the governing body and shall hold office at its pleasure. The board shall adopt rules of procedure for conducting its business and shall render decisions and findings in writing to the appellant with a duplicate copy to the Authority Having Jurisdiction.

107.2 Limitations of Authority. The Board of Appeals shall have no authority relative to interpretation of the administrative provisions of this code, nor shall the board be empowered to waive requirements of this code.

TABLE 104.5
MECHANICAL PERMIT FEES

Permit Issuance

1. For the issuance of each permit .... ____[1]
2. For issuing each supplemental permit for which the original permit has not expired or been canceled or finalized .... ____[1]

Unit Fee Schedule

1. Furnaces:

   For the installation or relocation of each forced-air or gravity-type furnace or burner, including ducts and vents attached to such appliance, not exceeding 100 000 British thermal units per hour (Btu/h) .... ____[1]

   For the installation or relocation of each forced-air or gravity-type furnace or burner, including ducts and vents attached to such appliance, exceeding 100 000 Btu/h .... ____[1]

   For the installation or relocation of each floor furnace, including vent .... ____[1]

   For the installation or relocation of each suspended heater, recessed wall heater, or floor-mounted unit heater .... ____[1]

2. Appliance Vents:

   For the installation, relocation, or replacement of each appliance vent installed and not included in an appliance permit .... ____[1]

3. Repairs or Additions:

   For the repair of, alteration of, or addition to each heating appliance, refrigeration unit, cooling unit, absorption unit, or each heating, cooling, absorption, or evaporative cooling, system including installation of controls regulated by this code .... ____[1]

4. Boilers, Compressors, and Absorption Systems:

   For the installation or relocation of each boiler or compressor, not exceeding 3 horsepower (hp), or each absorption system not exceeding 100 000 Btu/h .... ____[1]

   For the installation or relocation of each boiler or compressor exceeding 3 hp, not exceeding 15 hp, or each absorption system exceeding 100 000 Btu/h and including 500 000 Btu/h .... ____[1]

   For the installation or relocation of each boiler or compressor exceeding 15 hp, not exceeding 30 hp, or each absorption system exceeding 500 000 Btu/h, not exceeding 1 000 000 Btu/h .... ____[1]

   For the installation or relocation of each boiler or compressor exceeding 30 hp, not exceeding 50 hp, or for each absorption system exceeding 1 000 000 Btu/h, not exceeding 1 750 000 Btu/h .... ____[1]

   For the installation or relocation of each boiler or compressor exceeding 50 hp, or each absorption system exceeding 1 750 000 Btu/h .... ____[1]

5. Air Handlers:

   For each air-handling unit not exceeding 10 000 cubic feet per minute (cfm), including ducts attached thereto .... ____[1, 2]

6. Evaporative Coolers:

   For each air-handling unit exceeding 10 000 cfm .... ____[1]

   For each evaporative cooler other than portable type .... ____[1]

7. Ventilation and Exhaust:

   For each ventilation fan connected to a single duct .... ____[1]

   For each ventilation system that is not a portion of a heating or air-conditioning system authorized by a permit .... ____[1]

   For the installation of each hood that is served by mechanical exhaust, including the ducts for such hood .... ____[1]

8. Incinerators:

   For the installation or relocation of each domestic-type incinerator .... ____[1]

   For the installation or relocation of each commercial or industrial-type incinerator .... ____[1]

TABLE 104.5 (continued)
MECHANICAL PERMIT FEES

9. Miscellaneous:
   For each appliance or piece of equipment regulated by this code, but not classed in other appliance categories, or for which no other fee is listed in this table .... _____[1]
10. Fuel Gas Piping:
   Where Chapter 13 or Appendix B is applicable (See Section 101.2), permit fees for fuel-gas piping shall be as follows:
   For each gas piping system of one to five outlets .... _____[1]
   For each additional gas piping system, per outlet .... _____[1]
11. Process Piping:
   For each hazardous process piping system (HPP) of one to four outlets .... _____[1]
   For each HPP piping system of five or more outlets, per outlet .... _____[1]
   For each nonhazardous process piping system (NPP) of one to four outlets .... _____[1]
   For each NPP piping system of five or more outlets, per outlet .... _____[1]

Other Inspections and Fees

1. Inspections outside of normal business hours, per hour (minimum charge – 2 hours) .... _____[1]
2. Reinspection fees assessed under provisions of Section 105.2.6, per inspection .... _____[1]
3. Inspections for which no fee is specifically indicated, per hour (minimum charge – ½ hour) .... _____[1]
4. Additional plan review required by changes, additions, or revisions to plans or to plans for which an initial review has been completed, per hour (minimum charge – ½ hour) .... _____[1]

For SI units: 1000 British thermal units per hour = 0.293 kW, 1 horsepower = 0.746 kW, 1 cubic foot per minute = 0.00047 m^3/s

Notes:

[1] Jurisdiction will indicate their fees here.

[2] This fee shall not apply to an air-handling unit that is a portion of a factory-assembled appliance, cooling unit, evaporative cooler, or absorption unit for which a permit is required elsewhere in this code.

CALIFORNIA MECHANICAL CODE – MATRIX ADOPTION TABLE
CHAPTER 2 – DEFINITIONS

(Matrix Adoption Tables are non-regulatory, intended only as an aid to the code user. See Chapter 1 for state agency authority and building applications.)

Adopting Agency	BSC	BSC-CG	SFM	HCD			DSA			OSHPD						BSCC	DPH	AGR	DWR	CEC	CA	SL	SLC
				1	2	1-AC	AC	SS	SS/CC	1	1R	2	3	4	5								
Adopt Entire Chapter																							
Adopt Entire Chapter as amended (amended sections listed below)	X		X	X	X		X	X	X	X		X	X	X									
Adopt only those sections that are listed below																							
Chapter/Section																							
203.0			X	X	X					X	X	X	X	X	X								
204.0			X	X	X					X	X	X	X	X	X								
206.0				X	X																		
207.0			X	X	X					X	X	X	X	X	X								
208.0			X	X				X	X														
209.0	X		X				X	X	X	X	X	X	X	X	X								
210.0										X	X	X	X	X	X								
214.0			X	X	X																		
215.0			X																				
216.0				X	X																		
217.0			X	X	X					X		X	X	X									
218.0				X	X																		
222.0				X	X																		
223.0			X	X	X					X	X	X	X	X	X								

This state agency does not adopt sections identified with the following symbol: †

The Office of the State Fire Marshal's adoption of this chapter or individual sections is applicable to structures regulated by other state agencies pursuant to Section 1.11.0.

CHAPTER 2
DEFINITIONS

201.0 General.

201.1 Applicability. For the purpose of this code, the following terms have the meanings indicated in this chapter.

No attempt is made to define ordinary words, which are used in accordance with their established dictionary meanings, except where a word has been used loosely, and it is necessary to define its meaning as used in this code to avoid misunderstanding.

202.0 Definition of Terms.

202.1 General. The definitions of terms are arranged alphabetically according to the first word of the term.

203.0 – A –

Absorption Unit. An absorption refrigeration system that has been factory-assembled and tested prior to its installation.

Accepted Engineering Practice. That which conforms to technical or scientific-based principles, test, or standards that are accepted by the engineering profession.

Access Panel. A closure device used to cover an opening into a duct, an enclosure, equipment, or an appurtenance. [NFPA 96:3.3.1]

Accessible. Where applied to a device, appliance, or equipment, "accessible" means having access thereto, but which first may require the removal of an access panel, door, or similar obstruction.

Accessible, Readily. Having a direct access without the necessity of removing a panel, door, or similar obstruction.

Air, Class 1. Air with low contaminant concentration, low sensory-irritation intensity, and inoffensive odor. [ASHRAE 62.1:5.16.1]

Air, Class 2. Air with moderate contaminant concentration, mild sensory-irritation intensity, or mildly offensive odors. Class 2 air also includes air that is not necessarily harmful or objectionable, but that is inappropriate for transfer or recirculation to spaces used for different purposes. [ASHRAE 62.1:5.16.1]

Air, Class 3. Air with significant contaminant concentration, significant sensory-irritation intensity, or offensive odor. [ASHRAE 62.1:5.16.1]

Air, Class 4. Air with highly objectionable fumes or gases or with potentially dangerous particles, bioaerosols, or gases, at concentrations high enough to be considered harmful. [ASHRAE 62.1:5.16.1]

Air, Combustion. See Combustion Air.

Air, Conditioned. Air that has been treated to achieve a desired level of temperature, humidity, or cleanliness.

Air, Dilution. Air that enters a draft hood or draft regulator and mixes with the flue gases. [NFPA 54:3.3.2.2]

Air, Exhaust. Air being removed from any space or piece of equipment and conveyed directly to the atmosphere by means of openings or ducts.

Air, Makeup. Air that is provided to replace air being exhausted.

Air, Outside. Air from outside the building intentionally conveyed by openings or ducts to rooms or to conditioning equipment.

Air, Relief. [OSHPD 1, 1R, 2, 3, 4 & 5] *Air being exhausted directly from a building or a return duct system which is not contaminated by odors or other contaminants and could otherwise be used as return air if not exhausted from the building.*

Air, Return. Air from the conditioned area that is returned to the conditioning equipment for reconditioning.

Air, Supply. Air being conveyed to a conditioned area through ducts or plenums from a heat exchanger of a heating, cooling, absorption, or evaporative cooling system.

Air Dispersion Systems. Materials intended for use in air handling systems in exposed locations operating under positive pressure.

Air-Handling Unit. A blower or fan used for the purpose of distributing supply air to a room, space, or area.

Air Intakes. An opening in a building's envelope whose purpose is to allow outside air to be drawn into the structure to replace inside air that is removed by exhaust systems or to improve the quality of the inside air by providing a source of air having a lower concentration of odors, suspended particles, or heating content. [NFPA 96:3.3.2]

Air-Moving System. A system designed to provide heating, cooling, or ventilation in which one or more air-handling units are used to supply air to a common space or are drawing air from a common plenum or space.

Air Pollution Control Devices. Equipment and devices used for the purpose of cleaning air passing through them or by them in such a manner as to reduce or remove the impurities contained therein. [NFPA 96:3.3.3]

Anodeless Riser. An assembly of steel-cased plastic pipe used to make the transition between plastic piping installed underground and metallic piping installed aboveground. [NFPA 54:3.3.4]

Appliance. A device that utilizes an energy source to produce light, heat, power, refrigeration, air conditioning, or compressed fuel gas. This definition also shall include a vented decorative appliance.

Appliance, Fan-Assisted Combustion. An appliance equipped with an integral mechanical means to either draw or force products of combustion through the combustion chamber or heat exchanger. [NFPA 54:3.3.5.4]

Appliance, Low-Heat. A fuel-burning appliance that produces a continuous flue gas temperature, at the point of entrance to the flue, of not more than 1000°F (538°C).

Appliance, Medium-Heat. A fuel-burning appliance that produces a continuous flue gas temperature, at the point of entrance to the flue, of more than 1000°F (538°C) and less than 2000°F (1093°C).

Appliance Categorized Vent Diameter/Area. The minimum vent diameter/area permissible for Category I appliances to maintain a nonpositive vent static pressure where tested in accordance with nationally recognized standards. [NFPA 54:3.3.6]

Appliance Flue Outlet. The opening or openings in a cooking device where vapors, combustion gases, or both leave the cooking device. [NFPA 96:3.3.4] There might or might not be ductwork attached to this opening.

Appliance Fuel Connector. An assembly of listed semirigid or flexible tubing and fittings to carry fuel between a fuel-piping outlet and a fuel-burning appliance.

Approved. Acceptable to the Authority Having Jurisdiction.

Exception: [HCD 1 & HCD 2] *"Approved" means meeting the approval of the Enforcing Agency, except as otherwise provided by law, when used in connection with any system, material, type of construction, fixture or appliance as the result of investigations and tests conducted by the agency, or by reason of accepted principles or tests by national authorities, or technical, health, or scientific organizations or agencies.*

Notes:

(1) See Health and Safety Code Section 17920 for "Approved" as applied to residential construction and buildings or structures accessory thereto, as referenced in Sections 1.8.2.1.1 and 1.8.2.1.2.

(2) See Health and Safety Code Section 17921.1 for "Approved" as applied to the use of hotplates in residential construction referenced in Sections 1.8.2.1.1 and 1.8.2.1.2.

(3) See Health and Safety Code Section 19966 for "Approved" as applied to Factory-Built Housing as referenced in Sections 1.8.2.1.1 and 1.8.2.1.2.

(4) See Health and Safety Code Section 18201 for "Approved" as applied to Mobilehome Parks as referenced in Section 1.8.2.1.3.

(5) See Health and Safety Code Section 18862.1 for "Approved" as applied to Special Occupancy Parks as referenced in Section 1.8.2.1.3.

Approved Testing Agency. An organization primarily established for purposes of testing to approved standards and approved by the Authority Having Jurisdiction. ***[HCD 1 & HCD 2]*** *"Approved Testing Agency" is any agency which is determined by the Enforcing Agency, except as otherwise provided by statute, to have adequate personnel and expertise to carry out the testing of systems, materials, and various types of construction, fixtures or appliances.*

Appurtenance. An accessory or a subordinate part that enables the primary device to perform or improve its intended function. [NFPA 96:3.3.5]

Assembly Building. A building or a portion of a building used for the gathering together of 50 or more persons for such purposes as deliberation, education, instruction, worship, entertainment, amusement, drinking, dining, or awaiting transportation. ***[HCD 1, HCD 2 and SFM]*** *Refer to the California Building Code, Title 24, Part 2 for use and occupancy classification.*

Authority Having Jurisdiction. The organization, office, or individual responsible for enforcing the requirements of a code or standard, or for approving equipment, materials, installations, or procedures. The Authority Having Jurisdiction shall be a federal, state, local, or other regional department or an individual such as a plumbing official, mechanical official, labor department official, health department official, building official, or others having statutory authority. In the absence of statutory authority, the Authority Having Jurisdiction may be some other responsible party. This definition shall include the Authority Having Jurisdiction's duly authorized representative. ***[HCD 1, HCD 2, OSHPD 1, 2, 3 & 4 and SFM]*** *"Authority Having Jurisdiction" shall mean "Enforcing Agency" as defined in Section 207.0 of this code.*

Automatic. That which provides a function without the necessity of human intervention.

Automatic Boiler. A boiler equipped with certain controls and limit devices.

Azeotrope. A refrigerant blend containing two or more refrigerants whose equilibrium vapor and liquid phase compositions are the same at a given pressure. At this pressure, the slope of the temperature vs. composition curve equals zero, which mathematically is expressed as $(dt/dx)_p = 0$, which, in turn, implies the occurrence of a maximum, minimum, or saddle point temperature. Azeotropic blends exhibit some segregation of components at other conditions. The extent of the segregation depends on the particular azeotrope and the application. [ASHRAE 34:3]

204.0 – B –

Baffle Plate. An object placed in or near an appliance to change the direction or retard the flow of air, air-fuel mixtures, or flue gases.

Boiler. A closed vessel used for heating water or liquid, or for generating steam or vapor by direct application of heat from combustible fuels or electricity.

Boiler, High-Pressure. A boiler for generating steam at gauge pressures in excess of 15 psi (103 kPa), or for heating water to a temperature in excess of 250°F (121°C) or at a gauge pressure in excess of 160 psi (gauge pressure of 1103 kPa). [NFPA 211:3.3.14.2]

Boiler Room. A room where boilers are installed.

Bonding Conductor or Jumper. A reliable conductor to ensure the required electrical conductivity between metal parts required to be electrically connected. [NFPA 70:100(I)]

Breathing Zone. The region within an occupiable space between planes 3 inches and 72 inches (76 mm and 1829 mm) above the floor and exceeds 2 feet (610 mm) from the walls or fixed air-conditioning equipment. [ASHRAE 62.1:3]

Breathing Zone Outdoor Airflow. The outdoor airflow required in the breathing zone of the occupiable space or spaces in a ventilation zone. [ASHRAE 62.1:6.2.2.1]

Breeching. A metal connector for medium- and high-heat appliances.

Broiler. A general term including broilers, salamanders, barbecues, and other devices cooking primarily by radiated heat, excepting toasters. [NFPA 54:3.3.15]

BTU/H. The listed maximum capacity of any appliance, absorption unit, or burner expressed in British thermal units input per hour, unless otherwise noted.

Building. [HCD 1 & HCD 2] *Any structure used or intended for supporting or sheltering any use or occupancy.*

Exceptions: *"Building" shall not include the following:*

(1) Any mobilehome as defined in Health and Safety Code Section 18008.

(2) Any manufactured home as defined in Health and Safety Code Section 18007.

(3) Any commercial modular as defined in Health and Safety Code Section 18001.8 or any special purpose commercial modular as defined in Health and Safety Code Section 18012.5.

(4) Any recreational vehicle as defined in Health and Safety Code Section 18010.

(5) Any multifamily manufactured home, as defined in Health and Safety Code Section 18008.7.

For additional information, see Health and Safety Code Section 18908.

Building Code. The building code that is adopted by this || jurisdiction. ***[HCD 1, HCD 2, OSHPD 1, 1R, 2, 3, 4 & 5 and SFM]*** *"Building Code" shall mean the California Building Code, Title 24, Part 2.*

Building Official. See Authority Having Jurisdiction. *For applications listed in Section 1.11.0 regulated by the Office of the State Fire Marshal "Building Official" is the officer charged with the administration and enforcement of this code, or a regular deputy. See "Enforcing Agency". For the State of California, "Building Official" shall be the "Enforcing Agency" as specified in Section 1.11.0.*

Burner, Automatic Boiler. A device to convey fuel into the combustion chamber in proximity to its combustion air supply so as to permit a stable, controlled heat release compatible with the burner design and that is equipped with an ignition system to reliably ignite the entire heat release surface of the burner assembly.

205.0 – C –

CAS Number. The Chemical Abstract System registry number.

Ceiling Radiation Damper. A listed device installed in a ceiling membrane of a fire-resistance-rated floor-ceiling or roof-ceiling assembly to automatically limit the radiative heat transfer through an air inlet/outlet opening. [NFPA 5000:3.3.139.1]

Central Heating Plant or Heating Plant. Environmental heating equipment installed in a manner to supply heat by means of ducts or pipes to areas other than the room or space in which the equipment is located.

Certified. A formally stated recognition and approval of an acceptable level of competency, acceptable to the Authority Having Jurisdiction. [NFPA 96:3.3.10]

Certified Person. A person trained and certified by the equipment manufacturer, or by a recognized organization through a formal certification program for the system to be serviced or cleaned; that is acceptable to the Authority Having Jurisdiction.

Chimney. One or more passageways, vertical or nearly so, for conveying flue or vent gases to the outdoors. [NFPA 54:3.3.18]

Chimney, Factory-Built. A chimney composed of listed factory-built components assembled in accordance with the manufacturer's installation instructions to form the completed chimney. [NFPA 54:3.3.18.2]

Chimney, Masonry. A field-constructed chimney of solid masonry units, bricks, stones, listed masonry chimney units, or reinforced portland cement concrete, lined with approved chimney flue liners. [NFPA 54:3.3.18.3]

Chimney, Metal. A field-constructed chimney of metal. [NFPA 54:3.3.18.4]

Chimney Classifications:

Chimney, High-Heat Appliance-Type. A factory-built, masonry, or metal chimney suitable for removing the products of combustion from fuel-burning high-heat appliances producing combustion gases in excess of 2000°F (1093°C), measured at the appliance flue outlet.

Chimney, Low-Heat Appliance-Type. A factory-built, masonry, or metal chimney suitable for removing the products of combustion from fuel-burning low-heat appliances producing combustion gases not in excess of 1000°F (538°C) under normal operating conditions, but capable of producing combustion gases of 1400°F (760°C) during intermittent forced firing for periods up to one hour. All temperatures are measured at the appliance flue outlet.

Chimney, Medium-Heat Appliance-Type. A factory-built, masonry, or metal chimney suitable for removing the products of combustion from fuel-burning medium-heat appliances producing combustion gases, not in excess of 2000°F (1093°C), measured at the appliance flue outlet.

Chimney, Residential Appliance-Type. A factory-built or masonry chimney suitable for removing products of combustion from residential-type appliances producing combustion gases not in excess of 1000°F (538°C), measured at the appliance flue outlet. Factory-built Type HT chimneys have high-temperature thermal shock resistance.

Chimney Connector. The pipe that connects a fuel-burning appliance to a chimney. [NFPA 211:3.3.48.1]

Circulators (Circulating Pump). A device that circulates liquids within a closed circuit for an intended purpose.

Classified. See Listed (Third Party Certified).

Clean(ing). For kitchen exhaust systems and cooking equipment, the act of removing grease, oil deposits, and other residue. [NFPA 96:3.3.12]

Clearly Identified. Capable of being recognized by a person of normal vision without causing uncertainty and indecisiveness about the location or operating process of the identified item. [NFPA 96:3.3.13]

Closed Combustible Construction. Combustible building construction, including walls, structural framing, roofs, roof ceilings, floors, and floor-ceiling assemblies, continuously enclosing a grease duct on four sides where one or more sides require protection in accordance with Section 507.4.

Closed Combustion Solid-Fuel-Burning Appliance. A heat-producing appliance that employs a combustion chamber that has no openings other than the flue collar, fuel-charging door, and adjustable openings provided to control the amount of combustion air that enters the combustion chamber.

Closet. See Confined Space.

Clothes Dryer. An appliance used to dry wet laundry by means of heat. [NFPA 54:3.3.19]

Clothes Dryer, Type 1. Primarily used in family living environment. May or may not be coin-operated for public use. [NFPA 54:3.3.19.1]

Clothes Dryer, Type 2. Used in business with direct intercourse of the function with the public. May or may not be operated by public or hired attendant. May or may not be coin-operated. [NFPA 54:3.3.19.2]

Coastal High Hazard Areas. An area within the flood hazard area that is subject to high-velocity wave action, and shown on a Flood Insurance Rate Map or other flood hazard map as Zone V, VO, VE, or V1-30.

Code. A standard that is an extensive compilation of provisions covering broad subject matter or that is suitable for adoption into law independently of other codes and standards.

Combination Fire and Smoke Damper. A device that meets both the fire damper and smoke damper requirements. [NFPA 5000:3.3.139.2]

Combustible Material. A material that, in the form in which it is used and under the conditions anticipated, will ignite and burn; a material that does not meet the definition of noncombustible. [NFPA 54:3.3.64.1]

Combustion Air. The total amount of air provided to the space that contains fuel-burning equipment. Includes air for fuel combustion, draft hood dilution, and ventilation of the equipment enclosure.

Combustion Chamber. The portion of an appliance within which combustion occurs. [NFPA 54:3.3.21]

Commercial Food Heat-Processing Equipment. Equipment used in a food establishment for heat-processing food or utensils and that produces grease vapors, steam, fumes, smoke, or odors that are required to be removed through a local exhaust ventilation system.

Compensating Hood. A hood for commercial food heat-processing equipment that has an outside-air supply with air delivered below or within the hood. Where makeup air is diffused directly into the exhaust within the hood cavity, it becomes a short-circuit hood.

Compressor, Positive Displacement. A compressor in which increase in pressure is attained by changing the internal volume of the compression chamber.

Compressor, Refrigerant. A machine, with or without accessories, for compressing a refrigerant vapor.

Concealed Spaces. That portion(s) of a building behind walls, over suspended ceilings, in pipe chases, attics, and elsewhere whose size might normally range from 1¾ inch (44 mm) stud spaces to 8 foot (2438 mm) interstitial truss spaces and that might contain combustible materials such as building structural members, thermal, electrical insulation, or both, and ducting. Such spaces have sometimes been used as HVAC plenum chambers.

Condensate. The liquid phase produced by condensation of a particular gas or vapor.

Condenser. The part of the system designed to liquefy refrigerant vapor by removal of heat.

Condensing Appliance. An appliance that condenses part of the water vapor generated by the burning of hydrogen in fuels.

Condensing Unit. A mechanical refrigeration system, consisting of one or more power-driven compressors, condensers, liquid receivers where provided, and the regularly furnished accessories that have been factory assembled and tested prior to its installation.

Conditioned Space. An area, room, or space normally occupied and being heated or cooled for human habitation by any equipment.

Confined Space. A room or space having a volume less than 50 cubic feet per 1000 British thermal units per hour (Btu/h) (4.83 m^3/kW) of the aggregate input rating of all fuel-burning appliances installed in that space.

Construction Documents. Plans, specifications, written, graphic, and pictorial documents prepared or assembled for describing the design, location, and physical characteristics of the elements of a project necessary for obtaining a permit.

Continuous Enclosure. A recognized architectural or mechanical component of a building having a fire resistance rating as required for the structure and whose purpose is to enclose the vapor removal duct for its full length to its termination point outside the structure without any portion of the enclosure having a fire resistance rating less than the required value. [NFPA 96:3.3.22.1]

Continuous Pilot. A pilot that burns without turndown throughout the entire period that the boiler is in service, whether or not the main burner is firing.

Continuous Weld. A metal-joining method that produces a product without visible interruption or variation in quality. [NFPA 96:3.3.15] For the purpose of the definition, it specifically includes the exhaust compartment of hoods and welded joints of exhaust ducts, yet specifically does not include filter support frames or appendages inside hoods.

Conversion Burner. A unit consisting of a burner and its controls utilizing gaseous fuel for installation in an appliance originally utilizing another fuel. [NFPA 54:3.3.17.2]

Cooling. Air cooling to provide a room or space temperature of 68°F (20°C) or above.

Cooling System. All of the equipment, including associated refrigeration, intended or installed for the purpose of cooling air by mechanical means and discharging such air into any room or space. This definition shall not include an evaporative cooler.

Cooling Unit. A self-contained refrigeration system that has been factory assembled tested, and installed with or without conditioned air and ducts, without connecting any refrigerant-containing parts. This definition shall not include a portable cooling unit or an absorption unit.

Copper Alloy. A homogenous mixture of two or more metals in which copper is the primary component, such as brass and bronze.

Crawl Space. In a building, an area accessible by crawling, having a clearance less than human height, for access to plumbing or wiring, storage, etc.

CSST. An acronym for corrugated stainless steel tubing.

206.0 – D –

Damper. A valve or plate for controlling draft or the flow of gases, including air. [NFPA 211:3.3.52]

Fire Damper. An automatic-closing metal assembly consisting of one or more louvers, blades, slats, or vanes that closes upon detection of heat so as to restrict the passage of flame and is listed to the applicable recognized standards.

Smoke Damper. A damper arranged to seal off airflow automatically through a part of an air duct system so as to restrict the passage of smoke and is listed to the applicable recognized standard.

Volume Damper. A device that, when installed, will restrict, retard, or direct the flow of air in any duct, or the products of combustion in any heat-producing equipment, its vent connector, vent, or chimney.

Department. [HCD 1 & HCD 2] *"Department" means the Department of Housing and Community Development.*

Design Flood Elevation. The elevation of the "design flood," including wave height, relative to the datum specified on the community's legally designated flood hazard map. In areas designated as Zone AO, the design flood elevation is the elevation of the highest existing grade of the building's perimeter plus the depth number (in feet) specified on the flood hazard map. In areas designated as Zone AO where a depth number is not specified on the map, the depth number is taken as being equal to 2 feet (610 mm).

Detection Devices. Electrical, pneumatic, thermal, mechanical, or optical sensing instruments, or subcomponents of such instruments, whose purpose is to cause an automatic action upon the occurrence of some preselected event. [NFPA 96:3.3.17] In the context of this document, the event in question could be excessive temperature or flame, and the action could be the operation of a fire-extinguishing system.

Dips. Depression or cup like places in horizontal duct runs in which liquids could accumulate.

Direct Gas-Fired Non-Recirculating Industrial Air Heater. A non-recirculating industrial air heater in which all the products of combustion generated by the appliance are released into the outdoor airstream being heated. [NFPA 54:3.3.56.1]

Direct Gas-Fired Recirculating Industrial Air Heater. An air recirculating heater in which all of the products of combustion generated by the appliance are released into the airstream being heated. [NFPA 54:3.3.56.2]

Direct-Vent Appliances. Appliances that are constructed and installed so that all air for combustion is derived directly from the outdoors and all flue gases are discharged to the outdoors. [NFPA 54:3.3.5.3]

Discharge. The final portion of a duct or pipe where the product being conveyed is emptied or released from confinement; the termination point of the pipe or duct. [NFPA 96:3.3.18]

Discrete Products in Plenums. Individual, distinct products which are non-continuous such as pipe hangers, duct registers, duct fittings, and duct straps.

District Heating Plant. A power boiler plant designed to distribute hot water or steam to users located off the premises.

Draft Hood. A nonadjustable device built into an appliance, or made a part of the vent connector from an appliance, that is designed to:

(1) Provide for the ready escape of the flue gases from the appliance in the event of no draft, backdraft, or stoppage beyond the draft hood.

(2) Prevent a backdraft from entering the appliance.

(3) Neutralize the effect of stack action of the chimney or gas vent upon the operation of the appliance. [NFPA 54:3.3.31]

Duct. A tube or conduit for transmission of air, fumes, vapors, or dust. This definition shall not include:

(1) A vent, vent connector, or chimney connector.

(2) A tube or conduit wherein the pressure of the air exceeds 1 psi (7 kPa).

(3) The air passages of listed self-contained systems.

Duct Furnace. A furnace normally installed in distribution ducts of air-conditioning systems to supply warm air for heating. This definition applies only to an appliance that, for air circulation, depends on a blower not furnished as part of the furnace. [NFPA 54:3.3.45.3]

Duct System. A continuous passageway for the transmission of air and vapors that, in addition to the containment components themselves, might include duct fittings, dampers, plenums, other items, and air-handling equipment. [NFPA 96:3.3.20]

Ductless Mini-Split System. A heating and cooling equipment that includes one or multiple indoor evaporator, air handler, or both units, an outdoor condensing unit that is connected by refrigerant piping, and electrical wiring. A ductless mini-split system is capable of cooling or heating one or more rooms without the use of traditional ductwork.

Dwelling. A building or portion thereof that contains not more than two dwelling units.

Dwelling Unit. A building or portion thereof that contains living facilities, including provisions for sleeping, eating, cooking, and sanitation, as required by this code, for not more than one family.

207.0 – E –

Easily Accessible. See Accessible, Readily.

Effective Ground-Fault Current Path. An intentionally constructed, low-impedance electrically conductive path designed and intended to carry current under ground-fault conditions from the point of a ground fault on a wiring system to the electrical supply source and that facilitates the operation of the overcurrent protective device or ground-fault detectors on high-impedance grounded systems. [NFPA 54:3.3.34]

Electric Duct Heaters. A heater located in the airstream of a forced-air system where the air-moving unit is not provided as an integral part of the equipment.

Electric Heating Appliance. A device that produces heat energy to create a warm environment by the application of electric power to resistance elements, refrigerant compressors, or dissimilar material junctions.

Electrical Code. The National Electrical Code promulgated by the National Fire Protection Association, as adopted by this jurisdiction. ***[HCD 1 & HCD 2]*** *Whenever the term "Electrical Code" is used in this code, it shall mean the California Electrical Code, Title 24, Part 3.*

Emergency Alarm System. A system intended to provide the indication and warning of abnormal conditions and summon appropriate aid.

Emergency Control Station. An approved location on the premises where signals from emergency equipment are received.

Enforcing Agency. [HCD 1, HCD 2, SFM and OSHPD 1, 1R, 2, 3, 4 & 5] *"Enforcing Agency" is the designated department or agency as specified by statute and regulation.*

Environmental Air Duct. Ducting used for conveying air at temperatures not exceeding 250°F (121°C) to or from occupied areas of any occupancy through other than heating or air-conditioning systems, such as ventilation for human usage, domestic kitchen range exhaust, bathroom exhaust ducts, and domestic-type clothes dryer exhaust ducts.

Equipment. A general term including materials, fittings, devices, appliances, and apparatus used as part of or in connection with installations regulated by this code.

Evaporative Cooler. A device used for reducing the sensible heat of air for cooling by the process of evaporation of water into an airstream.

Evaporative Cooling System. Equipment intended or installed for the purpose of environmental cooling by an evaporative cooler from which the conditioned air is distributed through ducts or plenums to the conditioned area.

Evaporator. Part of a refrigeration system in which liquid refrigerant is vaporized to produce refrigeration.

Excess Flow Valve (EFV). A valve designed to activate when the fuel gas passing through it exceeds a prescribed flow rate. [NFPA 54:3.3.99.3]

208.0 – F –

Fabrication Area (Fab Area). An area within a Group H Occupancy semiconductor fabrication facility and related research and development areas in that there are processes involving hazardous production materials. Such areas are allowed to include ancillary rooms or areas such as dressing rooms and offices that are directly related to the fab area processes.

Factory-Built Grease Duct Enclosures. A listed factory-built grease duct system evaluated as an enclosure system for reduced clearances to combustibles and as an alternative to a duct with its fire-rated enclosure. [NFPA 96:3.3.22.2.1]

Family. [HCD 1] *"Family" is an individual or two or more persons who are related by blood or marriage, or otherwise, live together in a dwelling unit.*

Field-Applied Grease Duct Enclosures. A listed system evaluated for reduced clearances to combustibles and as an alternative to a duct with its fire-rated enclosure. [NFPA 96:3.3.22.2.2]

Fire Code. *Whenever the term "Fire Code" is used in this code, it shall mean the California Fire Code, Title 24, Part 9.*

Fire Partition. An interior wall or partition of a building that separates two areas and serves to restrict the spread of fire but does not qualify as a fire wall.

Fire Resistance Rating. The time, in minutes or hours, that materials or assemblies have withstood a fire exposure as established in accordance with ASTM E119 or UL 263.

Fire-Resistive Construction. Construction in accordance with the requirements of the building code for the time period specified.

Fire Wall. A wall separating buildings or subdividing a building to prevent the spread of the fire and having a fire resistance rating and structural stability. [NFPA 96:3.3.26]

Fireplace Stove. A chimney-connected, solid-fuel-burning stove (appliance) having part of its fire chamber open to the room.

Flammable Vapor or Fumes. The concentration of flammable constituents in air that exceeds 25 percent of its Lower Flammability Limit (LFL).

Flood Hazard Area. The greater of the following two areas:

(1) The area within a floodplain subject to a 1 percent or greater chance of flooding in any given year.

(2) The area designated as a flood hazard area on a community's flood hazard map, or otherwise legally designated.

Floor Furnace. A completely self-contained unit furnace suspended from the floor of the space being heated, taking air for combustion from outside this space. [NFPA 54:3.3.45.5] With means for observing flames and lighting the appliance from such space.

Forced-Air Furnace. A furnace equipped with a fan or blower that provides the primary means for circulation of air. [NFPA 54:3.3.45.6]

Downflow-Type Furnace. A forced-air-type furnace designed with airflow essentially in a vertical path, discharging air at or near the bottom of the furnace. [NFPA 211:3.3.79.2]

Enclosed Furnace. A specific heating or heating and ventilating furnace incorporating an integral total enclosure and using only outside air for combustion.

Horizontal-Type Furnace. A forced-air-type furnace designed with airflow through the furnace, essentially in a horizontal path. [NFPA 211:3.3.79.3]

Upflow-Type Furnace. A forced-air-type furnace designed with airflow essentially in a vertical path, discharging air at or near the top of the furnace. [NFPA 211:3.3.79.5]

Fractionation. A change in composition of a blend by preferential evaporation of the more volatile component or condensation of the less volatile component.

Fuel Gas. Natural, manufactured, liquefied petroleum, or a mixture of these.

Fume Incinerators. Devices utilizing intense heat or fire to break down, oxidize, or both vapors and odors contained in gases or air being exhausted into the atmosphere. [NFPA 96:3.3.27]

Furnace, Central. A self-contained appliance for heating air by transfer of heat of combustion through metal to the air and designed to supply heated air through ducts to spaces remote from or adjacent to the appliance location. [NFPA 54:3.3.45.1]

Fusible Link. A form of fixed-temperature heat-detecting device sometimes employed to restrain the operation of an electrical or mechanical control until its designed temperature is reached. [NFPA 96:3.3.28] Such devices are to be replaced following each operation.

Fusible Plug. A device arranged to relieve pressure by operation of a fusible member at a predetermined temperature.

209.0 – G –

Galvanized Steel. A steel that has been coated with a thin layer of zinc for corrosion protection; *conforming to the requirements of ASTM A653/A653M Standard Specification for Steel Sheet, Zinc-Coat (Galvanized) or Zinc-Iron Alloy-Coat (Galvanized) by the Hot Dip Process.*

Gas Convenience Outlet. A permanently mounted, hand-operated device providing a means for connecting and disconnecting an appliance or an appliance connector to the gas supply piping. The device includes an integral, manually operated gas valve with a nondisplaceable valve member so that disconnection can be accomplished only where the manually operated gas valve is in the closed position. [NFPA 54:3.3.48]

Gas Piping. An installation of pipe, valves, or fittings that are used to convey fuel gas, installed on any premises or in a building, but shall not include:

(1) A portion of the service piping.

(2) An approved piping connection 6 feet (1829 mm) or less in length between an existing gas outlet and a gas appliance in the same room with the outlet.

Gas Piping System. An arrangement of gas piping or regulators after the point of delivery and each arrangement of gas piping serving a building, structure, or premises, whether individually metered or not.

Generator. A device equipped with a means of heating used in an absorption system to drive refrigerant out of solution.

Gravity Heating System. A heating system consisting of a gravity-type warm air furnace, together with all air ducts or pipes and accessory apparatus installed in connection therewith.

Gravity-Type Floor Furnace. A floor furnace depending primarily on circulation of air by gravity. This classification also includes floor furnaces equipped with booster-type fans that do not materially restrict free circulation of air by gravity flow when such fans are not in operation. [NFPA 211:3.3.79.12.2]

Grease. Rendered animal fat, vegetable shortening, and other such oily matter used for the purposes of and resulting from cooking, preparing foods, or both. [NFPA 96:3.3.29] Grease might be liberated and entrained with exhaust air or might be visible as a liquid or solid.

Grease Ducts. A containment system for the transportation of air and grease vapors that is designed and installed to reduce the possibility of the accumulation of combustible condensation and the occurrence of damage if a fire occurs within the system. [NFPA 96:3.3.20.2]

Grease Filter. A removable component of the grease removal system designed to capture grease and direct it to a safe collection point. [NFPA 96:3.3.24.1]

Grease Filter, Mesh-Type. A filter construction consisting of a net made from intersecting strands with a space between each strand. [NFPA 96:3.3.24.2]

Grease Removal Devices. A system of components designed and intended to process vapors, gases, or air as it is drawn through such devices by collecting the airborne grease particles and concentrating them for further action at some future time, leaving the exiting air with a lower amount of combustible matter.

Greasetight. Constructed and performing in such a manner as not to permit the passage of grease under normal cooking conditions. [NFPA 96:3.3.31]

Grounding Electrode. A conducting object through which a direct connection to earth is established. [NFPA 70:100(I)]

210.0 – H –

Hazardous Location. An area or space where combustible dust, ignitable fibers, flammable liquids, volatile liquids, gases, vapors, or mixtures are or may be present in the air in quantities sufficient to produce explosive or ignitable mixtures.

Hazardous Process Piping (HPP). A process material piping or tubing conveying a liquid or gas that has a degree-of-hazard rating in health, flammability, or reactivity of Class 3 or 4, as ranked by the fire code.

|| ***Health Facilities. [OSHPD 1, 1R, 2, 3, 4 & 5]*** *Buildings specified within the statutory authority of the Office of Statewide Health Planning and Development.*

Heat Pump. A refrigeration system that extracts heat from one substance and transfers it to another portion of the same substance or to a second substance at a higher temperature for a beneficial purpose.

Heat (Energy) Recovery Ventilator. A device intended to remove air from buildings, replace it with outside air, and in the process transfer heat from the warmer to the colder airstreams.

Heating Degree Day. A unit, based upon temperature difference and time, used in estimating fuel consumption and specifying nominal annual heating load of a building. For any one day when the mean temperature is less than 65°F (18°C), there exist as many degree days as there is Fahrenheit degrees difference in temperature between mean temperature for the day and 65°F (18°C).

Heating Equipment. Includes warm air furnaces, warm air heaters, combustion products vents, heating air-distribution ducts and fans, and all steam and hot water piping, together with all control devices and accessories installed as part of, or in connection with, any environmental heating system or appliance regulated by this code.

Heating System. A warm air heating plant consisting of a heat exchanger enclosed in a casing, from which the heated air is distributed through ducts to various rooms and areas. A heating system includes the outside air, return air, and supply air system, and all accessory apparatus and equipment installed in connection therewith.

High Limit Control Device. An operating device installed and serving as an integral component of a deep-fat fryer that provides secondary limitation of the grease temperature by automatically disconnecting the thermal energy input when the temperature limit is exceeded. [NFPA 96:3.3.32]

High Purity Piping. A form of process piping but is usually specified for critical clean applications in the semiconductor, pharmaceutical, biotechnology, chemical, fiber optics, food, and dairy industries.

Highside. The parts of a refrigeration system subjected to approximately condenser pressure.

Hood. An air-intake device connected to a mechanical exhaust system for collecting and removing grease-laden vapors, fumes, smoke, steam, heat, or odors from commercial food heat-processing equipment.

Fixed Baffle. A listed unitary exhaust hood design where the grease removal device is a nonremovable assembly that contains an integral fire-activated water-wash fire-extinguishing system listed for this purpose. [NFPA 96:3.3.33.1]

Type I. A kitchen hood for collecting and removing grease and smoke.

Type II. A general kitchen hood for collecting and removing steam, vapor, heat, or odors.

Hot-Water-Heating Boiler. A boiler having a volume exceeding 120 gallons (454 L), a heat input exceeding 200 000 Btu/h (58.6 kW), or an operating temperature exceeding 210°F (99°C) that provides hot water to be used externally to itself.

HPM Storage Room. A room used for the storage or dispensing of hazardous production material (HPM) and that is classified as a Group H, Division 1 or Division 2 Occupancy.

Hydronics. Of or relating to a heating or cooling system that transfers energy by circulating a fluid through a system of pipes or tubing.

211.0 – I –

IDLH (Immediately Dangerous to Life and Health). A concentration of airborne contaminant's, normally expressed in parts per million (ppm) or milligrams per cubic meter (mg/m^3), that represents the maximum level from which one is capable of escaping within 30 minutes without escape-impairing symptoms or irreversible health effects. This level is established by the National Institute of Occupational Safety and Health (NIOSH).

Incinerator. An appliance or combustion chamber for the reduction, by burning, of rubbish, garbage, and other wastes. [NFPA 211:3.3.91]

Industrial Heating Equipment. Includes appliances, devices, or equipment used, or intended to be used, in an industrial, manufacturing, or commercial occupancy for applying heat to any material being processed, but shall not include water heaters, boilers, or portable equipment used by artisans in pursuit of a trade.

Insanitary Location. An area, space, or room where the air is unfit or undesirable for circulation to occupiable parts of a building.

Interconnected. Mutually assembled to another component in such a manner that the operation of one directly affects the other or that the contents of one specific duct system are allowed to encounter or contact the products being moved by another duct system. [NFPA 96:3.3.34]

Interlock. A device that senses a limit or off-limit condition or improper sequence of events and shuts down the offending or related piece of equipment or prevents proceeding in an improper sequence in order to prevent a hazardous condition from developing.

Intermittent Pilot. A pilot that burns during light-off and while the main burner is firing, and that is shut off with the main burner.

Interrupted Pilot. A pilot that burns during light-off and that is shut off during normal operation of the main burner.

212.0 – J –

Joint, Brazed. A joint obtained by joining of metal parts with alloys that melt at temperatures exceeding 840°F (449°C) but less than the melting temperature of the parts being joined.

Joint, Compression. A multipiece joint with cup-shaped threaded nuts that, when tightened, compress tapered sleeves so that they form a tight joint on the periphery of the tubing they connect.

Joint, Flanged. One made by bolting together a pair of flanged ends.

Joint, Flared. A metal-to-metal compression joint in which a conical spread is made on the end of a tube that is compressed by a flare nut against a mating flare.

Joint, Mechanical. General form for gastight or liquid-tight joints obtained by the joining of parts through a positive holding mechanical construction.

Joint, Press-Connect. A permanent mechanical joint consisting of an elastomeric seal or an elastomeric seal and corrosion-resistant grip ring. The joint is made with a pressing tool and jaw or ring approved by the fitting manufacturer.

Joint, Soldered. A joint obtained by the joining of metal parts with metallic mixtures or alloys that melt at a temperature up to and including 840°F (449°C).

Joint, Welded. A gastight joint obtained by the joining of metal parts in the plastic molten state.

213.0 – K –

No definitions.

214.0 – L –

Labeled. Equipment or materials bearing a label of a listing agency (accredited conformity assessment body). See Listed (Third Party Certified). ***[HCD 1 & HCD 2]*** *"Labeled" means equipment or materials to which has been attached a label, symbol or other identifying mark of an organization, approved by the Department, that maintains a periodic inspection program of production of labeled products, installations, equipment, or materials and by whose labeling the manufacturer indicates compliance with appropriate standards or performance in a specified manner.*

LEL (Lower Explosive Limit). See LFL.

LFL (Lower Flammable Limit or Lower Limit of Flammability). The minimum concentration of a substance that propagates a flame through a homogeneous mixture of the substance and air under the specified test conditions. The LFL is sometimes referred to as LEL (Lower Explosive Limit). For the purposes of this definition, LFL and LEL are identical.

Limited-Combustible Material. Refers to a building construction material that does not comply with the definition of noncombustible material that, in the form in which it is used, has a potential heat value not exceeding 3500 British thermal units per pound-force (Btu/lb) (8141 kJ/kg), where tested in accordance with NFPA 259, and includes either of the following:

(1) Materials having a structural base of noncombustible material, with a surfacing not exceeding a thickness of ⅛ of an inch (3.2 mm), that has a flame-spread index not greater than 50.

(2) Materials, in the form and thickness used, having neither a flame-spread index greater than 25 nor evidence of continued progressive combustion, and of such composition that surfaces that would be exposed by cutting through the material on any plane would have neither a flame-spread index greater than 25 nor evidence of continued progressive combustion, where tested in accordance with ASTM E84.

Line Contact Installation. An installation in which a furnace is installed so that building joists, studs, or framing are contacted by the furnace jacket upon the lines formed by the intersection of the jacket sides with the top surface.

Lineset. A set of two refrigerant pipes that extends from the condenser to the evaporator (cooling coil) in direct systems, consisting of a suction line and a liquid line.

Liquefied Petroleum Gas (LP-Gas). Means and includes a material composed predominantly of any of the following hydrocarbons or mixtures of them: propane, propylene, butanes (normal butane or isobutane), and butylenes. When reference is made to liquefied petroleum gas in this code, it shall refer to liquefied petroleum gases in either the liquid or gaseous state.

Liquefied Petroleum Gas (LP-Gas) Facilities. Liquefied petroleum gas (LP-Gas) facilities include tanks, containers, container valves, regulating equipment, meters, appurtenances, or any combination thereof for the storage and supply of liquefied petroleum gas for a building, structure, or premises.

Liquid-Tight. Constructed and performing in such a manner as not to permit the passage of liquid at any temperature. [NFPA 96:3.3.35]

Listed (Third Party Certified). Equipment or materials included in a list published by a listing agency (accredited conformity assessment body) that maintains periodic inspection of current production of listed equipment or materials and whose listing states either that the equipment or material complies with approved standards or has been tested and found suitable for use in a specified manner. ***[HCD 1 & HCD 2]*** *"Listed" means all products that appear in a list published by an approved testing or listing agency. For additional information, see Health and Safety Code Section 17920(h).*

Listed and Listing. [SFM] *"Listed" and "Listing" are terms referring to equipment or materials included in a list published by an approved testing laboratory, inspection agency, or other organization concerned with product evaluation that maintains periodic inspection of current productions of listed equipment or materials and which listing states that the material or equipment complies with approved nationally recognized codes, standards, or tests and has been tested or evaluated and found suitable for use in a specific manner. These terms shall also mean equipment or materials accepted by the State Fire Marshal as conforming to the provisions of these regulations and which are included in a list published by the State Fire Marshal.*

Listing Agency. An agency accredited by an independent and authoritative conformity assessment body to operate a material and product listing and labeling (certification) system and that are accepted by the Authority Having Jurisdiction, which is in the business of listing or labeling. The system includes initial and ongoing product testing, a periodic inspection on current production of listed (certified) products, and that makes available a published report of such listing in which specific information is included that the material or product is in accordance with applicable standards and found safe for use in a specific manner. ***[HCD 1 & HCD 2]*** *"Listing Agency" means an agency approved by the Department that is in the business of listing and labeling products, materials, equipment, and installations tested by an approved testing agency, and that maintains a periodic inspection program on current production of listed products, equipment, and installations, and that at least annually makes available a published report of these listings. For additional information, see Health and Safety Code Section 17920(i).*

Low-Pressure Hot-Water-Heating Boiler. A boiler furnishing hot water at pressures not exceeding 160 psi (1103 kPa) and at temperatures not exceeding 250°F (121°C).

Low-Pressure Steam-Heating Boiler. A boiler furnishing steam at pressures not exceeding 15 psi (103 kPa).

Lowside. Refers to the parts of a refrigeration system subjected to approximate evaporator pressure.

215.0 – M –

Machinery. The refrigeration equipment forming a part of the refrigeration system, including, but not limited to, the following: compressors, condensers, liquid receivers, evaporators, and connecting piping.

Machinery Room. [SFM] *Machinery Room is a room in which a refrigeration system is permanently installed and operated but not including evaporators located in a cold storage room, refrigerator box, air-cooled space or other enclosed space. Closets solely contained within, and opening only into, a room shall not be considered machinery rooms, but shall be considered a part of the machinery rooms in which they are contained or open into. It is not the intent of this definition to cause the space in which unit or self-contained systems of Group I refrigerants are locate to be classified as machinery rooms.*

Mechanical Ventilation. Ventilation provided by mechanically powered equipment, such as motor-driven fans and blowers, but not by devices such as wind-driven turbine ventilators and mechanically operated windows. [ASHRAE 62.1:3]

Miniature Boiler. A power boiler having an internal shell diameter of 16 inches (406 mm) or less, a gross volume of 5 cubic feet (0.14 m^3) or less, a heating surface of 20 square feet (1.86 m^2) or less (not applicable to electric boilers), and not exceeding 100 psi (689 kPa).

216.0 – N –

Natural Ventilation. Ventilation provided by thermal, wind, or diffusion effects through doors, windows, or other intentional openings in the building. [ASHRAE 62.1:3]

Noncombustible Material. As applied to building construction material, means a material that in the form in which it is used is either one of the following:

(1) A material that, in the form in which it is used and under the conditions anticipated, will not ignite, burn, support combustion, or release flammable vapors when subjected to fire or heat. Materials that are reported as passing ASTM E136 are considered noncombustible material.

(2) Material having a structural base of noncombustible material as defined in item 1 above, with a surfacing material not over ⅛ of an inch (3.2 mm) thick that has a flame-spread index not higher than 50.

Noncombustible does not apply to surface finish materials. Material required to be noncombustible for reduced clearances to flues, heating appliances, or other sources of high temperature shall refer to material in accordance with item 1 above. No material shall be classed as noncombustible that is subject to increase in combustibility or flame-spread index beyond the limits herein established, through the effects of age, moisture, or other atmospheric condition.

Nonhazardous Process Piping (NPP). Production material piping or tubing conveying a liquid or gas that is not classified as hazardous production material piping.

Nuisance. [HCD 1 & HCD 2] *"Nuisance" shall mean any nuisance as defined in Health and Safety Code Section 17920(l).* ||

Notes:

1. *For applications subject to the Mobilehome Parks Act as referenced in Section 1.8.3.2.2 of this code, refer to California Code of Regulations, Title 25, Division 1, Chapter 2 for the definition of "Nuisance".*
2. *For applications subject to the Special Occupancy Parks Act as referenced in Section 1.8.3.2.3 of this code, refer to California Code of Regulations, Title 25, Division 1, Chapter 2.2 for the definition of "Nuisance".*

217.0 – O –

Occupancy. The purpose for which a building or part thereof is used or intended to be used.

Occupancy Classification. Classifications are defined in the building code. ***[HCD 1, HCD 2 & SFM]*** *Whenever the term "Building Code" is used in this code, it shall mean the California Building Code, Title 24, Part 2.*

Occupational Exposure Limit (OEL). The time-weighted average (TWA) concentration for a normal 8-hour workday and a 40-hour workweek to which nearly all workers can be repeatedly exposed without adverse effect, based on the OSHA PEL, ACGIH TLV-TWA, TERA OARS-WEEL, or consistent value. [ASHRAE 34:3]

Occupiable Space. An enclosed space intended for human activities excluding those spaces intended primarily for other purposes such as storage rooms and equipment rooms that are only occupied occasionally and for short periods of time. [ASHRAE 62.1:3]

Open Combustible Construction. Combustible building construction, including wall, structural framing, roof, roof ceiling, floor, and floor-ceiling assemblies, adjacent to a grease duct on three or fewer sides where one or more sides require protection in accordance with Section 507.4.

218.0 – P –

Package Boiler. A class of boiler defined herein and shall be a boiler equipped and shipped complete with fuel-burning equipment, automatic controls and accessories, and mechanical draft equipment.

PE. Polyethylene.

PE-AL-PE. Polyethylene-aluminum-polyethylene.

PE-RT. Polyethylene of raised temperature.

PEL (Permissible Exposure Limit). The time-weighted average concentration [set by the U.S. Occupational Safety and Health Administration (OSHA)] for a normal 8-hour workday and a 40-hour workweek to which nearly all workers can be repeatedly exposed without adverse effect. Chemical manufacturers publish similar recommendations [e.g., acceptable exposure level (AEL), industrial exposure limit (IEL), or occupational exposure limit (OEL), depending on the company], generally for substances for which PEL has not been established. [ASHRAE 34:3] The maximum permitted time-weighted average exposures to be utilized are those published in 29 CFR 1910.1000.

PEX. Cross-linked polyethylene.

PEX-AL-PEX. Cross-linked polyethylene-aluminum-cross-linked polyethylene.

Pilot. A burner smaller than the main burner that is ignited by a spark or other independent and stable ignition source, and that provides ignition energy required to immediately light off the main burner.

Piping. The pipe or tube mains for interconnecting the various parts of a system. Piping includes pipe, tube, flanges, bolting, gaskets, valves, fittings the pressure-containing parts of other components such as expansion joints, strainers, and devices that serve such purposes as mixing, separating, snubbing, distributing, metering, or controlling flow, pipe-supporting fixtures and structural attachments.

Pitched. To be fixed or set at a desired angle or inclination. [NFPA 96:3.3.39]

Plenum. An air compartment or chamber including uninhabited crawl space areas above a ceiling or below a floor, including air spaces below raised floors of computer/data processing centers or attic spaces, to one or more ducts are connected and that forms part of either the supply-air, return-air, or exhaust-air system, other than the occupiable space being conditioned.

Plumbing Code. The Uniform Plumbing Code promulgated by the International Association of Plumbing and Mechanical Officials, as adopted by this jurisdiction. ***[HCD 1 & HCD 2]*** *Whenever the term "Plumbing Code" is used in this code, it shall mean the California Plumbing Code, Title 24, Part 5.*

Portable Cooling Unit. A self-contained refrigerating system, not over 3 horsepower (hp) (2.2 kW) rating that has been factory assembled and tested, installed without supply-air ducts and without connecting any refrigerant-containing parts. This definition shall not include an absorption unit.

Portable Evaporative Cooler. An evaporative cooler that discharges the conditioned air directly into the conditioned area without the use of ducts and can be readily transported from place to place without dismantling any portion thereof.

Portable Heating Appliance. A heating appliance designed for environmental heating that may have a self-contained fuel supply and is not secured or attached to a building by any means other than by a factory-installed power supply cord.

Portable Ventilating Equipment. Ventilating equipment that can be readily transported from place to place without dismantling a portion thereof and that is not connected to a duct.

Power Boiler. A boiler in which steam is generated at pressures exceeding 15 psi (103 kPa).

Power Boiler Plant. One or more power steam boilers or power hot water boilers and connecting piping and vessels within the same premises.

Power Hot Water Boiler (High Temperature Water Boiler). A boiler used for heating water or liquid to a pressure exceeding 160 psi (1103 kPa) or to a temperature exceeding 250°F (121°C).

PP. Polypropylene.

Pressure, Design. The maximum working pressure for which a specific part of a refrigeration system is designed.

Pressure, Field Test. A test performed in the field to prove system tightness.

Pressure-Imposing Element. A device or portion of the equipment used for the purpose of increasing the pressure of the refrigerant vapor.

Pressure-Limiting Device. A pressure-responsive mechanism designed to automatically stop the operation of the pressure-imposing element at a predetermined pressure.

Pressure-Relief Device. A pressure-actuated valve or rupture member or fusible plug designed to automatically relieve excessive pressure.

Pressure Test. The minimum gauge pressure to which a specific system component is subjected under test condition.

Pressure Vessel (Unfired). A closed container, having a nominal internal diameter exceeding 6 inches (152 mm) and a volume exceeding 1½ cubic feet (0.04 m^3), for liquids, gases, vapors subjected to pressures exceeding 15 psi (103 kPa), or steam under a pressure.

Pressure Vessel, Refrigerant. A refrigerant-containing receptacle that is a portion of a refrigeration system, but shall not include evaporators, headers, or piping of certain limited size and capacity.

Process Piping. Piping or tubing that conveys liquid or gas, which is used directly in research, laboratory, or production processes.

Product-Conveying Duct. Ducting used for conveying solid particulates, such as refuse, dust, fumes, and smoke; liquid particulate matter, such as spray residue, mists, and fogs; vapors, such as vapors from flammable or corrosive liquids; noxious and toxic gases; and air at temperatures exceeding 250°F (121°C).

Purge. The acceptable method of scavenging the combustion chamber, boiler passes, and breeching to remove combustible gases.

PVC. Polyvinyl chloride.

219.0 – Q –

Qualified. A competent and capable person or company that has met the requirements and training for a given field acceptable to the Authority Having Jurisdiction.

Quick-Disconnect Device, Fuel Gas. A hand-operated device that provides a means for connecting and disconnecting an appliance or an appliance connector to a gas supply and that is equipped with an automatic means to shut off the gas supply when the device is disconnected. [NFPA 54:3.3.28.3]

220.0 – R –

Radiant Room Heater. A room heater designed to transfer heat primarily by direct radiation. [NFPA 211:3.3.88.2.2]

Receiver, Liquid. A vessel permanently connected to a refrigeration system by inlet and outlet pipes for storage of liquid.

Recirculating Systems. Systems for control of smoke or grease-laden vapors from commercial cooking equipment that do not exhaust to the outside. [NFPA 96:3.3.41]

Reclaimed Refrigerants. Refrigerants reprocessed to the same specifications as new refrigerants by any means, including distillation. Such refrigerants have been chemically analyzed to verify that those specifications have been met. [ASHRAE 15:3]

Recovered Refrigerants. Refrigerants removed from a system in any condition without necessarily testing or processing them. [ASHRAE 15:3]

Recycled Refrigerants. Refrigerants for which contaminants have been reduced by oil separation, removal of noncondensible gases, and single or multiple passes through filter driers or other devices that reduce moisture, acidity, and particulate matter. [ASHRAE 15:3]

Refrigerant Designation. The unique identifying alphanumeric value assigned to an individual refrigerant.

Refrigerant Safety Classifications. Made up of a letter (A or B), that indicates the toxicity class, followed by a number (1, 2, or 3), that indicates the flammability class. Refrigerant blends are similarly classified, based on the compositions at their worst cases of fractionation, as separately determined for toxicity and flammability. In some cases, the worst case of fractionation is the original formulation.

Flammability Classification. Refrigerants shall be classified for flammability in accordance with one of the following:

Class 1. Refrigerants that do not show flame propagation where tested in air at 14.7 pound-force per square inch absolute (psia) (101 kPa) and 140°F (60°C).

Class 2. Refrigerants having a lower flammability limit (LFL) of more than 0.00625 pound per cubic foot (lb/ft^3) (0.10012 kg/m^3) at 140°F (60°C), 14.7 psia (101 kPa), and a heat of combustion of less than 8169 British thermal units per pound (Btu/lb) (1.8988 E+07 J/kg).

Class 3. Refrigerants that are highly flammable having a LFL of not more than 0.00625 lb/ft^3 (0.10012 kg/m^3) at 140°F (60°C) and 14.7 psia (101 kPa) or a heat of combustion not less than 8169 Btu/lb (1.8988 E+07 J/kg).

Toxicity Classification. Refrigerants shall be classified for the toxicity in accordance with one of the following:

Class A. Refrigerants have an occupational exposure limit (OEL) of not less than 400 parts per million (ppm).

Class B. Refrigerants have an OEL of less than 400 ppm.

Refrigeration Machinery Room. A room designed to house compressors and refrigerant pressure vessels.

Refrigeration Room or Space. A room or space in which an evaporator or brine coil is located for the purpose of reducing or controlling the temperature within the room or space to less than 68°F (20°C).

Refrigeration System, Absorption. A heat-operated closed refrigeration cycle in which a secondary fluid, the absorbent, absorbs a primary fluid, the refrigerant that has been vaporized in the evaporator.

Refrigeration System, Direct. A system in which the evaporator or condenser of the refrigerating system is in direct contact with the air or other substances to be cooled or heated. [ASHRAE 15:5.1.1]

Refrigeration System, Indirect. A system in which a secondary coolant cooled or heated by the refrigerating system is circulated to the air or other substance to be cooled or heated. Indirect systems are distinguished by the method of application given below. [ASHRAE 15:5.1.2]

ASHRAE
ASHRAE
ASHRAE
ASHRAE
ASHRAE

ASHRAE **Indirect Open Spray System.** A system in which a secondary coolant is in direct contact with the air or other substance to be cooled or heated. [ASHRAE 15:5.1.2.1]

ASHRAE **Double Indirect Open Spray System.** A system in which the secondary substance for an indirect open spray system is heated or cooled by the secondary coolant circulated from a second enclosure. [ASHRAE 15:5.1.2.2]

ASHRAE **Indirect Closed System.** A system in which a secondary coolant passes through a closed circuit in the air or other substance to be cooled or heated. [ASHRAE 15:5.1.2.3]

Refrigeration System, Mechanical. A combination of interconnected refrigerant-containing parts constituting one closed refrigerant circuit in which a refrigerant is circulated for the purpose of extracting heat and in which a compressor is used for compressing the refrigerant vapor.

Refrigeration System, Self-Contained. A complete factory-assembled and tested system that is shipped in one or more sections and has no refrigerant-containing parts that are joined in the field by other than companion or block valves.

Registered Design Professional. An individual who is registered or licensed by the laws of the state to perform such design work in the jurisdiction.

Relief Valve, Vacuum. A device which automatically opens or closes for relieving a vacuum with the system, depending on whether the vacuum is above or below a predetermined value.

Removable. Capable of being transferred to another location with a limited application of effort and tools. [NFPA 96:3.3.42]

Replacement Air. See Air, Makeup.

Residential Building. A building or portion thereof designed or used for human habitation.

Riser Heat Pipe. A duct that extends at an angle of 45 degrees (0.79 rad) from the horizontal. This definition shall not include any boot connection.

Room Heater. A freestanding, nonrecessed, environmental heating appliance installed in the space being heated and not connected to ducts.

Room Heater, Unvented. An unvented, self-contained, freestanding, nonrecessed, fuel- gas-burning appliance for furnishing warm air by gravity or fan circulation to the space in which installed, directly from the heater without duct connection. [NFPA 54:3.3.56.6]

Rupture Member. A pressure-relief device that operates by the rupture of a diaphragm within the device on a rise to a predetermined pressure.

221.0 – S –

Seam, Welded. See Joint, Welded.

Secondary Filtration. Fume incinerators, thermal recovery units, air pollution control devices or other filtration media installed in ducts or hoods located in the path of travel of exhaust products after the initial filtration.

Self-Contained. Having all essential working parts, except energy and control connections, so contained in a case or framework that they do not depend on appliances or fastenings outside of the machine.

Service Corridor. A fully enclosed passage used for transporting hazardous production materials and purposes other than required exiting.

Service Piping. The piping and equipment between the street gas main and the gas piping system inlet that is installed by, and is under the control and maintenance of, the serving gas supplier.

Shaft. An interior space enclosed by walls or construction extending through one or more stories or basements that connect openings in successive floors, or floors and roof, to accommodate elevators, dumbwaiters, mechanical equipment, or similar devices to transmit light or ventilation air.

Shaft Enclosure. The walls or construction forming the boundaries of a shaft.

Shall. Indicates a mandatory requirement.

Should. Indicates a recommendation or that which is advised but not required.

Single Hazard Area. Where two or more hazards can be simultaneously involved in fire by reason of their proximity, as determined by the Authority Having Jurisdiction. [NFPA 96:3.3.44]

Smoke Detector. An approved device that senses visible or invisible particles of combustion.

Solid Cooking Fuel. A solid, organic, consumable fuel such as briquettes, mesquite, hardwood, or charcoal. [NFPA 96:3.3.45]

Solid-Fuel Cooking Equipment. Cooking equipment that utilizes solid fuel. [NFPA 96:3.3.23.2] This equipment includes ovens, tandoori charcoal pots, grills, broilers, rotisseries, barbecue pits, or other type of cooking equipment that derives all or part of its heat source from the burning of solid cooking fuel.

Solvent. A substance (usually liquid) capable of dissolving or dispersing another substance; a chemical compound designed and used to convert solidified grease into a liquid or semiliquid state in order to facilitate a cleaning operation. [NFPA 96:3.3.46]

Spark Arrester. A device or method that minimizes the passage of airborne sparks and embers into a plenum, duct, and flue. [NFPA 96:3.3.48]

Standard. A document, the main text of which contains only mandatory provisions using the word "shall" to indicate requirements and that is in a form generally suitable for mandatory reference by another standard or code or for adoption into law. Nonmandatory provisions shall be located in an appendix, footnote, or fine-print note and are not to be considered a part of the requirements of a standard.

Stationary Fuel Cell Power Plant. A self-contained package or factory-matched packages that constitute an automatically operated assembly of integrated systems for generating useful electrical energy and recoverable energy that is permanently connected and fixed in place.

Steam-Heating Boiler. A boiler operated at pressures not exceeding 15 psi (103 kPa) for steam.

Strength, Ultimate. The highest stress level that the component can tolerate without rupture.

System Outdoor Airflow. The rate of outdoor airflow required at the ventilation system outdoor air intake.

222.0 –T–

Termination, Duct. The final or intended end portion of a duct system that is designed and functions to fulfill the obligations of the system in a satisfactory manner. [NFPA 96:3.3.19]

Testing Agency. [HCD 1 & HCD 2] *See "Approved Testing Agency".*

Thermal Recovery Unit. A device or series of devices whose purpose is to reclaim only the heat content of air, vapors, gases, or fluids that are being expelled through the exhaust system and to transfer the thermal energy so reclaimed to a location whereby a useful purpose can be served. [NFPA 96:3.3.49]

Trained. A person who has become proficient in performing a skill reliably and safely through instruction and practice/field experience acceptable to the Authority Having Jurisdiction. [NFPA 96:3.3.50]

Transition Gas Riser. A listed or approved section or sections of pipe and fittings used to convey fuel gas and installed in a gas piping system for the purpose of providing a transition from belowground to aboveground.

Trap. A cuplike or u-shaped configuration located on the inside of a duct system component where liquids can accumulate. [NFPA 96:3.3.51]

Type B Gas Vent. A factory-made gas vent listed by a nationally recognized testing agency for venting listed or approved appliances equipped to burn only gas.

Type B-W Gas Vent. A factory-made gas vent listed by a nationally recognized testing agency for venting listed or approved gas-fired vented wall furnaces.

Type L Gas Vent. A venting system consisting of listed vent piping and fittings for use with oil-burning appliances listed for use with Type L or with listed gas appliances.

223.0 –U–

UMC. [HCD 1, HCD 2, OSHPD 1, 1R, 2, 3, 4 & 5] *The most recent edition of Uniform Mechanical Code published by the International Association of Plumbing and Mechanical Officials.*

Unit Heater. A heating appliance designed for nonresidential space heating and equipped with an integral means for circulation of air.

Unusually Tight Construction. Construction where:

(1) Walls and ceilings exposed to the outdoors have a continuous water vapor retarder with a rating of 1 perm or less with openings gasketed or sealed.

(2) Weatherstripping is on openable windows and doors.

(3) Caulking or sealants are applied to areas such as joints around window and door frames, between sole plates and floors, between wall-ceiling joints, between wall panels, and at penetrations for plumbing, electrical, and gas lines and at other openings.

Use (Material). The placing in action or making available for service by opening or connecting a container utilized for confinement of material, whether a solid, liquid, or gas.

224.0 –V–

Vacuum. A pressure less than that exerted by the atmosphere.

Valve, Pressure-Relief. A pressure-actuated valve held closed by a spring or other means and designed to automatically relieve pressure in excess of its setting.

Valve, Stop. A device in a piping system to shut off the flow of the fluid.

Valve, Three-Way-Type Stop. A manually operated valve with one inlet that alternately can stop flow to either of two outlets.

Valves, Companion or Block. Pairs of mating stop valves valving off sections of refrigeration systems and arranged so that these sections may be joined before opening these valves or separated after closing them.

Vent, Gas. A passageway composed of listed factory-built components assembled in accordance with the manufacturer's installation instructions for conveying vent gases from appliances or their vent connectors to the outdoors. [NFPA 54:3.3.53]

Vent Connector, Gas. That portion of a gas-venting system that connects a listed gas appliance to a gas vent and is installed within the space or area in which the appliance is located.

Vent Offset. An arrangement of two or more fittings and pipe installed for the purpose of locating a vertical section of vent pipe in a different but parallel plane with respect to an adjacent section of a vertical vent pipe. [NFPA 54:3.3.102]

Vented Appliance Categories.

Category I. An appliance that operates with a nonpositive vent static pressure and with a vent gas temperature that avoids excessive condensate production in the vent. [NFPA 54:3.3.5.11.1]

Category II. An appliance that operates with a nonpositive vent static pressure and with a vent gas temperature that can cause excessive condensate production in the vent. [NFPA 54:3.3.5.11.2]

Category III. An appliance that operates with a positive vent static pressure and with a vent gas temperature that avoids excessive condensate production in the vent. [NFPA 54:3.3.5.11.3]

Category IV. An appliance that operates with a positive vent static pressure and with a vent gas temperature that can cause excessive condensate production in the vent. [NFPA 54:3.3.5.11.4]

Vented Decorative Appliance. A vented appliance whose only function is providing an aesthetic effect of flames.

Vented Wall Furnace. A self-contained, vented, fuel gas-burning appliance complete with grilles or equivalent, designed for incorporation in or permanent attachment to the structure of a building and furnishing heated air, circulated by gravity or by a fan, directly into the space to be heated through openings in the casing. [NFPA 54:3.3.45.7]

Ventilating Ceiling. A suspended ceiling containing many small apertures through which air, at low pressure, is forced downward from an overhead plenum dimensioned by the concealed space between the suspended ceiling and the floor or roof above.

Ventilation System. All of that equipment intended or installed for the purpose of supplying air to or removing air from, any room or space by mechanical means, other than equipment that is a portion of an environmental heating, cooling, absorption, or evaporative cooling system.

Venting Collar. The outlet opening of an appliance provided for connection of the vent system.

Venting System. The vent or chimney and its connectors, assembled to form a continuous open passageway from an appliance to the outdoors for the purpose of removing products of combustion. This definition also shall include a venting assembly that is an integral part of an appliance.

Venting System, Gravity-Type. A system that depends entirely on the heat from the fuel being used to provide the energy required to vent an appliance.

Venting System, Power-Type. A system that depends on a mechanical device to provide a positive draft within the venting system.

Volume, Internal Gross. The volume as determined from internal dimensions of the container, with no allowance for the volume of the internal parts.

225.0 – W –

Wall Heater. See Vented Wall Furnace.

Warm Air Furnace. An environmental heating appliance designed or arranged to discharge heated air through any duct or ducts. This definition shall not include a unit heater.

Water Heater or Hot-Water-Heating Boiler. An appliance designed primarily to supply hot water for domestic or commercial purposes and equipped with automatic controls limiting water temperature to a maximum of 210°F (99°C).

226.0 – X –

No definitions.

227.0 – Y –

No definitions.

228.0 – Z –

Zeotropic. Blends comprising multiple components of different volatilities that, when used in refrigeration cycles, change volumetric composition and saturation temperatures as they evaporate or condense at constant pressure. [ASHRAE 34:3]

CALIFORNIA MECHANICAL CODE – MATRIX ADOPTION TABLE
CHAPTER 3 – GENERAL REGULATIONS

(Matrix Adoption Tables are non-regulatory, intended only as an aid to the code user. See Chapter 1 for state agency authority and building applications.)

Adopting Agency	BSC	BSC-CG	SFM	HCD			DSA			OSHPD						BSCC	DPH	AGR	DWR	CEC	CA	SL	SLC
				1	2	1-AC	AC	SS	SS/CC	1	1R	2	3	4	5								
Adopt Entire Chapter	X																						
Adopt Entire Chapter as amended (amended sections listed below)			X	X	X			X	X	X	X	X	X	X	X								
Adopt only those sections that are listed below																							
Chapter/Section																							
303.2, Exception										X	X	X	X	X	X								
303.7			X					X	X														
303.7.1				X	X																		
303.13 *Exception*				X	X																		
311.2				X	X																		
312.1				X	X			X	X														
316.2				X	X																		
316.5				X	X																		
316.9				X	X																		
318.0										X	X	X	X	X	X								
319.1										X	X			X	X								
319.2												X		X									
320.1										X				X									
320.2												X		X									
320.3													X										
320.4										X				X									
320.5										X	X	X		X	X								
321.0										X	X	X	X	X	X								
322.0										X	X	X	X	X	X								
322.1										X		X	X	X									

This state agency does not adopt sections identified with the following symbol: †

The Office of the State Fire Marshal's adoption of this chapter or individual sections is applicable to structures regulated by other state agencies pursuant to Section 1.11.0.

CHAPTER 3
GENERAL REGULATIONS

301.0 General.

301.1 Applicability. This chapter covers general requirements for heating, ventilating, air-conditioning, refrigeration, miscellaneous heat-producing, and energy-utilizing equipment or appliances. Such equipment or appliances shall comply with the requirements of this code.

301.2 Approval. Equipment or appliance shall be approved by the Authority Having Jurisdiction for safe use or comply with applicable nationally recognized standards as evidenced by the listing and label of an approved agency. A list of accepted standards is included in Chapter 17. Defective materials or parts shall be replaced in such a manner as not to invalidate an approval.

301.3 Design of Equipment. Installers shall furnish satisfactory evidence that the appliance is constructed in accordance with the requirements of this code. The permanently attached label of an approved agency shall be permitted to be accepted as such evidence.

301.4 Electrical Connections. For equipment regulated by this code:

(1) Equipment requiring electrical connections of more than 50 volts shall have a positive means of disconnect adjacent to and in sight from the equipment served.

Exception: Other power disconnect means shall be acceptable where in accordance with *the California Electrical Code.*

(2) A 120 volt receptacle shall be located within 25 feet (7620 mm) of the equipment for service and maintenance purposes. The receptacle outlet shall be on the supply side of the disconnect switch. The receptacle need not be located on the same level as the equipment.

(3) Electrical wiring, controls, and connections to equipment and appliances regulated by this code shall be in accordance with *the California Electrical Code.*

301.5 Oil-Burning Appliances. The tank, piping, and valves for appliances burning oil shall be installed in accordance with the requirements of NFPA 31.

301.6 Personnel Protection. A metal guard shall be provided around exposed flywheels, fans, pulleys, belts, and moving machinery that are portions of a heating, ventilating, or refrigerating system.

302.0 Materials – Standards and Alternates.

302.1 Minimum Standards. Listed pipe, pipe fittings, appliances, appurtenances, equipment, materials, and devices used in a mechanical system shall be listed (third-party certified) by a listing agency (accredited conformity assessment body) as complying with the approved applicable recognized standards referenced in this code, and shall be free from defects. Unless otherwise provided for in this code, materials, appurtenances, or devices used or entering into the construction of mechanical systems, or parts thereof, shall be submitted to the Authority Having Jurisdiction for approval.

302.1.1 Marking. Each length of pipe and each pipe fitting, material, and device used in a mechanical system shall have cast, stamped, or indelibly marked on it any markings required by the applicable referenced standards and listing agency, and the manufacturer's mark or name, which shall readily identify the manufacturer to the end user of the product. Where required by the approved standard that applies, the product shall be marked with the weight and the quality of the product. Materials and devices used or entering into the construction of mechanical systems, or parts thereof, shall be marked and identified in a manner satisfactory to the Authority Having Jurisdiction. Such marking shall be done by the manufacturer. Field markings shall not be acceptable.

Exception: Markings shall not be required on nipples created from cutting and threading of approved pipe.

302.1.2 Standards. Standards listed or referred to in this chapter or other chapters cover materials that will conform to the requirements of this code, where used in accordance with the limitations imposed in this or other chapters thereof and their listing. Where a standard covers materials of various grades, weights, quality, or configurations, the portion of the listed standard that is applicable shall be used. Design and materials for special conditions or materials not provided for herein shall be permitted to be used by special permission of the Authority Having Jurisdiction after the Authority Having Jurisdiction has been satisfied as to their adequacy. A list of mechanical standards that appear in specific sections of this code is referenced in Table 1701.1. Standards referenced in Table 1701.1 shall be applied as indicated in the applicable referenced section. A list of additional standards, publications, practices and guides that are not referenced in specific sections of this code appear in Table 1701.2. The documents indicated in Table 1701.2 shall be permitted in accordance with Section 302.2.

302.1.3 Existing Buildings. In existing buildings or premises in which mechanical installations are to be altered, repaired, or renovated, the Authority Having Jurisdiction has discretionary powers to permit deviation from the provisions of this code, provided that such proposal to deviate is first submitted for proper determination in order that health and safety requirements, as they pertain to mechanical systems, shall be observed.

302.2 Alternate Materials and Methods of Construction Equivalency. Nothing in this code is intended to prevent the use of systems, methods, or devices of equivalent or superior quality, strength, fire resistance, effectiveness, durability, and safety over those prescribed by this code. Technical documentation shall be submitted to the Authority Having Jurisdiction to demonstrate equivalency. The Authority Having Jurisdiction shall have the authority to approve or disapprove the system, method, or device for the intended purpose.

However, the exercise of this discretionary approval by the Authority Having Jurisdiction shall have no effect beyond the jurisdictional boundaries of said Authority Having Jurisdiction. An alternate material or method of construction so approved shall not be considered as in accordance with the requirements, intent, or both of this code for a purpose other than that granted by the Authority Having Jurisdiction where the submitted data does not prove equivalency.

302.2.1 Testing. The Authority Having Jurisdiction shall have authority to require tests, as proof of equivalency.

302.2.1.1 Tests. Tests shall be made in accordance with approved testing standards, by an approved testing agency at the expense of the applicant. In the absence of such standards, the Authority Having Jurisdiction shall have the authority to specify the test procedure.

302.2.1.2 Request by the Authority Having Jurisdiction. The Authority Having Jurisdiction shall have the authority to require tests to be made or repeated where there is reason to believe that a material or device no longer is in accordance with the requirements on which its approval was based.

302.3 Alternative Engineered Design. An alternative engineered design shall comply with the intent of the provisions of this code and shall provide an equivalent level of quality, strength, effectiveness, fire resistance, durability, and safety. Material, equipment, or components shall be designed and installed in accordance with the manufacturer's installation instructions.

302.3.1 Permit Application. The registered design professional shall indicate on the design documents that the mechanical system, or parts thereof, is an alternative engineered design so that it is noted on the construction permit application. The permit and permanent permit records shall indicate that an alternative engineered design was part of the approved installation.

302.3.2 Technical Data. The registered design professional shall submit sufficient technical data to substantiate the proposed alternative engineered design and to prove that the performance meets the intent of this code.

302.3.3 Design Documents. The registered design professional shall provide two complete sets of signed and sealed design documents for the alternative engineered design for submittal to the Authority Having Jurisdiction. The design documents shall include floor plans of the work. Where appropriate, the design documents shall indicate location, sizing, and loading of appurtenances, equipment, appliances, and devices.

302.3.4 Design Approval. An approval of an alternative engineered design shall be at the discretion of the Authority Having Jurisdiction. The exercise of this discretionary approval by the Authority Having Jurisdiction shall have no effect beyond the jurisdictional boundaries of said Authority Having Jurisdiction. An alternative engineered design so approved shall not be considered as in accordance with the requirements, intent, or both of this code for a purpose other than that granted by the Authority Having Jurisdiction.

302.3.5 Design Review. The Authority Having Jurisdiction shall have the authority to require testing of the alternative engineered design in accordance with Section 302.2.1, including the authority to require an independent review of the design documents by a registered design professional selected by the Authority Having Jurisdiction and at the expense of the applicant.

302.3.6 Inspection and Testing. The alternative engineered design shall be tested and inspected in accordance with the submitted testing and inspection plan and the requirements of this code.

303.0 Installation.

303.1 Listed Appliances. The installation of equipment and appliances regulated by this code shall be in accordance with the conditions of the listing, the manufacturer's installation instructions and this code. The manufacturer's installation and operating instructions shall be attached to the appliance. Clearances of listed equipment and appliances from combustible materials shall be as specified in the listing or on the rating plate.

303.2 Closet or Alcove Installations. Central heating furnaces and boilers installed in closets or alcoves shall be listed for such installation. Central heating furnaces not listed for closet or alcove installation shall be installed in a room or space having a volume not less than 12 times the total volume of the furnace. Central heating boilers not listed for closet or alcove installation shall be installed in a room or space having a volume 16 times the volume of the boiler. ***[OSHPD 1, 1R, 2, 3, 4 & 5]*** *The total volume of the boilers shall be based on the total number of central-heating boilers that can operate at the same time.*

Exception: [OSHPD 1, 1R, 2, 3, 4 & 5] *A 25 percent reduction in the boiler room volume is allowed with forced-draft boilers and approved ventilation of the boiler room. In no case shall boiler room volume or clearances be reduced below those required by the conditions of the boiler listing. The boiler and the boiler room ventilation system, including fans, controls, and damper motors shall be on emergency power when required by Section 321.0. The ventilation system shall either operate continuously, or, if interlocked with the boiler(s) it shall not interfere with the proper boiler operation.*

Where the ceiling height of the room or space exceeds 8 feet (2438 mm), the volume shall be calculated on the basis of an 8 foot (2438 mm) height.

The installation clearances shall be in accordance with the appliance listing, shall not be reduced, and shall be installed in accordance with Section 904.1.

303.3 Unlisted Appliances. Except as otherwise permitted in this code, unlisted equipment and appliances shall be approved by the Authority Having Jurisdiction. Unlisted equipment and appliances shall be installed in accordance with the manufacturer's installation instructions and with clearances from combustible materials in accordance with Section 303.10 or Section 303.10.1.

303.4 Anchorage of Appliances. Appliances designed to be fixed in position shall be securely fastened in place in accordance with the manufacturer's installation instructions. Supports for appliances shall be designed and constructed to sustain vertical and horizontal loads within the stress limitations specified in the building code.

303.5 Movement. Movement of appliances with casters shall be limited by a restraining device installed in accordance with the connector and appliance manufacturer's installation instructions.

303.6 Identification of Equipment. Where more than one heating, cooling, ventilating, or refrigerating system is installed on the roof of a building or within a building, it shall be permanently identified as to the area or space served by the equipment.

303.7 Liquefied Petroleum Gas Facilities. Containers, container valves regulating equipment, and appurtenances for the storage and supply of liquefied petroleum gas shall be installed in accordance with NFPA 58 *and the California Fire Code.*

303.7.1 Liquefied Petroleum Gas [LP-Gas] Appliances. [HCD 1 & HCD 2] *Liquefied petroleum gas-burning appliances shall not be installed in a pit, basement, or similar location where heavier-than-air gas might collect. Appliances so fueled shall not be installed in an above-grade under-floor space or basement unless such location is provided with an approved means for removal of unburned gas.*

303.8 Appliances on Roofs. Appliances on roofs shall be designed or enclosed so as to withstand climatic conditions in the area in which they are installed. Where enclosures are provided, each enclosure shall permit easy entry and movement, shall be of reasonable height, and shall have at least a 30 inch (762 mm) clearance between the entire service access panel(s) of the appliance and the wall of the enclosure. [NFPA 54:9.4.1.1]

303.8.1 Load Capacity. Roofs on which appliances are to be installed shall be capable of supporting the additional load or shall be reinforced to support the additional load. [NFPA 54:9.4.1.2]

303.8.2 Fasteners. Access locks, screws, and bolts shall be of corrosion-resistant material. [NFPA 54:9.4.1.3]

303.8.3 Installation of Appliances on Roofs. Appliances shall be installed in accordance with the manufacturer's installation instructions. [NFPA 54:9.4.2.1]

303.8.4 Clearance. Appliances shall be installed on a well-drained surface of the roof. At least 6 feet (1829 mm) of clearance shall be available between any part of the appliance and the edge of a roof or similar hazard, or rigidly fixed rails, guards, parapets, or other building structures at least 42 inches (1067 mm) in height shall be provided on the exposed side. [NFPA 54:9.4.2.2]

303.8.5 Electrical Power. All appliances requiring an external source of electrical power for its operation shall be provided with the following:

(1) A readily accessible electrical disconnecting means within sight of the appliance that completely de-energizes the appliance.

(2) A 120-V ac grounding-type receptacle outlet on the roof adjacent to the appliance on the supply side of the disconnect switch. [NFPA 54:9.4.2.3]

303.8.6 Platform or Walkway. Where water stands on the roof at the appliance or in the passageways to the appliance, or where the roof is of a design having a water seal, a suitable platform, walkway, or both shall be provided above the waterline. Such platform(s) or walkway(s) shall be located adjacent to the appliance and control panels so that the appliance can be safely serviced where water stands on the roof. [NFPA 54:9.4.2.4]

303.9 Avoiding Strain on Gas Piping. Appliances shall be supported and connected to the piping so as not to exert undue strain on the connections. [NFPA 54:9.1.17]

303.10 Clearances. Appliances and their vent connectors shall be installed with clearances from combustible material so their operation does not create a hazard to persons or property. Minimum clearances between combustible walls and the back and sides of various conventional types of appliances and their vent connectors are specified in Chapter 8 and Chapter 9. [NFPA 54:9.2.2] Where not provided in this code, listed and unlisted equipment or appliances shall be installed to maintain the required clearances for servicing and to combustible construction in accordance with the listing and the manufacturer's installation instructions.

303.10.1 Clearance Reduction. Reduce clearances to combustible construction for listed equipment and appliances shall comply with the listing and Table 303.10.1. Where permitted by the manufacturer, and not provided in this code, reduce clearances to combustible construction for unlisted equipment and appliances shall comply with Table 303.10.1.

303.10.1.1 Type I Hood Exhaust System. Reduce clearances for Type I exhaust systems shall be in accordance with Section 507.4.2 through Section 507.4.2.3. Clearances from the duct or the exhaust fan to the interior surface of enclosures of combustible construction shall be in accordance with Section 510.7.3 and clearances shall not be reduced.

303.10.1.2 Product Conveying Ducts. Reduce clearances to combustibles construction for product conveying ducts shall be in accordance with Section 506.10.3 through Section 506.11.6.3.

303.10.1.3 Solid-Fuel Burning Appliances. For solid-fuel burning appliances, the clearance, after reduction, shall not be less than 12 inches (305 mm) to combustible walls and not less than 18 inches (457 mm) to combustible ceilings. The clearance, after reduction, shall be permitted to be less than 12 inches (305 mm) to combustible walls and less than 18 inches (457 mm) to combustible ceilings where the solid-fuel burning appliances is listed for lesser clearance.

303.11 Installation in Commercial Garages. Appliances installed in enclosed, basement, and underground parking structures shall be installed in accordance with NFPA 88A. [NFPA 54:9.1.11.1]

303.11.1 Repair Garages. Appliances installed in repair garages shall be installed in accordance with NFPA 30A. [NFPA 54:9.1.11.2]

303.12 Installation in Aircraft Hangars. Heaters in aircraft hangars shall be installed in accordance with NFPA 409. [NFPA 54:9.1.12]

303.13 Pit Location. Where excavation is necessary to install an appliance, it shall extend to a depth of 6 inches (152 mm) below and 12 inches (305 mm) on all sides of the appliance, except on the service side, which shall have 30 inches (762 mm). Where the depth of the excavation for either the appliance or passageway exceeds 12 inches (305 mm), walls shall be lined with concrete or masonry 4 inches (102 mm) above the adjoining ground level.

Exception: [HCD 1 & HCD 2] *Liquefied petroleum gas (LP-Gas) appliances as described in Section 303.7.1.*

304.0 Accessibility for Service.

304.1 General. Appliances shall be located with respect to building construction and other equipment so as to permit access to the appliance. Sufficient clearance shall be maintained to permit cleaning of heating surfaces; the replacement of filters, blowers, motors, burners, controls, and vent connections; the lubrication of moving parts where necessary; the adjustment and cleaning of burners and pilots; and the proper functioning of explosion vents, where provided. For attic installation, the passageway and servicing area adjacent to the appliance shall be floored. [NFPA 54:9.2.1]

Unless otherwise specified, not less than 30 inches (762 mm) in depth, width, and height of working space shall be provided.

Exception: A platform shall not be required for unit heaters or room heaters.

304.2 Sloped Roof. Where equipment or appliances that require service are installed on a roof having a slope of 4 units vertical in 12 units horizontal (33 percent slope) or more, a level platform of not less than 30 inches by 30 inches (762 mm by 762 mm) shall be provided at the service side of the equipment or appliance.

304.3 Access to Appliances on Roofs. Appliances located on roofs or other elevated locations shall be accessible. [NFPA 54:9.4.3.1]

304.3.1 Access. Buildings exceeding 15 feet (4572 mm) in height shall have an inside means of access to the roof unless other means acceptable to the Authority Having Jurisdiction are used. [NFPA 54:9.4.3.2]

304.3.1.1 Access Type. The inside means of access shall be a permanent, or foldaway inside stairway or ladder, terminating in an enclosure, scuttle, or trap door. Such scuttles or trap doors shall be not less than 22 inches by 24 inches (559 mm by 610 mm) in size, shall open easily and safely under all conditions, especially snow; and shall be constructed so as to permit access from the roof side unless deliberately locked on the inside.

Not less than 6 feet (1829 mm) of clearance shall be between the access opening and the edge of the roof or similar hazard or rigidly fixed rails or guards not less than 42 inches (1067 mm) in height shall be provided on the exposed side. Where parapets or other building structures are utilized in lieu of guards or rails, they shall be not less than 42 inches (1067 mm) in height. [NFPA 54:9.4.3.3]

304.3.1.2 Permanent Ladders. Permanent ladders required by Section 304.3.1.1 shall be constructed in accordance with the following:

(1) Side railings shall extend not less than 30 inches (762 mm) above the roof or parapet wall.

(2) Landings shall not exceed 18 feet (5486 mm) apart measured from the finished grade.

(3) Width shall be not less than 14 inches (356 mm) on center.

(4) Rungs spacing shall not exceed 12 inches (305 mm) on center, and each rung shall be capable of supporting a 300 pound (136.1 kg) load.

(5) Toe space shall be not less than 6 inches (152 mm).

304.3.2 Permanent Lighting. Permanent lighting shall be provided at the roof access. The switch for such lighting shall be located inside the building near the access means leading to the roof. [NFPA 54:9.4.3.4]

304.4 Appliances in Attics and Under-Floor Spaces. An attic or under-floor space in which an appliance is installed shall be accessible through an opening and passageway not less than the largest component of the appliance, and not less than 22 inches by 30 inches (559 mm by 762 mm).

304.4.1 Length of Passageway. Where the height of the passageway is less than 6 feet (1829 mm), the distance from the passageway access to the appliance shall not exceed 20 feet (6096 mm) measured along the centerline of the passageway. [NFPA 54:9.5.1.1]

304.4.2 Width of Passageway. The passageway shall be unobstructed and shall have solid flooring not less than 24 inches (610 mm) wide from the entrance opening to the appliance. [NFPA 54:9.5.1.2]

304.4.3 Work Platform. A level working platform not less than 30 inches by 30 inches (762 mm by 762 mm) shall be provided in front of the service side of the appliance. [NFPA 54:9.5.2]

Exception: A working platform need not be provided where the furnace is capable of being serviced from the required access opening. The furnace service side shall not exceed 12 inches (305 mm) from the access opening.

304.4.4 Lighting and Convenience Outlet. A permanent 120V receptacle outlet and a lighting fixture shall be installed near the appliance. The switch controlling the lighting fixture shall be located at the entrance to the passageway. [NFPA 54:9.5.3]

305.0 Location.

305.1 Installation in Garages. Appliances in residential garages and in adjacent spaces that open to the garage and are not part of the living space of a dwelling unit shall be installed so that all burners and burner-ignition devices are located not less than 18 inches (457 mm) above the floor unless listed as flammable vapor ignition resistant. [NFPA 54:9.1.10.1]

305.1.1 Physical Damage. Appliances installed in garages, warehouses, or other areas subject to mechanical damage shall be guarded against such damage by being installed behind protective barriers or by being elevated or located out of the normal path of vehicles.

305.1.2 Access from the Outside. Where appliances are installed in a separate, enclosed space having access only from outside of the garage, such appliances shall be permitted to be installed at floor level, providing the required combustion air is taken from the exterior of the garage. [NFPA 54:9.1.10.3]

305.1.3 Cellulose Nitrate Plastic Storage. Heating equipment located in rooms where cellulose nitrate plastic is stored or processed shall be in accordance with the fire code.

305.2 Flood Hazard Areas. For buildings located in flood hazard areas, heating, ventilating, air-conditioning, refrigeration, miscellaneous heat-producing, and energy-utilizing equipment and appliances shall be elevated at or above the elevation in accordance with the building code for utilities and attendant equipment or the elevation of the lowest floor, whichever is higher.

Exception: Equipment and appliances shall be permitted to be located below the elevation in accordance with the building code for utilities and attendant equipment or the elevation of the lowest floor, whichever is higher, provided that the systems are designed and installed to prevent water from entering or accumulating within their components and the systems are constructed to resist hydrostatic and hydrodynamic loads and stresses, including the effects of buoyancy, during the occurrence of flooding to such elevation.

305.2.1 Coastal High Hazard Areas. Mechanical systems in buildings located in coastal high hazard areas shall be in accordance with the requirements of Section 305.2, and mechanical systems, pipes, and appurtenances shall not be mounted on or penetrate through walls that are intended to breakaway under flood loads in accordance with the building code.

305.2.2 Air Exhaust and Intake Openings. Outside air exhaust openings and air intake openings shall be located at or above the elevation required by the building code for utilities and attendant equipment or the elevation of the lowest floor, whichever is higher.

305.3 Elevator Shaft. Unless required for the functionality and safety of the elevator system, mechanical systems shall not be located in an elevator shaft.

305.4 Drainage Pan. Where a water heater is located in an attic, in or on an attic ceiling assembly, floor-ceiling assembly, or floor-subfloor assembly where damage results from a leaking water heater, a watertight pan of corrosion-resistant materials shall be installed beneath the water heater with not less than ¾ of an inch (20 mm) diameter drain to an approved location. Such pan shall be not less than 1½ inches (38 mm) in depth.

306.0 Automatic Control Devices.

306.1 General. Heating appliances shall be equipped with a listed device or devices that will shut off the fuel supply to the main burner or burners in the event of pilot or ignition failure. Liquefied petroleum gas-air-burning heating appliances shall be equipped with a listed automatic device or devices that will shut off the flow of gas to the pilot in the event of ignition failure.

Exception: The listed shutoff devices shall not be required on range or cooking tops, log lighters, lights, or other open-burner manually operated appliances, or listed appliances not requiring such devices and specific industrial appliances as approved by the Authority Having Jurisdiction.

Heating appliances whose manual fuel controls are not readily accessible from the main portion of the building being heated shall be equipped with remote controls.

Forced-air and gravity-type warm air furnaces shall be equipped with a listed air outlet temperature limit control that cannot be set for temperatures exceeding 250°F (121°C). Such controls shall be located in the bonnet or plenum, within 2 feet (610 mm) of the discharge side of the heating element of gravity furnaces or in accordance with the conditions of listing.

Electric duct heaters shall be equipped with an approved automatic reset air outlet temperature limit control that will limit the outlet air temperature to not exceed 200°F (93°C). The electric elements of the heater shall be equipped with fusible links or a manual reset temperature limit control that will prevent outlet air temperature in excess of 250°F (121°C).

307.0 Labeling.

307.1 Fuel-Burning Appliances. Fuel-burning heating appliances shall bear a permanent and legible factory-applied nameplate on which shall appear:

(1) The name or trademark of the manufacturer.

(2) The approved fuel input rating of the appliance, expressed in Btu/h (kW).

(3) The model number or equivalent.

(4) The serial number.

(5) Instructions for the lighting, operation, and shutdown of the appliance.

(6) The type of fuel approved for use with the appliance.

(7) The symbol of an approved agency certifying compliance of the equipment with recognized standards.

(8) Required clearances from combustible surfaces on which or adjacent to which it is permitted to be mounted.

307.2 Electric Heating Appliances. Electric heating appliances shall bear a permanent and legible factory-applied nameplate on which shall appear:

(1) The name or trademark of the manufacturer.

(2) The model number or equivalent.

(3) The serial number.

(4) The electrical rating in volts, amperes (or watts), and, for other than single phase, the number of phases.

(5) The output rating in Btu/h (kW).

(6) The electrical rating in volts, amperes, or watts of each field-replaceable electrical component.

(7) The symbol of an approved agency certifying compliance of equipment with recognized standards.

(8) Required clearances from combustible surfaces on which or adjacent to which it is permitted to be mounted.

An appliance shall be accompanied by clear and complete installation instructions, including required clearances from combustibles other than mounting or adjacent surfaces, and temperature rating of field-installed wiring connections exceeding 140°F (60°C).

307.3 Heat Pump and Electric Cooling Appliances. Heat pumps and electric cooling appliances shall bear a permanent and legible factory-applied nameplate on which shall appear:

(1) The name or trademark of the manufacturer.

(2) The model number or equivalent.

(3) The serial number.

(4) The amount and type of refrigerant.

(5) The factory test pressures or pressures applied.

(6) The electrical rating in volts, amperes, and, for other than single phase, the number of phases.

(7) The output rating in Btu/h (kW).

(8) The electrical rating in volts, amperes, or watts of each field replaceable electrical component.

(9) The symbol of an approved agency certifying compliance of the equipment with recognized standards.

(10) Required clearances from combustible surfaces on which or adjacent to which it is permitted to be mounted.

An appliance shall be accompanied by clear and complete installation instructions, including required clearances from combustible other than mounting or adjacent surfaces, and temperature rating of field-installed wiring connections exceeding 140°F (60°C).

307.4 Absorption Units. Absorption units shall bear a permanent and legible factory-applied nameplate on which shall appear:

(1) The name or trademark of the manufacturer.

(2) The model number or equivalent.

(3) The serial number.

(4) The amount and type of refrigerant.

(5) Hourly rating in Btu/h (kW).

(6) The type of fuel approved for use with the unit.

(7) Cooling capacity Btu/h (kW).

(8) Required clearances from combustible surfaces on which or adjacent to which it is permitted to be mounted.

(9) The symbol of an approved agency certifying compliance of the equipment with recognized standards.

308.0 Improper Location.

308.1 General. Piping or equipment shall not be so located as to interfere with the normal use thereof or with the normal operation and use of windows, doors, or other required facilities.

309.0 Workmanship.

309.1 Engineering Practices. Design, construction, and workmanship shall comply with accepted engineering practices and shall be of such character as to secure the results sought to be obtained by this code.

309.2 Concealing Imperfections. It shall be unlawful to conceal cracks, holes, or other imperfections in materials by welding, brazing, or soldering, by using therein or thereon paint, wax, tar, solvent cement, other leak-sealing or repair agent.

309.3 Installation Practices. Mechanical systems shall be installed in a manner that is in accordance with this code, applicable standards, and the manufacturer's installation instructions.

310.0 Condensate Wastes and Control.

310.1 Condensate Disposal. Condensate from air washers, air-cooling coils, condensing appliances, and the overflow from evaporative coolers and similar water-supplied equipment or similar air-conditioning equipment shall be collected and discharged to an approved plumbing fixture or disposal area. Where discharged into the drainage system, equipment shall drain by means of an indirect waste pipe. The waste pipe shall have a slope of not less than ⅛ inch per foot (10.4 mm/m) or 1 percent slope and shall be of approved corrosion-resistant material not smaller than the outlet size in accordance with Section 310.3 or Section 310.4 for air-cooling coils or condensing appliances, respectively. Condensate or wastewater shall not drain over a public way.

310.1.1 Condensate Pumps. Where approved by the Authority Having Jurisdiction, condensate pumps shall be installed in accordance with the manufacturer's installation instructions. Pump discharge shall rise vertically to a point where it is possible to connect to a gravity condensate drain and discharged to an approved disposal point. Each condensing unit shall be provided with a separate sump and interlocked with the equipment to prevent the equipment from operating during a failure. Separate pumps shall be permitted to connect to a single gravity indirect waste where equipped with check valves and approved by the Authority Having Jurisdiction.

310.2 Condensate Control. Where an equipment or appliance is installed in a space where damage is capable of resulting from condensate overflow, other than damage to replaceable lay-in ceiling tiles, a drain line shall be provided and shall be drained in accordance with Section 310.1. An additional protection method for condensate overflow shall be provided in accordance with one of the following:

(1) A water level detecting device that will shut off the equipment or appliance in the event the primary drain is blocked.

(2) An additional watertight pan of corrosion-resistant material, with a separate drain line, installed beneath the cooling coil, unit, or the appliance to catch the overflow condensate due to a clogged primary condensate drain.

(3) An additional drain line at a level that is higher than the primary drain line connection of the drain pan.

(4) An additional watertight pan of corrosion-resistant material with a water level detection device installed beneath the cooling coil, unit, or the appliance to catch the overflow condensate due to a clogged primary condensate drain and to shut off the equipment.

The additional pan or the additional drain line connection shall be provided with a drain pipe of not less than ¾ of an inch (20 mm) nominal pipe size, discharging at a point that is readily observed.

310.2.1 Protection of Appurtenances. Where insulation or appurtenances are installed where damage is capable of resulting from a condensate drain pan overfill, such installations shall occur above the rim of the drain pan with supports. Where the supports are in contact with the condensate waste, the supports shall be of approved corrosion-resistant material.

310.3 Condensate Waste Pipe Material and Sizing. Condensate waste pipes from air-cooling coils shall be sized in accordance with the equipment capacity as specified in Table 310.3. The material of the piping shall comply with the pressure and temperature rating of the appliance or equipment, and shall be approved for use with the liquid being discharged.

TABLE 310.3
MINIMUM CONDENSATE PIPE SIZE

EQUIPMENT CAPACITY IN TONS OF REFRIGERATION	MINIMUM CONDENSATE PIPE DIAMETER (inches)
Up to 20	¾
21 – 40	1
41 – 90	1¼
91 – 125	1½
126 – 250	2

For SI units: 1 ton of refrigeration = 3.52 kW, 1 inch = 25 mm

The size of condensate waste pipes is for one unit or a combination of units, or as recommended by the manufacturer. The capacity of waste pipes assumes a ⅛ inch per foot (10.4 mm/m) or 1 percent slope, with the pipe running three-quarters full at the following pipe conditions:

Outside Air – 20%		Room Air – 80%	
DB	WB	DB	WB
90°F	73°F	75°F	62.5°F

For SI units: °C = (°F-32)/1.8

Condensate drain sizing for other slopes or other conditions shall be approved by the Authority Having Jurisdiction.

310.3.1 Cleanouts. Condensate drain lines shall be configured or provided with a cleanout to permit the clearing of blockages and for maintenance without requiring the drain line to be cut.

310.4 Appliance Condensate Drains. Condensate drain lines from individual condensing appliances shall be sized as required by the manufacturer's instructions. Condensate drain lines serving more than one appliance shall be approved by the Authority Having Jurisdiction prior to installation.

310.5 Point of Discharge. Air-conditioning condensate waste pipes shall connect indirectly, except where permitted in Section 310.6, to the drainage system through an air gap or air break to trapped and vented receptors, dry wells, leach pits, or the tailpiece of plumbing fixtures. A condensate drain shall be trapped in accordance with the appliance manufacturer's instructions or as approved.

310.6 Condensate Waste From Air-Conditioning Coils. Where the condensate waste from air-conditioning coils discharges by direct connection to a lavatory tailpiece or to an approved accessible inlet on a bathtub overflow, the connection shall be located in the area controlled by the same person controlling the air-conditioned space.

310.7 Plastic Fittings. Female plastic screwed fittings shall be used with plastic male fittings and plastic male threads.

311.0 Heating or Cooling Air System.

311.1 Source. A heating or cooling air system shall be provided with return air, outside air, or both. A heating or cooling air system regulated by this code and designed to replace required ventilation shall be arranged to discharge into a conditioned space not less than the amount of outside air specified in Chapter 4.

311.2 Air Filters. Air filters shall be installed in a heating, cooling, or makeup air system. Media-type air filters shall comply with UL 900. Electrostatic and high efficiency particulate filters shall comply with Section 936.0.

Exceptions:

(1) Systems serving single guest rooms or dwelling units shall not require a listed filter.

(2) Air filters used in listed appliances and in accordance with the manufacturer's instructions.

[HCD 1 & HCD 2] *In mechanically ventilated residential buildings, air filters shall be provided to clean outdoor and return air prior to its delivery to occupied spaces. The air filters shall comply with the California Energy Code requirements for labeling, ventilation system design and installation, air filter efficiency, and air filter pressure drop, as applicable.*

See California Energy Code Subchapters 3, 6, 7, and 9 for requirements related to air filtration for residential occupancies.

311.3 Prohibited Source. Outside or return air for a heating or cooling air system shall not be taken from the following locations:

(1) Less than 10 feet (3048 mm) in distance from an appliance vent outlet, a vent opening of a plumbing drainage system, or the discharge outlet of an exhaust fan, unless the outlet is 3 feet (914 mm) above the outside-air inlet.

(2) Less than 10 feet (3048 mm) above the surface of an abutting public way, sidewalk, street, alley, or driveway.

(3) A hazardous or insanitary location, or a refrigeration machinery room as defined in this code.

(4) An area, the volume of which is less than 25 percent of the entire volume served by such system, unless there is a permanent opening to an area the volume of which is equal to 25 percent of the entire volume served.

Exception: Such openings where used for a heating or cooling air system in a dwelling unit shall be permitted to be reduced to not less than 50 percent of the required area, provided the balance of the required return air is taken from a room or hall having not less than three doors leading to other rooms served by the furnace.

(5) A closet, bathroom, toilet room, or kitchen.

(6) Rooms or spaces containing a fuel-burning appliance therein. Where such room or space serves as source of return-air.

Exceptions:

(1) This shall not apply to fireplaces, fireplace appliances, residential cooking appliances, direct-vent appliances, enclosed furnaces, and domestic-type clothes dryers installed within the room or space.

(2) This shall not apply to a gravity-type or listed vented wall heating or cooling air system.

(3) This shall not apply to a blower-type heating or cooling air system installed in accordance with the following requirements:

(a) Where the return air is taken from a room or space having a volume exceeding 1 cubic foot (0.03 m^3) for each 10 Btu/h (0.003 kW) fuel input rating of fuel-burning appliances therein.

(b) Not less than 75 percent of the supply air is discharged back into the same room or space.

(c) Return-air inlets shall not be located within 10 feet (3048 mm) from an appliance firebox or draft diverter in the same enclosed room or confined space.

311.4 Return-Air Limitations. Return air from one dwelling unit shall not discharge into another dwelling unit through the heating or cooling air system.

312.0 Plumbing Connections.

312.1 General. Water supply, sanitary drainage, and backflow protection shall be in accordance with the *California Plumbing Code.*

313.0 Hangers and Supports.

313.1 General. Piping, tubing, appliances, and appurtenances shall be supported in accordance with this code, the manufacturer's installation instructions, and in accordance with the Authority Having Jurisdiction.

313.2 Material. Hangers and anchors shall be of sufficient strength to support the weight of the pipe or tubing and its contents. Piping or tubing shall be isolated from incompatible materials.

313.3 Suspended Piping. Suspended piping or tubing shall be supported at intervals not to exceed those shown in Table 313.3.

313.4 Alignment. Piping or tubing shall be supported in such a manner as to maintain its alignment and prevent sagging.

313.5 Underground Installation. Piping or tubing in the ground shall be laid on a firm bed for its entire length; where other support is otherwise provided, it shall be approved in accordance with Section 302.0.

313.6 Hanger Rod Sizes. Hanger rod sizes shall be not smaller than those shown in Table 313.6.

TABLE 313.6
HANGER ROD SIZES

PIPE AND TUBE SIZE (inches)	ROD SIZES (inches)
½ - 4	⅜
5 - 8	½
10 - 12	⅝

For SI units: 1 inch = 25.4 mm

313.7 Gas Piping. Gas piping shall be supported by metal straps or hooks at intervals not to exceed those shown in Table 1310.2.4.1.

314.0 Balancing.

314.1 General. Heating, ventilating, and air-conditioning systems (including hydronic systems) shall be balanced in accordance with one of the following methods:

(1) AABC National Standards for Total System Balance

(2) ACCA Manual B

(3) ASHRAE 111

(4) NEBB Procedural Standards for Testing Adjusting Balancing of Environmental Systems

(5) SMACNA HVAC Systems Testing, Adjusting, and Balancing

315.0 Louvers in Hurricane Prone Regions.

315.1 General. Louvers located in areas within hurricane-prone regions that are within 1 mile (2 km) of the coastal mean high water line where the basic wind speed is 110 miles per hour (mi/h) (49.2 m/s) or more; or portions of hurricane-prone regions where the basic wind speed is 120 mi/h (53.6 m/s) or more; or Hawaii, as described in ASCE 7 shall be tested in accordance with Section 315.1.1 and Section 315.1.2.

315.1.1 Testing. Louvers that protect air intake or exhaust openings shall be tested in accordance with AMCA 550 for resistance to wind-driven rain.

315.1.2 Impact Resistance Test. Upon request by the Authority Having Jurisdiction, louvers protecting intake and exhaust ventilation ducts that are not fixed in the open position and located within 30 feet (9144 mm) of the grade shall be tested for impact resistance in accordance with AMCA 540.

316.0 Protection of Piping, Tubing, Materials, and Structures.

316.1 General. Piping or tubing passing under or through walls shall be protected from breakage. Piping passing through or under cinders or other corrosive materials shall be protected from external corrosion in an approved manner. Approved provisions shall be made for expansion of hot water piping. Voids around piping or tubing passing through concrete floors on the ground shall be sealed.

316.2 Installation. Piping or tubing shall be installed so that the piping, tubing, or connections will not be subject to undue strains or stresses, and provisions shall be made for expansion, contraction, and structural settlement. No piping or tubing, unless designed and listed for such use, shall be directly embedded in concrete or masonry. No structural member shall be seriously weakened or impaired by cutting, notching, or otherwise as defined in the *California Building Code or California Residential Code.*

316.3 Corrosion, Erosion, and Mechanical Damage. Piping or tubing subject to corrosion, erosion, or mechanical damage shall be protected in an approved manner.

316.4 Protectively Coated Pipe. Protectively coated pipe or tubing shall be inspected and tested, and a visible void, damage, or imperfection to the pipe coating shall be repaired in an approved manner.

316.5 Fire-Resistant Construction. Piping, tubing, and duct system penetrations of fire-resistance-rated walls, partitions, floors, floor/ceiling assemblies, roof/ceiling assemblies, or shaft enclosures shall be protected in accordance with the requirements of the *California Building Code or California Residential Code.*

316.6 Steel Nail Plates. Plastic piping or tubing, copper or copper alloy piping or tubing, and ducts penetrating framing members to within 1 inch (25.4 mm) of the exposed framing shall be protected by steel nail plates not less than No. 18 gauge (0.0478 inches) (1.2141 mm) in thickness. The steel nail plate shall extend along the framing member not less than 1½ inches (38 mm) beyond the outside diameter of the pipe or tubing.

Exception: See Section 1310.3.3.

316.7 Sleeves. Sleeves shall be provided to protect piping through concrete and masonry walls and concrete floors.

Exception: Sleeves shall not be required where openings are drilled or bored.

316.7.1 Building Loads. Piping or tubing through concrete or masonry walls shall not be subject to a load from building construction.

316.7.2 Exterior Walls. In exterior walls, annular space between sleeves and pipes or tubing shall be sealed and made watertight, as approved by the Authority Having Jurisdiction. A penetration through fire-resistive construction shall be in accordance with Section 316.5.

316.8 Firewalls. A pipe sleeve through a firewall shall have the space around the pipe or tubing completely sealed with an approved fire-resistive material in accordance with other codes.

316.9 Structural Members. A structural member weakened or impaired by cutting, notching, or otherwise shall be reinforced, repaired, or replaced so as to be left in a safe structural condition in accordance with the requirements of the *California Building Code or California Residential Code.*

316.10 Rodentproofing. Mechanical system shall be constructed in such a manner as to restrict rodents or vermin from entering a building by following the ductwork from the outside into the building.

316.11 Metal Collars. In or on buildings where openings have been made in walls, floors, or ceilings for the passage of ductwork or pipes, such openings shall be closed and protected by the installation of approved metal collars securely fastened to the adjoining structure.

317.0 Trenching, Excavation, and Backfill.

317.1 Trenches. Trenches deeper than the footings of a building or structure, and paralleling the same, shall be located not less than 45 degrees (0.79 rad) from the bottom exterior edge of the footing, or as approved in accordance with Section 302.0.

317.2 Tunneling and Driving. Tunneling and driving shall be permitted to be done in yards, courts, or driveways of a building site. Where sufficient depth is available to permit, tunnels shall be permitted to be used between open-cut trenches. Tunnels shall have a clear height of 2 feet (610 mm) above the pipe and shall be limited in length to one-half the depth of the trench, with a maximum length of 8 feet (2438 mm). Where pipes are driven, the drive pipe shall be not less than one size larger than the pipe to be laid.

317.3 Open Trenches. Excavations required to be made for the installation of a mechanical system or part thereof, within the walls of a building, shall be open trench work and shall be kept open until it has been inspected, tested, and accepted.

317.4 Excavations. Excavations shall be completely backfilled as soon after inspection as practicable. Precaution shall be taken to ensure compactness of backfill around piping without damage to such piping. Trenches shall be backfilled in thin layers to 12 inches (305 mm) above the top of the piping with clean earth, which shall not contain stones, boulders, cinderfill, frozen earth, construction debris, or other materials that will damage or break the piping or cause corrosive action. Mechanical devices such as bulldozers, graders, etc., shall be permitted to then be used to complete backfill to grade. Fill shall be properly compacted. Precautions shall be taken to ensure permanent stability for pipe laid in filled or made ground.

318.0 Scope.

318.1 Applicability. *This part is applicable to health facilities regulated by OSHPD (See Adoption Tables for application for specific sections).*

Note: *This section has no corresponding provisions in the UMC. For the scope and authority of each state agency, refer to Chapter 1.*

318.2 Services/Systems and Utilities. *Refer to Section 1224.4.1 of the California Building Code.*

319.0 Steam and Hot-Water Systems.

319.1 Requirements for Hospitals and Optional Services Provided in Correctional Treatment Centers. [OSHPD 1, 1R, 4 & 5]

319.1.1 *Boilers shall have the capacity, based upon the rest ratings published by the Hydronics Institute or another acceptable national standard to supply the normal operating requirements of all connected systems and equipment.*

319.1.2 *A minimum of two boilers shall be provided. The arrangement of boilers shall be based on the capacity and capability of a boiler or boilers to operate all systems during-periods of breakdown or maintenance of any one boiler.*

319.1.3 *Boiler systems providing space heating shall be designed to maintain a minimum temperature of 60°F (15.6°C) in general patient areas and the temperatures specified in Table 4-A for sensitive areas during periods of breakdown or maintenance of any one boiler.*

319.1.4 *Boiler feed pumps, condensate return pumps, fuel oil pumps, and heating circulating pumps shall be connected and installed to provide standby service in the event of pump failure. Installation of duplex pumps or provision of a spare pump will meet this requirement.*

319.1.5 *At least two sources of heat (e.g. two pieces of equipment) shall be provided for supplying essential services such as sterilizers, hot water for dishwashing, and domestic hot water for minimum patient service, such as handwashing and baths. Booster heaters for dishwashing providing 125°F to 180°F (52°C to 82°C) water may be counted as the second source of heat for that service.*

319.2 Requirements for Skilled Nursing, Intermediate Care Facilities and Basic Services Provided in Correctional Treatment Centers. [For OSHPD 2 & 4]

319.2.1 *Boilers, if provided, shall accommodate Section 319.1.*

319.2.2 *Two or more interconnected water heaters are an acceptable means to provide two sources of heat for hot water (See Section 319.1.5).*

320.0 Air Conditioning and Heating Systems.

320.1 Requirements for Hospitals and Optional Services Provided in Correctional Treatment Centers. [OSHPD 1, 1R, 4 & 5]

320.1.1 *The systems shall be designed to provide the temperatures and relative humidity for sensitive areas or rooms shown in Table 4-A. When outdoor humidity and internal moisture sources are not sufficient to meet the requirements of sensitive areas or rooms in Table 4-A, humidification shall be provided by means of the healthcare facility air-handling systems. Temperature shall be individually controlled for each operating and delivery room. Burn unit patient rooms that require humidifiers to comply with the requirements of sensitive areas or rooms in Table 4-A shall be provided with individual humidity control. All humidifiers shall use dry steam. Humidifiers shall be located within air handling systems or ductwork to avoid moisture accumulation in downstream components, including filters and insulation.*

320.1.2 *Heating systems shall be designed based on the "Heating DB 99.6%" column of the Climatic Design Data in ASHRAE Handbook-Fundamentals. The systems shall be thermostatically controlled with appropriate zoning to achieve the above conditions.*

320.1.3 *Cooling systems shall be designed based on the 0.4% columns of the four Annual Design Conditions titled Cooling, Evaporation, Dehumidification, and Enthalpy shown by the Climate Design Data in ASHRAE Handbook-Fundamentals. The systems shall be thermostatically controlled with appropriate zoning to achieve the above conditions.*

320.2 Requirements for Skilled Nursing, Intermediate Care Facilities and Basic Services Provided in Correctional Treatment Centers. [For OSHPD 2 & 4]

320.2.1 *Systems shall accommodate the provisions of Sections 320.1.2 through 320.1.3.*

320.2.2 *Where air conditioning is provided, the system shall be thermostatically controlled in one or more zones.*

320.3 Requirements for Outpatient Facilities and Licensed Clinics. [For OSHPD 3]

320.3.1 *The system shall be designed to provide the temperature and humidity's for sensitive areas for rooms shown in Table 4-A.*

320.4 Telephone and Data Equipment Rooms. [OSHPD 1 & 4] *Where telecommunications service entrance rooms, technology equipment centers, or technology distribution rooms are provided in accordance with Section 1224.5 of the California Building Code, the following requirements shall apply:*

320.4.1 *Power for HVAC systems serving the room(s) shall be supplied by the Equipment Branch pursuant to the California Electrical Code. Where redundant systems are provided, only one shall be required to be supplied by the Equipment Branch.*

320.4.2 *Mechanical equipment or fixtures that are not directly related to the support of the room shall not be installed in or pass through the room.*

Exception: *Unrelated ductwork may be installed and shall be not less than 10 feet (3048 mm) above the finished floor.*

320.4.3 *HVAC systems shall be provided to maintain environmental conditions recommended in ASHRAE's Thermal Guidelines for Data Processing Environment and the requirements of the specific equipment installed.*

320.5 Psychiatric Services. [OSHPD 1, 1R, 2, 4 & 5] *For projects associated with provision of psychiatric services in acute psychiatric hospitals, general acute-care hospitals, and special treatment program service units in skilled nursing facilities, psychiatric, seclusion, and holding-patient rooms shall be designed with security diffusers, grilles, and registers.*

321.0 Essential Mechanical Provisions. [OSHPD 1, 1R, 2, 3 (Surgical Clinics only) 4 & 5] *During periods of power outages essential electrical power shall be provided for the following equipment:*

321.1 *(Does not apply to OSHPD 3 surgical clinic.) All heating equipment necessary to maintain a minimum temperature of 60°F (15.6°) in patient areas which are not specified in Section 322.0.*

321.2 *All heating equipment necessary to maintain the minimum temperatures listed in Table 4-A for sensitive areas specified in Section 322.0.*

321.3 *Equipment necessary for humidification of the areas listed in Section 322.0.*

321.4 *All supply, return, and exhaust fans required to maintain the positive and negative air balances as required in Table 4-A.*

321.5 *All control components and control systems necessary for the normal operation of equipment required to have essential electrical power.*

321.6 *Alarms for airborne infection isolation rooms and protective environment rooms.*

322.0 Sensitive Areas or Rooms. [OSHPD 1, 1R, 2, 3 (Surgical Clinics) 4 & 5] *The following are sensitive areas or rooms:*

(1) Operating room, hybrid operation room

(2) Cystoscopy

(3) Cardiac catheterization lab

(4) Trauma/cardiac room

(5) Delivery room, cesarean operating room

(6) Gastrointestinal endoscopy procedure room

(7) Post-anesthesia care unit

(8) Newborn nursery

(9) Newborn intensive-care nursery unit

(10) Intensive care

(11) Burn unit

322.1 *The following conditions shall be met for sensitive areas or rooms:*

(1) Thermostats and humidistats shall be either locally resettable and of the non-locking type or remotely resettable and of the locking type.

(2) Systems shall be capable of maintaining the rooms within the temperature range in Table 4-A during normal operation. Lower or higher temperature shall be permitted when patients' comfort and/or medical conditions require those conditions.

(3) The humidity ranges listed in Table 4-A are the minimum and maximum limits where control is specifically needed.

(4) Types of intensive care service spaces are listed in the California Building Code.

TABLE 303.10.1
REDUCTION OF CLEARANCES WITH SPECIFIED FORMS OF PROTECTION[1, 2, 3, 4, 5, 6, 7, 8, 9, 10, 11]
[NFPA 54: TABLE 10.2.3]

TYPE OF PROTECTION APPLIED TO AND COVERING ALL SURFACES OF COMBUSTIBLE MATERIAL WITHIN THE DISTANCE SPECIFIED AS THE REQUIRED CLEARANCE WITH NO PROTECTION	WHERE THE REQUIRED CLEARANCE WITH NO PROTECTION FROM APPLIANCE, VENT CONNECTOR, OR SINGLE-WALL METAL PIPE IS:									
	36 (INCHES)		18 (INCHES)		12 (INCHES)		9 (INCHES)		6 (INCHES)	
	ALLOWABLE CLEARANCES WITH SPECIFIED PROTECTION (INCHES)									
	USE COLUMN 1 FOR CLEARANCES ABOVE APPLIANCE OR HORIZONTAL CONNECTOR. USE COLUMN 2 FOR CLEARANCES FROM APPLIANCES, VERTICAL CONNECTOR, AND SINGLE-WALL METAL PIPE.									
	ABOVE (COLUMN 1)	SIDES AND REAR (COLUMN 2)	ABOVE (COLUMN 1)	SIDES AND REAR (COLUMN 2)	ABOVE (COLUMN 1)	SIDES AND REAR (COLUMN 2)	ABOVE (COLUMN 1)	SIDES AND REAR (COLUMN 2)	ABOVE (COLUMN 1)	SIDES AND REAR (COLUMN 2)
(1) 3½ inch thick masonry wall without ventilated air space	—	24	—	12	—	9	—	6	—	5
(2) ½ of an inch insulation board over 1 inch glass fiber or mineral wool batts	24	18	12	9	9	6	6	5	4	3
(3) 0.024 inch (nominal 24 gauge) sheet metal over 1 inch glass fiber or mineral wool batts reinforced with wire on rear face with ventilated air space	18	12	9	6	6	4	5	3	3	3
(4) 3½ inch thick masonry wall with ventilated air space	—	12	—	6	—	6	—	6	—	6
(5) 0.024 inch (nominal 24 gauge) sheet metal with ventilated air space	18	12	9	6	6	4	5	3	3	2
(6) ½ of an inch thick insulation board with ventilated air space	18	12	9	6	6	4	5	3	3	3
(7) 0.024 inch (nominal 24 gauge) sheet metal with ventilated air space over 0.024 inch (nominal 24 gauge) sheet metal with ventilated air space	18	12	9	6	6	4	5	3	3	3
(8) 1 inch glass fiber or mineral wool batts sandwiched between two sheets 0.024 inch (nominal 24 gauge) sheet metal with ventilated air space	18	12	9	6	6	4	5	3	3	3

For SI units: 1 inch = 25.4 mm, °C = (°F-32)/1.8

Notes:

1 Reduction of clearances from combustible materials shall not interfere with combustion air, draft hood clearance and relief, and accessibility of servicing.

2 Clearances shall be measured from the outer surface of the combustible material to the nearest point on the surface of the appliance, disregarding an intervening protection applied to the combustible material.

3 Spacers and ties shall be of noncombustible material. No spacer or tie shall be used directly opposite the appliance or connector.

4 Where clearance reduction systems use a ventilated air space, a provision for air circulation shall be provided as described. [See Figure 303.10.1(2) and Figure 303.10.1(3)]

5 There shall be not less than 1 inch (25.4 mm) between clearance reduction systems and combustible walls and ceilings for reduction systems using a ventilated air space.

6 Where a wall protector is mounted on a single flat wall away from corners, it shall have not less than a 1 inch (25.4 mm) air gap. To provide air circulation, the bottom and top edges, or the side and top edges, or edges shall be left open.

7 Mineral wool batts (blanket or board) shall have a density of not less than 8 pounds per cubic foot (lb/ft^3) (128 kg/m^3) and a minimum melting point of 1500°F (816°C).

8 Insulation material used as part of a clearance reduction system shall have a thermal conductivity of 1 British thermal unit inch per hour square foot degree Fahrenheit [Btu•in/(h•ft^2•°F)] [0.1W/(m•K)] or less.

9 There shall be not less than 1 inch (25.4 mm) between the appliance and the protector. In no case shall the clearance between the appliance and the combustible surface be reduced below that allowed in this table.

10 Clearances and thicknesses are minimum; larger clearances and thicknesses are acceptable.

11 Listed single-wall connectors shall be installed in accordance with the terms of their listing and the manufacturer's installation instructions.

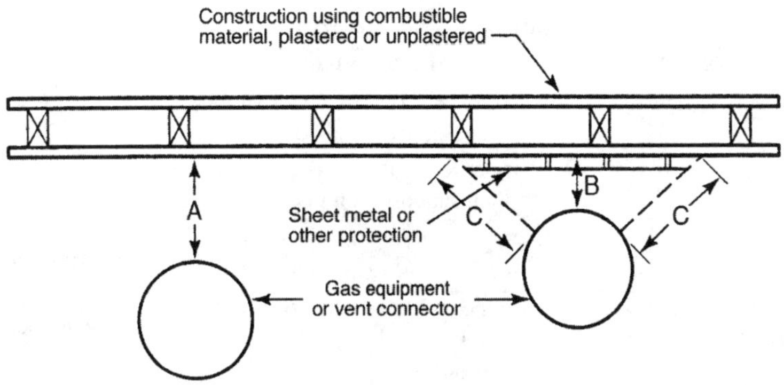

Notes:

[1] A – Equals the clearance with no protection specified in Table 802.7.3.3 and Table 904.2.2 and in the sections applying to various types of appliances.

[2] B – Equals the reduced clearance permitted in accordance with Table 303.10.1.

[3] The protection applied to the construction using combustible material shall extend far enough in each direction to make C equal to A.

FIGURE 303.10.1(1)[1, 2, 3]
EXTENT OF PROTECTION NECESSARY TO REDUCE CLEARANCES FROM GAS APPLIANCES OR VENT CONNECTORS
[NFPA 54: FIGURE 10.3.2.3(a)]

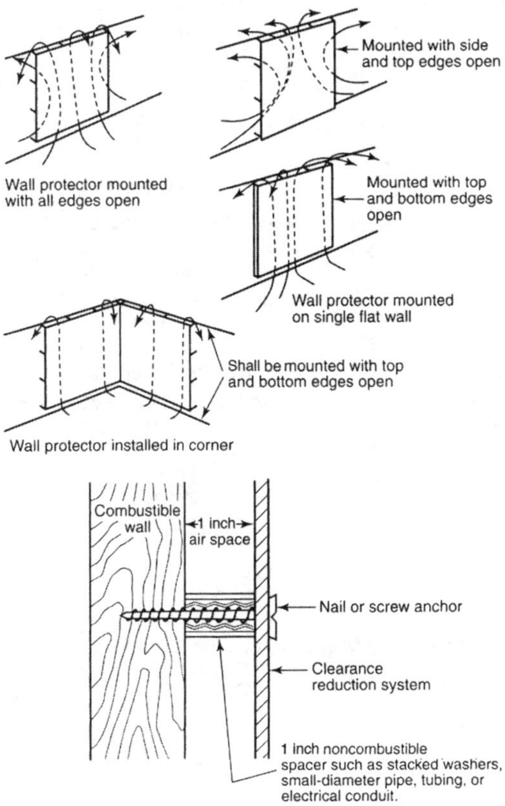

For SI units: 1 inch = 25.4 mm

Note: Masonry walls shall be permitted to be attached to combustible walls using wall ties. Spacers shall not be used directly behind appliance or connector.

FIGURE 303.10.1(2)
WALL PROTECTOR CLEARANCE REDUCTION SYSTEM
[NFPA 54: FIGURE 10.3.2.3(b)]

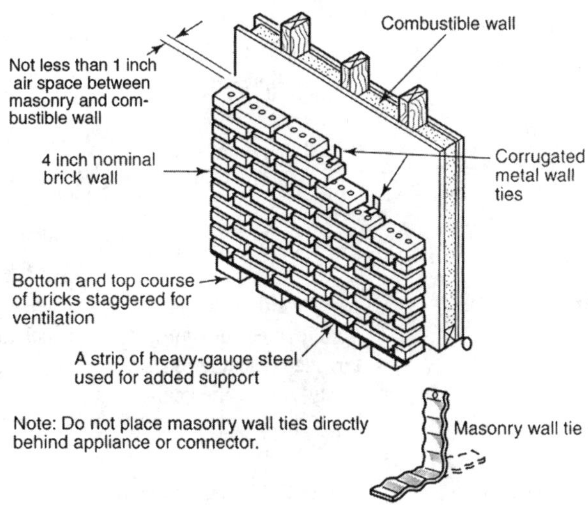

For SI units: 1 inch = 25.4 mm

FIGURE 303.10.1(3)
MASONRY CLEARANCE REDUCTION SYSTEM
[NFPA 54: FIGURE 10.3.2.3(c)]

TABLE 313.3
HANGERS AND SUPPORTS

MATERIALS	TYPES OF JOINTS	HORIZONTAL	VERTICAL
Cast	Lead and Oakum	5 feet, except 10 feet where 10 foot lengths are installed[1, 2, 3]	Base and each floor, not to exceed 15 feet
	Compression Gasket	Every other joint, unless over 4 feet then support each joint[1, 2, 3]	Base and each floor, not to exceed 15 feet
Cast-Iron Hubless	Shielded Coupling	Every other joint, unless over 4 feet then support each joint[1, 2, 3, 4]	Base and each floor, not to exceed 15 feet
Copper & Copper Alloys	Soldered, Brazed, Threaded, or Mechanical	1½ inches and smaller, 6 feet; 2 inches and larger, 10 feet	Each floor, not to exceed 10 feet[5]
Steel Pipe for Water DWV	Threaded or Welded	¾ inch and smaller, 10 feet; 1 inch and larger, 12 feet	Every other floor, not to exceed 25 feet[5]
Steel Pipe for Gas	Threaded or Welded	½ inch, 6 feet; ¾ inch and 1 inch, 8 feet; 1¼ inches and larger, 10 feet	½ inch, 6 feet; ¾ inch and 1 inch, 8 feet; 1¼ inches every floor level
Schedule 40 PVC and ABS	Solvent Cemented	All sizes, 4 feet; allow for expansion every 30 feet[3]	Base and each floor; provide mid-story guides; provide for expansion every 30 feet
CPVC	Solvent Cemented	1 inch and smaller, 3 feet; 1¼ inches and larger, 4 feet	Base and each floor; provide mid-story guides
Lead	Wiped or Burned	Continuous Support	Not to exceed 4 feet
Steel	Mechanical	In accordance with standards acceptable to the Authority Having Jurisdiction	
PEX	Cold Expansion, Insert and Compression	1 inch and smaller, 32 inches; 1¼ inches and larger, 4 feet	Base and each floor; provide mid-story guides
PEX-AL-PEX	Metal insert and metal compression	½ inch, ¾ inch, 1 inch } All sizes 98 inches	Base and each floor; provide mid-story guides
PE-AL-PE	Metal insert and metal compression	½ inch, ¾ inch, 1 inch } All sizes 98 inches	Base and each floor; provide mid-story guides
PE-RT	Insert and Compression	1 inch and smaller, 32 inches; 1¼ inches and larger, 4 feet	Base and each floor; provide mid-story guides
Polypropylene (PP)	Fusion weld (socket, butt, saddle, electrofusion), threaded (metal threads only), or mechanical	1 inch and smaller, 32 inches; 1¼ inches and larger, 4 feet	Base and each floor; provide mid-story guides

For SI unit: 1 inch = 25.4 mm, 1 foot = 304.8 mm

Notes:

1 Support adjacent to joint, not to exceed 18 inches (457 mm).

2 Brace not to exceed 40 feet (12 192 mm) intervals to prevent horizontal movement.

3 Support at each horizontal branch connection.

4 Hangers shall not be placed on the coupling.

5 Vertical water lines shall be permitted to be supported in accordance with recognized engineering principles with regard to expansion and contraction, where first approved by the Authority Having Jurisdiction.

CALIFORNIA MECHANICAL CODE – MATRIX ADOPTION TABLE
CHAPTER 4 – VENTILATION AIR

(Matrix Adoption Tables are non-regulatory, intended only as an aid to the code user. See Chapter 1 for state agency authority and building applications.)

Adopting Agency	BSC	BSC-CG	SFM	HCD			DSA			OSHPD						BSCC	DPH	AGR	DWR	CEC	CA	SL	SLC
				1	2	1-AC	AC	SS	SS/CC	1	1R	2	3	4	5								
Adopt Entire Chapter																							
Adopt Entire Chapter as amended (amended sections listed below)	X		X	X	X			X	X														
Adopt only those sections that are listed below										X	X	X	X	X	X								
Chapter/Section																							
401.0			X							X		X	X	X									
401.1											X	X	X	X	X								
401.2 with exception		X						X	X														
401.2.1		X						X	X														
402.0										X	X	X	X	X	X								
402.1	X							X	X	X													
402.1.3											X	X	X	X	X								
402.3				X	X																		
402.5				X	X																		
403.0										X	X	X	X	X	X								
403.7.2.1 – 403.7.2.4			X	X	X																		
405.0										X	X	X	X	X	X								
407.0 and subsections										X	X	X	X	X	X								
408.0 and subsections										X	X	X	X	X	X								
409.0 and subsections										X	X	X	X	X	X								
410.0 and subsections										X	X	X	X	X	X								
411.0 and subsections										X	X	X	X	X	X								
412.0 and subsections										X	X	X	X	X	X								
413.0 and subsections										X	X	X	X	X	X								
414.0 and subsections										X	X	X	X	X	X								
415.0 and subsections										X	X	X	X	X	X								
416.0 - 416.2										X	X	X	X	X	X								
416.3														X									
416.4										X	X	X	X	X	X								
417.0										X	X	X	X	X	X								
418.0 and subsections										X	X	X	X	X	X								
419.0 and subsections										X													
Table 4-A										X	X	X	X	X	X								
Table 4-B										X	X		X	X									
Table 4-C												X		X									
Table 402.1	X							X	X														
Table 403.7 *Note 11 & 12*				X	X																		

This state agency does not adopt sections identified with the following symbol: †

The Office of the State Fire Marshal's adoption of this chapter or individual sections is applicable to structures regulated by other state agencies pursuant to Section 1.11.0.

CHAPTER 4

VENTILATION AIR

401.0 General.

401.1 Applicability. This chapter contains requirements for ventilation air supply, exhaust, and makeup air requirements for occupiable spaces within a building. ***[OSHPD 1, 1R, 2, 3, 4 & 5]*** *See Sections 404.0 through 418.0.* ***[SFM]*** *Air filters shall comply with all requirements of Part 12, Title 24, Chapter 12-71, SFM Standard 12-71-1.*

401.2 Filters. [BSC-CG], [DSA-SS & DSA-SS/CC] *In mechanically ventilated buildings, provide regularly occupied areas of the building with air filtration media for outside and return air that provides at least a Minimum Efficiency Reporting Value (MERV) of 13. MERV 13 filters shall be installed prior to occupancy, and recommendations for maintenance with filters of the same value shall be included in the operation and maintenance manual in compliance with Chapter 5, Division 5.5. of the California Green Building Standards Code (CALGreen).*

Exception: *Existing mechanical equipment.*

401.2.1 Labeling. *Installed filters shall be clearly labeled by the manufacturer indicating the MERV rating.*

402.0 Ventilation Air. *[Not permitted for OSHPD 1, 2, 3 & 4]*

402.1 Occupiable Spaces. Occupiable spaces listed in Table 402.1 shall be designed to have ventilation (outdoor) air for occupants in accordance with this chapter. *Ventilation air requirements for occupancies regulated by the California Energy Commission are found in the California Energy Code.*

402.1.1 Construction Documents. The outdoor air ventilation rate and air distribution assumptions made in the design of the ventilation system shall be clearly identified on the construction documents.

402.1.2 Dwelling. Requirements for ventilation air rate for single-family dwellings shall be in accordance with this chapter or ASHRAE 62.2.

402.1.3 Ventilation in Health Care Facilities. Mechanical ventilation for health care facilities shall be designed and installed in accordance with this code and ASHRAE 170 ***[OSHPD 1, 1R, 2, 3, 4 & 5]****-2013, through Addendum ae, as published with "Guidelines for Design and Construction of Hospitals and Outpatient Facilities," 2014 edition (published by The Facility Guidelines Institute). All supply-air, return air, and exhaust-air systems shall comply with ASHRAE 170. The text of ASHRAE 170 shall be modified as follows:*

(1) ASHRAE 170. Section 6.1.2.1 -- Not adopted.

(2) ASHRAE 170. Section 6.3.2 -- Not adopted.

(3) ASHRAE 170. Table 6.4 -- Not adopted.

(4) ASHRAE 170. Section 6.4-6.4.4 -- Not adopted.

(5) ASHRAE 170. Section 6.9 -- Not adopted.

(6) ASHRAE 170. Section 7.1a -- Modify as follows:

Replace reference to Table 7.1 with reference to Table 4-A.

(7) ASHRAE 170. Section 7.2.1a through e -- Not adopted.

(8) ASHRAE 170. Section 7.2.2 a through c, and e -- Not adopted.

(9) ASHRAE 170. Section 7.2.3 -- Not adopted.

(10) ASHRAE 170. Section 7.3.1 -- Modify as follows:

Replace reference to Table 7.1 with reference to Table 4-A.

(11) ASHRAE 170. Section 7.4.1 -- Modify as follows:

Delete the Exception that allows for high return grilles.

402.2 Natural Ventilation. Natural ventilation systems shall be designed in accordance with this section and shall include mechanical ventilation systems designed in accordance with Section 403.0, Section 404.0, or both.

Exceptions:

(1) An engineered natural ventilation system where approved by the Authority Having Jurisdiction need not comply with Section 402.2.

(2) A mechanical ventilation system is not required where:

(a) natural ventilation openings comply with the requirements of Section 402.2 and are permanently open or have controls that prevent the openings from being closed during periods of expected occupancy or

(b) the zone is not served by heating or cooling equipment. [ASHRAE 62.1:6.4]

402.2.1 Floor Area to Be Ventilated. Spaces, or portions of spaces, to be naturally ventilated shall be located within a distance based on the ceiling height, in accordance with Section 402.2.1.1, Section 402.2.1.2, or Section 402.2.1.3, from operable wall openings in accordance with Section 402.2.2. For spaces with ceilings which are not parallel to the floor, the ceiling height shall be determined in accordance with Section 402.2.1.4. [ASHRAE 62.1:6.4.1]

402.2.1.1 Single Side Opening. For spaces with operable openings on one side of the space, the distance from the operable openings shall be not more than 2*H*, where *H* is the ceiling height. [ASHRAE 62.1:6.4.1.1]

402.2.1.2 Double Side Opening. For spaces with operable openings on two opposite sides of the space, the distance from the operable openings shall be not more than 5*H*, where *H* is the ceiling height. [ASHRAE 62.1:6.4.1.2]

402.2.1.3 Corner Openings. For spaces with operable openings on two adjacent sides of a space, such as two sides of a corner, the distance from the operable openings shall be not more than $5H$ along a line drawn between the two openings that are farthest apart. Floor area outside that line shall comply with Section 402.2.1.1. [ASHRAE 62.1:6.4.1.3]

402.2.1.4 Ceiling Height. The ceiling height, H, to be used in Section 402.2.1.1 through Section 402.2.1.3 shall be the minimum ceiling height in the space.

Exception: For ceilings that are increasing in height as distance from the openings is increased, the ceiling height shall be determined as the average height of the ceiling within 20 feet (6096 mm) from the operable openings. [ASHRAE 62.1:6.4.1.4]

402.2.2 Location and Size of Openings. Spaces, or portions of spaces, to be naturally ventilated shall be permanently open to operable wall openings directly to the outdoors, the openable area of which is a minimum of 4 percent of the net occupiable floor area. Where openings are covered with louvers or otherwise obstructed, openable area shall be based on the net free unobstructed area through the opening. Where interior rooms, or portions of rooms, without direct openings to the outdoors are ventilated through adjoining rooms, the opening between rooms shall be permanently unobstructed and shall have a free area of not less than 8 percent of the area of the interior room nor less than 25 square feet (2.3 m^2). [ASHRAE 62.1:6.4.2]

402.2.3 Control and Accessibility. The means to open required operable openings shall be readily accessible to building occupants where the space is occupied. Controls shall be designed to coordinate operation of the natural and mechanical ventilation systems. [ASHRAE 62.1:6.4.3]

402.3 Mechanical Ventilation. Where natural ventilation is not permitted by this section or the *California Building Standards Code,* mechanical ventilation systems shall be designed, constructed, and installed to provide a method of supply air and exhaust air. Mechanical ventilation systems shall include controls, manual or automatic, that enable the fan system to operate wherever the spaces served are occupied. The system shall be designed to maintain minimum outdoor airflow as required by Section 403.0 under any load conditions.

402.4 Outdoor Air Intake Protection. Required outdoor-air intakes shall be covered with a screen having not less than ¼ of an inch (6.4 mm) openings, and shall have not more than ½ of an inch (12.7 mm) openings.

402.4.1 Weather Protections. Outdoor air intakes that are part of the mechanical ventilation system shall be designed to manage rain entrainment, to prevent rain intrusion, and manage water from snow in accordance with ASHRAE 62.1.

402.5 Bathroom Exhaust Fans. [HCD 1 & HCD 2] *Each bathroom shall be mechanically ventilated in accordance with Division 4.5 of the California Green Building Standards Code (CALGreen).*

403.0 Ventilation Rates. *[Not permitted for OSHPD 1, 1R, 2, 3, 4 & 5]*

403.1 General. The design outdoor air intake flow rate for a ventilation system shall be determined in accordance with Section 403.2 through Section 403.9.4.

403.2 Zone Calculations. Ventilation zone parameters shall be determined in accordance with Section 403.2.1 through Section 403.2.3 for each ventilation zone served by the ventilation system. [ASHRAE 62.1:6.2.2]

403.2.1 Breathing Zone Outdoor Airflow. The outdoor airflow required in the breathing zone of the occupiable space or spaces in a ventilation zone, i.e., the breathing zone outdoor airflow (V_{bz}), shall be not less than the value determined in accordance with Equation 403.2.1.

$$V_{bz} = R_p \cdot P_z + R_a \cdot A_z \qquad \text{(Equation 403.2.1)}$$

Where:

A_z = zone floor area: the net occupiable floor area of the ventilation zone, square feet (m^2).

P_z = zone population: The number of people in the ventilation zone during typical usage.

R_p = outdoor airflow rate required per person as determined from Table 402.1.

R_a = outdoor airflow rate required per unit area as determined from Table 402.1. [ASHRAE 62.1:6.2.2.1]

403.2.2 Zone Air Distribution Effectiveness. The zone air distribution effectiveness (E_z) shall be not greater than the default value determined in accordance with Table 403.2.2. [ASHRAE 62.1:6.2.2.2]

403.2.3 Zone Outdoor Airflow. The zone outdoor airflow (V_{oz}), i.e., the outdoor airflow rate that shall be provided to the ventilation zone by the supply air distribution system, shall be determined in accordance with Equation 403.2.3. [ASHRAE 62.1:6.2.2.3]

$$V_{oz} = V_{bz}/E_z \qquad \text{(Equation 403.2.3)}$$

403.3 Single-Zone Systems. For ventilation systems where one or more air handlers supply a mixture of outdoor air and recirculated air to only one ventilation zone, the outdoor air intake flow (V_{ot}) shall be determined in accordance with Equation 403.3. [ASHRAE 62.1:6.2.3]

$$V_{ot} = V_{oz} \qquad \text{(Equation 403.3)}$$

403.4 One Hundred Percent Outdoor Air Systems. For ventilation systems where one or more air handlers supply only outdoor air to one or more ventilation zones, the outdoor air intake flow (V_{ot}) shall be determined in accordance with Equation 403.4. [ASHRAE 62.1:6.2.4]

$$V_{ot} = \Sigma_{\text{all zones}} V_{oz} \qquad \text{(Equation 403.4)}$$

403.5 Multiple-Zone Recirculating Systems. For ventilation systems where one or more air handlers supply a mixture of outdoor air and recirculated air to more than one ventilation zone, the outdoor air intake flow (V_{ot}) shall be determined in accordance with Section 403.5.1 through Section 403.5.4. [ASHRAE 62.1:6.2.5]

403.5.1 Primary Outdoor Air Fraction. The primary outdoor air fraction (Z_{pz}) shall be determined for ventilation zones in accordance with Equation 403.5.1. [ASHRAE 62.1:6.2.5.1]

$$Z_{pz} = V_{oz}/V_{pz} \qquad \text{(Equation 403.5.1)}$$

Where:

V_{pz} is the zone primary airflow, i.e., the primary airflow rate to the ventilation zone from the air handler, including outdoor air and recirculated air. [ASHRAE 62.1:6.2.5.1]

403.5.2 System Ventilation Efficiency. The system ventilation efficiency (E_v) shall be determined in accordance with Table 403.5.2 or Section 404.0. [ASHRAE 62.1:6.2.5.2]

403.5.3 Uncorrected Outdoor Air Intake. The uncorrected outdoor air intake (V_{ou}) flow shall be determined in accordance with Equation 403.5.3(1). [ASHRAE 62.1:6.2.5.3]

$$V_{ou} = D\,\Sigma_{\text{all zones}} (R_p \cdot P_z) + \Sigma_{\text{all zones}} (R_a \cdot A_z) \qquad \text{[Equation 403.5.3(1)]}$$

The occupant diversity ratio (D) shall be determined in accordance with Equation 403.5.3(2) to account for variations in population within the ventilation zones served by the system.

$$D = P_s/\Sigma_{\text{all zones}} P_z \qquad \text{[Equation 403.5.3(2)]}$$

Where the system population (P_s) is the total population in the area served by the system.

Exception: Alternative methods to account for occupant diversity shall be permitted, provided that the resulting (V_{ou}) value is not less than that determined in accordance with Equation 403.5.3(1). [ASHRAE 62.1:6.2.5.3.1]

403.5.4 Outdoor Air Intake. The design outdoor air intake flow (V_{ot}) shall be determined in accordance with Equation 403.5.4. [ASHRAE 62.1:6.2.5.4]

$$V_{ot} = V_{ou}/E_v \qquad \text{(Equation 403.5.4)}$$

403.6 Design for Varying Operating Conditions. Ventilation systems shall be designed to be capable of providing not less than the minimum ventilation rates required in the breathing zone where the zones served by the system are occupied, including all full and part-load conditions. The minimum outdoor air intake flow shall be permitted to be less than the design value at part-load conditions. [ASHRAE 62.1:6.2.6.1]

403.6.1 Short-Term Conditions. Where it is known that peak occupancy will be of short duration, or the ventilation will be varied or interrupted for a short period of time, the design shall be permitted to be based on the average conditions over a time period (T) determined in accordance with Equation 403.6.1.

$$T = 3v/V_{bz} \qquad \text{(Equation 403.6.1)}$$

Where:

T = averaging time period, minutes.

v = the volume of the ventilation zone for which averaging is being applied, cubic foot (m^3).

V_{bz} = the breathing zone outdoor airflow determined in accordance with Equation 403.2.1 and design value of the zone population (P_z), cubic foot per minute (cfm) (m^3/min).

Acceptable design adjustments based on this optional provision shall be in accordance with the following:

(1) Zones with fluctuating occupancy: The zone population (P_z) shall be permitted to be averaged over time (T).

(2) Zones with intermittent interruption of supply air: The average outdoor airflow supplied to the breathing zone over time (T) shall be not less than the breathing zone outdoor airflow (V_{bz}) calculated using Equation 403.2.1.

(3) Systems with intermittent closure of the outdoor air intake: The average outdoor air intake over time (T) shall be not less than the minimum outdoor air intake (V_{ot}) calculated using Equation 403.3, Equation 403.4, or Equation 403.5.4. [ASHRAE 62.1:6.2.6.2]

403.7 Exhaust Ventilation. Exhaust airflow shall be provided in accordance with the requirements in Table 403.7. Exhaust makeup air shall be permitted to be a combination of outdoor air, recirculated air, and transfer air.

403.7.1 Parking Garages. Exhaust rate for parking garages shall be in accordance with Table 403.7. Exhaust rate shall not be required for enclosed parking garages having a floor area of 1000 square feet (92.9 m^2) or less and used for the storage of 5 or less vehicles.

403.7.2 Enclosed Parking Garages. Mechanical ventilation systems for enclosed parking garages shall operate continuously.

Exceptions:

(1) Mechanical ventilation systems shall be permitted to operate intermittently where the system is designed to operate automatically upon detection of vehicle operation or the presence of occupants by approved automatic detection devices.

(2) Approved automatic carbon monoxide sensing devices shall be permitted to be employed to modulate the ventilation system to not exceed a maximum average concentration of carbon monoxide of 50 parts per million during an eight-hour period, with a concentration of not more than 200 parts per million for a period not exceeding one hour. Automatic carbon monoxide sensing devices installed to modulated parking garage ventilation systems shall be approved in accordance with Section 301.2.

403.7.2.1 Alternative Exhaust Ventilation for Enclosed Parking Garages.

403.7.2.2 Minimum Exhaust Rate. [HCD 1 & 2] *In lieu of the exhaust rates in Table 403.7, ventilation systems shall be capable of providing 14,000 cfm (6608 L/s) of exhaust air for each operating vehicle. Number of operating vehicles shall be determined based on 2.5 percent of all parking spaces (and not less than one vehicle).*

403.7.2.3 Exhaust Inlet Distribution. [HCD 1 & 2] *To ensure proper exhaust of contaminated air and fumes from parking garages, exhaust systems utilizing multiple exhaust inlets shall be designed so that exhaust inlets are distributed in such a manner that no portion of the parking garage is more than 50 feet (15 240 mm) from an exhaust inlet. Such exhaust inlets shall be installed so that the highest elevation of the exhaust inlet is no greater than 12 inches (305 mm) below the lowest ceiling level.*

Exception: *Garage exhaust systems designed without distributed exhaust inlets may have their exhaust inlets designed based on the principles of engineering and mechanics and shall provide the minimum required exhaust rate in Table 403.7.*

403.7.2.4 Exhaust System Operation. [HCD 1 & 2] *Exhaust systems shall operate continuously unless one of the exceptions to continuous operation of Section 403.7.2 is utilized.*

403.8 Dynamic Reset. The system shall be permitted to be designed to reset the outdoor air intake flow (V_{ot}), the space or ventilation zone airflow (V_{oz}) as operating conditions change. [ASHRAE 62.1:6.2.7]

403.9 Air Classification and Recirculation. Air shall be classified, and the recirculation or transfer shall be limited in accordance with Section 403.9.1 through Section 403.9.4. [ASHRAE 62.1:5.16] Recirculated air shall not be taken from prohibited locations in accordance with Section 311.3.

403.9.1 Class 1 Air. Recirculation or transfer of Class 1 air to other spaces shall be permitted. [ASHRAE 62.1:5.16.3.1]

403.9.2 Class 2 Air. Recirculation of Class 2 air within the space of origin shall be permitted. Recirculation or transfer of Class 2 air to other Class 2 or Class 3 spaces shall be permitted, provided the other spaces are used for the same or similar purpose or task and involve the same or similar pollutant sources as the Class 2 space. Transfer of Class 2 air to toilet rooms shall be permitted. Recirculation or transfer of Class 2 air to Class 4 spaces shall be permitted. Class 2 air shall not be recirculated or transferred to Class 1 spaces. Where using an energy recover device, recirculation from leakage, carryover, or transfer from the exhaust side of the energy recovery device shall be permitted and the recirculated Class 2 air shall not exceed 10 percent of the outdoor air intake flow. [ASHRAE 62.1:5.16.3.2]

403.9.3 Class 3 Air. Recirculation of Class 3 air within the space of origin shall be permitted. Class 3 air shall not be recirculated or transferred to other spaces. Where using an energy recover device, recirculation from leakage, carryover, or transfer from the exhaust side of the energy recovery device shall be permitted and the recirculated Class 3 air shall not exceed 5 percent of the outdoor air intake flow. [ASHRAE 62.1:5.16.3.3]

403.9.4 Class 4 Air. Class 4 air shall not be recirculated or transferred to other spaces or be recirculated within the space of origin. [ASHRAE 62.1:5.16.3.4]

404.0 Multiple-Zone Systems.

404.1 General. This section presents an alternative procedure for calculating the system ventilation efficiency (E_v) where values in Table 403.5.2 are not used. The system ventilation efficiency shall equal the lowest zone ventilation efficiency among the ventilation zones served by the air handler in accordance with Equation 404.1. [ASHRAE 62.1:A1.3]

$$E_v = \text{minimum } (E_{vz}) \qquad \text{(Equation 404.1)}$$

404.2 Average Outdoor Air Fraction. The average outdoor air fraction (X_s) for the ventilation system shall be determined in accordance with Equation 404.2.

$$X_s = V_{ou}/V_{ps} \qquad \text{(Equation 404.2)}$$

The uncorrected outdoor air intake (V_{ou}) shall be determined in accordance with Section 403.5.3, and the system primary airflow (V_{ps}) shall be determined at the condition analyzed. [ASHRAE 62.1:A1.1]

404.3 Zone Ventilation Efficiency. The zone ventilation efficiency (E_{vz}) shall be the efficiency with which a system distributes outdoor air from the intake to an individual breathing zone, and shall be determined in accordance with Section 404.3.1 or Section 404.3.2. [ASHRAE 62.1:A1.2]

404.3.1 Single Supply Systems. For single supply systems, where the air supplied to a ventilation zone is a mixture of outdoor air and system-level recirculated air, zone ventilation efficiency (E_{vz}) shall be determined in accordance with Equation 404.3.1. Examples of single supply systems include constant-volume reheat, single-duct VAV, single-fan dual-duct, and multizone systems.

$$E_{vz} = 1 + X_s - Z_{pz} \qquad \text{(Equation 404.3.1)}$$

The average outdoor air fraction for the system (X_s) shall be determined in accordance with Equation 404.2 and the primary outdoor air fraction for the zone (Z_{pz}) shall be determined in accordance with Section 403.5.1. [ASHRAE 62.1:A1.2.1]

404.3.2 Secondary-Recirculation Systems. For secondary-recirculation systems where the supply air or a portion thereof to each ventilation zone is recirculated air (air that has not been directly mixed with outdoor air) from other zones, zone ventilation efficiency (E_{vz}) shall be determined in accordance with Equation 404.3.2(1). Examples of secondary-recirculation systems include dual-fan dual-duct and fan-powered mixing-box systems, and systems that include transfer fans for conference rooms.

[Equation 404.3.2(1)]

$$E_{vz} = (F_a + X_s \bullet F_b - Z_{pz} \bullet E_p \bullet F_c)/F_a$$

The system air fractions Fa, Fb, and Fc shall be determined in accordance with Equation 404.3.2(2), Equation 404.3.2(3), and Equation 404.3.2(4). The zone primary air fraction (Ep) shall be determined in accordance with Equation 404.3.2(5). For single-zone and single-supply systems Ep shall equal to 1.0. The zone secondary recirculation fraction (Er) shall be determined by the designer based on system configuration. The zone air distribution effectiveness (Ez) shall be determined in accordance with Section 403.2.2. [ASHRAE 62.1:A1.2.2]

$$F_a = E_p + (1 - E_p) \bullet E_r \qquad \text{[Equation 404.3.2(2)]}$$

$$F_b = E_p \qquad \text{[Equation 404.3.2(3)]}$$

$$F_c = 1 - (1 - E_z) \bullet (1 - E_r) \bullet (1 - E_p) \qquad \text{[Equation 404.3.2(4)]}$$

$$E_p = V_{pz}/V_{dz} \qquad \text{[Equation 404.3.2(5)]}$$

Where:

E_p - Primary air fraction: The fraction of primary air in the discharge air to the ventilation zone.

E_r - Secondary recirculation fraction: In systems with secondary recirculation of return air, the fraction of secondary recirculated air to the zone that is representative of average system return air rather than air directly recirculated from the zone.

E_{vz} - Zone ventilation efficiency: The efficiency with which the system distributes air from the outdoor air intake to the breathing zone in any particular ventilation zone.

E_z - Zone air distribution effectiveness: A measure of the effectiveness of supply air distribution to the breathing zone. E_z is determined in accordance with Section 403.2.2.

F_a - Supply air fraction: The fraction of supply air to the ventilation zone from sources or air outside the zone.

F_b - Mixed air fraction: The fraction of supply air to the ventilation zone from fully mixed primary air.

F_c - Outdoor air fraction: The fraction of outdoor air to the ventilation zone from sources of air outside the zone.

V_{dz} - Zone discharge airflow: The expected discharge (supply) airflow to the zone that includes primary airflow and secondary recirculated airflow, cfm (m³/min).

V_{pz} - Zone primary airflow: Determine in accordance with Section 403.5.1.

X_s - Average outdoor air fraction: At the primary air handler, the fraction of outdoor air intake flow in the system primary airflow.

Z_{pz} - Primary outdoor air fraction: The outdoor air fraction required in the primary air supplied to the ventilation zone prior to the introduction of secondary recirculation air. [ASHRAE 62.1:A3]

405.0 Evaporative Cooling System for Health Care Facilities. [For OSHPD 1, 1R, 2, 3, 4 & 5] *Direct evaporative cooling systems where the air directly contacts the wetted surface or spray shall be limited in health facilities to nonpatient areas such as laundry rooms, food preparation areas, and boiler or machinery rooms. Similar rooms with high heating-producing equipment will be considered when specifically approved by the enforcing agency. The evaporative pads shall be a synthetic type. Filters shall be required in accordance with Tables 4-B and 4-C except utility rooms, i.e.: boiler or machinery rooms.*

406.0 Reserved.

407.0 Ventilation System Details. [OSHPD 1, 1R, 2, 3, 4 & 5]

407.1 General.

407.1.1 *All supply-air, return air, and exhaust-air systems shall be mechanically operated and such systems for areas listed in Table 4-A shall be operated continuously. Natural ventilation through windows or other openings such as louvers will be considered as supplemental to the required mechanical ventilation systems.*

Exceptions:

(1) Natural ventilation shall not be used in airborne infection isolation rooms and protective environment rooms.

(2) The number of air changes may be reduced to 25 percent of the indicated value in Table 4-A, when the

room is unoccupied, if provisions are made to ensure the following:

(1) The number of air changes per hour indicated is reestablished whenever the space is occupied.

(2) The pressure relationship with the surrounding rooms is maintained when the air changes per hour are reduced. In areas requiring no continuous directional control as identified in accordance with Table 4-A, ventilation systems may be shut down when the space is unoccupied and ventilation is not otherwise required. Ventilation shall not be reduced in rooms specifically used for airborne infection control, such as waiting rooms, triage rooms, corridors, reception areas, areas adjacent to waiting areas, airborne infection isolation rooms, negative pressure exam room, negative pressure x-ray treatment rooms, and protective environment rooms. All operating and delivery rooms shall maintain a minimum of six air changes per hour of total air when not in use.

407.1.2 *Fans serving exhaust systems shall be located at the discharge end of the system. The ventilation rates shown in Table 4-A shall be considered as minimum acceptable rates and shall not be construed as precluding the use of higher ventilation rates if they are required to meet design conditions.*

407.1.3 Services/Systems and Utilities. *(Refer to Section 1224.4.1 of the California Building Code).*

407.2 Outdoor Air Intakes and Exhaust Outlets.

407.2.1 Outdoor Air Intakes. *Outdoor air intakes shall be located at least 25 feet (7.62 m) from exhaust outlets of ventilating systems, combustion equipment stacks, medical-surgical vacuum systems, cooling towers, and areas that may collect vehicular exhaust or other noxious fumes. Plumbing vents shall be located in relation to outdoor air intakes per California Plumbing Code. The bottom of outdoor air intakes shall be located as high as practicable, but not less than 10 feet (3048 mm) above ground level. If installed above the roof, they shall be located 18 inches (457 mm) above roof level or 3 feet (914 mm) above a flat roof where heavy snowfall is anticipated.*

Exceptions:

(1) These dimensions may be reduced if it is demonstrated by the submission of details and calculations that location of intakes with respect to exhausts and their orientation, or the use of special filters, provides equal performance.

(2) The requirements regarding the bottom of outdoor air intakes and installation through the roof do not apply to skilled nursing facilities, intermediate-care facilities or nonsensitive areas in correctional treatment centers.

407.2.2 Exhaust Outlets. *Exhaust outlets shall be located a minimum of 10 feet (3048 mm) above adjoining grade and 10 feet (3048 mm) from doors, occupied areas, and operable windows.*

Exception: *Airborne infection isolation rooms shall comply with Section 414.1.*

407.3 Air Balance.

407.3.1 *The ventilation systems shall be designed and balanced to provide the general air balance relationship to adjacent areas, shown in Table 4-A. The ventilation systems shall be balanced in accordance with the latest edition of standards published by the Associated Air Balance Council (AABC), the National Environmental Balancing Bureau (NEBB), or the Testing, Adjusting and Balancing Bureau (TABB).*

407.4 Air Circulation.

407.4.1 *Design of the ventilation system shall provide air movement that is generally from clean to less clean areas.*

407.4.1.1 *Air supplied to operating rooms, cesarean operating rooms, cardiac catheterization labs, cystoscopy rooms, delivery rooms, and nurseries shall be delivered at or near the ceiling of the area served. In these areas and in morgues and autopsy rooms all air removed from the area shall be removed near floor level. Exhaust or recirculation inlets shall be located not less than 3 inches (76 mm) nor more than 8 inches (203 mm) above the finished floor, except in morgues and autopsy rooms where all of the exhaust air is removed through an autopsy table designed for this purpose. At least two exhaust or recirculation air inlets of equal capacity shall be used in all cardiac catheterization labs, cystoscopy rooms, operating rooms, and delivery rooms and shall be located not less than 3 inches (76 mm) nor more than 8 inches (203 mm) above the finished floor.*

Exception: *For airborne infection isolation rooms and protective environment rooms, see Sections 414.0 and 415.0.*

407.4.1.2 *Room supply air outlets and room recirculation and exhaust air inlets installed in nonsensitive areas shall be located not less than 3 inches (76 mm) above the floor.*

Exception: *For airborne infection isolation rooms and protective environment rooms, see Sections 414.0 and 415.0.*

407.4.1.3 *Corridors shall not be used to convey supply, return, transfer or exhaust air to or from any* ‖ *room if the corridor is required to be fire resistive construction per the California Building Code.*

Exceptions:

(1) Mechanically exhausted toilet rooms of 50 square feet (4.7 m²) or less and small rooms of 30 square feet (2.79 m²) or less such as janitor closets, housekeeping rooms, and electrical or telephone closets opening directly onto corridor.

(2) Air transfer caused by pressure differentials in rooms required to have a positive or negative air balance by Table 4-A.

407.4.1.4 *No space above a ceiling may be utilized as an outside-air, relief-air, supply-air, exhaust-air, or return-air plenum.*

Exception: *Designs specifically approved by the enforcing agency.*

407.4.1.5 *Air from a patient room, exam room, treatment room shall not be transferred to another similar room without first having passed through air filters as required by Table 4-B or Table 4-C.*

407.4.1.6 *Supply outlets and return and exhaust air inlets shall be located to prevent short-circuiting.*

407.5 Variable Air Volume.

407.5.1 Variable Air Volume Systems (VAV). *Variable air volume systems subjecting the patient to a fluctuating air movement are not acceptable for airborne infection isolation rooms, protective environment rooms or those critically sensitive areas listed in Section 322.0. For nonsensitive areas, variable air volume systems meeting the following criteria can be considered:*

407.5.1.1 *The VAV system shall comply with code requirements for outside air, total air, and pressure relationship through the full range of operation from minimum to maximum.*

407.5.1.2 *The central return or exhaust fan shall be controlled to accomplish the variable air volume requirements of the individual rooms served by the fan as described in Section 407.5.1.3.*

407.5.1.3 *Variable air volume for return or exhaust air shall be accomplished by utilizing an automatic modulating damper in the return or exhaust air for each zone. The damper will modulate from full open to minimum position in conjunction with the supply air VAV terminal box.*

|| *408.0 Filters. [OSHPD 1, 1R, 2, 3, 4 & 5]*

408.1 General. *Filter efficiencies shall be certified by the manufacturer and shall be based on ASHRAE 52.2 Method of Testing General Ventilation Air-Cleaning Devices for Removal Efficiency by Particle Size when specifically set forth in these standards.*

408.1.1 *A filter gauge shall be installed across each filter bank serving central air systems. The gauge shall be red lined or a filter alarm light installed to signal when the recommended maximum static pressure drop has been reached.*

408.1.2 *Central air-handling systems are defined as any unit requiring duct work on the supply or inlet side that serve more than one room.*

408.1.3 *Filter banks shall be visually inspected for torn media and bypass in filter frames by means of a flashlight or equivalent, both with fans in operation and stopped. Tears in media and bypass in filter frames shall be eliminated in accordance with the manufacturer's directions and the requirements of the enforcing agency prior to commencing operation of the system.*

408.1.4 *Central air-handling systems shall be maintained in a reasonably clean condition during construction and shall be cleaned as necessary prior to replacement of temporary filter used during construction to ensure that clean air will be delivered to the occupied spaces.*

408.1.5 *Filter bank No. 1 shall be located upstream of the air-conditioning equipment. Filter bank No. 2 and filter bank No. 3 shall be located downstream of the supply fan and all cooling and humidification equipment with efficiencies as indicated in Table 4-B or Table 4-C.*

Exception: *Dry steam-type humidifiers for local room humidity control may be installed in the supply air duct downstream of the final filter bank where designs are specifically approved by the enforcing agency. Dry steam is that which is defined in the ASHRAE HVAC Systems and Equipment Handbook.*

408.1.6 *Filter bank No. 2 and filter bank No. 3 media shall be rigid or supported (noncollapsing type) and shall operate on the principles of impingement, straining, and diffusion.*

408.2 Filters for Hospitals.

408.2.1 *All air-ventilation systems shall comply with code requirements of this section and shall have filter bank efficiencies as listed in Table 4-B.*

408.2.2 *Noncentral recirculating air systems providing cooling to high heat producing equipment located in nonsensitive areas shall have a filter with 30 percent average efficiency based on ASHRAE 52.2 or a minimum efficiency reporting value (MERV) of 8 based on ASHRAE 52.2.*

408.2.3 *Noncentral air systems serving any areas not listed in Table 4-B shall be provided with filter arrangement and efficiency specifically approved by the enforcing agency.*

408.2.4 *Noncentral recirculating air handling systems, for example, through-the-wall units, fan coil units, and heat pumps may be utilized for single patient rooms of one or more beds. Filtration for these units shall have a minimum weight arrestance value of 50 percent, based on ASHRAE 52.2 or a minimum efficiency reporting value (MERV) of 1, based on ASHRAE 52.2. The air ventilation system providing the minimum air changes of outdoor air shall comply with Table 4-B. These units may be used as recirculating units only. All outdoor air requirements shall be met by a separate central air handling systems.*

408.3 Filters for Skilled Nursing Facilities, Intermediate Care Facilities, and Correctional Treatment Centers.

408.3.1 *The air ventilation systems shall comply with code requirements of this section for skilled nursing facilities, intermediate care facilities and correctional treatment centers and shall have filter bank efficiencies as listed in Table 4-C.*

408.3.2 *Noncentral air systems serving single patient rooms of one or more beds shall comply with Table 4-C.*

408.3.3 *Noncentral recirculating air-handling systems, i.e. through the wall units, may be utilized for each patient room with one or more beds. Filtration for these units shall have a minimum weight arrestance value of 50 percent, based on ASHRAE 52.2 or a minimum efficiency reporting value (MERV) of 1, based on ASHRAE 52.2. The air ventilation system providing the minimum air changes of outdoor air shall comply with Table 4-C. These units may be used as recirculating units only. All outdoor air requirements shall be met by a separate central air handling system.*

408.3.4 *Airborne infection isolation rooms, protective environment rooms, and sensitive areas in correctional treatment centers shall comply with Section 408.2.*

408.4 Filters for Outpatient Facilities.

408.4.1 *The air ventilation systems shall comply with code requirements of this section for outpatient facilities and shall have filter bank efficiencies as listed in Table 4-B.*

408.4.2 *Noncentral air systems serving individual rooms shall comply with Table 4-B.*

409.0 Ducts. [OSHPD 1, 1R, 2, 3, 4 & 5]

409.1 *Ducts which penetrate construction, intended for X-ray or other radiation protection, shall not impair the effectiveness of the protection.*

409.2 *Duct linings and their use shall meet the requirements of Chapter 6, California Mechanical Code.*

409.3 Insulation of Ducts. *Cold air ducts shall be insulated wherever necessary or to prevent condensation.*

409.4 *The anchorage and supporting structural elements for airducts shall be designed to withstand the lateral forces as required by the California Building Code, Title 24, Part 2.*

410.0 Laboratory Ventilating Systems and Hoods. *[OSHPD 1, 1R, 2, 3, 4 & 5]*

410.1 Laboratory Ventilating Systems. *Laboratory ventilating systems shall comply with NFPA 99, as required by Section 1224.4.6.4 of the California Building Code.*

410.2 Exhaust Hoods and Safety Cabinets. *Hoods and safety cabinets may be used for normal exhaust of a space provided minimum air change rates are maintained. If air change standards in Table 4-A do not provide sufficient air for proper operation of exhaust hoods and safety cabinets (when in use), supplementary makeup air (filtered and preheated) shall be provided around these units to maintain the required airflow direction and exhaust velocity. Makeup systems for hoods shall be arranged to minimize "short circuiting" of air and to avoid reduction in air velocity at the point of contaminant capture.*

410.3 Laboratory Fume Hoods. *Laboratory fume hoods shall meet the following standards:*

410.3.1 General Standard. *Average face velocity shall be at least 75 feet per minute (0.38 meters per second). Exhaust system shall be separate from the building exhaust system. Exhaust fan shall be located at the discharge end of the system. Exhaust duct system shall be of noncombustible corrosion-resistant material as required to meet the planned usage of the hood.*

410.3.2 Special Standards for Use with Strong Oxidants. *Fume hoods and their associated equipment in the air stream intended for use with perchloric acid and other strong oxidants shall be constructed of stainless steel or other material consistent with special exposures. Hoods and equipment shall be provided with a water wash and drain system to permit periodic flushing of duct and hood. When perchloric acid or other strong oxidants are only transferred from one container to another, standard laboratory fume hoods and the associated equipment may be used in lieu of stainless steel construction.*

410.3.3 Special Standards for Use with Infectious or Radioactive Materials. *Each hood shall have a minimum face velocity of 90 to 110 feet per minute (0.45 to 0.56 meters per second) with suitable pressure-independent air-modulating devices and alarms to alert staff of fan shutdown or loss of airflow. Each hood shall have filters with a 99.97 percent efficiency (based on the DOP test method) in the exhaust stream and be designed and equipped to permit the safe removal, disposal, and replacement of contaminated filters. Filters shall be as close to the hood as practical to minimize duct contamination. Fume hoods intended for use with radioactive isotopes shall be constructed of stainless steel or other material suitable for the particular exposure.*

411.0 Kitchen and Dining Areas. [OSHPD 1, 1R, 2, 3, 4 & 5]

411.1 *The air from dining areas may be used to ventilate the food preparation areas only after it has passed through a filter with at least an 80 percent average efficiency based on ASHRAE 52.2 or a minimum efficiency reporting value (MERV) of 13, based on ASHRAE 52.2.*

Exception: *For skilled nursing facilities, intermediate care facilities and correctional treatment centers, the air from dining area may be used to ventilate food preparation areas only after it has passed through a filter with a 50 percent average efficiency based on ASHRAE 52.2 or a minimum efficiency reporting value (MERV) of 10, based on ASHRAE 52.2.*

412.0 Boiler, Mechanical, and Electrical Rooms. *[OSHPD 1, 1R, 2, 3, 4 & 5]*

412.1 *Boiler, heater and electrical equipment rooms shall be provided with outdoor air so as to maintain combustion rates of equipment and temperatures in the rooms and in adjoining areas as rated in this chapter.*

412.2 *Floor surfaces in occupied spaces above such rooms should not exceed a temperature of 85°F (29.4°C), and suitable insulation may be required.*

413.0 Odorous Rooms. [OSHPD 1, 1R, 2, 3, 4 & 5]

413.1 *Rooms in areas where excessive heat or moisture is generated, where objectional odors or dust are present, or where flammable or toxic gases may accumulate, which are used by health facility personnel or patients, shall be provided with exhaust ventilation to change the air a minimum of ten times per hour.*

413.2 *Kitchen, morgues and laundries located inside a hospital building or skilled nursing facility in which patients are accommodated, or treated, shall be ventilated with exhaust systems which will provide a minimum of ten air changes per hour and prevent odors from entering patient areas.*

414.0 Airborne Infection Isolation Rooms. [OSHPD 1, 1R, 2, 3, 4 & 5]

414.1 Exhaust Systems. *A separate, dedicated exhaust system shall be provided for airborne infection isolation rooms. The dedicated system may serve more than one airborne infection isolation room, adjoining toilet room and anteroom. The exhaust ducts shall be identified by appropriate labeling with the words "Caution Airborne Infection Isolation Rooms Exhaust" or similar terminology. Such labeling shall be in a manner which is not readily removable and shall appear on the exhaust duct at intervals of not more than 20 feet (6096 mm) and at least once near each room and each story traversed by the exhaust system. Exhaust fans shall comply with Section 407.1.2. The discharge from exhaust fans shall be located above the roof and shall be located a minimum of 25 feet (7620 mm) from areas that may be occupied, doors, operable windows, outdoor air intakes, or other openings into the building. The exhaust fan discharge shall be labeled in a manner which readily identifies the precautions which should be observed. To ensure that the airborne contaminates do not reenter the building, one of the following shall be provided:*

414.1.1 *Exhaust discharge from fan shall extend at least 7 feet (2134 mm) above the roof and discharge vertically upward. Self-draining stacks or equivalent shall be used for rain protection. Rain caps which divert the exhaust toward the roof shall be prohibited.*

414.1.2 *Exhaust shall discharge above roof level and through an accessible HEPA filter. The HEPA filter shall be located upstream of the exhaust fan and have a minimum efficiency of 99.97 percent based on the DOP method in accordance with Mil-Std. 282 or a minimum efficiency reporting value (MERV) of 17, based on ASHRAE 52.2. Filter gage shall be installed across the filter. For maintenance of air balance relationship, see Section 407.3.2. The 25-foot (7620 mm) dimension required by Section 414.1 may be reduced when a 99.97 percent HEPA filter or a minimum efficiency reporting value (MERV) of 17, based on ASHRAE 52.2 is used and the reduced dimension is specifically approved by the enforcing agency.*

414.2 Air Distribution. *The supply outlets and exhaust inlets shall be located to provide airflow patterns that prevent stagnation of the air and eliminate short circuiting of the supply to the exhaust, and minimize exposure of health care workers to airborne infectious particles. Supply-air outlets shall be located at or near the ceiling and at the end of the airborne infection isolation room which is opposite the head of the bed. Exhaust registers shall be located on the wall behind the patient's head, or as close to that wall as practical and shall be located not less than 3 inches (76 mm) nor more than 24 inches (610 mm) above the finished floor.*

Exception: *For correctional treatment centers, the location and design of the supply outlets an exhaust or return inlets shall not compromise the safety, security and protection of staff, inmates and property.*

415.0 Protective Environment Rooms. [OSHPD 1, 1R, 2, 3, 4 & 5]

415.1 Air Distribution. *The supply outlets and exhaust and return inlets shall be located to provide airflow patterns that prevent stagnation of the air and eliminate short circuiting of the supply to the exhaust or return. Supply air shall be delivered at or near the ceiling and near the patient's bed. All exhaust or return registers shall be located near the entrance to the protective environment room and not less than 3 inches (76 mm) nor more than 8 inches (203 mm) above the finished floor.*

Exception: *For correctional treatment centers, the location and design of the supply outlets and exhaust or return inlets shall not compromise the safety, security, and protection of staff, inmates and property.*

416.0 Alarms – Airborne Infection Isolation Rooms and Protective Environment Rooms. [OSHPD 1, 1R, 2, 3, 4 & 5]

416.1 *An alarm system which is based on static pressure control, volumetric control, or directional flow measurement shall be provided for each isolation room. The alarm system shall consist of a display monitor located on the corridor wall near the door to the room and a visual and audible alarm which annunciates at the room and at a nurses' station or other suitable location that will provide responsible surveillance. A time delay shall be provided to allow for routine openings of doors. The alarm shall annunciate when the supply, return, or exhaust fans are interrupted or when the minimum required pressure differential per ASHRAE 170 between the airborne infection isolation room and corridor or between the protective environment room and corridor is not being met during closed door conditions.*

416.2 *Other acceptable alarm systems will be allowed when designs are specifically approved by the enforcing agency.*

416.3 [For OSHPD 4] *For correctional treatment centers, the alarm system shall not create false alarms or security hazards.*

416.4 *Prior to acceptance of the rooms, the alarm system shall be tested and operated to demonstrate to the owner or designated representative that the installation and performance of the system conforms to design intent.*

417.0 Testing and Balancing Airborne Infection Isolation Rooms and Protective Environment Rooms. [OSHPD 1, 1R, 2, 3, 4 & 5] *Prior to acceptance of the rooms, all mechanical systems shall be tested, balanced, and operated to demonstrate to the owner or designated representative that the installation and performance of the systems conform to design intent. All testing and balancing shall be performed by a qualified independent agency certified by the Associated Air Balance Council (AABC): the National Environmental Balancing Bureau (NEBB); or the Testing, Adjusting and Balancing Bureau (TABB).*

418.0 Design Requirements for Ethylene Oxide (ETO) Sterilization Areas. [OSHPD 1, 1R, 2, 3, 4 & 5]

418.1 Air Changes. *The ETO sterilization equipment room shall be provided with minimum air changes per hour per Table 4-A and be maintained at a negative air balance.*

418.2 Exhaust Requirements.

418.2.1 *All air from the ETO sterilizer equipment room shall be exhausted to the outside by a dedicated system or other approved method.*

418.2.2 *The exhaust fan for the dedicated system shall be located at the discharge point of the system and identified as ETO Equipment Room Exhaust.*

418.2.3 Discharge Point. *The discharge point shall be a minimum of 25 feet (7620 mm) away from any outside intake, operable window or personnel passage.*

418.3 Ventilation Requirements.

418.3.1 Aeration Units. *The aeration units shall be ventilated through a nonrecirculating dedicated ventilation exhaust system.*

418.3.2 Capture Box. *When the drain is not located in the ETO sterilizer equipment room, ventilation is required by a capture box.*

418.3.3 Cylinder Change. *When not located in the ETO sterilizer equipment room, exhaust during cylinder change is required by installing a hood that is part of a dedicated ventilation exhaust system, positioned no more than 1 foot (305 mm) above or behind the point where the change of cylinders takes place.*

418.3.4 Sterilizer Relief Valve. *The ventilation of sterilizer relief valve is required through a pipe connected to the outlet of the relief valve exhausted directly to the outdoors at a point high enough to be away from passers by, and not near any windows that open, nor near any air-conditioning or ventilation air intakes.*

418.3.5 Ventilation of Sterilizer Door Area. *The system shall be designed to capture the ETO when the door is opened following the completion of the sterilization process. A hood or canopy closed on each end should be installed over the sterilization door. A hood or canopy shall be connected to a dedicated exhaust ventilation system.*

418.4 Gas Valves. *Installation of gas line hand valves at the connection to the supply cylinders are required to minimize leakage during cylinder change.*

418.5 Alarm Systems. *An Audible and visual alarm system shall be installed to alert sterilizer operating personnel if the air flow falls below design cubic feet per minute (L/s).*

419.0 Neonatal Intensive Care Units. [OSHPD 1]

419.1 Formula Preparation Area. *Air shall be supplied over the formulation preparation area by group E, nonaspirating supply diffusers. Air shall be returned or exhausted by registers located not less than 3 inches (76 mm) nor more than 8 inches (203 mm) above the finished floor in the cleanup area.*

419.2 Treatment Area/Room. *Air shall be supplied over the treatment surface by group E, nonaspirating supply diffusers. Air shall be returned or exhausted by registers located not less than 3 inches (76 mm) nor more than 8 inches (203 mm) above the finished floor, adjacent to the treatment surface.*

TABLE 4-A
PRESSURE RELATIONSHIP AND VENTILATION REQUIREMENTS FOR GENERAL ACUTE CARE HOSPITALS, SKILLED NURSING FACILITIES, INTERMEDIATE CARE FACILITIES, CORRECTIONAL TREATMENT CENTERS, OUTPATIENT FACILITIES, AND LICENSED CLINICS

FUNCTION OR SPACE	PRESSURE RELATIONSHIP TO ADJACENT AREAS (f) (n)	MINIMUM OUTDOOR ACH	MINIMUM TOTAL ACH	*MINIMUM TOTAL ACH IF 100% O.A.*	ALL ROOM AIR EXHAUSTED DIRECTLY TO OUTDOORS (j)	AIR RECIRCULATED BY MEANS OF ROOM UNITS (a)	DESIGN RELATIVE HUMIDITY(k), %	DESIGN TEMPERATURE (l),°F/°C
Airborne infection isolation anteroom (u)	(e)	NR	10	*10*	Yes	No	NR	NR
Airborne infection isolation room (u)	Negative	2	12	*12*	Yes	No	*max 60*	70-75/21-24
Airborne infection isolation treatment/exam room	*Negative*	*2*	*12*	*12*	*Yes*	*No*	*NR*	*NR*
Angiography room	*Positive*	*5*	*15*	*12*	*NR*	*NR*	*NR*	*NR*
Bathing room	Negative	NR	10	*NR*	Yes	No	NR	70-75/21-24
Bathroom	Negative	NR	10	*10*	Yes	No	NR	72-78/22-26
Bedpan room	Negative	NR	10	*10*	Yes	No	NR	NR
Blood bank/tissue storage	*NR*	*2*	*6*	*6*	*NR*	*NR*	*NR*	*NR*
Blood draw/phlebotomy	*NR*	*2*	*6*	*6*	*NR*	*NR*	*NR*	*NR*
Bronchoscopy, sputum collection, and pentamidine administration (n)	Negative	2	12	*12*	Yes	No	NR	68-73/20-23
Cardiac catheterization lab	*Positive*	*5*	*20*	*12*	*NR*	*No*	*max 60*	*70-75/21-24*
Clean linen storage	Positive	NR	2	*2*	NR	NR	NR	72-78/22-26
Clean workroom (central medical and surgical supply space)	Positive	2	4	*4*	NR	No	max 60	72-78/22-26
Clean workroom or clean holding (support)	Positive	2	4	*4*	NR	NR	NR	NR
Critical and intensive care	NR	2	6	*6*	NR	No	30-60	70-75/21-24
CT Scan	*NR*	*2*	*6*	*6*	*NR*	*NR*	*max 60*	*NR*
Darkroom (g)	Negative	2	10	*12*	Yes	No	NR	NR
Delivery room (caesarean) (m), (n), (o)	Positive	4	20	*12*	NR	No	20-60	68-75/20-24
Dialysis treatment area	NR	2	6	*6*	NR	NR	NR	72-78/22-26
Dialyzer reprocessing room	Negative	NR	10	*NR*	Yes	No	NR	NR
Dietary storage	NR	NR	2	*2*	NR	No	NR	72-78/22-26
Electroconvulsive therapy procedure room	*P*	*3*	*15*	*10*	*NR*	*NR*	*NR*	*NR*
Emergency department exam/treatment room (p)	NR	2	6	*6*	NR	NR	max 60	70-75/21-24
Endoscope cleaning	Negative	2	10	*10*	Yes	No	NR	NR
ER decontamination	Negative	2	12	*4*	Yes	No	NR	NR
ER waiting rooms	Negative	2	12	*12*	Yes (q)	NR	max 65	70-75/21-24
Examination room	NR	2	6	*6*	NR	NR	max 60	70-75/21-24
Fast track room	*NR*	*2*	*6*	*2*	*NR*	*NR*	*NR*	*NR*
Fluoroscopy room	*Negative*	*2*	*6*	*6*	*Yes*	*No*	*NR*	*NR*
Food preparation center (i)	NR	2	10	*10*	NR	No	NR	72-78/22-26
Gamma camera	*NR*	*2*	*6*	*6*	*NR*	*No*	*NR*	*NR*
Gastrointestinal endoscopy procedure room (x)	NR	2	6	*12*	NR	No	20-60	68-73/20-23
Hazardous material storage	Negative	2	10	*10*	Yes	No	NR	NR
Hydrotherapy	Negative	2	6	*6*	NR	NR	NR	72-80/22-27
Infusion room	*Positive*	*2*	*6*	6	*NR*	*NR*	*NR*	*NR*
Intermediate care (s)	NR	2	6	*6*	NR	NR	max 60	70-75/21-24

TABLE 4-A (continued)
PRESSURE RELATIONSHIP AND VENTILATION REQUIREMENTS FOR GENERAL ACUTE CARE HOSPITALS, SKILLED NURSING FACILITIES, INTERMEDIATE CARE FACILITIES, CORRECTIONAL TREATMENT CENTERS, OUTPATIENT FACILITIES, AND LICENSED CLINICS

FUNCTION OR SPACE	PRESSURE RELATION-SHIP TO ADJACENT AREAS (f) (n)	MINIMUM OUTDOOR ACH	MINIMUM TOTAL ACH	*MINIMUM TOTAL ACH IF 100% O.A.*	ALL ROOM AIR EXHAUSTED DIRECTLY TO OUTDOORS (j)	AIR RECIRCULATED BY MEANS OF ROOM UNITS (a)	DESIGN RELATIVE HUMIDITY(k), %	DESIGN TEMPERATURE (l),°F/°C
Interventional imaging procedure room	*Positive*	*5*	*15*	*12*	*NR*	*NR*	*NR*	*NR*
IV Prep. room	*Positive*	*2*	*6*	*6*	*NR*	*NR*	*NR*	*NR*
Janitor's closet, *housekeeping*	Negative	NR	10	*10*	Yes	No	NR	NR
Labor/delivery/recovery (LDR) (s)	NR	2	6	*2*	NR	NR	max 60	70-75/21-24
Labor/delivery/recovery/ postpartum (LDRP) (s)	NR	2	6	*2*	NR	NR	max 60	70-75/21-24
Laboratory, bacteriology (v)	Negative	2	6	*6*	Yes	NR	NR	70-75/21-24
Laboratory, biochemistry (v)	Negative	2	6	*6*	Yes	NR	NR	70-75/21-24
Laboratory, cytology (v)	Negative	2	6	*6*	Yes	NR	NR	70-75/21-24
Laboratory, general (v)	Negative	2	6	*6*	NR	NR	NR	70-75/21-24
Laboratory, glasswashing	Negative	2	10	*10*	Yes	NR	NR	NR
Laboratory, histology (v)	Negative	2	6	*6*	Yes	NR	NR	70-75/21-24
Laboratory, infectious disease and virus	Negative	2	6	*6*	Yes	NR	NR	70-75/21-24
Laboratory, media transfer (v)	Positive	2	4	*4*	NR	NR	NR	70-75/21-24
Laboratory, microbiology (v)	Negative	2	6	*6*	Yes	NR	NR	70-75/21-24
Laboratory, nuclear medicine (v)	Negative	2	6	*6*	Yes	NR	NR	70-75/21-24
Laboratory, pathology (v)	Negative	2	6	*6*	Yes	NR	NR	70-75/21-24
Laboratory, serology (v)	Negative	2	6	*6*	Yes	NR	NR	70-75/21-24
Laboratory, sterilizing	Negative	2	10	*10*	Yes	NR	NR	70-75/21-24
Lactation	*NR*	*2*	*6*	*2*	*NR*	*NR*	*NR*	*NR*
Laser eye room	Positive	3	15	*15*	NR	No	20-60	70-75/21-24
Laundry, general	Negative	2	10	*10*	Yes	No	NR	NR
Linen and trash chute room	Negative	NR	10	*10*	Yes	No	NR	NR
Medical/anesthesia gas storage (r)	Negative	NR	8	*8*	Yes	NR	NR	NR
Medication room	NR	2	4	*2*	NR	NR	max 60	70-75/21-24
Morgues and autopsy room (n)	Negative	2	12	*12*	Yes	No	NR	68-75/20-24
MRI room	*NR*	*2*	*6*	*6*	*NR*	*NR*	*NR*	*NR*
Negative-pressure x-ray room	*Negative*	*2*	*12*	*12*	*Yes*	*No*	*max 60*	*72-78/22-26*
Newborn intensive care	Positive	2	6	*6*	NR	No	30-60	72-78/22-26
Newborn intensive care formula room	*P*	*2*	*10*	*10*	*NR*	*No*	*NR*	*NR*
Newborn/*well baby* nursery suite	NR	2	6	*6*	NR	No	30-60	72-78/22-26
Nonrefrigerated body-holding room (h)	Negative	NR	10	*10*	Yes	No	NR	70-75/21-24
Nourishment area or room	NR	NR	2	*2*	NR	NR	NR	NR
Nuclear medicine (Gamma, PET, SPECT)	*Negative*	*2*	*6*	*6*	*Yes*	*No*	*NR*	*NR*
Nuclear medicine hot lab	Negative	NR	6	*6*	Yes	No	NR	70-75/21-24
Nuclear medicine treatment room	Negative	2	6	*6*	Yes	NR	NR	70-75/21-24
Nurse station (aa)	*(aa)*	*(aa)*	*2*	*2*	*(aa)*	*(aa)*	*(aa)*	*(aa)*
Observation	*NR*	*2*	*6*	*2*	*NR*	*NR*	*NR*	*70-75/21-24*

TABLE 4-A (continued)
PRESSURE RELATIONSHIP AND VENTILATION REQUIREMENTS FOR GENERAL ACUTE CARE HOSPITALS, SKILLED NURSING FACILITIES, INTERMEDIATE CARE FACILITIES, CORRECTIONAL TREATMENT CENTERS, OUTPATIENT FACILITIES, AND LICENSED CLINICS

FUNCTION OR SPACE	PRESSURE RELATIONSHIP TO ADJACENT AREAS (f) (n)	MINIMUM OUTDOOR ACH	MINIMUM TOTAL ACH	*MINIMUM TOTAL ACH IF 100% O.A.*	ALL ROOM AIR EXHAUSTED DIRECTLY TO OUTDOORS (j)	AIR RECIRCULATED BY MEANS OF ROOM UNITS (a)	DESIGN RELATIVE HUMIDITY(k), %	DESIGN TEMPERATURE (l),°F/°C
Occupational therapy	NR	*2*	*6*	*6*	NR	NR	NR	70-75/21-24
Operating room, *hybrid operating room* (m), (n), (o)	Positive	4	20	*12*	NR	No	20-60	68-75/20-24
Operating/surgical cystoscopic room (m), (n), (o)	Positive	4	20	*12*	NR	No	20-60	68-75/20-24
Patient corridor	NR	NR	2	*2*	NR	NR	NR	NR
Patient holding preparation	*NR*	*2*	*6*	*6*	*NR*	*No*	*NR*	*NR*
Patient room	NR	2	4 (y)	*2*	NR	NR	max 60	70-75/21-24
Pediatric play area	*NR*	*2*	*6*	*6*	*NR*	*NR*	*NR*	*NR*
Pharmacy (b)	Positive	2	4	*2*	NR	NR	NR	NR
Drug room	*NR*	*2*	*4*	*2*	*NR*	*NR*	*NR*	*NR*
HD ante room (b)	*Positive*	*NR*	*NR*	*NR*	*NR*	*NR*	*<60*	*≤68/≤20*
HD buffer room (b)	*Negative*	*NR*	*30*	*30*	*Yes*	*No*	*<60*	*≤68/≤20*
HD segregated compounding area (b)	*Negative*	*NR*	*12*	*12*	*Yes*	*NR*	*NR*	*NR*
HD storage (b)	*Negative*	*NR*	*12*	*12*	*Yes*	*NR*	*NR*	*NR*
Non-HD ante room (b)	*Positive*	*NR*	*30*	*30*	*NR*	*NR*	*<60*	*≤68/≤20*
Non-HD buffer room (b)	*Positive*	*NR*	*30*	*30*	*NR*	*No*	*<60*	*≤68/≤20*
Non-HD segregated compounding area (b)	*NR*	*NR*	*NR*	*NR*	*NR*	*NR*	*NR*	*NR*
Physical therapy (nursing facility)	Negative	2	6	*6*	NR	NR	NR	70-75/21-24
Physical therapy (diagnostic and treatment)	Negative	2	6	*6*	NR	NR	max 65	72-80/22-27
Post-anesthesia care unit	*NR*	*2*	*6*	*6*	*Yes*	*No*	*20-60*	*70-75/21-24*
Pre-screening area	*Negative*	*2*	*12*	*12*	*Yes (q)*	*NR*	*NR*	*NR*
Procedure room (o), (d)	Positive	3	15	*12*	NR	No	20-60	70-75/21-24
Protective environment anteroom (t)	(e)	NR	10	*15*	NR	No	NR	NR
Protective environment room (t)	Positive	2	12	*15*	NR	No	max 60	70-75/21-24
Radiology waiting rooms	Negative	2	12	*12*	Yes (q), (w)	NR	max 60	70-75/21-24
Recovery room	NR	2	6	*2*	NR	No	20-60	70-75/21-24
Recreation/activity room	*NR*	*2*	*6*	*6*	*NR*	*NR*	*NR*	*NR*
Resident gathering/activity/dining (nursing facility)	NR	4	4	*4*	NR	NR	NR	70-75/21-24
Resident room (nursing facility)	NR	2	2	*2*	NR	NR	NR	70-75/21-24
Resident unit corridor (nursing facility)	NR	NR	4	*2*	NR	NR	NR	NR
Seclusion room	*NR*	*2*	*6*	*2*	*NR*	*NR*	*NR*	*NR*
Semi-restricted corridor	*NR*	*2*	*4*	*2*	*NR*	*NR*	*NR*	*NR*
Shower room	*Negative*	*NR*	*10*	*NR*	*Yes*	*No*	*NR*	*NR*
Soiled linen sorting and storage	Negative	NR	10	*10*	Yes	No	NR	NR
Soiled or decontamination room	Negative	2	6	*4*	Yes	No	NR	72-78/22-26
Soiled workroom or soiled holding, *utility room*	Negative	2	10	*4*	Yes	No	NR	NR

TABLE 4-A (continued)
PRESSURE RELATIONSHIP AND VENTILATION REQUIREMENTS FOR GENERAL ACUTE CARE HOSPITALS, SKILLED NURSING FACILITIES, INTERMEDIATE CARE FACILITIES, CORRECTIONAL TREATMENT CENTERS, OUTPATIENT FACILITIES, AND LICENSED CLINICS

FUNCTION OR SPACE	PRESSURE RELATION-SHIP TO ADJACENT AREAS (f) (n)	MINIMUM OUTDOOR ACH	MINIMUM TOTAL ACH	*MINIMUM TOTAL ACH IF 100% O.A.*	ALL ROOM AIR EXHAUSTED DIRECTLY TO OUTDOORS (j)	AIR RECIRCULATED BY MEANS OF ROOM UNITS (a)	DESIGN RELATIVE HUMIDITY(k), %	DESIGN TEMPERATURE (l),°F/°C
Special purpose room (SNF & ICF only)	*NR*	*2*	*6*	*6*	*Yes*	*NR*	*NR*	*NR*
Speech therapy/audiology room	*NR*	*2*	*6*	*2*	*NR*	*NR*	*NR*	*NR*
Sterile storage	Positive	2	4	*4*	NR	NR	max 60	72-78/22-26
Sterilizer equipment room	Negative	NR	10	*10*	Yes	No	NR	NR
Substerile service area	NR	2	6	*6*	NR	No	NR	NR
Toilet room	Negative	NR	10	*10*	Yes	No	NR	NR
Trauma/*cardiac* room (crisis or shock) (c)	Positive	3	15	*12*	NR	No	20-60	70-75/21-24
Treatment room (surgery and critical care) (p)	NR	2	6	*6*	NR	NR	20-60	70-75/21-24
Treatment room (diagnostic and treatment) (x)	NR	2	6	*6*	NR	NR	max 60	70-75/21-24
Triage	Negative	2	12	*12*	Yes (q)	NR	max 60	70-75/21-24
Ultrasound room	*NR*	*2*	*6*	*6*	*NR*	*NR*	*NR*	*NR*
Unsterile supply	*NR*	*2*	*2*	*2*	*NR*	*NR*	*NR*	*NR*
Waiting area (nuclear medicine)	*Negative*	*2*	*12*	*12*	*Yes*	*No*	*NR*	*NR*
Waiting area primary care clinic	*Negative*	*2*	*10*	*10*	*Yes(q)*	*NR*	*NR*	*NR*
Warewashing	Negative	NR	10	*10*	Yes	No	NR	NR
Wound intensive care (burn unit)	NR	2	6	*6*	NR	No	40-60	70-75/21-24
X-ray (diagnostic and treatment)	NR	2	6	*6*	NR	NR	max 60	72-78/22-26
X-ray (surgery/critical care and catheterization)	Positive	3	15	*12*	NR	No	max 60	70-75/21-24

Note: NR = No requirement

Notes for Table 4-A:

a. Except where indicated by a "No" in this column, recirculating room HVAC units (with heating or cooling coils) are acceptable for providing that portion of the minimum total air changes per hour that is permitted by Section 7.1 (subparagraph [a][5]). Because of the cleaning difficulty and potential for buildup of contamination, recirculating room units shall not be used in areas marked "No." Recirculating devices with HEPA filters shall be permitted in existing facilities as interim, supplemental environmental controls to meet requirements for the control of airborne infectious agents. The design of either portable or fixed systems should prevent stagnation and short circuiting of airflow. The design of such systems shall also allow for easy access for scheduled preventative maintenance and cleaning.

b. *Additional air change, ISO class, differential pressure, continuous pressure monitoring and filtering requirements for compounding areas shall comply with California Board of Pharmacy regulations Title 16 §1735 & §1751, and USP <797> & <800>. Air supplied to the cleanroom suite must be introduced through 99.97% minimum HEPA filters located in the ceiling. At least 15 air changes per hour (ACPH) shall be provided to nonhazardous drug (non-HD) compounding buffer rooms through the ceiling. The HEPA filtered air from the PEC in the non-HD buffer room, when added to the HVAC-supplied HEPA-filtered air, shall increase the total HEPA-filtered ACPH to at least 30. If the PEC is used to meet the minimum total ACPH requirements, the PEC must not be turned off except for maintenance. All hazardous drug (HD) compounding areas and PECs shall be exhausted externally through 99.97% HEPA filtration. Minimum air changes shall be met under operating conditions. Returns and exhaust grilles shall be mounted low on the wall unless a visual smoke study demonstrates dilution of particles and sweeping out of particles from the entire room. One return/exhaust should be placed near the refrigerator compressor.*

c. The term *trauma room* as used herein is a first-aid room and/or emergency room used for general initial treatment of accident victims. The operating room within the trauma center that is routinely used for emergency surgery is considered to be an operating room by this standard.

d. Pressure relationships need not be maintained when the room is unoccupied.

e. See Section 7.2 of ASHRAE 170 and its subsections for pressure-relationship requirements.

f. *For operating rooms, cardiac catheterization labs, angiography rooms, cystoscopy rooms, delivery rooms, cesarean operating rooms, newborn intensive care, intensive care units, and nurseries provide approximately 15% excess supply air to the room or a sufficient quantity of excess supply air to maintain an appropriate positive air balance based on the room tightness and number of doors. For all rooms not listed in this footnote or not listed in Section 322.0 requiring either a positive or negative air balance, provide approximately 10% differential cfm between supply and return/exhaust airflow but not less than 25 cfm differential shall be provided regardless of room size. Room function, size, and tightness may be considered when determining the differential airflow required. Where continuous directional control is not required, variations between supply cfm and return or exhaust cfm shall be minimized.*

g. All air need not be exhausted if darkroom equipment has a scavenging exhaust duct attached and meets ventilation standards regarding NIOSH, OSHA, and local employee exposure limits.[2, 3]

h. A nonrefrigerated body-holding room is applicable only to facilities that do not perform autopsies on-site and use the space for short periods while waiting for the body to be transferred.

i. Minimum total air changes per hour (ach) shall be that required to provide proper makeup air to kitchen exhaust systems as specified in ANSI/ASHRAE Standard 154.[4] In some cases, excess exfiltration or infiltration to or from exit corridors compromises the exit corridor restrictions of NFPA 90A,[5] the pressure requirements of NFPA 96,[6] or the maximum defined in the table. During operation, a reduction to the number of air changes to any extent required for odor control shall be permitted when the space is not in use. (See FGI [2010] in Informative Appendix B.)

j. In some areas with potential contamination and/or odor problems, exhaust air shall be discharged directly to the outdoors and not recirculated to other areas. Individual circumstances may require special consideration for air exhausted to the outdoors. To satisfy exhaust needs, constant replacement air from the outdoors is necessary when the system is in operation.

k. The RH ranges listed are the minimum and/or maximum allowable at any point within the design temperature range required for that space.

l. Systems shall be capable of maintaining the rooms within the range during normal operation. Lower or higher temperature shall be permitted when patients' comfort and/or medical conditions require those conditions.

m. National Institute for Occupational Safety and Health (NIOSH) criteria documents regarding occupational exposure to waste anesthetic gases and vapors, and control of occupational exposure to nitrous oxide[7] indicate a need for both local exhaust (scavenging) systems and general ventilation of the areas in which the respective gases are utilized. Refer to NFPA 99 for other requirements.[8]

n. If pressure-monitoring device alarms are installed, allowances shall be made to prevent nuisance alarms. Short-term excursions from required pressure relationships shall be allowed while doors are moving or temporarily open. Simple visual methods such as smoke trail, ball-in-tube, or flutterstrip shall be permitted for verification of airflow direction.

o. Surgeons or surgical procedures may require room temperatures, ventilation rates, humidity ranges, and/or air distribution methods that exceed the minimum indicated ranges.

p. Treatment rooms used for bronchoscopy shall be treated as bronchoscopy rooms. Treatment rooms used for procedures with nitrous oxide shall contain provisions for exhausting anesthetic waste gases.

q. In a recirculating ventilation system, HEPA filters shall be permitted instead of exhausting the air from these spaces to the outdoors provided that the return air passes through the HEPA filters before it is introduced into any other spaces. The entire minimum total air changes per hour of recirculating airflow shall pass through HEPA filters. When these areas are open to larger, nonwaiting spaces, the exhaust air volume shall be calculated based on the seating area of the waiting area. (**Note:** The intent here is to not require the volume calculation to include a very large space [e.g., an atrium] just because a waiting area opens onto it.)

r. See NFPA 99 for further requirements.[8]

s. For intermediate care, labor/delivery/recovery rooms, and labor/delivery/recovery/postpartum rooms, four total ach shall be permitted when supplemental heating and/or cooling systems (radiant heating and cooling, baseboard heating, etc.) are used.

t. The protective environment airflow design specifications protect the patient from common environmental airborne infectious microbes (i.e., Aspergillus spores). *The anteroom shall have negative air pressure in relation to the protective environment room. A door louver, transfer grille, or other acceptable means may be provided to allow for airflow from the protective environment room to the anteroom. The protective environment room shall have positive-pressure in relation to the anteroom and adjoining toilet room.* Recirculation HEPA filters shall be permitted to increase the equivalent room air exchanges; however, the outdoor air changes are still required. Constant-volume airflow is required for consistent ventilation for the protected environment. The pressure relationship to adjacent areas shall remain unchanged if the PE room is utilized as a normal patient room. Rooms with reversible airflow provisions for the purpose of switching between protective environment and AII functions shall not be permitted.

u. The AII room described in this standard shall be used for isolating the airborne spread of infectious diseases, such as measles, varicella, or tuberculosis. *The airborne infection isolation room shall have negative pressure in relation to the anteroom, and the adjoining toilet room shall have negative pressure in relation to the airborne infection isolation room.* Supplemental recirculating devices using HEPA filters shall be permitted in the AII room to increase the equivalent room air exchanges; however, the minimum outdoor air changes of Table 4-A are still required. AII rooms that are retrofitted from standard patient rooms from which it is impractical to exhaust directly outdoors may be recirculated with air from the AII room, provided that air first passes through a HEPA filter. When the AII room is not utilized for airborne infection isolation, the pressure relationship to adjacent areas, when measured with the door closed, shall remain unchanged and the minimum total air change rate shall be 6 ach. Switching controls for reversible airflow provisions shall not be permitted. The anteroom shall have positive air pressure in relation to the airborne infection isolation room. A door louver, transfer grille, or other acceptable means may be provided to allow for airflow from the anteroom to the airborne infection isolation room.

v. When required, appropriate hoods and exhaust devices for the removal of noxious gases or chemical vapors shall be provided in accordance with NFPA 99.[8]

w. The requirement that all room air is exhausted directly to outdoors applies only to radiology waiting rooms programmed to hold patients who are waiting for chest x-rays for diagnosis of respiratory disease.

x. If the planned space is designated in the organization's operational plan to be utilized for both bronchoscopy and gastrointestinal endoscopy, the design parameters for "bronchoscopy, sputum collection, and pentamidine administration" shall be used.

y. For single-bed patient rooms using Group D diffusers, a minimum of six total ach shall be provided and calculated based on the volume from finished floor to 6 ft (1.83 m) above the floor.

z. *This* table *is based on Table 7.1 in ASHRAE 170, "Ventilation of Healthcare Facilities", and is used with expressed written permission from* ASHRAE.

aa. Nurse station pressure relationship and ventilation requirements shall match the area in which it is located.

ab. Air change per hour and ventilation rates for spaces not listed in Table 4-A may be per ASHRAE 62.1.

TABLE 4-B
FILTER EFFICIENCIES FOR CENTRAL VENTILATION AND AIR-CONDITIONING SYSTEMS IN GENERAL ACUTE CARE HOSPITALS, ACUTE PSYCHIATRIC HOSPITALS, OUTPATIENT FACILITIES, AND LICENSED CLINICS[1]

AREA DESIGNATION	MINIMUM NUMBER OF FILTER BANKS	FILTER EFFICIENCY % FILTER BANK (MINIMUM EFFICIENCY REPORTING VALUE MERV)[5]		
		NO. 1[1]	NO. 2[1]	NO. 3[1]
Orthopedic operating room, bone marrow transplant operating room, organ transplant operating room, NICU formula preparation room, NICU treatment area/room	*3*	*30%*	*90%*	*99.97%[3]*
		(8)	*(14)*	*(17)*
Protective environment rooms	*3*	*30%*	*90%*	*99.97%[4]*
		(8)	*(14)*	*(17)*
Angiography; cardiac catheterization labs; operating rooms; interventional imaging procedure rooms;delivery rooms nurseries; patient care, treatment, cystoscopy, cesarean operating room, diagnostic, and related areas; airborne infection isolation rooms; areas providing direct patient service or clean supplies such as sterile and clean processes, and patient area corridors	*2*	*30%*	*90%*	—
		(8)	*(14)*	—
Laboratories	*2*	*30%*	*80%*	—
		(8)	*(13)*	—
Administrative, med staff support areas, bulk storage, soiled holding areas, food preparation areas, public cafeterias, and laundries	*1*	*30%*	—	—
		(8)	—	—
Psychiatric hospitals intended for the care and treatment of inpatients who do not require acute medical services	*1*	30%	—	—
		(8)	—	—

1 Based on ASHRAE 52.2.
2 Based on DOP test in accordance with MIL-STD-282 or based on ASHRAE 52.2.
3 HEPA filters at air outlet or other locations when approved by the Authority Having Jurisdiction.
4 HEPA filter located in the supply duct which serves the positive-pressure isolation room or rooms may serve more than one supply outlet and more than one positive-pressure isolation room. HEPA filter or a filter with minimum efficiency reporting value (MERV) of 17 installation shall be designed and equipped to permit safe removal, disposal and replacement of filters.
5 The numbers in parentheses represent MERV rating based on ASHRAE 52.2.

TABLE 4-C
FILTER EFFICIENCIES FOR CENTRAL VENTILATION AND AIR-CONDITIONING SYSTEMS IN SKILLED NURSING FACILITIES AND INTERMEDIATE CARE FACILITIES AND CORRECTIONAL TREATMENT CENTERS[1]

AREA DESIGNATION	MINIMUM NUMBER OF FILTER BANKS	FILTER EFFICIENCY % FILTER BANK (MINIMUM EFFICIENCY REPORTING VALUE MERV)[3]	
		NO. 1[1]	NO. 2[1]
All areas for inpatient care, treatment and/or diagnosis, and those areas providing direct service or cleaning supplies)	*2*	*30%*	*80%*
		(8)	*(13)*
Administrative, bulk storage, soiled holding, laundries and food prep areas	*1*	*30%[2]*	—
		(8)	—

1 Based on ASHRAE 52.2.
2 Filters are not required for evaporative coolers serving laundries and food preparation areas.
3 The numbers in parentheses represent MERV rating based on ASHRAE 52.2.

TABLE 402.1
MINIMUM VENTILATION RATES IN BREATHING ZONE[1, 2]
[ASHRAE 62.1: TABLE 6.2.2.1]

Note: *Ventilation air supply requirements for occupancies regulated by the California Energy Commission are found in the California Energy Code.*

OCCUPANCY CATEGORY[4]	PEOPLE OUTDOOR Air Rate R_p (cfm/person)	AREA OUTDOOR Air Rate R_A (cfm/ft^2)	DEFAULT OCCUPANT DENSITY[3] (people/1000 ft^2)	AIR CLASS
CORRECTIONAL FACILITIES				
Booking/waiting	7.5	0.06	50	2
Cell	5	0.12	25	2
Day room	5	0.06	30	1
Guard stations	5	0.06	15	1
EDUCATIONAL FACILITIES				
Art classroom	10	0.18	20	2
Classrooms (ages 5-8)	10	0.12	25	1
Classrooms (age 9 plus)	10	0.12	35	1
Computer lab	10	0.12	25	1
Daycare (through age 4)	10	0.18	25	2
Daycare sickroom	10	0.18	25	3
Lecture classroom	7.5	0.06	65	1
Lecture hall (fixed seats)	7.5	0.06	150	1
Media center[a]	10	0.12	25	1
Multi-use assembly	7.5	0.06	100	1
Music/theater/dance	10	0.06	35	1
Science laboratories	10	0.18	25	2
University/college laboratories	10	0.18	25	2
Wood/metal shop	10	0.18	20	2
FOOD AND BEVERAGE SERVICE				
Bars, cocktail lounges	7.5	0.18	100	2
Cafeteria/fast food dining	7.5	0.18	100	2
Kitchen (cooking)	7.5	0.12	20	2
Restaurant dining rooms	7.5	0.18	70	2
GENERAL				
Break rooms	5	0.06	25	1
Coffee stations	5	0.06	20	1
Conference/meeting	5	0.06	50	1
Corridors	–	0.06	–	1
Occupiable storage rooms for liquids or gels[b]	5	0.12	2	2
HOTELS, MOTELS, RESORTS, DORMITORIES				
Barracks sleeping areas	5	0.06	20	1
Bedroom/living room	5	0.06	10	1
Laundry rooms, central	5	0.12	10	2
Laundry rooms within dwelling units	5	0.12	10	1
Lobbies/pre-function	7.5	0.06	30	1
Multipurpose assembly	5	0.06	120	1
OFFICE BUILDINGS				
Break Rooms	5	0.12	50	1
Main entry lobbies	5	0.06	10	1
Occupiable storage rooms for dry materials	5	0.06	2	1
Office space	5	0.06	5	1
Reception areas	5	0.06	30	1
Telephone/data entry	5	0.06	60	1
MISCELLANEOUS SPACES				
Bank or bank lobbies	7.5	0.06	15	1
Bank vaults/safe deposit	5	0.06	5	2
Computer (not printing)	5	0.06	4	1
Freezer and refrigerated spaces (<50°F)[e]	10	–	–	2
General manufacturing (excludes heavy industrial and processes using chemicals)	10	0.18	7	3
Pharmacy (prep. area)	5	0.18	10	2
Photo studios	5	0.12	10	1

TABLE 402.1 (continued)
MINIMUM VENTILATION RATES IN BREATHING ZONE[1, 2]
[ASHRAE 62.1: TABLE 6.2.2.1]

OCCUPANCY CATEGORY[4]	PEOPLE OUTDOOR Air Rate *Rp* (cfm/person)	AREA OUTDOOR Air Rate R_A (cfm/ft²)	DEFAULT OCCUPANT DENSITY[3] (people/1000 ft²)	AIR CLASS
Shipping/receiving[b]	10	0.12	2	2
Sorting, packing, light assembly	7.5	0.12	7	2
Telephone closets	–	–	–	1
Transportation waiting	7.5	0.06	100	1
Warehouses[b]	10	0.06	–	2
PUBLIC ASSEMBLY SPACES				
Auditorium seating area	5	0.06	150	1
Courtrooms	5	0.06	70	1
Legislative chambers	5	0.06	50	1
Libraries	5	0.12	10	1
Lobbies	5	0.06	150	1
Museums (children's)	7.5	0.12	40	1
Museums/galleries	7.5	0.06	40	1
Places of religious worship	5	0.06	120	1
RESIDENTIAL				
Common corridors	–	0.06	–	1
Dwelling unit[f, g]	5	0.06	See footnote[f]	1
RETAIL				
Sales (except as below)	7.5	0.12	15	2
Barber shop	7.5	0.06	25	2
Beauty and nail salons	20	0.12	25	2
Coin-operated laundries	7.5	0.12	20	2
Mall common areas	7.5	0.06	40	1
Pet shops (animal areas)	7.5	0.18	10	2
Supermarket	7.5	0.06	8	1
SPORTS AND ENTERTAINMENT				
Bowling alley (seating)	10	0.12	40	1
Disco/dance floors	20	0.06	100	2
Gambling casinos	7.5	0.18	120	1
Game arcades	7.5	0.18	20	1
Gym, sports arena (play area)[e]	20	0.18	7	2
Health club/aerobics room	20	0.06	40	2
Health club/weight rooms	20	0.06	10	2
Spectator areas	7.5	0.06	150	1
Stages, studios[d]	10	0.06	70	1
Swimming (pool & deck)[c]	–	0.48	–	2

For SI units: 1 cubic foot per minute = 0.0283 m³/min, 1 square foot = 0.0929 m²

Notes:

1 This table applies to no-smoking areas. Rates for smoking-permitted spaces shall be determined using other methods.

2 Volumetric airflow rates are based on an air density of 0.075 pounds of dry air per cubic foot (lbda/ft³) (1.201 kgda/m³), which corresponds to dry air at a barometric pressure of 1 atm (101 kPa) and an air temperature of 70°F (21°C). Rates shall be permitted to be adjusted for actual density, but such adjustment is not required for compliance with this chapter.

3 The default occupant density shall be used where actual occupant density is not known.

4 Where the occupancy category for a proposed space or zone is not listed, the requirements for the listed occupancy category that is most similar in terms of occupant density, activities, and building construction shall be used.

ITEM-SPECIFIC NOTES FOR TABLE 402.1

a For high school and college libraries, use values shown for Public Assembly Spaces – Libraries.

b Rate is capable of not being sufficient where stored materials include those having potentially harmful emissions.

c Rate does not allow for humidity control. Additional ventilation or dehumidification shall be permitted to remove moisture. "Deck area" refers to the area surrounding the pool that would be expected to be wetted during normal pool use, i.e., where the pool is occupied. Deck area that is not expected to be wetted shall be designated as a space type (for example, "spectator area").

d Rate does not include special exhaust for stage effects, e.g., dry ice vapors, smoke.

e Where combustion equipment is intended to be used on the playing surface or in the space, additional dilution ventilation, source control, or both shall be provided.

f Default occupancy for dwelling units shall be two persons for studio and one-bedroom units, with one additional person for each additional bedroom.

g Air from one residential dwelling shall not be recirculated or transferred to other spaces outside of that dwelling.

TABLE 403.2.2
ZONE AIR DISTRIBUTION EFFECTIVENESS[1, 2, 3, 4, 5]
[ASHRAE 62.1: TABLE 6.2.2.2]

AIR DISTRIBUTION CONFIGURATION	E_z
Ceiling supply of cool air.	1.0
Ceiling supply of warm air and floor return.	1.0
Ceiling supply of warm air 15°F or more above space temperature and ceiling return.	0.8
Ceiling supply of warm air less than 15°F above space temperature and ceiling return provided that the 150 feet per minute (fpm) supply air jet reaches to within 4.5 feet of floor level.	1.0[6]
Floor supply of cool air and ceiling return provided that the vertical throw is more than 50 fpm at a height of 4.5 feet or more above the floor.	1.0
Floor supply of cool air and ceiling return, provided low-velocity displacement ventilation achieves unidirectional flow and thermal stratification, or underfloor air distribution systems where the vertical throw is 50 fpm or less at a height of 4.5 feet above the floor.	1.2
Floor supply of warm air and floor return.	1.0
Floor supply of warm air and ceiling return.	0.7
Makeup supply drawn in on the opposite side of the room from the exhaust, return, or both.	0.8
Makeup supply drawn in near to the exhaust, return, or both locations.	0.5

For SI units: °C = (°F-32)/1.8, 1 foot per minute = 0.005 m/s, 1 foot = 304.8 mm

Notes:

1 "Cool air" is air cooler than space temperature.

2 "Warm air" is air warmer than space temperature.

3 "Ceiling supply" includes any point above the breathing zone.

4 "Floor supply" includes any point below the breathing zone.

5 As an alternative to using the above values, E_z shall be permitted to be regarded as equal to air change effectiveness determined in accordance with ASHRAE 129 for air distribution configurations except unidirectional flow.

6 For lower velocity supply air, E_z = 0.8

TABLE 403.5.2
SYSTEM VENTILATION EFFICIENCY[1, 2, 3]
[ASHRAE 62.1: TABLE 6.2.5.2]

MAX (Z_{pz})	E_v
≤ 0.15	1.0
≤ 0.25	0.9
≤ 0.35	0.8
≤ 0.45	0.7
≤ 0.55	0.6
> 0.55	Use Section 404.0

Notes:

1 "Max *Zpz*" refers to the largest value of *Zpz*, calculated in accordance with Equation 403.5.1, among the ventilation zones served by the system.

2 For values of Max (*Zpz)* between 0.15 and 0.55, the corresponding value of *Ev* shall be permitted to be determined by interpolating the values in the table.

3 The values of *Ev* in this table are based on a 0.15 average outdoor air fraction for the system (i.e., the ratio of the uncorrected outdoor air intake (*Vou)* to the total zone primary airflow for the zones served by the air handler). For systems with higher values of the average outdoor air fraction, this table is capable of resulting in unrealistically low values of E_V and the use of Section 404.0 is capable of yielding more practical results.

TABLE 403.7
MINIMUM EXHAUST RATES
[ASHRAE 62.1: TABLE 6.5]

OCCUPANCY CATEGORY[8]	EXHAUST RATE (cfm/unit)	EXHAUST RATE (cfm/ft^2)	AIR CLASS
Arenas[2]	–	0.50	1
Art classrooms	–	0.70	2
Auto repair rooms[1]	–	1.50	2
Barber shops	–	0.50	2
Bathroom[11,12]	*20/50*	–	2
Beauty and nail salons	–	0.60	2
Cells with toilet	–	1.00	2
Copy, printing rooms	–	0.50	2
Darkrooms	–	1.00	2
Educational science laboratories	–	1.00	2
Janitor closets, trash rooms, recycling	–	1.00	3
Kitchens – commercial	–	0.70	2
Kitchenettes	–	0.30	2
Locker rooms for athletic, industrial and health care facilities	–	0.50	2
Other locker rooms	–	0.25	2
Shower rooms[7,10]	20/50	–	2
Paint spray booths	–	–	4
Parking garages[3]	–	0.75	2
Pet shops (animal areas)	–	0.90	2
Refrigerating machinery rooms[6]	–	–	3
Residential – kitchens[7]	50/100	–	2
Soiled laundry storage rooms	–	1.00	3
Storage rooms, chemical	–	1.50	4
Toilets – private[5, 9]	25/50	–	2
Toilets – public[4, 9]	50/70	–	2
Woodwork shop/classrooms	–	0.50	2

For SI units: 1 cubic foot per minute = 0.0283 m^3/min, 1 square foot = 0.0929 m^2

Notes:

1 Stands where engines are run shall have exhaust systems that directly connect to the engine exhaust and prevent escape of fumes.

2 Where combustion equipment is intended to be used on the playing surface, additional dilution ventilation, source control, or both shall be provided.

3 Exhaust rate is not required for open parking garages as defined in accordance with the building code.

4 Rate is per water closet, urinal, or both. Provide the higher rate where periods of heavy use are expected to occur, e.g., toilets in theatres, schools, and sports facilities. Otherwise the lower rate shall be permitted to be used.

5 Rate is for a toilet room intended to be occupied by one person at a time. For continuous system operation during normal hours of use, the lower rate shall be permitted to be used. Otherwise the higher rate shall be used.

6 For refrigeration machinery rooms, the exhaust rate shall comply with Chapter 11.

7 For continuous system operation, the lower rates shall be permitted. Otherwise the higher rate shall be used.

8 For unlisted occupancies for a proposed space not listed in the table, the requirements for the listed occupancy that is most similar in terms of occupant density and occupancy type shall be used.

9 Exhaust air that has been cleaned in accordance with the criteria of Class 1 shall be permitted to be recirculated.

10 Rate is per shower head.

11 ***[HCD 1 & HCD 2]*** *A bathroom is any room containing a bathtub, a shower, a spa, or a similar source of moisture.*

12 ***[HCD 1 & HCD 2]*** *ANSI/ASHRAE 62.2: Tables 5.1 and 5.2.*

CALIFORNIA MECHANICAL CODE – MATRIX ADOPTION TABLE
CHAPTER 5 – EXHAUST SYSTEMS
(Matrix Adoption Tables are non-regulatory, intended only as an aid to the code user. See Chapter 1 for state agency authority and building applications.)

Adopting Agency	BSC	BSC-CG	SFM	HCD			DSA			OSHPD						BSCC	DPH	AGR	DWR	CEC	CA	SL	SLC
				1	2	1-AC	AC	SS	SS/CC	1	1R	2	3	4	5								
Adopt Entire Chapter				X	X								X										
Adopt Entire Chapter as amended (amended sections listed below)	X		X					X	X	X	X	X		X	X		X						
Adopt only those sections that are listed below		X																			X		
Chapter/Section																							
504.1.1										X	X	X		X	X								
505.7																					X		
505.7.1																					X		
508.5										X	X	X		X	X								
509.2.3			X																				
509.2.3.4																	X						
511.1.6		X						X	X														
512.2.5								X	X														
513.2.2			X					X	X														
513.2.2 Exception 2								X	X														
513.3.2			X					X	X														
513.11			X					X	X														
513.11.1			X					X	X														
516.2.7 Exception			X					X	X														
516.2.9			X					X	X														

This state agency does not adopt sections identified with the following symbol: †
The Office of the State Fire Marshal's adoption of this chapter or individual sections is applicable to structures regulated by other state agencies pursuant to Section 1.11.0.

CHAPTER 5
EXHAUST SYSTEMS

501.0 General.

501.1 Applicability. This chapter includes requirements for environmental air ducts, product-conveying systems, and commercial hoods and kitchen ventilation. Part I addresses environmental air ducts and product conveying systems. Part II addresses commercial hoods and kitchen ventilation.

502.0 Termination.

502.1 Exhaust Opening Protection. Exhaust openings terminating to the outdoors shall be covered with a corrosion-resistant screen having not less than ¼ of an inch (6.4 mm) openings, and shall have not more than ½ of an inch (12.7 mm) openings.

Exception: Clothes dryers.

502.2 Termination of Exhaust Ducts. Exhaust ducts shall terminate in accordance with Section 502.2.1 through Section 502.2.3.

502.2.1 Environmental Air Ducts. Environmental air duct exhaust shall terminate not less than 3 feet (914 mm) from a property line, 10 feet (3048 mm) from a forced air inlet, and 3 feet (914 mm) from openings into the building. Environmental exhaust ducts shall not discharge onto a public walkway.

502.2.2 Product Conveying Ducts. Ducts conveying explosive or flammable vapors, fumes, or dusts shall terminate not less than 30 feet (9144 mm) from a property line, 10 feet (3048 mm) from openings into the building, 6 feet (1829 mm) from exterior walls or roofs, 30 feet (9144 mm) from combustible walls or openings into the building that are in the direction of the exhaust discharge, and 10 feet (3048 mm) above adjoining grade.

Other product-conveying outlets shall terminate not less than 10 feet (3048 mm) from a property line, 3 feet (914 mm) from exterior walls or roofs, 10 feet (3048 mm) from openings into the building, and 10 feet (3048 mm) above adjoining grade.

502.2.3 Commercial Kitchen Ducts. Commercial kitchens exhaust ducts shall terminate in accordance with Section 510.9 or Section 519.5.

Part I – Environmental Air Ducts and Product-Conveying Systems.

503.0 Motors, Fans, and Filters.

503.1 General. Motors and fans shall be sized to provide the required air movement. Motors in areas that contain flammable vapors or dusts shall be of a type approved for such environments. A manually operated remote control installed at an approved location shall be provided to shut off fans or blowers in flammable vapor or dust systems. Equipment used in operations that generate explosive or flammable vapors, fumes, or dusts shall be interlocked with the ventilation system so that the equipment cannot be operated unless the ventilation fans are in operation. Motors for fans used to convey flammable vapors or dusts shall be located outside the duct or shall be protected with approved shields and dustproofing. Where belts are used, they shall not enter the duct unless the belt and pulley within the duct are enclosed. Motors and fans shall be accessible for servicing and maintenance.

503.2 Fans. Parts of fans in contact with explosive or flammable vapors, fumes, or dusts shall be of nonferrous or nonsparking materials, or their casing shall be lined or constructed of such material. Where the size and hardness of materials passing through a fan are capable of producing a spark, both the fan, and the casing shall be of nonsparking materials. Where fans are required to be spark-resistant, their bearings shall not be within the airstream, and parts of the fan shall be grounded. Fans in systems handling materials that are likely to clog the blades, and fans in buffing or woodworking exhaust systems, shall be of the radial-blade or tube-axial type.

Equipment used to exhaust explosive or flammable vapors, fumes, or dusts shall bear an identification plate stating the ventilation rate for which the system was designed.

Fans located in systems conveying corrosives shall be of materials that are resistant to the corrosive or shall be coated with corrosion-resistant materials.

504.0 Environmental Air Ducts.

504.1 General. Where not specified in this chapter, exhaust ducts shall be constructed and installed in accordance with Chapter 6 and shall be airtight as approved by the Authority Having Jurisdiction. Environmental air ducts that have an alternate function as a part of an approved smoke-control system do not require design as Class 1 product-conveying ducts.

Exceptions:

(1) Ductless range hoods where installed in accordance with the manufacturer's installation instructions.

(2) Condensing clothes dryers where installed in accordance with the manufacturer's installation instructions.

504.1.1 Backdraft Protection. Exhaust ducts shall terminate outside the building and shall be equipped with backdraft dampers or with motorized dampers that automatically shut where the systems or spaces served are not in use. ***[OSHPD 1, 1R, 2, 4 & 5] Exception:*** *Backdraft dampers are not required when the exhaust fan must operate continuously.*

Exception: Where the exhaust duct does not discharge into a common exhaust plenum and one of the following:

(1) The exhaust fan runs continuously.

(2) The exhaust duct serves space(s) that are not mechanically heated or cooled.

(3) The space served is maintained at positive pressure.

504.2 Independent Exhaust Systems. Single or combined mechanical exhaust systems shall be independent of other exhaust systems.

504.3 Domestic Range. Ducts used for domestic kitchen range or cooktop ventilation shall be of metal and shall have smooth interior surfaces.

Exception: Ducts for domestic kitchen downdraft grill-range ventilation installed under a concrete slab floor shall be permitted to be of approved Schedule 40 PVC provided:

(1) The under-floor trench in which the duct is installed shall be completely backfilled with sand or gravel.

(2) Not more than 1 inch (25.4 mm) of 6 inch diameter (152 mm) PVC coupling shall be permitted to protrude above the concrete floor surface.

(3) PVC pipe joints shall be solvent cemented to provide an air and greasetight duct.

(4) The duct shall terminate above grade outside the building and shall be equipped with a back-draft damper.

504.4 Clothes Dryers. A clothes dryer exhaust duct shall not be connected to a vent connector, gas vent, chimney, and shall not terminate into a crawl space, attic, or other concealed space. Exhaust ducts shall not be assembled with screws or other fastening means that extend into the duct and that are capable of catching lint, and that reduce the efficiency of the exhaust system. Exhaust ducts shall be constructed of rigid metallic material. Transition ducts used to connect the dryer to the exhaust duct shall be listed and labeled in accordance with UL 2158A, or installed in accordance with the clothes dryer manufacturer's installation instructions. Clothes dryer exhaust ducts shall terminate to the outside of the building in accordance with Section 502.2.1 and shall be equipped with a backdraft damper. Screens shall not be installed at the duct termination. Devices, such as fire or smoke dampers that will obstruct the flow of the exhaust shall not be used. Where joining of ducts, the male end shall be inserted in the direction of airflow.

504.4.1 Provisions for Makeup Air. Makeup air shall be provided in accordance with the following:

(1) Makeup air shall be provided for Type 1 clothes dryers in accordance with the manufacturer's instructions. [NFPA 54:10.4.3.1] Where a closet is designed for the installation of a clothes dryer, an opening of not less than 100 square inches (0.065 m^2) for makeup air shall be provided in the door or by other approved means.

(2) Provision for makeup air shall be provided for Type 2 clothes dryers, with a free area of not less than 1 square inch (0.0006 m^2) for each 1000 British thermal units per hour (Btu/h) (0.293 kW) total input rating of the dryer(s) installed. [NFPA 54:10.4.3.2]

504.4.2 Domestic Clothes Dryers. Where a compartment or space for a Type 1 clothes dryer is provided, not less than a 4 inch diameter (102 mm) exhaust duct of approved material shall be installed in accordance with Section 504.0.

Type 1 clothes dryer exhaust ducts shall be of rigid metal and shall have smooth interior surfaces. The diameter shall be not less than 4 inches nominal (100 mm), and the thickness shall be not less than 0.016 of an inch (0.406 mm).

504.4.2.1 Length Limitation. Unless otherwise permitted or required by the dryer manufacturer's instructions and approved by the Authority Having Jurisdiction, domestic dryer moisture exhaust ducts shall not exceed a total combined horizontal and vertical length of 14 feet (4267 mm), including two 90 degree (1.57 rad) elbows. A length of 2 feet (610 mm) shall be deducted for each 90 degree (1.57 rad) elbow in excess of two.

Exception: Where an exhaust duct power ventilator, in accordance with Section 504.4.2.3, is used, the maximum length of the dryer exhaust duct shall be permitted to be in accordance with the dryer exhaust duct power ventilator manufacturer's installation instructions.

504.4.2.2 Transition Ducts. Listed clothes dryer transition ducts not more than 6 feet (1829 mm) in length shall be permitted to be used to connect the Type 1 dryer to the exhaust ducts. Transition ducts and flexible clothes dryer transition ducts shall not be concealed within construction, and shall be installed in accordance with the manufacturer's installation instructions.

504.4.2.3 Exhaust Duct Power Ventilators. Dryer exhaust duct power ventilators for single residential clothes dryers shall be listed and labeled in accordance with UL 705 and installed in accordance with the manufacturer's installation instructions.

504.4.3 Commercial Clothes Dryers. Commercial dryer exhaust ducts shall be installed in accordance with their listings. The installation of commercial clothes dryer exhaust ducts shall comply with the appliance manufacturer's installation instructions.

504.4.3.1 Exhaust Ducts for Type 2 Clothes Dryers. Exhaust ducts for Type 2 clothes dryers shall comply with the following:

(1) Exhaust ducts for Type 2 clothes dryers shall comply with Section 504.4. [NFPA 54:10.4.5.1]

(2) Exhaust ducts for Type 2 clothes dryers shall be constructed of sheet metal or other noncombustible material. Such ducts shall be equivalent in strength and corrosion resistance to ducts made of galvanized sheet steel not less than 0.0195 of an inch (0.4953 mm) thick. [NFPA 54:10.4.5.2]

(3) Type 2 clothes dryers shall be equipped or installed with lint-controlling means. [NFPA 54:10.4.5.3]

(4) Exhaust ducts for Type 2 clothes dryers shall be installed with a clearance of not less than 6 inches (152 mm) from adjacent combustible

material. Where exhaust ducts for Type 2 clothes dryers are installed with reduced clearances, the adjacent combustible material shall be protected in accordance with Table 303.10.1. [NFPA 54:10.4.5.4]

(5) Where ducts pass through walls, floors, or partitions, the space around the duct shall be sealed with noncombustible material. [NFPA 54:10.4.5.5]

(6) Multiple installations of Type 2 clothes dryers shall be made in a manner to prevent adverse operation due to back pressures that are capable of being created in the exhaust systems. [NFPA 54:10.4.5.6] The exhaust fan shall operate continuously or shall be interlocked to exhaust air where a clothes dryer is in operation.

504.4.4 Common Exhaust. Where permitted by the clothes dryer manufacturer's installation instructions, multiple clothes dryers shall be permitted to be installed with a common exhaust. The common exhaust duct shall be constructed of rigid metal and shall be installed in a fire-resistant rated enclosure in accordance with the building code. The duct material shall be of rigid metal with a thickness of not less than 0.020 of an inch (0.508 mm) (24 gauge). The duct enclosure shall be provided with a cleanout opening at the base of not less than 12 inches by 12 inches (305 mm by 305 mm). The exhaust fan shall be located downstream of branch connections and operated continuously and shall be monitored by an approved means.

504.4.5 Duct Supports. Ducts shall be supported in accordance with SMACNA HVAC Duct Construction Standard – Metal and Flexible.

504.5 Heat (Energy) Recovery Ventilators. Heat (energy) recovery ventilators shall be installed in accordance with their listings and comply with the appliance manufacturer's installation instructions. Non-ducted heat recovery ventilators shall comply with UL 1815. Ducted heat recovery ventilators shall comply with UL 1812. Heat (energy) recovery ventilator ducts shall comply with Chapter 6.

504.6 Gypsum Wallboard Ducts. Bathroom and laundry room exhaust ducts shall be permitted to be of gypsum wallboard subject to the limitations of Section 602.5.

505.0 Product-Conveying Systems.

505.1 General. A mechanical ventilation or exhaust system shall be installed to control, capture, and remove emissions generated from product use or handling where required in accordance with the building code or fire code and where such emissions result in a hazard to life or property. The design of the system shall be such that the emissions are confined to the area in which they are generated by air currents, hoods, or enclosures and shall be exhausted by a duct system to a safe location or treated by removing contaminants. Ducts conveying explosives or flammable vapors, fumes, or dusts shall extend directly to the exterior of the building without entering other spaces and shall not extend into or through ducts and plenums.

Exception: Ducts conveying vapor or fumes having flammable constituents less than 25 percent of their Lower Flammability Limit (LFL) shall be permitted to pass through other spaces.

505.1.1 Mechanical Ventilation. A mechanical ventilation system shall be interlocked to operate with the equipment used to produce vapors, fumes, or dusts that are flammable or hazardous.

505.2 Incompatible Materials. Incompatible materials shall not be conveyed in the same system. [NFPA 91:4.2.2]

505.3 Flammability Limit. Unless the circumstances stipulated in Section 505.3.1, Section 505.3.2, or Section 505.3.3 exist, in systems conveying flammable vapors, gases, or mists, the concentration shall not exceed 25 percent of the lower flammability limit (LFL). [NFPA 91:4.2.3]

505.3.1 Higher Concentrations. Higher concentrations shall be permitted where the exhaust system is designed and protected in accordance with NFPA 69 using one or more of the following techniques:

(1) Combustible concentration reduction

(2) Oxidant concentration reduction

(3) Deflagration suppression

(4) Deflagration pressure containment [NFPA 91:4.2.3.1]

Contaminated air shall not be recirculated to occupied areas unless contaminants have been removed. Air contaminated with explosive or flammable vapors, fumes, or dusts; flammable or toxic gases; or radioactive material shall not be recirculated.

505.3.2 Ovens and Furnaces. Higher concentrations shall be permitted for ovens and furnaces designed and protected in accordance with NFPA 86. [NFPA 91:4.2.3.2]

505.3.3 Deflagration. Higher concentrations shall be permitted where deflagration venting is provided in accordance with NFPA 68. [NFPA 91:4.2.3.3]

505.4 Air-Moving Devices. Air-moving devices shall be sized to establish the velocity required to capture, control, and convey materials through the exhaust system. [NFPA 91:4.2.5]

505.5 Generating Flames, Sparks, or Hot Materials. Operations generating flames, sparks, or hot material such as from grinding wheels and welding shall not be manifolded into an exhaust system that air conveys flammable or combustible materials. [NFPA 91:4.2.6]

505.6 Fire Dampers. Fire dampers shall be permitted to be installed in exhaust systems in accordance with the following:

(1) Where ducts pass through fire barriers

(2) Where a collection system installed on the end of the system is protected with an automatic extinguishing system

(3) Where the duct system is protected with an automatic extinguishing system

(4) Where ducts have been listed with interrupters

(5) Where necessary to facilitate the control of smoke pursuant to the applicable NFPA standards [NFPA 91:4.2.9]

505.6.1 Prohibited. Fire dampers shall not be installed if the material being exhausted is toxic and if a risk evaluation indicates that the toxic hazard is greater than the fire hazard. [NFPA 91:4.2.10]

505.7 Fire Detection and Alarm Systems. Unless the conditions in Section 505.7.1 or Section 505.7.2 exist, fire detection and alarm systems shall not be interlocked to shut down air-moving devices. [NFPA 91:4.2.14]

505.7.1 Automatic Extinguishing System. Where shutdown is necessary for the effective operation of an automatic extinguishing system, it shall be permitted to interlock fire detection and alarm systems to shut down air-moving devices. [NFPA 91:4.2.14.1]

505.7.2 Shut Down Permitted. Where a documented risk analysis acceptable to the Authority Having Jurisdiction shows that the risk of damage from fire and the products of combustion would be higher with air-moving devices operating, it shall be permitted to interlock fire detection and alarm systems to shut down air-moving devices. [NFPA 91:4.2.14.2]

505.8 Product-Conveying Ducts Classification. Product-conveying ducts shall be classified according to their use, as follows:

Class 1 - Ducts conveying nonabrasives, such as smoke, spray, mists, fogs, noncorrosive fumes and gases, light fine dusts, or powders.

Class 2 - Ducts conveying moderately abrasive particulate in light concentrations, such as sawdust and grain dust, and buffing and polishing dust.

Class 3 - Ducts conveying Class 2 materials in high concentrations and highly abrasive materials in low concentrations, such as manganese, steel chips, and coke.

Class 4 - Ducts conveying highly abrasive material in high concentrations.

Class 5 - Ducts conveying corrosives, such as acid vapors.

505.9 Minimum Velocities and Circulation. The velocity and circulation of air in work areas shall be such that contaminant's are captured by an airstream at the area where the emissions are generated and conveyed into a product-conveying duct system. Mixtures within work areas where contaminants are generated shall be diluted to be accordance with Section 505.3 with air that does not contain other contaminants. The velocity of air within the duct shall be not less than set forth in Table 505.9.

Systems conveying particulate matter shall be designed by employing the constant velocity method. Systems conveying explosive or radioactive materials shall be pre-balanced through duct sizing. Other systems shall be permitted to be designed with balancing devices such as dampers. Dampers provided to balance airflow shall be provided with securely fixed minimum-position blocking devices to prevent restricting flow below the required volume or velocity.

505.10 Makeup Air. Makeup air shall be provided to replenish air exhausted by the ventilation system. Makeup air intakes shall be located so as to avoid recirculation of contaminated air within enclosures.

505.11 Hoods and Enclosures. Hoods and enclosures shall be used where contaminants originate in a concentrated area. The design of the hood or enclosure shall be such that air currents created by the exhaust systems will capture the contaminants and transport them directly to the exhaust duct. The volume of air shall be sufficient to dilute explosive or flammable vapors, fumes, or dusts in accordance with Section 505.9.

TABLE 505.9
MINIMUM DUCT DESIGN VELOCITIES*
[NFPA 91: TABLE A.4.1.5]

NATURE OF CONTAMINANTS	EXAMPLES	FEET PER MINUTE DESIGN VELOCITY (feet per minute)
Vapors, gases, smoke	Vapors, gases, and smoke	Any
Fumes	Welding	2000
Fine light dusts	Cotton lint, wood flour, litho powder	2500
Dry dusts and powders	Fine rubber dust, molding powder dust, jute lint, cotton dust, shavings (light), soap dust, leather shavings	3000
Average industrial dusts	Grinding dust, buffing lint (dry), wool jute dust (shaker waste), coffee beans, shoe dust, granite dust, silica flour, general material handling, brick cutting, clay dust, foundry (general), limestone dust, packaging and weighing asbestos dust in textile industries	3500
Heavy dusts	Sawdust (heavy and wet), metal turnings, foundry tumbling barrels and shake-out, sandblast dust, wood blocks, hog waste, brass turning, cast-iron boring dust, lead dust	4000
Heavy or moist dusts	Lead dust with chips, moist cement dust, asbestos chunks from transite pipe cutting machines, buffing lint (sticky), quick-lime dust	4500

For SI units: 1 foot per minute = 0.005 m/s

* Systems that are handling combustible particulate solids shall be in accordance with NFPA 654.

Hoods of steel shall have a base metal thickness not less than 0.027 of an inch (0.686 mm) (No. 22 gauge) for Class 1 and Class 5 metal duct systems; 0.033 of an inch (0.838 mm) (No. 20 gauge) for hoods serving a Class 2 duct system; 0.044 of an inch (1.118 mm) (No. 18 gauge) for hoods serving a Class 3 duct system; and 0.068 of an inch (1.727 mm) (No. 14 gauge) for hoods serving a Class 4 duct system.

Approved nonmetallic hoods and duct systems shall be permitted to be used for Class 5 corrosive systems where the corrosive mixture is nonflammable. Metal hoods used with Class 5 duct systems shall be protected with an approved corrosion-resistant material. Edges of hoods shall be rounded. The minimum clearance between hoods and combustible construction shall be the clearance required by the duct system.

505.12 Pharmacies – Compounding Area of Parenteral Solutions. [CA – Board of Pharmacy] *The pharmacy shall have a designated area for the preparation of sterile products for dispensing which shall be ventilated in a manner not interfering with laminar air flow.*

Note: *For additional pharmacy building standard requirements, see Chapter 12, California Building Code.*

505.12.1 Pharmacies – Laminar Flow Biological Safety Cabinet. [CA – Board of Pharmacy] *In all pharmacies preparing parenteral cytotoxic agents, all compounding shall be conducted within a certified Class II Type A or Class II Type B vertical laminar airflow hood with bag in – bag out design. The pharmacy must ensure that contaminated air plenums that are under positive air pressure are leak tight.*

Note: *For additional pharmacy building standard requirements, see Chapter 12, California Building Code.*

506.0 Product-Conveying Ducts.

506.1 Materials. Materials used in product-conveying duct systems shall be suitable for the intended use and shall be of metal.

Exceptions:

(1) Asbestos-cement, concrete, clay, or ceramic materials shall be permitted to be used where it is shown that these materials will be equivalent to metal ducts installed in accordance with this chapter.

(2) Ducts serving a Class 5 system shall be permitted to be constructed of approved nonmetallic material where the corrosive characteristics of the material being conveyed make a metal system unsuitable and where the mixture being conveyed is nonflammable.

Approved nonmetallic material shall be either a listed product having a flame-spread index not exceeding 25 and a smoke-developed rating of 50 or less on both inside and outside surfaces without evidence of continued progressive combustion, or shall have a flame-spread index not exceeding 25 and shall be installed with an automatic fire-sprinkler protection system inside the duct.

(3) Ducts used in central vacuum cleaning systems within a dwelling unit shall be constructed of materials in accordance with the applicable standards referenced in Chapter 17. Penetrations of fire walls or floor-ceiling or roof-ceiling assemblies shall be in accordance with the building code.

Copper or ferrous pipes or conduits extending from within the separation between a garage and dwelling unit to the central vacuuming unit shall be permitted to be used.

Aluminum ducts shall not be used in systems conveying flammable vapors, fumes, or explosive dusts, nor in Class 2, 3, or 4 systems. Galvanized steel and aluminum ducts shall not be used where the temperature of the material being conveyed exceeds 400°F (204°C).

Metal ducts used in Class 5 systems that are not resistant to the corrosiveness of the product shall be protected with an approved corrosion-resistant material.

506.2 Construction. Ducts used for conveying products shall be airtight construction as approved by the Authority Having Jurisdiction, and shall not have openings other than those required for operation and maintenance of the system. Ducts constructed of steel shall comply with Table 506.2(1) or Table 506.2(2).

Exceptions:

(1) Class 1 product-conveying ducts that operate at less than 4 inches water column (0.9 kPa) negative pressure and convey noncorrosive, nonflammable and nonexplosive materials at temperatures not exceeding 250°F (121°C) shall be permitted to be constructed in accordance with SMACNA HVAC Duct Construction Standards–Metal and Flexible.

(2) Ducts used in central vacuuming systems within a dwelling unit shall be constructed of materials in accordance with the applicable standards referenced in Chapter 17. Penetrations of fire-resistive walls or floor-ceiling or roof-ceiling assemblies shall be in accordance with the building code. Copper or ferrous pipes or conduit extending from within the separation between a garage and dwelling unit to the central vacuum unit shall be permitted to be used.

The use of rectangular ducts conveying particulates shall be subject to approval of the Authority Having Jurisdiction. The design of rectangular ducts shall consider the adhesiveness and buildup of products being conveyed within the duct.

Aluminum construction shall be permitted to be used in Class 1 duct systems. The thickness of aluminum ducts shall be not less than two Brown and Sharpe gauges thicker than the gauges required for steel ducts set forth in Table 506.2(1) and Table 506.2(2).

506.3 Penetrations. Exhaust ducts shall not pass through fire walls, as defined by NFPA 221. [NFPA 91:4.2.11]

506.3.1 Fire Barriers. Exhaust ducts passing through a fire barrier having a fire resistance rating of 2 hours or more shall meet one of the following specifications:

(1) Wrapped or encased with listed or approved materials having a fire resistance rating equal to the fire barrier for 10 feet (3048 mm) of the duct on each

TABLE 506.2(1)
MINIMUM SHEET METAL THICKNESS FOR ROUND DUCTS

NEGATIVE PRESSURE (inches water column)	REINF. SPACING (inches)	CLASS 1 (inches)								
		Up to 7	8 to 11	12 to 15	16 to 19	20 to 23	24 to 35	36 to 47	48 to59	60
To 7	0	0.021 (24 ga.)	0.021 (24 ga.)	0.033 (20 ga.)	0.044 (18 ga.)	0.055 (16 ga.)	0.068 (14 ga.)	0.127 (10 ga.)	—	—
	96	0.021 (24 ga.)	0.021 (24 ga.)	0.021 (24 ga.)	0.027 (22 ga.)	0.033 (20 ga.)	0.044 (18 ga.)	0.055 (16 ga.)	0.068 (14 ga.)	0.068 (14 ga.)
	48	0.021 (24 ga.)	0.021 (24 ga.)	0.021 (24 ga.)	0.021 (24 ga.)	0.021 (24 ga.)	0.027 (22 ga.)	0.033 (20 ga.)	0.044 (18 ga.)	0.055 (16 ga.)
	24	0.021 (24 ga.)	0.021 (24 ga.)	0.021 (24 ga.)	0.021 (24 ga.)	0.021 (24 ga.)	0.021 (24 ga.)	0.027 (22 ga.)	0.033 (20 ga.)	0.044 (18 ga.)
8 to 11	0	0.021 (24 ga.)	0.027 (22 ga.)	0.044 (18 ga.)	0.055 (16 ga.)	0.068 (14 ga.)	0.097 (12 ga.)	—	—	—
	96	0.021 (24 ga.)	0.027 (22 ga.)	0.027 (22 ga.)	0.044 (18 ga.)	0.044 (18 ga.)	0.044 (18 ga.)	0.068 (14 ga.)	0.097 (12 ga.)	0.097 (12 ga.)
	48	0.021 (24 ga.)	0.021 (24 ga.)	0.021 (24 ga.)	0.027 (22 ga.)	0.033 (20 ga.)	0.033 (20 ga.)	0.055 (16 ga.)	0.068 (14 ga.)	0.068 (14 ga.)
	24	0.021 (24 ga.)	0.021 (24 ga.)	0.021 (24 ga.)	0.021 (24 ga.)	0.027 (22 ga.)	0.027 (22 ga.)	0.044 (18 ga.)	0.055 (16 ga.)	0.055 (16 ga.)
12 to 15	0	0.021 (24 ga.)	0.033 (20 ga.)	0.055 (16 ga.)	0.068 (14 ga.)	0.097 (12 ga.)	0.097 (12 ga.)	—	—	—
	96	0.021 (24 ga.)	0.027 (22 ga.)	0.044 (18 ga.)	0.044 (18 ga.)	0.055 (16 ga.)	0.055 (16 ga.)	0.097 (12 ga.)	0.112 (11 ga.)	0.112 (11 ga.)
	48	0.021 (24 ga.)	0.027 (22 ga.)	0.027 (22 ga.)	0.033 (20 ga.)	0.044 (18 ga.)	0.044 (18 ga.)	0.068 (14 ga.)	0.068 (14 ga.)	0.097 (12 ga.)
	24	0.021 (24 ga.)	0.021 (24 ga.)	0.021 (24 ga.)	0.027 (22 ga.)	0.027 (22 ga.)	0.027 (22 ga.)	0.055 (16 ga.)	0.055 (16 ga.)	0.055 (16 ga.)
16 to 20	0	0.021 (24 ga.)	0.044 (18 ga.)	0.068 (14 ga.)	0.097 (12 ga.)	0.112 (11 ga.)	—	—	—	—
	96	0.021 (24 ga.)	0.033 (20 ga.)	0.055 (16 ga.)	0.055 (16 ga.)	0.068 (14 ga.)	0.068 (14 ga.)	0.112 (11 ga.)	0.112 (11 ga.)	0.112 (11 ga.)
	48	0.021 (24 ga.)	0.027 (22 ga.)	0.033 (20 ga.)	0.044 (18 ga.)	0.055 (16 ga.)	0.055 (16 ga.)	0.068 (14 ga.)	0.097 (12 ga.)	0.112 (11 ga.)
	24	0.021 (24 ga.)	0.021 (24 ga.)	0.027 (22 ga.)	0.033 (20 ga.)	0.044 (18 ga.)	0.044 (18 ga.)	0.055 (16 ga.)	0.068 (14 ga.)	0.097 (12 ga.)
		CLASS 2 (inches)								
To 7	0	0.027 (22 ga.)	0.027 (22 ga.)	0.033 (20 ga.)	0.044 (18 ga.)	0.055 (16 ga.)	0.068 (14 ga.)	0.127 (10 ga.)	—	—
	96	0.027 (22 ga.)	0.027 (22 ga.)	0.033 (20 ga.)	0.033 (20 ga.)	0.044 (18 ga.)	0.044 (18 ga.)	0.055 (16 ga.)	0.068 (14 ga.)	0.068 (14 ga.)
	48	0.027 (22 ga.)	0.027 (22 ga.)	0.033 (20 ga.)	0.033 (20 ga.)	0.044 (18 ga.)	0.044 (18 ga.)	0.055 (16 ga.)	0.055 (16 ga.)	0.055 (16 ga.)
	24	0.027 (22 ga.)	0.027 (22 ga.)	0.033 (20 ga.)	0.033 (20 ga.)	0.044 (18 ga.)	0.044 (18 ga.)	0.055 (16 ga.)	0.055 (16 ga.)	0.055 (16 ga.)
8 to 11	0	0.027 (22 ga.)	0.027 (22 ga.)	0.044 (18 ga.)	0.055 (16 ga.)	0.068 (14 ga.)	0.097 (12 ga.)	—	—	—
	96	0.027 (22 ga.)	0.027 (22 ga.)	0.033 (20 ga.)	0.044 (18 ga.)	0.044 (18 ga.)	0.044 (18 ga.)	0.068 (14 ga.)	0.097 (12 ga.)	0.097 (12 ga.)
	48	0.027 (22 ga.)	0.027 (22 ga.)	0.033 (20 ga.)	0.033 (20 ga.)	0.044 (18 ga.)	0.044 (18 ga.)	0.055 (16 ga.)	0.068 (14 ga.)	0.068 (14 ga.)
	24	0.027 (22 ga.)	0.027 (22 ga.)	0.033 (20 ga.)	0.033 (20 ga.)	0.044 (18 ga.)	0.044 (18 ga.)	0.055 (16 ga.)	0.055 (16 ga.)	0.055 (16 ga.)
12 to 15	0	0.027 (22 ga.)	0.033 (20 ga.)	0.055 (16 ga.)	0.068 (14 ga.)	0.097 (12 ga.)	0.097 (12 ga.)	—	—	—
	96	0.027 (22 ga.)	0.033 (20 ga.)	0.044 (18 ga.)	0.044 (18 ga.)	0.055 (16 ga.)	0.055 (16 ga.)	0.097 (12 ga.)	0.112 (11 ga.)	0.112 (11 ga.)
	48	0.027 (22 ga.)	0.027 (22 ga.)	0.033 (20 ga.)	0.033 (20 ga.)	0.044 (18 ga.)	0.044 (18 ga.)	0.068 (14 ga.)	0.068 (14 ga.)	0.097 (12 ga.)
	24	0.027 (22 ga.)	0.027 (22 ga.)	0.033 (20 ga.)	0.033 (20 ga.)	0.044 (18 ga.)	0.044 (18 ga.)	0.055 (16 ga.)	0.055 (16 ga.)	0.055 (16 ga.)
16 to 20	0	0.027 (22 ga.)	0.044 (18 ga.)	0.068 (14 ga.)	0.097 (12 ga.)	0.112 (11 ga.)	—	—	—	—
	96	0.027 (22 ga.)	0.033 (20 ga.)	0.055 (16 ga.)	0.055 (16 ga.)	0.068 (14 ga.)	0.068 (14 ga.)	0.112 (11 ga.)	0.112 (11 ga.)	0.112 (11 ga.)
	48	0.027 (22 ga.)	0.033 (20 ga.)	0.033 (20 ga.)	0.044 (18 ga.)	0.055 (16 ga.)	0.055 (16 ga.)	0.068 (14 ga.)	0.068 (14 ga.)	0.097 (12 ga.)
	24	0.027 (22 ga.)	0.033 (20 ga.)	0.033 (20 ga.)	0.033 (20 ga.)	0.044 (18 ga.)	0.044 (18 ga.)	0.055 (16 ga.)	0.055 (16 ga.)	0.055 (16 ga.)
		CLASS 3 (inches)								
To 7	0	0.033 (20 ga.)	0.033 (20 ga.)	0.044 (18 ga.)	0.044 (18 ga.)	0.055 (16 ga.)	0.068 (14 ga.)	0.127 (10 ga.)	—	—
	96	0.033 (20 ga.)	0.033 (20 ga.)	0.044 (18 ga.)	0.044 (18 ga.)	0.055 (16 ga.)	0.055 (16 ga.)	0.068 (14 ga.)	0.068 (14 ga.)	0.068 (14 ga.)
	48	0.033 (20 ga.)	0.033 (20 ga.)	0.044 (18 ga.)	0.044 (18 ga.)	0.055 (16 ga.)	0.055 (16 ga.)	0.068 (14 ga.)	0.068 (14 ga.)	0.068 (14 ga.)
	24	0.033 (20 ga.)	0.033 (20 ga.)	0.044 (18 ga.)	0.044 (18 ga.)	0.055 (16 ga.)	0.055 (16 ga.)	0.068 (14 ga.)	0.068 (14 ga.)	0.068 (14 ga.)
8 to 11	0	0.033 (20 ga.)	0.033 (20 ga.)	0.044 (18 ga.)	0.055 (16 ga.)	0.068 (14 ga.)	0.097 (12 ga.)	—	—	—
	96	0.033 (20 ga.)	0.033 (20 ga.)	0.044 (18 ga.)	0.044 (18 ga.)	0.055 (16 ga.)	0.055 (16 ga.)	0.068 (14 ga.)	0.097 (12 ga.)	0.097 (12 ga.)
	48	0.033 (20 ga.)	0.033 (20 ga.)	0.044 (18 ga.)	0.044 (18 ga.)	0.055 (16 ga.)	0.055 (16 ga.)	0.068 (14 ga.)	0.068 (14 ga.)	0.068 (14 ga.)
	24	0.033 (20 ga.)	0.033 (20 ga.)	0.044 (18 ga.)	0.044 (18 ga.)	0.055 (16 ga.)	0.055 (16 ga.)	0.068 (14 ga.)	0.068 (14 ga.)	0.068 (14 ga.)

TABLE 506.2(1) (continued)
MINIMUM SHEET METAL THICKNESS FOR ROUND DUCTS

NEGATIVE PRESSURE (inches water column)	REINF. SPACING (inches)	CLASS 3 (inches)								
		Up to 7	8 to 11	12 to 15	16 to 19	20 to 23	24 to 35	36 to 47	48 to59	60
12 to 15	0	0.133 (20 ga.)	0.133 (20 ga.)	0.155 (16 ga.)	0.168 (14 ga.)	0.197 (12 ga.)	0.197 (12 ga.)	—	—	—
	96	0.133 (20 ga.)	0.133 (20 ga.)	0.144 (18 ga.)	0.144 (18 ga.)	0.155 (16 ga.)	0.155 (16 ga.)	0.197 (12 ga.)	0.112 (11 ga.)	0.112 (11 ga.)
	48	0.133 (20 ga.)	0.133 (20 ga.)	0.144 (18 ga.)	0.144 (18 ga.)	0.155 (16 ga.)	0.155 (16 ga.)	0.168 (14 ga.)	0.168 (14 ga.)	0.168 (14 ga.)
	24	0.133 (20 ga.)	0.133 (20 ga.)	0.144 (18 ga.)	0.144 (18 ga.)	0.155 (16 ga.)	0.155 (16 ga.)	0.168 (14 ga.)	0.168 (14 ga.)	0.168 (14 ga.)
16 to 20	0	0.133 (20 ga.)	0.144 (18 ga.)	0.168 (14 ga.)	0.197 (12 ga.)	0.112 (11 ga.)	—	—	—	—
	96	0.133 (20 ga.)	0.133 (20 ga.)	0.155 (16 ga.)	0.155 (16 ga.)	0.168 (14 ga.)	0.168 (14 ga.)	0.112 (11 ga.)	0.112 (11 ga.)	0.112 (11 ga.)
	48	0.133 (20 ga.)	0.133 (20 ga.)	0.144 (18 ga.)	0.144 (18 ga.)	0.155 (16 ga.)	0.155 (16 ga.)	0.168 (14 ga.)	0.168 (14 ga.)	0.197 (12 ga.)
	24	0.133 (20 ga.)	0.133 (20 ga.)	0.144 (18 ga.)	0.144 (18 ga.)	0.155 (16 ga.)	0.155 (16 ga.)	0.168 (14 ga.)	0.168 (14 ga.)	0.168 (14 ga.)
		CLASS 4 (inches)								
To 7	0	0.155 (16 ga.)	0.155 (16 ga.)	0.155 (16 ga.)	0.155 (16 ga.)	0.155 (16 ga.)	0.168 (14 ga.)	0.168 (14 ga.)	—	—
	96	0.155 (16 ga.)	0.155 (16 ga.)	0.155 (16 ga.)	0.155 (16 ga.)	0.168 (14 ga.)	0.168 (14 ga.)	0.197 (12 ga.)	0.197 (12 ga.)	0.197 (12 ga.)
	48	0.155 (16 ga.)	0.155 (16 ga.)	0.155 (16 ga.)	0.155 (16 ga.)	0.168 (14 ga.)	0.168 (14 ga.)	0.197 (12 ga.)	0.197 (12 ga.)	0.197 (12 ga.)
	24	0.155 (16 ga.)	0.155 (16 ga.)	0.155 (16 ga.)	0.155 (16 ga.)	0.168 (14 ga.)	0.168 (14 ga.)	0.197 (12 ga.)	0.197 (12 ga.)	0.197 (12 ga.)
8 to 11	0	0.155 (16 ga.)	0.155 (16 ga.)	0.155 (16 ga.)	0.155 (16 ga.)	0.168 (14 ga.)	0.197 (12 ga.)	—	—	—
	96	0.155 (16 ga.)	0.155 (16 ga.)	0.155 (16 ga.)	0.155 (16 ga.)	0.168 (14 ga.)	0.168 (14 ga.)	0.197 (12 ga.)	0.197 (12 ga.)	0.197 (12 ga.)
	48	0.155 (16 ga.)	0.155 (16 ga.)	0.155 (16 ga.)	0.155 (16 ga.)	0.168 (14 ga.)	0.168 (14 ga.)	0.197 (12 ga.)	0.197 (12 ga.)	0.197 (12 ga.)
	24	0.155 (16 ga.)	0.155 (16 ga.)	0.155 (16 ga.)	0.155 (16 ga.)	0.168 (14 ga.)	0.168 (14 ga.)	0.197 (12 ga.)	0.197 (12 ga.)	0.197 (12 ga.)
12 to 15	0	0.155 (16 ga.)	0.155 (16 ga.)	0.155 (16 ga.)	0.168 (14 ga.)	0.197 (12 ga.)	0.197 (12 ga.)	—	—	—
	96	0.155 (16 ga.)	0.155 (16 ga.)	0.155 (16 ga.)	0.155 (16 ga.)	0.168 (14 ga.)	0.168 (14 ga.)	0.197 (12 ga.)	0.112 (11 ga.)	0.112 (11 ga.)
	48	0.155 (16 ga.)	0.155 (16 ga.)	0.155 (16 ga.)	0.155 (16 ga.)	0.168 (14 ga.)	0.168 (14 ga.)	0.197 (12 ga.)	0.197 (12 ga.)	0.097 (12 ga.)
	24	0.055 (16 ga.)	0.055 (16 ga.)	0.055 (16 ga.)	0.055 (16 ga.)	0.068 (14 ga.)	0.068 (14 ga.)	0.097 (12 ga.)	0.097 (12 ga.)	0.097 (12 ga.)
16 to 20	0	0.055 (16 ga.)	0.055 (16 ga.)	0.068 (14 ga.)	0.097 (12 ga.)	0.112 (11 ga.)	—	—	—	—
	96	0.055 (16 ga.)	0.055 (16 ga.)	0.055 (16 ga.)	0.055 (16 ga.)	0.068 (14 ga.)	0.068 (14 ga.)	0.112 (11 ga.)	0.112 (11 ga.)	0.112 (11 ga.)
	48	0.055 (16 ga.)	0.055 (16 ga.)	0.055 (16 ga.)	0.055 (16 ga.)	0.068 (14 ga.)	0.068 (14 ga.)	0.097 (12 ga.)	0.097 (12 ga.)	0.097 (12 ga.)
	24	0.055 (16 ga.)	0.055 (16 ga.)	0.055 (16 ga.)	0.055 (16 ga.)	0.068 (14 ga.)	0.068 (14 ga.)	0.097 (12 ga.)	0.097 (12 ga.)	0.097 (12 ga.)

For SI units: 1 inch = 25.4 mm, 1 inch water column = 0.249 kPa

side of the fire barrier including duct supports within this span.

(2) Constructed of materials and supports having a minimum fire resistance rating equal to the fire barrier.

(3) Enclosed with a shaft that is constructed of material having a fire resistance rating equal to the fire barrier for 10 feet (3048 mm) of the duct on each side of the fire barrier with no inlets to the duct within this distance, and the duct entry into and exit from the shaft is protected in accordance with Section 506.3.2. [NFPA 91:4.2.12]

506.3.2 Protection. Exhaust ducts passing through fire barriers of any fire resistance rating shall be protected by sealing the space around the duct with listed or approved fire stopping having a fire resistance rating equal to the fire resistance rating of the fire barrier. [NFPA 91:4.2.13]

506.4 Condensate. Joints in duct construction shall be liquidtight when the conveying system contains condensable vapors or liquids in suspension. [NFPA 91:4.3.6.1]

506.4.1 Drainage. Provisions shall be made for drainage of condensate at low points in the duct. [NFPA 91:4.3.6.2]

506.5 Fittings. Fittings in Class 2, 3, and 4 systems shall be not less than two gauges thicker than the thickness required for straight runs. Flexible metallic duct shall be permitted to be used for connecting ductwork to vibrating equipment. Duct systems subject to wide temperature fluctuations shall be provided with expansion joints.

Branches shall connect to main ducts at the large end of transitions at an angle not exceeding 45 degrees (0.79 rad).

Except for ducts used to convey noncorrosive vapors with no particulate, accessible cleanouts shall be provided at 10 foot (3048 mm) intervals and at changes in direction. Access openings shall also be provided for access to sprinklers and other equipment within the duct that require servicing.

506.6 Explosion Venting. Ducts conveying explosive dusts shall have explosion vents, openings protected by anti-flashback swing valves, or rupture diaphragms. Openings to relieve

TABLE 506.2(2)
MINIMUM SHEET METAL THICKNESS FOR RECTANGULAR DUCTS

NEGATIVE PRESSURE (inches water column)	REINF. SPACING (inches)	LONGEST SIDE OF DUCT									
		CLASS 1 (inches)					CLASS 2 (inches)				
		Up to 12	13 to 24	25 to 36	37 to 48	49 to 60	Up to 12	13 to 24	25 to 36	37 to 48	49 to 60
To 7	48	0.021 (24 ga.)	0.033 (20 ga.)	0.055 (16 ga.)	0.068 (14 ga.)	—	0.033 (20 ga.)	0.044 (18 ga.)	0.055 (16 ga.)	0.068 (14 ga.)	—
	24	0.021 (24 ga.)	0.027 (22 ga.)	0.033 (20 ga.)	0.033 (20 ga.)	0.033 (20 ga.)	0.033 (20 ga.)	0.044 (18 ga.)	0.055 (16 ga.)	0.055 (16 ga.)	0.055 (16 ga.)
	12	0.021 (24 ga.)	0.021 (24 ga.)	0.021 (24 ga.)	0.021 (24 ga.)	0.021 (24 ga.)	0.033 (20 ga.)	0.044 (18 ga.)	0.055 (16 ga.)	0.055 (16 ga.)	0.055 (16 ga.)
8 to 11	48	0.027 (22 ga.)	0.068 (14 ga.)	0.097 (12 ga.)	0.097 (12 ga.)	—	0.033 (20 ga.)	0.068 (14 ga.)	0.097 (12 ga.)	0.097 (12 ga.)	—
	24	0.027 (22 ga.)	0.055 (16 ga.)	0.055 (16 ga.)	0.068 (14 ga.)	0.068 (14 ga.)	0.033 (20 ga.)	0.055 (16 ga.)	0.055 (16 ga.)	0.068 (14 ga.)	0.068 (14 ga.)
	12	0.021 (24 ga.)	0.027 (22 ga.)	0.027 (22 ga.)	0.027 (22 ga.)	0.027 (22 ga.)	0.033 (20 ga.)	0.044 (18 ga.)	0.055 (16 ga.)	0.055 (16 ga.)	0.055 (16 ga.)
12 to 15	48	0.044 (18 ga.)	0.097 (12 ga.)	—	—	—	0.044 (18 ga.)	0.044 (18 ga.)	—	—	—
	24	0.044 (18 ga.)	0.055 (16 ga.)	0.097 (12 ga.)	0.097 (12 ga.)	0.097 (12 ga.)	0.044 (18 ga.)	0.044 (18 ga.)	0.097 (12 ga.)	0.097 (12 ga.)	0.097 (12 ga.)
	12	0.027 (22 ga.)	0.044 (18 ga.)	0.044 (18 ga.)	0.044 (18 ga.)	0.044 (18 ga.)	0.033 (20 ga.)	0.044 (18 ga.)	0.055 (16 ga.)	0.055 (16 ga.)	0.055 (16 ga.)
16 to 20	48	0.068 (14 ga.)	0.112 (11 ga.)	—	—	—	0.068 (14 ga.)	0.112 (11 ga.)	—	—	—
	24	0.068 (14 ga.)	0.068 (14 ga.)	0.112 (11 ga.)	0.112 (11 ga.)	0.112 (11 ga.)	0.068 (14 ga.)	0.068 (14 ga.)	0.112 (11 ga.)	0.112 (11 ga.)	0.112 (11 ga.)
	12	0.033 (20 ga.)	0.068 (14 ga.)	0.068 (14 ga.)	0.068 (14 ga.)	0.068 (14 ga.)	0.033 (20 ga.)	0.044 (18 ga.)	0.068 (14 ga.)	0.068 (14 ga.)	0.068 (14 ga.)
		CLASS 3 (inches)					CLASS 4 (inches)				
To 7	48	0.044 (18 ga.)	0.055 (16 ga.)	0.068 (14 ga.)	0.068 (14 ga.)	—	0.055 (16 ga.)	0.068 (14 ga.)	0.097 (12 ga.)	0.097 (12 ga.)	—
	24	0.044 (18 ga.)	0.055 (16 ga.)	0.068 (14 ga.)	0.068 (14 ga.)	0.068 (14 ga.)	0.055 (16 ga.)	0.068 (14 ga.)	0.097 (12 ga.)	0.097 (12 ga.)	0.097 (12 ga.)
	12	0.044 (18 ga.)	0.055 (16 ga.)	0.068 (14 ga.)	0.068 (14 ga.)	0.068 (14 ga.)	0.055 (16 ga.)	0.068 (14 ga.)	0.097 (12 ga.)	0.097 (12 ga.)	0.097 (12 ga.)
8 to 11	48	0.044 (18 ga.)	0.068 (14 ga.)	0.097 (12 ga.)	0.097 (12 ga.)	—	0.055 (16 ga.)	0.068 (14 ga.)	0.097 (12 ga.)	0.097 (12 ga.)	—
	24	0.044 (18 ga.)	0.055 (16 ga.)	0.068 (14 ga.)	0.068 (14 ga.)	0.068 (14 ga.)	0.055 (16 ga.)	0.068 (14 ga.)	0.097 (12 ga.)	0.097 (12 ga.)	0.097 (12 ga.)
	12	0.044 (18 ga.)	0.055 (16 ga.)	0.068 (14 ga.)	0.068 (14 ga.)	0.068 (14 ga.)	0.055 (16 ga.)	0.068 (14 ga.)	0.097 (12 ga.)	0.097 (12 ga.)	0.097 (12 ga.)
12 to 15	48	0.044 (18 ga.)	0.097 (12 ga.)	—	—	—	0.055 (16 ga.)	0.097 (12 ga.)	—	—	—
	24	0.044 (18 ga.)	0.055 (16 ga.)	0.097 (12 ga.)	0.097 (12 ga.)	0.097 (12 ga.)	0.055 (16 ga.)	0.068 (14 ga.)	0.097 (12 ga.)	0.097 (12 ga.)	0.097 (12 ga.)
	12	0.044 (18 ga.)	0.055 (16 ga.)	0.068 (14 ga.)	0.068 (14 ga.)	0.068 (14 ga.)	0.055 (16 ga.)	0.068 (14 ga.)	0.097 (12 ga.)	0.097 (12 ga.)	0.097 (12 ga.)
16 to 20	48	0.068 (14 ga.)	0.112 (11 ga.)	—	—	—	0.055 (16 ga.)	0.112 (11 ga.)	—	—	—
	24	0.068 (14 ga.)	0.068 (14 ga.)	0.112 (11 ga.)	0.112 (11 ga.)	0.112 (11 ga.)	0.055 (16 ga.)	0.068 (14 ga.)	0.112 (11 ga.)	0.112 (11 ga.)	0.112 (11 ga.)
	12	0.044 (18 ga.)	0.068 (14 ga.)	0.068 (14 ga.)	0.068 (14 ga.)	0.068 (14 ga.)	0.055 (16 ga.)	0.068 (14 ga.)	0.097 (12 ga.)	0.097 (12 ga.)	0.097 (12 ga.)

For SI units: 1 inch = 25.4 mm, 1 inch water column = 0.249 kPa

explosive forces shall be located outside the building. Where relief devices cannot provide sufficient pressure relief, ductwork shall be designed to withstand an internal pressure of not less than 100 pounds-force per square inch (psi) (689 kPa).

Where a room or building contains a dust explosion hazard that is external to protected equipment, as defined in NFPA 654, such areas shall be provided with deflagration venting to a safe outside location.

506.7 Supports. Supports shall be of noncombustible materials, and the spacing shall not exceed 12 feet (3658 mm) for 8 inch (203 mm) ducts and 20 feet (6096 mm) for larger ducts.

506.8 Fire Protection. Sprinklers or other fire-protection devices shall be installed within ducts having a cross-sectional dimension exceeding 10 inches (254 mm) where the duct conveys flammable vapors or fumes. Sprinklers shall be installed at 12 foot (3658 mm) intervals in horizontal ducts and at changes in direction. In vertical runs, sprinklers shall be installed at the top and at alternate floor levels.

506.8.1 Loads. Duct supports shall be designed to carry the weight of the duct half filled with material. Where sprinkler protection is provided or cleaning of duct will be performed, the hanger's design shall include the weight of the expected liquid accumulation. Duct supports shall be designed to prevent placing loads on connected equipment. [NFPA 91:4.6.1 – 4.6.3]

506.8.2 Corrosion. Hangers and supports exposed to corrosive atmospheres shall be corrosion resistant. [NFPA 91:4.6.4]

506.8.3 Vibration and Stress. To prevent vibration and stress on the duct, hangers and supports shall be securely fastened to the building or structure. [NFPA 91:4.6.5]

506.8.4 Expansion and Contraction. Hangers and supports shall be designed to allow for expansion and contraction. [NFPA 91:4.6.6]

506.9 Protection from Physical Damage. Ducts installed in locations where they are subject to physical damage shall be protected by guards.

506.10 Duct Clearances. Unless the conditions in Section 506.10.1 or Section 506.10.2 exist, duct systems and system components shall have a clearance of at least 6 inches (152 mm) from stored combustible materials, and not less than ½ of an inch (13 mm) clearance from combustible construction. [NFPA 91:4.7.1]

506.10.1 Protection Provided. Where stored combustible material or combustible construction is protected from ductwork by the use of materials or products listed for protection purposes, clearance shall be maintained in accordance with those listings. [NFPA 91:4.7.1.1]

506.10.2 Systems Conveying Combustible Materials. Unless the conditions in Section 506.10.3 exist, duct systems and system components handling combustible material shall have a clearance of not less than 18 inches (457 mm) from combustible construction or a combustible material. [NFPA 91:4.7.2]

506.10.3 Reduced Clearance Permitted. When the ductwork system is operating at 140°F (60°C) or below and is equipped with an approved automatic extinguishing system designed for the specific hazard, the clearance shall be permitted to be reduced to 6 inches (152 mm) from combustible materials and ½ of an inch (12.7 mm) from combustible construction. [NFPA 91:4.7.2.1]

506.10.4 Clearance Increases. All duct systems and system components operating at temperatures above 140°F (60°C) shall have clearances from stored combustible materials or combustible construction not less than those listed in Table 506.10.4. [NFPA 91:4.7.3]

TABLE 506.10.4
BASIC MINIMUM CLEARANCES TO UNPROTECTED SURFACES
[NFPA 91: TABLE 4.7.3]

DUCT GAS TEMPERATURE	LARGEST DUCT DIMENSION (inches)	CLEARANCE (inches)
140°F – 600°F incl.	8	8
	>8	12
>600°F – 900°F incl.	8	18
	>8	24
>900°F	All ducts lined with refractory	24

For SI units: 1 inch = 25.4 mm, °C=(°F-32)/1.8

506.10.4.1 Temperatures Over 900°F. Ducts handling materials at temperatures in excess of 900°F (482°C) shall be lined with refractory material or the equivalent. [NFPA 91:4.7.3.1]

506.10.4.2 Clearance Reduction. When stored combustible materials or combustible construction are protected from ductwork in accordance with Section 506.11, the clearance established in Table 506.10.4 shall be permitted to be reduced in accordance with Table 506.11, but not to less than specified in Section 506.10. [NFPA 91:4.7.3.2]

506.11 Clearance Reduction Methods. It shall be permitted to protect stored combustible material or combustible construction from ductwork in accordance with Table 506.11 and Section 506.11.1 through Section 506.11.6. In no case shall the clearance between the duct and the combustible surface be reduced below that allowed in Table 506.11. [NFPA 91:4.7.4, 4.7.4.1]

506.11.1 Spacers and Ties. Spacers and ties for protection materials shall be of noncombustible material and shall not be installed on the duct side of the protection system. [NFPA 91:4.7.4.2]

506.11.2 Wool Batts Insulation. Mineral wool batts (blanket or board) shall have a density of not less than 8 pounds per cubic feet (lb/ft^3) (128 kg/m^3) and have a melting point of not less than 1500°F (816°C). [NFPA 91:4.7.4.3]

506.11.3 Insulation Board. Insulation board used as a part of a clearance-reduction system shall meet the following criteria:

(1) Have a thermal conductivity of 1 British thermal unit inch per hour square foot degree Fahrenheit [Btu•in/(h•ft^2•°F)] [0.14 W/(m•K)] or less.

(2) Be formed of noncombustible material. [NFPA 91:4.7.4.4]

506.11.4 Duct and Thermal Shield. With all clearance reduction systems, not less than 1 inch (25.4 mm) clear space shall be provided between the duct and the thermal shield. [NFPA 91:4.7.4.5]

506.11.5 Thermal Shield and Combustible Surface. When using clearance reduction systems that include an air gap, not less than 1 inch (25.4 mm) clear space shall be provided between the thermal shield and the combustible surface. [NFPA 91:4.7.4.6]

506.11.6 Reduce Clearance with Air Gaps. When using clearance reduction systems that include an air gap between the combustible surface and the selected means of protection, air circulation shall be provided by one of the methods in accordance with Section 506.11.6.1 through Section 506.11.6.2. [NFPA 91:4.7.4.7]

506.11.6.1 Air Circulation. Air circulation shall be permitted to be provided by leaving all edges of the wall protecting system open with at least a 1 inch (25.4 mm) air gap. [NFPA 91:4.7.4.7.1]

506.11.6.2 Single Flat Wall. If the means for protection is mounted on a single flat wall away from corners, air circulation shall be permitted to be provided by one of the following:

(1) Leaving only the top and bottom edges open to circulation by maintaining the 1 inch (25.4 mm) air gap.

(2) Leaving the top and both side edges open to circulation by maintaining the 1 inch (25.4 mm) air gap. [NFPA 91:4.7.4.7.2]

506.11.6.3 Thermal Shielding. Thermal shielding that covers two walls in a corner shall be permitted to be open at the top and bottom edges with not less than 1 inch (25.4 mm) air gap. [NFPA 91:4.7.4.7.3]

TABLE 506.11
REDUCTION OF DUCT CLEARANCE WITH SPECIFIED FORMS OF PROTECTION
[NFPA 91: TABLE 4.7.4]

FORM OF PROTECTION*	MAXIMUM ALLOWABLE REDUCTION IN CLEARANCE (percent)	
	AS WALL PROTECTOR OR VERTICAL SURFACE	AS CEILING PROTECTOR OR HORIZONTAL SURFACE
3½ inch thick masonry wall without ventilated air space	33	None
½ inch thick noncombustible insulation board over 1 inch glass fiber or mineral wool batts without ventilated air space	50	33
0.024 inch (24 gauge) sheet metal over 1 inch glass fiber or mineral wool batts reinforced with wire, or equivalent on rear face with at least a 1 inch air gap	66	66
3½ inch (90 mm) thick masonry wall with at least a 1 inch air gap	66	None
0.024 inch (24 gauge) sheet metal with at least a 1 inch air gap	66	50
½ inch thick noncombustible insulation board with at least a 1 inch air gap	66	50
0.024 inch (24 gauge) sheet metal with ventilated air space over at least 0.024 inch (24 gauge) sheet metal with at least a 1 inch air gap	66	50
1 inch glass fiber or mineral wool batts sandwiched between two sheets of 0.024 inch (24 guage) sheet metal with at least a 1 inch air gap	66	50

For SI units: 1 inch = 25.4 mm

* Clearance reduction applied to and covering all combustible surfaces within the distance specified as required clearance with no protection in Table 506.10.4.

Part II - Commercial Hoods and Kitchen Ventilation.

507.0 General Requirements.

507.1 Type I Hood Exhaust System. Exhaust systems serving Type I hoods shall comply with Section 507.0 through Section 518.0.

507.2 Exhaust System. Cooking equipment used in processes producing smoke or grease-laden vapors shall be equipped with an exhaust system that is in accordance with the equipment and performance requirements of this chapter. [NFPA 96:4.1.1] Such equipment and its performance shall be maintained in accordance with the requirements of this chapter during periods of operation of the cooking equipment. [NFPA 96:4.1.2] The following equipment shall be kept in working condition:

(1) Cooking equipment
(2) Hoods
(3) Ducts (where applicable)
(4) Fans
(5) Fire-extinguishing equipment
(6) Special effluent or energy control equipment [NFPA 96:4.1.3]

Maintenance and repairs shall be performed on components at intervals necessary to maintain good working conditions as follows:

(1) Airflows shall be maintained. [NFPA 96:4.1.4]
(2) The responsibility for inspection, testing, maintenance, and cleanliness of the ventilation control and fire protection of the commercial cooking operations shall ultimately be that of the owner of the system, provided that this responsibility has not been transferred in written form to a management company, tenant, or other party. [NFPA 96:4.1.5]
(3) Solid-fuel cooking equipment shall comply with the requirements of Section 517.0. [NFPA 96:4.1.6]
(4) Multitenant applications shall require the concerted cooperation of design, installation, operation, and maintenance responsibilities by tenants and by the building owner. [NFPA 96:4.1.7]
(5) Interior surfaces of the exhaust system shall be accessible for cleaning and inspection purposes. [NFPA 96:4.1.8]
(6) Cooking equipment used in fixed, mobile, or temporary concessions, such as trucks, buses, trailers, pavilions, tents, or a form of roofed enclosure, shall be in accordance with this chapter unless otherwise exempted by the Authority Having Jurisdiction. [NFPA 96:4.1.9]

507.3 Listed Devices. Penetrations shall be sealed with listed devices in accordance with the requirements of Section 507.3.1.

507.3.1 Penetration. Devices that require penetration of a Type I hood or grease duct, such as pipe and conduit penetration fittings and fasteners, shall be listed in accordance with UL 710 or UL 1978. Seams, joints, and penetrations of the hood enclosure shall comply with Section 508.3.2. Seams, joints, and penetrations of the ductwork shall comply with Section 510.5.3.

507.4 Clearance. Where enclosures are not required, hoods, grease removal devices, exhaust fans, and ducts shall have a clearance of not less than 18 inches (457 mm) to combustible material, 3 inches (76 mm) to limited-combustible material, and 0 inches (0 mm) to noncombustible material. [NFPA 96:4.2.1]

507.4.1 Listed. Where a hood, duct, or grease removal device is listed for clearances less than those in accordance with Section 507.4, the listing requirements shall be permitted. [NFPA 96:4.2.2]

507.4.2 Clearance Reduction. Where a clearance reduction system consisting of 0.013 of an inch (0.33 mm) (28 gauge) sheet metal spaced out 1 inch (25.4 mm) on noncombustible spacers is provided, there shall be not less than 9 inches (229 mm) clearance to combustible material. [NFPA 96:4.2.3.1]

507.4.2.1 Mineral Wool Batts or Ceramic Fiber Blanket. Where a clearance reduction system consisting of 0.027 of an inch (0.686 mm) (22 gauge) sheet metal on 1 inch (25.4 mm) mineral wool batts or ceramic fiber blanket reinforced with wire mesh or equivalent spaced out 1 inch (25.4 mm) on noncombustible spacers is provided, there shall be not less than 3 inches (76 mm) clearance to combustible material. [NFPA 96:4.2.3.2]

507.4.2.2 Field-Applied Grease Duct Enclosure. Where a clearance reduction system consisting of a listed and labeled field-applied grease duct enclosure material, system, product, or method of construction specifically evaluated for such purpose in accordance with ASTM E2336, the required clearance shall be in accordance with the listing. [NFPA 96:4.2.3.3]

507.4.2.3 Zero Clearance. Zero clearance to limited-combustible materials shall be permitted where protected by one of the following:

(1) Metal lath and plaster.
(2) Ceramic tile.
(3) Quarry tile.
(4) Other noncombustible materials or assembly of noncombustible materials that are listed for the purpose of reducing clearance.
(5) Other materials and products that are listed for the purpose of reducing clearance. [NFPA 96:4.2.3.4]

507.4.3 Clearance Integrity. In the event of damage, the material or product shall be repaired and restored to meet its intended listing or clearance requirements and shall be approved by the Authority Having Jurisdiction. [NFPA 96:4.2.4.1]

» **507.4.3.1 Fire.** In the event of a fire within a kitchen exhaust system, the duct and its enclosure (rated shaft, factory-built grease duct enclosure, or field-applied grease duct enclosure) shall be inspected by qualified personnel to determine whether the duct and protection method are structurally sound, capable of maintaining their fire protection function, and in accordance with this chapter for continued operation. [NFPA 96:4.2.4.2]

» **507.4.3.2 Required Protection.** Protection shall be provided on the wall from the bottom of the hood to the floor, or to the top of the noncombustible material extending to the floor, to the same level as required in Section 507.4. [NFPA 96:4.2.4.3]

» **507.4.3.3 Protection Methods.** The protection methods for ducts to reduce clearance shall be applied to the combustible or limited-combustible construction, not to the duct itself. [NFPA 96:4.2.4.4]

» **507.4.4 Factory Built.** Factory-built grease duct enclosures shall be protected with a through-penetration firestop system classified in accordance with ASTM E814 or UL 1479 having an "F" and "T" rating equal to the fire resistance rating of the assembly being penetrated from the point at which the duct penetrates a ceiling, wall, or floor to the outlet terminal. The factory-built grease duct protection system shall be listed in accordance with UL 2221. The factory-built grease duct protection system shall be installed in accordance with the manufacturer's installation instructions and the listing requirements. [NFPA 96:4.3.3]

» **507.4.5 Field Applied.** Field-applied grease duct enclosures shall be protected with a through penetration firestop system classified in accordance with ASTM E814 or UL 1479 having an "F" and "T" rating equal to the fire resistance rating of the assembly being penetrated. The surface of the field fabricated grease duct shall be continuously covered on sides from the point at which the duct enclosure penetrates a ceiling, wall, or floor to the outlet terminal. The field-applied grease duct shall be listed in accordance with ASTM E2336 and installed in accordance with the manufacturer's installation instructions and the listing requirements. [NFPA 96:4.3.1]

» **507.4.6 Both Field-Applied and Factory Built.** Field-applied grease duct enclosures and factory-built grease duct enclosures shall demonstrate that they provide mechanical and structural integrity, resiliency, and stability where subjected to expected building environmental conditions, duct movement under general operating conditions, and duct movement due to fire conditions. [NFPA 96:4.3.4]

» **507.4.6.1 Physical Damage.** Measures shall be taken to prevent physical damage to a material or product used for the purpose of reducing clearances.

Exception: Where the duct is protected with a field-applied grease duct enclosure or factory-built grease duct enclosure.

507.4.6.2 Specification. The specifications of material, gauge, and construction of the duct used in the testing and listing of field-applied grease duct enclosures and factory-built grease duct enclosures shall be included as minimum requirements in their listing and installation documentation. [NFPA 96:4.3.5] «

507.4.6.3 Clearance Options. The following clearance options for which field-applied grease duct enclosures and factory-built grease duct enclosures have been successfully evaluated shall be clearly identified in their listing and installation documentation and on their labels: «

(1) Open combustible construction clearance at manufacturer's requested dimensions.

(2) Closed combustible construction clearance at manufacturer's requested dimensions, with or without specified ventilation.

(3) Rated shaft clearance at manufacturer's requested dimensions, with or without specified ventilation. [NFPA 96:4.3.6]

507.4.7 Building and Structural Contact. A duct shall be permitted to contact noncombustible floors, interior walls, and other noncombustible structures or supports, but it shall not be in contact for more than 50 percent of its surface area for each lineal foot of contact length. [NFPA 96:4.4.1] «

507.4.7.1 Corrosion Protection. Where duct contact must exceed the requirements of Section 507.4.7, the duct shall be protected from corrosion. [NFPA 96:4.4.2] «

507.4.7.2 Zero Clearance. Where the duct is listed for zero clearance to combustibles or otherwise protected with a material or product listed for the purpose of reducing clearance to zero, the duct shall be permitted to exceed the contact limits of Section 507.4.7 without additional corrosion protection. [NFPA 96:4.4.3] «

507.4.8 Clearance Between Duct and Interior Surfaces. Clearances between the duct and interior surfaces of enclosures shall be in accordance with the requirements of Section 507.4. [NFPA 96:4.5] «

507.5 Drawings. A drawing(s) of the exhaust system installation along with a copy of operating instructions for subassemblies and components used in the exhaust system, including electrical schematics, shall be on the premises. [NFPA 96:4.6] «

507.6 Notification of Change. Where required by the Authority Having Jurisdiction, notification in writing shall be given of an alteration, replacement, or relocation of an exhaust, extinguishing system or part thereof or cooking equipment. [NFPA 96:4.7] «

Satisfaction shall be provided to the Authority Having Jurisdiction that the complete exhaust system as addressed in this chapter is installed and operable in accordance with the approved design and the manufacturer's installation instructions.

508.0 Type I Hoods.

508.1 Where Required. Type I hoods shall be installed at or above commercial-type deep-fat fryers, broilers, grills, hot-top ranges, ovens, barbecues, rotisseries, and similar equipment that emits comparable amounts of smoke or grease in a food-processing establishment. For the purpose of this section, a food-processing establishment shall include a building or portion thereof used for the processing of food, but shall not include a dwelling unit.

Exceptions:

(1) Cooking appliance that is in accordance with UL 710B for reduced emissions where the grease discharge does not exceed 2.9 E-09 ounces per cubic inch (oz/in^3) (5.0 E-06 kg/m^3) where operated with a total airflow of 500 cubic feet per minute (cfm) (0.236 m^3/s).

(2) Recirculating systems listed in accordance with UL 710B and installed in accordance with Section 516.0.

508.2 Listed Type I Hood Assemblies. Listed hood assemblies shall be installed in accordance with the terms of their listing and the manufacturer's installation instructions. Listed hood assemblies shall be tested in accordance with UL 710. [NFPA 96:5.4]

508.2.1 Listed Ultraviolet Hoods. Listed ultraviolet hoods shall be installed and maintained in accordance with the terms of their listing and the manufacturer's installation instructions. Duct systems connected to ultraviolet hoods shall comply with Section 510.0. Ultraviolet hoods shall be tested and listed in accordance with UL 710 and UL 710C. [NFPA 96:5.5]

508.2.2 Construction of Listed Exhaust Hoods. Listed exhaust hoods with or without exhaust dampers shall be permitted to be constructed of materials required by the listing. [NFPA 96:5.1.6]

508.2.3 Assembly of Listed Exhaust Hoods. Listed exhaust hoods with or without exhaust dampers shall be permitted to be assembled in accordance with the listing requirements. [NFPA 96:5.1.7]

508.3 Construction of Type I Hoods. The hood or that portion of a primary collection means designed for collecting cooking vapors and residues constructed of steel shall be not less than 0.048 of an inch (1.219 mm) (No. 18 MSG), stainless steel not less than 0.036 of an inch (0.914 mm) (No. 20 MSG) in thickness, or other approved material of equivalent strength and fire and corrosion resistance. [NFPA 96:5.1.1]

Exception: Listed exhaust hoods.

508.3.1 Grease Vapor. Wall-mounted exhaust hood assemblies shall be tight fitting against the back wall as to not permit passage of grease vapor behind the hood, or between the back wall and the hood assembly. [NFPA 96:5.1.13]

508.3.2 Seams, Joints, and Penetrations. Seams, joints, and penetrations of the hood enclosure that direct and capture grease-laden vapors and exhaust gases shall have a liquid-tight continuous external weld to the hood's lower outermost perimeter. [NFPA 96:5.1.2]

Exceptions:

(1) Seams, joints, and penetrations of the hood shall be permitted to be internally welded, provided that the weld is formed smooth or ground smooth, so as to not trap grease, and is cleanable. [NFPA 96:5.1.3]

(2) Penetrations shall be permitted to be sealed by devices that are listed for such use and whose presence does not detract from the hood's or duct's structural integrity. [NFPA 96:5.1.5]

508.3.2.1 Sealed. Internal hood joints, seams, filter support frames, and appurtenances attached inside the hood shall be sealed or otherwise made greasetight. [NFPA 96:5.1.4]

508.3.3 Eyebrow-Type Hoods. Eyebrow-type hoods over gas or electric ovens shall be permitted to have a duct constructed as required in Section 510.0 from the oven flue(s) connected to the hood canopy upstream of the exhaust plenum, as shown in Figure 508.3.3. [NFPA 96:5.1.8.1]

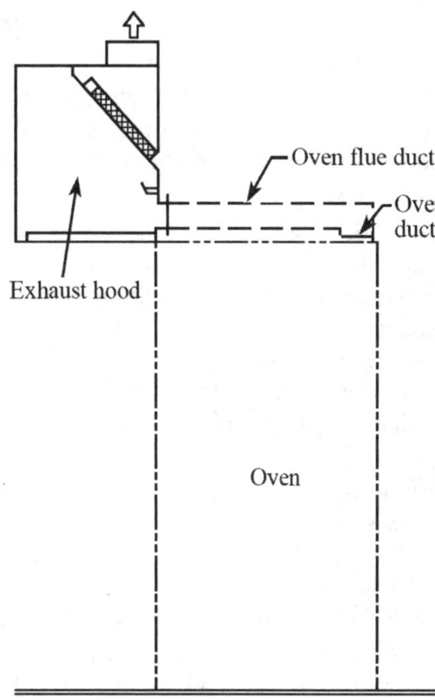

FIGURE 508.3.3
TYPICAL SECTION OF EYEBROW-TYPE HOOD
[NFPA 96: FIGURE 5.1.8.1]

508.3.3.1 Duct Connection. The duct connecting the oven flue(s) to the hood canopy shall be connected with a continuous weld or have a duct-to-duct connection. [See Figure 511.1.2(2) through Figure 511.1.2(4)] [NFPA 96:5.1.8.2]

508.3.4 Insulation. Insulation materials other than electrical insulation shall have a flame spread index of not more than 25, where tested in accordance with ASTM E84 or UL 723. Adhesives or cements used in the installation of insulating materials shall be in accordance with this section where tested with the specific insulating material. [NFPA 96:5.1.9, 5.1.10]

»**508.3.5 Exhaust Hood Assemblies with Integrated Supply-Air Plenums.** The construction and size of exhaust hood assemblies with integrated supply air plenums shall be in accordance with the requirements of Section 508.1 through Section 508.5. [NFPA 96:5.3.1]

»**508.3.5.1 Outer Shell.** The construction of the outer shell or the inner exhaust shell shall be in accordance with Section 508.1 through Section 508.3.4. [NFPA 96:5.3.2]

»**508.3.5.2 Inner Shell.** Where the outer shell is welded, the inner shell shall be of greasetight construction. [NFPA 96:5.3.3]

»**508.3.5.3 Fire Dampers.** A fire-actuated damper shall be installed in the supply air plenum at each point where a supply air duct inlet or a supply air outlet penetrates the continuously welded shell of the assembly. [NFPA 96:5.3.4.1]

»**508.3.5.3.1 Listing.** The fire damper shall be listed for such use or be part of a listed exhaust hood with or without exhaust damper. [NFPA 96:5.3.4.2]

»**508.3.5.3.2 Actuating Temperature.** The actuation device shall have a temperature rating not to exceed 286°F (141°C). [NFPA 96:5.3.4.3]

»**508.3.5.3.3 Exemption.** Supply air plenums that discharge air from the face rather than from the bottom or into the exhaust hood and that are isolated from the exhaust hood by the continuously welded shell extending to the lower outermost perimeter of the entire hood assembly shall not require a fire-actuated damper. [NFPA 96:5.3.4.4]

»**508.4 Supports.** Hoods shall be secured in place ***[OSHPD 1,*** || ***1R, 2, 4 & 5]*** *to resist the lateral loads given in the California Building Code, Title 24, Part 2* by noncombustible supports. The supports shall be capable of supporting the expected weight of the hood and plus 800 pounds (362.9 kg).

»**508.5 Hood Size.** Hoods shall be sized in accordance with the airflow capacity in accordance with Section 508.5.1.1 and installed to provide for the removal of heat, and capture and removal of grease-laden vapors in accordance with Section 511.2.2.

»**508.5.1 Canopy Size and Location.** For canopy type commercial cooking hoods, the inside edge thereof shall overhang or extend a horizontal distance of not less than 6 inches (152 mm) beyond the edge of the cooking surface on open sides, and the vertical distance between the lip of the hood and the cooking surface shall not exceed 4 feet (1219 mm).

Exception: Listed exhaust hoods are to be installed in accordance with the terms of their listings and the manufacturer's installation instructions.

»**508.5.1.1 Capacity of Hoods.** Canopy-type commercial cooking hoods shall exhaust through the hood with a quantity of air not less than determined by the application in accordance with Section 508.5.1.2 through Section 508.5.1.5. The exhaust quantity shall be the net exhaust from the hood determined in accordance with Equation 508.5.1.1. The duty level for the hood shall be the duty level of the appliance that has the highest (heaviest) duty level of appliances installed underneath the hood.

Exception: Listed exhaust hoods installed in accordance with the manufacturer's installation instructions.

$$E_{NET} = E_{HOOD} - MA_{ID} \qquad \text{(Equation 508.5.1.1)}$$

Where:

E_{NET} = net hood exhaust

E_{HOOD} = total hood exhaust

MA_{ID} = makeup air, internal discharge

508.5.1.2 Extra-Heavy-Duty Cooking Appliances. The minimum net airflow for hoods used for solid fuel cooking appliances such as charcoal, briquette, and mesquite to provide the heat source for cooking shall be in accordance with Table 508.5.1.2. «

TABLE 508.5.1.2
EXTRA-HEAVY-DUTY COOKING APPLIANCE AIRFLOW «

TYPE OF HOOD	AIRFLOW (cubic foot per minute per linear foot of hood)
Backshelf/pass-over	Not permitted
Double island canopy (per side)	550
Eyebrow	Not permitted
Single island canopy	700
Wall-mounted canopy	550

For SI units: 1 cubic foot per minute = 0.00047 m³/s, 1 foot = 304.8 mm

508.5.1.3 Heavy-Duty Cooking Appliances. The minimum net airflow for hoods used for cooking appliances such as gas under-fired broilers, gas chain (conveyor) broilers, electric and gas wok ranges, and electric and gas over-fired (upright) broilers shall be in accordance with Table 508.5.1.3. «

TABLE 508.5.1.3
HEAVY-DUTY COOKING APPLIANCE AIRFLOW «

TYPE OF HOOD	AIRFLOW (cubic foot per minute per linear foot of hood)
Backshelf/pass-over	400
Double island canopy (per side)	400
Eyebrow	Not permitted
Single island canopy	600
Wall-mounted canopy	400

For SI units: 1 cubic foot per minute = 0.00047 m³/s, 1 foot = 304.8 mm

»**508.5.1.4 Medium-Duty Cooking Appliances.** The minimum net airflow for hoods used for cooking appliances such as electric and gas hot-top ranges, gas open-burner ranges (with or without oven), electric and gas flat griddles, electric and gas double-sided griddles, electric and gas fryers (including open deep fat fryers, donut fryers, kettle fryers, and pressure fryers), and electric and gas conveyor pizza ovens shall be in accordance with Table 508.5.1.4.

» **TABLE 508.5.1.4**
MEDIUM-DUTY COOKING APPLIANCE AIRFLOW

TYPE OF HOOD	AIRFLOW (cubic foot per minute per linear foot of hood)
Backshelf/pass-over	300
Double island canopy (per side)	300
Eyebrow	250
Single island canopy	500
Wall-mounted canopy	300

For SI units: 1 cubic foot per minute = 0.00047 m^3/s, 1 foot = 304.8 mm

»**508.5.1.5 Light-Duty Cooking Appliances.** The minimum net airflow for hoods used for cooking appliances such as gas and electric ovens (including standard, bake, roasting, revolving, retherm, convection, combination convection/steamer, rotisserie, countertop conveyorized baking/finishing, deck, and pastry), discrete element ranges (with or without oven), electric and gas steam-jacketed kettles less than 20 gallons (76 L), electric and gas pasta cookers, electric and gas compartment steamers (both pressure and atmospheric), electric and gas cheese melters, electric and gas tilting skillets (braising pans) electric and gas rotisseries, and electric and gas salamanders shall be in accordance with Table 508.5.1.5.

» **TABLE 508.5.1.5**
LIGHT-DUTY COOKING APPLIANCE AIRFLOW

TYPE OF HOOD	AIRFLOW (cubic foot per minute per linear foot of hood)
Backshelf/pass-over	250
Double island canopy (per side)	250
Eyebrow	250
Single island canopy	400
Wall-mounted canopy	200

For SI units: 1 cubic foot per minute = 0.00047 m^3/s, 1 foot = 304.8 mm

»**508.5.2 Noncanopy-Type Hoods.** Noncanopy-type commercial cooking hoods shall be installed and sized in accordance with the manufacturer's installation instructions, and Section 508.5.2.1 and Section 508.5.2.2.

Exception: Listed hood assemblies designed and installed specifically for the intended use.

508.5.2.1 Installation. Noncanopy-type commercial cooking hoods shall be installed with the edge of the hood set back not more than 1 foot (305 mm) from the edge of the cooking surface, and the vertical distance between the lip of the hood and the cooking surface shall not exceed 3 feet (914 mm). «

508.5.2.2 Capacity. In addition to other requirements for hoods specified in this section, the volume of air exhausting through a noncanopy-type hood to the duct system shall be not less than 300 cubic feet per minute per lineal foot [$(ft^3/min)/ft$)] [0.464 $(m^3/s)/m$] of cooking equipment. Listed noncanopy exhaust hoods and filters shall be sized and installed in accordance with the terms of their listing and the manufacturer's installation instructions. «

508.5.3 Labeling. Type I hoods shall bear a label indicating the exhaust flow rate in cubic feet per minute per lineal foot [$(m^3/s)/m$]. «

508.6 Solid-Fuel Hood Assemblies. Where solid-fuel cooking equipment is to be used, the solid-fuel hood assembly shall be in accordance with Section 517.0. «

508.7 Exhaust Outlets. An exhaust outlet within an unlisted hood shall be located so as to optimize the capture of particulate matter. Each outlet shall serve not more than a 12 foot (3658 mm) section of an unlisted hood. «

509.0 Grease Removal Devices in Hoods.

509.1 Grease Removal Devices. Listed grease filters or other listed grease removal devices intended for use with commercial cooking operations shall be provided. Listed grease filters and grease removal devices that are removable, but not an integral component of a specific listed exhaust hood, shall be listed in accordance with UL 1046. [NFPA 96:6.1.1, 6.1.2]

509.1.1 Grease Filters, Mesh-Type. Mesh filters shall not be used unless evaluated as an integral part of a listed exhaust hood or listed in conjunction with a primary filter in accordance with UL 1046. [NFPA 96:6.1.3]

509.2 Installation. The distance between the grease removal device and the cooking surface shall be not less than 18 inches (457 mm). [NFPA 96:6.2.1.1]

509.2.1 Vertical Distance. Where grease removal devices are used in conjunction with charcoal or charcoal-type broilers, including gas or electrically heated charbroilers, a vertical distance of not less than 4 feet (1219 mm) shall be maintained between the lower edge of the grease removal device and the cooking surface. [NFPA 96:6.2.1.2]

Exceptions:

(1) For cooking equipment without exposed flame and where flue gases bypass grease removal devices, the minimum vertical distance shall be permitted to be reduced to not less than 6 inches (152 mm). [NFPA 96:6.2.1.3]

(2) Where a grease removal device is listed for separation distances less than those required in Section 509.2 and Section 509.2.1, the listing requirements shall be permitted. [NFPA 96:6.2.1.4]

(3) Grease removal devices supplied as part of listed hood assemblies shall be installed in accordance with the terms of the listing and the manufacturer's installation instructions. [NFPA 96:6.2.1.5]

509.2.2 Grease Removal Device Protection. Where the distance between the grease removal device and the appliance flue outlet (heat source) is less than 18 inches (457 mm), grease removal devices shall be protected from combustion gas outlets and from direct flame impingement occurring during normal operation of cooking appliances producing high flue gas temperatures. [NFPA 96:6.2.2.1]

509.2.2.1 Installation. This protection shall be permitted to be accomplished by the installation of a steel or stainless steel baffle plate between the heat source and the grease removal device. [NFPA 96:6.2.2.2]

509.2.2.2 Size and Location. The baffle plate shall be sized and located so that flames or combustion gases shall travel a distance not less than 18 inches (457 mm) from the heat source to the grease removal device. [NFPA 96:6.2.2.3]

509.2.2.3 Clearance. The baffle shall be located not less than 6 inches (152 mm) from the grease removal devices. [NFPA 96:6.2.2.4]

509.2.3 Grease Filters. Grease filters shall be listed and constructed of steel or other non-combustible material, and shall be of rigid construction that will not distort or crush under normal operation, handling, cleaning, or replacement. ***[SFM]*** *For applications listed in Section 1.11.0 regulated by the Office of the State Fire Marshal, grease filters shall be Class I when tested in accordance with the test method in Title 24, Part 12, Chapter 12-71, SFM Standard 12-71-1.*

509.2.3.1 Arrangement. Grease filters shall be arranged so that exhaust air passes through the grease filters. [NFPA 96:6.2.3.4]

509.2.3.2 Accessibility. Grease filters shall be easily accessible for removal. [NFPA 96:6.2.3.5]

509.2.3.3 Angled Installation. Grease filters shall be installed at an angle not less than 45 degrees (0.79 rad) from the horizontal. [NFPA 96:6.2.3.6]

509.2.3.4 [DPH] *Grease filters shall be Class I when tested in accordance with the test method in SFM 12-71-1.*

509.2.4 Grease Drip Trays. Grease filters shall be equipped with a grease drip tray beneath their lower edges. [NFPA 96:6.2.4.1]

509.2.4.1 Size and Pitch. Grease drip trays shall be kept to the minimum size needed to collect grease and shall be pitched to drain into an enclosed metal container having a capacity not exceeding 1 gallon (4 L). [NFPA 96:6.2.4.2, 6.2.4.3]

509.2.5 Grease Filter Orientation. Grease filters that require a specific orientation to drain grease shall be clearly so designated, or the hood shall be constructed so that filters cannot be installed in the wrong orientation. [NFPA 96:6.2.5]

509.3 Solid-Fuel Grease Removal Devices. Where solid-fuel cooking equipment is provided with grease removal devices, these devices shall be in accordance with Section 517.0.

510.0 Exhaust Duct Systems.

510.1 General. Ducts shall not pass through fire walls. [NFPA 96:7.1.1]

510.1.1 Fire Hazards. Ducts shall lead as directly to the exterior of the building, so as not to unduly increase a fire hazard. [NFPA 96:7.1.2]

510.1.2 Interconnection. Duct systems shall not be interconnected with a building ventilation or exhaust system. [NFPA 96:7.1.3]

510.1.3 Duct Installation. Ducts shall be installed with not less than 2 percent slope on horizontal runs up to 75 feet (22 860 mm) and not less than 8 percent slope on horizontal runs more than 75 feet (22 860 mm). Factory-built grease ducts shall be permitted to be installed in accordance with the listing and the manufacturer's installation instructions. Horizontal ducts shall be provided with access in accordance with Section 510.3.3.

Drains shall be provided at low points in horizontal ducts. Where provided, drains shall be continuously welded to the exhaust duct or listed grease duct drains in accordance with the terms of the listing and the manufacturer's installation instructions.

Ducts shall be installed without forming dips or traps. In manifold (common duct) systems, the lowest end of the main duct shall be connected flush on the bottom with the branch duct. [NFPA 96:7.1.4 – 7.1.4.5]

510.1.4 Accessibility. Openings required for accessibility shall be in accordance with Section 510.3 through Section 510.3.2. [NFPA 96:7.1.5]

510.1.5 Sign. A sign shall be placed on access panels stating the following:

ACCESS PANEL – DO NOT OBSTRUCT [NFPA 96:7.1.6]

510.1.6 Bracing and Supports. Duct bracing and supports shall be of noncombustible material, securely attached to the structure and designed to carry gravity and lateral loads within the stress limitations of the building code. Bolts, screws, rivets, and other mechanical fasteners shall not penetrate duct walls.

510.1.7 Type I Exhaust Duct Systems. Listed grease ducts shall be installed in accordance with the terms of their listings and manufacturer's installation instructions. [NFPA 96:7.1.7] «

510.2 Clearance. Clearance between ducts and combustible materials shall be provided in accordance with the requirements of Section 507.4. [NFPA 96:7.2]

510.3 Openings. Openings shall be provided at the sides or at the top of the duct, whichever is more accessible, and at changes of direction. Openings shall be protected by approved access constructed and installed in accordance with the requirements of Section 510.3.7. [NFPA 96:7.3.1, 7.3.2]

Exception: Openings shall not be required in portions of the duct that are accessible from the duct entry or discharge. [NFPA 96:7.3.3]

510.3.1 Access Panel. For hoods with dampers in the exhaust or supply collar, an access panel for cleaning and inspection shall be provided in the duct or the hood within 18 inches (457 mm) of the damper. [NFPA 96:7.3.4]

Exception: Dampers that are accessible from under the hood.

510.3.2 Access for Cleaning and Inspection. Exhaust fans with ductwork connected to both sides shall have access for cleaning and inspection within 3 feet (914 mm) of each side of the fan. Wall-mounted exhaust fans shall have access for cleaning and inspection within 3 feet (914 mm) of the exhaust fan. [NFPA 96:7.3.7, 7.3.8]

510.3.3 Horizontal Ducts. On horizontal ducts, not less than one 20 inch by 20 inch (508 mm by 508 mm) opening shall be provided for personnel entry. [NFPA 96:7.4.1.1]

510.3.3.1 Cleaning. Where an opening of the size specified in Section 510.3.3 is not possible, openings large enough to permit thorough cleaning shall be provided at 12 feet (3658 mm) intervals. [NFPA 96:7.4.1.2]

510.3.3.2 Safe Access and Work Platform. Where not easily accessible from a 10 foot (3048 mm) stepladder, openings on horizontal grease duct systems shall be provided with safe access and a work platform. [NFPA 96:7.4.1.3]

510.3.3.3 Support. Support systems for horizontal grease duct systems 24 inches (610 mm) and larger in a cross-sectional dimension shall be designed for the weight of the ductwork plus 800 pounds (362.9 kg) at a point in the duct systems. [NFPA 96:7.4.1.4]

510.3.4 Vertical Ducts. On vertical ductwork where personnel entry is possible, access shall be provided at the top of the vertical riser to accommodate descent. [NFPA 96:7.4.2.1]

510.3.4.1 Access. Where personnel entry is not possible, an access for cleaning shall be provided on each floor. [NFPA 96:7.4.2.2]

510.3.4.2 Safe Access and Work Platform. Where not easily accessible from a 10 foot (3048 mm) stepladder, openings on vertical grease ducts shall be provided with safe access and a work platform. [NFPA 96:7.4.2.3]

510.3.5 Nonlisted Ductwork. On nonlisted ductwork, the edge of the opening shall be not less than 1½ inches (38 mm) from all outside edges of the duct or welded seams. [NFPA 96:7.4.1.5]

510.3.6 Access Panels. Access panels shall be of the same material and thickness as the duct. Access panels shall have a gasket or sealant that is rated for 1500°F (816°C) and shall be greasetight. Fasteners, such as bolts, weld studs, latches, or wing nuts, used to secure the access panels shall be carbon steel or stainless steel and shall not penetrate duct walls.

Exception: Listed grease duct access door assemblies (access panels) shall be installed in accordance with the terms of the listings and the manufacturer's installation instructions. [NFPA 96:7.4.3]

510.3.6.1 Within an Enclosure. Where openings are located in ducts within an enclosure, the access panel including its components shall be of the same fire rating as the enclosure.

510.3.7 Fire Protection System Devices. Openings for installation, servicing, and inspection of listed fire protection system devices and for duct cleaning shall be provided in ducts and enclosures and shall be in accordance with the requirements of Section 510.3 through Section 510.3.2 and Section 510.7.7. Enclosure openings required to reach access panels in the ductwork shall be large enough for removal of the access panel through the enclosure opening. [NFPA 96:7.4.4]

510.4 Listed Grease Ducts. Listed grease ducts shall be installed in accordance with the terms of the listing and the manufacturer's installation instructions. [NFPA 96:7.1.7]

510.4.1 Factory-Built Grease Ducts. Factory-built grease ducts in accordance with UL 1978 shall be permitted to incorporate non-welded joints in accordance with their listings.

510.5 Other Grease Ducts. Other grease ducts shall comply with the requirements of Section 510.5.1 through Section 510.5.5. [NFPA 96:7.5]

510.5.1 Materials. Ducts shall be constructed of and supported by carbon steel not less than 0.060 of an inch (1.524 mm) (No. 16 MSG) in thickness or stainless steel not less than 0.048 of an inch (1.219 mm) (No. 18 MSG) in thickness. [NFPA 96:7.5.1.1]

510.5.2 Factory-Built Grease Ducts. Factory-built grease ducts listed in accordance with UL 1978 shall be permitted to use materials in accordance with their listing. [NFPA 96:7.5.1.2]

510.5.3 Installation. Seams, joints, penetrations, and duct-to-hood collar connections shall have a liquid-tight continuous external weld. [NFPA 96:7.5.2.1]

Exceptions:

(1) Factory-built grease duct listed in accordance with UL 1978 shall be permitted to incorporate non-welded joint construction in accordance with their listing. [NFPA 96:7.5.2.1.1]

(2) Duct-to-hood collar connections as shown in Figure 510.5.3 shall not require a liquid-tight continuous external weld. [NFPA 96:7.5.2.2]

(3) Penetrations shall be permitted to be sealed by other listed devices that are tested to be greasetight and are evaluated under the same conditions of fire severity as the hood or enclosure of listed grease extractors and whose presence does not detract from the hood or the duct's structural integrity. [NFPA 96:7.5.2.3]

(4) Internal welding shall be permitted, provided the joint is formed or ground smooth and is readily accessible for inspection. [NFPA 96:7.5.2.4]

510.5.3.1 Duct Leakage Test. Prior to the use of or concealment of a portion of a grease duct system, a leakage test shall be performed to determine that all welded joints and seams are liquid tight. [NFPA 96:7.5.2.1.2]

510.5.3.2 Welded Duct Connection. Duct-to-duct connection shall be as follows:

(1) Telescoping joint, as shown in Figure 510.5.3.2(1).

(2) Bell-type joint, as shown in Figure 510.5.3.2(2).

(3) Flange with edge weld, as shown in Figure 510.5.3.2(3).

(4) Flange with filled weld, as shown in Figure 510.5.3.2(4). [NFPA 96:7.5.5.1]

510.5.4 Butt Welded Connections. Butt welded connections shall not be permitted. [NFPA 96:7.5.5.2]

510.5.5 Telescoping and Bell-Type Connections. For telescoping and bell-type connections, the inside duct section shall be uphill of the outside duct section. [NFPA 96:7.5.5.3]

510.5.6 Duct Leakage Test. Prior to the use of or concealment of a grease duct system, a leakage test shall be performed to determine that welded joints and seams are liquid tight. The leakage test shall consist of a light test, water pressure test, or an approved equivalent test. The permit holder shall be responsible for providing the necessary equipment and for performing the test. Such test shall be conducted in accordance with ASHRAE 154.

510.6 Exterior Installations. The exterior portion of the ductwork shall be vertical where possible and shall be installed and supported on the exterior of a building. Bolts, screws, rivets, and other mechanical fasteners shall not penetrate duct walls. Clearance of a duct shall be in accordance with Section 507.4. [NFPA 96:7.6.1 – 7.6.3]

510.6.1 Weather Protection. Ducts shall be protected on the exterior by paint or other suitable weather-protective coating. Ducts constructed of stainless steel shall not be required to have additional paint or weather-protective coatings. Ductwork subject to corrosion shall have minimal contact with the building surface. [NFPA 96:7.6.4 – 7.6.6]

510.7 Interior Installations. In buildings more than one story in height, and in one-story buildings where the roof-ceiling assembly is required to have a fire resistance rating, the ducts shall be enclosed in a continuous enclosure extending from the lowest fire-rated ceiling or floor above the hood, through concealed spaces, to or through the roof, to maintain the integrity of the fire separations required by the applicable building code provisions. The enclosure shall be sealed around the duct at the point of penetration of the first fire-rated barrier after the hood, to maintain the fire resistance rating of the enclosure. The enclosure shall be vented to the exterior of the building through weather-protected openings. [NFPA 96:7.7.1.2 – 7.7.1.4]

Exception: The continuous enclosure provisions shall not be required where a field-applied grease duct enclosure or a factory-built grease duct enclosure (see Section 507.4.4 through Section 507.4.6) is protected with a listed duct-through-penetration protection system equivalent to the fire resistance rating of the assembly being penetrated, and where the materials are installed in accordance with the conditions of the listings and the manufacturer's installation instructions and are acceptable to the Authority Having Jurisdiction. [NFPA 96:7.7.1.5]

510.7.1 Less than Four Stories. Buildings less than four stories in height shall have an enclosure with a fire resistance rating of not less than 1 hour. [NFPA 96:7.7.2.1.1]

510.7.2 Four Stories or More. Buildings four stories or more in height shall have an enclosure with a fire resistance rating of not less than 2 hours. [NFPA 96:7.7.2.1.2]

510.7.3 Clearance. Clearance from the duct or the exhaust fan to the interior surface of enclosures of combustible construction shall be not less than 18 inches (457 mm), and clearance from the duct to the interior surface of enclosures of noncombustible or limited-combustible construction shall be not less than 6 inches (152 mm). Provisions for reducing clearances as described in Section 507.4 through Section 507.4.3.3 shall not be applicable to enclosures. [NFPA 96:7.7.2.2.1 – 7.7.2.2.3]

Exception: Clearance from the outer surfaces of field-applied grease duct enclosures and factory-built grease duct enclosures to the interior surfaces of construction installed around them shall be permitted to be reduced where the field-applied grease duct enclosure materials and the factory-built grease duct enclosures are installed in accordance with the conditions of the listings and the manufacturer's installation instructions and are acceptable to the Authority Having Jurisdiction. [NFPA 96:7.7.2.2.4]

510.7.4 Mechanical and Structural Integrity. Field-applied grease duct enclosures and factory-built grease duct enclosures shall provide mechanical and structural integrity, resiliency, and stability where subjected to expected building environmental conditions, duct movement under general operating conditions, and duct movement as a result of interior and exterior fire conditions. [NFPA 96:7.7.2.2.5]

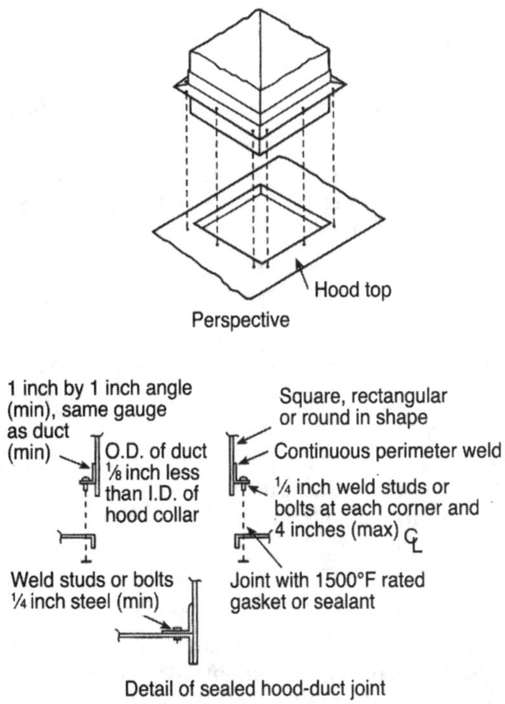

For SI units: 1 inch = 25.4 mm, °C = (°F-32)/1.8

FIGURE 510.5.3
PERMITTED DUCT-TO HOOD COLLAR CONNECTION
[NFPA 96: FIGURE 7.5.2.2]

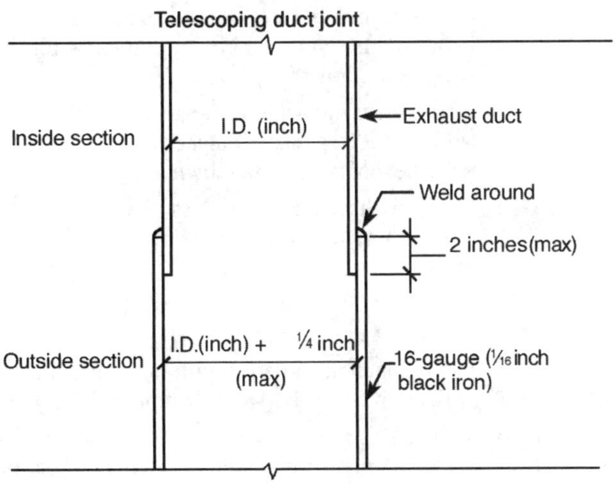

For SI units: 1 inch = 25.4 mm

Notes:

1. Duct size decreases (going upward) with each telescope.
2. Smaller (inside) duct section shall be above or uphill (on sloped duct), to be self draining into larger (outside) duct.

FIGURE 510.5.3.2(1)
TELESCOPING -TYPE DUCT CONNECTION
[NFPA 96: FIGURE 7.5.5.1(a)]

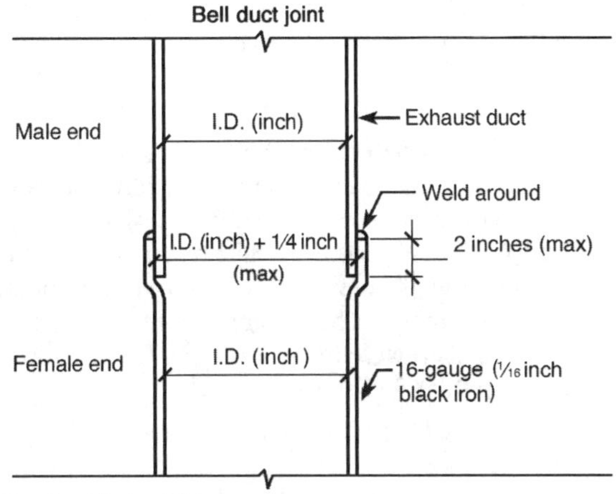

For SI units: 1 inch = 25.4 mm

Notes:

1. Duct size stays the same throughout the duct system.
2. Smaller (inside) male duct end is always above or uphill (on sloped duct), to be self draining into larger (outside) female duct end.

FIGURE 510.5.3.2(2)
BELL-TYPE DUCT CONNECTION
[NFPA 96: FIGURE 7.5.5.1(b)]

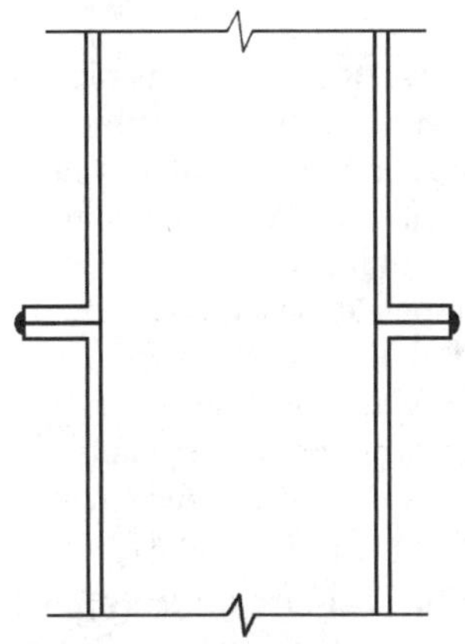

FIGURE 510.5.3.2(3)
FLANGE WITH EDGE WELD DUCT CONNECTION
[NFPA 96: FIGURE 7.5.5.1(c)]

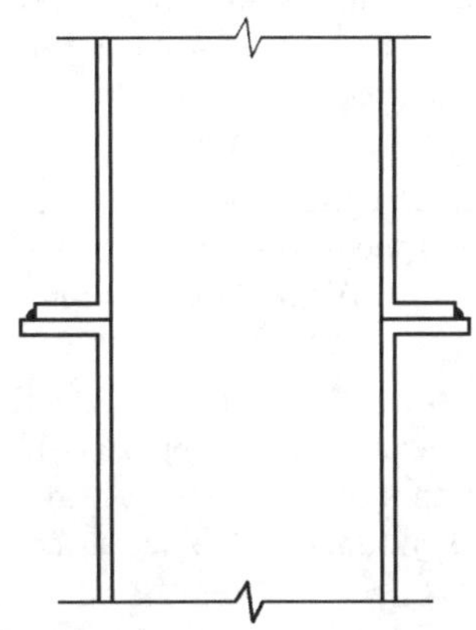

FIGURE 510.5.3.2(4)
FLANGE WITH FILLED WELD DUCT CONNECTION
[NFPA 96: FIGURE 7.5.5.1(d)]

510.7.5 Materials. For field-applied grease duct enclosures and factory-built grease duct enclosures, the materials and products shall be provided in accordance with Section 510.7.5.1 and Section 510.7.5.2.

510.7.5.1 Protection from Physical Damage. Measures shall be taken to prevent physical damage to a covering or enclosure material. Damage to the covering or enclosure shall be repaired, and the covering or enclosure shall be restored in accordance with its intended listing and fire-resistance rating, and be acceptable to the Authority Having Jurisdiction. [NFPA 96:7.7.3.1, 7.7.3.2]

510.7.5.2 Inspection. In the event of a fire within a kitchen exhaust system, the duct, the enclosure, and the covering directly applied to the duct shall be inspected by qualified personnel to determine whether the duct, the enclosure, and the covering directly applied to the duct are structurally sound, capable of maintaining their fire protection functions, approved for continued operation, and acceptable to the Authority Having Jurisdiction. [NFPA 96:7.7.3.3]

510.7.6 Listed. For listed grease ducts, see Section 510.4.

510.7.7 Fire Doors. Where openings in the enclosure walls are provided, they shall be protected by listed fire doors of proper rating. Fire doors shall be installed in accordance with NFPA 80. Openings on other listed materials or products shall be clearly identified and labeled according to the terms of the listing and the manufacturer's instructions and shall be acceptable to the Authority Having Jurisdiction. [NFPA 96:7.7.4.1 – 7.7.4.3] The fire door shall be readily accessible, aligned and of a size to allow access to the rated access panels on the ductwork. [NFPA 96:7.7.4.4]

510.7.8 Ducts with Enclosure(s). A duct system shall constitute an individual system serving exhaust hoods in one fire zone on one floor. Multiple ducts shall not be permitted in a single enclosure unless acceptable to the Authority Having Jurisdiction. [NFPA 96:7.7.5]

510.8 Underground Installations. Grease ducts installed underground shall be approved for underground installation. The material of the grease duct shall be corrosion-resistant and shall comply with Section 510.5.1.

510.8.1 Grease Receptacle. The grease duct shall be sloped to drain the grease back to an approved grease collection device. A grease collection device shall be located at the base of the vertical riser.

510.8.2 Cleanouts. For horizontal installations, cleanouts for cleaning and maintenance shall be provided on the top portion of the grease duct in accordance with Section 510.3 and shall be labeled at the interior portion of the duct.

510.9 Termination of Type I Hood Exhaust System. The exhaust system shall terminate as follows:

(1) Outside the building with a fan or duct.

(2) Through the roof or to the roof from outside in accordance with Section 510.9.1, or through a wall in accordance with Section 510.9.2. [NFPA 96:7.8.1]

510.9.1 Rooftop Terminations. Rooftop terminations shall be arranged with or provided with the following:

(1) Not less than 10 feet (3048 mm) of horizontal clearance from the outlet to adjacent buildings, property lines, and air intakes.

(2) Not less than 5 feet (1524 mm) of horizontal clearance from the outlet (fan housing) to a combustible structure.

(3) A vertical separation of 3 feet (914 mm) below an exhaust outlet for air intakes within 10 feet (3048 mm) of the exhaust outlet.

(4) The ability to drain grease out of traps or low points formed in the fan or duct near the termination of the system into a collection container that is noncombustible, closed, rainproof, and structurally sound for the service to which it is applied, and that will not sustain combustion.

(5) A grease collection device that is applied to exhaust systems that does not inhibit the performance of a fan.

(6) A listed grease collection system that is in accordance with Section 510.9.1(4) and Section 510.9.1(5).

(7) A listed grease duct in accordance with Section 507.4.7 or ductwork in accordance with Section 507.4.8.

(8) A hinged upblast fan supplied with flexible weatherproof electrical cable and service hold-open retainer to permit inspection and cleaning that is listed for commercial cooking equipment with the following conditions:
   (a) Where the fan attaches to the ductwork, the ductwork is not less than 18 inches (457 mm) away from the roof surface, as shown in Figure 510.9.1.
   (b) The fan discharges not less than 40 inches (1016 mm) away from the roof surface, as shown in Figure 510.9.1.

(9) Other approved fan, provided it is in accordance with the following criteria:
   (a) The fan is in accordance with the requirements of Section 510.9.1(3) and Section 511.1.3.
   (b) Its discharge or its extended duct discharge is in accordance with the requirements of Section 510.9.1(2). (See Section 511.1.3)
   (c) Exhaust fan discharge is directed up and away from the roof surface. [NFPA 96:7.8.2.1]

510.9.1.1 Listed Flexible Connectors. Listed flexible connectors shall be permitted to be used on exterior roof locations where required for proper equipment vibration isolation.

510.9.1.2 Inspection and Cleaning. Fans shall be provided with safe access and a work surface for inspection and cleaning. [NFPA 96:7.8.2.2]

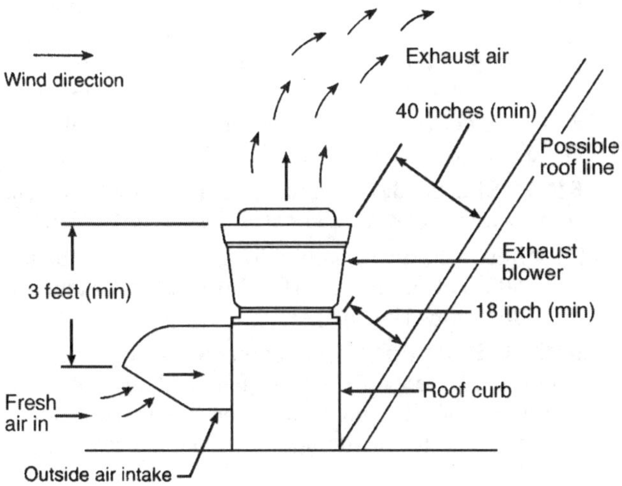

For SI units: 1 inch = 25.4 mm, 1 foot = 304.8 mm

FIGURE 510.9.1
UPBLAST FAN CLEARANCES
[NFPA 96: FIGURE 7.8.2.1]

510.9.2 Wall Terminations. Wall terminations shall be arranged with or provided with the following properties:

(1) Through a noncombustible wall with not less than 10 feet (3048 mm) of clearance from the outlet to adjacent buildings, property lines, grade level, combustible construction, electrical equipment or lines, and the closest point of an air intake or operable door or window at or below the plane of the exhaust termination. The closest point of an air intake or operable door or window above the plane of the exhaust termination shall be not less than 10 feet (3048 mm) in distance, plus 3 inches (76 mm) for each 1 degree (0.017 rad) from horizontal, the angle of degree being measured from the center of the exhaust termination to the center of the air intake, operable door or window, as indicated in Figure 510.9.2.

Exception: A wall termination in a secured area shall be permitted to be at a lower height above grade where acceptable to the Authority Having Jurisdiction.

(2) The exhaust flow shall be directed perpendicularly outward from the wall face or upward.

(3) The ductwork shall be pitched to drain the grease back into the hood(s) or with a drain provided to bring the grease back into a container within the building or into a remote grease trap.

(4) A listed grease duct shall comply with Section 510.3.3 through Section 510.3.7; other ducts shall comply with Section 510.5.

(5) An approved fan shall comply with the requirements of Section 510.9.2(3), and Section 511.1.1 or Section 511.1.3. [NFPA 96:7.8.3]

» **510.10 Solid-Fuel Duct Systems.** Where solid-fuel cooking equipment is to be vented, the duct system shall be in accordance with Section 517.0.

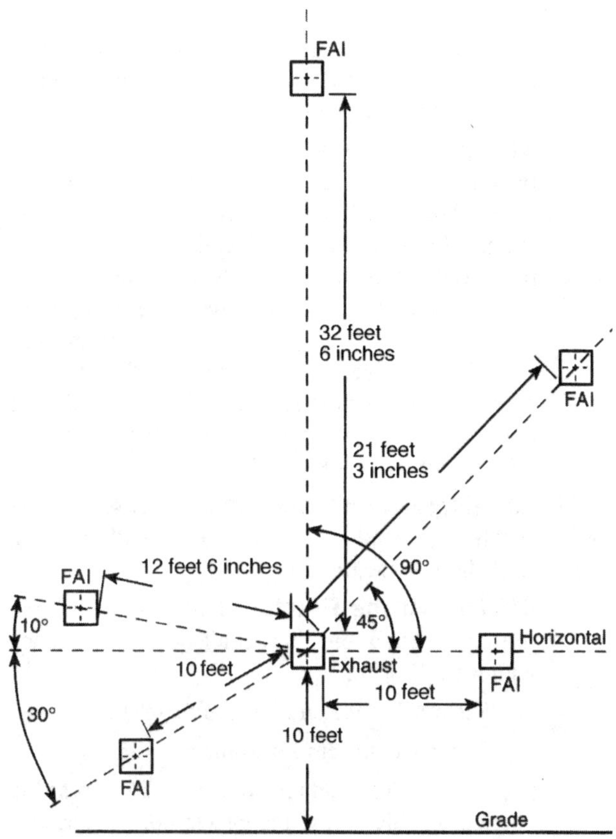

For SI units: 1 inch = 25.4 mm, 1 foot = 304.8 mm

Notes:

1. Fresh air intake (FAI) applies to an air intake, including an operable door or window.
2. Example:
   FAI is same plane as exhaust or lower: 10 feet (min.) between closest edges.
   FAI above plane of exhaust: 10 feet + 3 inches.

FIGURE 510.9.2
EXHAUST TERMINATION DISTANCE FROM FRESH AIR INTAKE (FAI) OR OPERABLE DOOR OR WINDOW
[NFPA 96: FIGURE 7.8.3]

511.0 Air Movement.

511.1 Exhaust Fans for Commercial Cooking Operations. Exhaust fans shall be installed in accordance with Section 511.1.1 through Section 511.1.6. Exhaust fans shall comply with UL 762 and be installed in accordance with the manufacturer's installation instructions.

511.1.1 Upblast Fans. Upblast fans with motors surrounded by the airstream shall be hinged and supplied with flexible weatherproof electrical cable, and service hold-open retainers. Installation shall comply with the requirements of Section 510.9. Upblast fans shall have a drain directed to a readily accessible and visible grease receptacle not to exceed 1 gallon (4 L). [NFPA 96:8.1.2]

511.1.2 In-Line Exhaust Fans. In-line fans shall be of the type with the motor located outside the airstream and with belts and pulleys protected from the airstream by a greasetight housing. In-line fans shall be connected to the exhaust duct by flanges securely bolted as shown in Figure 511.1.2(1) through Figure 511.1.2(4), or by a system specifically listed for such use. Flexible connectors shall not be used. [NFPA 96:8.1.3.1 – 8.1.3.3]

511.1.2.1 Accessibility. Where the design or positioning of the fan allows grease to be trapped, a drain directed to a readily accessible and visible grease receptacle, not exceeding 1 gallon (4 L), shall be provided. In-line exhaust fans shall be located in an easily accessible area of approved size to allow for service or removal. Where the duct system connected to the fan is in an enclosure, the space or room in which the exhaust fan is located shall have the same fire resistance rating as the enclosure. [NFPA 96:8.1.3.4 – 8.1.3.6]

511.1.3 Utility Set Exhaust Fans. Utility set exhaust fans shall be installed in accordance with Section 511.1.3.1 through Section 511.1.3.3.

511.1.3.1 At the Rooftop. Fans installed at the rooftop termination point shall be in accordance with the following:

(1) Section 510.9.1 and Section 510.9.1.2.

(2) Flexible connectors shall be permitted.

(3) A drain shall be directed to a readily accessible and visible grease receptacle not to exceed 1 gallon (4 L).

511.1.3.2 Within the Building. Fans installed within the building shall be in accordance with the following:

(1) Located in an accessible area of a size to allow for service or removal. [NFPA 96:8.1.4.2]

(2) Flexible connectors shall be prohibited. [NFPA 96:8.1.4.5]

(3) A drain shall be directed to a readily accessible and visible grease receptacle not to exceed 1 gallon (4 L). [NFPA 96:8.1.4.6]

511.1.3.3 Duct Systems. Duct systems connected to fans in an enclosure shall be in accordance with the following:

(1) The space or room in which the exhaust fan is located shall have the same fire resistance rating as the enclosure.

(2) The fan shall be connected to the exhaust duct by flanges securely bolted as shown in Figure 511.1.2(1) through Figure 511.1.2(4) or by a system specifically listed for such use. [NFPA 96:8.1.4.3, 8.1.4.4]

511.1.4 Construction. Exhaust fan housings shall be constructed of carbon steel not less than 0.060 of an inch (1.524 mm) (No. 16 MSG) in thickness, of stainless steel not less than 0.048 of an inch (1.219 mm) (No. 18 MSG) in thickness, or, where listed, in accordance with the terms of the listing. [NFPA 96:8.1.5]

511.1.5 Openings. Openings for cleaning, servicing, and inspection shall be in accordance with the requirements of Section 510.3.2. Clearances shall be in accordance with the requirements of Section 507.4, or Section 510.7.3 and Section 510.7.4 where installed within an enclosure. [NFPA 96:8.1.6.1, 8.1.6.2]

511.1.6 Standard. Wiring and electrical equipment shall comply with *California Electrical Code*. [NFPA 96:8.1.7]

511.2 Airflow. The air velocity through a duct shall be not less than 500 feet per minute (ft/min) (2.54 m/s) and not exceed 2500 ft/min (12.7 m/s).

511.2.1 Exceptions. Transition duct sections that do not exceed 3 feet (914 mm) in length and do not contain grease traps shall be permitted to be connected to hoods and exhaust fans that do not meet this velocity. [NFPA 96:8.2.1.2]

511.2.2 Exhaust-Air Volumes. Exhaust air volumes for hoods shall be of sufficient level to provide for capture and removal of grease-laden cooking vapors. Test data, performance tests approved by the Authority Having Jurisdiction, or both shall be displayed, provided on request, or both. [NFPA 96:8.2.2.1, 8.2.2.2] Lower exhaust air volumes shall be permitted during no-load and partial load cooking conditions, provided they are sufficient to capture and remove flue gases and cooking effluent from cooking equipment.

511.2.2.1 Performance Test. A performance test shall be conducted upon completion and before final approval of the installation of a ventilation system serving commercial cooking appliances. The test shall verify the rate of exhaust airflow in accordance with Section 508.5.1.2 through Section 508.5.1.5. The permit holder shall furnish the necessary test equipment and devices required to perform the tests. [ASHRAE 154:4.7.1]

511.2.2.2 Capture and Containment Test. The permit holder shall verify the capture and containment performance of Type I hoods. A field test shall be conducted with all appliances under the hood at operating temperatures, all the hoods operating at design airflows, and with all sources of replacement air operating at design airflows for the restaurant. Capture and containment shall be verified visually by observing smoke or steam produced by actual cooking operation or by simulating cooking using devices such as smoke candles or smoke puffers. Smoke bombs shall not be used. [ASHRAE 154:4.7.2]

511.2.3 Operation. A hood exhaust fan(s) shall continue to operate after the extinguishing system has been activated, unless fan shutdown is required by a listed component of the ventilation system, or by the design of the extinguishing system. The hood exhaust fan shall not be required to start upon activation of the extinguishing system where the exhaust fan and cooking equipment served by the fan have been shut down. The exhaust fan shall be provided with a means so that the fan is activated when an appliance under the hood is turned on. [NFPA 96:8.2.3]

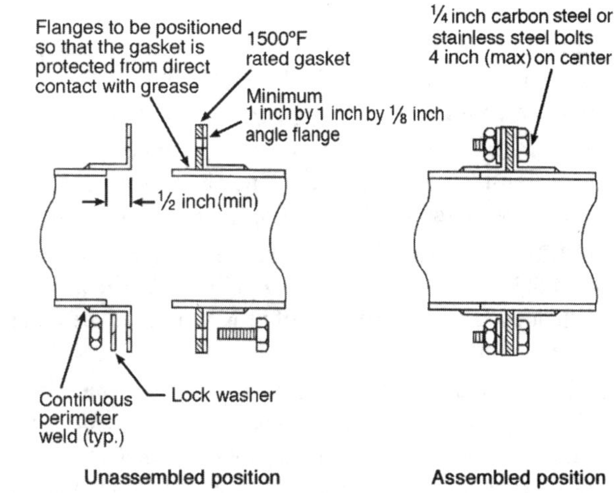

For SI units: 1 inch = 25.4 mm, °C = (°F-32)/1.8

**FIGURE 511.1.2(1)
TYPICAL SECTION OF DUCT-TO-FAN CONNECTION-BUTT JOINT METHOD
[NFPA 96: FIGURE 8.1.3.2(a)]**

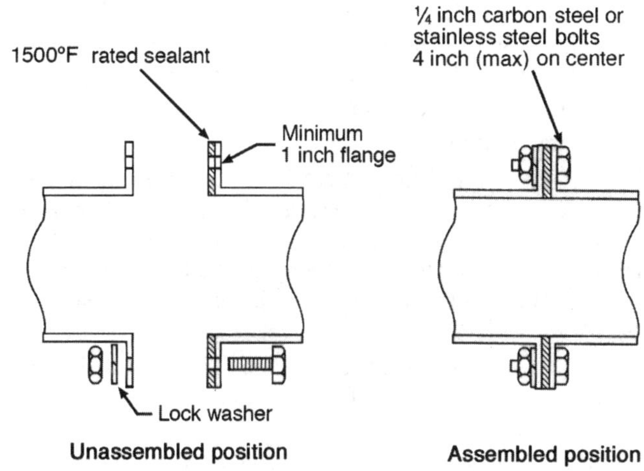

For SI units: 1 inch = 25.4 mm, °C = (°F-32)/1.8)

**FIGURE 511.1.2(3)
TYPICAL SECTION OF DUCT-TO-FAN CONNECTION–SEALANT METHOD
[NFPA 96: FIGURE 8.1.3.2(c)]**

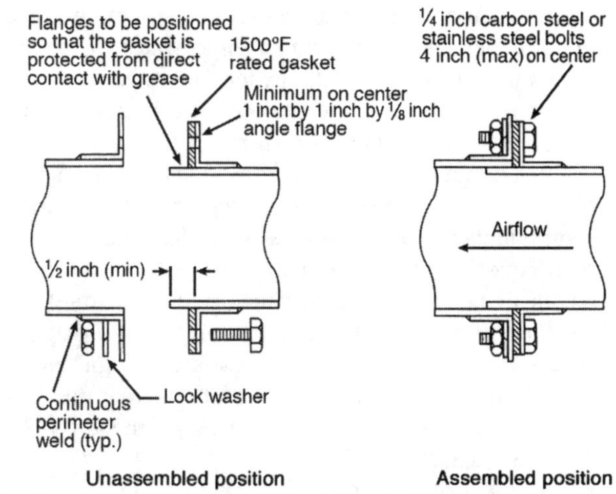

For SI units: 1 inch = 25.4 mm, °C = (°F-32)/1.8

**FIGURE 511.1.2(2)
TYPICAL SECTION OF DUCT-TO-FAN CONNECTION-OVERLAPPING METHOD
[NFPA 96: FIGURE 8.1.3.2(b)]**

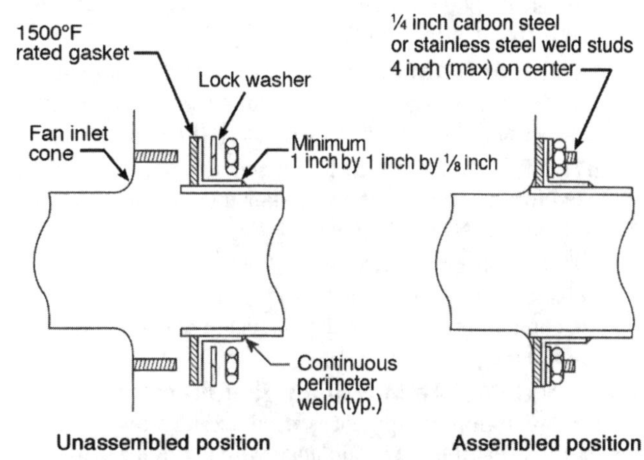

For SI units: 1 inch = 25.4 mm, °C = (°F-32)/1.8

**FIGURE 511.1.2(4)
TYPICAL SECTION OF DUCT-TO-FAN CONNECTION-DIRECT TO FAN INLET CONE METHOD
[NFPA 96: FIGURE 8.1.3.2(d)]**

511.3 Makeup Air. The makeup air quantity shall prevent negative pressures in the commercial cooking area(s) from exceeding 0.02 inch water column (0.005 kPa). Where the fire-extinguishing system activates, makeup air supplied internally to a hood shall be shut off.

For compensating hoods, where a Type I or Type II hood has an internal discharge of makeup air, the makeup air flow shall not exceed 10 percent of the exhaust airflow, the exhaust airflow shall be the net exhaust from the hood in accordance with Section 508.5.1.2 through Section 508.5.1.5. The total hood exhaust shall be determined in accordance with Equation 511.3.

$$E_{NET} = E_{HOOD} - MA_{ID} \quad \text{(Equation 511.3)}$$

Where:

E_{NET} = net hood exhaust

E_{HOOD} = total hood exhaust

MA_{ID} = makeup air, internal discharge

511.3.1 Air Balance. Design plans for a facility with a commercial kitchen ventilation system shall include a schedule or diagram indicating the design outdoor air balance. The design outdoor air balance shall indicate the exhaust and replacement air for the facility and the net exfiltration where applicable. The total replacement airflow rate shall equal the total exhaust airflow rate and the net exfiltration.

511.4 Common Duct (Manifold) Systems. Master kitchen exhaust ducts that serve multiple tenants shall include provision to bleed air from outdoors or from adjacent spaces into the master exhaust duct where required to maintain the necessary minimum air velocity in the master exhaust duct. [NFPA 96:8.4.1]

511.4.1 Connections. The bleed-air ducts shall connect to the top or side of the master exhaust duct. [NFPA 96:8.4.2]

511.4.2 Fire Damper. The bleed-air duct shall have a fire damper not less than 12 inches (305 mm) from the exhaust duct connection. [NFPA 96:8.4.3]

511.4.3 Construction and Clearance. The bleed-air duct shall have the same construction and clearance requirements as the main exhaust duct from the connection to the exhaust duct to not less than 12 inches (305 mm) on both sides of the fire damper. [NFPA 96:8.4.4]

511.4.4 Adjustment. Each bleed air duct shall have a means of adjusting (e.g., using volume dampers) the bleed air quantity. [NFPA 96:8.4.5]

511.4.5 Adjustment Location. Means to adjust the bleed air quantity shall be installed between the fire damper and the source of bleed air. [NFPA 96:8.4.6]

511.4.6 Bleed Air Duct. A bleed air duct shall not be used for the exhaust of grease-laden vapors and shall be so labeled. [NFPA 96:8.4.7]

511.4.7 Disconnect. Unused tenant exhaust connections to the master exhaust duct that are not used as bleed air connections shall be disconnected and sealed at the main duct. [NFPA 96:8.4.8]

511.5 Solid-Fuel Air Movement Requirements. Where solid-fuel cooking equipment is used, exhaust and replacement air also shall be in accordance with Section 517.0.

512.0 Auxiliary Equipment.

512.1 Dampers. Dampers shall not be installed in exhaust ducts or exhaust duct systems. [NFPA 96:9.1.1]

512.1.1 Use. Where specifically listed for such use or where required as part of a listed device or system, dampers in exhaust ducts or exhaust duct systems shall be permitted. [NFPA 96:9.1.2]

512.2 Electrical Equipment. Wiring systems shall not be installed in ducts. [NFPA 96:9.2.1]

512.2.1 Device Installation in Ducts. Motors, lights, and other electrical devices shall be permitted to be installed in ducts or hoods or to be located in the path of travel of exhaust products where specifically listed for such use. [NFPA 96:9.2.2]

512.2.2 Lighting Units. Lighting units in hoods shall not be located in concealed spaces except as permitted by Section 512.2.3 and Section 512.2.4. [NFPA 96:9.2.3.2]

512.2.3 Concealed Spaces. Lighting units shall be permitted in concealed spaces where such units are part of a listed exhaust hood. [NFPA 96:9.2.3.3]

512.2.4 Listed Lighting Units. Listed lighting units specifically listed for such use and installed in accordance with the terms of the listing shall be permitted to be installed in concealed spaces. [NFPA 96:9.2.3.4]

512.2.5 Standard. Electrical equipment shall be installed in accordance with *California Electrical Code*, with due regard to the effects of heat, vapor, and grease on the equipment.

512.3 Other Equipment. Fume incinerators, thermal recovery units, air pollution control devices, or other devices shall be permitted to be installed in ducts, hoods or to be located in the path of travel of exhaust products where specifically listed for such use. Downgrading other parts of the exhaust system due to the installation of these approved devices, whether listed or not, shall not be permitted. [NFPA 96:9.3.1, 9.3.2]

512.3.1 Fire-Extinguishing System. An equipment, listed or otherwise, that provides secondary filtration or air pollution control and that is installed in the path of travel of exhaust products shall be provided with an approved automatic fire-extinguishing system, installed in accordance with fire-extinguishing system manufacturer's installation instructions, for the protection of the component sections of the equipment, and shall include protection of the ductwork downstream of the equipment, whether or not the equipment is provided with a damper. Filter media used in secondary filtration or air pollution control units and not in accordance with Section 509.2.3 shall have fire protection that is adequate for the filter media being used in accordance with the fire-extinguishing system manufacturer's installation instructions. Where the equipment provides a source of ignition, it shall be provided with a detection to operate the fire-extinguishing system protecting the equipment. [NFPA 96:9.3.3 – 9.3.4]

512.3.2 Air Recirculation. Where a cooking exhaust system employs an air pollution control device that recirculates air into the building, the requirements of Section 516.0 shall apply. [NFPA 96:9.3.5]

512.4 Solid-Fuel Auxiliary Equipment. Where solid fuel cooking comprises a part of a cooking operation, additional provisions, and equipment as described in Section 517.0 shall be used where required.

513.0 Fire-Extinguishing Equipment.

513.1 General. Fire-extinguishing equipment for the protection of grease removal devices, hood exhaust plenums, and exhaust duct systems shall be provided. [NFPA 96:10.1.1]

513.1.1 Protection. Cooking equipment that produces grease-laden vapors and is capable of being a source of ignition of grease in the hood, grease removal device, or duct shall be protected by fire-extinguishing equipment. [NFPA 96:10.1.2]

513.2 Types of Equipment. Fire-extinguishing equipment shall include both automatic fire-extinguishing systems as primary protection and portable fire extinguishers as secondary backup. [NFPA 96:10.2.1]

513.2.1 Identification. A placard shall be conspicuously placed near the fire extinguisher that states that the fire protection system shall be activated prior to using the fire-extinguisher. [NFPA 96:10.2.2]

513.2.2 Standard. *Commercial cooking equipment that produce grease laden vapors shall be provided with a Type I Hood, in accordance with this Code, and an automatic fire extinguishing system that is listed and labeled for its intended use as follows:*

(1) Wet chemical extinguishing system (NFPA 17A, Standard for Wet Chemical Extinguishing Systems), complying with UL 300, Fire Testing of Fire Extinguishing Systems for Protection of Restaurant Cooking Equipment. All existing dry chemical and wet chemical extinguishing systems shall comply with UL 300, Fire Testing of Fire Extinguishing Systems for Protection of Restaurant Cooking Equipment.

(2) Carbon dioxide extinguishing systems (NFPA 12, Standard on Carbon Dioxide Extinguishing Systems).

(3) Automatic fire sprinkler systems (NFPA 13, Standard for the Installation of Sprinkler Systems).

All existing dry chemical and wet chemical extinguishing systems shall comply with UL 300, no later than the second required servicing of the system following the effective date of this section.

|| **Exceptions:**

(1) Automatic fire-extinguishing equipment provided as part of listed recirculating systems in accordance with UL 710B. [NFPA 96:10.2.5]

|| *(2)* ***[DSA-SS and DSA-SS/CC]*** *Public school kitchens, without deep-fat fryers, shall be upgraded to a UL 300, Fire Testing of Fire Extinguishing Systems for Protection of Restaurant Cooking Equipment, compliant system during state-funded modernization projects that are under the jurisdiction of the Division of the State Architect.*

513.2.3 Installation. Automatic fire-extinguishing systems shall be installed in accordance with the terms of their listing, the manufacturer's installation instructions, and the following standards where applicable:

(1) NFPA 12

(2) NFPA 13

(3) NFPA 17

(4) NFPA 17A

513.2.4 Modification of Existing Hood Systems. An abandoned pipe or conduit from a previous installation shall be removed from within the hood, plenum, and exhaust duct. [NFPA 96:10.2.7.1]

513.2.4.1 Sealing. Penetrations and holes resulting from the removal of conduit or piping shall be sealed with listed or equivalent liquid-tight sealing devices. [NFPA 96:10.2.7.2]

513.2.4.2 Obstructions. The addition of obstructions to spray patterns from the cooking appliance nozzle(s) such as baffle plates, shelves, or a modification shall not be permitted. [NFPA 96:10.2.7.3]

513.2.4.3 System Re-evaluation. Changes or modifications to the hazard after installation of the fire-extinguishing systems shall result in re-evaluation of the system design by a properly trained, qualified, and certified person(s). [NFPA 96:10.2.7.4]

513.2.5 Fixed Baffle Hoods with Water Wash. Grease removal devices, hood exhaust plenums, and exhaust ducts requiring protection in accordance with Section 513.1 shall be permitted to be protected by a listed fixed baffle hood containing a constant or fire-actuated water wash system that is listed and in accordance with UL 300 or other equivalent standards and shall be installed in accordance with the requirements of their listing. [NFPA 96:10.2.8.1]

513.2.5.1 Domestic Water Supply. The water for listed, fixed baffle hood assemblies shall be permitted to be supplied from the domestic water supply where the minimum water pressure and flow are provided in accordance with the terms of the listing. [NFPA 96:10.2.8.3]

513.2.5.2 Control Valve. The water supply shall be controlled by a supervised water supply control valve. [NFPA 96:10.2.8.4]

513.2.5.3 Activation. The water wash in the fixed baffle hood specifically listed to extinguish a fire shall be activated by the cooking equipment extinguishing system. [NFPA 96:10.2.8.5]

513.2.5.4 Water-Wash System. A water-wash system approved to be used for protection of the grease removal device(s), hood exhaust plenum(s), exhaust duct(s), or combination thereof shall include instruction and electrical interface for simultaneous activation of the water-wash system from an automatic fire-extinguishing system, where the automatic fire-extinguishing system is used for cooking equipment protection. [NFPA 96:10.2.8.6]

513.2.5.5 Exception. Where the fire-extinguishing system provides protection for the cooking equipment, hood, and duct, activation of the water-wash shall not be required. [NFPA 96:10.2.8.7]

513.2.5.6 Water Supply. The water required for listed automatic fire-extinguishing systems shall be permitted to be supplied from the domestic water supply where the minimum water pressure and flow are provided in accordance with the terms of the listing. The water supply shall be controlled by a supervised water supply control valve. Where the water supply is from a dedicated fire protection water supply in a building with one or more fire sprinkler systems, separate indicating control valves and drains shall be provided and arranged so that the hood system and sprinkler system are capable of being controlled individually. [NFPA 96:10.2.9]

513.2.6 Water Valve Supervision. Valves controlling the water supply to listed fixed baffle hood assemblies, automatic fire-extinguishing systems, or both shall be listed indicating type of valve and shall be supervised open by one of the following methods:

(1) Central station, proprietary, or remote station alarm service.

(2) Local alarm service that will cause the sounding of an audible signal at a constantly attended point.

(3) Locking valves open.

(4) Sealing of valves and approved weekly recorded inspection. [NFPA 96:10.2.10]

513.3 Simultaneous Operation. Fixed pipe extinguishing systems in a single hazard area shall be arranged for simultaneous automatic operation upon actuation of any one of the systems. [NFPA 96:10.3.1]

513.3.1 Automatic Sprinkler System. Simultaneous operation shall not be required where the one fixed pipe extinguishing system is an automatic sprinkler system. Where an automatic sprinkler system is used in conjunction with a water-based fire-extinguishing system served by the same water supply, hydraulic calculations shall consider both systems operating simultaneously. [NFPA 96:10.3.2, 10.3.2.1]

513.3.2 Dry or Wet Chemical Systems. Simultaneous operation shall be required where a dry or wet chemical system is used to protect common exhaust ductwork by NFPA 17A, *Wet Chemical Extinguishing Systems*.

513.4 Fuel and Electric Power Shutoff. Upon activation of a fire-extinguishing system for a cooking operation, sources of fuel and electric power that produce heat to equipment requiring protection by that system shall automatically shut off. [NFPA 96:10.4.1]

Exception: Solid-fuel cooking operations.

513.4.1 Steam. Steam supplied from an external source shall not be required to automatically shut off. [NFPA 96:10.4.2]

513.4.2 Protection Not Required. A gas appliance not requiring protection, but located under ventilating equipment where protected appliances are located, shall be automatically shut off upon activation of the extinguishing system. [NFPA 96:10.4.3]

513.4.3 Manual Reset. Shutoff devices shall require manual reset. [NFPA 96:10.4.4]

513.5 Manual Activation. A readily accessible means for manual activation shall be located between 42 inches and 48 inches (1067 mm and 1219 mm) above the floor, be accessible in the event of a fire, be located in a path of egress, and clearly identify the hazard protected. Not less than one manual actuation device shall be located not less than 10 feet (3048 mm) and not more than 20 feet (6096 mm) from the protected exhaust system(s) within the path of egress or at an alternative location acceptable to the Authority Having Jurisdiction. Manual actuation using a cable-operated pull station shall not require more than 40 pounds-force (lbf) (178 N) of force, with a pull movement not to exceed 14 inches (356 mm) to activate the automatic fire extinguishing system. The automatic and manual means of system activation external to the control head or releasing device shall be separate and independent of each other so that failure of one will not impair the operation of the other except as permitted in Section 513.5.1. [NFPA 96:10.5.1 – 10.5.2]

513.5.1 Location of Manual Activation Device. The manual means of system activation shall be permitted to be common with the automatic means where the manual activation device is located between the control head or releasing device and the first fusible link. [NFPA 96:10.5.3]

513.5.2 Automatic Sprinkler System. An automatic sprinkler system shall not require a manual means of system activation. [NFPA 96:10.5.4]

513.5.3 Manual Actuator(s). The means for manual activation shall be mechanical or rely on electrical power for activation in accordance with Section 513.5.4. [NFPA 96:10.5.5]

513.5.4 Standby Power Supply. Electrical power shall be permitted to be used for manual activation where a standby power supply is provided or where supervision is provided in accordance with Section 513.7. [NFPA 96:10.5.6]

513.6 System Annunciation. Upon activation of an automatic fire-extinguishing system, an audible alarm or visual indicator shall be provided to show that the system has activated. [NFPA 96:10.6.1]

513.6.1 Signaling. Where a fire alarm signaling system is serving the occupancy where the extinguishing system is located, the activation of the automatic fire-extinguishing system shall activate the fire alarm signaling system. [NFPA 96:10.6.2]

513.7 System Supervision. Where electrical power is required to operate the fixed automatic fire-extinguishing system, the system shall be provided with a reserve power supply and be monitored by a supervisory alarm except as permitted in accordance with Section 513.7.1. [NFPA 96:10.7.1]

513.7.1 Automatic Fire-Extinguishing System. Where a fixed automatic fire-extinguishing system includes automatic mechanical detection and actuation as a backup detection system, electrical power monitoring, and reserve power supply shall not be required. [NFPA 96:10.7.2]

513.7.2 Supervision. System supervision shall not be required where a fire-extinguishing system(s) is interconnected or interlocked with the cooking equipment power source(s) so that where the fire-extinguishing system becomes inoperable due to power failure, sources of fuel or electric power that produce heat to cooking equipment serviced by that hood shall automatically shut off. [NFPA 96:10.7.3]

513.7.3 Listed Water Wash System. System supervision shall not be required where an automatic fire-extinguishing system, including automatic mechanical detection and actuation, is electrically connected to a listed fire-actuated water-wash system for simultaneous operation of both systems. [NFPA 96:10.7.4]

513.8 Special Design and Application. Hoods containing automatic fire-extinguishing systems are protected areas; therefore, these hoods are not considered obstructions to overhead sprinkler systems and shall not require floor coverage underneath. [NFPA 96:10.8.1]

513.8.1 Single Device. A single detection device, listed with the extinguishing system, shall be permitted for more than one appliance where installed in accordance with the terms of the listing. [NFPA 96:10.8.2]

513.9 Review and Certification. Where required, complete drawings of the system installation, including the hood(s), exhaust duct(s), and appliances, along with the interface of the fire-extinguishing system detectors, piping, nozzles, fuel and electric power shutoff devices, agent storage container(s), and manual actuation device(s), shall be submitted to the Authority Having Jurisdiction. [NFPA 96:10.9.1]

513.10 Installation Requirements. Installation of systems shall be performed by persons properly trained and qualified to install the specific system being provided. The installer shall provide certification to the Authority Having Jurisdiction that the installation is in agreement with the terms of the listing and the manufacturer's installation instructions, approved design, or both. [NFPA 96:10.9.2]

513.11 Portable Fire Extinguishers. Portable fire extinguishers shall be selected and installed in kitchen cooking areas in accordance with *Chapter 3, Title 19 California Code of Regulations* and shall be specifically listed for such use. Class K fire extinguishers shall be provided for cooking appliances hazards that involve combustible cooking media such as vegetable oils, animal oils, and fats. [NFPA 96:10.10.1, 10.10.2]

513.11.1 Other Fire Extinguishers. Portable fire extinguishers shall be provided for other hazards in kitchen areas and shall be selected and installed in accordance with *Chapter 3, Title 19 California Code of Regulations*. [NFPA 96:10.10.3]

513.12 Maintenance. Portable fire extinguishers shall be maintained in accordance with NFPA 10. [NFPA 96:10.10.4]

513.12.1 Permitted Use. Portable fire extinguishers listed specifically for use in the kitchen cooking areas shall also be permitted.

513.13 Solid-Fuel Fire-Extinguishing Equipment. Where solid-fuel cooking equipment is served by fire extinguishing equipment, the provisions of Section 517.0 shall apply.

514.0 Procedures for the Use, Inspection, Testing, and Maintenance of Equipment.

514.1 Operating Procedures. Exhaust systems shall be operated where cooking equipment is turned on. [NFPA 96:11.1.1]

514.1.1 Filters. Filter-equipped exhaust systems shall not be operated with filters removed. [NFPA 96:11.1.2]

514.1.2 Openings. Openings provided for replacing air exhausted through ventilating equipment shall not be restricted by covers, dampers, or other means that would reduce the operating efficiency of the exhaust system. [NFPA 96:11.1.3]

514.1.3 Posting of Instructions. Instructions for manually operating the fire-extinguishing system shall be posted conspicuously in the kitchen and shall be reviewed with employees by the management. [NFPA 96:11.1.4]

514.1.4 Listing and Manufacturer's Instructions. Listed exhaust hoods shall be operated in accordance with the terms of their listings and the manufacturer's instructions. [NFPA 96:11.1.5]

514.1.5 Nonoperational. Cooking equipment shall not be operated while its fire-extinguishing system or exhaust system is nonoperational or impaired. [NFPA 96:11.1.6]

514.1.6 Secondary Control Equipment. Secondary filtration and pollution control equipment shall be operated in accordance with the terms of its listing and the manufacturer's instructions. [NFPA 96:11.1.7]

514.1.7 Inspection Frequency. Inspection and maintenance of "other equipment" as allowed in Section 512.3 shall be conducted by trained and qualified persons at a frequency determined by the manufacturer's instructions or the equipment listing. [NFPA 96:11.1.8]

514.2 Inspection, Testing, and Maintenance. Maintenance of the fire-extinguishing systems and listed exhaust hoods containing a constant or fire-activated water system that is listed to extinguish a fire in the grease removal devices, hood exhaust plenums, and exhaust ducts shall be made by trained, qualified, and certified person(s) acceptable to the Authority Having Jurisdiction not less than every 6 months. [NFPA 96:11.2.1]

514.2.1 Requirements. Actuation and control components, including remote manual pull stations, mechanical and electrical devices, detectors, and actuators shall be tested for proper operation during the inspection in accordance with the manufacturer's instructions. The specific inspection and maintenance requirements of the extinguishing system standards as well as the applicable installation and maintenance manuals for the listed system and service bulletins shall be followed. [NFPA 96:11.2.2, 11.2.3]

514.2.2 Fusible Links and Sprinklers. Fusible links of the metal alloy type and automatic sprinklers of the metal alloy type shall be replaced not less than semiannually except as permitted by Section 514.2.3 and Section 514.2.4. [NFPA 96:11.2.4]

514.2.3 Inspection Tag. The year of manufacture and the date of installation of the fusible links shall be marked on the system inspection tag. The tag shall be signed or initialed by the installer.

Detection devices that are bulb-type automatic sprinklers and fusible links other than the metal alloy type shall be examined and cleaned or replaced annually. [NFPA 96:11.2.5, 11.2.5.1, 11.2.6]

514.2.4 Temperature-Sensing Elements. Fixed temperature-sensing elements other than the fusible metal alloy type shall be permitted to remain continuously in service, provided they are inspected and cleaned, or replaced where necessary in accordance with the manufacturer's instructions every 12 months or more frequently to ensure operation of the system. [NFPA 96:11.2.7]

514.2.5 Certification. Where required, certificates of inspection and maintenance shall be forwarded to the Authority Having Jurisdiction. [NFPA 96:11.2.8]

514.3 Inspection for Grease Buildup. The entire exhaust system shall be inspected for grease buildup by a trained, qualified, and certified person(s) acceptable to the Authority Having Jurisdiction and in accordance with Table 514.3. [NFPA 96:11.4]

TABLE 514.3
SCHEDULE OF INSPECTION FOR GREASE BUILDUP
[NFPA 96: TABLE 11.4]

TYPE OR VOLUME OF COOKING	INSPECTION
Systems serving solid-fuel cooking operations.	Monthly
Systems serving high-volume cooking operations such as 24-hour cooking, charbroiling, or wok cooking.	Quarterly
Systems serving moderate-volume cooking operations.	Semiannually
Systems serving low-volume cooking operations, such as churches, day camps, seasonal businesses, or senior centers.	Annually

514.4 Cleaning of Exhaust Systems. Where, upon inspection, the exhaust system is found to be contaminated with deposits from grease-laden vapors, the contaminated portions of the exhaust system shall be cleaned by a trained, qualified, and certified person(s) acceptable to the Authority Having Jurisdiction. [NFPA 96:11.6.1]

514.4.1 Removal of Contaminants. Hoods, grease removal devices, fans, ducts, and other appurtenances shall be cleaned to remove combustible contaminants prior to surfaces becoming heavily contaminated with grease or oily sludge. [NFPA 96:11.6.2]

514.4.2 Electrical Switches. At the start of the cleaning process, electrical switches that could be activated accidentally shall be locked out. [NFPA 96:11.6.3]

514.4.3 Fire Suppression System. Components of the fire suppression system shall not be rendered inoperable during the cleaning process. [NFPA 96:11.6.4]

514.4.4 Inoperable. Fire-extinguishing systems shall be permitted to be rendered inoperable during the cleaning process where serviced by trained and qualified persons. [NFPA 96:11.6.5]

514.4.5 Solvents/Cleaning Aids. Flammable solvents or other flammable cleaning aids shall not be used. [NFPA 96:11.6.6]

514.4.6 Cleaning Chemicals. Cleaning chemicals shall not be applied on fusible links or other detection devices of the automatic extinguishing system. [NFPA 96:11.6.7]

514.4.7 Coating. After the exhaust system is cleaned, it shall not be coated with powder or other substance. [NFPA 96:11.6.8]

514.4.8 Access Panels and Cover Plates. Where cleaning procedures are completed, access panels (doors) and cover plates shall be restored to their normal operational condition. [NFPA 96:11.6.9]

514.4.9 Date of Inspection. Where an access panel is removed, a service company label or tag preprinted with the name of the company and giving the date of inspection or cleaning shall be affixed near the affected access panels. [NFPA 96:11.6.10]

514.4.10 Airflow. Dampers and diffusers shall be positioned for proper airflow. [NFPA 96:11.6.11]

514.4.11 Operable State. Where cleaning procedures are completed, electrical switches and system components shall be returned to an operable state. [NFPA 96:11.6.12]

514.4.12 Certification of Service. Where an exhaust cleaning service is used, a certificate showing the name of the servicing company, the name of the person performing the work, and the date of inspection or cleaning shall be maintained on the premises. [NFPA 96:11.6.13]

514.4.13 Report Provided. After cleaning or inspection is completed, the exhaust cleaning company and the person performing the work at the location shall provide the owner of the system with a written report that also specifies areas that were inaccessible or not cleaned. [NFPA 96:11.6.14]

514.4.14 Unclean Area. Where required, certificates of inspection and cleaning and reports of areas not cleaned shall be submitted to the Authority Having Jurisdiction. [NFPA 96:11.6.15]

514.5 Cooking Equipment Maintenance. Inspection and servicing of the cooking equipment shall be made not less than annually by properly trained and qualified persons. [NFPA 96:11.7.1]

514.5.1 Cleaning. Cooking equipment that collects grease below the surface, behind the equipment, or in cooking equipment flue gas exhaust, such as griddles or charbroilers, shall be inspected and, where found with grease accumulation, cleaned by a properly trained, qualified, and certified person acceptable to the Authority Having Jurisdiction. [NFPA 96:11.7.2]

515.0 Minimum Safety Requirements for Cooking Equipment.

515.1 Cooking Equipment. Cooking equipment shall be approved based on one of the following criteria:

(1) Listings by a testing laboratory.

(2) Test data acceptable to the Authority Having Jurisdiction. [NFPA 96:12.1.1]

515.1.1 Installation. Listed appliances shall be installed in accordance with the terms of their listings and the manufacturer's installation instructions. Solid fuel used for flavoring within a gas-operated appliance shall be in a solid fuel holder (smoker box) that is listed with the equipment. [NFPA 96:12.1.2.1, 12.1.2.1.1]

515.1.1.1 Re-evaluation. Cooking appliances requiring protection shall not be moved, modified, or rearranged without prior re-evaluation of the fire-extinguishing system by the system installer or servicing agent, unless otherwise allowed by the design of the fire-extinguishing system. A solid fuel holder shall not be added to an existing appliance until the fire-extinguishing system has been evaluated by the fire-extinguishing system service provider. [NFPA 96:12.1.2.2, 12.1.2.2.1]

515.1.1.2 Prior Location. The fire-extinguishing system shall not require re-evaluation where the cooking appliances are moved for the purpose of maintenance and cleaning, provided the appliances are returned to approved design location prior to cooking operations, and disconnected fire-extinguishing system nozzles attached to the appliances are reconnected in accordance with the manufacturer's instructions and listing. [NFPA 96:12.1.2.3]

515.1.1.3 Minimum Space. Deep-fat fryers shall be installed with not less than a 16 inch (406 mm) space between the fryer and surface flames from adjacent cooking equipment. [NFPA 96:12.1.2.4]

515.1.1.4 Space Not Required. Where a steel or tempered glass baffle plate is installed not less than 8 inches (203 mm) in height between the fryer and surface flames of the adjacent appliance, the requirement for a 16 inch (406 mm) space shall not apply. [NFPA 96:12.1.2.5]

515.1.1.5 Minimum Height. Where the fryer and the surface flames are at different horizontal planes, a height of not less than 8 inches (203 mm) shall be measured from the higher of the two. [NFPA 96:12.1.2.5.1]

515.2 Operating Controls. Deep-fat fryers shall be equipped with a separate high-limit control in addition to the adjustable operating control (thermostat) to shut off fuel or energy where the fat temperature reaches 475°F (246°C) at 1 inch (25.4 mm) below the surface. [NFPA 96:12.2]

516.0 Recirculating Systems.

516.1 General Requirements. Recirculating systems containing or for use with appliances used in processes producing smoke or grease-laden vapors shall be equipped with components in accordance with the following:

(1) The clearance requirements of Section 507.4.

(2) The hood shall comply with the requirements of Section 508.0.

(3) Grease removal devices shall comply with Section 509.0.

(4) The air movement requirements of Section 511.2.1 and Section 511.2.2.

(5) Auxiliary equipment (such as particulate and odor removal devices) shall comply with Section 512.0.

(6) Fire-extinguishing equipment shall comply with the requirements of Section 513.0.

Exception: Fire-extinguishing equipment in accordance with Section 513.1 and Section 513.5.

(7) The use and maintenance requirements of Section 514.0.

(8) The minimum safety requirements of Section 515.0.

(9) The requirements of Section 516.0. [NFPA 96:13.1]

(10) Provisions shall be provided for latent heat and excessive moisture acceptable to the Authority Having Jurisdiction.

516.2 Design Restrictions. Recirculating systems shall comply with Section 516.2.1 through Section 516.2.9. [NFPA 96:13.2]

516.2.1 Gas/Electrically Fueled Cooking Appliances. Gas-fueled or electrically fueled cooking appliances shall be used. Listed gas-fueled equipment designed for use with specific recirculating systems shall have the flue outlets connected in the intended manner. Gas-fueled appliances shall have not less than 18 inches (457 mm) of clearance from the flue outlet to the filter inlet in accordance with Section 509.2.2 through Section 509.2.2.3 and shall be in accordance with the installation requirements of NFPA 54 or NFPA 58. [NFPA 96:13.2.1 – 13.2.3]

516.2.2 Recirculation. Recirculating systems shall be listed with a testing laboratory in accordance with UL 710B or equivalent. [NFPA 96:13.2.4]

516.2.3 Protection. Cooking appliances that require protection and that are under a recirculating hood shall be protected by either the integral fire protection system in accordance with UL 710B or Section 513.0. [NFPA 96:13.2.4.2]

516.2.4 Maximum Limits. A recirculating system shall not use cooking equipment that exceeds that recirculating system's labeled maximum limits for that type of equipment, stated in maximum energy input, maximum cooking temperature, and maximum square area of cooking surface or cubic volume of cooking cavity. [NFPA 96:13.2.6]

516.2.5 Label. The listing label shall show the type(s) of cooking equipment tested and the maximum limits specified in Section 516.2.4. [NFPA 96:13.2.7]

516.2.6 Fire Damper. A fire-actuated damper shall be installed at the exhaust outlet of the system. [NFPA 96:13.2.8] The actuation device for the fire damper shall have a maximum temperature rating of 375°F (191°C). [NFPA 96:13.2.10]

516.2.7 Installation of Electrical Wiring. No electrical wiring shall be installed in the interior sections of the hood plenum that is capable of becoming exposed to grease.

Exception: As permitted by *the California Electrical Code.*

516.2.8 Power Supply. The power supply of an electrostatic precipitator (ESP) shall be of the "cold spark," ferroresonant type in which the voltage falls off as the current draw of a short increases. [NFPA 96:13.2.11]

516.2.9 Listing Evaluation. Listing evaluation shall include the following:

(1) Capture and containment of vapors at published and labeled airflows.

(2) Grease discharge at the exhaust outlet of the system not to exceed an average of 2.9 E-09 (oz/in^3) (5.0 E-06 kg/m^3) of exhausted air sampled from that equipment at maximum amount of product that is capable of being processed over a continuous 8 hour test with the system operating at its minimum listed airflow.

(3) Listing and labeling of clearance to combustibles from the sides, top, and bottom.

(4) Electrical connection in the field in accordance with *the California Electrical Code.*

(5) Interlocks on removable components that lie in the path of airflow within the unit to ensure that they are in place during operation of the cooking appliance. [NFPA 96:13.2.12]

516.3 Interlocks. The recirculating system shall be provided with interlocks of critical components and operations as indicated in Section 516.3.1 through Section 516.3.3.1 such that, where an interlock is interrupted, the cooking appliance shall not be able to operate. [NFPA 96:13.3.1]

516.3.1 Airflow Sections. Closure panels encompassing airflow sections shall have interlocks to ensure the panels are in place and fully sealed. [NFPA 96:13.3.2]

516.3.2 Filter Component. Each filter component (grease and odor) shall have an interlock to prove the component is in place. [NFPA 96:13.3.3]

516.3.3 ESP Interlocks. Each ESP shall have a sensor to prove its performance is as designed, with no interruption of the power to exceed 2 minutes. [NFPA 96:13.3.4.1]

516.3.3.1 Manual Reset. The sensor shall be a manual reset device or circuit. [NFPA 96:13.3.4.2]

516.3.4 Airflow Switch or Transducer. An airflow switch or transducer shall be provided after the last filter component to ensure that a minimum airflow is maintained. The airflow switch or transducer shall open the interlock circuit where the airflow falls 25 percent below the system's normal operating flow or 10 percent below its listed minimum rating, whichever is lower. The airflow switch or transducer shall be a manual reset device or circuit. [NFPA 96:13.3.5.1 – 13.3.5.3]

516.4 Location and Application Restrictions. The location of recirculating systems shall be approved by the Authority Having Jurisdiction. Items to be reviewed in the fire risk assessment shall include, but not be limited to, life safety, combustibility of surroundings, proximity to air vents, and total fuel load. [NFPA 96:13.4]

516.5 Additional Fire Safety Requirements. In addition to the appliance nozzle(s), a recirculating system shall be listed with the fire protection for grease filters, grease filtration, odor filtration units, and ductwork, where applicable. [NFPA 96:13.5.1]

516.5.1 Installation Downstream. In addition to other fire-extinguishing system activation device, there shall be a fire-extinguishing system activation device installed downstream of an ESP. [NFPA 96:13.5.2]

516.5.2 Locations. The requirements of Section 513.6 shall also apply to recirculating system locations. [NFPA 96:13.5.3]

516.6 Use and Maintenance. Automatic or manual covers on cooking appliances, especially fryers, shall not interfere with the application of the fire suppression system. [NFPA 96:13.6.1]

516.6.1 Manufacturer's Instructions. Filters shall be cleaned or replaced in accordance with the manufacturer's instructions. [NFPA 96:13.6.2]

516.6.2 Cleaning Schedule. ESPs shall be cleaned a minimum of once per week and in accordance with the manufacturer's instructions. [NFPA 96:13.6.3]

516.6.3 Hood Plenum and Blower Section Cleaning Schedule. The entire hood plenum and the blower section shall be cleaned not less than once every 3 months. [NFPA 96:13.6.4]

516.6.4 Inspection of Safety Interlocks. Inspection and testing of the total operation and safety interlocks in accordance with the manufacturer's instructions shall be performed by qualified service personnel not less than once every 6 months, or more frequently where required. [NFPA 96:13.6.5]

516.6.5 Inspection. Fire-extinguishing equipment shall be inspected in accordance with Section 514.2. [NFPA 96:13.6.6]

516.6.6 Maintenance Log. A signed and dated log of maintenance as performed in accordance with Section 516.6.3 and Section 516.6.4 shall be available on the premises for use by the Authority Having Jurisdiction. [NFPA 96:13.6.7]

517.0 Solid-Fuel Cooking Operations.

517.1 Venting Application. Venting requirements of solid-fuel cooking operations shall be determined in accordance with Section 517.1.1 through Section 517.1.6. [NFPA 96:14.1]

517.1.1 Natural Draft. Where solid-fuel cooking equipment is required by the manufacturer to have a natural draft, the vent shall be in accordance with Section 517.4. [NFPA 96:14.1.1]

517.1.2 System Compliance. Where the solid-fuel cooking equipment has a self-contained top, is the appliance to be vented in an isolated space (except for a single water heater with its own separate vent), has a separate makeup air system, and is provided with supply and return air (not supplied or returned from other spaces), the system shall be in accordance with Section 517.4 and Section 517.6. [NFPA 96:14.1.2]

517.1.3 Makeup Air System. Where the solid-fuel cooking equipment is located in a space with other vented equipment, the vented equipment shall have an exhaust system interlocked with a makeup air system for the space in accordance with Section 517.6. [NFPA 96:14.1.3]

517.1.4 Natural Draft Ventilation Systems. Natural draft ventilation systems and power-exhausted ventilation systems shall comply with Section 517.3, Section 517.4, and Section 517.6. [NFPA 96:14.1.4]

517.1.5 Opening Requirements. Where a solid-fuel cooking appliance allows effluent to escape from the appliance opening, this opening shall be covered by a hood and an exhaust system that meets the requirements of Section 517.3, Section 517.4, and Section 517.6. [NFPA 96:14.1.5]

517.1.6 Spark Arresters. Solid-fuel cooking operations shall have spark arresters to minimize the passage of airborne sparks and embers into plenums and ducts. Where the solid-fuel cooking operation is not located under a hood, a spark arrester shall be provided to minimize the passage of sparks and embers into flues and chimneys. [NFPA 96:14.1.6, 14.1.7]

517.2 Location of Appliances. Appliances shall be located with respect to building construction and other equipment so as to permit access to the appliance. [NFPA 96:14.2.1]

517.2.1 Prohibited Location. Solid-fuel cooking appliances shall not be installed in confined spaces. [NFPA 96:14.2.2]

Exception: Solid-fuel cooking appliances listed for installation in confined spaces such as alcoves shall be installed in accordance with the terms of the listing and the manufacturer's installation instructions. [NFPA 96:14.2.3]

517.2.2 Flammable Vapors. Solid-fuel cooking appliances shall not be installed in a location where gasoline or other flammable vapors or gases are present. [NFPA 96:14.2.4]

517.3 Hoods for Solid-Fuel Cooking. Hoods shall be sized and located in a manner capable of capturing and containing the effluent discharging from the appliances. The hood and its exhaust system shall be in accordance with the provisions of Section 508.0 through Section 513.0. [NFPA 96:14.3.1, 14.3.2]

517.3.1 Separation. Except as permitted in Section 517.3.1.1, exhaust systems serving solid-fuel cooking equipment, including gas or electrically operated equipment, shall be separate from other exhaust systems. [NFPA 96:14.3.3]

Exception: Cooking equipment not requiring automatic fire-extinguishing equipment (in accordance with Section 513.0) shall be permitted to be installed under a common hood with solid-fuel cooking equipment that is served by a duct system separate from other exhaust systems. [NFPA 96:14.3.5]

517.3.1.1 Equipment with Solid Fuel for Flavoring. Gas-operated equipment utilizing solid fuel for flavoring that is in accordance with the following conditions shall not be required to have a separate exhaust system:

(1) The solid fuel holder (smoker box) shall be listed with the gas-operated equipment.

(2) The solid fuel holder shall be located underneath the gas burner.

(3) Spark arresters that are in accordance with Section 517.1.6 shall be provided.

(4) The maximum quantity of solid fuel consumed shall not exceed 4.5 pounds (2.04 kg) per hour per 100 000 Btu/hr (29 kW) of gas burner capacity.

(5) The gas-operated equipment shall be protected by a fire suppression system listed for the equipment, including the solid fuel holder.

(6) Gas-operated equipment with integral solid fuel holder(s) intended for flavoring, such as a radiant charbroiler(s), shall comply with the requirements of UL 300 that address the gas radiant charbroiler(s) and mesquite wood charbroiler(s).

(7) A fire suppression system nozzle(s) shall be installed to protect the solid fuel holder.

(8) The fire suppression system shall be designed and installed to protect the entire cooking operation.

(9) The solid fuel holder shall be limited to a size of 2000 cubic inches (0.0328 m^3), with no dimension to exceed 20 inches (508 mm).

(10) A maximum of one solid fuel holder for each 100 000 Btu/hr (29 kW), or portion thereof, of burner capacity shall be permitted.

(11) The inspection frequency shall be the same as for solid fuel cooking operations in Table 514.3. [NFPA 96:14.3.4]

517.4 Exhaust Systems for Solid-Fuel Cooking. Where a hood is not required, in buildings where the duct system does not exceed three stories in height, a duct complying with Section 510.0 shall be provided. [NFPA 96:14.4]

517.4.1 Hood. Where a hood is used in buildings where the duct system does not exceed three stories in height, the duct system shall be in accordance with Section 510.0. [NFPA 96:14.4.1]

517.4.2 Building Exceeding Four Stories. A listed or approved grease duct system that is four stories in height or more shall be provided for solid-fuel cooking exhaust systems. [NFPA 96:14.4.2]

517.4.3 Prohibited. Wall terminations of solid-fuel exhaust systems shall be prohibited. [NFPA 96:14.4.4]

517.5 Grease Removal Devices for Solid-Fuel Cooking. Grease removal devices shall be constructed of steel or stainless steel or be approved for solid-fuel cooking. [NFPA 96:14.5.1]

517.5.1 Spark Arrester Devices. Where airborne sparks and embers can be generated by the solid-fuel cooking operation, spark arrester devices shall be used prior to using the grease removal device, to minimize the entrance of these sparks and embers into the grease removal device and into the hood and duct system. [NFPA 96:14.5.2]

517.5.2 Filters. Filters shall be not less than 4 feet (1219 mm) above the appliance cooking surface. [NFPA 96:14.5.3]

517.6 Air Movement for Solid-Fuel Cooking. Exhaust system requirements shall comply with Section 511.0 for hooded operation or shall be installed in accordance with the manufacturer's installation instructions for unhooded applications. [NFPA 96:14.6.1]

517.6.1 Replacement Air. A replacement or makeup air system shall be provided to ensure a positive supply of replacement air at times during cooking operations. [NFPA 96:14.6.2]

517.6.2 Operation. Makeup air systems serving solid-fuel cooking operations shall be interlocked with the exhaust air system and powered, where necessary, to prevent the space from attaining a negative pressure while the solid-fuel appliance is in operation. [NFPA 96:14.6.3]

517.7 Fire-Extinguishing Equipment for Solid-Fuel Cooking. Solid-fuel cooking appliances that produce grease-laden vapors shall be protected by listed fire-extinguishing equipment.

Exception: Where acceptable to the Authority Having Jurisdiction, solid-fuel cooking appliances constructed of solid masonry or reinforced portland or refractory cement concrete and vented in accordance with NFPA 211 shall not require fixed automatic fire-extinguishing equipment. [NFPA 96:14.7.1, 14.7.2]

517.7.1 Grease Removal Devices, Hoods, and Duct Systems. Listed fire-extinguishing equipment shall be provided for the protection of grease removal devices, hoods, and duct systems. [NFPA 96:14.7.3]

Exception: Where acceptable to the Authority Having Jurisdiction, solid-fuel cooking appliances constructed of solid masonry or reinforced portland or refractory cement concrete and vented in accordance with NFPA 211 shall not require automatic fire-extinguishing equipment for the protection of grease removal devices, hoods, and duct systems. [NFPA 96:14.7.4]

517.7.2 Listed Fire-Extinguishing Equipment. Listed fire-extinguishing equipment for solid-fuel-burning cooking appliances, where required, shall be in accordance with Section 513.0 and shall use water-based agents. [NFPA 96:14.7.5]

517.7.3 Rating and Design. Fire-extinguishing equipment shall be rated and designed to extinguish solid-fuel cooking fires. The fire-extinguishing equipment shall be of sufficient size to totally extinguish fire in the entire hazard area and prevent reignition of the fuel. [NFPA 96:14.7.6, 14.7.7]

517.7.4 Listing/Class. Solid-fuel appliances (whether or not under a hood) with fireboxes of 5 cubic feet (0.14 m^3) volume or less shall have not less than a listed 2-A rated water-spray fire extinguisher or a 1.6 gallon (6.1 L) wet chemical fire extinguisher listed for Class K fires in accordance with NFPA 10 with a travel distance of not more than 20 feet (6096 mm) to the appliance. [NFPA 96:14.7.8]

517.7.5 Fixed-Water Pipe System. Solid-fuel appliances with fireboxes exceeding 5 cubic feet (0.14 m^3) shall be provided with a fixed-water pipe system with a hose in the kitchen capable of reaching the firebox. The hose shall be equipped with an adjustable nozzle capable of producing a fine to medium spray or mist. The nozzle shall be of the type that cannot produce a straight stream. The system shall have an operating pressure of not less than 40 psi (276 kPa) and shall provide not less than 5 gallons per minute (gpm) (0.3 L/s). [NFPA 96:14.7.9]

517.7.6 Fuel Storage. Fuel storage areas shall be provided with a sprinkler system in accordance with NFPA 13 except where permitted in accordance with the following:

(1) Where approved by the Authority Having Jurisdiction, fuel storage areas shall be permitted to be protected with a fixed water pipe system with a hose capable of reaching all parts of the area.

(2) In lieu of the sprinkler system outlined in Section 517.7.6, a listed 2-A rated water spray fire extinguisher or a 1.6 gallon (6.1 L) wet chemical fire extinguisher listed for Class K fires with a travel distance of not more than 20 feet (6096 mm) to the solid fuel piles shall be permitted to be used for a solid fuel pile, provided that the fuel pile does not exceed 5 cubic feet (0.14 m^3). [NFPA 96:14.9.2.8 – 9.2.8.2]

517.7.7 Auxiliary Fuel. In addition to the requirements of Section 517.7.4, Section 517.7.5 and Section 517.8, where a solid-fuel cooking appliance is also provided with auxiliary electric, gas, oil, or other fuel for ignition or supplemental heat and the appliance is also served by a portion of a fire-extinguishing system in accordance with Section 513.0, such auxiliary fuel shall be shut off upon actuation of the fire-extinguishing system. [NFPA 96:14.7.11]

517.8 Other Safety Requirements. Metal-fabricated solid-fuel cooking appliances shall be listed for the application where produced in practical quantities or shall be approved by the Authority Having Jurisdiction. Where listed, metal-fabricated solid fuel cooking appliances shall be installed in accordance with the terms of their listings and with the applicable requirements of this chapter. [NFPA 96:14.9.4.1, 14.9.4.2]

517.8.1 Site-Built Solid Fuel Cooling Appliances. Site-built solid-fuel cooking appliances shall be submitted for approval to the Authority Having Jurisdiction before being considered for installation. Units submitted to the Authority Having Jurisdiction shall be installed, operated, and maintained in accordance with the approved terms of the manufacturer's instructions and additional requirements in accordance with the Authority Having Jurisdiction. [NFPA 96:14.9.4.3]

517.8.2 Additional Devices. Except for the spark arrester required in Section 517.1.6, there shall be no additional devices in a portion of the appliance, flue pipe, and chimney of a natural draft solid-fuel operation. [NFPA 96:14.9.4.4]

517.8.3 Prohibited. No solid-fuel cooking device shall be permitted for deep-fat frying involving more than 1 quart (qt) (1 L) of liquid shortening, and solid-fuel cooking device shall not be permitted within 3 feet (914 mm) of a deep-fat frying unit. [NFPA 96:14.9.4.5]

518.0 Downdraft Appliances.

518.1 General. A downdraft appliance ventilation system containing, or for use with appliances used in processes that produce, smoke or grease-laden vapors shall be equipped with components that are in accordance with the following:

(1) The clearance requirements in accordance with Section 507.4.

(2) The primary collection means designed for collecting cooking vapors and residues in accordance with the requirements of Section 508.0.

(3) Grease removal devices that comply with Section 509.0.

(4) Special-purpose filters as listed in accordance with UL 1046.

(5) Exhaust ducts that comply with Section 510.0.

(6) The air movement requirements in accordance with Section 511.2.1 and Section 511.2.2.

(7) Auxiliary equipment (such as particulate and odor removal devices) are in accordance with Section 512.0.

(8) Fire-extinguishing equipment that is in accordance with the requirements of Section 513.0, and as specified in Section 518.3.

(9) The use and maintenance requirements in accordance with Section 514.0.

(10) The minimum safety requirements in accordance with Section 515.0. [NFPA 96:15.1.1]

518.2 Ventilation System. The downdraft appliance ventilation system shall be capable of capturing and containing the effluent discharge from the appliance(s) it is serving. [NFPA 96:15.1.2]

518.3 Fire-Extinguishing Equipment. Fire-extinguishing equipment on a downdraft appliance ventilation system shall comply with the following:

(1) Cooking surface, duct, and plenum protection shall be provided.

(2) Not less than one fusible link or heat detector shall be installed within an exhaust duct opening in accordance with the manufacturer's listing.

(3) A fusible link or heat detector shall be provided above the protected cooking appliance and in accordance with the extinguishing system manufacturer's listing.

(4) A manual activation device shall be provided as part of the appliance at a height approved by the Authority Having Jurisdiction.

(5) Portable fire extinguishers shall be installed in accordance with Section 513.11. [NFPA 96:15.2]

518.3.1 Integral Fire-Extinguishing System. A listed downdraft appliance ventilation system employing an integral fire-extinguishing system including detection systems that has been evaluated for grease and smoke capture, fire extinguishing, and detection shall be considered to be in accordance with Section 518.3. [NFPA 96:15.2.1]

518.3.2 Interlocks. The downdraft appliance ventilation system shall be provided with interlocks such that the cooking fuel supply will not be activated unless the exhaust and supply air system have been activated. [NFPA 96:15.2.2]

518.4 Airflow Switch or Transducer. An airflow switch or transducer shall be provided after the last filter component to ensure that a minimum airflow is maintained. [NFPA 96:15.3.1]

518.4.1 Interlocks. The airflow switch or transducer shall open the interlock circuit where the airflow is less than 25 percent the system's normal operating flow or less than 10 percent its listed minimum rating, whichever is less. [NFPA 96:15.3.2]

518.4.2 Manual Reset. The airflow switch or transducer shall be a manual reset device or circuit. [NFPA 96:15.3.3]

518.5 Surface Materials. Surfaces located directly above the cooking appliance shall be of noncombustible or limited-combustible materials. [NFPA 96:15.4]

519.0 Type II Hood Exhaust System Requirements.

519.1 Where Required. Type II hoods shall be installed above equipment and dishwashers that generate steam, heat, and products of combustion, and where grease or smoke is not present.

Exceptions:

(1) Dishwashing machines connected to a Type II duct system and exhausted directly to the outdoors.

(2) Dishwashing machines with a self-contained condensing system listed in accordance with UL 921 and installed in a space where the HVAC system has been engineered to accommodate the latent and sensible heat load emitted from such appliances as approved by the Authority Having Jurisdiction. Such equipment shall be provided with an interlocking device to prevent opening of the appliance prior to completion of its cycle.

519.2 Construction of Type II Hoods. Type II hoods constructed of steel shall be not less than 0.024 of an inch (0.61 mm) (No. 24 gauge). Hoods constructed of copper shall be of copper sheets weighing not less than 0.17 ounces per square inch (oz/in^2) (7.47 kg/m^2). Joints and seams shall be substantially tight. Solder shall not be used except for sealing a joint or seam. «

519.3 Dishwashing Appliances. The net airflow for Type II hoods used for dishwashing equipment shall be not less than «

200 cubic feet per minute (0.094 m^3/s) per linear foot (m) of hood length.

519.4 Type II Exhaust Duct Systems. Ducts and plenums serving Type II hoods shall be constructed of rigid metallic materials in accordance with Chapter 6. Duct bracing and supports shall comply with Chapter 6. Ducts subject to positive pressure shall be adequately sealed.

519.5 Termination of Type II Hood Exhaust System. The exhaust system shall terminate as follows:

(1) Rooftop terminations shall terminate not less than 10 feet (3048 mm) from a property line, and the exhaust flow shall be directed away from the roof surface of the roof, not less than 40 inches (1016 mm).

(2) Horizontal terminations shall terminate not less than 10 feet (3048 mm) from adjacent buildings, property lines, operable openings, and from grade level.

(3) The termination outlet shall not be directed onto a public way.

519.6 Makeup Air. Makeup air shall be provided in accordance with Section 511.3.

CALIFORNIA MECHANICAL CODE – MATRIX ADOPTION TABLE
CHAPTER 6 – DUCT SYSTEMS

(Matrix Adoption Tables are non-regulatory, intended only as an aid to the code user. See Chapter 1 for state agency authority and building applications.)

Adopting Agency	BSC	BSC-CG	SFM	HCD			DSA			OSHPD						BSCC	DPH	AGR	DWR	CEC	CA	SL	SLC
				1	2	1-AC	AC	SS	SS/CC	1	1R	2	3	4	5								
Adopt Entire Chapter	X																						
Adopt Entire Chapter as amended (amended sections listed below)			X	X	X			X	X	X	X	X	X	X	X								
Adopt only those sections that are listed below																							
Chapter/Section																							
601.2, *Exception*				X	X																		
602.1										X	X	X	X	X	X								
602.2.1				X	X																		
602.1, *Exception*										X	X	X	X	X	X								
602.6.1										X	X	X	X	X	X								
603.3.3				X	X																		
603.4.1										X	X	X	X	X	X								
603.10.1				†	†																		
603.10.1.1				X	X																		
604.1										X	X	X	X	X	X								
604.2										X	X	X	X	X	X								
604.3										X	X	X		X	X								
605.1				X	X																		
605.8			X					X	X														
606.1				X	X																		
606.1.1										X	X	X	X	X	X								
608.1			X					X	X														

This state agency does not adopt sections identified with the following symbol: †

The Office of the State Fire Marshal's adoption of this chapter or individual sections is applicable to structures regulated by other state agencies pursuant to Section 1.11.0.

CHAPTER 6
DUCT SYSTEMS

601.0 General.

601.1 Applicability. Ducts and plenums that are portions of a heating, cooling, ventilation, or exhaust system shall comply with the requirements of this chapter.

601.2 Sizing Requirements. Duct systems shall be sized in accordance with ACCA Manual D listed in Table 1701.1, or by other approved methods.

***Exception: [HCD 1 & HCD 2]** Duct sizing calculations are not required for existing duct systems.*

602.0 Material.

602.1 General. Materials used for duct systems shall comply with Section 602.2 through Section 602.8 as applicable.

***Exception: [OSHPD 1, 1R, 2, 3, 4 & 5]** See Section 407.4.1.3.*

[Not permitted for OSHPD 1, 1R, 2, 3, 4 & 5] Concealed building spaces or independent construction within buildings shall be permitted to be used as ducts or plenums. Gypsum board shall not be used for positive pressure ducts.

Exception: In healthcare facilities, concealed spaces shall not be permitted to be used as ducts or plenums.

602.2 Combustibles Within Ducts or Plenums. Materials exposed within ducts or plenums shall be noncombustible or shall have a flame spread index not to exceed 25 and a smoke-developed index not to exceed 50, where tested as a composite product in accordance with ASTM E84 or UL 723.

Exceptions:

(1) Return-air and outside-air ducts, plenums, or concealed spaces that serve a dwelling unit.

(2) Air filters in accordance with the requirements of Section 311.2.

(3) Water evaporation media in an evaporative cooler.

(4) Charcoal filters where protected with an approved fire suppression system.

(5) Products listed and labeled for installation within plenums in accordance with Section 602.2.1 through Section 602.2.3.

(6) Smoke detectors.

(7) Duct insulation, coverings, and linings and other supplementary materials installed in accordance with Section 604.0.

(8) Materials in a hazardous fabrication area including the areas above and below the fabrication area sharing a common air recirculation path with the fabrication area.

602.2.1 Electrical. Electrical wiring in plenums shall comply with *the California Electrical Code*. Electrical wires and cables and optical fiber cables shall be listed and labeled for use in plenums and shall have a flame spread distance not exceeding 5 feet (1524 mm), an average optical density not exceeding 0.15, and a peak optical density not exceeding 0.5, where tested in accordance with NFPA 262.

602.2.2 Fire Sprinkler Piping. Nonmetallic fire sprinkler piping in plenums shall be listed and labeled for use in plenums and shall have a flame spread distance not exceeding 5 feet (1524 mm), an average optical density not exceeding 0.15 and, a peak optical density not exceeding 0.5, where tested in accordance with UL 1887.

602.2.3 Pneumatic Tubing. Nonmetallic pneumatic tubing in plenums shall be listed and labeled for use in plenums and shall have a flame spread distance not exceeding 5 feet (1524 mm), an average optical density not exceeding 0.15, and a peak optical density not exceeding 0.5, where tested in accordance with UL 1820.

602.2.4 Discrete Products in Plenums. Discrete plumbing, mechanical, and electrical products that are located in a plenum and have exposed combustible material shall be listed and labeled in accordance with UL 2043.

602.3 Metal. Ducts, plenums, or fittings of metal shall comply with SMACNA HVAC Duct Construction Standards – Metal and Flexible.

602.4 Phenolic. Ducts, plenums, or fittings of phenolic shall be constructed in accordance with SMACNA Phenolic Duct Construction Standards.

602.5 Gypsum. Where gypsum products are exposed in ducts or plenums, the air temperature shall be restricted to a range from 50°F (10°C) to 125°F (52°C), and moisture content shall be controlled so that the material is not adversely affected. All gypsum products shall have a mold or mildew resistant surface. For the purpose of this section, gypsum products shall not be exposed in supply ducts.

602.6 Factory-Made Air Ducts. Factory-made air ducts shall be approved for the use intended or shall be in accordance with the requirements of UL 181. Each portion of a factory-made air duct system shall be identified by the manufacturer with a label or other identification indicating compliance with its class designation.

***602.6.1 Flexible Ducts. [OSHPD 1, 1R, 2, 3, 4 & 5]** In hospital building projects and all other health-care facilities, including clinics and correctional treatment centers, flexible ducts of not more than 10 feet (3048 mm) in length may be used to connect supply, return or exhaust-air terminal devices to rigid duct systems. Where constant volume, variable volume or mixing boxes are utilized, flexible duct of not more than 10 feet (3048 mm), may be used on the inlet side for alignment. An internal impervious liner shall be provided to isolate insulation material from conditioned air.*

602.7 Vibration Isolators. Vibration isolation connectors installed between mechanical equipment and metal ducts (or casings) shall be made of an approved material and shall not exceed 10 inches (254 mm) in length.

602.8 Corridors. Corridors shall not be used to convey air to or from rooms where the corridor is required to be of fire-resistive construction in accordance with the building code except where permitted by the building code.

603.0 Installation of Ducts.

603.1 General. The pressure classification of ducts shall be not less than the design operating pressure of the air distribution in which the duct is utilized.

603.2 Under Floor or Crawl Space. Air ducts installed under a floor in a crawl space shall be installed in accordance with the following:

(1) Shall not prevent access to an area of the crawl space.

(2) Where it is required to move under ducts for access to areas of the crawl space, a vertical clearance of not less than 18 inches (457 mm) shall be provided.

603.3 Metal Ducts. Ducts shall be supported at each change of direction and in accordance with SMACNA HVAC Duct Construction Standards – Metal and Flexible. Riser ducts shall be held in place by means of metal straps or angles and channels to secure the riser to the structure.

Metal ducts shall be installed with not less than 4 inches (102 mm) separation from earth. Ducts shall be installed in a building with clearances that will retain the full thickness of fireproofing on structural members.

603.3.1 Rectangular Ducts. Supports for rectangular ducts shall be installed on two opposite sides of each duct and shall be riveted, bolted, or metal screwed to each side of the duct at intervals specified.

603.3.2 Horizontal Round Ducts. Horizontal round ducts not more than 40 inches (1016 mm) in diameter where suspended from above shall be supported with one hanger per interval and in accordance with Section 603.3.2.1 through Section 603.3.2.3.

603.3.2.1 Tight-Fitting Around the Perimeter. Ducts shall be equipped with tight-fitting circular bands extending around the entire perimeter of the duct at each specified support interval.

603.3.2.2 Size of Circular Bands. Circular bands shall be not less than 1 inch (25.4 mm) wide nor less than equivalent to the gauge of the duct material it supports.

Exception: Ducts not more than 10 inches (254 mm) in diameter shall be permitted to be supported by No. 18 gauge galvanized steel wire.

603.3.2.3 Connection. Each circular band shall be provided with means of connecting to the suspending support.

603.3.3 Earthquake Loads. Ducts located in structures that are installed in areas classified as seismic design category C, D, E, or F shall be in accordance with the *California Building Code or California Residential Code.*

603.4 Factory-Made Air Ducts. Factory-made air ducts shall be listed and labeled in accordance with UL 181 and installed in accordance with the terms of their listing, the manufacturer's installation instructions, and SMACNA HVAC Duct Construction Standards – Metal and Flexible.

Factory-made air ducts shall not be used for vertical risers in air-duct systems serving more than two stories and shall not penetrate a fire-resistance-rated assembly or construction.

Factory-made air ducts shall be installed with not less than 4 inches (102 mm) of separation from earth, except where installed as a liner inside of concrete, tile, or metal pipe and shall be protected from physical damage.

The temperature of the air to be conveyed in a duct shall not exceed 250°F (121°C). Flexible air connectors shall not be permitted.

603.4.1 Length Limitation. ***[Not permitted for OSHPD 1, 1R, 2, 3, 4 & 5]*** Factory-made flexible air ducts and connectors shall be not more than 5 feet (1524 mm) in length and shall not be used in lieu of rigid elbows or fittings. Flexible air ducts shall be permitted to be used as an elbow at a terminal device. TIA TIA TIA TIA

Exception: Residential occupancies.

603.5 Flexible Air Ducts. Flexible air ducts shall comply with UL 181, and shall be installed in accordance with the manufacturer's installation instructions and SMACNA HVAC Duct Construction Standards – Metal and Flexible.

Flexible air duct installations shall comply with the following:

(1) Ducts shall be installed using the minimum required length to make the connection.

(2) Horizontal duct runs shall be supported at not more than 4 feet (1219 mm) intervals.

(3) Vertical risers shall be supported at not more than 6 feet (1829 mm) intervals.

(4) Sag between support hangers shall not exceed ½ inch (12.7 mm) per foot (305 mm) of support spacing.

(5) Supports shall be rigid and shall be not less than 1½ inches (38 mm) wide at point of contact with the duct surface.

(6) Duct bends shall be not less than one duct diameter bend radius.

(7) Screws shall not penetrate the inner liner of non-metallic flexible ducts unless permitted in accordance with the manufacturer's installation instructions.

(8) Fittings for attaching non-metallic ducts shall be beaded and have a collar length of not less than 2 inches (51 mm) for attaching the duct.

Exception: A bead shall not be required where metal worm-gear clamps are used or where attaching metallic ducts using screws in accordance with the manufacturer's installation instructions.

(9) Duct inner liner shall be installed at not less than 1 inch (25.4 mm) on the collar and past the bead prior to the application of the tape and mechanical fastener. Where mastic is used instead of tape, the mastic shall be applied in accordance the mastic manufacturer's instructions.

(10) Duct outer vapor barriers shall be secured using two wraps of approved tape. A mechanical fastener shall be permitted to be used in place of, or in combination with, the tape.

(11) Flexible air ducts shall not penetrate a fire-resistance-rated assembly or construction.

(12) The temperature of the air to be conveyed in a flexible air duct shall not exceed 250°F (121°C).

(13) Flexible Air ducts shall be sealed in accordance with Section 603.10.

603.6 Plastic Ducts. Plastic air ducts and fittings shall be permitted where installed underground and listed for such use.

603.7 Protection of Ducts. Ducts installed in locations where they are exposed to mechanical damage by vehicles or from other causes shall be protected by approved barriers.

603.8 Support of Ducts. Installers shall provide the manufacturer's field fabrication and installation instructions.

Factory-made air ducts that are in accordance with UL 181 shall be supported in accordance with the manufacturer's installation instructions. Other ducts shall comply with SMACNA HVAC Duct Construction Standards – Metal and Flexible.

603.9 Protection Against Flood Damage. In flood hazard areas, ducts shall be located above the elevation required by the building code for utilities and attendant equipment or the elevation of the lowest floor, whichever is higher, or shall be designed and constructed to prevent water from entering or accumulating within the ducts during floods up to such elevation. Where the ducts are located below that elevation, the ducts shall be capable of resisting hydrostatic and hydrodynamic loads and stresses, including the effects of buoyancy, during the occurrence of flooding to such elevation.

603.10 Joints and Seams of Ducts. Joints and seams for duct systems shall comply with SMACNA HVAC Duct Construction Standards – Metal and Flexible. Joints of duct systems shall be made substantially airtight by means of tapes, mastics, gasketing, or other means. Crimp joints for round ducts shall have a contact lap of not less than 1½ inches (38 mm) and shall be mechanically fastened by means of not less than three sheet-metal screws equally spaced around the joint, or an equivalent fastening method.

Joints and seams and reinforcements for factory-made air ducts and plenums shall comply with the conditions of prior approval in accordance with the installation instructions that shall accompany the product. Closure systems for sealing factory made air ducts and plenums shall be listed and labeled in accordance with UL 181A or UL 181B, and marked in accordance with Table 603.10.

TABLE 603.10
CLOSURE MARKINGS

TYPE OF DUCTWORK	STANDARD	TYPE OF CLOSURE SYSTEM	MARKING
Rigid Metallic or Rigid Fiberglass	UL 181A	Pressure Sensitive Tape	181A-P
Rigid Metallic or Rigid Fiberglass	UL 181A	Mastic Tape	181A-M
Rigid Metallic or Rigid Fiberglass	UL 181A	Heat Sensitive Tape	181A–H
Flexible Air Ducts	UL 181B	Pressure Sensitive Tape*	181B-FX*
Flexible Air Ducts	UL 181B	Mastic*	181B-M*

* Mechanical fasteners shall be used in conjunction with a listed pressure sensitive tape or mastic in accordance with UL181. Nonmetallic mechanical fasteners shall be listed and labeled in accordance with UL 181B and labeled "181B-C."

603.10.1 Duct Leakage Tests. *[Not adopted by HCD]* Ductwork shall be leak-tested in accordance with the SMACNA HVAC Air Duct Leakage Test Manual. Representative sections totaling not less than 10 percent of the total installed duct area shall be tested. Where the tested 10 percent fail to comply with the requirements of this section, then 40 percent of the total installed duct area shall be tested. Where the tested 40 percent fail to comply with the requirements of this section, then 100 percent of the total installed duct area shall be tested. Sections shall be selected by the building owner or designated representative of the building owner. Positive pressure leakage testing shall be permitted for negative pressure ductwork. The permitted duct leakage shall be not more than the following:

$$L_{max} = C_L P^{0.65} \quad \text{(Equation 603.10.1)}$$

Where:

L_{max} = maximum permitted leakage, (ft³/min)/100 square feet [0.0001 (m³/s)/m²] duct surface area.

C_L = six, duct leakage class, (ft³/min)/100 square feet [0.0001 (m³/s)/m²] duct surface area at 1 inch water column (0.2 kPa).

P = test pressure, which shall be equal to the design duct pressure class rating, inch water column (kPa).

603.10.1.1 Duct Leakage Tests for Residential Buildings. [HCD 1 & HCD 2] *See California Energy Code Section 150.0(m)(11) for low-rise residential; and Section 140.4(l) for duct leakage tests for other residential buildings.*

603.11 Cross Contamination. Exhaust ducts and venting systems under positive pressure shall not extend into or pass through ducts or plenums.

603.12 Underground Installation. Ducts installed underground shall be approved for the installation and shall have a slope of not less than 1/8 inch per foot (10.4 mm/m) back to the main riser. Ducts, plenums, and fittings shall be permitted to be constructed of concrete, clay, or ceramics where installed in the ground or in a concrete slab, provided the joints are sealed and duct is secured in accordance with SMACNA HVAC Duct Construction Standards – Metal and Flexible. Metal ducts where installed in or under a concrete slab shall be encased in not less than 2 inches (51 mm) of concrete, secured in accordance with SMACNA HVAC Duct Construction Standards-Metal and Flexible.

603.13 Air Dispersion Systems. Where installed, air dispersion systems shall be completely in exposed locations in duct systems under positive pressure, and not pass through or penetrate fire-resistant-rated construction. Air dispersion systems shall be listed and labeled in accordance with UL 2518.

604.0 Insulation of Ducts.

604.1 General. Air ducts conveying air at temperatures exceeding 140°F (60°C) shall be insulated to maintain an insulation surface temperature of not more than 140°F (60°C). Factory-made air ducts and insulations intended for installation on the exterior of ducts shall be legibly printed with the name of the manufacturer, the thermal resistance (R) value at installed thickness, flame-spread index, and smoke developed index of the composite material. Internal duct liners and insulation shall be installed in accordance with SMACNA HVAC Duct Construction Standards – Metal and Flexible. ***[OSHPD 1, 1R, 2, 3, 4 & 5]*** *Cold air ducts shall be insulated wherever necessary or to prevent condensation.*

Exceptions:

(1) Factory-installed plenums, casings, or ductwork furnished as a part of HVAC equipment tested and rated in accordance with approved energy efficiency standards.

(2) Ducts or plenums located in conditioned spaces where heat gain or heat loss will not increase energy use.

(3) For runouts less than 10 feet (3048 mm) in length to air terminals or air outlets, the rated R-value of insulation need not exceed R-3.5.

(4) Backs of air outlets and outlet plenums exposed to unconditioned or indirectly conditioned spaces with face areas exceeding 5 square feet (0.5 m^2) need not exceed R-2; those 5 square feet (0.5 m^2) or smaller need not be insulated.

(5) Ducts and plenums used exclusively for evaporative cooling systems.

604.1.1 Within Ducts or Plenums. Materials installed within ducts and plenums for insulating, sound deadening, or other purposes shall have a mold, humidity, and erosion-resistant surface where tested in accordance with UL 181. Duct liners in systems operating with air velocities exceeding 2000 feet per minute (10.16 m/s) shall be fastened with both adhesive and mechanical fasteners, and exposed edges shall have approved treatment to withstand the operating velocity. Where the internal insulation is capable of being in contact with condensates or other liquids, the material shall be water-resistant.

604.1.2 Duct Coverings and Linings. Insulation applied to the surface of ducts, including duct coverings, linings, tapes, and adhesives, located in buildings shall have a flame-spread index not to exceed 25 and a smoke-developed index not to exceed 50, where tested in accordance with ASTM E84 or UL 723. The specimen preparation and mounting procedures of ASTM E2231 shall be used. Air duct coverings and linings shall not flame, glow, smolder, or smoke where tested in accordance with ASTM C411 at the temperature to which they are exposed in service. In no case shall the test temperature be less than 250°F (121°C). Coverings shall not penetrate a fire-resistance-rated assembly.

604.2 [OSHPD 1, 1R, 2, 3 (surgical clinics), 4 & 5] *Thermal acoustical lining materials shall not be installed within ducts, terminal boxes, sound traps, and other in-duct systems serving areas such as operating, cesarean operating rooms, delivery rooms, post anesthesia care units, cystoscopy, cardiac catheterization labs, nurseries, intensive care units, newborn intensive care units, and airborne infection isolation rooms unless terminal filters with 90 percent average efficiency based on ASHRAE Standard 52.2 or minimum efficiency rating value (MERV) of 14 are installed downstream of the duct lining.*

604.3 [OSHPD 1, 1R, 2, 4 & 5] *Thermal or acoustical lining materials shall not be installed within ducts which are downstream of the 99.97 percent high-efficiency particulate air (HEPA) filter or with minimum efficiency rating value (MERV) of 17 required in Section 408.2.1 for protective environment rooms.*

605.0 Smoke Dampers, Fire Dampers, and Ceiling Dampers.

605.1 Smoke Dampers. Smoke dampers shall comply with UL 555S, and shall be installed in accordance with the manufacturer's installation instructions where required by the *California Building Code or California Residential Code.*

605.2 Fire Dampers. Fire dampers shall comply with UL 555, and shall be installed in accordance with the manufacturer's installation instructions where required by the building code. Fire dampers shall have been tested for closure under airflow conditions and shall be labeled for both maximum airflow permitted and direction of flow. Where more than one damper is installed at a point in a single air path, the entire airflow shall be assumed to be passing through the smallest damper area.

Ductwork shall be connected to damper sleeves or assemblies in accordance with the fire damper manufacturer's installation instructions.

605.3 Ceiling Radiation Dampers. Ceiling radiation dampers shall comply with UL 555C, and shall be installed in accordance with the manufacturer's installation instructions in the fire-resistive ceiling membrane of floor-ceiling and roof-ceiling assemblies where required by the building code. Fire dampers not meeting the temperature limitation of ceiling radiation dampers shall not be used as a substitute.

605.4 Multiple Arrangements. Where size requires the use of multiple dampers, each damper shall be listed for use in multiple arrangements and installed in accordance with the manufacturer's installation instructions.

605.5 Access and Identification. Fire and smoke dampers shall be provided with an approved means of access large enough to allow inspection and maintenance of the damper and its operating parts. The access shall not affect the integrity of the fire-resistance-rated assembly. The access openings shall not reduce the fire-resistance rating of the assembly.

Access shall not require the use of tools. Access doors in ducts shall be tight fitting and approved for the required duct construction. Access points shall be permanently identified on the exterior by a label with letters not less than ½ of an inch (12.7 mm) in height reading as one of the following:

(1) Smoke Damper

(2) Fire Damper

(3) Fire/Smoke Damper

605.6 Freedom from Interference. Dampers shall be installed in a manner to ensure positive closing or opening as required by function. Interior liners or insulation shall be held back from portions of a damper, its sleeve, or adjoining duct that would interfere with the damper's proper operation. Exterior materials shall be installed so as to not interfere with the operation or maintenance of external operating devices needed for the function of the damper.

605.7 Temperature Classification of Operating Elements. Fusible links, thermal sensors, and pneumatic or electric operators shall have a temperature rating or classification as in accordance with the building code.

605.8 *When the automatic activation of a smoke damper or a combination smoke-fire damper occurs, the HVAC system serving such dampers shall immediately shut down.*

Exceptions:

(1) HVAC systems that are part of an engineered smoke evacuation system.

(2) Where the automatic activation causes all the smoke dampers and combination smoke-fire dampers to close in the enclosed space having a common atmosphere where openings are required to be protected.

(3) Where analysis demonstrates shutoff would create a greater hazard.

The HVAC system shall not be restarted again until all the dampers are reset and fully opened.

606.0 Ventilating Ceilings.

606.1 General. Perforated ceilings shall be permitted to be used for air supply within the limitations of this section. Exit corridors, where required to be of fire-resistive construction by the *California Building Code*, shall not have ventilating ceilings.

606.1.1 [OSHPD 1, 1R, 2, 3, 4 & 5] *Ventilating ceilings are not permitted in health facilities.*

606.2 Requirements. Ventilating ceilings shall comply with the following:

(1) Suspended ventilating ceiling material shall have a Class 1 flame spread classification on both sides, determined in accordance with the building code. Suspended ventilating ceiling supports shall be of noncombustible materials.

(2) Lighting fixtures recessed into ventilating ceilings shall be of a type approved for that purpose.

607.0 Use of Under-Floor Space as Supply Plenum for Dwelling Units.

607.1 General. An under-floor space shall be permitted to be used as a supply plenum.

607.2 Dwelling Units. The use of under-floor space shall be limited to dwelling units not more than two stories in height. Except for the floor immediately above the under-floor plenum, supply ducts shall be provided extending from the plenum to registers on other floor levels.

Exception: In flood hazard areas, under-floor spaces shall not be used as supply plenums unless the flood opening requirements in the building code are met.

607.3 Enclosed. Such spaces shall be cleaned of all loose combustible scrap material and shall be tightly enclosed.

607.4 Flammable Materials. The enclosing material of the under-floor space, including the sidewall insulation, shall be not more flammable than 1 inch (25.4 mm) (nominal) wood boards (flame-spread index of 200). Installation of foam plastics is regulated by the building code.

607.5 Access. Access shall be through an opening in the floor and shall be not less than 24 inches by 24 inches (610 mm by 610 mm).

607.6 Automatic Control. A furnace supplying warm air to under-floor space shall be equipped with an automatic control that will start the air-circulating fan where the air in the furnace bonnet reaches a temperature not exceeding 150°F (66°C). Such control shall be one that cannot be set to exceed 150°F (66°C).

607.7 Temperature Limit. A furnace supplying warm air to such space shall be equipped with an approved temperature limit control that will limit outlet air temperature to 200°F (93°C).

607.8 Noncombustible Receptacle. A noncombustible receptacle shall be placed below each floor opening into the air chamber, and such receptacle shall comply with Section 607.8.1 through Section 607.8.3.

607.8.1 Location. The receptacle shall be securely suspended from the floor members and shall be not more than 18 inches (457 mm) below the floor opening.

607.8.2 Area. The area of the receptacle shall extend 3 inches (76 mm) beyond the opening on all sides.

607.8.3 Perimeter. The perimeter of the receptacle shall have a vertical lip not less than 1 inch (25.4 mm) high at the open sides where it is at the level of the bottom of the joists, or 3 inches (76 mm) high where the receptacle is suspended.

607.9 Floor Registers. Floor registers shall be designed for easy removal in order to give access for cleaning the receptacles.

607.10 Exterior Wall and Interior Stud Partitions. Exterior walls and interior stud partitions shall be fire blocked at the floor.

607.11 Wall Register. Each wall register shall be connected to the air chamber by a register box or boot.

607.12 Distance from Combustible. A duct complying with Section 602.0 shall extend from the furnace supply outlet not less than 6 inches (152 mm) below combustible framing.

607.13 Vapor Barrier. The entire ground surface of the under-floor space shall be covered with a vapor barrier having a thickness not less than 4 mils (0.1 mm) and a flame-spread index of not more than 200.

607.14 Prohibited. Fuel gas lines and plumbing waste cleanouts shall not be located within the space.

608.0 Automatic Shutoffs.

608.1 Air-Moving Systems and Smoke Detectors. Air-moving systems supplying air in excess of 2000 cubic feet per minute (ft^3/min) (0.9439 m^3/s) to enclosed spaces within buildings shall be equipped with an automatic shutoff. Automatic shutoff shall be accomplished by interrupting the power source of the air-moving equipment upon detection of smoke in the main supply-air duct served by such equipment. Duct smoke detectors shall comply with UL 268A*, shall be labeled by an approved agency, approved and listed by California State Fire Marshal,* and shall be installed in accordance with the manufacturer's installation instructions. Such devices shall be compatible with the operating velocities, pressures, temperatures, and humidities of the system. Where fire-detection or alarm systems are provided for the building, the smoke detectors shall be supervised by such systems in an approved manner, *and installed in accordance with NFPA 72 and the California Building and Fire Codes.*

Exceptions:

(1) Where the space supplied by the air-moving equipment is served by a total coverage smoke-detection system in accordance with *the California Fire Code*, interconnection to such system shall be permitted to be used to accomplish the required shutoff.

(2) Automatic shutoff is not required where occupied rooms served by the air-handling equipment have direct exit to the exterior, and the travel distance does not exceed 100 feet (30 480 mm).

(3) Automatic shutoff is not required for Group R, Division 3 and Group U Occupancies.

(4) Automatic shutoff is not required for approved smoke-control systems or where analysis demonstrates shutoff would create a greater hazard, such as shall be permitted to be encountered in air-moving equipment supplying specialized portions of Group H Occupancies. Such equipment shall be required to have smoke detection with remote indication and manual shutoff capability at an approved location.

(5) Smoke detectors that are factory installed in listed air-moving equipment shall be permitted to be used in lieu of smoke detectors installed in the main supply-air duct served by such equipment.

CALIFORNIA MECHANICAL CODE – MATRIX ADOPTION TABLE
CHAPTER 7 – COMBUSTION AIR

(Matrix Adoption Tables are non-regulatory, intended only as an aid to the code user. See Chapter 1 for state agency authority and building applications.)

Adopting Agency	BSC	BSC-CG	SFM	HCD			DSA			OSHPD						BSCC	DPH	AGR	DWR	CEC	CA	SL	SLC
				1	2	1-AC	AC	SS	SS/CC	1	1R	2	3	4	5								
Adopt Entire Chapter	X		X	X	X			X	X	X	X	X	X	X	X		X						
Adopt Entire Chapter as amended (amended sections listed below)																							
Adopt only those sections that are listed below																							
Chapter/Section																							

This state agency does not adopt sections identified with the following symbol: †

The Office of the State Fire Marshal's adoption of this chapter or individual sections is applicable to structures regulated by other state agencies pursuant to Section 1.11.0.

CHAPTER 7
COMBUSTION AIR

701.0 General.

701.1 Applicability. Air for combustion, ventilation, and dilution of flue gases for appliances installed in buildings shall be obtained by application of one of the methods covered in Section 701.4 through Section 701.9.3. Where the requirements of Section 701.4 are not met, outdoor air shall be introduced in accordance with methods covered in Section 701.6 through Section 701.9.3.

Exceptions:

(1) This provision shall not apply to direct-vent appliances.

(2) Type 1 clothes dryers that are provided with makeup air in accordance with Section 504.4.1. [NFPA 54:9.3.1.1]

701.1.1 Other Types of Appliances. Appliances of other than natural draft design, appliances not designated as Category I vented appliances, and appliances equipped with power burners shall be provided with combustion, ventilation, and dilution air in accordance with the appliance manufacturer's instructions. [NFPA 54:9.3.1.2]

701.2 Pressure Difference. Where used, a draft hood or a barometric draft regulator shall be installed in the same room or enclosure as the appliance served so as to prevent a difference in pressure between the hood or regulator and the combustion-air supply. [NFPA 54:9.3.1.4]

701.3 Makeup Air. Where exhaust fans, clothes dryers, and kitchen ventilation systems interfere with the operation of appliances, makeup air shall be provided. [NFPA 54:9.3.1.5]

701.4 Indoor Combustion Air. The required volume of indoor air shall be determined in accordance with the method in Section 701.4.1 or Section 701.4.2, except that where the air infiltration rate is known to be less than 0.40 ACH (air change per hour), the method in Section 701.4.2 shall be used. The total required volume shall be the sum of the required volume calculated for appliances located within the space. Rooms communicating directly with the space in which the appliances are installed through openings not furnished with doors, and through combustion air openings sized and located in accordance with Section 701.5, are considered a part of the required volume. [NFPA 54:9.3.2]

701.4.1 Standard Method. The required volume shall be not less than 50 cubic feet per 1000 British thermal units per hour (Btu/h) (4.83 m^3/kW). [NFPA 54:9.3.2.1]

701.4.2 Known Air Infiltration Rate Method. Where the air infiltration rate of a structure is known, the minimum required volume shall be determined as follows [NFPA 54:9.3.2.2]:

(1) For appliances other than fan-assisted, calculate using the following Equation 701.4.2(1). [NFPA 54:9.3.2.2(1)]

(2) For fan-assisted appliances, calculate using the following Equation 701.4.2(2). [NFPA 54: 9.3.2.2(2)]

(3) For purposes of these calculations, an infiltration rate greater than 0.60 ACH shall not be used in the equations in Section 701.4.2(1) and Section 701.4.2(2). [NFPA 54:9.3.2.2(3)]

[Equation 701.4.2(1)]

$$\text{Required Volume}_{other} \geq (21\ ft^3/ACH) \times (I_{other}/1000\ Btu/h)$$

[Equation 701.4.2(2)]

$$\text{Required Volume}_{fan} \geq (15\ ft^3/ACH) \times (I_{fan}/1000\ Btu/h)$$

Where:

I_{other} = All appliances other than fan-assisted input (Btu/h)

I_{fan} = Fan-assisted appliance input (Btu/h)

ACH = Air change per hour (percent of volume of space exchanged per hour, expressed as a decimal)

For SI units: 1 cubic foot = 0.0283 m^3, 1000 British thermal units per hour = 0.293 kW

701.5 Indoor Opening Size and Location. Openings used to connect indoor spaces shall be sized and located in accordance with the following:

(1) Each opening shall have a free area of not less than 1 square inch per 1000 Btu/h (0.002 m^2/kW) of the total input rating of appliances in the space, but not less than 100 square inches (0.065 m^2). One opening shall commence within 12 inches (305 mm) of the top of the enclosure, and one opening shall commence within 12 inches (305 mm) of the bottom of the enclosure (see Figure 701.5). The dimension of air openings shall be not less than 3 inches (76 mm).

(2) The volumes of spaces in different stories shall be considered as communicating spaces where such spaces are connected by one or more openings in doors or floors having a total free area of not less than 2 square inches per 1000 Btu/h (0.004 m^2/kw) of total input rating of appliances. [NFPA 54:9.3.2.3]

701.6 Outdoor Combustion Air. Outdoor combustion air shall be provided through opening(s) to the outdoors in accordance with the methods in Section 701.6.1 or Section 701.6.2. The dimension of air openings shall be not less than 3 inches (76 mm). [NFPA 54:9.3.3]

701.6.1 Two Permanent Openings Method. Two permanent openings, one commencing within 12 inches (305 mm) of the top of the enclosure and one commencing within 12 inches (305 mm) of the bottom of the enclosure, shall be provided. The openings shall communicate directly, or by ducts, with the outdoors or spaces that freely communicate with the outdoors as follows:

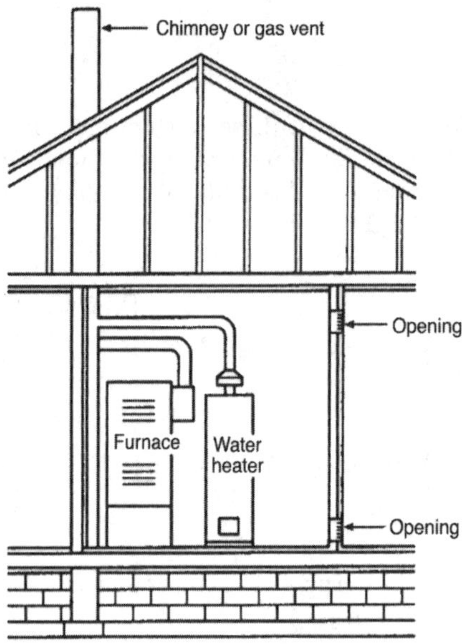

FIGURE 701.5
COMBUSTION AIR FROM ADJACENT INDOOR SPACES THROUGH INDOOR COMBUSTION AIR OPENINGS
[NFPA 54: FIGURE A.9.3.2.3(1)]

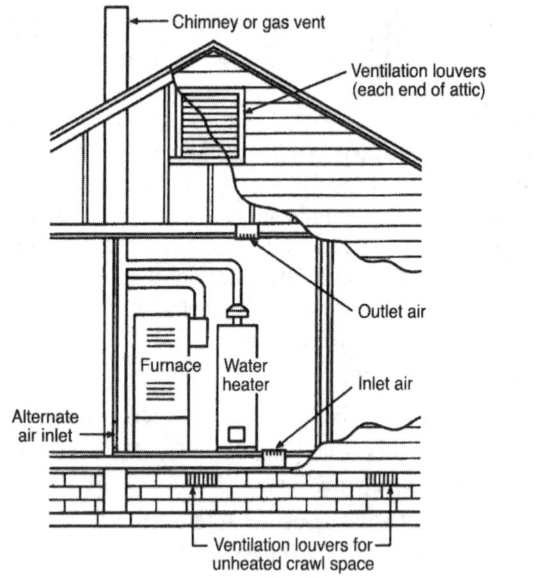

FIGURE 701.6.1(1)
COMBUSTION AIR FROM OUTDOORS
INLET AIR FROM VENTILATED CRAWL SPACE AND OUTLET AIR TO VENTILATED ATTIC
[NFPA 54: FIGURE A.9.3.3.1(1)(a)]

(1) Where directly communicating with the outdoors or where communicating to the outdoors through vertical ducts, each opening shall have a free area of not less than 1 square inch per 4000 Btu/h (0.0005 m^2/kW) of total input rating of appliances in the enclosure. [See Figure 701.6.1(1) and Figure 701.6.1(2)]

(2) Where communicating with the outdoors through horizontal ducts, each opening shall have a free area of not less than 1 square inch per 2000 Btu/h (0.001 m^2/kW) of total input rating of appliances in the enclosure. [See Figure 701.6.1(3)] [NFPA 54:9.3.3.1]

701.6.2 One Permanent Opening Method. One permanent opening, commencing within 12 inches (305 mm) of the top of the enclosure, shall be provided. The appliance shall have clearances of at least 1 inch (25.4 mm) from the sides and back and 6 inches (152 mm) from the front of the appliance. The opening shall directly communicate with the outdoors or shall communicate through a vertical or horizontal duct to the outdoors or spaces that freely communicate with the outdoors (see Figure 701.6.2) and shall have a minimum free area of the following:

(1) One square inch per 3000 Btu/h (0.0007 m^2/kW) of the total input rating of all appliances located in the enclosure.

(2) Not less than the sum of the areas of all vent connectors in the space. [NFPA 54:9.3.3.2]

701.7 Combination Indoor and Outdoor Combustion Air. The use of a combination of indoor and outdoor combustion air shall be in accordance with Section 701.7.1 through Section 701.7.3. [NFPA 54:9.3.4] (see Appendix F for example calculations)

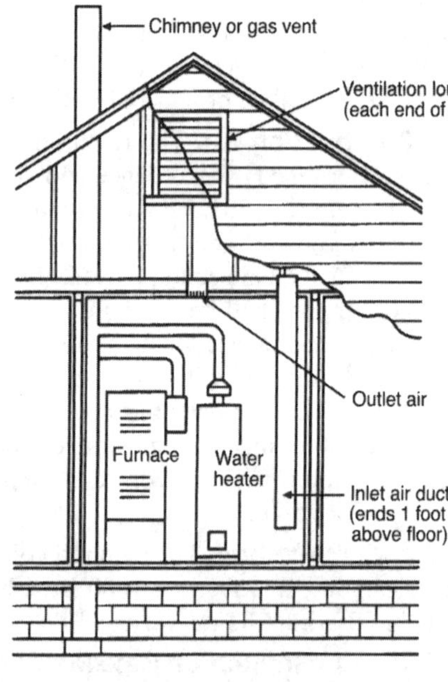

FIGURE 701.6.1(2)
COMBUSTION AIR FROM OUTDOORS THROUGH VENTILATED ATTIC
[NFPA 54: FIGURE A.9.3.3.1(1)(b)]

701.7.1 Indoor Openings. Where used, openings connecting the interior spaces shall be in accordance with Section 701.5. [NFPA 54:9.3.4(1)]

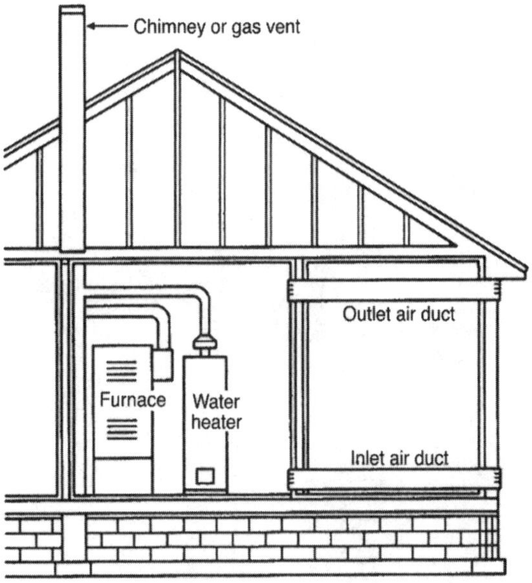

FIGURE 701.6.1(3)
COMBUSTION AIR FROM OUTDOORS THROUGH HORIZONTAL DUCTS
[NFPA 54: FIGURE A.9.3.3.1(2)]

701.7.2 Outdoor Openings. Outdoor openings shall be located in accordance with Section 701.6. [NFPA 54: 9.3.4(2)]

701.7.3 Outdoor Opening(s) Size. The outdoor opening(s) size shall be calculated in accordance with the following:

(1) The ratio of interior spaces shall be the available volume of all communicating spaces divided by the required volume.

(2) The outdoor size reduction factor shall be one minus the ratio of interior spaces.

(3) The minimum size of outdoor opening(s) shall be the full size of outdoor opening(s) calculated in accordance with Section 701.6, multiplied by the reduction factor. The minimum dimension of air openings shall not be less than 3 inches (76 mm). [NFPA 54:9.3.4(3)]

701.8 Engineered Installations. Engineered combustion air installations shall provide an adequate supply of combustion, ventilation, and dilution air and shall be approved by the Authority Having Jurisdiction. [NFPA 54:9.3.5]

701.9 Mechanical Combustion Air Supply. Where combustion air is provided by a mechanical air supply system, the combustion air shall be supplied from outdoors at the rate of not less than 0.35 cubic feet per minute per 1000 Btu/h [0.034 (m^3/min)/kW] for appliances located within the space. [NFPA 54:9.3.6]

701.9.1 Exhaust Fans. Where exhaust fans are installed, additional air shall be provided to replace the exhausted air. [NFPA 54:9.3.6.1]

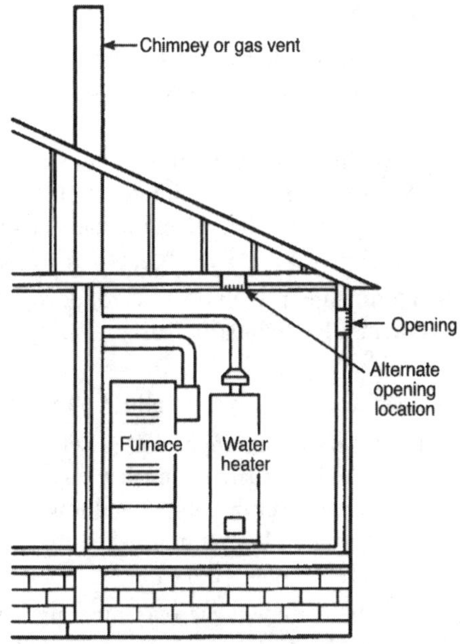

FIGURE 701.6.2
COMBUSTION AIR FROM OUTDOORS THROUGH SINGLE COMBUSTION AIR OPENING
[NFPA 54: FIGURE A.9.3.3.2]

701.9.2 Interlock. Each of the appliances served shall be interlocked to the mechanical air supply system to prevent main burner operation where the mechanical air supply system is not in operation. [NFPA 54:9.3.6.2]

701.9.3 Specified Combustion Air. Where combustion air is provided by the building's mechanical ventilation system, the system shall provide the specified combustion air rate in addition to the required ventilation air. [NFPA 54:9.3.6.3]

701.10 Louvers, Grilles, and Screens. The required size of openings for combustion, ventilation, and dilution air shall be based on the net free area of each opening. Where the free area through a design of louver, grille, or screen is known, it shall be used in calculating the size opening required to provide the free area specified. Where the louver and grille design and free area are not known, it shall be assumed that wood louvers have 25 percent free area and metal louvers and grilles have 75 percent free area. Nonmotorized louvers and grilles shall be fixed in the open position. [NFPA 54:9.3.7.1]

701.10.1 Minimum Screen Mesh Size. Screens shall be not less than ¼ of an inch (6.4 mm) mesh. [NFPA 54:9.3.7.2]

701.10.2 Motorized Louvers. Motorized louvers shall be interlocked with the appliance so they are proven in the full open position prior to main burner ignition and during main burner operation. Means shall be provided to prevent the main burner from igniting where the louver fail to open during burner start-up and to shut down the main burner where the louvers close during burner operation. [NFPA 54:9.3.7.3]

701.11 Combustion Air Ducts. Combustion air ducts shall comply with the following [NFPA 54:9.3.8]:

(1) Ducts shall be constructed of galvanized steel or a material having equivalent corrosion resistance, strength, and rigidity.

Exception: Within dwellings units, unobstructed stud and joist spaces shall not be prohibited from conveying combustion air, provided that not more than one fireblock is removed. [NFPA 54:9.3.8.1]

(2) Ducts shall terminate in an unobstructed space, allowing free movement of combustion air to the appliances. [NFPA 54:9.3.8.2]

(3) Ducts shall serve a single space. [NFPA 54:9.3.8.3]

(4) Ducts shall not serve both upper and lower combustion air openings where both such openings are used. The separation between ducts serving upper and lower combustion air openings shall be maintained to the source of combustion air. [NFPA 54:9.3.8.4]

(5) Ducts shall not be screened where terminating in an attic space. [NFPA 54:9.3.8.5]

(6) Combustion air intake openings located on the exterior of the building shall have the lowest side of the combustion air intake openings located at least 12 inches (305 mm) vertically from the adjoining finished ground level. [NFPA 54:9.3.8.8]

(7) Horizontal upper combustion air ducts shall not slope downward toward the source of combustion air. [NFPA 54:9.3.8.6]

(8) The remaining space surrounding a chimney liner, gas vent, special gas vent, or plastic piping installed within a masonry, metal, or factory-built chimney shall not be used to supply combustion air.

Exception: Direct-vent appliances designed for installation in a solid fuel-burning fireplace where installed in accordance with the manufacturer's installation instructions. [NFPA 54:9.3.8.7]

701.12 Dampers Prohibited. Combustion air ducts or plenums shall not be installed so as to require openings in or penetrations through construction where fire dampers are required. Manually operated dampers shall not be installed in combustion air openings. With prior approval, power-actuated movable louvers admitting combustion air shall be permitted to be used and, where installed, shall be electrically interlocked with the main burner fuel-supply valve so as to prevent fuel delivery unless the louvers are in the fully open position.

702.0 Extra Device or Attachment.

702.1 General. No device or attachment shall be installed on an appliance that is capable of impairing the combustion of gas. [NFPA 54:9.1.15]

CALIFORNIA MECHANICAL CODE – MATRIX ADOPTION TABLE
CHAPTER 8 – CHIMNEYS AND VENTS

(Matrix Adoption Tables are non-regulatory, intended only as an aid to the code user. See Chapter 1 for state agency authority and building applications.)

Adopting Agency	BSC	BSC-CG	SFM	HCD			DSA			OSHPD						BSCC	DPH	AGR	DWR	CEC	CA	SL	SLC
				1	2	1-AC	AC	SS	SS/CC	1	1R	2	3	4	5								
Adopt Entire Chapter	X		X	X	X			X	X	X	X	X	X	X	X								
Adopt Entire Chapter as amended (amended sections listed below)																							
Adopt only those sections that are listed below																							
Chapter/Section																							

This state agency does not adopt sections identified with the following symbol: †

The Office of the State Fire Marshal's adoption of this chapter or individual sections is applicable to structures regulated by other state agencies pursuant to Section 1.11.0.

CHAPTER 8
CHIMNEYS AND VENTS

801.0 General.

801.1 Applicability. The requirements of this chapter shall govern the venting of fuel-burning appliances.

801.2 Venting of Gas Appliances. Low-heat and medium-heat gas appliances shall be vented in accordance with this chapter. Other gas appliances shall be vented in accordance with NFPA 211 or other applicable standards.

801.3 Appliances Fueled by Other Fuels. Appliances fueled by fuels other than gas shall be vented in accordance with NFPA 211 and the appliance manufacturer's instructions.

802.0 Venting of Appliances.

802.1 Listing. Type B and Type B-W gas vents shall comply with UL 441, Type L gas vents shall comply with UL 641.

802.1.1 Installation. Listed vents shall be installed in accordance with this chapter and the manufacturer's installation instructions. [NFPA 54:12.2.1]

802.1.2 Prohibited Discharge. Appliance vents shall not discharge into a space enclosed by screens having openings less than ¼ of an inch (6.4 mm) mesh.

802.2 Connection to Venting Systems. Except as permitted in Section 802.2.1 through Section 802.2.8, all appliances shall be connected to venting systems. [NFPA 54:12.3.1]

802.2.1 Appliances Not Required to be Vented. The following appliances shall not be required to be vented:

(1) Listed ranges.

(2) Built-in domestic cooking units listed and marked for optional venting.

(3) Listed hot plates and listed laundry stoves.

(4) Listed Type 1 clothes dryers exhausted in accordance with Section 504.4.

(5) A single listed booster-type (automatic instantaneous) water heater, when designed and used solely for the sanitizing rinse requirements of a dishwashing machine, provided that the appliance is installed with the draft hood in place and unaltered, if a draft hood is required, in a commercial kitchen having a mechanical exhaust system; where installed in this manner, the draft hood outlet shall not be less than 36 inches (914 mm) vertically and 6 inches (152 mm) horizontally from any surface other than the appliance.

(6) Listed refrigerators.

(7) Counter appliances.

(8) Room heaters listed for unvented use.

(9) Direct gas-fired makeup air heaters.

(10) Other appliances listed for unvented use and not provided with flue collars.

(11) Specialized appliances of limited input such as laboratory burners or gas lights. [NFPA 54:12.3.2]

802.2.2 Maximum Input Rating. Where any or all of the appliances in Section 802.2.1(5) through Section 802.2.1(11) are installed so the aggregate input rating exceeds 20 Btu/hr/ft^3 (207 W/m^3) of room or space in which it is installed, one or more shall be provided with venting systems or other approved means for conveying the vent gases to the outdoors so the aggregate input rating of the remaining unvented appliances does not exceed 20 Btu/hr/ft^3 (207 W/m^3). [NFPA 54:12.3.2.1]

802.2.3 Adjacent Room or Space. Where the calculation includes the volume of an adjacent room or space, the room or space in which the appliances are installed shall be directly connected to the adjacent room or space by a doorway, archway, or other opening of comparable size that cannot be closed. [NFPA 54:12.3.2.2]

802.2.4 Ventilating Hoods. Ventilating hoods and exhaust systems shall be permitted to be used to vent appliances installed in commercial applications and to vent industrial appliances, particularly where the process itself requires fume disposal. [NFPA 54:12.3.3]

802.2.5 Well-Ventilated Spaces. The operation of industrial appliances such that its flue gases are discharged directly into a large and well-ventilated space shall be permitted. [NFPA 54:12.3.4]

802.2.6 Direct-Vent Appliances. Listed direct-vent appliances shall be installed in accordance with the manufacturer's installation instructions and Section 802.8.2. [NFPA 54:12.3.5]

802.2.7 Appliances with Integral Vents. Appliances incorporating integral venting means shall be installed in accordance with the manufacturer's installation instructions, Section 802.8, and Section 802.8.1. [NFPA 54:12.3.6]

802.2.8 Incinerators. Commercial industrial-type incinerators shall be vented in accordance with NFPA 82. [NFPA 54:12.3.7]

802.3 Design and Construction. Venting systems shall be designed and constructed to convey flue and vent gases to the outdoors. [NFPA 54:12.1]

802.3.1 Appliance Draft Requirements. A venting system shall satisfy the draft requirements of the appliance in accordance with the manufacturer's instructions. [NFPA 54:12.4.1]

802.3.2 Appliance Venting Requirements. Appliances required to be vented shall be connected to a venting system designed and installed in accordance with the provisions of Section 802.4 through Section 802.15.1. [NFPA 54:12.4.2]

802.3.3 Mechanical Draft Systems. Mechanical draft systems shall be listed and installed in accordance with both the appliance and the mechanical draft system manufacturer's installation instructions. [NFPA 54:12.4.3.1]

802.3.3.1 Venting. Appliances requiring venting shall be permitted to be vented by means of mechanical draft systems of either forced or induced draft design. [NFPA 54:12.4.3.2]

802.3.3.2 Leakage. Forced draft systems and portions of induced draft systems under positive pressure during operation shall be designed and installed so as to prevent leakage of flue or vent gases into a building. [NFPA 54:12.4.3.3]

802.3.3.3 Vent Connectors. Vent connectors serving appliances vented by natural draft shall not be connected into any portion of mechanical draft systems operating under positive pressure. [NFPA 54:12.4.3.4]

802.3.3.4 Operation. Where a mechanical draft system is employed, provision shall be made to prevent the flow of gas to the main burners where the draft system is not performing so as to satisfy the operating requirements of the appliance for safe performance. [NFPA 54:12.4.3.5]

802.3.3.5 Exit Terminals. The exit terminals of mechanical draft systems shall be not less than 7 feet (2134 mm) above finished ground level where located adjacent to public walkways and shall be located as specified in Section 802.8 and Section 802.8.1. [NFPA 54:12.4.3.6]

802.3.4 Ventilating Hoods and Exhaust Systems. Ventilating hoods and exhaust systems shall be permitted to be used to vent appliances installed in commercial applications. [NFPA 54:12.4.4.1]

802.3.4.1 Automatically Operated Appliance. Where automatically operated appliances, other than commercial cooking appliances, are vented through a ventilating hood or exhaust system equipped with a damper or with a power means of exhaust, provisions shall be made to allow the flow of gas to the main burners where the damper is open to a position to properly vent the appliance and where the power means of exhaust is in operation. [NFPA 54:12.4.4.2]

802.3.5 Circulating Air Ducts and Furnace Plenums. Venting systems shall not extend into or pass through a fabricated air duct or furnace plenum. [NFPA 54:12.4.5.1]

802.3.6 Above-Ceiling or Nonducted Air Handling System. Where a venting system passes through an above-ceiling air space or other nonducted portion of an air-handling system, it shall conform to one of the following requirements:

(1) The venting system shall be a listed special gas vent, other system serving a Category III or Category IV appliance, or other positive pressure vent, with joints sealed in accordance with the appliance or vent manufacturer's instructions.

(2) The vent system shall be installed such that no fittings or joints between sections are installed in the above-ceiling space.

(3) The venting system shall be installed in a conduit or enclosure with joints between the interior of the enclosure and the ceiling space sealed. [NFPA 54:12.4.5.2]

802.4 Type of Venting System to be Used. The type of venting system to be used shall be in accordance with Table 802.4. [NFPA 54:12.5.1]

802.4.1 Plastic Piping. Where plastic piping is used to vent an appliance, the appliance shall be listed for use with such venting materials and the appliance manufacturer's installation instructions shall identify the specific plastic piping material. [NFPA 54:12.5.2]

802.4.2 Plastic Vent Joints. Plastic pipe and fittings used to vent appliances shall be installed in accordance with the appliance manufacturer's installation instructions. Where primer is required, it shall be of a contrasting color. [NFPA 54:12.5.3]

802.4.3 Special Gas Vent. Special gas vents shall be listed and installed in accordance with the special gas vent manufacturer's installation instructions. [NFPA 54:12.5.4]

802.5 Masonry, Metal, and Factory-Built Chimneys. Chimneys shall be installed in accordance with Section 802.5.1 through Section 802.5.3.

802.5.1 Factory-Built Chimneys. Factory-built chimneys shall be installed in accordance with the manufacturer's installation instructions. Factory-built chimneys used to vent appliances that operate at positive vent pressure shall be listed for such application. [NFPA 54:12.6.1.1]

802.5.1.1 Decorative Shrouds. Decorative shrouds addressed in Section 802.5.4.3 shall be listed or labeled in accordance with UL 103 for factory-built residential chimneys, UL 127 for factory-built fireplaces, or UL 1482 for solid-fuel room heaters.

802.5.1.2 Listing Requirements. Factory-built chimneys shall comply with the requirements of UL 103 or of UL 959. Factory-built chimneys for use with wood-burning appliances shall comply with the Type HT requirements of UL 103. [NFPA 211:6.1.3.1, 6.1.3.2]

802.5.2 Metal Chimneys. Metal chimneys shall be built and installed in accordance with NFPA 211. [NFPA 54:12.6.1.2]

802.5.3 Masonry Chimneys. Masonry chimneys shall be built and installed in accordance with NFPA 211 and lined with approved clay flue lining, a listed chimney lining system, or other approved material that resists cor-

TABLE 802.4
TYPE OF VENTING SYSTEM TO BE USED
[NFPA 54: TABLE 12.5.1]

APPLIANCES	TYPE OF VENTING SYSTEM	LOCATION OF REQUIREMENTS
Listed Category I appliances	Type B gas vent	Section 802.6
Listed appliances equipped with draft hood	Chimney	Section 802.5
Appliances listed for use with Type B gas vent	Single-wall metal pipe Listed chimney lining system for gas venting Special gas vent listed for these appliances	Section 802.7 Section 802.5.3 Section 802.4.3
Listed vented wall furnaces	Type B-W gas vent	Section 802.6, Section 907.0
Category II appliances Category III appliances Category IV appliances	As specified or furnished by manufacturers of listed appliances	Section 802.4.1 and Section 802.4.3
Incinerators	Single-wall metal pipe	NFPA 82
Appliances that can be converted to use of solid fuel Unlisted combination gas- and oil-burning appliances Combination gas- and solid-fuel-burning appliances Appliances listed for use with chimneys only Unlisted appliances	Chimney	Section 802.5
Listed combination gas- and oil-burning appliances	Type L vent Chimney	Section 802.6 Section 802.5
Decorative appliances in vented fireplace	Chimney	Section 911.2
Gas-fired toilets	Single-wall metal pipe	Section 802.7, Section 929.3
Direct-vent appliances	—	Section 802.2.6
Appliances with integral vents	—	Section 802.2.7

rosion, erosion, softening, or cracking from vent gases at temperatures not exceeding 1800°F (982°C).

Exception: Masonry chimney flues lined with a chimney lining system specifically listed for use with listed appliances with draft hoods, Category I appliances, and other appliances listed for use with Type B vents shall be permitted. The liner shall be installed in accordance with the liner manufacturer's installation instructions. A permanent identifying label shall be attached at the point where the connection is to be made to the liner. The label shall read: "This chimney liner is for appliances that burn gas only. Do not connect to solid-or liquid-fuel-burning appliances or incinerators." [NFPA 54:12.6.1.3]

802.5.4 Termination. A chimney for a residential-type or low-heat appliance shall extend not less than 3 feet (914 mm) above the highest point where it passes through a roof of a building and not less than 2 feet (610 mm) higher than a portion of a building within a horizontal distance of 10 feet (3048 mm). (See Figure 802.5.4) [NFPA 54:12.6.2.1]

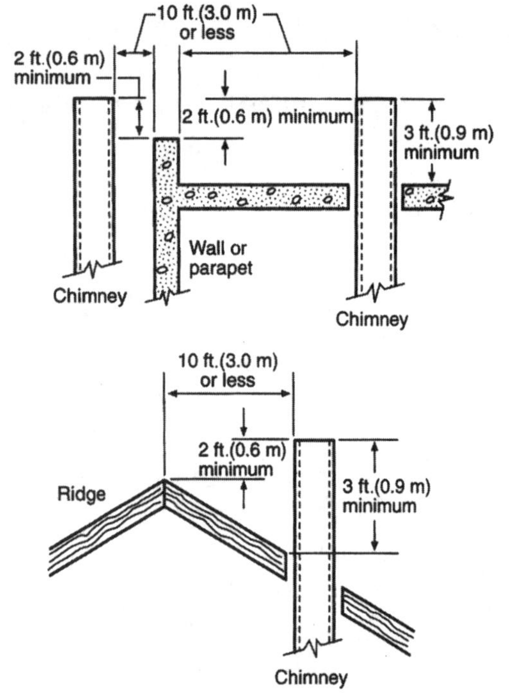

(a) Termination 10 ft.(3.0 m) or Less from Ridge, Wall, or Parapet

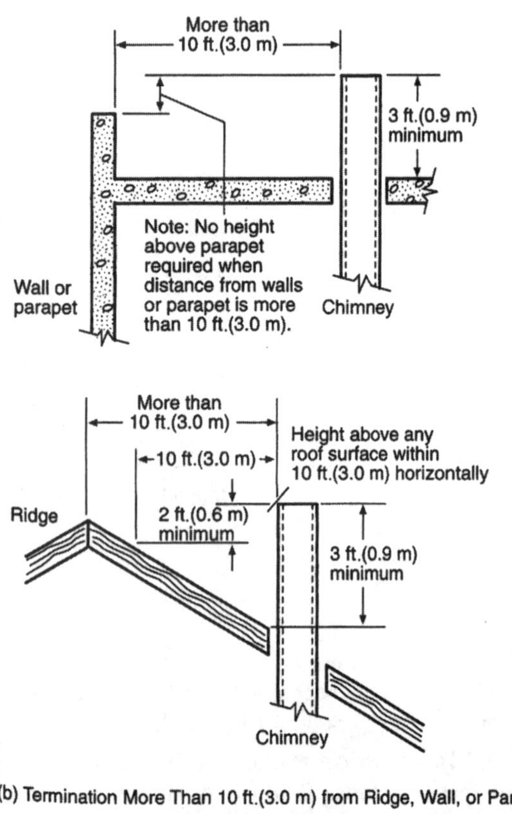

(b) Termination More Than 10 ft.(3.0 m) from Ridge, Wall, or Parapet

FIGURE 802.5.4
TYPICAL TERMINATION LOCATIONS FOR CHIMNEYS AND SINGLE-WALL METAL PIPES SERVING RESIDENTIAL-TYPE AND LOW-HEAT APPLIANCE [NFPA 54: FIGURE A.12.6.2.1]

802.5.4.1 Medium-Heat Gas Appliances. A chimney for medium-heat appliance shall extend at least less than 10 feet (3048 mm) higher than any portion of any building within 25 feet (7620 mm). [NFPA 54:12.6.2.2]

802.5.4.2 Chimney Height. A chimney shall extend not less than 5 feet (1524 mm) above the highest connected appliance draft hood outlet or flue collar. [NFPA 54:12.6.2.3]

802.5.4.3 Decorative Shrouds. Decorative shrouds shall not be installed at the termination of factory-built chimneys except where such shrouds are listed and labeled for use with the specific factory-built chimney system and are installed in accordance with the manufacturer's installation instructions. [NFPA 54:12.6.2.4]

802.5.5 Size of Chimneys. The effective area of a chimney venting system serving listed appliances with draft hoods, Category I appliances, and other appliances listed for use with Type B vents shall be in accordance with one of the following methods:

(1) Those listed in Section 803.0.

(2) For sizing an individual chimney venting system for a single appliance with a draft hood, the effective areas of the vent connector and chimney flue shall be not less than the area of the appliance flue collar or draft hood outlet or greater than seven times the draft hood outlet area.

(3) For sizing a chimney venting system connected to two appliances with draft hoods, the effective area of the chimney flue shall be not less than the area of the larger draft hood outlet plus 50 percent of the area of the smaller draft hood outlet or greater than seven times the smaller draft hood outlet area.

(4) Chimney venting systems using mechanical draft shall be sized in accordance with approved engineering methods.

(5) Other approved engineering methods. [NFPA 54:12.6.3.1]

802.5.6 Inspection of Chimneys or Vents. This inspection shall be made after chimneys, vents, or parts thereof, authorized by the permit, have been installed and before such vent or part thereof has been covered or concealed.

802.5.7 Inspection of Chimneys. Before replacing an existing appliance or connecting a vent connector to a chimney, the chimney passageway shall be examined to ascertain that it is clear and free of obstructions and shall be cleaned where previously used for venting solid- or liquid-fuel-burning appliances or fireplaces. [NFPA 54:12.6.4.1]

802.5.7.1 Standard. Chimneys shall be lined in accordance with NFPA 211.

Exception: Existing chimneys shall be permitted to have their use continued when an appliance is replaced by an appliance of similar type, input rating, and efficiency, where the chimney complies with Section 802.5.7 through Section 802.5.7.3 and the sizing of the chimney is in accordance with Section 802.5.5. [NFPA 54:12.6.4.2]

802.5.7.2 Cleanouts. Cleanouts shall be examined to determine that they will remain tightly closed where not in use. [NFPA 54:12.6.4.3]

802.5.7.3 Existing Chimney. Where inspection reveals that an existing chimney is not safe for the intended application, it shall be repaired, rebuilt, lined, relined, or replaced with a vent or chimney in accordance with NFPA 211, and shall be approved for the appliances to be attached. [NFPA 54:12.6.4.4]

802.5.8 Chimney Serving Appliances Burning Other Fuels. An appliance shall not be connected to a chimney flue serving a separate appliance designed to burn solid fuel. [NFPA 54:12.6.5.1]

802.5.8.1 Gas and Liquid-Fuel-Burning Appliances. Where one chimney serves gas appliances and liquid fuel-burning appliances, the appliances shall be connected through separate openings or shall be connected through a single opening where joined by a fitting located as close as practical to the chimney. Where two or more openings are provided into one chimney flue, they shall be at different levels. Where the gas appliance is automatically controlled, it shall be equipped with a safety shutoff device. [NFPA 54:12.6.5.2]

802.5.8.2 Gas and Solid-Fuel-Burning Appliances. A listed combination gas-and solid-fuel-burning appliance connected to a single chimney flue shall be equipped with a manual reset device to shut off gas to the main burner in the event of sustained backdraft or flue gas spillage. The chimney flue shall be sized to properly vent the appliance. [NFPA 54:12.6.5.3]

802.5.8.3 Combination Gas- and Oil-Burning Appliances. A single chimney flue serving a listed combination gas- and oil-burning appliance shall be sized to properly vent the appliance. [NFPA 54:12.6.5.4]

802.5.9 Support of Chimneys. All portions of chimneys shall be supported for the design and weight of the materials employed. Listed factory-built chimneys shall be supported and spaced in accordance with the manufacturer's installation instructions. [NFPA 54:12.6.6]

802.5.10 Cleanouts. Where a chimney that formerly carried flue products from liquid or solid fuel-burning appliances is used with an appliance using fuel gas, an accessible cleanout shall be provided. The cleanout shall have a tight-fitting cover and be installed so its upper edge is not less than 6 inches (152 mm) below the lower edge of the lowest chimney inlet opening. [NFPA 54:12.6.7]

802.5.11 Space Surrounding Lining or Vent. The remaining space surrounding a chimney liner, gas vent, special gas vent, or plastic piping installed within a masonry chimney shall not be used to vent another appliance.

Exception: The insertion of another liner or vent within the chimney as provided in this code and the liner or vent manufacturer's instructions. [NFPA 54:12.6.8.1]

802.5.11.1 Combustion Air. The remaining space surrounding a chimney liner, gas vent, special gas vent, or plastic piping installed within a masonry, metal or factory-built chimney flue shall not be used to supply combustion air.

Exception: Direct-vent appliances designed for installation in a solid-fuel-burning fireplace where installed in accordance with the manufacturer's installation instructions. [NFPA 54:12.6.8.2]

802.6 Gas Vents. The installation of gas vents shall meet the following requirements:

(1) Gas vents shall be installed in accordance with the manufacturer's installation instructions.
(2) A Type B-W gas vent shall have a listed capacity not less than that of the listed vented wall furnace to which it is connected.
(3) Gas vents installed within masonry chimneys shall be installed in accordance with the manufacturer's installation instructions. Gas vents installed within masonry chimneys shall be identified with a permanent label installed at the point where the vent enters the chimney. The label shall contain the following language: "This gas vent is for appliances that burn gas. Do not connect to solid or liquid fuel–burning appliances or incinerators."
(4) Screws, rivets, and other fasteners shall not penetrate the inner wall of double-wall gas vents, except at the transition from the appliance draft hood outlet, flue collar, or single-wall metal connector to a double-wall vent. [NFPA 54:12.7.1]

802.6.1 Termination Requirements. A gas vent shall terminate in accordance with one of the following:

(1) Gas vents that are 12 inches (300 mm) or less in size and located not less than 8 feet (2438 mm) from a vertical wall or similar obstruction shall terminate above the roof in accordance with Figure 802.6.1 and Table 802.6.1.
(2) Gas vents that are over 12 inches (300 mm) in size or are located less than 8 feet (2438 mm) from a vertical wall or similar obstruction shall terminate not less than 2 feet (610 mm) above the highest point where they pass through the roof and not less than 2 feet (610 mm) above a portion of a building within 10 feet (3048 mm) horizontally.
(3) Industrial appliances as provided in Section 802.2.5.
(4) Direct-vent systems as provided in Section 802.2.6.
(5) Appliance with integral vents as provided in Section 802.2.7.
(6) Mechanical draft systems as provided in Section 802.3.3 through Section 802.3.3.5.
(7) Ventilating hoods and exhaust systems as provided in Section 802.3.4 and Section 802.3.4.1. [NFPA 54:12.7.2(1)]

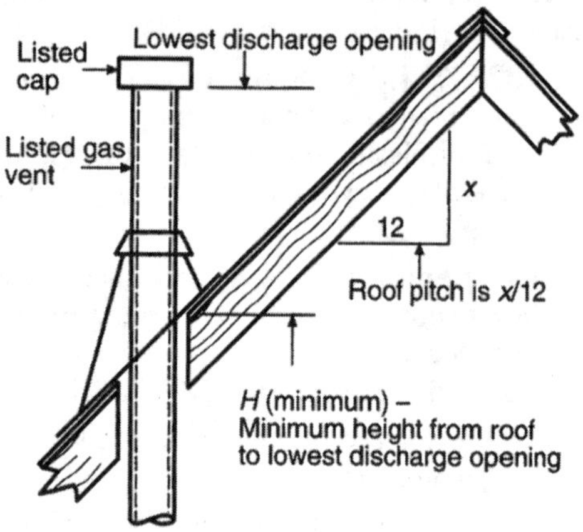

FIGURE 802.6.1
GAS VENT TERMINATION LOCATIONS FOR LISTED CAPS 12 INCHES OR LESS IN SIZE NOT LESS THAN 8 FEET FROM A VERTICAL WALL
[NFPA 54: FIGURE 12.7.2]

TABLE 802.6.1
ROOF PITCH HEIGHT
[NFPA 54: TABLE 12.7.2]

ROOF PITCH	H (minimum) (feet)
Flat to 6/12	1.0
Over 6/12 to 7/12	1.25
Over 7/12 to 8/12	1.5
Over 8/12 to 9/12	2.0
Over 9/12 to 10/12	2.5
Over 10/12 to 11/12	3.25
Over 11/12 to 12/12	4.0
Over 12/12 to 14/12	5.0
Over 14/12 to 16/12	6.0
Over 16/12 to 18/12	7.0
Over 18/12 to 20/12	7.5
Over 20/12 to 21/12	8.0

For SI units: 1 inch = 25.4 mm, 1 foot = 304.8 mm

802.6.1.1 Type B and L Vents. A Type B or a Type L gas vent shall terminate not less than 5 feet (1524 mm) in vertical height above the highest connected appliance draft hood or flue collar. [NFPA 54:12.7.2(2)]

802.6.1.2 Type B-W Vents. A Type B-W gas vent shall terminate not less than 12 feet (3658 mm) in vertical height above the bottom of the wall furnace. [NFPA 54:12.7.2(3)]

802.6.1.3 Exterior Wall Termination. A gas vent extending through an exterior wall shall not terminate adjacent to the wall or below eaves or parapets, except as provided in Section 802.2.6 and Section 802.3.3 through Section 802.3.3.5. [NFPA 54:12.7.2(4)]

802.6.1.4 Decorative Shrouds. Decorative shrouds shall not be installed at the termination of gas vents except where such shrouds are listed for use with the specific gas venting system and are installed in accordance with the manufacturer's installation instructions. [NFPA 54:12.7.2(5)]

802.6.1.5 Termination Cap. A gas vent shall extend through the roof flashing, roof jack, or roof thimble and terminate with a listed cap or listed roof assembly. [NFPA 54:12.7.2(6)]

802.6.1.6 Forced Air Inlet. A gas vent shall terminate not less than 3 feet (914 mm) above a forced air inlet located within 10 feet (3048 mm). [NFPA 54:12.7.2(7)]

802.6.1.7 Insulation Shield. Where a vent passes through an insulated assembly, an approved metal shield shall be installed between the vent and insulation. The shield shall extend not less than 2 inches (51 mm) above the insulation and be secured to the structure in accordance with the manufacturer's installation instructions.

802.6.2 Size of Gas Vents. Venting systems shall be sized and constructed in accordance with Section 803.0 or other approved engineering methods and the gas vent and the appliance manufacturer's instructions. [NFPA 54:12.7.3]

802.6.2.1 Category I Appliances. The sizing of natural draft venting systems serving one or more listed appliances equipped with a draft hood or appliances listed for use with a Type B gas vent, installed in a single story of a building, shall be in accordance with one of the following:

(1) The provisions of Section 803.0.

(2) Vents serving fan-assisted combustion system appliances, or combinations of fan-assisted combustion system and draft hood-equipped appliances shall be sized in accordance with Section 803.0 or other approved engineering methods.

(3) For sizing an individual gas vent for a single, draft hood-equipped appliance, the effective area of the vent connector and the gas vent shall be not less than the area of the appliance draft hood outlet or exceeding seven times the draft hood outlet area.

(4) For sizing a gas vent connected to two appliances, with draft hoods, the effective area of the vent shall be not less than the area of the larger draft hood outlet plus 50 percent of the area of the smaller draft hood outlet or exceeding seven times the smaller draft hood outlet area.

(5) Approved engineering practices. [NFPA 54:12.7.3.1]

802.6.2.2 Vent Offsets. Type B and Type L vents sized in accordance with Section 802.6.2.1(3) or Section 802.6.2.1(4) shall extend in a generally vertical direction with offsets not exceeding 45 degrees except that a vent system having not more than one

60 degree offset shall be permitted. Any angle greater than 45 degrees from the vertical is considered horizontal. The total horizontal distance of a vent plus the horizontal vent connector serving draft hood-equipped appliances shall not be greater than 75 percent of the vertical height of the vent. [NFPA 54:12.7.3.2]

802.6.2.3 Category II, Category III, and Category IV Appliances. The sizing of gas vents for Category II, Category III, and Category IV appliances shall be in accordance with the appliance manufacturer's instructions. The sizing of plastic pipe specified by the appliance manufacturer as a venting material for Category II, Category III, and Category IV appliances shall be in accordance with the appliance manufacturers' instructions. [NFPA 54:12.7.3.3]

802.6.2.4 Sizing. Chimney venting systems using mechanical draft shall be sized in accordance with approved engineering methods. [NFPA 54:12.7.3.4]

802.6.3 Gas Vents Serving Appliances on More than One Floor. A common vent shall be permitted in multistory installations to vent Category I appliances located on more than one floor level, provided the venting system is designed and installed in accordance with approved engineering methods.

For the purpose of this section, crawl spaces, basements, and attics shall be considered as floor levels. [NFPA 54:12.7.4.1]

802.6.3.1 Occupiable Space. Appliances connected to the common vent shall be located in rooms separated from an occupiable space. Each of these rooms shall have provisions for an adequate supply of combustion, ventilation, and dilution air that is not supplied from an occupiable space. (See Figure 802.6.3.1) [NFPA 54:12.7.4.2]

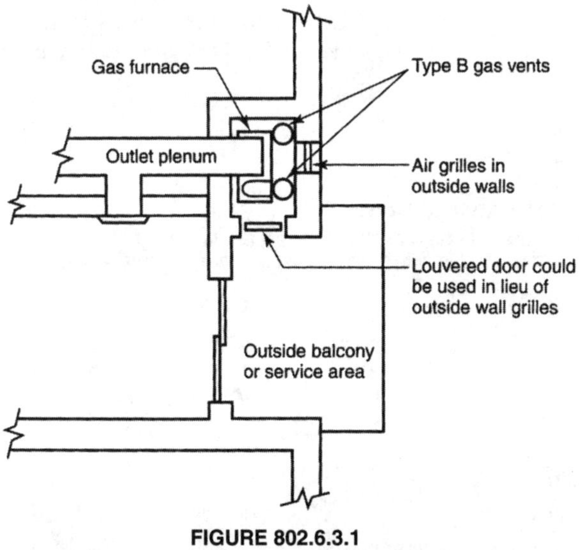

FIGURE 802.6.3.1
PLAN VIEW OF PRACTICAL SEPARATION METHOD FOR MULTISTORY GAS VENTING
[NFPA 54: FIGURE A.12.7.4.2]

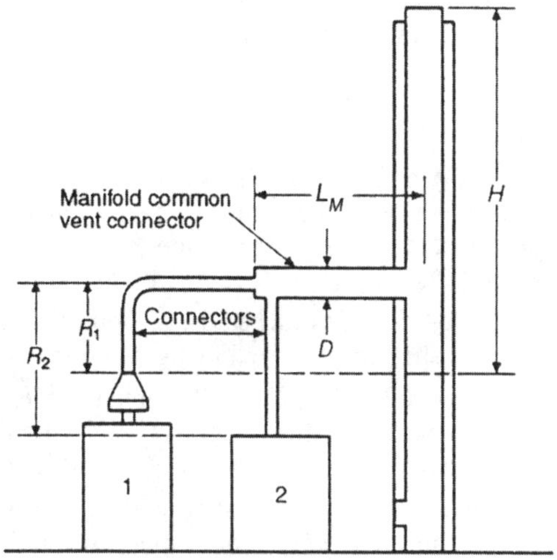

FIGURE 802.6.3.2
USE OF MANIFOLDED COMMON VENT CONNECTOR
[NFPA 54: FIGURE F.1(k)]

802.6.3.2 Multistory Venting System. The size of the connectors and common segments of multistory venting systems for appliances listed for use with a Type B double-wall gas vent shall be in accordance with Table 803.2(1), provided all of the following apply:

(1) The available total height (*H*) for each segment of a multistory venting system is the vertical distance between the level of the highest draft hood outlet or flue collar on that floor and the centerline of the next highest interconnection tee.

(2) The size of the connector for a segment is determined from the appliance's gas input rate and available connector rise, and shall not be smaller than the draft hood outlet or flue collar size.

(3) The size of the common vertical vent segment, and of the interconnection tee at the base of that segment, is based on the total appliance's gas input rate entering that segment and its available total height. [NFPA 54:12.7.4.3]

802.6.4 Support of Gas Vents. Gas vents shall be supported and spaced in accordance with the manufacturer's installation instructions. [NFPA 54:12.7.5]

802.6.5 Marking. In those localities where solid and liquid fuels are used extensively, gas vents shall be permanently identified by a label attached to the wall or ceiling at a point where the vent connector enters the gas vent. The label shall read: "This gas vent is for appliances that burn gas. Do not connect to solid or liquid fuel-burning appliances or incinerators." The Authority Having Jurisdiction shall determine whether its area constitutes such a locality. [NFPA 54:12.7.6]

802.7 Single-Wall Metal Pipe. Single-wall metal pipe shall be constructed of galvanized sheet steel not less than 0.0304 of an inch (0.7722 mm) thick or of other approved, noncombustible, corrosion-resistant material. [NFPA 54:12.8.1]

802.7.1 Cold Climate. Uninsulated single-wall metal pipe shall not be used outdoors for venting appliances in regions where the 99 percent winter design temperature is below 32°F (0°C). [NFPA 54:12.8.2]

802.7.2 Termination. The termination of single-wall metal pipe shall meet the following requirements:

(1) Single-wall metal pipe shall terminate at least 5 feet (1524 mm) in vertical height above the highest connected appliance draft hood outlet or flue collar.

(2) Single-wall metal pipe shall extend at least 2 feet (610 mm) above the highest point where it passes through a roof of a building and at least 2 feet (610 mm) higher than any portion of a building within a horizontal distance of 10 feet (3048 mm). (See Figure 802.5.4)

(3) An approved cap or roof assembly shall be attached to the terminus of a single-wall metal pipe. [NFPA 54:12.8.3]

802.7.3 Installation with Appliances Permitted by Section 802.4. Single-wall metal pipe shall not be used as a vent in dwellings and residential occupancies. [NFPA 54:12.8.4.1]

802.7.3.1 Limitations. Single-wall metal pipe shall be used for runs directly from the space in which the appliance is located through the roof or exterior wall to the outer air. A pipe passing through a roof shall extend without interruption through the roof flashing, roof jacket, or roof thimble. [NFPA 54:12.8.4.2]

802.7.3.2 Attic or Concealed Space. Single-wall metal pipe shall not originate in an unoccupied attic or concealed space and shall not pass through an attic, inside wall, concealed space, or floor. [NFPA 54:12.8.4.3]

802.7.3.3 Clearances. Minimum clearances from single-wall metal pipe to combustible material shall be in accordance with Table 802.7.3.3. Reduced clearances from single-wall metal pipe to combustible material shall be as specified for vent connectors in Table 303.10.1. [NFPA 54:12.8.4.4]

802.7.3.4 Combustible Exterior Wall. Single-wall metal pipe shall not pass through a combustible exterior wall unless guarded at the point of passage by a ventilated metal thimble not smaller than the following:

(1) For listed appliances with draft hoods and appliances listed for use with Type B gas vents, the thimble shall be a minimum of 4 inches (102 mm) larger in diameter than the metal pipe. Where there is a run of not less than 6 feet (1829 mm) of metal pipe in the opening between the draft hood outlet and the thimble, the thimble shall be a minimum of 2 inches (51 mm) larger in diameter than the metal pipe.

(2) For unlisted appliances having draft hoods, the thimble shall be a minimum of 6 inches (152 mm) larger in diameter than the metal pipe.

(3) For residential and low-heat appliances, the thimble shall be a minimum of 12 inches (305 mm) larger in diameter than the metal pipe.

Exception: In lieu of thimble protection, combustible material in the wall shall be removed a sufficient distance from the metal pipe to provide the specified clearance from such metal pipe to combustible material. Any material used to close up such opening shall be noncombustible. [NFPA 54:12.8.4.6]

802.7.3.5 Roof Thimble. Where a single-wall metal pipe passes through a roof constructed of combustible material, a noncombustible, nonventilating thimble shall be used at the point of passage. The thimble shall extend not less than 18 inches (457 mm) above and 6 inches (152 mm) below the roof

TABLE 802.7.3.3
CLEARANCE FOR CONNECTORS*
[NFPA 54: TABLE 12.8.4.4]

APPLIANCE	MINIMUM DISTANCE FROM COMBUSTIBLE MATERIAL (inches)			
	LISTED TYPE B GAS VENT MATERIAL	LISTED TYPE L VENT MATERIAL	SINGLE-WALL METAL PIPE	FACTORY-BUILT CHIMNEY SECTIONS
Listed appliance with draft hoods and appliance listed for use with Type B gas vents	As listed	As listed	6	As listed
Residential boilers and furnaces with listed gas conversion burner and with draft hood	6	6	9	As listed
Residential appliances listed for use with Type L vents	Not permitted	As listed	9	As listed
Listed gas-fired toilets	Not permitted	As listed	As listed	As listed
Unlisted residential appliances with draft hood	Not permitted	6	9	As listed
Residential and low-heat appliance other than those above	Not permitted	9	18	As listed
Medium-heat appliance	Not permitted	Not permitted	36	As listed

For SI units: 1 inch = 25.4 mm

* These clearances shall apply unless the installation instructions of a listed appliance or connector specify different clearances, in which case the listed clearances shall apply.

with the annular space open at the bottom and closed at the top. The thimble shall be sized in accordance with Section 802.7.3.4. [NFPA 54:12.8.4.5]

802.7.4 Size of Single-Wall Metal Pipe. Single-wall metal piping shall comply with Section 802.7.4.1 through Section 802.7.4.3. [NFPA 54:12.8.5]

802.7.4.1 Sizing of Venting System. A venting system of a single-wall metal pipe shall be sized in accordance with one of the following methods and the appliance manufacturer's instructions:

(1) For a draft hood-equipped appliance, in accordance with Section 803.0.

(2) For a venting system for a single appliance with a draft hood, the areas of the connector and the pipe each shall not be less than the area of the appliance flue collar or draft hood outlet, whichever is smaller. The vent area shall not exceed seven times the draft hood outlet area.

(3) Other approved engineering methods. [NFPA 54:12.8.5(1)]

802.7.4.2 Non-Round Metal Pipe. Where a single-wall metal pipe is used and has a shape other than round, it shall have an effective area equal to the effective area of the round pipe for which it is substituted, and the internal dimension of the pipe shall be not less than 2 inches (51 mm). [NFPA 54:12.8.5(2)]

802.7.4.3 Venting Capacity. The vent cap or a roof assembly shall have a venting capacity not less than that of the pipe to which it is attached. [NFPA 54:12.8.5(3)]

802.7.5 Support of Single-Wall Metal Pipe. Portions of single-wall metal pipe shall be supported for the design and weight of the material employed. [NFPA 54:12.8.6]

802.7.6 Marking. Single-wall metal pipe shall comply with the marking provisions of Section 802.6.5. [NFPA 54:12.8.7]

802.8 Through-the-Wall Vent Termination. A mechanical draft venting system shall terminate at least 3 feet (914 mm) above any forced air inlet located within 10 feet (3048 mm). (See Figure 802.8)

Exceptions:

(1) This provision shall not apply to the combustion-air intake of a direct-vent appliance.

(2) This provision shall not apply to the separation of the integral outdoor-air inlet and flue gas discharge of listed outdoor appliances. [NFPA 54:12.9.1]

802.8.1 Mechanical Draft Venting System. A mechanical draft venting system of other than direct-vent type shall terminate not less than 4 feet (1219 mm) below, 4 feet (1219 mm) horizontally from, or 1 foot (305 mm) above a door, operable window, or gravity air inlet into a building. The bottom of the vent terminal shall be located not less than 12 inches (305 mm) above finished ground level. [NFPA 54:12.9.2]

802.8.2 Direct-Vent Appliance. The vent terminal of a direct-vent appliance with an input of 10 000 Btu/h (3 kW) or less shall be located at least 6 inches (152 mm) from any air opening into a building, an appliance with an input over 10 000 Btu/h (3 kW) but not over 50 000 Btu/h (14.7 kW) shall be installed with a 9 inch (229 mm) vent termination clearance, and an appliance with an input exceeding 50 000 Btu/h (14.7 kW) shall have at least a 12 inch (305 mm) vent termination clearance. The bottom of the vent terminal and the air intake shall be located at least 12 inches (305 mm) above finished ground level. [NFPA 54:12.9.3]

802.8.3 Category I through Category IV and Non-categorized Appliances. Through-the-wall vents for Category II and Category IV appliances and noncategorized condensing appliances shall not terminate over public walkways or over an area where condensate or vapor could create a nuisance or hazard or could be detrimental to the operation of regulators, relief valves, or other equipment. Where local experience indicates that condensate is a problem with Category I and Category III appliances, this provision shall also apply.

Drains for condensate shall be installed in accordance with the appliance and the vent manufacturer's installation instructions. [NFPA 54:12.9.4]

802.8.4 Annular Spaces. Where vents, including those for direct-vent appliances or combustion air intake pipes, penetrate outside walls of buildings, the annular spaces around such penetrations shall be permanently sealed using approved materials to prevent entry of combustion products into the building. [NFPA 54:12.9.5]

802.8.5 Vent Terminals. Vent systems for Category IV appliances that terminate through an outside wall of a building and discharge flue gases perpendicular to the adjacent wall shall be located not less than 10 feet (3048 mm) horizontally from an operable opening in an adjacent building.

Exception: This shall not apply to vent terminals that are 2 feet (610 mm) or more above or 25 feet (7620 mm) or more below operable openings. [NFPA 54:12.9.6]

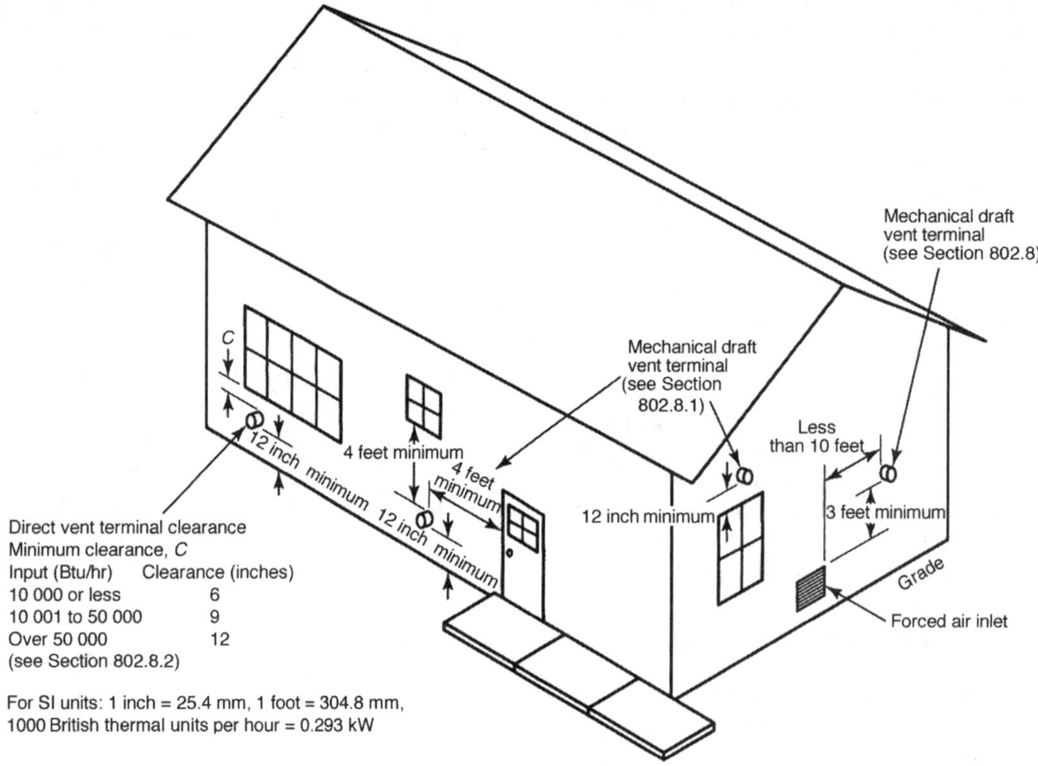

FIGURE 802.8
EXIT TERMINALS OF MECHANICAL DRAFT AND DIRECT-VENT VENTING SYSTEMS
[NFPA 54: FIGURE A.12.9]

802.9 Condensation Drain. Provision shall be made to collect and dispose of condensate from venting systems serving Category II and Category IV appliances and noncategorized condensing appliances in accordance with Section 802.8.3. [NFPA 54:12.10.1]

802.9.1 Local Experience. Where local experience indicates that condensation is a problem, provision shall be made to drain off and dispose of condensate from venting systems serving Category I and Category III appliances in accordance with Section 802.8.3. [NFPA 54:12.10.2]

802.10 Vent Connectors for Category I Appliances. A vent connector shall be used to connect an appliance to a gas vent, chimney, or single-wall metal pipe, except where the gas vent, chimney, or single-wall metal pipe is directly connected to the appliance. [NFPA 54:12.11.1]

802.10.1 Materials. A vent connector shall be made of noncombustible, corrosion resistant material capable of withstanding the vent gas temperature produced by the appliance and of a thickness to withstand physical damage. [NFPA 54:12.11.2.1]

802.10.1.1 Unconditioned Area. Where the vent connector used for an appliance having a draft hood or a Category I appliance is located in or passes through an unconditioned area, attic or crawl space, that portion of the vent connector shall be listed Type B, Type L, or listed vent material having equivalent insulation qualities.

Exception: Single-wall metal pipe located within the exterior walls of the building and located in an unconditioned area other than an attic or a crawl space having a local 99 percent winter design temperature of 5°F (-15°C) or higher. [NFPA 54:12.11.2.2]

802.10.1.2 Residential Type Appliances. Vent connectors for residential-type appliances shall comply with the following:

(1) Vent connectors for listed appliances having draft hoods, appliances having draft hoods and equipped with listed conversion burners, and Category I appliances that are not installed in attics, crawl spaces, or other unconditioned areas shall be one of the following:

(a) Type B or Type L vent material.

(b) Galvanized sheet steel not less than 0.018 of an inch (0.457 mm) thick.

(c) Aluminum (1100 or 3003 alloy or equivalent) sheet not less than 0.027 of an inch (0.686 mm) thick.

(d) Stainless steel sheet not less than 0.012 of an inch (0.305 mm) thick.

(e) Smooth interior wall metal pipe having resistance to heat and corrosion equal to or greater than that of Section 802.10.1.2(1)(b), Section 802.10.1.2(1)(c), or Section 802.10.1.2(1)(d).

(f) A listed vent connector.

(2) Vent connectors shall not be covered with insulation.

Exception: Listed insulated vent connectors shall be installed in accordance with the manufacturer's installation instructions. [NFPA 54:12.11.2.3]

802.10.1.3 Non-Residential Low-Heat Appliances. A vent connector for a nonresidential low-heat appliance shall be a factory-built chimney section or steel pipe having resistance to heat and corrosion equivalent to that for the appropriate galvanized pipe as specified in Table 802.10.1.3. Factory-built chimney sections shall be joined together in accordance with the chimney manufacturer's instructions. [NFPA 54:12.11.2.4]

TABLE 802.10.1.3
MINIMUM THICKNESS FOR GALVANIZED STEEL VENT CONNECTORS FOR LOW-HEAT APPLIANCES
[NFPA 54: TABLE 12.11.2.4]

DIAMETER OF CONNECTOR (inches)	MINIMUM THICKNESS (inches)
Less than 6	0.019
6 to less than 10	0.023
10 to 12 inclusive	0.029
14 to 16 inclusive	0.034
Over 16	0.056

For SI units: 1 inch = 25.4 mm, 1 square inch = 0.000645 m^2

802.10.1.4 Medium-Heat Appliances. Vent connectors for medium-heat appliances shall be constructed of factory-built, medium-heat chimney sections or steel of a thickness not less than that specified in Table 802.10.1.4 and shall comply with the following:

(1) A steel vent connector for an appliance with a vent gas temperature in excess of 1000°F (538°C) measured at the entrance to the connector shall be lined with medium-duty fire brick or the equivalent.

(2) The lining shall be at least 2½ inches (64 mm) thick for a vent connector having a diameter or greatest cross-sectional dimension of 18 inches (457 mm) or less.

(3) The lining shall be at least 4½ inches (114 mm) thick laid on the 4½ inches (114 mm) bed for a vent connector having a diameter or greatest cross-sectional dimension greater than 18 inches (457 mm).

(4) Factory-built chimney sections, if employed, shall be joined together in accordance with the chimney manufacturer's instructions. [NFPA 54:12.11.2.5]

TABLE 802.10.1.4
MINIMUM THICKNESS FOR STEEL VENT CONNECTORS FOR MEDIUM-HEAT APPLIANCES
[NFPA 54: TABLE 12.11.2.5]

VENT CONNECTOR SIZE		
DIAMETER (inches)	AREA (square inches)	MINIMUM THICKNESS (inches)
Up to 14	Up to 154	0.053
Over 14 to 16	154 to 201	0.067
Over 16 to 18	201 to 254	0.093
Over 18	Larger than 254	0.123

For SI units: 1 inch = 25.4 mm, 1 square inch = 0.000645 m^2

802.10.2 Size of Vent Connector. A vent connector for an appliance with a single draft hood or for a Category I fan-assisted combustion system appliance shall be sized and installed in accordance with Section 803.0 or other approved engineering methods. [NFPA 54:12.11.3.1]

802.10.2.1 Manifold. For a single appliance having more than one draft hood outlet or flue collar, the manifold shall be constructed according to the instructions of the appliance manufacturer. Where there are no instructions, the manifold shall be designed and constructed in accordance with approved engineering practices. As an alternative method, the effective area of the manifold shall equal the combined area of the flue collars or draft hood outlets, and the vent connectors shall have a minimum 1 foot (305 mm) rise. [NFPA 54:12.11.3.2]

802.10.2.2 Size. Where two or more appliances are connected to a common vent or chimney, each vent connector shall be sized in accordance with Section 803.0 or other approved engineering methods. [NFPA 54:12.11.3.3]

As an alternative method applicable where the appliances are draft hood-equipped, each vent connector shall have an effective area not less than the area of the draft hood outlet of the appliance to which it is connected. [NFPA 54:12.11.3.4]

802.10.2.3 Height. Where two or more appliances are vented through a common vent connector or vent manifold, the common vent connector or vent manifold shall be located at the highest level consistent with available headroom and clearance to combustible material and sized in accordance with Section 803.0 or other approved engineering methods. [NFPA 54:12.11.3.5]

As an alternative method applicable only where there are two draft hood-equipped appliances, the effective area of the common vent connector or vent manifold and all junction fittings shall be not less than the area of the larger vent connector plus 50 percent of the area of the smaller flue collar outlet. [NFPA 54:12.11.3.6]

802.10.2.4 Size Increase. Where the size of a vent connector is increased to overcome installation limitations and obtain connector capacity equal to the appliance input, the size increase shall be made at the appliance draft hood outlet. [NFPA 54:12.11.3.7]

802.10.3 Two or More Appliances Connected to a Single Vent. Where two or more openings are provided into one chimney flue or vent, either of the following shall apply:

(1) The openings shall be at different levels.

(2) The connectors shall be attached to the vertical portion of the chimney or vent at an angle of 45 degrees or less relative to the vertical. [NFPA 54:12.11.4.1]

802.10.3.1 Height of Connector. Where two or more vent connectors enter a common vent, chimney flue, or single-wall metal pipe, the smaller connector shall enter at the highest level consistent with the available headroom or clearance to combustible material. [NFPA 54:12.11.4.2]

802.10.3.2 Pressure. Vent connectors serving Category I appliances shall not be connected to a portion of a mechanical draft system operating under positive static pressure, such as those serving Category III or Category IV appliances. [NFPA 54:12.11.4.3]

802.10.4 Clearance. Minimum clearances from vent connectors to combustible material shall comply with Table 802.7.3.3.

Exception: The clearance between a vent connector and combustible material shall be permitted to be reduced where the combustible material is protected as specified for vent connectors in Table 303.10.1. [NFPA 54:12.11.5]

802.10.5 Joints. Joints between sections of connector piping and connections to flue collars or draft hood outlets shall be fastened in accordance with one of the following methods:

(1) Sheet metal screws.

(2) Vent connectors of listed vent material assembled and connected to flue collars or draft hood outlets in accordance with the manufacturer's instructions.

(3) Other approved means. [NFPA 54:12.11.6]

802.10.6 Slope. A vent connector shall be installed without any dips or sags and shall slope upward toward the vent or chimney at least $\frac{1}{4}$ inch per foot (20.8 mm/m).

Exception: Vent connectors attached to a mechanical draft system installed in accordance with appliance and the draft system manufacturers' instructions. [NFPA 54:12.11.7]

802.10.7 Length of Vent Connector. The length of vent connectors shall comply with Section 802.10.7.1 or Section 802.10.7.2.

802.10.7.1 Single Wall Connector. The maximum horizontal length of a single-wall connector shall be 75 percent of the height of the chimney or vent except for engineered systems. [NFPA 54:12.11.8.1]

802.10.7.2 Type B Double Wall Connector. The maximum horizontal length of a Type B double-wall connector shall be 100 percent of the height of the chimney or vent, except for engineered systems. The maximum length of an individual connector for a chimney or vent system serving multiple appliances, from the appliance outlet to the junction with the common vent or another connector, shall be 100 percent of the height of the chimney or vent. [NFPA 54:12.11.8.2]

802.10.8 Support. A vent connector shall be supported for the design and weight of the material employed to maintain clearances and prevent physical damage and separation of joints. [NFPA 54:12.11.9]

802.10.9 Chimney Connection. Where entering a flue in a masonry or metal chimney, the vent connector shall be installed above the extreme bottom to avoid stoppage. Where a thimble or slip joint is used to facilitate removal of the connector, the connector shall be attached to or inserted into the thimble or slip joint to prevent the connector from falling out. Means shall be employed to prevent the connector from entering so far as to restrict the space between its end and the opposite wall of the chimney flue. [NFPA 54:12.11.10]

802.10.10 Inspection. The entire length of a vent connector shall be readily accessible for inspection, cleaning, and replacement. [NFPA 54:12.11.11]

802.10.11 Fireplaces. A vent connector shall not be connected to a chimney flue serving a fireplace unless the fireplace flue opening is permanently sealed. [NFPA 54:12.11.12]

802.10.12 Passage through Ceilings, Floors, or Walls. A vent connector shall not pass through a ceiling, floor, or fire-resistance-rated wall. A single-wall metal pipe connector shall not pass through an interior wall.

Exception: Vent connectors made of listed Type B or Type L vent material and serving listed appliances with draft hoods and other appliances listed for use with Type B gas vents that pass through walls or partitions constructed of combustible material shall be installed with not less than the listed clearance to combustible material.

802.10.12.1 Medium-Heat Appliances. Vent connectors for medium-heat appliances shall not pass through walls or partitions constructed of combustible material. [NFPA 54:12.11.13.2]

802.11 Vent Connectors for Category II, Category III, and Category IV Appliances. The vent connectors for Category II, Category III, and Category IV appliances shall be in accordance with Section 802.4 through Section 802.4.3. [NFPA 54:12.12]

802.12 Draft Hoods and Draft Controls. Vented appliances shall be installed with draft hoods.

Exception: Dual oven-type combination ranges; incinerators; direct-vent appliances; fan-assisted combustion system appliances; appliances requiring chimney draft for operation; single firebox boilers equipped with conversion burners with inputs exceeding 400 000 Btu/h (117 kW); appliances equipped with blast, power, or pressure burners that are not listed for use with draft hoods; and appliances designed for forced venting.

802.12.1 Installation. A draft hood supplied with or forming a part of listed vented appliances shall be installed without alteration, exactly as furnished and specified by the appliance manufacturer. [NFPA 54:12.13.2]

Where a draft hood is not supplied by the appliance manufacturer where one is required, a draft hood shall be installed, be of a listed or approved type, and, in the absence of other instructions, be of the same size as the appliance flue collar. Where a draft hood is required with a conversion burner, it shall be of a listed or approved type. [NFPA 54:12.13.2.1]

Where a draft hood of special design is needed or preferable, the installation shall be approved and in accordance with the recommendations of the appliance manufacturer. [NFPA 54:12.13.2.2]

802.12.2 Draft Control Devices. Where a draft control device is part of the appliance or is supplied by the appliance manufacturer, it shall be installed in accordance with the manufacturer's instructions. In the absence of manufacturer's instructions, the device shall be attached to the flue collar of the appliance or as near to the appliance as practical. [NFPA 54:12.13.3]

802.12.3 Additional Devices. Appliances requiring controlled chimney draft shall be permitted to be equipped with listed double-acting barometric draft regulators installed and adjusted in accordance with the manufacturer's installation instructions. [NFPA 54:12.13.4]

802.12.4 Location. Draft hoods and barometric draft regulators shall be installed in the same room or enclosure as the appliance in such a manner as to prevent a difference in pressure between the hood or regulator and the combustion air supply. [NFPA 54:12.13.5]

802.12.5 Positioning. Draft hoods and draft regulators shall be installed in the position for which they were designed with reference to the horizontal and vertical planes and shall be located so that the relief opening is not obstructed by a part of the appliance or adjacent construction. The appliance and its draft hood shall be located so that the relief opening is accessible for checking vent operation. [NFPA 54:12.13.6]

802.12.6 Clearance. A draft hood shall be located so that its relief opening is not less than 6 inches (152 mm) from a surface except that of the appliance it serves and the venting system to which the draft hood is connected. Where a greater or lesser clearance is indicated on the appliance label, the clearance shall not be less than that specified on the label. Such clearances shall not be reduced. [NFPA 54:12.13.7]

802.13 Manually Operated Dampers. A manually operated damper shall not be placed in an appliance vent connector. Fixed baffles shall not be classified as manually operated dampers. [NFPA 54:12.14]

802.14 Obstructions. Devices that retard the flow of vent gases shall not be installed in a vent connector, chimney, or vent. The following shall not be considered as obstructions:

(1) Draft regulators and safety controls specifically listed for installation in venting systems and installed in accordance with the manufacturer's installation instructions.

(2) Approved draft regulators and safety controls designed and installed in accordance with approved engineering methods.

(3) Listed heat reclaimers and automatically operated vent dampers installed in accordance with the manufacturer's installation instructions.

(4) Vent dampers serving listed appliances installed in accordance with Section 803.1 or Section 803.2 or other approved engineering methods.

(5) Approved economizers, heat reclaimers, and recuperators installed in venting systems of appliances not required to be equipped with draft hoods, provided the appliance manufacturer's installation instructions cover the installation of such a device in the venting system and performance in accordance with Section 802.3 and Section 802.3.1 is obtained. [NFPA 54:12.16]

802.15 Automatically Operated Vent Dampers. An automatically operated vent damper shall be of a listed type. [NFPA 54:12.15]

802.15.1 Listing. Automatically operated vent dampers for oil-fired appliances shall comply with UL 17. The automatic damper control shall comply with UL 378.

803.0 Sizing of Category I Venting Systems.

803.1 Single Appliance Vent Table 803.1.2(1) through Table 803.1.2(6). Table 803.1.2(1) through Table 803.1.2(6) shall not be used where obstructions are installed in the venting system. The installation of vents serving listed appliances with vent dampers shall be in accordance with the appliance manufacturer's installation instructions or in accordance with the following:

(1) The maximum capacity of the vent system shall be determined using the NAT Max column.

(2) The minimum capacity shall be determined as though the appliance were a fan-assisted appliance, using the FAN Min column to determine the minimum capacity of the vent system. Where the corresponding "FAN Min" is "NA", the vent configuration shall not be permitted and an alternative venting configuration shall be utilized. [NFPA 54:13.1.1]

803.1.1 Vent Downsizing. Where the vent size determined from the tables is smaller than the appliance draft hood outlet or flue collar, the use of the smaller size shall be permitted provided that the installation is in accordance with the following requirements:

(1) The total vent height (*H*) is not less than 10 feet (3048 mm).

(2) Vents for appliance draft hood outlets or flue collars 12 inches (305mm) in diameter or smaller are not reduced more than one table size.

(3) Vents for appliance draft hood outlets or flue collars exceeding 12 inches (305 mm) in diameter are not reduced more than two table sizes.

(4) The maximum capacity listed in the tables for a fan-assisted appliance is reduced by 10 percent (0.90 x maximum table capacity).

(5) The draft hood outlet exceeds 4 inches (102 mm) in diameter. A 3 inch (76 mm) diameter vent shall not be connected to a 4 inch (102 mm) diameter draft hood outlet. This provision shall not apply to fan-assisted appliances. [NFPA 54:13.1.2]

803.1.2 Elbows. Single-appliance venting configurations with zero (0) lateral lengths in Table 803.1.2(1), Table 803.1.2(2), and Table 803.1.2(5) shall not have elbows in the venting system. Single-appliance venting with lateral length, include two 90 degree elbows. For each additional elbow up to and including 45 degrees, the maximum capacity listed in the venting tables shall be reduced by 5 percent. For each additional elbow greater than 45 degrees up to and including 90 degrees, the maximum capacity listed in the venting tables shall be reduced by 10 percent. Where multiple offsets occur in a vent, the total lateral length of all offsets combined shall not exceed that specified in Table 803.1.2(1) through Table 803.1.2(5). [NFPA 54:13.1.3]

803.1.3 Zero Lateral. Zero lateral (*L*) shall apply to a straight vertical vent attached to a top outlet draft hood or flue collar. [NFPA 54:13.1.4]

803.1.4 High-Altitude Installations. Sea level input ratings shall be used where determining maximum capacity for high-altitude installation. Actual input (derated for altitude) shall be used for determining minimum capacity for high-altitude installation. [NFPA 54:13.1.5]

803.1.5 Multiple Input Ratings. For appliances with more than one input rate, the minimum vent capacity (FAN Min) determined from the tables shall be less than the lowest appliance input rating, and the maximum vent capacity (FAN Max/NAT Max) determined from the tables shall exceed the highest appliance rating input. [NFPA 54:13.1.6]

803.1.6 Corrugated Chimney Liner Reduction. Listed corrugated metallic chimney liner systems in masonry chimneys shall be sized by using Table 803.1.2(1) or Table 803.1.2(2) for Type B vents, with the maximum capacity reduced by 20 percent (0.80 x maximum capacity) and the minimum capacity as shown in Table 803.1.2(1) or Table 803.1.2(2).

Corrugated metallic liner systems installed with bends or offsets shall have their maximum capacity further reduced in accordance with Section 803.1.2. The 20 percent reduction for corrugated metallic chimney liner systems includes an allowance for one long radius 90 degree (1.57 rad) turn at the bottom of the liner. [NFPA 54:13.1.7]

803.1.7 Connection to Chimney Liners. Connections between chimney liners and listed double-wall connectors shall be made with listed adapters designed for such purpose. [NFPA 54:13.1.8]

803.1.8 Vertical Vent Upsizing Using 7 x Rule. Where the vertical vent has a larger diameter than the vent connector, the vertical vent diameter shall be used to determine the minimum vent capacity, and the connector diameter shall be used to determine the maximum vent capacity. The flow area of the vertical vent shall not exceed seven times the flow area of the listed appliance categorized vent area, flue collar area, or draft hood outlet area unless designed in accordance with approved engineering methods. [NFPA 54:13.1.9]

803.1.9 Draft Hood Conversion Accessories. Draft hood conversion accessories for use with masonry chimneys venting listed Category I fan-assisted appliances shall be listed and installed in accordance with the listed accessory manufacturer's installation instructions. [NFPA 54:13.1.10]

803.1.10 Chimney and Vent Locations. Table 803.1.2(1) through Table 803.1.2(5) shall be used only for chimneys and vents not exposed to the outdoors below the roof line. A Type B vent or listed chimney lining system passing through an unused masonry chimney flue shall not be considered to be exposed to the outdoors. Where vents extend outdoors above the roof more than 5 feet (1524 mm) higher than required by Table 802.6.1, and where vents terminate in accordance with Section 802.6.1(2), the outdoor portion of the vent shall be enclosed as required by this paragraph for vents not considered to be exposed to the outdoors, or such venting system shall be engineered. A Type B vent passing through an unventilated enclosure or chase insulated to a value of not less than R8 shall not be considered to be exposed to the outdoors. Table 803.1.2(3) in combination with Table 803.1.2(6) shall be used for clay tile-lined exterior masonry chimneys, provided all of the following requirements are met:

(1) The vent connector is Type B double wall.

(2) The vent connector length is limited to 18 in./in. (18 mm/mm) of vent connector diameter.

(3) The appliance is draft hood equipped.

(4) The input rating is less than the maximum capacity given in Table 803.1.2(3).

(5) For a water heater, the outdoor design temperature shall be not less than 5°F (-15°C).

(6) For a space-heating appliance, the input rating is greater than the minimum capacity given by Table 803.1.2(6). [NFPA 54:13.1.11]

803.1.11 Residential and Low-Heat Appliances. Flue lining system for residential and low heat appliance shall be in accordance with Section 803.1.11.1 and Section 803.1.11.2.

803.1.11.1 Clay Flue Lining. Clay flue lining shall be manufactured in accordance with ASTM C315 or other approved standard.

803.1.11.2 Chimney Lining. Chimney lining shall be listed in accordance with UL 1777.

803.1.12 Corrugated Vent Connector Size. Corrugated vent connectors shall not be smaller than the listed appliance categorized vent diameter, flue collar diameter, or draft hood outlet diameter. [NFPA 54:13.1.12]

803.1.13 Upsizing. Vent connectors shall not be upsized more than two sizes exceeding the listed appliance categorized vent diameter, flue collar diameter, or draft hood outlet diameter. [NFPA 54:13.1.13]

803.1.14 Single Run of Vent. In a single run of vent or vent connector, more than one diameter and type shall be permitted to be used, provided that the sizes and types are permitted by the tables. [NFPA 54:13.1.14]

803.1.15 Interpolation. Interpolation shall be permitted in calculating capacities for vent dimensions that fall between table entries. [NFPA 54:13.1.15]

803.1.16 Extrapolation. Extrapolation beyond the table entries shall not be permitted. [NFPA 54:13.1.16]

803.1.17 Engineering Methods. For vent heights lower than 6 feet (1829 mm) and higher than shown in the tables, engineering methods shall be used to calculate vent capacities. [NFPA 54:13.1.17]

803.1.18 Height Entries. Where the actual height of a vent falls between entries in the height column of the applicable table in Table 803.1.2(1) through Table 803.1.2(6), either of the following shall be used:

(1) Interpolation.

(2) The lower appliance input rating shown in the table entries for FAN MAX and NAT MAX column values, and the higher appliance input rating for the FAN MIN column values. [NFPA 54:13.1.18]

803.2 Multiple Appliance Vent Table 803.2(1) through Table 803.2(9). Venting Table 803.2(1) through Table 803.2(9) shall not be used where obstructions are installed in the venting system. The installation of vents serving listed appliances with vent dampers shall be in accordance with the appliance manufacturer's instructions or in accordance with the following:

(1) The maximum capacity of the vent connector shall be determined using the NAT Max column.

(2) The maximum capacity of the vertical vent or chimney shall be determined using the FAN + NAT column when the second appliance is a fan-assisted appliance, or the NAT + NAT column when the second appliance is equipped with a draft hood.

(3) The minimum capacity shall be determined as if the appliance were a fan-assisted appliance, as follows:

(a) The minimum capacity of the vent connector shall be determined using the FAN Min column.

(b) The FAN + FAN column shall be used when the second appliance is a fan-assisted appliance, and the FAN + NAT column shall be used when the second appliance is equipped with a draft hood, to determine whether the vertical vent or chimney configuration is not permitted (NA). Where the vent configuration is NA, the vent configuration shall not be permitted and an alternative venting configuration shall be utilized. [NFPA 54:13.2.1]

803.2.1 Vent Connector Maximum Length. The maximum vent connector horizontal length shall be 18 inches per inch (18 mm/mm) of connector diameter as shown in Table 803.2.1, or as permitted by Section 803.2.2. [NFPA 54:13.2.2]

803.2.2 Vent Connector Exceeding Maximum Length. The vent connector shall be routed to the vent utilizing the shortest possible route. Connectors with longer horizontal lengths than those listed in Table 803.2.1 are permitted under the following conditions:

(1) The maximum capacity (FAN Max or NAT Max) of the vent connector shall be reduced 10 percent for each additional multiple of the length listed in Table 803.2.1. For example, the maximum length listed for a 4 inch (100 mm) connector is 6 feet (1829 mm). With a connector length greater than 6 feet (1829 mm) but not exceeding 12 feet (3658 mm), the maximum capacity must be reduced by 10 percent (0.90 x maximum vent connector capacity). With a connector length greater than 12 feet (3658 mm) but not exceeding 18 feet (5486 mm), the maximum capacity must be reduced by 20 percent (0.80 x maximum vent capacity).

(2) For a connector serving a fan-assisted appliance, the minimum capacity (FAN Min) of the connector shall be determined by referring to the corresponding single appliance table. For Type B double-wall connectors, Table 803.1.2(1) shall be used. For single-wall connectors, Table 803.1.2(2) shall be used. The height (*H*) and lateral (*L*) shall be measured according to the procedures for a single-appliance vent, as if the other appliances were not present. [NFPA 54:13.2.3]

TABLE 803.2.1
VENT CONNECTOR MAXIMUM LENGTH
[NFPA 54: TABLE 13.2.2]

CONNECTOR DIAMETER (inches)	MAXIMUM CONNECTOR HORIZONTAL LENGTH (feet)
3	4½
4	6
5	7½
6	9
7	10½
8	12
9	13½
10	15
12	18
14	21
16	24
18	27
20	30
22	33
24	36

For SI units: 1 inch = 25.4 mm, 1 foot = 304.8 mm

803.2.3 Vent Connector Manifold. Where the vent connectors are combined prior to entering the vertical portion of the common vent to form a common vent manifold, the size of the common vent manifold and the common vent shall be determined by applying a 10 percent reduction (0.90 x maximum common vent capacity) to the common vent capacity part of the common vent tables. The length of the common vent manifold (*LM*) shall not exceed 18 inches per inch (18 mm/mm) of common vent diameter (*D*). (See Figure 802.6.3.2) [NFPA 54:13.2.4]

803.2.4 Vent Offset. Where the common vertical vent is offset, the maximum capacity of the common vent shall be reduced in accordance with Section 803.2.5, and the horizontal length of the common vent offset shall not exceed 18 inches per inch (18 mm/mm) of common vent diameter (*D*). Where multiple offsets occur in a common vent, the total horizontal length of offsets combined shall not exceed 18 inches per inch (18 mm/mm) of the common vent diameter. [NFPA 54:13.2.5]

803.2.5 Elbow Reduction. For each elbow up to and including 45 degrees (0.79 rad) in the common vent, the maximum common vent capacity listed in the venting tables shall be reduced by 5 percent. For each elbow exceeding 45 degrees (0.79 rad) up to and including 90 degrees (1.57 rad), the maximum common vent capacity listed in the venting tables shall be reduced by 10 percent. [NFPA 54:13.2.6]

803.2.6 Elbows in Connectors. The vent connector capacities listed in the common vent sizing tables include allowance for two 90 degree elbows. For each additional elbow up to and including 45 degrees, the maximum vent connector capacity listed in the venting tables shall be reduced by 5 percent. For each elbow greater than 45 degrees up to and including 90 degrees, the maximum vent connector capacity listed in the venting tables shall be reduced by 10 percent. [NFPA 54:13.2.7]

»» **803.2.7 Common Vent Minimum Size.** The cross-sectional area of the common vent shall be equal to or greater than the cross-sectional area of the largest connector. [NFPA 54:13.2.8]

»» **803.2.8 Tee and Wye Fittings.** Tee and wye fittings connected to a common gas vent shall be considered as part of the common gas vent and constructed of materials consistent with that of the common gas vent. [NFPA 54:13.2.9]

»» **803.2.9 Size of Fittings.** At the point where tee or wye fittings connect to a common gas vent, the opening size of the fitting shall be equal to the size of the common vent. Such fittings shall not be prohibited from having reduced size openings at the point of connection of appliance gas vent connectors. [NFPA 54:13.2.10]

»» **803.2.10 High-Altitude Installations.** Sea level input ratings shall be used where determining maximum capacity for high-altitude installation. Actual input (derated for altitude) shall be used for determining minimum capacity for high-altitude installation. [NFPA 54:13.2.11]

803.2.11 Connector Rise. The connector rise (*R*) for each appliance connector shall be measured from the draft hood outlet or flue collar to the centerline where the vent gas streams come together. [NFPA 54:13.2.12] «

803.2.12 Vent Height. For multiple appliances located on one floor, the total height (*H*) shall be measured from the highest draft hood outlet or flue collar up to the level of the outlet of the common vent. [NFPA 54:13.2.13] «

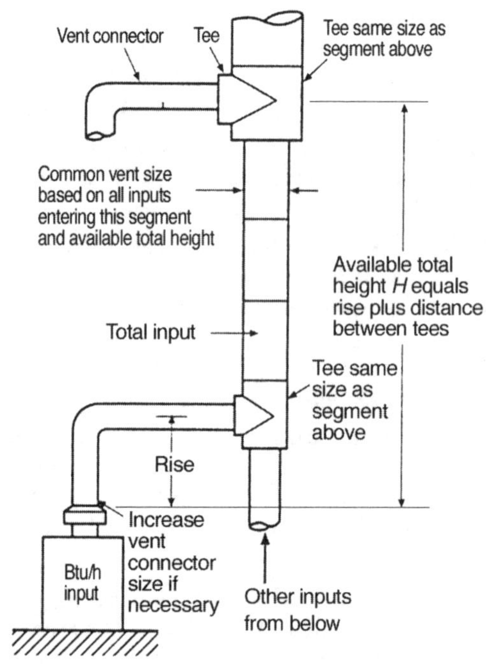

FIGURE 803.2.13
MULTISTORY GAS VENT DESIGN PROCEDURE FOR EACH SEGMENT OF SYSTEM
[NFPA 54: FIGURE F.1(m)] «

803.2.13 Multistory Installations. For multistory installations, the total height (*H*) for each segment of the system shall be the vertical distance between the highest draft hood outlet or flue collar entering that segment and the centerline of the next higher interconnection tee. (See Figure 803.2.13) [NFPA 54:13.2.14] «

803.2.14 Size of Vents for Multistory Installations. The size of the lowest connector and of the vertical vent leading to the lowest interconnection of a multistory system shall be in accordance with Table 803.1.2(1) or Table 803.1.2(2) for available total height (*H*) up to the lowest interconnection. (See Figure 803.2.13) [NFPA 54:13.2.15] «

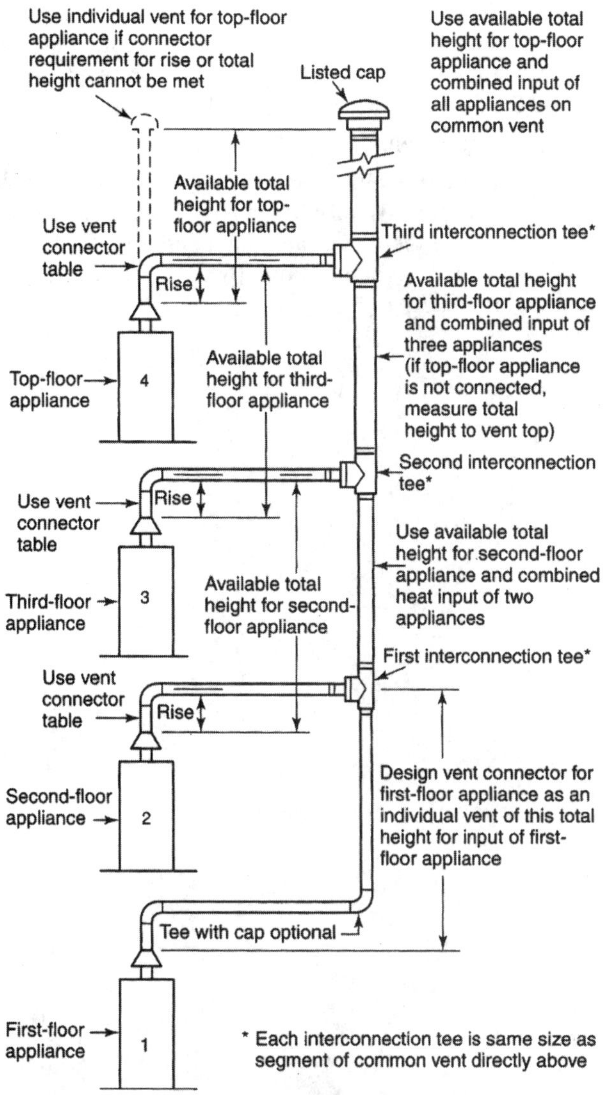

FIGURE 803.2.14
PRINCIPLES OF DESIGN OF MULTISTORY VENTS USING VENT CONNECTOR AND COMMON VENT DESIGN TABLES [NFPA 54: FIGURE F.1(n)]

803.2.15 Vent Type Multistory Installation. Where used in multistory systems, vertical common vents shall be Type B double-wall and shall be installed with a listed vent cap. [NFPA 54:13.2.16]

803.2.16 Offsets in Multistory Installations. Offsets in multistory common vent systems shall be limited to a single offset in each system, and systems with an offset shall comply with the following:

(1) The offset angle shall not exceed 45 degrees (0.79 rad) from vertical.

(2) The horizontal length of the offset shall not exceed 18 inches per inch (18 mm/mm) of common vent diameter of the segment in which the offset is located.

(3) For the segment of the common vertical vent containing the offset, the common vent capacity listed in the common venting tables shall be reduced by 20 percent (0.80 x maximum common vent capacity).

(4) A multistory common vent shall not be reduced in size above the offset. [NFPA 54:13.2.17]

803.2.17 Vertical Vent Size Limitation. Where two or more appliances are connected to a vertical vent or chimney, the flow area of the largest section of vertical vent or chimney shall not exceed seven times the smallest listed appliance categorized vent areas, flue collar area, or draft hood outlet area unless designed in accordance with approved engineering methods. [NFPA 54:13.2.18]

803.2.18 Multiple Input Ratings. For appliances with more than one input rate, the minimum vent connector capacity (FAN Min) determined from the tables shall be less than the lowest appliance input rating, and the maximum vent connector capacity (FAN Max or NAT Max) determined from the tables shall exceed the highest appliance input rating. [NFPA 54:13.2.19]

803.2.19 Corrugated Metallic Chimney Liner Reduction. Listed corrugated metallic chimney liner systems in masonry chimneys shall be sized by using Table 803.2(1) or Table 803.2(2) for Type B vents, with the maximum capacity reduced by 20 percent (0.80 x maximum capacity) and the minimum capacity as shown in Table 803.2(1) or Table 803.2(2). Corrugated metallic liner systems installed with bends or offsets shall have their maximum capacity further reduced in accordance with Section 803.2.5 and Section 803.2.6. The 20 percent reduction for corrugated metallic chimney liner systems includes an allowance for one long radius 90 degree (1.57 rad) turn at the bottom of the liner. [NFPA 54:13.2.20]

803.2.20 Chimneys and Vents. Table 803.2(1) through Table 803.2(5) shall be used only for chimneys and vents not exposed to the outdoors below the roof line. A Type B vent or listed chimney lining system passing through an unused masonry chimney flue shall not be considered to be exposed to the outdoors. A Type B vent passing through an unventilated enclosure or chase insulated to a value of not less than R-8 shall not be considered to be exposed to the outdoors. Where vents extend outdoors above the roof more than 5 feet (1524 mm) higher than required by Table 802.6.1, and where vents terminate in accordance with Section 802.6.1(1), the outdoor portion of the vent shall be enclosed as required by this section for vents not considered to be exposed to the outdoors or such venting system shall be engineered. Table 803.2(6) through Table 803.2(9) shall be used for clay-tile-lined exterior masonry chimneys, provided all the following conditions are met:

(1) The vent connector is Type B double-wall.

(2) At least one appliance is draft hood-equipped.

(3) The combined appliance input rating is less than the maximum capacity given by Table 803.2(6) (for NAT+NAT) or Table 803.2(8) (for FAN+NAT).

(4) The input rating of each space-heating appliance is greater than the minimum input rating given by Table 803.2(7) (for NAT+NAT) or Table 803.2(9) (for FAN+NAT).

(5) The vent connector sizing is in accordance with Table 803.2(3). [NFPA 54:13.2.22]

803.2.21 Vent Connector Sizing. Vent connectors shall not be increased more than two sizes greater than the listed appliance categorized vent diameter, flue collar diameter, or draft hood outlet diameter. Vent connectors for draft hood-equipped appliances shall not be smaller than the draft hood outlet diameter. Where a vent connector sizes determined from the tables for a fan-assisted appliance(s) is smaller than the flue collar diameter, the use of the smaller size(s) shall be permitted, provided that the installation complies with all of the following conditions:

(1) Vent connectors for fan-assisted appliance flue collars 12 inches (300 mm) in diameter or smaller are not reduced by more than one table size [e.g., 12 inches to 10 inches (300 mm to 250 mm) is a one-size reduction] and those larger than 12 inches (300 mm) in diameter are not reduced more than two table sizes [e.g., 24 inches to 20 inches (600 mm to 500 mm) is a two-size reduction].

(2) The fan-assisted appliance(s) is common vented with a draft hood-equipped appliance(s).

(3) The vent connector has a smooth interior wall. [NFPA 54:13.2.24]

803.2.22 Combination of Pipe Types and Sizes. All combinations of pipe sizes, single-wall metal pipe, and double-wall metal pipe shall be allowed within any connector run(s) or within the common vent, provided ALL of the appropriate tables permit ALL of the desired sizes and types of pipe, as if they were used for the entire length of the subject connector or vent. Where single-wall and Type B double-wall metal pipes are used for vent connectors within the same venting system, the common vent shall be sized using Table 803.2(2) or Table 803.2(4) as appropriate. [NFPA 54:13.2.25]

803.2.23 Multiple Connector and Vent Sizes. Where a table permits more than one diameter of pipe to be used for a connector or vent, all the permitted sizes shall be permitted to be used. [NFPA 54:13.2.26]

803.2.24 Interpolation. Interpolation shall be permitted in calculating capacities for vent dimensions that fall between table entries. [NFPA 54:13.2.27]

803.2.25 Extrapolation. Extrapolation beyond the table entries shall not be permitted. [NFPA 54:13.2.28]

803.2.26 Engineering Methods. For vent heights lower than 6 feet (1829 mm) and higher than shown in the tables, engineering methods shall be used to calculate vent capacities. [NFPA 54:13.2.29]

803.2.27 Height Entries. Where the actual height of a vent falls between entries in the height column of the applicable table in Table 803.2(1) through Table 803.2(9), one of the following shall be used:

(1) Interpolation.

(2) The lower appliance input rating shown in the table entries for FAN MAX and NAT MAX column values; and the higher appliance input rating for the FAN MIN column values. [NFPA 54:13.2.30]

99% Winter Design Temperatures for the Contiguous United States

This map is a necessarily generalized guide to temperatures in the contiguous United States. Temperatures shown for areas such as mountainous regions and large urban centers may not be accurate. The data used to develop this map are from the 1993 ASHRAE Handbook — Fundamentals (Chapter 24, Table 1: Climate Conditions for the United States).

For 99% winter design temperatures in Alaska, consult the ASHRAE Handbook — Fundamentals.

99% winter design temperatures for Hawaii are greater than 37°F.

For SI units: °C = (°F-32)/1.8

FIGURE 803.1.2(6)
RANGE OF WINTER DESIGN TEMPERATURES USED IN ANALYZING EXTERIOR MASONRY CHIMNEYS IN THE UNITED STATES
[NFPA 54: FIGURE F.2.4]

TABLE 803.1.2(1)
TYPE B DOUBLE-WALL GAS VENT [NFPA 54: TABLE 13.1(a)]*

						NUMBER OF APPLIANCES:			SINGLE							
						APPLIANCE TYPE:			CATEGORY I							
						APPLIANCE VENT CONNECTION:			CONNECTED DIRECTLY TO VENT							
		VENT DIAMETER – *D* (inch)														
		3			4			5			6			7		
		APPLIANCE INPUT RATING IN THOUSANDS OF BTU PER HOUR														
HEIGHT *H* (feet)	LATERAL *L* (feet)	FAN		NAT	FAN		NAT	FAN		NAT	FAN		NAT	FAN		NAT
		Min	Max	Max	Min	Max	Max	Min	Max	Max	Min	Max	Max	Min	Max	Max
6	0	0	78	46	0	152	86	0	251	141	0	375	205	0	524	285
	2	13	51	36	18	97	67	27	157	105	32	232	157	44	321	217
	4	21	49	34	30	94	64	39	153	103	50	227	153	66	316	211
	6	25	46	32	36	91	61	47	149	100	59	223	149	78	310	205
8	0	0	84	50	0	165	94	0	276	155	0	415	235	0	583	320
	2	12	57	40	16	109	75	25	178	120	28	263	180	42	365	247
	5	23	53	38	32	103	71	42	171	115	53	255	173	70	356	237
	8	28	49	35	39	98	66	51	164	109	64	247	165	84	347	227
10	0	0	88	53	0	175	100	0	295	166	0	447	255	0	631	345
	2	12	61	42	17	118	81	23	194	129	26	289	195	40	402	273
	5	23	57	40	32	113	77	41	187	124	52	280	188	68	392	263
	10	30	51	36	41	104	70	54	176	115	67	267	175	88	376	245
15	0	0	94	58	0	191	112	0	327	187	0	502	285	0	716	390
	2	11	69	48	15	136	93	20	226	150	22	339	225	38	475	316
	5	22	65	45	30	130	87	39	219	142	49	330	217	64	463	300
	10	29	59	41	40	121	82	51	206	135	64	315	208	84	445	288
	15	35	53	37	48	112	76	61	195	128	76	301	198	98	429	275
20	0	0	97	61	0	202	119	0	349	202	0	540	307	0	776	430
	2	10	75	51	14	149	100	18	250	166	20	377	249	33	531	346
	5	21	71	48	29	143	96	38	242	160	47	367	241	62	519	337
	10	28	64	44	38	133	89	50	229	150	62	351	228	81	499	321
	15	34	58	40	46	124	84	59	217	142	73	337	217	94	481	308
	20	48	52	35	55	116	78	69	206	134	84	322	206	107	464	295
30	0	0	100	64	0	213	128	0	374	220	0	587	336	0	853	475
	2	9	81	56	13	166	112	14	283	185	18	432	280	27	613	394
	5	21	77	54	28	160	108	36	275	176	45	421	273	58	600	385
	10	27	70	50	37	150	102	48	262	171	59	405	261	77	580	371
	15	33	64	NA	44	141	96	57	249	163	70	389	249	90	560	357
	20	56	58	NA	53	132	90	66	237	154	80	374	237	102	542	343
	30	NA	NA	NA	73	113	NA	88	214	NA	104	346	219	131	507	321
50	0	0	101	67	0	216	134	0	397	232	0	633	363	0	932	518
	2	8	86	61	11	183	122	14	320	206	15	497	314	22	715	445
	5	20	82	NA	27	177	119	35	312	200	43	487	308	55	702	438
	10	26	76	NA	35	168	114	45	299	190	56	471	298	73	681	426
	15	59	70	NA	42	158	NA	54	287	180	66	455	288	85	662	413
	20	NA	NA	NA	50	149	NA	63	275	169	76	440	278	97	642	401
	30	NA	NA	NA	69	131	NA	84	250	NA	99	410	259	123	605	376
100	0	NA	NA	NA	0	218	NA	0	407	NA	0	665	400	0	997	560
	2	NA	NA	NA	10	194	NA	12	354	NA	13	566	375	18	831	510
	5	NA	NA	NA	26	189	NA	33	347	NA	40	557	369	52	820	504
	10	NA	NA	NA	33	182	NA	43	335	NA	53	542	361	68	801	493
	15	NA	NA	NA	40	174	NA	50	321	NA	62	528	353	80	782	482
	20	NA	NA	NA	47	166	NA	59	311	NA	71	513	344	90	763	471
	30	NA	NA	NA	NA	NA	NA	78	290	NA	92	483	NA	115	726	449
	50	NA	NA	NA	NA	NA	NA	NA	NA	NA	147	428	NA	180	651	405

For SI units: 1 inch = 25.4 mm, 1 foot = 304.8 mm, 1000 British thermal units per hour = 0.293 kW, 1 square inch = 0.000645 m^2

* NA: Not applicable.

TABLE 803.1.2(1)
TYPE B DOUBLE-WALL GAS VENT [NFPA 54: TABLE 13.1(a)] (continued)

										NUMBER OF APPLIANCES:	SINGLE					
										APPLIANCE TYPE:	CATEGORY I					
										APPLIANCE VENT CONNECTION:	CONNECTED DIRECTLY TO VENT					
		VENT DIAMETER – *D* (inch)														
		8			9			10			12			14		
		APPLIANCE INPUT RATING IN THOUSANDS OF BTU PER HOUR														
HEIGHT *H* (feet)	LATERAL *L* (feet)	FAN		NAT	FAN		NAT	FAN		NAT	FAN		NAT	FAN		NAT
		Min	Max	Max	Min	Max	Max	Min	Max	Max	Min	Max	Max	Min	Max	Max
6	0	0	698	370	0	897	470	0	1121	570	0	1645	850	0	2267	1170
	2	53	425	285	63	543	370	75	675	455	103	982	650	138	1346	890
	4	79	419	279	93	536	362	110	668	445	147	975	640	191	1338	880
	6	93	413	273	110	530	354	128	661	435	171	967	630	219	1330	870
8	0	0	780	415	0	1006	537	0	1261	660	0	1858	970	0	2571	1320
	2	50	483	322	60	619	418	71	770	515	98	1124	745	130	1543	1020
	5	83	473	313	99	607	407	115	758	503	154	1110	733	199	1528	1010
	8	99	463	303	117	596	396	137	746	490	180	1097	720	231	1514	1000
10	0	0	847	450	0	1096	585	0	1377	720	0	2036	1060	0	2825	1450
	2	48	533	355	57	684	457	68	852	560	93	1244	850	124	1713	1130
	5	81	522	346	95	671	446	112	839	547	149	1229	829	192	1696	1105
	10	104	504	330	122	651	427	142	817	525	187	1204	795	238	1669	1080
15	0	0	970	525	0	1263	682	0	1596	840	0	2380	1240	0	3323	1720
	2	45	633	414	53	815	544	63	1019	675	86	1495	985	114	2062	1350
	5	76	620	403	90	800	529	105	1003	660	140	1476	967	182	2041	1327
	10	99	600	386	116	777	507	135	977	635	177	1446	936	227	2009	1289
	15	115	580	373	134	755	491	155	953	610	202	1418	905	257	1976	1250
20	0	0	1057	575	0	1384	752	0	1756	930	0	2637	1350	0	3701	1900
	2	41	711	470	50	917	612	59	1150	755	81	1694	1100	107	2343	1520
	5	73	697	460	86	902	599	101	1133	738	135	1674	1079	174	2320	1498
	10	95	675	443	112	877	576	130	1105	710	172	1641	1045	220	2282	1460
	15	111	654	427	129	853	557	150	1078	688	195	1609	1018	248	2245	1425
	20	125	634	410	145	830	537	167	1052	665	217	1578	990	273	2210	1390
30	0	0	1173	650	0	1548	855	0	1977	1060	0	3004	1550	0	4252	2170
	2	33	826	535	42	1072	700	54	1351	865	74	2004	1310	98	2786	1800
	5	69	811	524	82	1055	688	96	1332	851	127	1981	1289	164	2759	1775
	10	91	788	507	107	1028	668	125	1301	829	164	1944	1254	209	2716	1733
	15	105	765	490	124	1002	648	143	1272	807	187	1908	1220	237	2674	1692
	20	119	743	473	139	977	628	160	1243	784	207	1873	1185	260	2633	1650
	30	149	702	444	171	929	594	195	1189	745	246	1807	1130	305	2555	1585
50	0	0	1297	708	0	1730	952	0	2231	1195	0	3441	1825	0	4934	2550
	2	26	975	615	33	1276	813	41	1620	1010	66	2431	1513	86	3409	2125
	5	65	960	605	77	1259	798	90	1600	996	118	2406	1495	151	3380	2102
	10	86	935	589	101	1230	773	118	1567	972	154	2366	1466	196	3332	2064
	15	100	911	572	117	1203	747	136	1536	948	177	2327	1437	222	3285	2026
	20	113	888	556	131	1176	722	151	1505	924	195	2288	1408	244	3239	1987
	30	141	844	522	161	1125	670	183	1446	876	232	2214	1349	287	3150	1910
100	0	0	1411	770	0	1908	1040	0	2491	1310	0	3925	2050	0	5729	2950
	2	21	1155	700	25	1536	935	30	1975	1170	44	3027	1820	72	4313	2550
	5	60	1141	692	71	1519	926	82	1955	1159	107	3002	1803	136	4282	2531
	10	80	1118	679	94	1492	910	108	1923	1142	142	2961	1775	180	4231	2500
	15	93	1095	666	109	1465	895	126	1892	1124	163	2920	1747	206	4182	2469
	20	105	1073	653	122	1438	880	141	1861	1107	181	2880	1719	226	4133	2438
	30	131	1029	627	149	1387	849	170	1802	1071	215	2803	1663	265	4037	2375
	50	197	944	575	217	1288	787	241	1688	1000	292	2657	1550	350	3856	2250

For SI units: 1 inch = 25.4 mm, 1 foot = 304.8 mm, 1000 British thermal units per hour = 0.293 kW, 1 square inch = 0.000645 m^2

TABLE 803.1.2(1)
TYPE B DOUBLE-WALL GAS VENT [NFPA 54: TABLE 13.1(a)] (continued)

						NUMBER OF APPLIANCES:	SINGLE									
						APPLIANCE TYPE:	CATEGORY I									
						APPLIANCE VENT CONNECTION:	CONNECTED DIRECTLY TO VENT									
		VENT DIAMETER – *D* (inch)														
		16			18			20			22			24		
		APPLIANCE INPUT RATING IN THOUSANDS OF BTU PER HOUR														
HEIGHT *H* (feet)	LATERAL *L* (feet)	FAN Min	FAN Max	NAT Max	FAN Min	FAN Max	NAT Max	FAN Min	FAN Max	NAT Max	FAN Min	FAN Max	NAT Max	FAN Min	FAN Max	NAT Max
6	0	0	2983	1530	0	3802	1960	0	4721	2430	0	5737	2950	0	6853	3520
	2	178	1769	1170	225	2250	1480	296	2782	1850	360	3377	2220	426	4030	2670
	4	242	1761	1160	300	2242	1475	390	2774	1835	469	3370	2215	555	4023	2660
	6	276	1753	1150	341	2235	1470	437	2767	1820	523	3363	2210	618	4017	2650
8	0	0	3399	1740	0	4333	2220	0	5387	2750	0	6555	3360	0	7838	4010
	2	168	2030	1340	212	2584	1700	278	3196	2110	336	3882	2560	401	4634	3050
	5	251	2013	1330	311	2563	1685	398	3180	2090	476	3863	2545	562	4612	3040
	8	289	2000	1320	354	2552	1670	450	3163	2070	537	3850	2530	630	4602	3030
10	0	0	3742	1925	0	4782	2450	0	5955	3050	0	7254	3710	0	8682	4450
	2	161	2256	1480	202	2868	1890	264	3556	2340	319	4322	2840	378	5153	3390
	5	243	2238	1461	300	2849	1871	382	3536	2318	458	4301	2818	540	5132	3371
	10	298	2209	1430	364	2818	1840	459	3504	2280	546	4268	2780	641	5099	3340
15	0	0	4423	2270	0	5678	2900	0	7099	3620	0	8665	4410	0	10 393	5300
	2	147	2719	1770	186	3467	2260	239	4304	2800	290	5232	3410	346	6251	4080
	5	229	2696	1748	283	3442	2235	355	4278	2777	426	5204	3385	501	6222	4057
	10	283	2659	1712	346	3402	2193	432	4234	2739	510	5159	3343	599	6175	4019
	15	318	2623	1675	385	3363	2150	479	4192	2700	564	5115	3300	665	6129	3980
20	0	0	4948	2520	0	6376	3250	0	7988	4060	0	9785	4980	0	11 753	6000
	2	139	3097	2000	175	3955	2570	220	4916	3200	269	5983	3910	321	7154	4700
	5	219	3071	1978	270	3926	2544	337	4885	3174	403	5950	3880	475	7119	4662
	10	273	3029	1940	334	3880	2500	413	4835	3130	489	5896	3830	573	7063	4600
	15	306	2988	1910	372	3835	2465	459	4786	3090	541	5844	3795	631	7007	4575
	20	335	2948	1880	404	3791	2430	495	4737	3050	585	5792	3760	689	6953	4550
30	0	0	5725	2920	0	7420	3770	0	9341	4750	0	11 483	5850	0	13 848	7060
	2	127	3696	2380	159	4734	3050	199	5900	3810	241	7194	4650	285	8617	5600
	5	206	3666	2350	252	4701	3020	312	5863	3783	373	7155	4622	439	8574	5552
	10	259	3617	2300	316	4647	2970	386	5803	3739	456	7090	4574	535	8505	5471
	15	292	3570	2250	354	4594	2920	431	5744	3695	507	7026	4527	590	8437	5391
	20	319	3523	2200	384	4542	2870	467	5686	3650	548	6964	4480	639	8370	5310
	30	369	3433	2130	440	4442	2785	540	5574	3565	635	6842	4375	739	8239	5225
50	0	0	6711	3440	0	8774	4460	0	11 129	5635	0	13 767	6940	0	16 694	8430
	2	113	4554	2840	141	5864	3670	171	7339	4630	209	8980	5695	251	10 788	6860
	5	191	4520	2813	234	5826	3639	283	7295	4597	336	8933	5654	394	10 737	6818
	10	243	4464	2767	295	5763	3585	355	7224	4542	419	8855	5585	491	10 652	6749
	15	274	4409	2721	330	5701	3534	396	7155	4511	465	8779	5546	542	10 570	6710
	20	300	4356	2675	361	5641	3481	433	7086	4479	506	8704	5506	586	10 488	6670
	30	347	4253	2631	412	5523	3431	494	6953	4421	577	8557	5444	672	10 328	6603
100	0	0	7914	4050	0	10 485	5300	0	13 454	6700	0	16 817	8600	0	20 578	10 300
	2	95	5834	3500	120	7591	4600	138	9577	5800	169	11 803	7200	204	14 264	8800
	5	172	5797	3475	208	7548	4566	245	9528	5769	293	11 748	7162	341	14 204	8756
	10	223	5737	3434	268	7478	4509	318	9447	5717	374	11 658	7100	436	14 105	8683
	15	252	5678	3392	304	7409	4451	358	9367	5665	418	11 569	7037	487	14 007	8610
	20	277	5619	3351	330	7341	4394	387	9289	5613	452	11 482	6975	523	13 910	8537
	30	319	5505	3267	378	7209	4279	446	9136	5509	514	11 310	6850	592	13 720	8391
	50	415	5289	3100	486	6956	4050	572	8841	5300	659	10 979	6600	752	13 354	8100

For SI units: 1 inch = 25.4 mm, 1 foot = 304.8 mm, 1000 British thermal units per hour = 0.293 kW, 1 square inch = 0.000645 m^2

TABLE 803.1.2(2)
TYPE B DOUBLE-WALL GAS VENT [NFPA 54: TABLE 13.1(b)]*

								NUMBER OF APPLIANCES:	SINGLE							
								APPLIANCE TYPE:	CATEGORY I							
								APPLIANCE VENT CONNECTION:	SINGLE-WALL METAL CONNECTOR							
		VENT DIAMETER – *D* (inch)														
		3			4			5			6			7		
		APPLIANCE INPUT RATING IN THOUSANDS OF BTU PER HOUR														
HEIGHT *H* (feet)	LATERAL *L* (feet)	FAN		NAT	FAN		NAT	FAN		NAT	FAN		NAT	FAN		NAT
		Min	Max	Max	Min	Max	Max	Min	Max	Max	Min	Max	Max	Min	Max	Max
6	0	38	77	45	59	151	85	85	249	140	126	373	204	165	522	284
	2	39	51	36	60	96	66	85	156	104	123	231	156	159	320	213
	4	NA	NA	33	74	92	63	102	152	102	146	225	152	187	313	208
	6	NA	NA	31	83	89	60	114	147	99	163	220	148	207	307	203
8	0	37	83	50	58	164	93	83	273	154	123	412	234	161	580	319
	2	39	56	39	59	108	75	83	176	119	121	261	179	155	363	246
	5	NA	NA	37	77	102	69	107	168	114	151	252	171	193	352	235
	8	NA	NA	33	90	95	64	122	161	107	175	243	163	223	342	225
10	0	37	87	53	57	174	99	82	293	165	120	444	254	158	628	344
	2	39	61	41	59	117	80	82	193	128	119	287	194	153	400	272
	5	52	56	39	76	111	76	105	185	122	148	277	186	190	388	261
	10	NA	NA	34	97	100	68	132	171	112	188	261	171	237	369	241
15	0	36	93	57	56	190	111	80	325	186	116	499	283	153	713	388
	2	38	69	47	57	136	93	80	225	149	115	337	224	148	473	314
	5	51	63	44	75	128	86	102	216	140	144	326	217	182	459	298
	10	NA	NA	39	95	116	79	128	201	131	182	308	203	228	438	284
	15	NA	NA	NA	NA	NA	72	158	186	124	220	290	192	272	418	269
20	0	35	96	60	54	200	118	78	346	201	114	537	306	149	772	428
	2	37	74	50	56	148	99	78	248	165	113	375	248	144	528	344
	5	50	68	47	73	140	94	100	239	158	141	363	239	178	514	334
	10	NA	NA	41	93	129	86	125	223	146	177	344	224	222	491	316
	15	NA	NA	NA	NA	NA	80	155	208	136	216	325	210	264	469	301
	20	NA	NA	NA	NA	NA	NA	186	192	126	254	306	196	309	448	285
30	0	34	99	63	53	211	127	76	372	219	110	584	334	144	849	472
	2	37	80	56	55	164	111	76	281	183	109	429	279	139	610	392
	5	49	74	52	72	157	106	98	271	173	136	417	271	171	595	382
	10	NA	NA	NA	91	144	98	122	255	168	171	397	257	213	570	367
	15	NA	NA	NA	115	131	NA	151	239	157	208	377	242	255	547	349
	20	NA	NA	NA	NA	NA	NA	181	223	NA	246	357	228	298	524	333
	30	NA	NA	NA	NA	NA	NA	NA	NA	NA	NA	NA	NA	389	477	305
50	0	33	99	66	51	213	133	73	394	230	105	629	361	138	928	515
	2	36	84	61	53	181	121	73	318	205	104	495	312	133	712	443
	5	48	80	NA	70	174	117	94	308	198	131	482	305	164	696	435
	10	NA	NA	NA	89	160	NA	118	292	186	162	461	292	203	671	420
	15	NA	NA	NA	112	148	NA	145	275	174	199	441	280	244	646	405
	20	NA	NA	NA	NA	NA	NA	176	257	NA	236	420	267	285	622	389
	30	NA	NA	NA	NA	NA	NA	NA	NA	NA	315	376	NA	373	573	NA
100	0	NA	NA	NA	49	214	NA	69	403	NA	100	659	395	131	991	555
	2	NA	NA	NA	51	192	NA	70	351	NA	98	563	373	125	828	508
	5	NA	NA	NA	67	186	NA	90	342	NA	125	551	366	156	813	501
	10	NA	NA	NA	85	175	NA	113	324	NA	153	532	354	191	789	486
	15	NA	NA	NA	132	162	NA	138	310	NA	188	511	343	230	764	473
	20	NA	NA	NA	NA	NA	NA	168	295	NA	224	487	NA	270	739	458
	30	NA	NA	NA	NA	NA	NA	231	264	NA	301	448	NA	355	685	NA
	50	NA	NA	NA	NA	NA	NA	NA	NA	NA	NA	NA	NA	540	584	NA

For SI units: 1 inch = 25.4 mm, 1 foot = 304.8 mm, 1000 British thermal units per hour = 0.293 kW, 1 square inch = 0.000645 m^2

* NA: Not applicable.

TABLE 803.1.2(2)
TYPE B DOUBLE-WALL GAS VENT [NFPA 54: TABLE 13.1(b)] (continued)*

		NUMBER OF APPLIANCES: SINGLE											
		APPLIANCE TYPE: CATEGORY I											
		APPLIANCE VENT CONNECTION: SINGLE-WALL METAL CONNECTOR											
		VENT DIAMETER – *D* (inch)											
		8			9			10			12		
		APPLIANCE INPUT RATING IN THOUSANDS OF BTU PER HOUR											
HEIGHT *H* (feet)	LATERAL *L* (feet)	FAN		NAT	FAN		NAT	FAN		NAT	FAN		NAT
		Min	Max	Max	Min	Max	Max	Min	Max	Max	Min	Max	Max
6	0	211	695	369	267	894	469	371	1118	569	537	1639	849
	2	201	423	284	251	541	368	347	673	453	498	979	648
	4	237	416	277	295	533	360	409	664	443	584	971	638
	6	263	409	271	327	526	352	449	656	433	638	962	627
8	0	206	777	414	258	1002	536	360	1257	658	521	1852	967
	2	197	482	321	246	617	417	339	768	513	486	1120	743
	5	245	470	311	305	604	404	418	754	500	598	1104	730
	8	280	458	300	344	591	392	470	740	486	665	1089	715
10	0	202	844	449	253	1093	584	351	1373	718	507	2031	1057
	2	193	531	354	242	681	456	332	849	559	475	1242	848
	5	241	518	344	299	667	443	409	834	544	584	1224	825
	10	296	497	325	363	643	423	492	808	520	688	1194	788
15	0	195	966	523	244	1259	681	336	1591	838	488	2374	1237
	2	187	631	413	232	812	543	319	1015	673	457	1491	983
	5	231	616	400	287	795	526	392	997	657	562	1469	963
	10	284	592	381	349	768	501	470	966	628	664	1433	928
	15	334	568	367	404	742	484	540	937	601	750	1399	894
20	0	190	1053	573	238	1379	750	326	1751	927	473	2631	1346
	2	182	708	468	227	914	611	309	1146	754	443	1689	1098
	5	224	692	457	279	896	596	381	1126	734	547	1665	1074
	10	277	666	437	339	866	570	457	1092	702	646	1626	1037
	15	325	640	419	393	838	549	526	1060	677	730	1587	1005
	20	374	616	400	448	810	526	592	1028	651	808	1550	973
30	0	184	1168	647	229	1542	852	312	1971	1056	454	2996	1545
	2	175	823	533	219	1069	698	296	1346	863	424	1999	1308
	5	215	806	521	269	1049	684	366	1324	846	524	1971	1283
	10	265	777	501	327	1017	662	440	1287	821	620	1927	1243
	15	312	750	481	379	985	638	507	1251	794	702	1884	1205
	20	360	723	461	433	955	615	570	1216	768	780	1841	1166
	30	461	670	426	541	895	574	704	1147	720	937	1759	1101
50	0	176	1292	704	220	1724	948	295	2223	1189	428	3432	1818
	2	168	971	613	209	1273	811	280	1615	1007	401	2426	1509
	5	204	953	602	257	1252	795	347	1591	991	496	2396	1490
	10	253	923	583	313	1217	765	418	1551	963	589	2347	1455
	15	299	894	562	363	1183	736	481	1512	934	668	2299	1421
	20	345	866	543	415	1150	708	544	1473	906	741	2251	1387
	30	442	809	502	521	1086	649	674	1399	848	892	2159	1318
100	0	166	1404	765	207	1900	1033	273	2479	1300	395	3912	2042
	2	158	1152	698	196	1532	933	259	1970	1168	371	3021	1817
	5	194	1134	688	240	1511	921	322	1945	1153	460	2990	1796
	10	238	1104	672	293	1477	902	389	1905	1133	547	2938	1763
	15	281	1075	656	342	1443	884	447	1865	1110	618	2888	1730
	20	325	1046	639	391	1410	864	507	1825	1087	690	2838	1696
	30	418	988	NA	491	1343	824	631	1747	1041	834	2739	1627
	50	617	866	NA	711	1205	NA	895	1591	NA	1138	2547	1489

For SI units: 1 inch = 25.4 mm, 1 foot = 304.8 mm, 1000 British thermal units per hour = 0.293 kW, 1 square inch = 0.000645 m^2

* NA: Not applicable.

TABLE 803.1.2(3)
MASONRY CHIMNEY [NFPA 54: TABLE 13.1(c)]*

							NUMBER OF APPLIANCES:	SINGLE								
							APPLIANCE TYPE:	CATEGORY I								
							APPLIANCE VENT CONNECTION:	TYPE B DOUBLE-WALL CONNECTOR								
		TYPE B DOUBLE-WALL CONNECTOR DIAMETER – *D* (inch) TO BE USED WITH CHIMNEY AREAS WITHIN THE SIZE LIMITS AT BOTTOM														
		3			4			5			6			7		
		APPLIANCE INPUT RATING IN THOUSANDS OF BTU PER HOUR														
HEIGHT *H* (feet)	LATERAL *L* (feet)	FAN Min	FAN Max	NAT Max	FAN Min	FAN Max	NAT Max	FAN Min	FAN Max	NAT Max	FAN Min	FAN Max	NAT Max	FAN Min	FAN Max	NAT Max
6	2	NA	NA	28	NA	NA	52	NA	NA	86	NA	NA	130	NA	NA	180
	5	NA	NA	25	NA	NA	49	NA	NA	82	NA	NA	117	NA	NA	165
8	2	NA	NA	29	NA	NA	55	NA	NA	93	NA	NA	145	NA	NA	198
	5	NA	NA	26	NA	NA	52	NA	NA	88	NA	NA	134	NA	NA	183
	8	NA	NA	24	NA	NA	48	NA	NA	83	NA	NA	127	NA	NA	175
10	2	NA	NA	31	NA	NA	61	NA	NA	103	NA	NA	162	NA	NA	221
	5	NA	NA	28	NA	NA	57	NA	NA	96	NA	NA	148	NA	NA	204
	10	NA	NA	25	NA	NA	50	NA	NA	87	NA	NA	139	NA	NA	191
15	2	NA	NA	35	NA	NA	67	NA	NA	114	NA	NA	179	53	475	250
	5	NA	NA	35	NA	NA	62	NA	NA	107	NA	NA	164	NA	NA	231
	10	NA	NA	28	NA	NA	55	NA	NA	97	NA	NA	153	NA	NA	216
	15	NA	NA	NA	NA	NA	48	NA	NA	89	NA	NA	141	NA	NA	201
20	2	NA	NA	38	NA	NA	74	NA	NA	124	NA	NA	201	51	522	274
	5	NA	NA	36	NA	NA	68	NA	NA	116	NA	NA	184	80	503	254
	10	NA	NA	NA	NA	NA	60	NA	NA	107	NA	NA	172	NA	NA	237
	15	NA	NA	NA	NA	NA	NA	NA	NA	97	NA	NA	159	NA	NA	220
	20	NA	NA	NA	NA	NA	NA	NA	NA	83	NA	NA	148	NA	NA	206
30	2	NA	NA	41	NA	NA	82	NA	NA	137	NA	NA	216	47	581	303
	5	NA	NA	NA	NA	NA	76	NA	NA	128	NA	NA	198	75	561	281
	10	NA	NA	NA	NA	NA	67	NA	NA	115	NA	NA	184	NA	NA	263
	15	NA	NA	NA	NA	NA	NA	NA	NA	107	NA	NA	171	NA	NA	243
	20	NA	NA	NA	NA	NA	NA	NA	NA	91	NA	NA	159	NA	NA	227
	30	NA	NA	NA	NA	NA	NA	NA	NA	NA	NA	NA	NA	NA	NA	188
50	2	NA	NA	NA	NA	NA	92	NA	NA	161	NA	NA	251	NA	NA	351
	5	NA	NA	NA	NA	NA	NA	NA	NA	151	NA	NA	230	NA	NA	323
	10	NA	NA	NA	NA	NA	NA	NA	NA	138	NA	NA	215	NA	NA	304
	15	NA	NA	NA	NA	NA	NA	NA	NA	127	NA	NA	199	NA	NA	282
	20	NA	NA	NA	NA	NA	NA	NA	NA	NA	NA	NA	185	NA	NA	264
	30	NA	NA	NA	NA	NA	NA	NA	NA	NA	NA	NA	NA	NA	NA	NA
Minimum internal area of chimney (square inches)		12			19			28			38			50		
Maximum internal area of chimney (square inches)		Seven times the listed appliance categorized vent area, flue collar area, or draft hood outlet areas.														

For SI units: 1 inch = 25.4 mm, 1 foot = 304.8 mm, 1000 British thermal units per hour = 0.293 kW, 1 square inch = 0.000645 m^2

* NA: Not applicable.

TABLE 803.1.2(3)
MASONRY CHIMNEY [NFPA 54: TABLE 13.1(c)] (continued)*

								NUMBER OF APPLIANCES:	SINGLE				
								APPLIANCE TYPE:	CATEGORY I				
								APPLIANCE VENT CONNECTION:	TYPE B DOUBLE-WALL CONNECTOR				
		TYPE B DOUBLE-WALL CONNECTOR DIAMETER – *D* (inch) TO BE USED WITH CHIMNEY AREAS WITHIN THE SIZE LIMITS AT BOTTOM											
		8			9			10			12		
		APPLIANCE INPUT RATING IN THOUSANDS OF BTU PER HOUR											
HEIGHT *H* (feet)	LATERAL *L* (feet)	FAN		NAT	FAN		NAT	FAN		NAT	FAN		NAT
		Min	Max	Max	Min	Max	Max	Min	Max	Max	Min	Max	Max
6	2	NA	NA	247	NA	NA	320	NA	NA	401	NA	NA	581
	5	NA	NA	231	NA	NA	298	NA	NA	376	NA	NA	561
8	2	NA	NA	266	84	590	350	100	728	446	139	1024	651
	5	NA	NA	247	NA	NA	328	149	711	423	201	1007	640
	8	NA	NA	239	NA	NA	318	173	695	410	231	990	623
10	2	68	519	298	82	655	388	98	810	491	136	1144	724
	5	NA	NA	277	124	638	365	146	791	466	196	1124	712
	10	NA	NA	263	155	610	347	182	762	444	240	1093	668
15	2	64	613	336	77	779	441	92	968	562	127	1376	841
	5	99	594	313	118	759	416	139	946	533	186	1352	828
	10	126	565	296	148	727	394	173	912	567	229	1315	777
	15	NA	NA	281	171	698	375	198	880	485	259	1280	742
20	2	61	678	375	73	867	491	87	1083	627	121	1548	953
	5	95	658	350	113	845	463	133	1059	597	179	1523	933
	10	122	627	332	143	811	440	167	1022	566	221	1482	879
	15	NA	NA	314	165	780	418	191	987	541	251	1443	840
	20	NA	NA	296	186	750	397	214	955	513	277	1406	807
30	2	57	762	421	68	985	558	81	1240	717	111	1793	1112
	5	90	741	393	106	962	526	125	1216	683	169	1766	1094
	10	115	709	373	135	927	500	158	1176	648	210	1721	1025
	15	NA	NA	353	156	893	476	181	1139	621	239	1679	981
	20	NA	NA	332	176	860	450	203	1103	592	264	1638	940
	30	NA	NA	288	NA	NA	416	249	1035	555	318	1560	877
50	2	51	840	477	61	1106	633	72	1413	812	99	2080	1243
	5	83	819	445	98	1083	596	116	1387	774	155	2052	1225
	10	NA	NA	424	126	1047	567	147	1347	733	195	2006	1147
	15	NA	NA	400	146	1010	539	170	1307	702	222	1961	1099
	20	NA	NA	376	165	977	511	190	1269	669	246	1916	1050
	30	NA	NA	327	NA	NA	468	233	1196	623	295	1832	984
Minimum internal area of chimney (square inches)		63			78			95			132		
Maximum internal area of chimney (square inches)		Seven times the listed appliance categorized vent area, flue collar area, or draft hood outlet areas.											

For SI units: 1 inch = 25.4 mm, 1 foot = 304.8 mm, 1000 British thermal units per hour = 0.293 kW, 1 square inch = 0.000645 m^2
* NA: Not applicable.

TABLE 803.1.2(4)
MASONRY CHIMNEY [NFPA 54: TABLE 13.1(d)]*

							NUMBER OF APPLIANCES:			SINGLE						
							APPLIANCE TYPE:			CATEGORY I						
							APPLIANCE VENT CONNECTION:			SINGLE-WALL METAL CONNECTOR						
		SINGLE-WALL METAL CONNECTOR DIAMETER – *D* (inch) TO BE USED WITH CHIMNEY AREAS WITHIN THE SIZE LIMITS AT BOTTOM														
		3			4			5			6			7		
		APPLIANCE INPUT RATING IN THOUSANDS OF BTU PER HOUR														
HEIGHT *H* (feet)	LATERAL *L* (feet)	FAN		NAT	FAN		NAT	FAN		NAT	FAN		NAT	FAN		NAT
		Min	Max	Max	Min	Max	Max	Min	Max	Max	Min	Max	Max	Min	Max	Max
6	2	NA	NA	28	NA	NA	52	NA	NA	86	NA	NA	130	NA	NA	180
	5	NA	NA	25	NA	NA	48	NA	NA	81	NA	NA	116	NA	NA	164
8	2	NA	NA	29	NA	NA	55	NA	NA	93	NA	NA	145	NA	NA	197
	5	NA	NA	26	NA	NA	51	NA	NA	87	NA	NA	133	NA	NA	182
	8	NA	NA	23	NA	NA	47	NA	NA	82	NA	NA	126	NA	NA	174
10	2	NA	NA	31	NA	NA	61	NA	NA	102	NA	NA	161	NA	NA	220
	5	NA	NA	28	NA	NA	56	NA	NA	95	NA	NA	147	NA	NA	203
	10	NA	NA	24	NA	NA	49	NA	NA	86	NA	NA	137	NA	NA	189
15	2	NA	NA	35	NA	NA	67	NA	NA	113	NA	NA	178	166	473	249
	5	NA	NA	32	NA	NA	61	NA	NA	106	NA	NA	163	NA	NA	230
	10	NA	NA	27	NA	NA	54	NA	NA	96	NA	NA	151	NA	NA	214
	15	NA	NA	NA	NA	NA	46	NA	NA	87	NA	NA	138	NA	NA	198
20	2	NA	NA	38	NA	NA	73	NA	NA	123	NA	NA	200	163	520	273
	5	NA	NA	35	NA	NA	67	NA	NA	115	NA	NA	183	NA	NA	252
	10	NA	NA	NA	NA	NA	59	NA	NA	105	NA	NA	170	NA	NA	235
	15	NA	NA	NA	NA	NA	NA	NA	NA	95	NA	NA	156	NA	NA	217
	20	NA	NA	NA	NA	NA	NA	NA	NA	80	NA	NA	144	NA	NA	202
30	2	NA	NA	41	NA	NA	81	NA	NA	136	NA	NA	215	158	578	302
	5	NA	NA	NA	NA	NA	75	NA	NA	127	NA	NA	196	NA	NA	279
	10	NA	NA	NA	NA	NA	66	NA	NA	113	NA	NA	182	NA	NA	260
	15	NA	NA	NA	NA	NA	NA	NA	NA	105	NA	NA	168	NA	NA	240
	20	NA	NA	NA	NA	NA	NA	NA	NA	88	NA	NA	155	NA	NA	223
	30	NA	NA	NA	NA	NA	NA	NA	NA	NA	NA	NA	NA	NA	NA	182
50	2	NA	NA	NA	NA	NA	91	NA	NA	160	NA	NA	250	NA	NA	350
	5	NA	NA	NA	NA	NA	NA	NA	NA	149	NA	NA	228	NA	NA	321
	10	NA	NA	NA	NA	NA	NA	NA	NA	136	NA	NA	212	NA	NA	301
	15	NA	NA	NA	NA	NA	NA	NA	NA	124	NA	NA	195	NA	NA	278
	20	NA	NA	NA	NA	NA	NA	NA	NA	NA	NA	NA	180	NA	NA	258
	30	NA	NA	NA	NA	NA	NA	NA	NA	NA	NA	NA	NA	NA	NA	NA
Minimum internal area of chimney (square inches)		12			19			28			38			50		
Maximum internal area of chimney (square inches)		Seven times the listed appliance categorized vent area, flue collar area, or draft hood outlet areas.														

For SI units: 1 inch = 25.4 mm, 1 foot = 304.8 mm, 1000 British thermal units per hour = 0.293 kW, 1 square inch = 0.000645 m^2

* NA: Not applicable.

TABLE 803.1.2(4)
MASONRY CHIMNEY [NFPA 54: TABLE 13.1(d)] (continued)*

				NUMBER OF APPLIANCES:				SINGLE					
				APPLIANCE TYPE:				CATEGORY I					
				APPLIANCE VENT CONNECTION:				SINGLE-WALL METAL CONNECTOR					
		SINGLE-WALL METAL CONNECTOR DIAMETER – *D* (inch) TO BE USED WITH CHIMNEY AREAS WITHIN THE SIZE LIMITS AT BOTTOM											
		8			9			10			12		
		APPLIANCE INPUT RATING IN THOUSANDS OF BTU PER HOUR											
HEIGHT *H* (feet)	LATERAL *L* (feet)	FAN		NAT	FAN		NAT	FAN		NAT	FAN		NAT
		Min	Max	Max	Min	Max	Max	Min	Max	Max	Min	Max	Max
6	2	NA	NA	247	NA	NA	319	NA	NA	400	NA	NA	580
	5	NA	NA	230	NA	NA	297	NA	NA	375	NA	NA	560
8	2	NA	NA	265	NA	NA	349	382	725	445	549	1021	650
	5	NA	NA	246	NA	NA	327	NA	NA	422	673	1003	638
	8	NA	NA	237	NA	NA	317	NA	NA	408	747	985	621
10	2	216	518	297	271	654	387	373	808	490	536	1142	722
	5	NA	NA	276	334	635	364	459	789	465	657	1121	710
	10	NA	NA	261	NA	NA	345	547	758	441	771	1088	665
15	2	211	611	335	264	776	440	362	965	560	520	1373	840
	5	261	591	312	325	755	414	444	942	531	637	1348	825
	10	NA	NA	294	392	722	392	531	907	504	749	1309	774
	15	NA	NA	278	452	692	372	606	873	481	841	1272	738
20	2	206	675	374	258	864	490	252	1079	625	508	1544	950
	5	255	655	348	317	842	461	433	1055	594	623	1518	930
	10	312	622	330	382	806	437	517	1016	562	733	1475	875
	15	NA	NA	311	442	773	414	591	979	539	823	1434	835
	20	NA	NA	292	NA	NA	392	663	944	510	911	1394	800
30	2	200	759	420	249	982	556	340	1237	715	489	1789	1110
	5	245	737	391	306	958	524	417	1210	680	600	1760	1090
	10	300	703	370	370	920	496	500	1168	644	708	1713	1020
	15	NA	NA	349	428	884	471	572	1128	615	798	1668	975
	20	NA	NA	327	NA	NA	445	643	1089	585	883	1624	932
	30	NA	NA	281	NA	NA	408	NA	NA	544	1055	1539	865
50	2	191	837	475	238	1103	631	323	1408	810	463	2076	1240
	5	NA	NA	442	293	1078	593	398	1381	770	571	2044	1220
	10	NA	NA	420	355	1038	562	447	1337	728	674	1994	1140
	15	NA	NA	395	NA	NA	533	546	1294	695	761	1945	1090
	20	NA	NA	370	NA	NA	504	616	1251	660	844	1898	1040
	30	NA	NA	318	NA	NA	458	NA	NA	610	1009	1805	970
Minimum internal area of chimney (square inches)		63			78			95			132		
Maximum internal area of chimney (square inches)		Seven times the listed appliance categorized vent area, flue collar area, or draft hood outlet areas.											

For SI units: 1 inch = 25.4 mm, 1 foot = 304.8 mm, 1000 British thermal units per hour = 0.293 kW, 1 square inch = 0.000645 m^2

* NA: Not applicable.

TABLE 803.1.2(5)
SINGLE-WALL METAL PIPE OR TYPE B ASBESTOS-CEMENT VENT [NFPA 54: TABLE 13.1(e)]*

					NUMBER OF APPLIANCES:	SINGLE			
					APPLIANCE TYPE:	DRAFT HOOD-EQUIPPED			
					APPLIANCE VENT CONNECTION:	CONNECTED DIRECTLY TO PIPE OR VENT			
		DIAMETER – *D* (inch) TO BE USED WITH CHIMNEY AREAS WITHIN THE SIZE LIMITS AT BOTTOM							
		3	4	5	6	7	8	10	12
HEIGHT *H* (feet)	LATERAL *L* (feet)	APPLIANCE INPUT RATING IN THOUSANDS OF BTU PER HOUR							
		MAXIMUM APPLIANCE INPUT RATING IN THOUSANDS OF BTU PER HOUR							
6	0	39	70	116	170	232	312	500	750
	2	31	55	94	141	194	260	415	620
	5	28	51	88	128	177	242	390	600
8	0	42	76	126	185	252	340	542	815
	2	32	61	102	154	210	284	451	680
	5	29	56	95	141	194	264	430	648
	10	24	49	86	131	180	250	406	625
10	0	45	84	138	202	279	372	606	912
	2	35	67	111	168	233	311	505	760
	5	32	61	104	153	215	289	480	724
	10	27	54	94	143	200	274	455	700
	15	NA	46	84	130	186	258	432	666
15	0	49	91	151	223	312	420	684	1040
	2	39	72	122	186	260	350	570	865
	5	35	67	110	170	240	325	540	825
	10	30	58	103	158	223	308	514	795
	15	NA	50	93	144	207	291	488	760
	20	NA	NA	82	132	195	273	466	726
20	0	53	101	163	252	342	470	770	1190
	2	42	80	136	210	286	392	641	990
	5	38	74	123	192	264	364	610	945
	10	32	65	115	178	246	345	571	910
	15	NA	55	104	163	228	326	550	870
	20	NA	NA	91	149	214	306	525	832
30	0	56	108	183	276	384	529	878	1370
	2	44	84	148	230	320	441	730	1140
	5	NA	78	137	210	296	410	694	1080
	10	NA	68	125	196	274	388	656	1050
	15	NA	NA	113	177	258	366	625	1000
	20	NA	NA	99	163	240	344	596	960
	30	NA	NA	NA	NA	192	295	540	890
50	0	NA	120	210	310	443	590	980	1550
	2	NA	95	171	260	370	492	820	1290
	5	NA	NA	159	234	342	474	780	1230
	10	NA	NA	146	221	318	456	730	1190
	15	NA	NA	NA	200	292	407	705	1130
	20	NA	NA	NA	185	276	384	670	1080
	30	NA	NA	NA	NA	222	330	605	1010

For SI units: 1 inch = 25.4 mm, 1 foot = 304.8 mm, 1000 British thermal units per hour = 0.293 kW, 1 square inch = 0.000645 m^2

* NA: Not applicable.

TABLE 803.1.2(6)
EXTERIOR MASONRY CHIMNEY [NFPA 54: TABLE 13.1(f)][1, 2]

			NUMBER OF APPLIANCES:		SINGLE			
			APPLIANCE TYPE:		NAT			
			APPLIANCE VENT CONNECTION:		TYPE B DOUBLE-WALL CONNECTOR			
MINIMUM ALLOWABLE INPUT RATING OF SPACE-HEATING APPLIANCE IN THOUSANDS OF BTU PER HOUR								
VENT HEIGHT *H* (feet)	INTERNAL AREA OF CHIMNEY (square inches)							
	12	19	28	38	50	63	78	113
	Local 99% winter design temperature: 37°F or greater							
6	0	0	0	0	0	0	0	0
8	0	0	0	0	0	0	0	0
10	0	0	0	0	0	0	0	0
15	NA	0	0	0	0	0	0	0
20	NA	NA	123	190	249	184	0	0
30	NA	NA	NA	NA	NA	393	334	0
50	NA	NA	NA	NA	NA	NA	NA	579
	Local 99% winter design temperature: 27°F to 36°F							
6	0	0	68	116	156	180	212	266
8	0	0	82	127	167	187	214	263
10	0	51	97	141	183	201	225	265
15	NA	NA	NA	NA	233	253	274	305
20	NA	NA	NA	NA	NA	307	330	362
30	NA	NA	NA	NA	NA	419	445	485
50	NA	NA	NA	NA	NA	NA	NA	763
	Local 99% winter design temperature: 17°F to 26°F							
6	NA	NA	NA	NA	NA	215	259	349
8	NA	NA	NA	NA	197	226	264	352
10	NA	NA	NA	NA	214	245	278	358
15	NA	NA	NA	NA	NA	296	331	398
20	NA	NA	NA	NA	NA	352	387	457
30	NA	NA	NA	NA	NA	NA	507	581
50	NA	NA	NA	NA	NA	NA	NA	NA
	Local 99% winter design temperature: 5°F to 16°F							
6	NA	NA	NA	NA	NA	NA	NA	416
8	NA	NA	NA	NA	NA	NA	312	423
10	NA	NA	NA	NA	NA	289	331	430
15	NA	NA	NA	NA	NA	NA	393	485
20	NA	NA	NA	NA	NA	NA	450	547
30	NA	NA	NA	NA	NA	NA	NA	682
50	NA	NA	NA	NA	NA	NA	NA	972
	Local 99% winter design temperature: -10°F to 4°F							
6	NA	NA	NA	NA	NA	NA	NA	484
8	NA	NA	NA	NA	NA	NA	NA	494
10	NA	NA	NA	NA	NA	NA	NA	513
15	NA	NA	NA	NA	NA	NA	NA	586
20	NA	NA	NA	NA	NA	NA	NA	650
30	NA	NA	NA	NA	NA	NA	NA	805
50	NA	NA	NA	NA	NA	NA	NA	1003
Local 99% winter design temperature: -11°F or lower Not recommended for any vent configurations								

For SI units: 1 inch = 25.4 mm, 1 foot = 304.8 mm, 1000 British thermal units per hour = 0.293 kW, 1 square inch = 0.000645 m^2, °C = (°F-32)/1.8

Notes:

[1] See Figure 803.1.2(6) for a map showing local 99 percent winter design temperatures in the United States.

[2] NA: Not applicable.

TABLE 803.2(1)
TYPE B DOUBLE-WALL VENT [NFPA 54: TABLE 13.2(a)]*

					NUMBER OF APPLIANCES:				TWO OR MORE							
					APPLIANCE TYPE:				CATEGORY I							
					APPLIANCE VENT CONNECTION:				TYPE B DOUBLE-WALL CONNECTOR							
		VENT CONNECTOR CAPACITY														
		TYPE B DOUBLE-WALL VENT AND CONNECTOR DIAMETER – *D* (inch)														
		3			4			5			6			7		
		APPLIANCE INPUT RATING LIMITS IN THOUSANDS OF BTU PER HOUR														
VENT HEIGHT *H* (feet)	CONNECTOR RISE *R* (feet)	FAN		NAT	FAN		NAT	FAN		NAT	FAN		NAT	FAN		NAT
		Min	Max	Max	Min	Max	Max	Min	Max	Max	Min	Max	Max	Min	Max	Max
6	1	22	37	26	35	66	46	46	106	72	58	164	104	77	225	142
	2	23	41	31	37	75	55	48	121	86	60	183	124	79	253	168
	3	24	44	35	38	81	62	49	132	96	62	199	139	82	275	189
8	1	22	40	27	35	72	48	49	114	76	64	176	109	84	243	148
	2	23	44	32	36	80	57	51	128	90	66	195	129	86	269	175
	3	24	47	36	37	87	64	53	139	101	67	210	145	88	290	198
10	1	22	43	28	34	78	50	49	123	78	65	189	113	89	257	154
	2	23	47	33	36	86	59	51	136	93	67	206	134	91	282	182
	3	24	50	37	37	92	67	52	146	104	69	220	150	94	303	205
15	1	21	50	30	33	89	53	47	142	83	64	220	120	88	298	163
	2	22	53	35	35	96	63	49	153	99	66	235	142	91	320	193
	3	24	55	40	36	102	71	51	163	111	68	248	160	93	339	218
20	1	21	54	31	33	99	56	46	157	87	62	246	125	86	334	171
	2	22	57	37	34	105	66	48	167	104	64	259	149	89	354	202
	3	23	60	42	35	110	74	50	176	116	66	271	168	91	371	228
30	1	20	62	33	31	113	59	45	181	93	60	288	134	83	391	182
	2	21	64	39	33	118	70	47	190	110	62	299	158	85	408	215
	3	22	66	44	34	123	79	48	198	124	64	309	178	88	423	242
50	1	19	71	36	30	133	64	43	216	101	57	349	145	78	477	197
	2	21	73	43	32	137	76	45	223	119	59	358	172	81	490	234
	3	22	75	48	33	141	86	46	229	134	61	366	194	83	502	263
100	1	18	82	37	28	158	66	40	262	104	53	442	150	73	611	204
	2	19	83	44	30	161	79	42	267	123	55	447	178	75	619	242
	3	20	84	50	31	163	89	44	272	138	57	452	200	78	627	272

	COMMON VENT CAPACITY											
	TYPE B DOUBLE-WALL COMMON VENT DIAMETER – *D* (inch)											
	4			5			6			7		
	COMBINED APPLIANCE INPUT RATING IN THOUSANDS OF BTU PER HOUR											
VENT HEIGHT *H* (feet)	FAN +FAN	FAN +NAT	NAT +NAT	FAN +FAN	FAN +NAT	NAT +NAT	FAN +FAN	FAN +NAT	NAT +NAT	FAN +FAN	FAN +NAT	NAT +NAT
6	92	81	65	140	116	103	204	161	147	309	248	200
8	101	90	73	155	129	114	224	178	163	339	275	223
10	110	97	79	169	141	124	243	194	178	367	299	242
15	125	112	91	195	164	144	283	228	206	427	352	280
20	136	123	102	215	183	160	314	255	229	475	394	310
30	152	138	118	244	210	185	361	297	266	547	459	360
50	167	153	134	279	244	214	421	353	310	641	547	423
100	175	163	NA	311	277	NA	489	421	NA	751	658	479

For SI units: 1 inch = 25.4 mm, 1 foot = 304.8 mm, 1000 British thermal units per hour = 0.293 kW, 1 square inch = 0.000645 m^2

* NA: Not applicable.

TABLE 803.2(1)
TYPE B DOUBLE-WALL VENT [NFPA 54: TABLE 13.2(a)] (continued)

		NUMBER OF APPLIANCES: TWO OR MORE								
		APPLIANCE TYPE: CATEGORY I								
		APPLIANCE VENT CONNECTION: TYPE B DOUBLE-WALL CONNECTOR								
		VENT CONNECTOR CAPACITY								
		TYPE B DOUBLE-WALL VENT AND CONNECTOR DIAMETER – *D* (inch)								
		8			9			10		
		APPLIANCE INPUT RATING LIMITS IN THOUSANDS OF BTU PER HOUR								
VENT HEIGHT *H* (feet)	CONNECTOR RISE *R* (feet)	FAN		NAT	FAN		NAT	FAN		NAT
		Min	Max	Max	Min	Max	Max	Min	Max	Max
6	1	92	296	185	109	376	237	128	466	289
	2	95	333	220	112	424	282	131	526	345
	3	97	363	248	114	463	317	134	575	386
8	1	100	320	194	118	408	248	138	507	303
	2	103	356	230	121	454	294	141	564	358
	3	105	384	258	123	492	330	143	612	402
10	1	106	341	200	125	436	257	146	542	314
	2	109	374	238	128	479	305	149	596	372
	3	111	402	268	131	515	342	152	642	417
15	1	110	389	214	134	493	273	162	609	333
	2	112	419	253	137	532	323	165	658	394
	3	115	445	286	140	565	365	167	700	444
20	1	107	436	224	131	552	285	158	681	347
	2	110	463	265	134	587	339	161	725	414
	3	113	486	300	137	618	383	164	764	466
30	1	103	512	238	125	649	305	151	802	372
	2	105	535	282	129	679	360	155	840	439
	3	108	555	317	132	706	405	158	874	494
50	1	97	627	257	120	797	330	144	984	403
	2	100	645	306	123	820	392	148	1014	478
	3	103	661	343	126	842	441	151	1043	538
100	1	91	810	266	112	1038	341	135	1285	417
	2	94	822	316	115	1054	405	139	1306	494
	3	97	834	355	118	1069	455	142	1327	555

	COMMON VENT CAPACITY								
	TYPE B DOUBLE-WALL COMMON VENT DIAMETER – *D* (inch)								
	8			9			10		
	COMBINED APPLIANCE INPUT RATING IN THOUSANDS OF BTU PER HOUR								
VENT HEIGHT *H* (feet)	FAN +FAN	FAN +NAT	NAT +NAT	FAN +FAN	FAN +NAT	NAT +NAT	FAN +FAN	FAN +NAT	NAT +NAT
6	404	314	260	547	434	335	672	520	410
8	444	348	290	602	480	378	740	577	465
10	477	377	315	649	522	405	800	627	495
15	556	444	365	753	612	465	924	733	565
20	621	499	405	842	688	523	1035	826	640
30	720	585	470	979	808	605	1209	975	740
50	854	706	550	1164	977	705	1451	1188	860
100	1025	873	625	1408	1215	800	1784	1502	975

For SI units: 1 inch = 25.4 mm, 1 foot = 304.8 mm, 1000 British thermal units per hour = 0.293 kW, 1 square inch = 0.000645 m^2

TABLE 803.2(1)
TYPE B DOUBLE-WALL VENT [NFPA 54: TABLE 13.2(a)] (continued)*

						NUMBER OF APPLIANCES:			TWO OR MORE				
						APPLIANCE TYPE:			CATEGORY I				
						APPLIANCE VENT CONNECTION:			TYPE B DOUBLE-WALL CONNECTOR				
		VENT CONNECTOR CAPACITY											
		TYPE B DOUBLE-WALL VENT AND CONNECTOR DIAMETER – *D* (inch)											
		12			14			16			18		
		APPLIANCE INPUT RATING LIMITS IN THOUSANDS OF BTU PER HOUR											
VENT HEIGHT *H* (feet)	CONNECTOR RISE *R* (feet)	FAN		NAT	FAN		NAT	FAN		NAT	FAN		NAT
		Min	Max	Max	Min	Max	Max	Min	Max	Max	Min	Max	Max
6	2	174	764	496	223	1046	653	281	1371	853	346	1772	1080
	4	180	897	616	230	1231	827	287	1617	1081	352	2069	1370
	6	NA	NA	NA	NA	NA	NA	NA	NA	NA	NA	NA	NA
8	2	186	822	516	238	1126	696	298	1478	910	365	1920	1150
	4	192	952	644	244	1307	884	305	1719	1150	372	2211	1460
	6	198	1050	772	252	1445	1072	313	1902	1390	380	2434	1770
10	2	196	870	536	249	1195	730	311	1570	955	379	2049	1205
	4	201	997	664	256	1371	924	318	1804	1205	387	2332	1535
	6	207	1095	792	263	1509	1118	325	1989	1455	395	2556	1865
15	2	214	967	568	272	1334	790	336	1760	1030	408	2317	1305
	4	221	1085	712	279	1499	1006	344	1978	1320	416	2579	1665
	6	228	1181	856	286	1632	1222	351	2157	1610	424	2796	2025
20	2	223	1051	596	291	1443	840	357	1911	1095	430	2533	1385
	4	230	1162	748	298	1597	1064	365	2116	1395	438	2778	1765
	6	237	1253	900	307	1726	1288	373	2287	1695	450	2984	2145
30	2	216	1217	632	286	1664	910	367	2183	1190	461	2891	1540
	4	223	1316	792	294	1802	1160	376	2366	1510	474	3110	1920
	6	231	1400	952	303	1920	1410	384	2524	1830	485	3299	2340
50	2	206	1479	689	273	2023	1007	350	2659	1315	435	3548	1665
	4	213	1561	860	281	2139	1291	359	2814	1685	447	3730	2135
	6	221	1631	1031	290	2242	1575	369	2951	2055	461	3893	2605
100	2	192	1923	712	254	2644	1050	326	3490	1370	402	4707	1740
	4	200	1984	888	263	2731	1346	336	3606	1760	414	4842	2220
	6	208	2035	1064	272	2811	1642	346	3714	2150	426	4968	2700

	COMMON VENT CAPACITY											
	TYPE B DOUBLE-WALL COMMON VENT DIAMETER – *D* (inch)											
	12			14			16			18		
	COMBINED APPLIANCE INPUT RATING IN THOUSANDS OF BTU PER HOUR											
VENT HEIGHT *H* (feet)	FAN +FAN	FAN +NAT	NAT +NAT	FAN +FAN	FAN +NAT	NAT +NAT	FAN +FAN	FAN +NAT	NAT +NAT	FAN +FAN	FAN +NAT	NAT +NAT
6	900	696	588	1284	990	815	1735	1336	1065	2253	1732	1345
8	994	773	652	1423	1103	912	1927	1491	1190	2507	1936	1510
10	1076	841	712	1542	1200	995	2093	1625	1300	2727	2113	1645
15	1247	986	825	1794	1410	1158	2440	1910	1510	3184	2484	1910
20	1405	1116	916	2006	1588	1290	2722	2147	1690	3561	2798	2140
30	1658	1327	1025	2373	1892	1525	3220	2558	1990	4197	3326	2520
50	2024	1640	1280	2911	2347	1863	3964	3183	2430	5184	4149	3075
100	2569	2131	1670	3732	3076	2450	5125	4202	3200	6749	5509	4050

For SI units: 1 inch = 25.4 mm, 1 foot = 304.8 mm, 1000 British thermal units per hour = 0.293 kW, 1 square inch = 0.000645 m^2

* NA: Not applicable.

TABLE 803.2(1)
TYPE B DOUBLE-WALL VENT [NFPA 54: TABLE 13.2(a)] (continued)*

		NUMBER OF APPLIANCES:	TWO OR MORE							
		APPLIANCE TYPE:	CATEGORY I							
		APPLIANCE VENT CONNECTION:	TYPE B DOUBLE-WALL CONNECTOR							
		VENT CONNECTOR CAPACITY								
		TYPE B DOUBLE-WALL VENT AND CONNECTOR DIAMETER – *D* (inch)								
		20			22			24		
		APPLIANCE INPUT RATING LIMITS IN THOUSANDS OF BTU PER HOUR								
VENT HEIGHT *H* (feet)	CONNECTOR RISE *R* (feet)	FAN		NAT	FAN		NAT	FAN		NAT
		Min	Max	Max	Min	Max	Max	Min	Max	Max
6	2	NA	NA	NA	NA	NA	NA	NA	NA	NA
	4	NA	NA	NA	NA	NA	NA	NA	NA	NA
	6	NA	NA	NA	NA	NA	NA	NA	NA	NA
8	2	NA	NA	NA	NA	NA	NA	NA	NA	NA
	4	471	2737	1800	560	3319	2180	662	3957	2590
	6	478	3018	2180	568	3665	2640	669	4373	3130
10	2	NA	NA	NA	NA	NA	NA	NA	NA	NA
	4	486	2887	1890	581	3502	2280	686	4175	2710
	6	494	3169	2290	589	3849	2760	694	4593	3270
15	2	NA	NA	NA	NA	NA	NA	NA	NA	NA
	4	523	3197	2060	624	3881	2490	734	4631	2960
	6	533	3470	2510	634	4216	3030	743	5035	3600
20	2	NA	NA	NA	NA	NA	NA	NA	NA	NA
	4	554	3447	2180	661	4190	2630	772	5005	3130
	6	567	3708	2650	671	4511	3190	785	5392	3790
30	2	NA	NA	NA	NA	NA	NA	NA	NA	NA
	4	619	3840	2365	728	4861	2860	847	5606	3410
	6	632	4080	2875	741	4976	3480	860	5961	4150
50	2	NA	NA	NA	NA	NA	NA	NA	NA	NA
	4	580	4601	2633	709	5569	3185	851	6633	3790
	6	594	4808	3208	724	5826	3885	867	6943	4620
100	2	NA	NA	NA	NA	NA	NA	NA	NA	NA
	4	523	5982	2750	639	7254	3330	769	8650	3950
	6	539	6143	3350	654	7453	4070	786	8892	4810

	COMMON VENT CAPACITY								
	TYPE B DOUBLE-WALL COMMON VENT DIAMETER – *D* (inch)								
	20			22			24		
	COMBINED APPLIANCE INPUT RATING IN THOUSANDS OF BTU PER HOUR								
VENT HEIGHT *H* (feet)	FAN +FAN	FAN +NAT	NAT +NAT	FAN +FAN	FAN +NAT	NAT +NAT	FAN +FAN	FAN +NAT	NAT +NAT
6	2838	2180	1660	3488	2677	1970	4206	3226	2390
8	3162	2439	1860	3890	2998	2200	4695	3616	2680
10	3444	2665	2030	4241	3278	2400	5123	3957	2920
15	4026	3133	2360	4971	3862	2790	6016	4670	3400
20	4548	3552	2640	5573	4352	3120	6749	5261	3800
30	5303	4193	3110	6539	5157	3680	7940	6247	4480
50	6567	5240	3800	8116	6458	4500	9837	7813	5475
100	8597	6986	5000	10 681	8648	5920	13 004	10 499	7200

For SI units: 1 inch = 25.4 mm, 1 foot = 304.8 mm, 1000 British thermal units per hour = 0.293 kW, 1 square inch = 0.000645 m^2

* NA: Not applicable.

TABLE 803.2(2)
TYPE B DOUBLE-WALL VENT [NFPA 54: TABLE 13.2(b)]*

							NUMBER OF APPLIANCES:				TWO OR MORE					
							APPLIANCE TYPE:				CATEGORY I					
							APPLIANCE VENT CONNECTION:				SINGLE-WALL METAL CONNECTOR					
		VENT CONNECTOR CAPACITY														
		SINGLE-WALL METAL VENT CONNECTOR DIAMETER – *D* (inch)														
		3			4			5			6			7		
		APPLIANCE INPUT RATING LIMITS IN THOUSANDS OF BTU PER HOUR														
VENT HEIGHT	CONNECTOR RISE	FAN		NAT	FAN		NAT	FAN		NAT	FAN		NAT	FAN		NAT
H (feet)	*R* (feet)	Min	Max	Max	Min	Max	Max	Min	Max	Max	Min	Max	Max	Min	Max	Max
6	1	NA	NA	26	NA	NA	46	NA	NA	71	NA	NA	102	207	223	140
	2	NA	NA	31	NA	NA	55	NA	NA	85	168	182	123	215	251	167
	3	NA	NA	34	NA	NA	62	121	131	95	175	198	138	222	273	188
8	1	NA	NA	27	NA	NA	48	NA	NA	75	NA	NA	106	226	240	145
	2	NA	NA	32	NA	NA	57	125	126	89	184	193	127	234	266	173
	3	NA	NA	35	NA	NA	64	130	138	100	191	208	144	241	287	197
10	1	NA	NA	28	NA	NA	50	119	121	77	182	186	110	240	253	150
	2	NA	NA	33	84	85	59	124	134	91	189	203	132	248	278	183
	3	NA	NA	36	89	91	67	129	144	102	197	217	148	257	299	203
15	1	NA	NA	29	79	87	52	116	138	81	177	214	116	238	291	158
	2	NA	NA	34	83	94	62	121	150	97	185	230	138	246	314	189
	3	NA	NA	39	87	100	70	127	160	109	193	243	157	255	333	215
20	1	49	56	30	78	97	54	115	152	84	175	238	120	233	325	165
	2	52	59	36	82	103	64	120	163	101	182	252	144	243	346	197
	3	55	62	40	87	107	72	125	172	113	190	264	164	252	363	223
30	1	47	60	31	77	110	57	112	175	89	169	278	129	226	380	175
	2	51	62	37	81	115	67	117	185	106	177	290	152	236	397	208
	3	54	64	42	85	119	76	122	193	120	185	300	172	244	412	235
50	1	46	69	34	75	128	60	109	207	96	162	336	137	217	460	188
	2	49	71	40	79	132	72	114	215	113	170	345	164	226	473	223
	3	52	72	45	83	136	82	119	221	123	178	353	186	235	486	252
100	1	45	79	34	71	150	61	104	249	98	153	424	140	205	585	192
	2	48	80	41	75	153	73	110	255	115	160	428	167	212	593	228
	3	51	81	46	79	157	85	114	260	129	168	433	190	222	603	256

	COMMON VENT CAPACITY											
	TYPE B DOUBLE-WALL COMMON VENT DIAMETER – *D* (inch)											
	4			5			6			7		
	COMBINED APPLIANCE INPUT RATING IN THOUSANDS OF BTU PER HOUR											
VENT HEIGHT *H* (feet)	FAN +FAN	FAN +NAT	NAT +NAT	FAN +FAN	FAN +NAT	NAT +NAT	FAN +FAN	FAN +NAT	NAT +NAT	FAN +FAN	FAN +NAT	NAT +NAT
6	NA	78	64	NA	113	99	200	158	144	304	244	196
8	NA	87	71	NA	126	111	218	173	159	331	269	218
10	NA	94	76	163	137	120	237	189	174	357	292	236
15	121	108	88	189	159	140	275	221	200	416	343	274
20	131	118	98	208	177	156	305	247	223	463	383	302
30	145	132	113	236	202	180	350	286	257	533	446	349
50	159	145	128	268	233	208	406	337	296	622	529	410
100	166	153	NA	297	263	NA	469	398	NA	726	633	464

For SI units: 1 inch = 25.4 mm, 1 foot = 304.8 mm, 1000 British thermal units per hour = 0.293 kW, 1 square inch = 0.000645 m^2

* NA: Not applicable.

TABLE 803.2(2)
TYPE B DOUBLE-WALL VENT [NFPA 54: TABLE 13.2(b)] (continued)

		NUMBER OF APPLIANCES: TWO OR MORE								
		APPLIANCE TYPE: CATEGORY I								
		APPLIANCE VENT CONNECTION: SINGLE-WALL METAL CONNECTOR								
		VENT CONNECTOR CAPACITY								
		SINGLE-WALL METAL VENT CONNECTOR DIAMETER – *D* (inch)								
		8			9			10		
		APPLIANCE INPUT RATING LIMITS IN THOUSANDS OF BTU PER HOUR								
VENT HEIGHT *H* (feet)	CONNECTOR RISE *R* (feet)	FAN		NAT	FAN		NAT	FAN		NAT
		Min	Max	Max	Min	Max	Max	Min	Max	Max
6	1	262	293	183	325	373	234	447	463	286
	2	271	331	219	334	422	281	458	524	344
	3	279	361	247	344	462	316	468	574	385
8	1	285	316	191	352	403	244	481	502	299
	2	293	353	228	360	450	292	492	560	355
	3	302	381	256	370	489	328	501	609	400
10	1	302	335	196	372	429	252	506	534	308
	2	311	369	235	381	473	302	517	589	368
	3	320	398	265	391	511	339	528	637	413
15	1	312	380	208	397	482	266	556	596	324
	2	321	411	248	407	522	317	568	646	387
	3	331	438	281	418	557	360	579	690	437
20	1	306	425	217	390	538	276	546	664	336
	2	317	453	259	400	574	331	558	709	403
	3	326	476	294	412	607	375	570	750	457
30	1	296	497	230	378	630	294	528	779	358
	2	307	521	274	389	662	349	541	819	425
	3	316	542	309	400	690	394	555	855	482
50	1	284	604	245	364	768	314	507	951	384
	2	294	623	293	376	793	375	520	983	458
	3	304	640	331	387	816	423	535	1013	518
100	1	269	774	249	345	993	321	476	1236	393
	2	279	788	299	358	1011	383	490	1259	469
	3	289	801	339	368	1027	431	506	1280	527

	COMMON VENT CAPACITY								
	TYPE B DOUBLE-WALL COMMON VENT DIAMETER – *D* (inch)								
	8			9			10		
	COMBINED APPLIANCE INPUT RATING IN THOUSANDS OF BTU PER HOUR								
VENT HEIGHT *H* (feet)	FAN +FAN	FAN +NAT	NAT +NAT	FAN +FAN	FAN +NAT	NAT +NAT	FAN +FAN	FAN +NAT	NAT +NAT
6	398	310	257	541	429	332	665	515	407
8	436	342	285	592	473	373	730	569	460
10	467	369	309	638	512	398	787	617	487
15	544	434	357	738	599	456	905	718	553
20	606	487	395	824	673	512	1013	808	626
30	703	570	459	958	790	593	1183	952	723
50	833	686	535	1139	954	689	1418	1157	838
100	999	846	606	1378	1185	780	1741	1459	948

For SI units: 1 inch = 25.4 mm, 1 foot = 304.8 mm, 1000 British thermal units per hour = 0.293 kW, 1 square inch = 0.000645 m^2

TABLE 803.2(3)
MASONRY CHIMNEY [NFPA 54: TABLE 13.2(c)]*

								NUMBER OF APPLIANCES:			TWO OR MORE					
								APPLIANCE TYPE:			CATEGORY I					
								APPLIANCE VENT CONNECTION:			TYPE B DOUBLE-WALL CONNECTOR					
		VENT CONNECTOR CAPACITY														
		TYPE B DOUBLE-WALL VENT CONNECTOR DIAMETER – *D* (inch)														
		3			4			5			6			7		
		APPLIANCE INPUT RATING LIMITS IN THOUSANDS OF BTU PER HOUR														
VENT HEIGHT *H* (feet)	CONNECTOR RISE *R* (feet)	FAN		NAT	FAN		NAT	FAN		NAT	FAN		NAT	FAN		NAT
		Min	Max	Max	Min	Max	Max	Min	Max	Max	Min	Max	Max	Min	Max	Max
6	1	24	33	21	39	62	40	52	106	67	65	194	101	87	274	141
	2	26	43	28	41	79	52	53	133	85	67	230	124	89	324	173
	3	27	49	34	42	92	61	55	155	97	69	262	143	91	369	203
8	1	24	39	22	39	72	41	55	117	69	71	213	105	94	304	148
	2	26	47	29	40	87	53	57	140	86	73	246	127	97	350	179
	3	27	52	34	42	97	62	59	159	98	75	269	145	99	383	206
10	1	24	42	22	38	80	42	55	130	71	74	232	108	101	324	153
	2	26	50	29	40	93	54	57	153	87	76	261	129	103	366	184
	3	27	55	35	41	105	63	58	170	100	78	284	148	106	397	209
15	1	24	48	23	38	93	44	54	154	74	72	277	114	100	384	164
	2	25	55	31	39	105	55	56	174	89	74	299	134	103	419	192
	3	26	59	35	41	115	64	57	189	102	76	319	153	105	448	215
20	1	24	52	24	37	102	46	53	172	77	71	313	119	98	437	173
	2	25	58	31	39	114	56	55	190	91	73	335	138	101	467	199
	3	26	63	35	40	123	65	57	204	104	75	353	157	104	493	222
30	1	24	54	25	37	111	48	52	192	82	69	357	127	96	504	187
	2	25	60	32	38	122	58	54	208	95	72	376	145	99	531	209
	3	26	64	36	40	131	66	56	221	107	74	392	163	101	554	233
50	1	23	51	25	36	116	51	51	209	89	67	405	143	92	582	213
	2	24	59	32	37	127	61	53	225	102	70	421	161	95	604	235
	3	26	64	36	39	135	69	55	237	115	72	435	180	98	624	260
100	1	23	46	24	35	108	50	49	208	92	65	428	155	88	640	237
	2	24	53	31	37	120	60	51	224	105	67	444	174	92	660	260
	3	25	59	35	38	130	68	53	237	118	69	458	193	94	679	285

	COMMON VENT CAPACITY														
	MINIMUM INTERNAL AREA OF MASONRY CHIMNEY FLUE (square inches)														
	12			19			28			38			50		
	COMBINED APPLIANCE INPUT RATING IN THOUSANDS OF BTU PER HOUR														
VENT HEIGHT *H* (feet)	FAN +FAN	FAN +NAT	NAT +NAT	FAN +FAN	FAN +NAT	NAT +NAT	FAN +FAN	FAN +NAT	NAT +NAT	FAN +FAN	FAN +NAT	NAT +NAT	FAN +FAN	FAN +NAT	NAT +NAT
6	NA	74	25	NA	119	46	NA	178	71	NA	257	103	NA	351	143
8	NA	80	28	NA	130	53	NA	193	82	NA	279	119	NA	384	163
10	NA	84	31	NA	138	56	NA	207	90	NA	299	131	NA	409	177
15	NA	NA	36	NA	152	67	NA	233	106	NA	334	152	523	467	212
20	NA	NA	41	NA	NA	75	NA	250	122	NA	368	172	565	508	243
30	NA	NA	NA	NA	NA	NA	NA	270	137	NA	404	198	615	564	278
50	NA	NA	NA	NA	NA	NA	NA	NA	NA	NA	NA	NA	NA	620	328
100	NA	NA	NA	NA	NA	NA	NA	NA	NA	NA	NA	NA	NA	NA	348

For SI units: 1 inch = 25.4 mm, 1 foot = 304.8 mm, 1000 British thermal units per hour = 0.293 kW, 1 square inch = 0.000645 m^2

* NA: Not applicable.

TABLE 803.2(3)
MASONRY CHIMNEY [NFPA 54: TABLE 13.2(c)] (continued)*

NUMBER OF APPLIANCES: TWO OR MORE
APPLIANCE TYPE: CATEGORY I
APPLIANCE VENT CONNECTION: TYPE B DOUBLE-WALL CONNECTOR

VENT CONNECTOR CAPACITY — TYPE B DOUBLE-WALL VENT CONNECTOR DIAMETER – *D* (inch) — APPLIANCE INPUT RATING LIMITS IN THOUSANDS OF BTU PER HOUR

VENT HEIGHT *H* (feet)	CONNECTOR RISE *R* (feet)	8			9			10		
		FAN		NAT	FAN		NAT	FAN		NAT
		Min	Max	Max	Min	Max	Max	Min	Max	Max
6	1	104	370	201	124	479	253	145	599	319
	2	107	436	232	127	562	300	148	694	378
	3	109	491	270	129	633	349	151	795	439
8	1	113	414	210	134	539	267	156	682	335
	2	116	473	240	137	615	311	160	776	394
	3	119	517	276	139	672	358	163	848	452
10	1	120	444	216	142	582	277	165	739	348
	2	123	498	247	145	652	321	168	825	407
	3	126	540	281	147	705	366	171	893	463
15	1	125	511	229	153	658	297	184	824	375
	2	128	558	260	156	718	339	187	900	432
	3	131	597	292	159	760	382	190	960	486
20	1	123	584	239	150	752	312	180	943	397
	2	126	625	270	153	805	354	184	1011	452
	3	129	661	301	156	851	396	187	1067	505
30	1	119	680	255	145	883	337	175	1115	432
	2	122	715	287	149	928	378	179	1171	484
	3	125	746	317	152	968	418	182	1220	535
50	1	115	798	294	140	1049	392	168	1334	506
	2	118	827	326	143	1085	433	172	1379	558
	3	121	854	357	147	1118	474	176	1421	611
100	1	109	907	334	134	1222	454	161	1589	596
	2	113	933	368	138	1253	497	165	1626	651
	3	116	956	399	141	1282	540	169	1661	705

COMMON VENT CAPACITY — MINIMUM INTERNAL AREA OF MASONRY CHIMNEY FLUE (square inches) — COMBINED APPLIANCE INPUT RATING IN THOUSANDS OF BTU PER HOUR

VENT HEIGHT *H* (feet)	63			78			113		
	FAN +FAN	FAN +NAT	NAT +NAT	FAN +FAN	FAN +NAT	NAT +NAT	FAN +FAN	FAN +NAT	NAT +NAT
6	NA	458	188	NA	582	246	1041	853	NA
8	NA	501	218	724	636	278	1144	937	408
10	606	538	236	776	686	302	1226	1010	454
15	682	611	283	874	781	365	1374	1156	546
20	742	668	325	955	858	419	1513	1286	648
30	816	747	381	1062	969	496	1702	1473	749
50	879	831	461	1165	1089	606	1905	1692	922
100	NA	NA	499	NA	NA	669	2053	1921	1058

For SI units: 1 inch = 25.4 mm, 1 foot = 304.8 mm, 1000 British thermal units per hour = 0.293 kW, 1 square inch = 0.000645 m^2

* NA: Not applicable.

TABLE 803.2(4)
MASONRY CHIMNEY [NFPA 54: TABLE 13.2(d)]*

		NUMBER OF APPLIANCES: TWO OR MORE														
		APPLIANCE TYPE: CATEGORY I														
		APPLIANCE VENT CONNECTION: SINGLE-WALL METAL CONNECTOR														
		VENT CONNECTOR CAPACITY														
		SINGLE-WALL METAL VENT CONNECTOR DIAMETER – *D* (inch)														
		3			4			5			6			7		
		APPLIANCE INPUT RATING LIMITS IN THOUSANDS OF BTU PER HOUR														
VENT HEIGHT *H* (feet)	CONNECTOR RISE *R* (feet)	FAN		NAT	FAN		NAT	FAN		NAT	FAN		NAT	FAN		NAT
		Min	Max	Max	Min	Max	Max	Min	Max	Max	Min	Max	Max	Min	Max	Max
6	1	NA	NA	21	NA	NA	39	NA	NA	66	179	191	100	231	271	140
	2	NA	NA	28	NA	NA	52	NA	NA	84	186	227	123	239	321	172
	3	NA	NA	34	NA	NA	61	134	153	97	193	258	142	247	365	202
8	1	NA	NA	21	NA	NA	40	NA	NA	68	195	208	103	250	298	146
	2	NA	NA	28	NA	NA	52	137	139	85	202	240	125	258	343	177
	3	NA	NA	34	NA	NA	62	143	156	98	210	264	145	266	376	205
10	1	NA	NA	22	NA	NA	41	130	151	70	202	225	106	267	316	151
	2	NA	NA	29	NA	NA	53	136	150	86	210	255	128	276	358	181
	3	NA	NA	34	97	102	62	143	166	99	217	277	147	284	389	207
15	1	NA	NA	23	NA	NA	43	129	151	73	199	271	112	268	376	161
	2	NA	NA	30	92	103	54	135	170	88	207	295	132	277	411	189
	3	NA	NA	34	96	112	63	141	185	101	215	315	151	286	439	213
20	1	NA	NA	23	87	99	45	128	167	76	197	303	117	265	425	169
	2	NA	NA	30	91	111	55	134	185	90	205	325	136	274	455	195
	3	NA	NA	35	96	119	64	140	199	103	213	343	154	282	481	219
30	1	NA	NA	24	86	108	47	126	187	80	193	347	124	259	492	183
	2	NA	NA	31	91	119	57	132	203	93	201	366	142	269	518	205
	3	NA	NA	35	95	127	65	138	216	105	209	381	160	277	540	229
50	1	NA	NA	24	85	113	50	124	204	87	188	392	139	252	567	208
	2	NA	NA	31	89	123	60	130	218	100	196	408	158	262	588	230
	3	NA	NA	35	94	131	68	136	231	112	205	422	176	271	607	255
100	1	NA	NA	23	84	104	49	122	200	89	182	410	151	243	617	232
	2	NA	NA	30	88	115	59	127	215	102	190	425	169	253	636	254
	3	NA	NA	34	93	124	67	133	228	115	199	438	188	262	654	279

	COMMON VENT CAPACITY														
	MINIMUM INTERNAL AREA OF MASONRY CHIMNEY FLUE (square inches)														
	12			19			28			38			50		
	COMBINED APPLIANCE INPUT RATING IN THOUSANDS OF BTU PER HOUR														
VENT HEIGHT *H* (feet)	FAN +FAN	FAN +NAT	NAT +NAT	FAN +FAN	FAN +NAT	NAT +NAT	FAN +FAN	FAN +NAT	NAT +NAT	FAN +FAN	FAN +NAT	NAT +NAT	FAN +FAN	FAN +NAT	NAT +NAT
6	NA	NA	25	NA	118	45	NA	176	71	NA	255	102	NA	348	142
8	NA	NA	28	NA	128	52	NA	190	81	NA	276	118	NA	380	162
10	NA	NA	31	NA	136	56	NA	205	89	NA	295	129	NA	405	175
15	NA	NA	36	NA	NA	66	NA	230	105	NA	335	150	NA	400	210
20	NA	NA	NA	NA	NA	74	NA	247	120	NA	362	170	NA	503	240
30	NA	NA	NA	NA	NA	NA	NA	NA	135	NA	398	195	NA	558	275
50	NA	NA	NA	NA	NA	NA	NA	NA	NA	NA	NA	NA	NA	612	325
100	NA	NA	NA	NA	NA	NA	NA	NA	NA	NA	NA	NA	NA	NA	NA

For SI units: 1 inch = 25.4 mm, 1 foot = 304.8 mm, 1000 British thermal units per hour = 0.293 kW, 1 square inch = 0.000645 m^2

* NA: Not applicable.

TABLE 803.2(4)
MASONRY CHIMNEY [NFPA 54: TABLE 13.2(d)] (continued)*

		NUMBER OF APPLIANCES: TWO OR MORE								
		APPLIANCE TYPE: CATEGORY I								
		APPLIANCE VENT CONNECTION: SINGLE-WALL METAL CONNECTOR								
		VENT CONNECTOR CAPACITY								
		SINGLE-WALL METAL VENT CONNECTOR DIAMETER – *D* (inch)								
		8			9			10		
		APPLIANCE INPUT RATING LIMITS IN THOUSANDS OF BTU PER HOUR								
VENT HEIGHT *H* (feet)	CONNECTOR RISE *R* (feet)	FAN Min	FAN Max	NAT Max	FAN Min	FAN Max	NAT Max	FAN Min	FAN Max	NAT Max
6	1	292	366	200	362	474	252	499	594	316
	2	301	432	231	373	557	299	509	696	376
	3	309	491	269	381	634	348	519	793	437
8	1	313	407	207	387	530	263	529	672	331
	2	323	465	238	397	607	309	540	766	391
	3	332	509	274	407	663	356	551	838	450
10	1	333	434	213	410	571	273	558	727	343
	2	343	489	244	420	640	317	569	813	403
	3	352	530	279	430	694	363	580	880	459
15	1	349	502	225	445	646	291	623	808	366
	2	359	548	256	456	706	334	634	884	424
	3	368	586	289	466	755	378	646	945	479
20	1	345	569	235	439	734	306	614	921	387
	2	355	610	266	450	787	348	627	986	443
	3	365	644	298	461	831	391	639	1042	496
30	1	338	665	250	430	864	330	600	1089	421
	2	348	699	282	442	908	372	613	1145	473
	3	358	729	312	452	946	412	626	1193	524
50	1	328	778	287	417	1022	383	582	1302	492
	2	339	806	320	429	1058	425	596	1346	545
	3	349	831	351	440	1090	466	610	1386	597
100	1	315	875	328	402	1181	444	560	1537	580
	2	326	899	361	415	1210	488	575	1570	634
	3	337	921	392	427	1238	529	589	1604	687

	COMMON VENT CAPACITY								
	MINIMUM INTERNAL AREA OF MASONRY CHIMNEY FLUE (square inches)								
	63			78			113		
	COMBINED APPLIANCE INPUT RATING IN THOUSANDS OF BTU PER HOUR								
VENT HEIGHT *H* (feet)	FAN +FAN	FAN +NAT	NAT +NAT	FAN +FAN	FAN +NAT	NAT +NAT	FAN +FAN	FAN +NAT	NAT +NAT
6	NA	455	187	NA	579	245	NA	846	NA
8	NA	497	217	NA	633	277	1136	928	405
10	NA	532	234	771	680	300	1216	1000	450
15	677	602	280	866	772	360	1359	1139	540
20	765	661	321	947	849	415	1495	1264	640
30	808	739	377	1052	957	490	1682	1447	740
50	NA	821	456	1152	1076	600	1879	1672	910
100	NA	NA	494	NA	NA	663	2006	1885	1046

For SI units: 1 inch = 25.4 mm, 1 foot = 304.8 mm, 1000 British thermal units per hour = 0.293 kW, 1 square inch = 0.000645 m^2

* NA: Not applicable.

TABLE 803.2(5)
SINGLE-WALL METAL PIPE OR TYPE B ASBESTOS-CEMENT VENT [NFPA 54: TABLE 13.2(e)]*

		NUMBER OF APPLIANCES:	TWO OR MORE				
		APPLIANCE TYPE:	DRAFT HOOD-EQUIPMENT				
		APPLIANCE VENT CONNECTION:	DIRECT TO PIPE OR VENT				
		VENT CONNECTOR CAPACITY					
		VENT CONNECTOR DIAMETER – *D* (inch)					
TOTAL VENT HEIGHT *H* (feet)	CONNECTOR RISE *R* (feet)	3	4	5	6	7	8
		APPLIANCE INPUT RATING IN THOUSANDS OF BTU PER HOUR					
6-8	1	21	40	68	102	146	205
	2	28	53	86	124	178	235
	3	34	61	98	147	204	275
15	1	23	44	77	117	179	240
	2	30	56	92	134	194	265
	3	35	64	102	155	216	298
30 and up	1	25	49	84	129	190	270
	2	31	58	97	145	211	295
	3	36	68	107	164	232	321

	COMMON VENT CAPACITY						
	COMMON VENT DIAMETER – *D* (inch)						
TOTAL VENT HEIGHT *H* (feet)	4	5	6	7	8	10	12
	COMBINED APPLIANCE INPUT RATING IN THOUSANDS OF BTU PER HOUR						
6	48	78	111	155	205	320	NA
8	55	89	128	175	234	365	505
10	59	95	136	190	250	395	560
15	71	115	168	228	305	480	690
20	80	129	186	260	340	550	790
30	NA	147	215	300	400	650	940
50	NA	NA	NA	360	490	810	1190

For SI units: 1 inch = 25.4 mm, 1 foot = 304.8 mm, 1000 British thermal units per hour = 0.293 kW, 1 square inch = 0.000645 m^2
* NA: Not applicable.

TABLE 803.2(6)
EXTERIOR MASONRY CHIMNEY [NFPA 54: TABLE 13.2(f)]*

				NUMBER OF APPLIANCES:	TWO OR MORE		
				APPLIANCE TYPE:	NAT + NAT		
				APPLIANCE VENT CONNECTION:	TYPE B DOUBLE-WALL CONNECTOR		
COMBINED APPLIANCE MAXIMUM INPUT RATING IN THOUSANDS OF BTU PER HOUR								
VENT HEIGHT *H* (feet)	INTERNAL AREA OF CHIMNEY (square inches)							
	12	19	28	38	50	63	78	113
6	25	46	71	103	143	188	246	NA
8	28	53	82	119	163	218	278	408
10	31	56	90	131	177	236	302	454
15	NA	67	106	152	212	283	365	546
20	NA	NA	NA	NA	NA	325	419	648
30	NA	NA	NA	NA	NA	NA	496	749
50	NA	NA	NA	NA	NA	NA	NA	922
100	NA	NA	NA	NA	NA	NA	NA	NA

For SI units: 1 inch = 25.4 mm, 1 foot = 304.8 mm, 1000 British thermal units per hour = 0.293 kW, 1 square inch = 0.000645 m^2
* NA: Not applicable.

TABLE 803.2(7)
EXTERIOR MASONRY CHIMNEY [NFPA 54: TABLE 13.2(g)][1, 2]

				NUMBER OF APPLIANCES:	TWO OR MORE			
				APPLIANCE TYPE:	NAT + NAT			
				APPLIANCE VENT CONNECTION:	TYPE B DOUBLE-WALL CONNECTOR			
MINIMUM ALLOWABLE INPUT RATING OF SPACE-HEATING APPLIANCE IN THOUSANDS OF BTU PER HOUR								
VENT HEIGHT *H* (feet)	INTERNAL AREA OF CHIMNEY (square inches)							
	12	19	28	38	50	63	78	113
Local 99% winter design temperature: 37°F or greater								
6	0	0	0	0	0	0	0	NA
8	0	0	0	0	0	0	0	0
10	0	0	0	0	0	0	0	0
15	NA	0	0	0	0	0	0	0
20	NA	NA	NA	NA	NA	184	0	0
30	NA	NA	NA	NA	NA	393	334	0
50	NA	NA	NA	NA	NA	NA	NA	579
100	NA	NA	NA	NA	NA	NA	NA	NA
Local 99% winter design temperature: 27°F to 36°F								
6	0	0	68	NA	NA	180	212	NA
8	0	0	82	NA	NA	187	214	263
10	0	51	NA	NA	NA	201	225	265
15	NA	NA	NA	NA	NA	253	274	305
20	NA	NA	NA	NA	NA	307	330	362
30	NA	NA	NA	NA	NA	NA	445	485
50	NA	NA	NA	NA	NA	NA	NA	763
100	NA	NA	NA	NA	NA	NA	NA	NA
Local 99% winter design temperature: 17°F to 26°F								
6	NA	NA	NA	NA	NA	NA	NA	NA
8	NA	NA	NA	NA	NA	NA	264	352
10	NA	NA	NA	NA	NA	NA	278	358
15	NA	NA	NA	NA	NA	NA	331	398
20	NA	NA	NA	NA	NA	NA	387	457
30	NA	NA	NA	NA	NA	NA	NA	581
50	NA	NA	NA	NA	NA	NA	NA	862
100	NA	NA	NA	NA	NA	NA	NA	NA
Local 99% winter design temperature: 5°F to 16°F								
6	NA	NA	NA	NA	NA	NA	NA	NA
8	NA	NA	NA	NA	NA	NA	NA	NA
10	NA	NA	NA	NA	NA	NA	NA	430
15	NA	NA	NA	NA	NA	NA	NA	485
20	NA	NA	NA	NA	NA	NA	NA	547
30	NA	NA	NA	NA	NA	NA	NA	682
50	NA	NA	NA	NA	NA	NA	NA	NA
100	NA	NA	NA	NA	NA	NA	NA	NA
Local 99% winter design temperature: 4°F or lower Not recommended for any vent configurations								

For SI units: 1 inch = 25.4 mm, 1 foot = 304.8 mm, 1000 British thermal units per hour = 0.293 kW, 1 square inch = 0.000645 m^2, °C = (°F-32)/1.8

Notes:

[1] See Figure 803.1.2(6) for a map showing local 99 percent winter design temperatures in the United States.

[2] NA: Not applicable.

TABLE 803.2(8)
EXTERIOR MASONRY CHIMNEY [NFPA 54: TABLE 13.2(h)]*

				NUMBER OF APPLIANCES:	TWO OR MORE			
				APPLIANCE TYPE:	FAN + NAT			
				APPLIANCE VENT CONNECTION:	TYPE B DOUBLE-WALL CONNECTOR			
COMBINED APPLIANCE MAXIMUM INPUT RATING IN THOUSANDS OF BTU PER HOUR								
VENT HEIGHT *H* (feet)	INTERNAL AREA OF CHIMNEY (square inches)							
	12	19	28	38	50	63	78	113
6	74	119	178	257	351	458	582	853
8	80	130	193	279	384	501	636	937
10	84	138	207	299	409	538	686	1010
15	NA	152	233	334	467	611	781	1156
20	NA	NA	250	368	508	668	858	1286
30	NA	NA	NA	404	564	747	969	1473
50	NA	NA	NA	NA	NA	831	1089	1692
100	NA	NA	NA	NA	NA	NA	NA	1921

For SI units: 1 inch = 25.4 mm, 1 foot = 304.8 mm, 1000 British thermal units per hour = 0.293 kW, 1 square inch = 0.000645 m^2

* NA: Not applicable.

TABLE 803.2(9)
EXTERIOR MASONRY CHIMNEY [NFPA 54: TABLE 13.2(i)][1, 2]

				NUMBER OF APPLIANCES:	TWO OR MORE		
				APPLIANCE TYPE:	FAN + NAT		
				APPLIANCE VENT CONNECTION:	TYPE B DOUBLE-WALL CONNECTOR		
MINIMUM ALLOWABLE INPUT RATING OF SPACE-HEATING APPLIANCE IN THOUSANDS OF BTU PER HOUR								
VENT HEIGHT *H* (feet)	INTERNAL AREA OF CHIMNEY (square inches)							
	12	19	28	38	50	63	78	113
Local 99% winter design temperature: 37°F or greater								
6	0	0	0	0	0	0	0	0
8	0	0	0	0	0	0	0	0
10	0	0	0	0	0	0	0	0
15	NA	0	0	0	0	0	0	0
20	NA	NA	123	190	249	184	0	0
30	NA	NA	NA	334	398	393	334	0
50	NA	NA	NA	NA	NA	714	707	579
100	NA	NA	NA	NA	NA	NA	NA	1600
Local 99% winter design temperature: 27°F to 36°F								
6	0	0	68	116	156	180	212	266
8	0	0	82	127	167	187	214	263
10	0	51	97	141	183	201	225	265
15	NA	111	142	183	233	253	274	305
20	NA	NA	187	230	284	307	330	362
30	NA	NA	NA	330	319	419	445	485
50	NA	NA	NA	NA	NA	672	705	763
100	NA	NA	NA	NA	NA	NA	NA	1554
Local 99% winter design temperature: 17°F to 26°F								
6	0	55	99	141	182	215	259	349
8	52	74	111	154	197	226	264	352
10	NA	90	125	169	214	245	278	358
15	NA	NA	167	212	263	296	331	398
20	NA	NA	212	258	316	352	387	457
30	NA	NA	NA	362	429	470	507	581
50	NA	NA	NA	NA	NA	723	766	862
100	NA	NA	NA	NA	NA	NA	NA	1669
Local 99% winter design temperature: 5°F to 16°F								
6	NA	78	121	166	214	252	301	416
8	NA	94	135	182	230	269	312	423
10	NA	111	149	198	250	289	331	430
15	NA	NA	193	247	305	346	393	485
20	NA	NA	NA	293	360	408	450	547
30	NA	NA	NA	377	450	531	580	682
50	NA	NA	NA	NA	NA	797	853	972
100	NA	NA	NA	NA	NA	NA	NA	1833
Local 99% winter design temperature: -10°F to 4°F								
6	NA	NA	145	196	249	296	349	484
8	NA	NA	159	213	269	320	371	494
10	NA	NA	175	231	292	339	397	513
15	NA	NA	NA	283	351	404	457	586
20	NA	NA	NA	333	408	468	528	650
30	NA	NA	NA	NA	NA	603	667	805
50	NA	NA	NA	NA	NA	NA	955	1003
100	NA	NA	NA	NA	NA	NA	NA	NA
Local 99% winter design temperature: -11°F or lower Not recommended for any vent configurations								

For SI units: 1 inch = 25.4 mm, 1 foot = 304.8 mm, 1000 British thermal units per hour = 0.293 kW, 1 square inch = 0.000645 m^2, °C = (°F-32)/1.8

Notes:

[1] See Figure 803.1.2(6) for a map showing local 99 percent winter design temperatures in the United States.

[2] NA: Not applicable.

CALIFORNIA MECHANICAL CODE – MATRIX ADOPTION TABLE
CHAPTER 9 – INSTALLATION OF SPECIFIC APPLIANCES

(Matrix Adoption Tables are non-regulatory, intended only as an aid to the code user. See Chapter 1 for state agency authority and building applications.)

Adopting Agency	BSC	BSC-CG	SFM	HCD			DSA			OSHPD						BSCC	DPH	AGR	DWR	CEC	CA	SL	SLC
				1	2	1-AC	AC	SS	SS/CC	1	1R	2	3	4	5								
Adopt Entire Chapter	X		X					X	X				X										
Adopt Entire Chapter as amended (amended sections listed below)				X	X					X	X	X		X	X								
Adopt only those sections that are listed below																	X						
Chapter/Section																							
902.2										X	X	X		X	X								
911.0																	X						
911.1										X	X	X		X	X								
911.2.2				†	†																		
912.1				†	†																		
912.2*(1)*				X	X																		
916.2.1				X	X																		
916.2.1.1				X	X																		
930.0				†	†																		
930.1				†	†																		

This state agency does not adopt sections identified with the following symbol: †

The Office of the State Fire Marshal's adoption of this chapter or individual sections is applicable to structures regulated by other state agencies pursuant to Section 1.11.0.

CHAPTER 9
INSTALLATION OF SPECIFIC APPLIANCES

901.0 General.

901.1 Applicability. This chapter addresses requirements for the design, construction, and installation of specific appliances. In addition to the requirements of this chapter, appliances shall comply with the general requirements of Chapter 3.

902.0 General.

902.1 Nonindustrial Appliance. This chapter is applicable primarily to nonindustrial-type appliances and installations and, unless specifically indicated, does not apply to industrial-type appliances and installations. Listed appliances shall be installed in accordance with their listing and the manufacturer's installation instructions or, as elsewhere specified in this chapter, as applicable to the appliance. Unlisted appliances shall be installed as specified in this part as applicable to the appliances. For additional information concerning particular appliances and accessories, including industrial types, reference can be made to the standards listed in Chapter 17.

902.2 Combustion Air from Bedroom or Bathroom. Appliances shall not be installed so their combustion, ventilation, and dilution air are obtained only from a bedroom or bathroom unless the bedroom or bathroom has the required volume in accordance with Section 701.4. [NFPA 54:10.1.2]

[OSHPD 1, 1R, 2, 4 & 5] *Warm air furnaces shall not be installed under openable windows on exterior walls.*

902.3 Added or Converted Appliances. When additional or replacement appliances or equipment is installed or an appliance is converted to gas from another fuel, the location in which the appliances or equipment is to be operated shall be checked to verify the following:

(1) Air for combustion and ventilation is provided where required, in accordance with the provisions of Section 701.0. Where existing facilities are not adequate, they shall be upgraded to meet Section 701.0 specifications.

(2) The installation components and appliances meet the clearances to combustible material provisions of Section 303.10. It shall be determined that the installation and operation of the additional or replacement appliances do not render the remaining appliances unsafe for continued operation.

(3) The venting system is constructed and sized in accordance with the provisions of Section 802.0. Where the existing venting system is not adequate, it shall be upgraded to comply with Section 802.0. [NFPA 54:9.1.2]

902.4 Type of Gas(es). The appliance shall be connected to the fuel gas for which it was designed. No attempt shall be made to convert the appliance from the gas specified on the rating plate for use with a different gas without consulting the manufacturer's installation instructions, the serving gas supplier, or the appliance manufacturer for complete instructions. [NFPA 54:9.1.3]

902.5 Safety Shutoff Devices for Unlisted LP-Gas Appliance Used Indoors. Unlisted appliances for use with undiluted LP-Gases and installed indoors, except attended laboratory equipment, shall be equipped with safety shutoff devices of the complete shutoff type. [NFPA 54:9.1.4]

902.6 Fuel Input Rate. The fuel input rate to the appliance shall not be increased or decreased in violation of the approved rating at the altitude where it is being used.

902.7 Use of Air or Oxygen Under Pressure. Where air or oxygen under pressure is used in connection with the gas supply, effective means such as a back pressure regulator and relief valve shall be provided to prevent air or oxygen from passing back into the gas piping. Where oxygen is used, installation shall be in accordance with NFPA 51. [NFPA 54:9.1.5]

902.8 Building Structural Members. Appliances and equipment shall be furnished either with load-distributing bases or with a sufficient number of supports to prevent damage to either the building structure or the appliance and the equipment. [NFPA 54:9.1.8.1]

902.8.1 Structural Capacity. At the locations selected for installation of appliances and equipment, the dynamic and static load-carrying capacities of the building structure shall be checked to determine whether they are adequate to carry the additional loads. The appliances and equipment shall be supported and shall be connected to the piping so as not to exert undue stress on the connections. [NFPA 54:9.1.8.2]

902.9 Flammable Vapors. Appliances shall not be installed in areas where the open use, handling, or dispensing of flammable liquids occurs, unless the design, operation, or installation reduces the potential of ignition of the flammable vapors. Appliances installed in compliance with Section 305.1 through Section 305.1.2, Section 303.11, or Section 303.12 shall be considered to comply with the intent of this provision. [NFPA 54:9.1.9]

902.10 Solid-Fuel Burning Appliances. Unless otherwise specified, solid-fuel burning appliances shall be installed in accordance with NFPA 211 and the manufacturer's installation instructions.

902.11 Combination of Appliances and Equipment. A combination of appliances, equipment, attachments, or devices used together in a manner shall be in accordance with the standards that apply to the individual appliance and equipment. [NFPA 54:9.1.21]

902.12 Protection of Gas Appliances from Fumes or Gases Other than Products of Combustion. Non-direct-vent appliances installed in beauty shops, barber shops, or other facilities where chemicals that generate corrosive or flammable products such as aerosol sprays are routinely used shall be located in a mechanical room separate or partitioned off from other areas with provisions for combustion and dilution air from outdoors. Direct-vent appliances in such facilities shall be in accordance with the appliance manufacturer's installation instructions. [NFPA 54:9.1.6.2]

902.13 Process Air. In addition to air needed for combustion in commercial or industrial processes, process air shall be provided as required for cooling of appliances, equipment, or material; for controlling dew point, heating, drying, oxidation, dilution, safety exhaust, odor control, and air for compressors; and for comfort and proper working conditions for personnel. [NFPA 54:9.1.7]

902.14 Gas Appliance Pressure Regulators. Where the gas supply pressure is higher than that at which the appliance is designed to operate or varies beyond the design pressure limits of the appliance, a gas appliance pressure regulator shall be installed. [NFPA 54:9.1.18]

902.15 Venting of Gas Appliance Pressure Regulators. Venting of gas appliance pressure regulators shall comply with the following requirements:

(1) Appliance pressure regulators requiring access to the atmosphere for successful operation shall be equipped with vent piping leading outdoors or, if the regulator vent is an integral part of the appliance, into the combustion chamber adjacent to a continuous pilot, unless constructed or equipped with a vent limiting means to limit the escape of gas from the vent opening in the event of diaphragm failure.

(2) Vent limiting means shall be employed on listed appliance pressure regulators only.

(3) In the case of vents leading outdoors, means shall be employed to prevent water from entering this piping and also to prevent blockage of vents by insects and foreign matter.

(4) Under no circumstances shall a regulator be vented to the appliance flue or exhaust system.

(5) In the case of vents entering the combustion chamber, the vent shall be located so the escaping gas is readily ignited by the pilot and the heat liberated thereby does not adversely affect the normal operation of the safety shutoff system. The terminus of the vent shall be securely held in a fixed position relative to the pilot. For manufactured gas, the need for a flame arrester in the vent piping shall be determined.

(6) A vent line(s) from an appliance pressure regulator and a bleed line(s) from a diaphragm-type valve shall not be connected to a common manifold terminating in a combustion chamber. Vent lines shall not terminate in positive-pressure-type combustion chambers. [NFPA 54:9.1.19]

902.16 Bleed Lines for Diaphragm-Type Valves. Bleed lines shall comply with the following requirements:

(1) Diaphragm-type valves shall be equipped to convey bleed gas to the outdoors or into the combustion chamber adjacent to a continuous pilot.

(2) In the case of bleed lines leading outdoors, means shall be employed to prevent water from entering this piping and also to prevent blockage of vents by insects and foreign matter.

(3) Bleed lines shall not terminate in the appliance flue or exhaust system.

(4) In the case of bleed lines entering the combustion chamber, the bleed line shall be located so the bleed gas is readily ignited by the pilot and the heat liberated thereby does not adversely affect the normal operation of the safety shutoff system. The terminus of the bleed line shall be securely held in a fixed position relative to the pilot. For manufactured gas, the need for a flame arrester in the bleed line piping shall be determined.

(5) A bleed line(s) from a diaphragm-type valve and a vent line(s) from an appliance pressure regulator shall not be connected to a common manifold terminating in a combustion chamber. Bleed lines shall not terminate in positive-pressure-type combustion chambers. [NFPA 54:9.1.20]

903.0 Air-Conditioning Appliances.

903.1 Electric Air Conditioners. Electric air conditioning systems designed for permanent installation shall comply with UL 1995 or UL 60335-2-40.

903.2 Gas-Fired Air Conditioners and Heat Pumps. Gas-fired air conditioners shall comply with Section 903.2.1 through Section 903.2.7.

903.2.1 Independent Gas Piping. Gas piping serving heating appliances shall be permitted to also serve cooling appliances where heating and cooling appliances cannot be operated simultaneously. [NFPA 54:10.2.1]

903.2.2 Connection of Gas Engine-Powered Air Conditioners. To protect against the effects of normal vibration in service, gas engines shall not be rigidly connected to the gas supply piping. [NFPA 54:10.2.2]

903.2.3 Clearances for Indoor Installation. The installation of air-conditioning appliances shall comply with the following requirements:

(1) Listed air-conditioning appliances shall be installed with clearances in accordance with the terms of their listing and the manufacturer's installation instructions.

(2) Unlisted air-conditioning appliances shall be installed with clearances from combustible material of not less than 18 inches (457 mm) above the appliance and at the sides, front, rear and in accordance with the manufacturer's installation instructions. [NFPA 54:10.2.3(2)]

(3) Listed and unlisted air-conditioning appliances shall be permitted to be installed with reduced clearances to combustible material, provided that the combustible material or appliance is protected as described in Table 303.10.1 and such reduction is allowed by the manufacturer's installation instructions. [NFPA 54:10.2.3(3)]

(4) Where the furnace plenum is adjacent to plaster on metal lath or noncombustible material attached to combustible material, the clearance shall be measured to the surface of the plaster or other noncombustible finish where the clearance specified is 2 inches (51 mm) or less. [NFPA 54:10.2.3(4)]

(5) Listed air-conditioning appliances shall have the clearance from supply ducts, within 3 feet (914 mm) of the furnace plenum be not less than that specified from the furnace plenum. No clearance is necessary beyond this distance. [NFPA 54:10.2.3(5)]

903.2.4 Assembly and Installation. Air-conditioning appliances shall be installed in accordance with the manufacturer's installation instructions. Unless the appliance is listed for installation on a combustible surface, such as a floor or roof, or unless the surface is protected in an approved manner, it shall be installed on a surface of noncombustible construction with noncombustible material and surface finish and with no combustible material against the underside thereof. [NFPA 54:10.2.4]

903.2.5 Furnace Plenums and Air Ducts. A furnace plenum supplied as a part of the air-conditioning appliance shall be installed in accordance with the manufacturer's instructions. Where a furnace plenum is not supplied with the appliance, any fabrication and installation instructions provided by the manufacturer shall be followed. The method of connecting supply and return ducts shall facilitate proper circulation of air. Where the air conditioner is installed within an enclosure, the installation shall comply with Section 904.7.3. [NFPA 54:10.2.5]

903.2.6 Refrigeration Coils. The installation of refrigeration coils shall be in accordance with Section 904.8 and Section 904.9. [NFPA 54:10.2.6]

903.2.7 Switches in Electrical Supply Line. Means for interrupting the electrical supply to the air-conditioning appliance and to its associated cooling tower (if supplied and installed in a location remote from the air conditioner) shall be provided within sight of and not over 50 feet (15 240 mm) from the air conditioner and the cooling tower. [NFPA 54:10.2.7]

904.0 Central Heating Boilers and Furnaces.

904.1 Location. Central heating furnace and low-pressure boiler installations in bedrooms or bathrooms shall comply with one of the following:

(1) Central heating furnaces and low-pressure boilers shall be permitted to be installed in a closet located in the bedroom or bathroom, provided the closet is equipped with a listed, gasketed door assembly, and a listed self-closing device. The self-closing door assembly shall comply with the requirements of Section 904.1.1. The door assembly shall be installed with a threshold and bottom door seal and shall comply with the requirements of Section 904.1.2. Combustion air for such installations shall be obtained from the outdoors. The closet shall be for the exclusive use of the central heating furnace or low-pressure boiler.

(2) Central heating furnaces and low-pressure boilers shall be of the direct-vent type.

904.1.1 Self-Closing Doors. Self-closing doors shall swing easily and freely, and shall be equipped with a self-closing device to cause the door to close and latch each time it is opened. The closing mechanism shall not have a hold-open feature.

904.1.2 Gasketing. Gasketing on gasketed doors or frames shall be furnished in accordance with the published listings of the door, frame, or gasketing material manufacturer.

Exception: Where acceptable to the Authority Having Jurisdiction, gasketing of noncombustible or limited-combustible material shall be permitted to be applied to the frame, provided closing and latching of the door are not inhibited.

904.2 Clearance. Central heating furnaces and low-pressure boilers shall be provided with clearances in accordance with Section 904.2.1 through Section 904.2.9.

904.2.1 Listed Units. Listed central heating furnaces and low-pressure boilers shall be installed with clearances in accordance with the terms of their listings and the manufacturer's installation instructions.

904.2.2 Unlisted Units. Unlisted central-heating furnaces and low pressure boilers shall be installed with clearances from combustible material not less than those specified in Table 904.2.2. [NFPA 54:10.3.2.2]

904.2.3 Listed and Unlisted Units. Listed and unlisted central heating furnaces and low-pressure boilers shall be permitted to be installed with reduced clearances to combustible material, provided that the combustible material or appliance is protected as described in Table 303.10.1 and Figure 303.10.1(1) through Figure 303.10.1(3), and such reduction is allowed by the manufacturer's installation instructions. [NFPA 54:10.3.2.3]

904.2.4 Front Clearance. Front clearance shall be sufficient for servicing the burner and the furnace or boiler. [NFPA 54:10.3.2.4]

904.2.5 Adjacent to Plaster or Noncombustible Materials. Where the furnace plenum is adjacent to plaster on metal lath or noncombustible material attached to combustible material, the clearance shall be measured to the surface of the plaster or other noncombustible finish where the clearance specified is 2 inches (51 mm) or less. [NFPA 54:10.3.2.5]

904.2.6 Interference. The clearances to these appliances shall not interfere with combustion air, draft hood clearance and relief, and accessibility for servicing. [NFPA 54:10.3.2.6]

904.2.7 Supply Air Ducts to Listed Furnaces. Supply air ducts connecting to listed central heating furnaces shall have the same minimum clearance to combustibles as required for the furnace supply plenum for a distance of not less than 3 feet (914 mm) from the supply plenum. Clearance shall not be required beyond the 3 feet (914 mm) distance. [NFPA 54:10.3.2.7]

904.2.8 Supply Air Ducts to Unlisted Furnaces. Supply air ducts connecting to unlisted central heating furnaces equipped with temperature limit controls with a maximum setting of 250°F (121°C) shall have a minimum clearance to combustibles of 6 inches (152 mm) for a distance of not less than 6 feet (1829 mm) from the furnace supply plenum. Clearance shall not be required beyond the 6 feet (1829 mm) distance. [NFPA 54:10.3.2.8]

904.2.9 Central Heating Furnaces. Central heating furnaces other than those listed in Section 904.2.7 or Section 904.2.8 shall have clearances from the supply ducts of not less than 18 inches (457 mm) from the furnace plenum for the first 3 feet (914 mm), then 6 inches (152 mm) for the next 3 feet (914 mm) and 1 inch (25.4 mm) beyond 6 feet (1829 mm). [NFPA 54:10.3.2.9]

904.3 Assembly and Installation. A central-heating boiler or furnace shall be installed in accordance with the manufacturer's instructions in one of the following manners:

(1) On a floor of noncombustible construction with noncombustible flooring, and surface finish and with no combustible material against the underside thereof.

(2) On fire-resistive slabs or arches having no combustible material against the underside thereof.

Exceptions:

(1) Appliances listed for installation on a combustible floor.

(2) Installation on a floor protected in an approved manner. [NFPA 54:10.3.3]

904.3.1 Under-Floor Installation. Furnaces installed in an under-floor area of the building shall be in accordance with the Section 904.3.1.1 through Section 904.3.1.3.

904.3.1.1 Supported by Ground. Where a furnace is supported by the ground, it shall be installed on a concrete slab not less than 3 inches (76 mm) above the adjoining ground level.

904.3.1.2 Supported from Above. Where a furnace is supported from above, a clearance of not less than 6 inches (152 mm) shall be provided from finished grade.

904.3.1.3 Excavation. Where excavation is necessary to install a furnace, it shall be installed in accordance with Section 303.11.

904.4 Temperature- or Pressure-Limiting Devices. Steam and hot water boilers, respectively, shall be provided with approved automatic limiting devices for shutting down the burner(s) to prevent boiler steam pressure or boiler water temperature from exceeding the maximum allowable working pressure or temperature. Safety limit controls shall not be used as operating controls. [NFPA 54:10.3.4]

904.5 Low-Water Cutoff. All water boilers and steam boilers shall be provided with an automatic means to shut off the fuel supply to the burner(s) if the boiler water level drops below the lowest safe water line. In lieu of the low-water cutoff, water tube or coil-type boilers that require forced circulation to prevent overheating and failure shall have an approved flow sensing device arranged to shut down the boiler when the flow rate is inadequate to protect the boiler against overheating. [NFPA 54:10.3.5]

904.6 Steam Safety and Pressure-Relief Valves. Steam and hot water boilers shall be equipped, respectively, with listed or approved steam safety or pressure-relief valves of appropriate discharge capacity and conforming with ASME requirements. A shutoff valve shall not be placed between the relief valve and the boiler or on discharge pipes between such valves and the atmosphere. [NFPA 54:10.3.6]

904.6.1 Discharge. Relief valves shall be piped to discharge near the floor. [NFPA 54:10.3.6.1]

904.6.2 Size. The entire discharge piping shall be not less than the same size as the relief valve discharge piping. [NFPA 54:10.3.6.2]

904.6.3 End Connections. Discharge piping shall not contain a threaded end connection at its termination point. [NFPA 54:10.3.6.3]

TABLE 904.2.2
CLEARANCES TO COMBUSTIBLE MATERIAL FOR UNLISTED FURNACES AND BOILERS*
[NFPA 54: TABLE 10.3.2.2]

APPLIANCE	MINIMUM CLEARANCE (inches)					
	ABOVE AND SIDES OF FURNACE PLENUM	TOP OF BOILER	JACKET SIDES AND REAR	FRONT	DRAFT HOOD AND BAROMETRIC DRAFT REGULATOR	SINGLE-WALL VENT CONNECTOR
1. Automatically fired, forced air or gravity system, equipped with temperature limit control that is not capable of being set to exceed 250°F.	6	–	6	18	6	18
2. Automatically fired heating boilers – steam boilers operating at not over 15 pounds-force per square inch (psi) and hot water boilers operating at 250°F or less.	6	6	6	18	18	18
3. Central heating boilers and furnaces, other than in 1 or 2.	18	18	18	18	18	18

For SI units: 1 inch = 25.4 mm, °C = (°F-32)/1.8, 1 pound-force per square inch = 6.8947 kPa

* See Section 904.1 for additional requirements for central heating boilers and furnaces.

904.7 Furnace Plenums and Air Ducts. Furnace plenums and air ducts shall be installed in accordance with NFPA 90A or NFPA 90B. [NFPA 54:10.3.7.1]

904.7.1 Supplied as a Part of Furnace. A furnace plenum supplied as a part of a furnace shall be installed in accordance with the manufacturer's instructions. [NFPA 54:10.3.7.2]

904.7.2 Not Supplied with the Furnace. Where a furnace plenum is not supplied with the furnace, any fabrication and installation instructions provided by the manufacturer shall be followed. The method of connecting supply and return ducts shall facilitate proper circulation of air. [NFPA 54:10.3.7.3]

904.7.3 Return Air. Where a furnace is installed so supply ducts carry air circulated by the furnace to areas outside the space containing the furnace, the return air shall also be handled by a duct(s) sealed to the furnace casing and terminating outside the space containing the furnace. [NFPA 54:10.3.7.4]

904.8 Refrigeration Coils. The installation of refrigeration coils shall comply with the following requirements:

(1) A refrigeration coil shall not be installed in conjunction with a forced-air furnace where circulation of cooled air is provided by the furnace blower, unless the blower has the capacity to overcome the external static pressure resistance imposed by the duct system and refrigeration coil at the air flow rate for heating or cooling, whichever is greater.

(2) Furnaces shall not be located upstream from refrigeration coils, unless the refrigeration coil is designed or equipped so as not to develop excessive temperature or pressure.

(3) Refrigeration coils shall be installed in parallel with or on the downstream side of central furnaces to avoid condensation in the heating element, unless the furnace has been specifically listed for downstream installation. With a parallel flow arrangement, the dampers or other means used to control flow of air shall be tight to prevent a circulation of cooled air through the furnace.

(4) Means shall be provided for disposal of condensate and to prevent dripping of condensate on the heating element. [NFPA 54:10.3.8]

904.9 Cooling Units Used with Heating Boilers. Boilers, where used in conjunction with refrigeration systems, shall be installed so that the chilled medium is piped in parallel with the heating boiler with appropriate valves to prevent the chilled medium from entering the heating boiler. [NFPA 54:10.3.9.1]

904.9.1 Exposed to Refrigerated Air Circulation. Where hot-water-heating boilers are connected to heating coils located in air-handling units where they can be exposed to refrigerated air circulation, such boiler piping systems shall be equipped with flow control valves or other automatic means to prevent gravity circulation of the boiler water during the cooling cycle. [NFPA 54:10.3.9.2]

904.10 Furnace (Upright and Horizontal). Upright furnaces shall be permitted to be installed in an attic, furred, or under-floor space exceeding 5 feet (1524 mm) in height, provided the required listings and furnace and duct clearances are observed. Horizontal furnaces shall be permitted to be installed in an attic, furred, or under-floor space, provided the required listings and furnace and duct clearances are observed.

904.11 Solid-Fuel-Fired Furnaces. Factory-built solid-fuel-fired furnaces shall comply with UL 391 and shall be installed in accordance with the manufacturer's installation instructions.

904.12 Oil-Fired Central Furnaces. Oil-fired central furnaces shall comply with UL 727 and shall be installed in accordance with the manufacturer's installation instructions.

904.13 Commercial or Industrial Gas Heaters. Commercial or industrial gas-fired heaters shall comply with UL 795 and shall be installed in accordance with the manufacturer's installation instructions.

904.14 Electric Central Furnaces. Electric central heating furnaces shall comply with UL 1995 and shall be installed in accordance with the manufacturer's installation instructions.

905.0 Duct Furnaces.

905.1 Clearances. The installation of duct furnaces shall comply with the following clearance requirements:

(1) Listed duct furnaces shall be installed with clearances of not less than 6 inches (152 mm) between adjacent walls, ceilings, and floors of combustible material and the furnace draft hood. Furnaces listed for installation at lesser clearances shall be installed in accordance with their listings and the manufacturer's installation instructions. In no case shall the clearance be such as to interfere with combustion air and accessibility.

(2) Unlisted duct furnaces shall be installed with clearances to combustible material in accordance with the clearances specified for unlisted furnaces and boilers in Table 904.2.2. Combustible floors under unlisted duct furnaces shall be protected in an approved manner.

905.2 Installation of Duct Furnaces. Duct furnaces shall be installed in accordance with the manufacturer's installation instructions. [NFPA 54:10.10.2]

905.3 Access Panels. The ducts connected to duct furnaces shall have removable access panels on both the upstream and downstream sides of the furnace. [NFPA 54:10.10.3]

905.4 Location of Draft Hoods and Controls. The controls, combustion-air inlet, and draft hoods for duct furnaces shall be located outside the ducts. The draft hood shall be located in the same enclosure from which combustion air is taken. [NFPA 54:10.10.4]

905.5 Circulating Air. Where a duct furnace is installed so that supply ducts carry air circulated by the furnace to areas outside the space containing the furnace, the return air shall also be handled by a duct(s) sealed to the furnace casing and terminating outside the space containing the furnace. The duct furnace shall be installed on the positive-pressure side of the circulating air blower. [NFPA 54:10.10.5]

905.6 Duct Furnaces Used with Refrigeration Systems. A duct furnace shall not be installed in conjunction with a refrigeration coil where circulation of cooled air is provided by the blower.

Exception: Where the blower has sufficient capacity to overcome the external static resistance imposed by the duct system, furnace, and the cooling coil and the air throughput necessary for heating or cooling, whichever is greater. [NFPA 54:10.10.6.1]

905.6.1 In Conjunction with Cooling Appliances. Duct furnaces used in conjunction with cooling appliances shall be installed in parallel with or on the upstream side of cooling coils to avoid condensation within heating elements. With a parallel flow arrangement, the dampers or other means used to control the flow of air shall be sufficiently tight to prevent any circulation of cooled air through the unit.

Exception: Where the duct furnace has been specifically listed for downstream installation. [NFPA 54:10.10.6.2]

905.6.2 Located Upstream from Cooling Coils. Where duct furnaces are to be located upstream from cooling units, the cooling unit shall be so designed or equipped as to not develop excessive temperatures or pressures. [NFPA 54:10.10.6.3]

905.6.3 Heat Exchangers. Where a duct furnace is installed downstream of an evaporative cooler or air washer, the heat exchanger shall be constructed of corrosion-resistant materials. Stainless steel, ceramic-coated steel, and an aluminum-coated steel in which the bond between the steel and the aluminum is an iron-aluminum alloy are considered to be corrosion resistant. Air washers operating with chilled water that deliver air below the dew point of the ambient air at the duct furnace shall be considered as refrigeration systems. [NFPA 54:10.10.6.4]

905.7 Installation in Commercial Garages and Aircraft Hangars. Duct furnaces installed in garages for more than three motor vehicles or in aircraft hangars shall be of a listed type and shall be installed in accordance with Section 303.11 and Section 303.12. [NFPA 54:10.10.7]

905.8 Electric Duct Heaters. Electric duct heaters installed within an air duct shall be listed and labeled in accordance with UL 1996 and designed for the maximum air temperature. The duct heater and fan shall be interlocked such that the electric duct heater operates when the fan is operating.

905.8.1 Installation. Duct heaters shall be installed in accordance with the manufacturer's installation instructions, and shall not create a hazard to persons or property. Where installed 4 feet (1219 mm) or less from a heat pump or air conditioner, the duct heater shall be listed for such installation.

905.8.2 Clearance. A working space clearance shall be maintained to permit replacement of controls and heating elements and for adjusting and cleaning of controls. The working space for energized equipment shall comply with *the California Electrical Code*.

906.0 Floor Furnaces.

906.1 Installation. The installation of floor furnaces shall comply with the following requirements:

(1) Listed floor furnaces shall be installed in accordance with their listing and the manufacturer's installation instructions.

(2) Unlisted floor furnaces shall not be installed on combustible floors.

(3) Thermostats controlling floor furnaces shall not be located in a room or space that is capable of being separated from the room or space in which the register of the floor furnace is located.

906.2 Temperature Limit Controls. Floor furnaces shall be provided with temperature limit controls in accordance with the following requirements:

(1) Listed automatically operated floor furnaces shall be equipped with temperature limit controls. [NFPA 54:10.11.2.1]

(2) Unlisted automatically operated floor furnaces shall be equipped with a temperature limit control arranged to shut off the flow of gas to the burner in the event the temperature at the warm air outlet register exceeds 350°F (177°C) above room temperature. [NFPA 54:10.11.2.2]

906.3 Combustion and Circulating Air. Combustion and circulating air shall be provided in accordance with Section 701.0. [NFPA 54:10.11.3]

906.4 Placement. The following provisions apply to furnaces that serve one story:

(1) Floors. Floor furnaces shall not be installed in the floor of any doorway, stairway landing, aisle, or passageway of any enclosure, public or private, or in an exitway from any such room or space.

(2) Walls and Corners. The register of a floor furnace with a horizontal warm air outlet shall not be placed closer than 6 inches (152 mm) from the nearest wall. A distance of at least 18 inches (457 mm) from two adjoining sides of the floor furnace register to walls shall be provided to eliminate the necessity of occupants walking over the warm air discharge. The remaining sides shall be a minimum of 6 inches (152 mm) from a wall. Wall register models shall not be placed closer than 6 inches (152 mm) to a corner.

(3) Draperies. The furnace shall be placed so that a door, drapery, or similar object cannot be nearer than 12 inches (305 mm) to any portion of the register of the furnace. [NFPA 54:10.11.4]

906.5 Bracing. The space provided for the furnace shall be framed with doubled joists and with headers not lighter than the joists. [NFPA 54:10.11.5]

906.6 Support. Means shall be provided to support the furnace where the floor register is removed. [NFPA 54:10.11.6]

906.7 Clearance. The lowest portion of the floor furnace shall have at least a 6 inch (152 mm) clearance from the general ground level. A reduced clearance to a minimum of 2 inches (51 mm) shall be permitted, provided the lower 6 inches (152 mm) portion of the floor furnace is sealed by the manufacturer to prevent entrance of water. Where these clearances are not present, the ground below and to the sides shall be excavated to form a "basin-like" pit under the furnace so that the required clearance is provided beneath the lowest portion of the furnace. A 12 inch (305 mm) clearance shall be provided on all sides except the control side, which has an 18 inch (457 mm) clearance. [NFPA 54:10.11.7]

906.8 Access. The space in which a floor furnace is installed shall be accessible by an opening in the foundation not less than 24 inches by 18 inches (610 mm by 457 mm) or by a trap door not less than 24 inches by 24 inches (610 mm by 610 mm) in a cross-section thereof, and a passageway not less than 24 inches by 18 inches (610 mm by 457 mm) in a cross-section thereof. [NFPA 54:10.11.8]

906.9 Seepage Pan. Where the excavation exceeds 12 inches (305 mm) in depth or water seepage is likely to collect, a watertight copper pan, concrete pit, or other approved material shall be used, unless adequate drainage is provided or the appliance is sealed by the manufacturer to meet this condition. A copper pan shall be made of not less than 16 ounces per square foot (oz/ft^2) (4.9 kg/m^2) sheet copper. The pan shall be anchored in place so as to prevent floating, and the walls shall extend at least 4 inches (102 mm) above the ground level with at least a 6 inches (152 mm) clearance on all sides, except the control side, which shall have at least an 18 inch (457 mm) clearance. [NFPA 54:10.11.9]

906.10 Wind Protection. Floor furnaces shall be protected, where necessary, against severe wind conditions. [NFPA 54:10.11.10]

906.11 Upper-Floor Installations. Listed floor furnaces shall be permitted to be installed in an upper floor, provided the furnace assembly projects below into a utility room, closet, garage, or similar nonhabitable space. In such installations, the floor furnace shall be enclosed completely (entirely separated from the nonhabitable space) with means for air intake to meet the provisions of Section 701.0, with access for servicing, minimum furnace clearances of 6 inches (152 mm) to all sides and bottom, and with the enclosure constructed of portland cement plaster or metal lath or other noncombustible material. [NFPA 54:10.11.11]

906.12 First Floor Installation. Listed floor furnaces installed in the first or ground floors of buildings shall not be required to be enclosed unless the basements of these buildings have been converted to apartments or sleeping quarters, in which case the floor furnace shall be enclosed as specified for upper floor installations and shall project into a nonhabitable space. [NFPA 54:10.11.12]

906.13 Oil-Fired Floor Furnaces. Oil-fired floor furnaces shall comply with UL 729 and installed in accordance with the manufacturer's installation instructions.

907.0 Wall Furnaces.

907.1 Installation. Listed wall furnaces shall be installed in accordance with their listings and the manufacturer's installation instructions. Wall furnaces installed in or attached to combustible material shall be listed for such installation.

907.1.1 Unlisted Wall Furnaces. Unlisted wall furnaces shall not be installed in or attached to combustible material. [NFPA 54:10.26.1.2]

907.1.2 Vented Wall Furnaces. Vented wall furnaces connected to a Type B-W gas vent system listed only for a single story shall be installed only in single-story buildings or the top story of multistory buildings. Vented wall furnaces connected to a Type B-W gas vent system listed for installation in multistory buildings shall be permitted to be installed in single-story or multistory buildings. Type B-W gas vents shall be attached directly to a solid header plate that serves as a firestop at that point and that shall be permitted to be an integral part of the vented wall furnace, as illustrated in Figure 907.1.2. The stud space in which the vented wall furnace is installed shall be ventilated at the first ceiling level by installation of the ceiling plate spacers furnished with the gas vent. Firestop spacers shall be installed at each subsequent ceiling or floor level penetrated by the vent. [NFPA 54:10.26.1.3]

907.1.3 Direct Vent Wall Furnaces. Direct-vent wall furnaces shall be installed with the vent-air intake terminal in the outdoors. The thickness of the walls on which the furnace is mounted shall be within the range of wall thickness marked on the furnace and covered in the manufacturer's installation instructions. [NFPA 54:10.26.1.4]

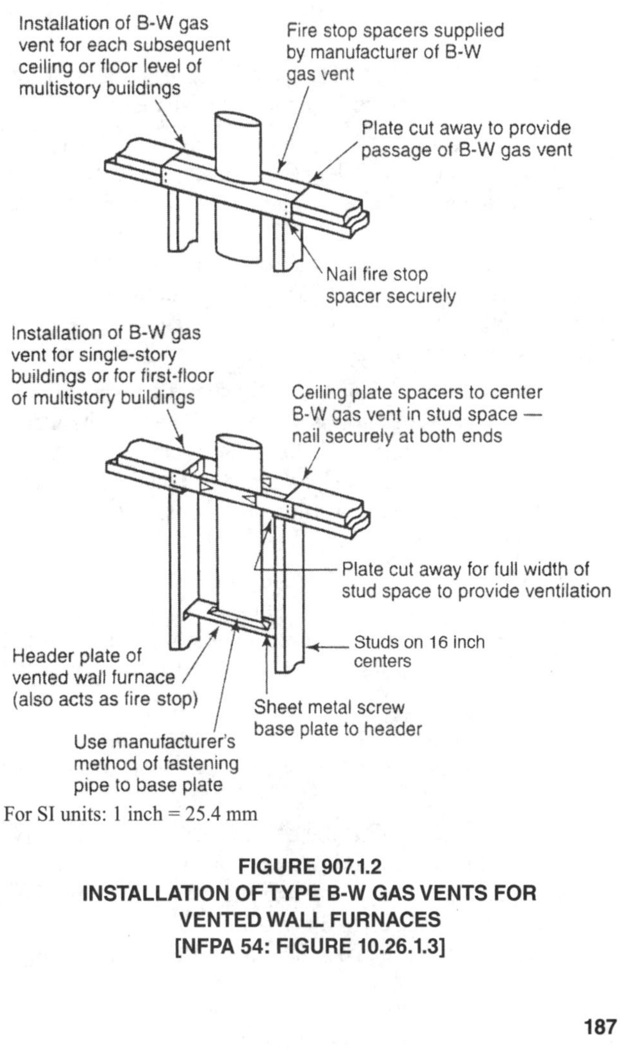

For SI units: 1 inch = 25.4 mm

FIGURE 907.1.2
INSTALLATION OF TYPE B-W GAS VENTS FOR VENTED WALL FURNACES
[NFPA 54: FIGURE 10.26.1.3]

907.1.4 Panels, Grilles, and Access Doors. Panels, grilles, and access doors that are required to be removed for normal servicing operations shall not be attached to the building. For additional information on the venting of wall furnaces, see Section 802.0. [NFPA 54:10.26.1.5]

907.2 Location. Wall furnaces shall be located so as not to cause a hazard to walls, floors, curtains, furniture, or doors. Wall furnaces installed between bathrooms and adjoining rooms shall not circulate air from bathrooms to other parts of the building. [NFPA 54:10.26.2]

907.3 Combustion and Circulating Air. Combustion and circulating air shall be provided in accordance with Section 701.0. [NFPA 54:10.26.3]

907.4 Oil-Fired Wall Furnaces. Oil-fired wall furnaces shall comply with UL 730 and installed in accordance with the manufacturer's installation instructions.

908.0 Clothes Dryers.

908.1 Electric Clothes Dryers. Commercial electric clothes dryers shall comply with UL 1240 and installed in accordance with the manufacturer's installation instructions. Residential and coin-operated electric clothes dryers shall comply with UL 2158 and installed in accordance with the manufacturer's installation instructions.

908.2 Gas-Fired Clothes Dryers. Gas-fired clothes dryers shall comply with Section 908.2.1 through Section 908.2.3.

908.2.1 Clearance. The installation of clothes dryers shall comply with the following requirements:

(1) Listed Type 1 clothes dryers shall be installed with a clearance of not less than 6 inches (152 mm) from adjacent combustible material. Clothes dryers listed for installation at reduced clearances shall be installed in accordance with their listing and the manufacturer's installation instructions. Type 1 clothes dryers installed in closets shall be listed for such installation.

(2) Listed Type 2 clothes dryers shall be installed with clearances of not less than that shown on the marking plate and in the manufacturer's instructions. Type 2 clothes dryers designed and marked, "For use only in noncombustible locations," shall not be installed elsewhere.

(3) Unlisted clothes dryers shall be installed with clearances to combustible material of not less than 18 inches (457 mm). Combustible floors under unlisted clothes dryers shall be protected in an approved manner.

908.2.2 Exhausting to the Outdoors. Type 1 and Type 2 clothes dryers shall be exhausted to the outside air in accordance with Section 504.4.

908.2.3 Multiple-Family or Public Use. Clothes dryers installed for multiple-family or public use shall be equipped with approved safety shutoff devices and shall be installed as specified for a Type 2 clothes dryer in accordance with Section 504.4.3.1. [NFPA 54:10.4.6]

909.0 Conversion Burners.

909.1 General. Installation of conversion burners shall comply with CSA Z21.8. [NFPA 54:10.5]

910.0 Burner Assemblies.

910.1 Oil Burners. Oil burners shall comply with UL 296 and installed in accordance with the manufacturer's installation instructions.

910.2 Gas Burners. Commercial gas burners shall comply with UL 295 and installed in accordance with the manufacturer's installation instructions.

911.0 Decorative Appliances for Installation in Vented Fireplaces.

911.1 Prohibited Installations. Decorative appliances for installation in vented fireplaces shall not be installed in bathrooms or bedrooms unless the appliance is listed and the bedroom or bathroom has the required volume in accordance with Section 701.4. [NFPA 54:10.6.1] ***[OSHPD 1, 1R, 2, 4 & 5]*** *A vented decorative appliance shall not be located in any hospital, skilled nursing facility, intermediate care facility, or correctional treatment center.*

911.2 Installation. A decorative appliance for installation in a vented fireplace shall be installed only in a vented fireplace having a working chimney flue and constructed of noncombustible materials. These appliances shall not be thermostatically controlled. [NFPA 54:10.6.2]

911.2.1 Listed Decorative Appliance. A listed decorative appliance for installation in a vented fireplace shall be installed in accordance with its listing and the manufacturer's installation instructions.

911.2.2 ***Reserved.***

911.2.3 Unlisted Decorative Appliance. An unlisted decorative appliance for installation in a vented fireplace shall be installed in a fireplace having a permanent free opening, based on appliance input rating and chimney height, equal to or greater than that specified in Table 911.2. [NFPA 54:10.6.2.3]

911.3 Fireplace Screens. A fireplace screen shall be installed with a decorative appliance for installation in a vented fireplace. [NFPA 54:10.6.3]

912.0 Gas Fireplaces, Vented.

912.1 Prohibited Installations. ***Reserved.***

912.2 Installation. The installation of vented gas fireplaces shall comply with the following requirements:

(1) ***[HCD 1 & HCD 2]*** *Any newly installed gas fireplace shall be a direct-vent sealed-combustion type.*

(2) Listed vented gas fireplaces shall be installed in accordance with their listing and the manufacturer's installation instructions and where installed in or attached to combustible material shall be specifically listed for such installation.

(3) Unlisted vented gas fireplaces shall not be installed in or attached to combustible material. They shall have a clearance at the sides and rear of not less than 18 inches (457 mm). Combustible floors under unlisted vented gas fireplaces shall be protected in an approved manner. Unlisted appliances of other than the direct-vent type shall be equipped with a draft hood and shall be vented in accordance with Section 802.0. Appliances that use metal, asbestos, or ceramic material to direct radiation to the front of the appliance shall have a clearance of 36 inches (914 mm) in front and, where constructed with a double back of metal or ceramic, shall be installed with a clearance of not less than 18 inches (457 mm) at the sides and 12 inches (305 mm) at the rear.

(4) Panels, grilles, and access doors that are required to be removed for normal servicing operations shall not be attached to the building.

(5) Direct-vent gas fireplaces shall be installed with the vent-air intake terminal in the outdoors and in accordance with the manufacturer's installation instructions.

912.3 Combustion and Circulating Air. Combustion and circulating air shall be provided in accordance with Section 701.0. [NFPA 54:10.7.3]

913.0 Factory-Built Fireplaces and Fireplace Stoves.

913.1 Factory-Built Fireplaces. Factory-built fireplaces shall comply with UL 127 and installed in accordance with the manufacturer's installation instructions.

913.1.1 Gasketed Fireplace Doors. A gasketed fireplace door shall not be installed on a factory-built fireplace, except where the fireplace system has been tested in accordance with UL 127.

913.2 Fireplace Stoves. Fireplace stoves shall comply with UL 737 and installed in accordance with the manufacturer's installation instructions.

913.3 Fireplace Accessories. Heat exchangers, glass doors assemblies, combustion air vents, and termination caps shall comply with UL 907 and installed in accordance with the manufacturer's installation instructions.

914.0 Non-Recirculating Direct Gas-Fired Industrial Air Heaters.

914.1 Application. Direct gas-fired industrial air heaters of the non-recirculating type shall be listed in accordance with CSA Z83.4. [NFPA 54:10.8.1]

914.2 Prohibited Installations. Non-recirculating direct gas-fired industrial air heaters shall not serve any area containing sleeping quarters. Non-recirculating direct gas-fired industrial air heaters shall not recirculate room air. [NFPA 54:10.8.2.1, 10.8.2.2]

914.3 Installation. Non-recirculating direct gas-fired industrial air heaters shall be installed in accordance with the manufacturer's instructions. [NFPA 54:10.8.3.1]

914.3.1 Industrial or Commercial Occupancies. Non-recirculating direct gas-fired industrial air heaters shall be installed only in industrial or commercial occupancies. [NFPA 54:10.8.3.2]

914.3.2 Fresh Air Ventilation. Non-recirculating direct gas-fired industrial air heaters shall be permitted to provide fresh air ventilation. [NFPA 54:10.8.3.3]

914.3.3 Access Required. Non-recirculating direct gas-fired industrial air heaters shall be provided with access for removal of burners; for replacement of motors, controls, filters, and other working parts; and for adjustment and lubrication of parts requiring maintenance. [NFPA 54:10.8.3.4]

914.4 Clearance from Combustible Materials. Non-recirculating direct gas-fired industrial air heaters shall be installed with a clearance from combustible materials of not less than that shown on the rating plate and the manufacturer's installation instructions. [NFPA 54:10.8.4]

914.5 Air Supply. All air to the non-recirculating direct gas-fired industrial air heater shall be ducted directly from outdoors. Where outdoor air dampers or closing louvers are used, they shall be verified to be in the open position prior to main burner operation. [NFPA 54:10.8.5]

914.6 Atmospheric Vents, Gas Reliefs, or Bleeds. Non-recirculating direct gas-fired industrial air heaters with valve train components equipped with atmospheric vents, gas

TABLE 911.2
FREE OPENING AREA OF CHIMNEY DAMPER FOR VENTING FLUE GASES FROM UNLISTED DECORATIVE APPLIANCES FOR INSTALLATION IN VENTED FIREPLACES
[NFPA 54: TABLE 10.6.2.3]

CHIMNEY HEIGHT (feet)	MINIMUM PERMANENT FREE OPENING (square inches)*						
	8	13	20	29	39	51	64
	APPLIANCE INPUT RATING (Btu/h)						
6	7800	14 000	23 200	34 000	46 400	62 400	80 000
8	8400	15 200	25 200	37 000	50 400	68 000	86 000
10	9000	16 800	27 600	40 400	55 800	74 400	96 400
15	9800	18 200	30 200	44 600	62 400	84 000	108 800
20	10 600	20 200	32 600	50 400	68 400	94 000	122 200
30	11 200	21 600	36 600	55 200	76 800	105 800	138 600

For SI units: 1 foot = 304.8 mm, 1000 British thermal units per hour = 0.293 kW, 1 square inch = 0.000645 m^2

* The first six minimum permanent free openings [8 square inches (0.005 m^2) to 51 square inches (0.03 m^2)] correspond to the cross-sectional areas of chimneys having diameters of 3 inches (76 mm) through 8 inches (203 mm), respectively. The 64 square inch (0.04 m^2) opening corresponds to the cross-sectional area of a standard 8 inch (203 mm) by 8 inch (203 mm) chimney tile.

reliefs, or bleeds shall have their vent lines, gas reliefs, or bleeds lead to a safe point outdoors. Means shall be employed on these lines to prevent water from entering and to prevent blockage from insects and foreign matter. An atmospheric vent line shall not be required to be provided on a valve train component equipped with a listed vent limiter. [NFPA 54:10.8.6]

914.7 Relief Openings. The design of the installation shall include adequate provisions to permit the non-recirculating direct gas-fired industrial air heater to operate at its rated airflow without overpressurizing the space served by the heater by taking into account the structure's designed infiltration rate, properly designed relief openings, or an interlocked powered exhaust system, or a combination of these methods. [NFPA 54:10.8.7]

914.7.1 Infiltration Rate. The structure's designed infiltration rate and the size of relief opening(s) shall be determined by approved engineering methods. [NFPA 54:10.8.7.1]

914.7.2 Louver or Gravity Dampers. Louver or counterbalanced gravity damper relief openings shall be permitted. Where motorized dampers or closeable louvers are used, they shall be proved to be in their open position prior to main burner operation. [NFPA 54:10.8.7.2]

914.8 Purging. Inlet ducting, where used, shall be purged with not less than four air changes prior to an ignition attempt. [NFPA 54:10.8.8]

915.0 Recirculating Direct Gas-Fired Industrial Air Heaters.

915.1 Application. Direct gas-fired industrial air heaters of the recirculating type shall be listed in accordance with CSA Z83.18. [NFPA 54:10.9.1]

915.2 Prohibited Installations. Recirculating direct gas-fired industrial air heaters shall not serve any area containing sleeping quarters. Recirculating direct gas-fired industrial air heaters shall not recirculate room air in buildings that contain flammable solids, liquids, or gases; explosive materials; or substances that can become toxic when exposed to flame or heat. [NFPA 54:10.9.2.1, 10.9.2.2]

915.3 Installation. Installation of direct gas-fired industrial air heaters shall comply with the following requirements:

(1) Recirculating direct gas-fired industrial air heaters shall be installed in accordance with the manufacturer's installation instructions.

(2) Recirculating direct gas-fired industrial air heaters shall be installed in industrial or commercial occupancies. [NFPA 54:10.9.3]

915.4 Clearance from Combustible Materials. Recirculating direct gas-fired industrial air heaters shall be installed with a clearance from combustible materials of not less than that shown on the rating plate and the manufacturer's installation instructions. [NFPA 54:10.9.4]

915.5 Air Supply. Ventilation air to the recirculating direct gas-fired industrial air heater shall be ducted directly from outdoors. Air to the recirculating direct gas-fired industrial air heater in excess of the minimum ventilation air specified on the heater's rating plate shall be taken from the building, ducted directly from outdoors, or a combination of both. Where outdoor air dampers or closing louvers are used, they shall be verified to be in the open position prior to main burner operation. [NFPA 54:10.9.5]

915.6 Atmospheric Vents, Gas Reliefs, or Bleeds. Recirculating direct gas-fired industrial air heaters with valve train components equipped with atmospheric vents, gas reliefs, or bleeds shall have their vent lines, gas reliefs, or bleeds lead to a safe point outdoors. Means shall be employed on these lines to prevent water from entering and to prevent blockage from insects and foreign matter. An atmospheric vent line shall not be required to be provided on a valve train component equipped with a listed vent limiter. [NFPA 54:10.9.6]

915.7 Relief Openings. The design of the installation shall include adequate provisions to permit the recirculating direct gas-fired industrial air heater to operate at its rated airflow without overpressurizing the space served by the heater by taking into account the structure's designed infiltration rate, properly designed relief openings or an interlocked powered exhaust system, or a combination of these methods. [NFPA 54:10.9.7]

915.7.1 Infiltration Rate. The structure's designed infiltration rate and the size of relief opening(s) shall be determined by approved engineering methods. [NFPA 54:10.9.7.1]

915.7.2 Louver or Gravity Dampers. Louver or counterbalanced gravity damper relief openings shall be permitted. Where motorized dampers or closeable louvers are used, they shall be proved to be in their open position prior to main burner operation. [NFPA 54:10.9.7.2]

915.8 Purging. Inlet ducting, where used, shall be purged with not less than four air changes prior to an ignition attempt. [NFPA 54:10.9.8]

916.0 Room Heaters.

916.1 Electric Room Heaters. Electric room heaters shall comply with UL 2021.

916.2 Gas-Fired Room Heaters. Gas-fired room heaters shall comply with Section 916.2.1 through Section 916.2.4.

916.2.1 Prohibited Installations. Unless specifically permitted by the Authority Having Jurisdiction, unvented room heaters shall not be installed as primary heat sources. Unvented room heaters shall not be permitted in spaces that do not have the required volume of indoor air as defined in Section 701.4.

[HCD 1 & HCD 2] *Unvented fuel-burning room heaters shall not be installed, used, maintained, or permitted to exist in a Group R Occupancy.*

916.2.1.1 Unvented Room Heaters. Unvented room heaters shall not be installed in bathrooms or bedrooms.

Exceptions:

(1) Where approved by the Authority Having Jurisdiction, one listed wall-mounted unvented room heater equipped with an oxygen depletion safety shutoff system shall be permitted to be installed in a bathroom, provided that the input rating does not exceed 6000 Btu/h (1.76 kW)

and combustion and ventilation air is provided as specified in Section 902.2.

(2) Where approved by the Authority Having Jurisdiction, one listed wall-mounted unvented room heater equipped with an oxygen depletion safety shutoff system shall be permitted to be installed in a bedroom, provided that the input rating does not exceed 10 000 Btu/h (3 kW) and combustion and ventilation air is provided as specified in Section 902.2. [NFPA 54:10.22.1]

(3) Portable oil fired unvented heating appliances used as supplemental heating in storage occupancies, utility occupancies, and in accordance with the fire code.

***[HCD 1 & HCD 2]** Unvented fuel-burning room heaters shall not be installed, used, maintained, or permitted to exist in a Group R Occupancy.*

916.2.2 Installations in Institutions. Room heaters shall not be installed in the following occupancies:

(1) Residential board and care

(2) Health care [NFPA 54:10.22.3]

916.2.3 Clearance. A room heater shall be placed so as not to cause a hazard to walls, floors, curtains, furniture, doors where open, and to the free movements of persons within the room. Heaters designed and marked, "For use in noncombustible fireplace only," shall not be installed elsewhere. Listed room heaters shall be installed in accordance with their listings and the manufacturer's installation instructions. In no case shall the clearances be such as to interfere with combustion air and accessibility. Unlisted room heaters shall be installed with clearances from combustible material not less than the following:

(1) Circulating type room heaters having an outer jacket surrounding the combustion chamber, arranged with openings at top and bottom so that air circulates between the inner and outer jacket, and without openings in the outer jacket to permit direct radiation, shall have clearance at sides and rear of not less than 12 inches (305 mm).

(2) Radiating type room heaters other than those of the circulating type described in Section 916.2.3(1) shall have clearance at sides and rear of not less than 18 inches (457 mm), except that heaters that make use of metal, asbestos, or ceramic material to direct radiation to the front of the heater shall have a clearance of 36 inches (914 mm) in front and, where constructed with a double back of metal or ceramic, shall be permitted to be installed with a clearance of 18 inches (457 mm) at sides and 12 inches (305 mm) at rear. Combustible floors under unlisted room heaters shall be protected in an approved manner.

916.2.4 Wall-Type Room Heaters. Wall-type room heaters shall not be installed in or attached to walls of combustible material unless listed for such installation. [NFPA 54:10.23.5]

916.3 Solid-Fuel-Type Room Heaters. Solid-fuel type room heaters shall comply with UL 1482.

917.0 Unit Heaters.

917.1 Support. Suspended-type unit heaters shall be safely and adequately supported, with due consideration given to their weight and vibration characteristics. Hangers and brackets shall be of noncombustible material. [NFPA 54:10.25.1]

917.2 Clearance. Suspended-type unit heaters shall comply with the following requirements:

(1) A listed unit heater shall be installed with clearances from combustible material of not less than 18 inches (457 mm) at the sides, 12 inches (305 mm) at the bottom, and 6 inches (152 mm) above the top where the unit heater has an internal draft hood or 1 inch (25.4 mm) above the top of the sloping side of a vertical draft hood. A unit heater listed for reduced clearances shall be installed in accordance with its listing and the manufacturer's installation instructions.

(2) Unlisted unit heaters shall be installed with clearances to combustible material of not less than 18 inches (457 mm).

(3) Clearances for servicing shall be in accordance with the manufacturer's installation instructions.

917.2.1 Floor-Mounted-Type Unit Heaters. Floor-mounted-type unit heaters shall comply with the following requirements:

(1) A listed unit heater shall be installed with clearances from combustible material at the back and one side of not less than 6 inches (152 mm). Where the flue gases are vented horizontally, the 6 inch (152 mm) clearance shall be measured from the draft hood or vent instead of the rear wall of the unit heater. A unit heater listed for reduced clearances shall be installed in accordance with its listing and the manufacturer's installation instructions.

(2) Floor-mounted-type unit heaters installed on combustible floors shall be listed for such installation.

(3) Combustible floors under unlisted floor-mounted unit heaters shall be protected in an approved manner.

(4) Clearances for servicing shall be in accordance with the manufacturer's instructions.

917.3 Combustion and Circulating Air. Combustion and circulating air shall be provided in accordance with Section 701.0. [NFPA 54:10.25.3]

917.4 Ductwork. A unit heater shall not be attached to a warm air duct system unless listed and marked for such installation. [NFPA 54:10.25.4]

917.5 Installation in Commercial Garages and Aircraft Hangars. Unit heaters installed in garages for more than three motor vehicles or in aircraft hangars shall be of a listed type and shall be installed in accordance with Section 303.11 and Section 303.12. [NFPA 54:10.25.5]

917.6 Oil-Fired Unit Heaters. Oil-fired unit heaters shall comply with UL 731 and installed in accordance with the manufacturer's installation instructions.

918.0 Food Service Appliance, Floor Mounted.

918.1 Clearance for Listed Appliances. Listed floor-mounted food service appliances, such as ranges for hotels and restaurants, deep-fat fryers, unit broilers, kettles, steam cookers, steam generators, and baking and roasting ovens, shall be installed not less than 6 inches (152 mm) from combustible material except that not less than a 2 inch (51 mm) clearance shall be maintained between a draft hood and combustible material. Floor-mounted food service appliances listed for installation at lesser clearances shall be installed in accordance with its listing and the manufacturer's installation instructions. Appliances designed and marked, "For use only in noncombustible locations," shall not be installed elsewhere.

918.2 Clearance for Unlisted Appliances. Unlisted floor-mounted food service appliances shall be installed to provide a clearance to combustible material of not less than 18 inches (457 mm) from the sides and rear of the appliance and from the vent connector and not less than 48 inches (1219 mm) above cooking tops and at the front of the appliance.

Clearances for unlisted appliances installed in partially enclosed areas such as alcoves shall not be reduced. Reduced clearances for unlisted appliances installed in rooms that are not partially enclosed shall be in accordance with Table 303.10.1. [NFPA 54:10.12.2]

918.3 Mounting on Combustible Floors. Listed floor-mounted food service appliances that are listed specifically for installation on floors constructed of combustible material shall be permitted to be installed on combustible floors in accordance with its listing and the manufacturer's installation instructions.

918.3.1 Not Listed for Mounting on Combustible Floors. Floor-mounted food service appliances that are not listed for mounting on a combustible floor shall be mounted in accordance with Section 918.4 or be mounted in accordance with one of the following:

(1) Where the appliance is set on legs that provide not less than 18 inches (457 mm) open space under the base of the appliance or where it has no burners and no portion of any oven or broiler within 18 inches (457 mm) of the floor, it shall be permitted to be mounted on a combustible floor without special floor protection, provided at least one sheet metal baffle is between the burner and the floor.

(2) Where the appliance is set on legs that provide not less than 8 inches (203 mm) open space under the base of the appliance, it shall be permitted to be mounted on combustible floors, provided the floor under the appliance is protected with not less than $^{3}/_{8}$ of an inch (9.5 mm) insulating millboard covered with sheet metal not less than 0.0195 of an inch (0.4953 mm) thick. The preceding specified floor protection shall extend not less than 6 inches (152 mm) beyond the appliance on all sides.

(3) Where the appliance is set on legs that provide not less than 4 inches (102 mm) under the base of the appliance, it shall be permitted to be mounted on combustible floors, provided the floor under the appliance is protected with hollow masonry not less than 4 inches (102 mm) in thickness covered with sheet metal not less than 0.0195 of an inch (0.4953 mm) thick. Such masonry courses shall be laid with ends unsealed and joints matched in such a way as to provide for free circulation of air through the masonry.

(4) Where the appliance does not have legs at least 4 inches (102 mm) high, it shall be permitted to be mounted on combustible floors, provided the floor under the appliance is protected by two courses of 4 inch (102 mm) hollow clay tile, or equivalent, with courses laid at right angles and with ends unsealed and joints matched in such a way as to provide for free circulation of air through such masonry courses, and covered with steel plate not less than $^{3}/_{16}$ of an inch (4.8 mm) in thickness. [NFPA 54:10.12.3.2]

918.4 Installation on Noncombustible Floors. Listed floor-installed food service appliances that are designed and marked "For use only in noncombustible locations" shall be installed on floors of noncombustible construction with noncombustible flooring and surface finish and with no combustible material against the underside thereof, or on noncombustible slabs or arches having no combustible material against the underside thereof. Such construction shall in all cases extend not less than 12 inches (305 mm) beyond the appliance on all sides. [NFPA 54:10.12.4.1, 10.12.4.2]

918.5 Combustible Material Adjacent to Cooking Top. Listed and unlisted food service ranges shall be installed to provide clearance to combustible material of not less than 18 inches (457 mm) horizontally for a distance of up to 2 feet (610 mm) above the surface of the cooking top where the combustible material is not completely shielded by high shelving, warming closet, or other system. Reduced combustible material clearances are permitted where protected in accordance with Table 303.10.1. [NFPA 54:10.12.5]

918.6 Use with Casters. Floor-mounted appliances with casters shall be listed for such construction and shall be installed in accordance with the manufacturer's installation instructions for limiting the movement of the appliance to prevent strain on the connection. [NFPA 54:10.12.6]

918.7 Level Installation. Floor-mounted food service appliances shall be installed level on a firm foundation. [NFPA 54:10.12.7]

918.8 Ventilation. Means shall be provided to ventilate the space in which food service appliance is installed to permit combustion of the gas. [NFPA 54:10.12.8]

919.0 Food Service Appliances, Counter Appliances.

919.1 Vertical Clearance. A vertical distance of not less than 48 inches (1219 mm) shall be provided between the top of food service hot plates and griddles and combustible material. [NFPA 54:10.13.1]

919.2 Clearance for Listed Appliances. Listed food service counter appliances such as hot plates and griddles, food and dish warmers, and coffee brewers and urns, where installed on combustible surfaces, shall be set on their own bases or legs and shall be installed with a horizontal clearance of not less than 6 inches (152 mm) from combustible material, except that not less than a 2 inches (51 mm) clearance shall be maintained between a draft hood and combustible material. Food service counter appliances listed for installation at lesser clearances shall be installed in accordance with their listing and the manufacturer's installation instructions.

919.3 Clearance for Unlisted Appliances. Unlisted food service hot plates and griddles shall be installed with a horizontal clearance from combustible material of not less than 18 inches (457 mm). Unlisted gas food service counter appliances, including coffee brewers and urns, waffle bakers, and hot water immersion sterilizers, shall be installed with a horizontal clearance from combustible material of not less than 12 inches (305 mm). Reduced clearances for gas food service counter appliances shall be in accordance with Table 303.10.1. Unlisted food and dish warmers shall be installed with a horizontal clearance from combustible material of not less than 6 inches (152 mm). [NFPA 54:10.13.3]

919.4 Mounting of Unlisted Appliances. Unlisted food service counter appliances shall not be set on combustible material unless they have legs that provide not less than 4 inches (102 mm) of open space below the burners and the combustible surface is protected with insulating millboard not less than ¼ of an inch (6.4 mm) thick covered with sheet metal not less than 0.0122 of an inch (0.3099 mm) thick, or with equivalent protection. [NFPA 54:10.13.4]

920.0 Household Cooking Appliances.

920.1 Electric Household Cooking Appliances. Electric household cooking appliances designed for permanent installations shall be installed in accordance with the manufacturer's installation instructions. Household electric ranges shall comply with UL 858.

920.2 Gas-Fired Household Cooking Appliances. Gas-fired household cooking appliances shall comply with Section 920.3 though Section 920.4.4.

920.3 Floor-Mounted Units. Floor mounted units shall be installed in accordance with Section 920.3.1 and Section 920.3.2.

920.3.1 Clearance from Combustible Material. The clearances specified as follows shall not interfere with combustion air, accessibility for operation, and servicing:

(1) Listed floor-mounted household cooking appliances, where installed on combustible floors, shall be set on their own bases or legs and shall be installed in accordance with their listing and the manufacturer's installation instructions.

(2) Listed household cooking appliances with listed gas room heater sections shall be installed so that the warm air discharge side shall have a clearance of not less than 18 inches (457 mm) from adjacent combustible material. A clearance of not less than 36 inches (914 mm) shall be provided between the top of the heater section and the bottom of cabinets.

(3) Listed household cooking appliances that include a solid or liquid fuel-burning section shall be spaced from combustible material and otherwise installed in accordance with their listing and the manufacturer's installation instructions for the supplementary fuel section of the appliance.

(4) Unlisted floor-mounted household cooking appliances shall be installed with not less than 6 inches (152 mm) clearance at the back and sides to combustible material. Combustible floors under unlisted appliances shall be protected in an approved manner.

920.3.2 Vertical Clearance Above Cooking Top. Household cooking appliances shall have a vertical clearance above the cooking top of not less than 30 inches (762 mm) to combustible material or metal cabinets. A minimum clearance of 24 inches (610 mm) is permitted where one of the following is installed:

(1) The underside of the combustible material or metal cabinet above the cooking top is protected with not less than ¼ of an inch (6.4 mm) insulating millboard covered with sheet metal not less than 0.0122 of an inch (0.3099 mm) thick.

(2) A metal ventilating hood of sheet metal not less than 0.0122 of an inch (0.3099 mm) thick is installed above the cooking top with a clearance of not less than ¼ of an inch (6.4 mm) between the hood and the underside of the combustible material or metal cabinet, and the hood is as wide as the appliance and is centered over the appliance.

(3) A listed cooking appliance or microwave oven installed over a listed cooking appliance shall be in accordance with the terms of the upper appliance's listing and the manufacturer's installation instructions. Microwave ovens shall comply with UL 923.

920.4 Built-In Units. Built-in units shall be installed in accordance with Section 920.4.1 through Section 920.4.4.

920.4.1 Installation. Listed built-in household cooking appliances shall be installed in accordance with their listing and the manufacturer's installation instructions. The installation shall not interfere with combustion air, accessibility for operation, and servicing. Unlisted built-in household cooking appliances shall not be installed in or adjacent to combustible material.

920.4.2 Vertical Clearance. Built-in top (or surface) cooking appliances shall have a vertical clearance above the cooking top of not less than 30 inches (762 mm) to combustible material or metal cabinets. A clearance of not less than 24 inches (610 mm) is permitted where one of the following is installed:

(1) The underside of the combustible material or metal cabinet above the cooking top is protected with not less than ¼ of an inch (6.4 mm) insulating millboard covered with sheet metal not less than 0.0122 of an inch (0.3099 mm) thick.

(2) A metal ventilating hood of sheet metal not less than 0.0122 of an inch (0.3099 mm) thick is installed

above the cooking top with a clearance of not less than ¼ of an inch (6.4 mm) between the hood and the underside of the combustible material or metal cabinet, and the hood not less than the width of the appliance and is centered over the appliance.

(3) A listed cooking appliance or microwave oven installed over a listed cooking appliance shall be in accordance with the terms of the upper appliance listing and the manufacturer's installation instructions.

920.4.3 Horizontal Clearance. The horizontal distance from the center of the burner head(s) of a listed top (or surface) cooking appliance to vertical combustible walls extending above the top panel shall be not less than that distance specified by the permanent marking on the appliance. [NFPA 54:10.15.2.3]

920.4.4 Level Installation. Built-in household cooking appliances shall be installed so that the cooking top, broiler pan, or oven racks are level. [NFPA 54:10.15.2.4]

921.0 Cooking Appliances Listing.

921.1 Commercial Electric Ranges. Commercial electric ranges shall comply with UL 197 and installed in accordance with the manufacturer's installation instructions.

921.2 Commercial Wood-Fired Baking Ovens. Commercial wood-fired baking ovens (refractory type) shall comply with UL 2162 and installed in accordance with the manufacturer's installation instructions.

921.3 Oil-Burning Ranges. Oil-burning ranges shall comply with UL 896 and installed in accordance with the manufacturer's installation instructions.

922.0 Open-Top Broiler Units.

922.1 Listed Units. Listed open-top broiler units shall be installed in accordance with the manufacturer's installation instructions. [NFPA 54:10.18.1]

922.2 Unlisted Units. Unlisted open-top broiler units shall be installed in accordance with the manufacturer's instructions but shall not be installed in combustible material. [NFPA 54:10.18.2]

922.3 Protection Above Domestic Units. Domestic open-top broiler units shall be provided with a metal ventilating hood not less than 0.0122 of an inch (0.3099 mm) thick with a clearance of not less than ¼ of an inch (6.4 mm) between the hood and the underside of combustible material or metal cabinets. A clearance of at least 24 inches (610 mm) shall be maintained between the cooking top and the combustible material or metal cabinet, and the hood shall be at least as wide as the open-top broiler unit and centered over the unit. Listed domestic open-top broiler units incorporating an integral exhaust system and listed for use without a ventilating hood shall not be required to be provided with a ventilating hood if installed in accordance with Section 920.3.2(1). [NFPA 54:10.18.3]

922.4 Commercial Units. Commercial open-top broiler units shall be provided with ventilation in accordance with NFPA 96. [NFPA 54:10.18.4]

923.0 Outdoor Cooking Appliances.

923.1 Listed Units. Listed outdoor cooking appliances shall be installed in accordance with their listing and the manufacturer's installation instructions.

923.2 Unlisted Units. Unlisted outdoor cooking appliances shall be installed outdoors with clearances to combustible material of not less than 36 inches (914 mm) at the sides and back and not less than 48 inches (1219 mm) at the front. In no case shall the appliance be located under overhead combustible construction. [NFPA 54:10.19.2]

924.0 Illuminating Appliances.

924.1 Clearances for Listed Appliances. Listed illuminating appliances shall be installed in accordance with their listing and the manufacturer's installation instructions.

924.2 Clearances for Unlisted Appliances. Clearances for unlisted illuminating appliances shall comply with the following:

(1) Unlisted enclosed illuminating appliances installed outdoors shall be installed with clearances in any direction from combustible material of not less than 12 inches (305 mm). [NFPA 54:10.15.2.1(1)]

(2) Unlisted enclosed illuminating appliances installed indoors shall be installed with clearances in any direction from combustible material of not less than 18 inches (457 mm). [NFPA 54:10.15.2.1(2)]

924.2.1 Open-Flame Type. Clearances shall comply with the following:

(1) Unlisted open-flame illuminating appliances installed outdoors shall have clearances from combustible material not less than that specified in Table 924.2.1. The distance from ground level to the base of the burner shall be a minimum of 7 feet (2134 mm) where installed within 2 feet (610 mm) of walkways. Lesser clearances shall be permitted to be used where acceptable to the Authority Having Jurisdiction.

(2) Unlisted open-flame illuminating appliances installed outdoors shall be equipped with a limiting orifice or other limiting devices that will maintain a flame height consistent with the clearance from combustible material, as given in Table 924.2.1.

(3) Appliances designed for flame heights in excess of 30 inches (762 mm) shall be permitted to be installed if acceptable to the Authority Having Jurisdiction. Such appliances shall be equipped with a safety shutoff device or automatic ignition.

(4) Unlisted open-flame illuminating appliances installed indoors shall have clearances from combustible material acceptable to the Authority Having Jurisdiction. [NFPA 54:10.15.2.2]

TABLE 924.2.1
CLEARANCES FOR UNLISTED OUTDOOR OPEN-FLAME ILLUMINATING APPLIANCES
[NFPA 54:TABLE 10.15.2.2]

FLAME HEIGHT ABOVE BURNER HEAD (inches)	MINIMUM CLEARANCE FROM COMBUSTIBLE MATERIAL (feet)*	
	HORIZONTAL	VERTICAL
12	2	6
18	3	8
24	3	10
30	4	12

For SI units: 1 inch = 25.4 mm, 1 foot = 304.8 mm
* Measured from the nearest portion of the burner head.

924.3 Mounting on Buildings. Illuminating appliances designed for wall or ceiling mounting shall be securely attached to substantial structures in such a manner that they are not dependent on the gas piping for support. [NFPA 54:10.15.3]

924.4 Mounting on Posts. Illuminating appliances designed for post mounting shall be securely and rigidly attached to a post. Posts shall be rigidly mounted. The strength and rigidity of posts greater than 3 feet (914 mm) in height shall be at least equivalent to that of a 2½ inch (64 mm) diameter post constructed of 0.064 of an inch (1.626 mm) thick steel or a 1 inch (25.4 mm) Schedule 40 steel pipe. Posts 3 feet (914 mm) or less in height shall not be smaller than a ¾ of an inch (19.1 mm) Schedule 40 steel pipe. Drain openings shall be provided near the base of posts where water collecting inside the posts is possible. [NFPA 54:10.15.4]

924.5 Appliance Pressure Regulators. Where an appliance pressure regulator is not supplied with an illuminating appliance and the service line is not equipped with a service pressure regulator, an appliance pressure regulator shall be installed in the line serving one or more illuminating appliances. [NFPA 54:10.15.5]

925.0 Incinerators and Crematories.

925.1 Field Constructed Commercial-Industrial Incinerators. Field constructed commercial-industrial incinerators shall be constructed and installed in accordance with NFPA 82.

925.2 Factory-Built Commercial Crematories. Factory-built commercial incinerators and crematories shall comply with UL 2790 and installed in accordance with the manufacturer's installation instructions.

925.3 Residential Incinerators. Residential incinerators shall comply with UL 791 and installed in accordance with the manufacturer's installation instructions.

926.0 Infrared Heaters.

926.1 Support. Suspended-type infrared heaters shall be fixed in position independent of gas and electric supply lines. Hangers and brackets shall be of noncombustible material. Heaters subject to vibration shall be provided with vibration-isolating hangers. [NFPA 54:10.17.1]

926.2 Clearance. The installation of infrared heaters shall comply with the following clearance requirements:

(1) Listed heaters shall be installed with clearances from combustible material in accordance with their listing and the manufacturer's installation instructions.

(2) Unlisted heaters shall be installed in accordance with clearances from combustible material acceptable to the Authority Having Jurisdiction.

(3) In locations used for the storage of combustible materials, signs shall be posted to specify the maximum permissible stacking height to maintain required clearances from the heater to the combustibles.

926.3 Combustion and Ventilation Air. Where unvented infrared heaters are used, natural or mechanical means shall be provided to supply and exhaust at least 4 ft³/min/1000 Btu/h (0.38 m³/min/kW) input of installed heaters. [NFPA 54:10.17.3.1]

926.3.1 Exhaust Openings. Exhaust openings for removing flue products shall be above the level of the heaters. [NFPA 54:10.17.3.2]

926.4 Installation in Commercial Garages and Aircraft Hangars. Overhead heaters installed in garages for more than three motor vehicles or in aircraft hangars shall be of a listed type and shall be installed in accordance with Section 303.11 and Section 303.12. [NFPA 54:10.17.4]

927.0 Pool Heaters.

927.1 Location. A pool heater shall be located or protected so as to minimize accidental contact of hot surfaces by persons. [NFPA 54:10.20.1]

927.2 Clearance. The installation of pool heaters shall comply with the following requirements:

(1) In no case shall the clearances be such as to interfere with combustion air, draft hood, or vent terminal clearance and relief, and accessibility for servicing.

(2) A listed pool heater shall be installed in accordance with its listing and the manufacturer's installation instructions.

(3) An unlisted pool heater shall be installed with a clearance of not less than 12 inches (305 mm) on the sides and the rear. A combustible floor under an unlisted pool heater shall be protected in an approved manner.

927.3 Temperature or Pressure-Limiting Devices. An unlisted pool heater shall be provided with overtemperature protection or overtemperature and overpressure protection by means of an approved device(s). [NFPA 54:10.20.3.1]

927.3.1 Pressure Relief Valve. Where a pool heater is provided with over-temperature protection only and is installed with any device in the discharge line of the heater that can restrict the flow of water from the heater to the pool (such as a check valve, shutoff valve, therapeutic pool valving, or flow nozzles), a pressure-relief valve shall be installed either in the heater or between the heater and the restrictive device. [NFPA 54:10.20.3.2]

927.4 Bypass Valves. Where an integral bypass system is not provided as a part of the pool heater, a bypass line and valve shall be installed between the inlet and outlet piping for use in adjusting the flow of water through the heater. [NFPA 54:10.20.4]

927.5 Venting. A pool heater listed for outdoor installation shall be installed with the venting means supplied by the manufacturer and in accordance with the manufacturer's installation instructions. [NFPA 54:10.20.5]

928.0 Refrigerators.

928.1 Clearance. Refrigerators shall be provided with clearances for ventilation at the top and back in accordance with the manufacturer's instructions. Where such instructions are not available, at least 2 inches (51 mm) shall be provided between the back of the refrigerator and the wall at least 12 inches (305 mm) above the top. [NFPA 54:10.21.1]

928.2 Venting or Ventilating Kits Approved for Use with a Refrigerator. Where an accessory kit is used for conveying air for burner combustion or unit cooling to the refrigerator from areas outside the room in which it is located, or for conveying combustion products diluted with air containing waste heat from the refrigerator to areas outside the room in which it is located, the kit shall be installed in accordance with the refrigerator manufacturer's instructions. [NFPA 54:10.21.2]

929.0 Gas-Fired Toilets.

929.1 Clearance. A listed gas-fired toilet shall be installed in accordance with its listing and the manufacturer's installation instructions, provided that the clearance shall be such to afford ready accessibility for use, cleanout, and necessary servicing.

929.2 Installation on Combustible Floors. Listed gas-fired toilets installed on combustible floors shall be listed for such installation. [NFPA 54:10.24.2]

929.3 Vents. Vents or vent connectors that are capable of being contacted during casual use of the room in which the toilet is installed shall be protected or shielded to prevent such contact. [NFPA 54:10.24.3]

930.0 Appliances for Installation in Manufactured Housing. *[Not adopted by HCD]*

930.1 General. ***[Not adopted by HCD]*** Appliances installed in manufactured housing after the initial sale shall be listed for installation in manufactured housing, or approved, and shall be installed in accordance with the requirements of this code and the manufacturer's installation instructions. Appliances installed in the living space of manufactured housing shall be in accordance with the requirements of Section 701.0. [NFPA 54:10.29]

931.0 Small Ceramic Kilns.

931.1 General. The provisions of this section apply to kilns used for ceramics that have a maximum interior volume of 20 cubic feet (0.57 m^3) and are used for hobby or noncommercial purposes.

931.2 Installation. Kilns shall be installed in accordance with the manufacturer's installation instructions and the provisions of this code.

931.3 Fuel-Gas Controls. Fuel-gas controls shall comply with Section 306.0 and Section 902.4. Standing pilots shall not be used with gas-fired kilns.

931.4 Electrical Equipment. All electrical equipment used as part of, or in connection with, the installation of a kiln shall be in accordance with the requirements in the electrical code. Electric kilns shall be listed and labeled in accordance with UL 499.

931.5 Installations Inside Buildings. In addition to other requirements specified in this section, interior installations shall comply with the requirements of Section 931.5.1 through Section 931.5.5.

931.5.1 Kiln Clearances. The sides and tops of kilns shall be located not less than 18 inches (457 mm) from a noncombustible wall surface and 3 feet (914 mm) from a combustible wall surface. Kilns shall be installed on noncombustible flooring consisting of not less than 2 inches (51 mm) of solid masonry or concrete extending not less than 12 inches (305 mm) beyond the base or supporting members of the kiln.

Exception: These clearances shall be permitted to be reduced, provided the kiln is installed in accordance with its listing.

In no case shall the clearance on the gas or electrical control side of a kiln be reduced to less than 30 inches (762 mm).

931.5.2 Hoods. A canopy-type hood shall be installed directly above each kiln. The face opening area of the hood shall be equal to or greater than the top horizontal surface area of the kiln. The hood shall be constructed of not less than 0.024 of an inch (0.61 mm) (No. 24 gauge) galvanized steel or equivalent and be supported at a height of between 12 inches (305 mm) and 30 inches (762 mm) above the kiln by noncombustible supports.

Exception: Electric kilns installed with listed exhaust blowers shall be permitted to be used where marked as being suitable for the kiln and installed in accordance with the manufacturer's installation instructions.

931.5.3 Gravity Ventilation Ducts. Each hood shall be connected to a gravity ventilation duct extending in a vertical direction to outside the building. This duct shall be of the same construction as the hood and shall have a minimum cross-sectional area of not less than one-fifteenth of the face opening area of the hood. The duct shall terminate not less than 12 inches (305 mm) above a portion of a building within 4 feet (1219 mm) and terminate not less than 4 feet (1219 mm) from an openable window or other opening into the building or adjacent property line. The duct opening to the outside shall be shielded, without reduction of duct area, to prevent entrance of rain into the duct. The duct shall be supported at each section by noncombustible supports.

931.5.4 Makeup Air. Provisions shall be made for air to enter the room in which a kiln is installed at a rate not less than the air being removed through the kiln hood.

931.5.5 Hood and Duct Clearances. A hood and duct serving a fuel-burning kiln shall have a clearance from combustible construction of not less than 18 inches (457 mm). This clearance shall be permitted to be reduced in accordance with Table 303.10.1.

931.6 Exterior Installations. Kilns shall be installed with minimum clearances as specified in Section 931.5.1. Wherever a kiln is located under a roofed area and is partially enclosed by more than two vertical wall surfaces, a hood and gravity ventilation duct shall be installed in accordance with Section 931.5.2, Section 931.5.3, and Section 931.5.5.

932.0 Outdoor Open Flame Decorative Appliances.

932.1 General. Permanently fixed in place outdoor open flame decorative appliances shall be installed in accordance with Section 932.1.1 through Section 932.1.3. [NFPA 54:10.31]

932.1.1 Listed Units. Listed outdoor open flame decorative appliances shall be installed in accordance with the manufacturer's installation instructions. [NFPA 54:10.31.1]

932.1.2 Unlisted Units. Unlisted outdoor open flame decorative appliances shall be installed outdoors in accordance with the manufacturer's installation instructions and with clearances to combustible material of not less than 36 inches (914 mm) from the sides. In no case shall the appliance be located under overhead combustible construction. [NFPA 54:10.31.2]

932.1.3 Connection to the Piping System. The connection to the gas piping system shall be in accordance with Section 1312.1(1), Section 1312.1(2), Section 1312.1(4), or Section 1312.1(5). [NFPA 54:10.31.3]

933.0 Evaporative Cooling Systems.

933.1 General. Evaporative cooling systems, including air ducts and fire dampers that are a portion of an evaporative cooling system, shall be in accordance with Section 933.2 through Section 933.4.3. Evaporative cooling systems shall be provided with outside air as specified for cooling systems in Section 403.0.

933.2 Location. Evaporative cooling systems shall be installed so as to minimize the probability of damage from an external source.

933.3 Access, Inspection, and Repair. Evaporative coolers shall be accessible for inspection, service, and replacement without removing permanent construction.

933.4 Installation. An evaporative cooler supported by the building structure shall be installed on a level base and shall be secured directly or indirectly to the building structure, to prevent displacement of the cooler.

933.4.1 Modifications to the Supporting Structure. Modifications made to the supporting framework of buildings as a result of the installation shall be in accordance with the requirements of the building code. Openings in exterior walls shall be flashed in an approved manner in accordance with the requirements of the building code.

933.4.2 On the Ground. An evaporative cooler supported directly by the ground shall be isolated from the ground by a level concrete slab extending not less than 3 inches (76 mm) above the adjoining ground level.

933.4.3 On a Platform. An evaporative cooler supported on an aboveground platform shall be elevated not less than 6 inches (152 mm) above adjoining ground level.

934.0 Refrigeration Appliances.

934.1 Self-Contained Refrigerators and Freezers. Factory-built commercial refrigerators and freezers shall comply with UL 471 and installed in accordance with the manufacturer's installation instructions.

934.2 Unit Coolers. Factory-built unit coolers for use in refrigerators, freezers, refrigerated warehouses, and walk-in coolers shall comply with UL 412 and installed in accordance with the manufacturer's installation instructions.

934.3 Self-Contained Mechanical Refrigeration Systems. Self-contained mechanical refrigeration systems for use in walk-in coolers shall comply with UL 427 and installed in accordance with the manufacturer's installation instructions.

935.0 Ductless Mini-Split Systems Installation.

935.1 General. A ductless mini-split system installation shall be installed in accordance with the manufacturer's installation instructions and Section 310.2 for condensate control.

936.0 Air Filter Appliances.

936.1 Electrostatic Air Cleaners. Electrostatic air cleaners shall comply with UL 867 and installed in accordance with the manufacturer's installation instructions.

936.2 High-Efficiency Particulate Air Filter Units. High-efficiency particulate air filter units for use in industrial and laboratory exhaust and ventilation systems shall be installed in accordance with the manufacturer's installation instructions.

937.0 Gaseous Hydrogen Systems.

937.1 General. Gaseous hydrogen systems shall be installed in accordance with NFPA 2.

938.0 Compressed Natural Gas (CNG) Vehicular Fuel Systems.

938.1 General. The installation of compressed natural gas (CNG) fueling (dispensing) systems shall conform to NFPA 52. [NFPA 54:10.28]

CALIFORNIA MECHANICAL CODE – MATRIX ADOPTION TABLE
CHAPTER 10 – BOILERS AND PRESSURE VESSELS

(Matrix Adoption Tables are non-regulatory, intended only as an aid to the code user. See Chapter 1 for state agency authority and building applications.)

Adopting Agency	BSC	BSC-CG	SFM	HCD			DSA			OSHPD						BSCC	DPH	AGR	DWR	CEC	CA	SL	SLC
				1	2	1-AC	AC	SS	SS/CC	1	1R	2	3	4	5								
Adopt Entire Chapter	X		X					X	X	X	X	X	X	X	X		X						
Adopt Entire Chapter as amended (amended sections listed below)				X	X																		
Adopt only those sections that are listed below																							
Chapter/Section																							
1001.2				X	X																		

This state agency does not adopt sections identified with the following symbol: †

The Office of the State Fire Marshal's adoption of this chapter or individual sections is applicable to structures regulated by other state agencies pursuant to Section 1.11.0.

CHAPTER 10

BOILERS AND PRESSURE VESSELS

1001.0 General.

1001.1 Applicability. The requirements of this chapter shall apply to the construction, installation, operation, repair, and alteration of boilers and pressure vessels. Low-pressure boilers shall comply with this chapter and Section 904.0.

Exceptions:

(1) Listed and approved potable water heaters with a nominal capacity not exceeding 120 gallons (454 L) and having a heat input not exceeding 200 000 British thermal units per hour (Btu/h) (58.6 kW) used for hot water supply at a pressure not exceeding 160 pounds-force per square inch (psi) (1103 kPa) and at temperatures not exceeding 210°F (99°C), in accordance with the plumbing code.

(2) Pressure vessels used for unheated water supply, including those containing air that serves as a cushion and is compressed by the introduction of water and tanks connected to sprinkler systems.

(3) Portable unfired pressure vessels and Interstate Commerce Commission (I.C.C.) containers.

(4) Containers for liquefied petroleum gases, bulk oxygen, and medical gas that are regulated by the fire code.

(5) Unfired pressure vessels in business, factory, hazardous, mercantile, residential, storage, and utility occupancies having a volume not exceeding 5 cubic feet (0.14 m^3) and operating at pressures not exceeding 250 psi (1724 kPa).

(6) Pressure vessels used in refrigeration systems shall comply with Chapter 11.

(7) Pressure tanks used in conjunction with coaxial cables, telephone cables, power cables, and other similar humidity control systems.

(8) A boiler or pressure vessel subject to regular inspection by federal inspectors or licensed by federal authorities.

1001.2 Boiler Rooms and Enclosures. Boiler rooms and enclosures shall comply with the *California Building Code*.

1001.3 Air for Combustion and Ventilation. Air for combustion and ventilation shall be provided in accordance with Chapter 7.

1001.4 Drainage. For heating or hot-water-supply boiler applications, the boiler room shall be equipped with a floor drain or other approved means for disposing of the accumulation of liquid wastes incident to cleaning, recharging, and routine maintenance. No steam pipe shall be directly connected to a part of a plumbing or drainage system, nor shall a water having a temperature above 140°F (60°C) be discharged under pressure directly into a part of a drainage system. Pipes from boilers shall discharge by means of indirect waste piping as determined by the Authority Having Jurisdiction or the boiler manufacturer's instructions.

1001.5 Mounting. Equipment shall be set or mounted on a level base capable of supporting and distributing the weight contained thereon. Boilers, tanks, and equipment shall be securely anchored to the structure. Equipment requiring vibration isolation shall be installed as designed by a registered design professional and approved by the Authority Having Jurisdiction.

1001.5.1 Floors. Boilers shall be mounted on floors of noncombustible construction unless listed for mounting on combustible flooring.

1001.6 Chimneys or Vents. Boilers shall be connected to a chimney or vent, as provided for other fuel-burning equipment in Chapter 8 of this code.

1002.0 Standards.

1002.1 General. Pressure vessels shall be constructed and designed in accordance with the ASME Boiler & Pressure Vessel Code (BPVC) Section VIII. Boilers shall be constructed, designed, and installed in accordance with one of the following:

(1) ASME BPVC Section I

(2) ASME BPVC Section IV

(3) NFPA 85

1002.2 Oil-Burning Boilers. Oil-burning boilers shall comply with Section 1002.2.1 and Section 1002.2.2.

1002.2.1 Listing & Labeling. Oil-burning boilers shall be listed and labeled in accordance with UL 726.

1002.2.2 Installation. Tanks, piping, and valves for oil-burning boilers shall be installed in accordance with NFPA 31.

1002.3 Electric Boilers. Electric boilers shall be listed and labeled in accordance with UL 834.

1002.4 Solid-Fuel-Fired Boilers. Solid-fuel-fired boilers shall comply with UL 2523 and shall be installed in accordance with the manufacturer's installation instructions.

1002.5 Dual Purpose Water Heater. Water heaters utilized for combined space- and water-heating applications shall be listed or labeled in accordance with the standards referenced in Table 1203.2, and shall be installed in accordance with the manufacturer's installation instructions.

1003.0 Detailed Requirements.

1003.1 Safety Requirements. The construction of boilers and pressure vessels and the installation thereof shall be in accordance with minimum requirements for safety from structural and mechanical failure and excessive pressures as established by the Authority Having Jurisdiction in accordance with nationally recognized standards.

1003.2 Controls. Required electrical, mechanical, safety, and operating controls shall carry the approval of an approved testing agency or be accepted by the Authority Having Jurisdiction. Electrical controls shall be of such design and construction as to be suitable for installation in the environment in which they are located.

1003.2.1 Automatic Boilers. Automatic boilers shall be equipped with controls and limit devices in accordance with ASME CSD-1 or Table 1003.2.1.

The Authority Having Jurisdiction shall have the authority to approve solid-fuel-fired boilers that comply with the safety requirements for automatic gas fired boilers or oil fired boilers.

1003.3 Gauges. Steam boilers shall be provided with a pressure gauge and a water level glass. Water boilers shall be provided with a pressure gauge and a temperature gauge. Automatic boilers shall be equipped with the following gauges, as applicable:

(1) Oil temperature

(2) Oil suction pressure

(3) High and low gas pressure

(4) Stack temperature

(5) Windbox pressure

1003.4 Stack Dampers. Stack dampers on boilers fired with oil or solid fuel shall not close off more than 80 percent of the stack area where closed, except on automatic boilers with prepurge, automatic draft control, and interlock. Operative dampers shall not be placed within a stack, flue, or vent of a gas-fired boiler, except on an automatic boiler with prepurge, automatic draft control, and interlock.

1003.5 Welding. Welding on pressure vessels shall be done by certified welders in accordance with nationally recognized standards.

1004.0 Expansion Tanks.

1004.1 General. An expansion tank shall be installed in a hot-water-heating system as a means for controlling increased pressure caused by thermal expansion. Expansion tanks shall be of the closed or open type and securely fastened to the structure. Tanks shall be rated for the pressure of the system. Supports shall be capable of carrying twice the weight of the tank filled with water without placing a strain on connecting piping.

Hot-water-heating systems incorporating hot water tanks or fluid relief columns shall be installed to prevent freezing under normal operating conditions.

1004.2 Open-Type Expansion Tanks. Open type expansion tanks shall be located not less than 3 feet (914 mm) above the highest point of the system. Such tanks shall be sized based on the capacity of the system. An overflow with a diameter of not less than one-half the size of the supply or not less than 1 inch (25 mm) in diameter shall be installed at the top of the tank. The overflow shall discharge through an air gap into the drainage system.

1004.3 Closed-Type Systems. Closed-type systems shall have an airtight tank or other approved air cushion that will be consistent with the volume and capacity of the system, and shall be designed for a hydrostatic test pressure of two and one-half times the allowable working pressure of the system. Expansion tanks for systems designed to operate at more than 30 pounds-force per square inch (psi) (207 kPa) shall comply with ASME BPVC Section VIII. Provisions shall be made for draining the tank without emptying the system.

1004.4 Minimum Capacity of Closed-Type Tank. The minimum capacity for a gravity-type hot water system expansion tank shall be in accordance with Table 1004.4(1). The minimum capacity for a forced-type hot water system expansion tank shall be in accordance with Table 1004.4(2), or Equation 1004.4. Equation 1004.4 shall not be used for diaphragm-type expansion tanks.

$$V_t = \frac{(0.00041t - 0.0466)\,V_s}{\left(\frac{P_a}{P_f} - \frac{P_a}{P_o}\right)} \qquad \text{(Equation 1004.4)}$$

Where:

V_t = Minimum volume of expansion tank, gallons

V_s = Volume of system, not including expansion tank, gallons

t = Average operating temperature, °F

P_a = Atmospheric pressure, feet H_2O absolute

P_f = Fill pressure, feet H_2O absolute

P_o = Maximum operating pressure, feet H_2O absolute

For SI units: 1 gallon = 3.785 L, °C = (°F-32)/1.8, 1 foot of water = 2.99 kPa

TABLE 1004.4(1)
EXPANSION TANK CAPACITIES FOR GRAVITY HOT WATER SYSTEMS[1]

INSTALLED EQUIVALENT DIRECT RADIATION[2] (square feet)	TANK CAPACITY (gallons)
Up to 350	18
Up to 450	21
Up to 650	24
Up to 900	30
Up to 1100	35
Up to 1400	40
Up to 1600	2 to 30
Up to 1800	2 to 30
Up to 2000	2 to 35
Up to 2400	2 to 40

For SI units: 1 gallon = 3.785 L, 1 square foot = 0.0929 m²

Notes:

[1] Based on a two-pipe system with an average operating water temperature of 170°F (77°C), using cast-iron column radiation with a heat emission rate of 150 British thermal units per square foot hour [Btu/(ft²•h)] (473 W/m²) equivalent direct radiation.

[2] For systems that exceed 2400 square feet (222.9 m²) of installed equivalent direct water radiation, the required capacity of the cushion tank shall be increased on the basis of 1 gallon (4 L) tank capacity per 33 square feet (3.1 m²) of additional equivalent direct radiation.

TABLE 1004.4(2)
EXPANSION TANK CAPACITIES FOR FORCED HOT WATER SYSTEMS[1]

SYSTEM VOLUME[2] (gallons)	TANK CAPACITY DIAPHRAGM TYPE (gallons)	TANK CAPACITY NON-DIAPHRAGM TYPE (gallons)
100	9	15
200	17	30
300	25	45
400	33	60
500	42	75
1000	83	150
2000	165	300

For SI units: 1 gallon = 3.785 L

Notes:

[1] Based on an average operating water temperature of 195°F (91°C), a fill pressure of 12 psig (83 kPa), and an operating pressure of not more than 30 psig (207 kPa).

[2] Includes volume of water in boiler, radiation, and piping, not including expansion tank.

1005.0 Safety or Relief Valve Discharge.

1005.1 General. Pressurized vessels or boilers shall be provided with overpressure protection by means of a listed pressure relief valve installed in accordance with the manufacturer's installation instructions.

1005.2 Discharge Piping. The discharge piping serving a temperature relief valve, pressure relief valve, or combination of both shall have no valves, obstructions, or means of isolation and provided with the following:

(1) Equal to the size of the valve outlet and shall discharge full size to the flood level of the area receiving the discharge and pointing down.

(2) Materials shall be rated at not less than the operating temperature of the system and approved for such use.

(3) Discharge pipe shall discharge independently by gravity through an air gap into the drainage system or outside of the building with the end of the pipe not exceeding 2 feet (610 mm) and not less than 6 inches (152 mm) above the ground and pointing downwards.

(4) Discharge in such a manner that does not cause personal injury or structural damage.

(5) No part of such discharge pipe shall be trapped or subject to freezing.

(6) The terminal end of the pipe shall not be threaded.

(7) Discharge from a relief valve into a water heater pan shall be prohibited.

1005.3 Splash Shield. Where the operating temperature exceeds 212°F (100°C), the discharge pipe shall be installed with a splash shield or centrifugal separator.

1005.4 Hazardous Discharge. Where the discharge from safety valves is capable of being hazardous, discharge of steam inside the boiler room, such discharge shall be discharged to the outside of the boiler room. Discharges from relief valves on industrial boilers shall be discharged to an approved location.

1005.5 Vacuum Relief Valve. Hot-water heating systems that are subjected to a vacuum while in operation or during shutdown shall be protected with a vacuum relief valve. Where the piping configuration, equipment location, and valve outlets are located below the boiler elevation, the system shall be equipped with a vacuum relief valve at the highest point.

1006.0 Shutoff Valves.

1006.1 General. An approved manual shutoff valve shall be installed upstream of all control devices on the main burner of a gas-fired boiler. The takeoff point for the gas supply to the pilot shall be upstream of the gas shutoff valve of the main burner and shall be valved separately. A union or other approved means of disconnect shall be provided immediately downstream of these shutoff valves.

1007.0 Gas-Pressure Regulators.

1007.1 General. An approved gas-pressure regulator shall be installed on gas-fired boilers where the gas supply pressure is exceeding that at which the main burner is designed to operate. A separate approved gas-pressure regulator shall be installed to regulate the gas pressure to the pilot or pilots.

1008.0 Low-Water Cutoff.

1008.1 General. Hot water boilers and steam boilers shall be installed with a low-water cutoff. A coil-type boiler or a water-tube boiler that requires forced circulation to prevent overheating of the coils or tubes shall be installed with a flow-sensing device in the outlet piping in lieu of the low-water cutoff. The low-water cutoff or the flow sensing device shall be installed so as to prevent damage to the boiler and to permit testing of the fuel-supply cutoff without draining the heating system. The low-water cutoff shall shut off the combustion at a water level setpoint that is in accordance with the boiler manufacturer's instructions.

1009.0 Combustion Regulators – Safety Valves.

1009.1 General. The following requirements shall be retroactive:

(1) Hot-water-heating boilers, other than manually fired, shall be equipped with two temperature combustion regulators in series. Steam-heating boilers, other than manually fired, shall be equipped with a pressure combustion regulator and a low-water cutoff. (See Section 1008.0)

(2) Boilers and pressure vessels shall be provided with the required number, size, and capacity of safety or relief valves to ensure positive relief of overpressure in accordance with nationally recognized standards, as applicable. Valves so employed shall be constructed, sealed, and installed in accordance with nationally recognized standards, as applicable.

1010.0 Clearance for Access.

1010.1 General. Where boilers are installed or replaced, clearance shall be provided to allow access for inspection, maintenance, and repair. Passageways around all sides of boilers shall have an unobstructed width of not less than 18 inches (457 mm). Clearance for repair and cleaning shall be permitted to be provided through a door or access panel into another area, provided the opening is of sufficient size.

Exception: Subject to the approval of the Authority Having Jurisdiction, boilers shall be permitted to be installed with a side clearance of less than 18 inches (457 mm), provided that the lesser clearance does not inhibit inspection, maintenance, or repair.

1010.2 Power Boilers. Power boilers having a steam-generating capacity in excess of 5000 pounds per hour (lb/h) (0.6299 kg/s) or having a heating surface in excess of 1000 square feet (92.9 m^2) or input in excess of 5 000 000 Btu/h (1464 kW) shall have a clearance of not less than 7 feet (2134 mm) from the top of the boiler to the ceiling.

1010.3 Steam-Heating Boilers, Hot Water Boilers, and Power Boilers. Steam-heating boilers and hot-water-heating boilers that exceed one of the following limits:

(1) 5 000 000 Btu/h input (1464 kW)

(2) 5000 pounds steam per hour (0.6299 kg/s) capacity

(3) 1000 square foot (92.9 m^2) heating surface

Power boilers that do not exceed one of the following limits:

(1) 5 000 000 Btu/h input (1464 kW)

(2) 5000 pounds steam per hour (0.6299 kg/s) capacity

(3) 1000 square foot (92.9 m^2) heating surface

Boilers with manholes on top of the boiler, except those described in Section 1010.2 and Section 1010.4, shall have a clearance of not less than 3 feet (914 mm) from the top of the boiler to the ceiling.

1010.4 Package Boilers, Steam-Heating Boilers, and Hot-Water-Heating Boilers. Package boilers, steam-heating boilers, and hot-water-heating boilers with no manhole on top of the shell and not exceeding one of the above limits shall have a clearance of not less than 2 feet (610 mm) from the ceiling.

1011.0 Boilers, Stokers, and Steam Generators.

1011.1 General. The design, installation, and operation of single burner boilers, multiple burner boilers, stokers, and atmospheric fluidized-bed boilers with not less than a fuel input rating of 12.5 E+06 Btu/h (3.663 MW) to pulverized fuel systems, fired or unfired steam generators used to recover heat from combustion turbines and to other combustion turbine exhaust systems shall be in accordance with NFPA 85. That portion of the oil-burning system supplied on boilers and covered within the scope of NFPA 85 shall be installed in accordance with NFPA 85.

1012.0 Operating Adjustments and Instructions.

1012.1 General. Hot water boiler installations, upon completion, shall have controls set, adjusted, and tested by the installing contractor. A complete control diagram of a permanent legible type, together with complete boiler operating instructions, shall be furnished by the installer for each installation.

1013.0 Inspections and Tests.

1013.1 General. An installation for which a permit is required shall not be put into service until it has been inspected and approved by the Authority Having Jurisdiction.

It shall be the duty of the owner or his authorized representative to notify the Authority Having Jurisdiction that the installation is ready for inspection and test. It also shall be the duty of the owner or his authorized representative to post in a conspicuous position on the installation a notice in substantially the following form: "Warning! This installation has not been inspected and approved by the Authority Having Jurisdiction and shall not be covered or concealed until so inspected and approved," and it shall be unlawful for anyone other than the Authority Having Jurisdiction to remove such notice. The Authority Having Jurisdiction shall require such tests as it deems necessary to determine that the installation is in accordance with the provision of this section. Such tests shall be made by the owner or his authorized representative in the presence of the Authority Having Jurisdiction.

Exception: On installations designed and supervised by a registered design professional, the Authority Having Jurisdiction shall have the authority to permit inspection and testing by such registered design professional.

Where the owner or his authorized representative requests inspection of a boiler prior to its installation, the Authority Having Jurisdiction shall make such inspection.

1013.2 Operating Permit. It shall be unlawful to operate a boiler or pressure vessel without first obtaining a valid operating permit to do so from the Authority Having Jurisdiction. Such permit shall be displayed in a conspicuous place adjacent to the boiler or vessel. The operating permit shall not be issued until the equipment has been inspected and approved by the Authority Having Jurisdiction.

Exception: The operation of steam-heating boilers, low-pressure hot-water-heating boilers, hot water supply boilers, and pressure vessels in residential occupancies of less than six dwelling units and utility occupancies.

1013.3 Maintenance Inspection. The Authority Having Jurisdiction shall inspect boilers and pressure vessels operated under a permit in accordance with ASHRAE/ACCA 180 at such intervals as deemed necessary, but not less frequently than in accordance with Section 1013.4 through Section 1013.7.

1013.4 Power and Miniature Boilers. Power boilers and miniature boilers shall be inspected externally annually. Where construction and operating conditions permit, they shall be subject to inspection internally annually.

1013.5 Steam-Heating and Water-Heating Boilers. Steam-heating boilers and hot-water-heating boilers shall be inspected externally annually. Where construction and operating conditions permit, they shall also be subject to inspection internally annually.

1013.6 Automatic Steam-Heating Boilers. Automatic steam-heating boilers shall be inspected externally biennially. Where construction and operating conditions permit, they shall be subject to inspection internally biennially.

1013.7 Unfired Pressure Vessels. Unfired pressure vessels shall be inspected externally biennially. Where subject to corrosion and construction permits, they shall be subject to inspection internally biennially.

Inspection of boilers and pressure vessels covered by insurance shall be permitted to be made by employees of the insuring company holding commissions from the National Board of Boiler and Pressure Vessel Inspectors, subject to approval of the Authority Having Jurisdiction. Approved insuring company inspectors shall make reports on prescribed forms on inspections authorized by the Authority Having Jurisdiction. The reports shall be filed in the Authority Having Jurisdiction office. Company inspectors shall notify the Authority Having Jurisdiction of suspension of insurance because of dangerous conditions, new insurance in effect, and discontinuance of insurance coverage.

1014.0 Operation and Maintenance of Boilers and Pressure Vessels.

1014.1 General. Boilers and pressure vessels shall be operated and maintained in accordance with requirements for protection of the public established by the Authority Having Jurisdiction in accordance with nationally recognized standards.

The Authority Having Jurisdiction shall notify the owner or authorized representative of defects or deficiencies and properly corrected. Where such corrections are not made, or where the operation of the boiler or pressure vessel is deemed unsafe by the Authority Having Jurisdiction, they shall have the authority to revoke the permit to operate the boiler or pressure vessel. Where the operation of a boiler or pressure vessel is deemed by the Authority Having Jurisdiction to constitute an immediate danger, the pressure on such boiler or pressure vessel shall be permitted to be relieved at the owner's cost and the boiler or pressure vessel shall not thereafter be operated without the approval of the Authority Having Jurisdiction.

TABLE 1003.2.1
CONTROLS AND LIMIT DEVICES FOR AUTOMATIC BOILERS

BOILER GROUP	FUEL	FUEL INPUT RANGE[1] (INCLUSIVE), BTU/H	TYPE OF PILOT[2]	SAFETY CONTROL TIMING (NOMINAL MAXIMUM TIME IN SECONDS)				ASSURED FUEL SUPPLY CONTROL[4]	ASSURED AIR SUPPLY CONTROL[5]	LOW FIRE START UP CONTROL[6]	PRE-PURGING CONTROL[7]	HOT WATER TEMPERATURE AND LOW WATER LIMIT CONTROLS[8]	STEAM PRESSURE AND LOW WATER LIMIT CONTROLS[9]	APPROVED FUEL SHUT-OFF[10]	CONTROL AND LIMIT DEVICE SYSTEM DESIGN[11]
				TRIAL FOR PILOT	TRIAL FOR MAIN BURNER FRAME		MAIN BURNER FLAME FAILURE[3]								
					DIRECT ELECTRIC IGNITION	FLAME PILOT									
A	Gas	0 - 400 000	Any type	90	Not Required	90	90	Not Required	Required	Not Required	Not Required	Required	Required	Not Required	Required
B	Gas	400 001 - 2 500 000	Interrupted or Intermittent	15	15	15	2-4	Not Required	Required	Not Required	Not Required	Required	Required	Not Required	Required
C	Gas	2 500 001 - 5 000 000	Interrupted or Intermittent	15	15	15	2-4	Required	Required	Required	Required	Required	Required	Required	Required
D	Gas	Over 5 000 000	Interrupted	15	15	15	2-4	Required	Required	Required	Required	Required	Required	Required	Required
E	Oil	0 - 400 000	Any type	Not Required	90	90	90	Not Required	Required	Not Required	Not Required	Required	Required	Not Required	Required
F	Oil	400 001 - 1 000 000	Interrupted	Not Required	30	30	2-4	Required	Required	Not Required	Not Required	Required	Required	Not Required	Required
G	Oil	1 000 001 - 3 000 000	Interrupted	Not Required	15	15	2-4	Required	Required	Not Required	Not Required	Required	Required	Not Required	Required
H	Oil	Over 3 000 000	Interrupted	15	15	60	2-4	Required	Required	Required	Required	Required	Required	Required	Required
K	Electric	All	Not required	Not Required	Not Required	Not Required	Not Required	Not Required	Not Required	Not Required	Not Required	Required	Required	Not Required	Required
L	Gas, Oil and/or Coal	12 500 000 or more	Any	10 sec Per NFPA 85	Per NFPA 85	Per NFPA 85	Per NFPA 85	Per NFPA 85	Per NFPA 85	Per NFPA 85	Per NFPA 85	Per ASME Power Boiler Code, Section 1 and NFPA 85	Per ASME Power Boiler Code, Section 1 and NFPA 85	Per NFPA 85	Per NFPA 85
M	Heat Recovery Steam Generator	Any	None	Per NFPA 85	Per NFPA 85	Per NFPA 85	Per NFPA 85	Per NFPA 85	Per NFPA 85	Per NFPA 85	Per NFPA 85	Per ASME Boiler & Pressure Code & NFPA 85	Per ASME Boiler & Pressure Code & NFPA 85	Per NFPA 85	Per NFPA 85

For SI units: 1000 British thermal units per hour = 0.293 kW

FOOTNOTES FOR TABLE 1003.2.1 (continued)

[1] Fuel input shall be determined by one of the following:
The burner input shall not exceed the input shown on the burner nameplate or as otherwise identified by the manufacturer.
The nominal boiler rating, as determined by the building official, plus 25 percent.

[2] Automatic boilers shall have one flame failure device on each burner, which shall prove the presence of an ignition source at the point where it will ignite the main burner, except that boiler groups A, B, E, F, and G, which are equipped with direct electric ignition, shall monitor the main burner, and boiler groups using interrupted pilots shall monitor the main burner after the prescribed limited trial and ignition periods. Boiler group A, equipped with continuous pilot, shall accomplish 100 percent shutoff within 90 seconds upon pilot flame failure. The use of intermittent pilots in boiler group C is limited to approved burner units.

[3] In boiler groups B, C, and D a 90 second main burner flame failure limit shall be permitted to be applied where continuous pilots are provided on manufacturer assembled boiler-burner units that have been approved by an approved testing agency in accordance with nationally recognized standards approved by the building official. Boiler groups F and G equipped to re-energize their ignition systems within 0.8 second after main burner flame failure will be permitted 30 seconds for group F or 15 seconds for group G to re-establish their main burner flames.

[4] Boiler groups C and D shall have controls interlocked to accomplish a non-recycling fuel shutoff upon high or low gas pressure, and boiler groups F, G, and H using steam or air for fuel atomization shall have controls interlocked to accomplish a nonrecycling fuel shutoff upon low atomizing steam or air pressure. Boiler groups F, G, and H equipped with a preheated oil system shall have controls interlocked to provide fuel shutoff upon low oil temperature.

[5] Automatic boilers shall have controls interlocked to shut off the fuel supply in the event of draft failure where forced or induced draft fans are used or, in the event of low combustion airflow, where a gas power burner is used. Where a single motor directly driving both the fan and the oil pump is used, a separate control is not required.

[6] Boiler groups C, D, and H, where firing in excess of 400 000 Btu/h (117 kW) per combustion chamber, shall be provided with low fire start of its main burner system to permit smooth light-off. This will normally be a rate of one-third of its maximum firing rate.

[7] Boiler groups C, D, and H shall not permit pilot or main burner trial for ignition operation before a purging operation of sufficient duration to permit not less than four complete air changes through the furnace, including a combustion chamber and the boiler passes. Where this is not readily determinable, five complete air changes of the furnace, including combustion chamber up to the first pass, will be considered equivalent. An atmospheric gas burner with no mechanical means of creating air movement or an oil burner that obtains two-thirds or more of the air required for combustion without mechanical means of creating air movement shall not require purge by means of four air changes, so long as its secondary air openings are not provided with means of closing. Where such burners have means of closing secondary air openings, a time delay shall be provided that puts these closures in a normally open position for four minutes before an attempt for ignition. An installation with a trapped combustion chamber shall, in every case, be provided with a mechanical means of creating air movement for purging.

[8] An automatic hot-water-heating boiler, low-pressure hot-water-heating boiler, and power hot water boiler shall be equipped with two high-temperature limit controls with a manual reset on the control, with the higher setting interlocked to shut off the main fuel supply, except that manual reset on the high-temperature limit control shall not be required on an automatic package boiler not exceeding 400 000 Btu/h (117 kW) input and that has been approved by an approved testing agency. An automatic hot-water heating, power boiler, and package hot-water supply boiler shall be equipped with one low-water level limit control with a manual reset interlocked to shut off the fuel supply, so installed as to prevent damage to the boiler and to permit testing of the control without draining the heating system, except on boilers used in Group R Occupancies of less than six units and in Group U Occupancies and further, except that the low-water level limit control is not required on package hot-water supply boilers approved by a nationally recognized testing agency. However, a low- water flow limit control installed in the circulating water line shall be permitted to be used instead of the low-water level limit control for the same purpose on coil-type boilers.

[9] An automatic low-pressure steam-heating boiler, small power boiler, and power steam boiler shall be equipped with two high-steam pressure limit controls interlocked to shut off the fuel supply to the main burner with manual reset on the control, with the higher setting and two low-water-level limit controls, one of which shall be provided with a manual reset device and independent of the feed water controller. Coil-type flash steam boilers shall be permitted to use two high-temperature limit controls, one of which shall be manually reset in the hot water coil section of the boiler instead of the low-water level limit control.

[10] Boiler groups C, D, and H shall use an approved automatic reset safety shutoff valve for the main burner fuel shutoff, which shall be interlocked to the programming control devices required. On oil burners where the safety shutoff valve will be subjected to pressures in excess of 10 psi (69 kPa) where the burner is not firing, a second safety shutoff valve shall be provided in series with the first. Boiler groups C and D using gas in excess of 1 psi (7 kPa) pressure or having a trapped combustion chamber or employing horizontal fire tubes shall be equipped with two approved safety shutoff valves, one of which shall be an automatic reset type, one of which shall be permitted to be used as an operating control, and both of which shall be interlocked to the limit-control devices required. Boiler groups C and D using gas in excess of 1 psi (7 kPa) pressure shall be provided with a permanent and ready means for making periodic tightness checks of the main fuel safety shutoff valves.

[11] Control and limit device systems shall be grounded with operating voltage not to exceed 150 volts, except that, upon approval by the building official, existing control equipment to be reused in an altered boiler control system shall be permitted to use 220 volts single phase with one side grounded, provided such voltage is used for all controls. Control and limit devices shall interrupt the ungrounded side of the circuit. A readily accessible means of manually disconnecting the control circuit shall be provided with controls so arranged that where they are de-energized, the burner shall be inoperative.

CALIFORNIA MECHANICAL CODE – MATRIX ADOPTION TABLE
CHAPTER 11 – REFRIGERATION

(Matrix Adoption Tables are non-regulatory, intended only as an aid to the code user. See Chapter 1 for state agency authority and building applications.)

Adopting Agency	BSC	BSC-CG	SFM	HCD			DSA			OSHPD						BSCC	DPH	AGR	DWR	CEC	CA	SL	SLC
				1	2	1-AC	AC	SS	SS/CC	1	1R	2	3	4	5								
Adopt Entire Chapter	X			X	X												X						
Adopt Entire Chapter as amended (amended sections listed below)			X					X	X	X	X	X	X	X	X								
Adopt only those sections that are listed below																							
Chapter/Section																							
1108.4			X					X	X														
Table 1104.1										X	X	X	X	X	X								

This state agency does not adopt sections identified with the following symbol: †

The Office of the State Fire Marshal's adoption of this chapter or individual sections is applicable to structures regulated by other state agencies pursuant to Section 1.11.0.

CHAPTER 11
REFRIGERATION

1101.0 General.

1101.1 Applicability. Part I governs the design, installation, and construction of refrigeration systems, equipment, refrigerant piping, pressure vessels, safety devices, replacement of parts, alterations, and substitution of different refrigerants. Part II governs the installation and construction of cooling towers.

1101.2 Equipment. Equipment for refrigerant recovery, recycling, or both shall comply with UL 1963.

Part I – Refrigeration Systems.

1102.0 Refrigeration Systems.

1102.1 General. Refrigeration systems using a refrigerant other than ammonia shall comply with this chapter and ASHRAE 15.

1102.2 Ammonia Refrigeration Systems. Refrigeration systems using ammonia as the refrigerant shall comply with IIAR 2, IIAR 3, IIAR 4, and IIAR 5 and shall not be required to comply with this chapter.

›› **1102.3 Refrigerants.** The refrigerant used shall be of a type listed in Table 1102.3 or in accordance with ASHRAE 34 where approved by the Authority Having Jurisdiction.

Exception: Lithium bromide absorption systems using water as the refrigerant.

1103.0 Classification.

1103.1 Classification of Refrigerants. Refrigerants shall be classified in accordance with Table 1102.3 or in accordance with ASHRAE 34 where approved by the Authority Having Jurisdiction.

1103.1.1 Safety Group. Table 1102.3 classifies refrigerants by toxicity and flammability, and assigns safety groups using combinations of toxicity class and flammability class. For the purposes of this chapter, the refrigerant Groups A1, A2L, A2, A3, B1, B2L, B2, and B3 shall be considered to be individual and distinct safety groups. Each refrigerant is assigned into not more than one group.

1103.2 Classification of Refrigeration Systems. Refrigeration systems shall be classified according to the degree of probability that a leakage of refrigerant will enter an occupancy-classified area in accordance with Section 1103.2.1 and Section 1103.2.2. [ASHRAE 15:5.2]

1103.2.1 High-Probability System. Systems in which the basic design, or the location of components, is such that a leakage of refrigerant from a failed connection, seal, or component will enter the occupied space shall be classified as high-probability systems. A high-probability system shall be a direct system or an indirect open spray system in which the refrigerant is capable of producing pressure that is more than the secondary coolant. [ASHRAE 15:5.2.1]

1103.2.2 Low-Probability System. Systems in which the basic design, or the location of the components, is such that a leakage of refrigerant from a failed connection, seal, or component is not capable of entering the occupied space shall be classified as low-probability systems. A low-probability system shall be an indirect closed system, double indirect system, or an indirect open spray system. In a low-probability indirect open spray system, the secondary coolant pressure remains more than the refrigerant pressure in operating and standby conditions. [ASHRAE 15:5.2.2]

1103.3 Higher Flammability Refrigerants. Group A3 and B3 refrigerants shall not be used except where approved by the Authority Having Jurisdiction.

Exceptions:

(1) Laboratories with more than 100 square feet (9.29 m^2) of space per person.

(2) Industrial occupancies.

(3) Listed portable-unit systems containing not more than 0.331 pounds (0.150 kg) of Group A3 refrigerant, provided that the equipment is installed in accordance with the listing and the manufacturer's installation instructions. [ASHRAE 15:7.5.3]

1104.0 Requirements for Refrigerant and Refrigeration System Use.

1104.1 System Selection. Refrigeration systems shall be limited in application in accordance with Table 1104.1, and the requirements of Section 1104.0.

1104.2 Refrigerant Concentration Limit. The concentration of refrigerant in a complete discharge of an independent circuit of high-probability systems shall not exceed the amounts shown in Table 1102.3, except as provided in Section 1104.3 and Section 1104.4. The volume of occupied space shall be determined in accordance with Section 1104.2.1 through Section 1104.2.3.

Exceptions:

(1) Listed equipment containing not more than 6.6 pounds (2.99 kg) of refrigerant, regardless of the refrigerant safety classification, provided the equipment is installed in accordance with the listing and with the manufacturer's installation instructions.

(2) Listed equipment for use in laboratories with more than 100 square feet (9.29 m^2) of space per person, regardless of the refrigerant safety classification, provided that the equipment is installed in accordance with the listing and the manufacturer's installation instructions. [ASHRAE 15:7.2]

1104.2.1 Volume Calculations. The volume used to convert from refrigerant concentration limits to refrigerating system quantity limits for refrigerants in Section 1104.2 shall be based on the volume of space to which refrigerant disperses in the event of a refrigerant leak. [ASHRAE 15:7.3]

1104.2.2 Nonconnecting Spaces. Where a refrigerating system or part thereof is located in one or more enclosed occupied spaces that do not connect through permanent openings or HVAC ducts, the volume of the smallest occupied space shall be used to determine the refrigerant quantity limit in the system. Where different stories and floor levels connect through an open atrium or mezzanine arrangement, the volume to be used in calculating the refrigerant quantity limit shall be determined by multiplying the floor area of the lowest space by 8.2 feet (2499 mm). [ASHRAE 15:7.3.1]

1104.2.3 Ventilated Spaces. Where a refrigerating system or a part thereof is located within an air handler, in an air distribution duct system, or in an occupied space served by a mechanical ventilation system, the entire air distribution system shall be analyzed to determine the worst-case distribution of leaked refrigerant. The worst case or the smallest volume in which the leaked refrigerant disperses shall be used to determine the refrigerant quantity limit, subject to the criteria in accordance with Section 1104.2.3.1 through Section 1104.2.3.3. [ASHRAE 15:7.3.2]

1104.2.3.1 Closures. Closures in the air distribution system shall be considered. Where one or more spaces of several arranged in parallel are capable of being closed off from the source of the refrigerant leak, their volume(s) shall not be used in the calculation.

Exceptions: The following closure devices shall not be considered:

(1) Smoke dampers, fire dampers, and combination smoke and fire dampers that close only in an emergency not associated with a refrigerant leak.

(2) Dampers, such as variable-air-volume (VAV) boxes, that provide limited closure where airflow is not reduced below 10 percent of its maximum with the fan running. [ASHRAE 15:7.3.2.1]

1104.2.3.2 Plenums. The space above a suspended ceiling shall not be included in calculating the refrigerant quantity limit in the system unless such space is part of the air supply or return system. [ASHRAE 15:7.3.2.2]

1104.2.3.3 Supply and Return Ducts. The volume of the supply and return ducts and plenums shall be included where calculating the refrigerant quantity limit in the system. [ASHRAE 15:7.3.2.3]

1104.3 Institutional Occupancies. The RCL value required in Section 1104.2 shall be reduced by 50 percent for the areas of institutional occupancies. The total of Group A2, B2, A3, and B3 refrigerants shall not exceed 550 pounds (249.5 kg) in the occupied areas and machinery rooms of institutional occupancies.

Exception: The total of all Group A2L refrigerants shall not be limited in machinery rooms of institutional occupancies.

1104.4 Industrial Occupancies and Refrigerated Rooms. Section 1104.2 shall not apply in industrial occupancies and refrigerated rooms where in accordance with the following:

(1) The space(s) containing the machinery is (are) separated from other occupancies by tight construction with tight-fitting doors.

(2) Access is restricted to authorized personnel.

(3) The floor area per occupant is not less than 100 square feet (9.29 m^2).

Exception: The minimum floor area shall not apply where the space is provided with egress directly to the outdoors or into approved building exits.

(4) Refrigerant detectors are installed with the sensing location and alarm level as required in refrigeration machinery rooms in accordance with Section 1106.2.2.2.

(5) Open flames and surfaces exceeding 800°F (427°C) shall not be permitted where a Group A2, B2, A3, or B3 refrigerant, is used.

(6) Electrical equipment that is in accordance with Class 1, Division 2, of the California Electrical Code where the quantity of a Group A2, B2, A3, or B3 refrigerant in an independent circuit is capable of exceeding 25 percent of the lower flammability limit (LFL) upon release to the space based on the volume determined in accordance with Section 1104.2.1 through Section 1104.2.3.

(7) Refrigerant containing parts in systems exceeding 100 horsepower (74.6 kW) compressor drive power, except evaporators used for refrigeration or dehumidification, condensers used for heating, control and pressure-relief valves for either, and connecting piping, are located in a machinery room or outdoors. [ASHRAE 15:7.2.2]

1104.5 Flammable Refrigerants. The total of Group A2, B2, A3, and B3 refrigerants, other than Group A2L and B2L refrigerants shall not exceed 1100 pounds (498.9 kg) without approval by the Authority Having Jurisdiction. Institutional Occupancies shall comply with Section 1104.3.

1104.6 Applications for Human Comfort and for Nonindustrial Occupancies. In nonindustrial occupancies, Group A2, A3, B1, B2, and B3 refrigerants shall not be used in high-probability systems for human comfort.

1104.7 Refrigerant Type and Purity. Refrigerants shall be of a type specified by the equipment manufacturer. Unless otherwise specified by the equipment manufacturer, refrigerants used in new equipment shall be of purity in accordance with AHRI 700.

1104.7.1 Recovered Refrigerants. Recovered refrigerants shall not be reused except in the system from which they were removed or as provided in Section 1104.7.2 or Section 1104.7.3. Where contamination is evident by discoloration, odor, acid test results, or system history, recovered refrigerants shall be reclaimed in accordance with Section 1104.7.3. [ASHRAE 15:7.5.1.4]

1104.7.2 Recycled Refrigerants. Recycled refrigerants shall not be reused except in systems using the same refrigerant and lubricant designation and belonging to the same owner as the systems from which they were removed. Where contamination is evident by discoloration, odor, acid test results, or system history, recycled refrigerants shall be reclaimed in accordance with Section 1104.7.3.

Exception: Drying shall not be required in order to use recycled refrigerants where water is the refrigerant, is used as an absorbent or is a deliberate additive. [ASHRAE 15:7.5.1.5]

1104.7.3 Reclaimed Refrigerants. Used refrigerants shall not be reused in a different owner's equipment unless tested and found to be in accordance with the requirements of AHRI 700. Contaminated refrigerants shall not be used unless reclaimed and is in accordance with AHRI 700. [ASHRAE 15:7.5.1.6]

1104.7.4 Mixing. Refrigerants, including refrigerant blends, with different designations as in accordance with Table 1102.3 shall not be mixed in a system.

Exception: Addition of a second refrigerant shall be permitted where specified by the equipment manufacturer to improve oil return at low temperatures. The refrigerant and amount added shall be in accordance with the manufacturer's instructions. [ASHRAE 15:7.5.1.7]

1104.8 Changing Refrigerants. A change in the type of refrigerant in a system shall not be made without notifying the Authority Having Jurisdiction, the user, and due observance of safety requirements. The refrigerant being considered shall be evaluated for suitability. [ASHRAE 15:5.3]

1105.0 General Requirements.

1105.1 Human Comfort. Cooling systems used for human comfort shall be in accordance with the return-air and outside-air provisions for furnaces in Section 904.7 and Section 904.8. Cooling equipment used for human comfort in dwelling units shall be selected to satisfy the calculated loads determined in accordance with the reference standards in Chapter 17 or other approved methods. Refrigerants used for human comfort shall be in accordance with Section 1104.6.

1105.2 Supports and Anchorage. Supports and anchorage for refrigeration equipment and piping shall be designed in accordance with the building code as Occupancy Category H (hazardous facilities). Supports shall be made of noncombustible materials.

Exceptions:

(1) Equipment containing Group A1 refrigerants shall be permitted to be supported by the same materials permitted for the building type.

(2) The use of approved vibration isolators specifically designed for the normal, wind, and seismic loads encountered, shall be permitted.

A compressor or portion of a condensing unit supported from the ground shall rest on a concrete or other approved base extending not less than 3 inches (76 mm) above the adjoining ground level.

1105.3 Access. An unobstructed readily accessible opening and passageway not less than 36 inches (914 mm) in width and 80 inches (2032 mm) in height shall be provided and maintained to the compressor, valves required by this chapter, or other portions of the system requiring routine maintenance.

Exceptions:

(1) Refrigerant evaporators, suspended overhead, shall be permitted to use portable means of access.

(2) Air filters, brine control or stop valves, fan motors or drives, and remotely de-energized electrical connections shall be permitted to be provided access to an unobstructed space not less than 30 inches (762 mm) in depth, width, and height. Where an access opening is immediately adjacent to these items and the equipment is capable of being serviced, repaired, and replaced from this opening, the dimensions shall be permitted to be reduced to 22 inches (559 mm) by 30 inches (762 mm) provided the largest piece of equipment is removed through the opening.

(3) Cooling equipment, using Group A1 refrigerants or brine, located in an attic or furred space shall be permitted to be provided access by a minimum opening and passageway thereto of not less than 22 inches (559 mm) by 30 inches (762 mm).

(4) Cooling or refrigeration equipment, using Group A1 or B1 refrigerants or brine, located on a roof or on an exterior wall of a building, shall be permitted to be provided access as for furnaces in Section 304.3.

1105.4 Illumination and Service Receptacles. In addition to the requirements of Section 301.4, permanent lighting fixtures shall be installed for equipment required by this code to be accessible or readily accessible. Such fixtures shall provide illumination to perform the required tasks for which access is provided. Control of the illumination source shall be provided at the access entrance.

Exceptions:

(1) Lighting fixtures shall be permitted to be omitted where the fixed lighting of the building will provide the required illumination.

(2) Equipment located on the roof or on the exterior walls of a building.

1105.5 Ventilation of Rooms Containing Condensing Units. Where not in a refrigerant machinery room, rooms or spaces in which a refrigerant-containing portion of a condensing unit is installed shall be provided with ventilation in accordance with Section 1105.5.1 or Section 1105.5.2. Ventilation for machinery rooms shall comply with Section 1106.0.

1105.5.1 Permanent Gravity Ventilation Openings. Permanent gravity ventilation openings of not less than 2 square feet (0.2 m^2) net free area opening shall be terminated directly to the outside of the building or extend to the outside of the building by continuous ducts.

1105.5.2 Mechanical Exhaust System. A mechanical exhaust system shall be designed to provide a complete change of air not less than every 20 minutes in such room or space and shall discharge to the outside of the building.

Exceptions:

(1) A condensing unit in a room or space where the cubical content exceeds 1000 cubic feet per horsepower (ft^3/hp) (37.95 m^3/kW) of the unit.

(2) A condensing unit in a room or space that has permanent gravity ventilation having an area of 2 square feet (0.2 m^2) or more to other rooms or openings exceeding 1000 ft^3/hp (37.95 m^3/kW).

1105.6 Prohibited Locations. Refrigeration systems or portions thereof shall not be located within a required exit enclosure. Refrigeration compressors exceeding 5 horsepowers (3.7 kW) rating shall be located not less than 10 feet (3048 mm) from an exit opening in a Group A; Group B; Group E; Group F; Group I; Group R, Division 1; or Group S Occupancy, unless separated by a one-hour fire-resistive occupancy separation.

1105.7 Condensate. Condensate from air-cooling coils shall be collected and drained to an approved location. Drain pans and coils shall be arranged to allow thorough drainage and access for cleaning. Where temperatures drop below freezing, heat tracing and insulation of condensate drains shall be installed.

1105.8 Defrost. Where defrost cycles are required for portions of the system, provisions shall be made for collection and disposal of the defrost liquid in a safe and sanitary manner.

1105.9 Overflows. Where condensate or defrost liquids are generated in an attic or furred space, and structural damage will result from overflow, provisions for overflow shall be provided.

1105.10 Condensate, Defrost, and Overflow Disposal. Disposal of condensate, defrost, or overflow discharges shall comply with Section 310.0.

1105.11 Refrigerant Port Protection. Air conditioning refrigerant circuit access ports located outdoors shall be protected from unauthorized access with locking-type tamper-resistant caps or in a manner approved by the Authority Having Jurisdiction.

Exception: Refrigerant ports in secure locations protected by walls or fencing and requiring key access.

1105.12 Storage. Refrigerants and refrigerant oils not charged within the refrigeration system shall be stored in accordance with Section 1105.12.1 and the fire code. Storage of materials in a refrigeration machinery room shall comply with the fire code.

1105.12.1 Quantity. The amount of refrigerant stored in a machinery room in containers not provided with relief valves and piping in accordance with Section 1113.0 shall not exceed 330 pounds (149.7 kg). Refrigerant shall be stored in approved storage containers. Additional quantities of refrigerant shall be stored in an approved storage facility. [ASHRAE 15:11.5]

1106.0 Refrigeration Machinery Rooms.

1106.1 Where Required. Refrigeration systems shall be provided with a refrigeration machinery room where the conditions as outlined in Section 1106.1.1 through Section 1106.1.4 exist.

Exception: Refrigeration equipment shall be permitted to be located outdoors in accordance with ASHRAE 15.

1106.1.1 Quantity. The quantity of refrigerant in a single, independent refrigerant circuit of a system exceeds the amounts of Table 1102.3.

1106.1.2 Equipment. Direct- and indirect-fired absorption equipment is used.

Exception: Direct and indirect-fired lithium bromide absorption systems using water as the refrigerant.

1106.1.3 A1 System. An A1 system having an aggregate combined compressor horsepower of 100 (74.6 kW) or more is used.

1106.1.4 A1 Refrigerant. The system contains other than a Group A1 refrigerant.

Exceptions:

(1) Lithium bromide absorption systems using water as the refrigerant.

(2) Systems containing less than 300 pounds (136.1 kg) of refrigerant R-123 and located in an approved exterior location.

Refrigeration machinery rooms shall house refrigerant-containing portions of the system other than the piping and evaporators permitted by Section 1104.4, discharge piping required of this chapter, and cooling towers regulated by Part II of this chapter, and their essential piping.

1106.2 Refrigeration Machinery Room, General Requirements. Where a refrigeration system is located indoors and a machinery room is required in accordance with Section 1106.1, the machinery room shall be in accordance with Section 1106.2.1 through Section 1106.2.5.2.

1106.2.1 Access. Machinery rooms shall not be prohibited from housing other mechanical equipment unless specifically prohibited elsewhere in this chapter. A machinery room shall be so dimensioned that parts are accessible with space for service, maintenance, and operations. There shall be clear head room of not less than 7.25 feet (2210 mm) below equipment situated over passageways. [ASHRAE 15:8.11.1]

1106.2.2 Openings. Each refrigeration machinery room shall have a tight-fitting door or doors opening outward, self-closing where they open into the building and adequate in number to ensure freedom for persons to escape in an emergency. With the exception of access doors and panels in air ducts and air-handling units in accordance with Section 1106.6, there shall be no openings that will permit passage of escaping refrigerant to other parts of the building. [ASHRAE 15:8.11.2]

1106.2.2.1 Detectors and Alarms. Each refrigeration machinery room shall contain one or more refrigerant detectors in accordance with Section 1106.2.2.2, located in areas where refrigerant from a leak will concentrate, that actuate an alarm and mechanical ventilation in accordance with Section 1106.2.4 at a set point not more than the corresponding Occupational Exposure Limit, OEL, in accordance with Table 1102.3, a set point determined in accordance with the OEL as defined in Chapter 2 shall be approved by the Authority Having Jurisdiction. The alarm shall annunciate visual and audible alarms inside the refrigeration machinery room and outside each entrance to the refrigeration machinery room. The alarms required in this section shall be of the manual reset type with the reset located inside the refrigeration machinery room. Alarms set at other levels, such as IDLH, and automatic reset alarms shall be permitted in addition to those required in accordance with this section. The meaning of each alarm shall be clearly marked by signage near the annunciator.

Exception: Refrigerant detectors are not required where only systems using R-718 (water) are located in the refrigeration machinery room.

1106.2.2.2 Refrigerant Detectors. Refrigerant detectors required in accordance with Section 1106.2.2.1 or Section 1107.1.7 shall meet all of the following conditions:

(1) The refrigerant detector shall perform automatic self-testing of sensors. Where a failure is detected, a trouble signal shall be activated.

(2) The refrigerant detector shall have one or more set points to activate responses in accordance with Section 1106.2.2.1 or Section 1107.1.7.

(3) The refrigerant detector as installed, including any sampling tubes, shall activate responses within a time not to exceed 30 seconds after exposure to refrigerant concentration exceeding the set point value specified in Section 1106.2.2.1 or Section 1107.1.7.

1106.2.3 Mechanical Ventilation. Machinery rooms shall be vented to the outdoors, utilizing mechanical ventilation in accordance with Section 1106.2.4 and Section 1106.2.5.

1106.2.4 Ventilation. Mechanical ventilation referred to in Section 1106.2.3 shall be by one or more power-driven fans capable of exhausting air from the machinery room at not less than the amount shown in accordance with Section 1106.2.5.

To obtain a reduced airflow for normal ventilation, multiple fans or multispeed fans shall be used. Provision shall be made to supply make-up air to replace that being exhausted. Ducts for supply and exhaust to the machinery room shall serve no other area. The makeup air supply locations shall be positioned relative to the exhaust air locations to avoid short-circuiting. Inlets to the exhaust ducts shall be located in an area where refrigerant from a leak will concentrate, in consideration of the location of the replacement supply air paths, refrigerating machines, and the density of the refrigerant relative to air.

Inlets to exhaust ducts shall be within 1 foot (305 mm) of the lowest point of the machinery room for refrigerants that are heavier than air, and shall be within 1 foot (305 mm) of the highest point for refrigerants that are lighter than air. The discharge of the exhaust air shall be to the outdoors in such a manner as not to cause a nuisance or danger.

1106.2.5 Emergency Ventilation-Required Airflow. An emergency ventilation system shall be required to exhaust an accumulation of refrigerant due to leaks or a rupture of the system. The emergency ventilation required shall be capable of removing air from the machinery room in not less than the airflow quantity in Section 1106.2.5.1 or Section 1106.2.5.2. Where multiple refrigerants are present, then the highest airflow quantity shall apply.

1106.2.5.1 Ventilation - A1, A2, A3, B1, B2L, B2 and B3 Refrigerants. The emergency ventilation for A1, A2, A3, B1, B2L, B2 and B3 refrigerants shall have the capacity to provide mechanical exhaust at a rate as determined in accordance with Equation 1106.2.5.1:

$$Q = 100\sqrt{G} \qquad \text{(Equation 1106.2.5.1)}$$

Where:

Q = Air flow rate, cubic feet per minute.

G = Refrigerant mass in largest system, pounds.

For SI units: 1 cubic foot per minute = 0.00047 m^3/s, 1 pound = 0.453 kg

1106.2.5.2 Ventilation - Group A2L Refrigerants. The emergency ventilation for A2L refrigerants shall have the capacity to provide mechanical exhaust at a rate determined in accordance with Table 1106.2.5.2:

TABLE 1106.2.5.2
REQUIRED AIRFLOW FOR GROUP A2L REFRIGERANTS

REFRIGERANT	MINIMUM AIR FLOW* (CFM)
R32	32 500
R143a	28 600
R444A	13 700
R444B	22 400
R445A	16 400
R446A	50 500
R447A	50 200
R447B	29 600
R451A	14 900
R451B	14 900
R452B	31 500
R454A	4290
R454B	6650
R454C	32 800
R455A	4770
R457A	31 400
R1234yf	16 500
R1234zeE	12 600

For SI units: 1 cubic foot per minute = 0.00047 m^3/s

* The values were tabulated from the following equation:

$Q_{A2L} \geq [(\rho \cdot v \cdot A)/(LFL \cdot 0.50)]$ (Equation 1106.2.5.2)

Where:

ρ = Refrigerant density, pounds per cubic feet (kg/m^3).

v = Refrigerant velocity equal to the refrigerant acoustic velocity (speed of sound), feet per second (m/s).

A = Cross-section flow area of refrigerant leak, square feet (m^2), A = 0.00136 ft^2 (0.000126 m^2).

LFL = Lower Flammability Limit, or $ETFL_{60}$ where no LFL exist, published value in accordance with ASHRAE 34.

Q_{A2L} = Minimum required air flow rate, conversion to other units of measures is permitted, cubic feet per second (m^3/s).

For exact ventilation rates and for refrigerants not listed, the ventilation rate shall be calculated using this equation.

1106.3 Normal Operation. A part of the refrigeration machinery room mechanical ventilation shall be in accordance with the following:

(1) Operated, where occupied, to supply not less than 0.5 cfm/ft^2 (2.54 L/s/m^2) of machinery room area or 20 cubic feet per minute (9.44 L/s) per person.

(2) Operable, where occupied at a volume required to not exceed the higher of a temperature rise of 18°F (10°C) above inlet air temperature or a maximum temperature of 122°F (50°C).

1106.4 Natural Ventilation. Where a refrigerating system is located outdoors more than 20 feet (6096 mm) from buildings opening and is enclosed by a penthouse, lean-to, or other open structure, natural or mechanical ventilation shall be provided. The requirements for such natural ventilation shall be in accordance with the following:

(1) The free-aperture cross section for the ventilation of a machinery room shall be not less than as determined in accordance with Equation 1106.4.

$$F = \sqrt{G} \quad \text{[Equation 1106.4]}$$

Where:

F = The free opening area, square feet.

G = The mass of refrigerant in the largest system, any part of which is located in the machinery room, pounds.

For SI units: 1 cubic foot per minute = 0.00047 m^3/s, 1 pound = 0.453 kg

(2) The location of the gravity ventilation openings shall be based on the relative density of the refrigerant to air. [ASHRAE 15:8.11.5(a), (b)]

1106.5 Combustion Air. No open flames that use combustion air from the machinery room shall be installed where refrigerant is used. Combustion equipment shall not be installed in the same machinery room with refrigerant-containing equipment except under one of the following conditions:

(1) Combustion air shall be ducted from outside the machinery room and sealed in such a manner as to prevent refrigerant leakage from entering the combustion chamber.

(2) A refrigerant detector, that is in accordance with Section 1106.2.2.1, shall be installed to automatically shut down the combustion process in the event of refrigerant leakage.

Exception: Machinery rooms where carbon dioxide (R-744) or water (R-718) is the refrigerant.

1106.6 Airflow. There shall be no airflow to or from an occupied space through a machinery room unless the air is ducted and sealed in such a manner as to prevent a refrigerant leakage from entering the airstream. Access doors and panels in ductwork and air-handling units shall be gasketed and tight fitting. [ASHRAE 15:8.11.7]

1106.7 Ventilation Intake. Makeup air intakes to replace the exhaust air shall be provided to the refrigeration machinery room directly from outside the building. Intakes shall be located as required by other sections of the code and fitted with backdraft dampers or other approved flow-control means to prevent reverse flow. Distribution of makeup air shall be arranged to provide thorough mixing within the refrigeration machinery room to prevent short circuiting of the makeup air directly to the exhaust.

1106.8 Maximum Temperature. Ventilation or mechanical cooling systems shall be provided to maintain a temperature of not more than 104°F (40°C) in the refrigerant machinery room under design load and weather conditions.

1106.9 Refrigerant Parts in Air Duct. Joints and refrigerant-containing parts of a refrigerating system located in an air duct carrying conditioned air to and from an occupied space shall be constructed to withstand a temperature of 700°F (371°C) without leakage into the airstream. [ASHRAE 15:8.8]

1106.10 Dimensions. Refrigeration machinery rooms shall be of such dimensions that system parts are readily accessible with approved space for maintenance and operations. An unobstructed walking space not less than 36 inches (914 mm) in width and 80 inches (2032 mm) in height shall be maintained throughout, allowing free access to not less than two sides of moving machinery and approaching each stop valve. Access to refrigeration machinery rooms shall be restricted to authorized personnel and posted with a permanent sign.

1106.11 Restricted Access. Access to the refrigeration machinery room shall be restricted to authorized personnel. Doors shall be clearly marked or permanent signs shall be posted at each entrance to indicate this restriction. [ASHRAE 15:8.11.8]

1106.12 Exits. Exits shall comply with the building code for special hazards.

1107.0 Machinery Room, Special Requirements.

1107.1 General. In cases specified in the rules of Section 1106.1, a refrigeration machinery room shall comply with the special requirements in accordance with Section 1107.1.1 through Section 1107.1.10, in addition to Section 1106.2.

1107.1.1 Flame-Producing Devices. There shall be no flame-producing device or continuously operating hot surface over 800°F (427°C) permanently installed in the room.

1107.1.2 Doors. Doors communicating with the building shall be approved, self-closing, tight-fitting fire doors.

1107.1.3 Walls, Floors, and Ceilings. Walls, floor, and ceiling shall be tight and of noncombustible construction. Walls, floor, and ceiling separating the refrigeration machinery room from other occupied spaces shall be not less than one-hour fire-resistive construction.

1107.1.4 Machinery Rooms. The refrigeration machinery room shall have a door that opens directly to the outdoors or through a vestibule equipped with self-closing, tight-fitting doors.

1107.1.5 Exterior Openings. Exterior openings, where present, shall not be under a fire escape or an open stairway.

1107.1.6 Sealing. All pipes piercing the interior walls, ceiling, or floor of such rooms shall be tightly sealed to the walls, ceiling, or floor through which they pass.

1107.1.7 Group A2L and B2L Refrigerants. Where refrigerant of Groups A2L or B2L are used, the requirements of Class 1, Division 2, of *the California Electrical Code,* shall not apply to the machinery room provided that the conditions in Section 1107.1.7.1 through Section 1107.1.7.3 are met.

1107.1.7.1 Mechanical Ventilation. The mechanical ventilation system in the machinery room is run continuously in accordance with Section 1106.2.5 and failure of the mechanical ventilation system actuates an alarm, or the mechanical ventilation system in the machinery room is activated by one or more refrigerant detectors, in accordance with the requirements of Section 1106.2.2.1 and Section 1106.2.2.2.

1107.1.7.2 Refrigeration Detectors. For the refrigerant detection required in Section 1106.2.2.1, detection of refrigerant concentration that exceeds 25 percent of the LFL or the upper detection limit of the refrigerant detector, whichever is lower, shall automatically de-energize the following equipment in the machinery room:

(a) refrigerant compressors

(b) refrigerant pumps

(c) normally-closed automatic refrigerant valves

1107.1.7.3 Machinery Rooms. The machinery room shall comply with Section 1107.1.8.

1107.1.8 Group A2, A3, B2, or B3 Refrigerants. Where any refrigerant of Groups A2, A3, B2, or B3 are used, the machinery room shall comply with Class 1, Division 2, of *the California Electrical Code.*

1107.1.9 Refrigeration Systems. As part of the mechanical ventilation system in accordance with Section 1106.2.4, refrigeration systems that contain more than 110 pounds (50 kg) of any Group A2L, A2, A3, B2L, B2, or B3, refrigerant shall have not less than one exhaust air inlet located adjacent to each system not more than 9 feet (3 m) away.

1107.1.10 Remote Control. Remote control of the mechanical equipment in the refrigeration machinery room shall be provided immediately outside the machinery room door solely for the purpose of shutting down the equipment in an emergency. Ventilation fans shall be on a separate electrical circuit and have a control switch located immediately outside the machinery room door.

1108.0 Refrigeration Machinery Room Equipment and Controls.

1108.1 General. Equipment, piping, ducts, vents, or similar devices that are not essential for the refrigeration process, maintenance of the equipment, or for the illumination, ventilation, or fire protection of the room shall not be placed in or pass through a refrigeration machinery room.

1108.2 Electrical. Electrical equipment and installations shall comply with the electrical code. The refrigeration machinery room shall not be classified as a hazardous location except as provided in Section 1107.1.7 or Section 1107.1.8.

1108.3 Emergency Shut-off. A clearly identified emergency shut-off switch of the break-glass type or with an approved tamper-resistant cover shall be provided immediately adjacent to and outside of the principal refrigeration machinery room entrance. The switch shall provide off-only control of refrigerant compressors, refrigerant pumps, and normally-closed automatic refrigerant valves located in the machinery room. For other than A1 and B1 refrigerants, emergency shutoff shall be automatically activated by refrigerant Alarm 2 in accordance with Section 1106.2.2.1.

1108.4 Installation, Maintenance, and Testing. Detection and alarm systems in accordance with Section 1106.2.2.1 shall be installed, maintained, and tested in accordance with the fire code *and with the equipment manufacturer's specification.*

1108.5 Emergency Pressure Control System. Where required by the fire code, an emergency pressure control system shall be installed in accordance with applicable fire code requirements.

1109.0 Refrigeration Piping, Containers, and Valves.

1109.1 Materials. Materials used in the construction and installation of refrigerating systems shall be compatible with the conveying refrigerant used. Materials shall not be used that will deteriorate due to the chemical action of the refrigerant, lubricant, or combination of both where exposed to air or moisture to a degree that poses a safety hazard. [ASHRAE 15:9.1.1] Refrigerant piping shall be metallic.

1109.1.1 Copper and Copper Alloy Pipe. Copper and copper alloy refrigeration piping, valves, fittings, and related parts used in the construction and installation of refrigeration systems shall be approved for the intended use. Refrigeration piping shall comply with ASME B31.5.

1109.1.2 Copper Linesets. Copper linesets shall comply with ASTM B280 or ASTM B1003.

» **1109.1.3 Iron and Steel.** Iron and steel refrigeration piping, valves, fittings, and related parts shall be approved for the intended use. Pipe exceeding 2 inches (50 mm) iron pipe size shall be electric-resistance welded or seamless pipe. Refrigeration piping shall comply with ASME B31.5.

» **1109.1.4 Prohibited Contact.** Aluminum, zinc, magnesium, or their alloys shall not be used in contact with methyl chloride. Magnesium alloys shall not be used where in contact with halogenated refrigerants. [ASHRAE 15:9.1.2]

1109.2 Joints. Iron or steel pipe joints shall be of approved threaded, flanged, or welded types. Exposed threads shall be tinned or coated with an approved corrosion inhibitor. Copper or copper alloy pipe joints of iron pipe size shall be of approved threaded, flanged, press-connect or brazed types. Copper tubing joints and connections shall be connected by approved flared, lapped, swaged, or brazed joints, soldered joints, or mechanical joints that comply with UL 207 either individually or as part of an assembly or a system by an approved nationally recognized laboratory. Piping and tubing shall be installed so as to prevent vibration and strains at joints and connections.

1109.3 Penetration of Piping. Refrigerant piping shall not penetrate floors, ceilings, or roofs.

Exceptions:

(1) Penetrations connecting the basement and the first floor.

(2) Penetrations connecting the top floor and a machinery penthouse or roof installation.

(3) Penetrations connecting adjacent floors served by the refrigeration system.

(4) Penetrations of a direct system where the refrigerant concentration does not exceed that listed in Table 1102.3 for the smallest occupied space through which the refrigerant piping passes.

(5) In other than industrial occupancies and where the refrigerant concentration exceeds that listed in Table 1102.3 for the smallest occupied space, penetrations that connect separate pieces of equipment that are in accordance with one of the following:

  (a) Enclosed by an approved gastight, fire-resistive duct or shaft with openings to those floors served by the refrigerating system.

  (b) Located on the exterior wall of a building where vented to the outdoors or to the space served by the system and not used as an air shaft, closed court, or similar space. [ASHRAE 15:8.10.3]

1109.4 Location of Refrigeration Piping. Refrigerant piping crossing an open space that affords passageway in a building shall be not less than 7.25 feet (2210 mm) above the floor unless the piping is located against the ceiling of such space and is permitted by the Authority Having Jurisdiction. [ASHRAE 15:8.10.1]

1109.4.1 Protection from Mechanical Damage. Passages shall not be obstructed by refrigerant piping. Refrigerant piping shall not be located in an elevator, dumbwaiter, or other shaft containing a moving object, or in a shaft that has openings to living quarters, or to means of egress. Refrigerant piping shall not be installed in an enclosed public stairway, stair landing, or means of egress. [ASHRAE 15:8.10.2]

1109.5 Underground Piping. Refrigerant piping placed underground shall be protected against corrosion.

1109.5.1 Piping in Concrete Floors. Refrigerant piping installed in concrete floors shall be encased in a pipe duct. Refrigerant piping shall be isolated and supported to prevent damaging vibration, stress, or corrosion. [ASHRAE 15:8.10.4]

1109.6 Support. In addition to the requirements of Section 1105.2, piping and tubing shall be securely fastened to a permanent support within 6 feet (1829 mm) following the first bend in such tubing from the compressor and within 2 feet (610 mm) of each subsequent bend or angle. Piping and tubing shall be supported at points not more than 15 feet (4572 mm) apart.

1109.7 Pipe Enclosure. Refrigerant piping and tubing shall be installed so that it is not subject to damage from an external source. Soft annealed copper tubing shall not exceed 1⅜ inches (35 mm) nominal size. Mechanical joints, other than approved press-connect joints, shall not be made on tubing exceeding ¾ of an inch (20 mm) nominal size. Soft annealed copper tubing conveying refrigerant shall be enclosed in iron or steel piping and fittings, or in conduit, molding, or raceway that will protect the tubing against mechanical injury from an exterior source.

Exceptions:

(1) Tubing entirely within or tubing within 5 feet (1524 mm) of a refrigerant compressor where so located that it is not subject to external injury.

(2) Copper tubing serving a dwelling unit, where such tubing contains Group A1 refrigerant and is placed in locations not subject to damage from an external source.

1109.8 Visual Inspection. Refrigerant piping and joints erected on the premises shall be exposed to view for visual inspection prior to being covered or enclosed.

Exception: Copper tubing enclosed in iron or steel piping conduit, molding, or raceway, provided there are no fittings or joints concealed therein.

1109.9 Condensation. Piping and fittings that convey brine, refrigerant, or coolants that during normal operation are capable of reaching a surface temperature below the dew point of the surrounding air and that are located in spaces or areas where condensation will cause a hazard to the building occupants or damage to the structure, electrical or other equipment shall be protected to prevent such damage.

1109.10 Identification. Piping shall be in accordance with the reference standard for identification. The type of refrigerant, function and pressure shall be indicated.

1110.0 Valves.

1110.1 More than 6.6 Pounds of Refrigerant. Systems containing more than 6.6 pounds (2.99 kg) of refrigerant shall have stop valves installed at the following locations:

(1) The suction inlet of a compressor, compressor unit, or condensing unit.

(2) The discharge of a compressor, compressor unit, or condensing unit.

(3) The outlet of a liquid receiver.

Exceptions:

(1) Systems that have a refrigerant pumpout function capable of storing the refrigerant charge, or are equipped with the provisions for pumpout of the refrigerant.

(2) Self-contained systems. [ASHRAE 15:9.12.4]

1110.2 More than 110 Pounds of Refrigerant. Systems containing more than 110 pounds (49.9 kg) of refrigerant shall have stop valves installed at the following locations:

(1) The suction inlet of a compressor, compressor unit, or condensing unit.

(2) The discharge outlet of a compressor, compressor unit, or condensing unit.

(3) The inlet of a liquid receiver, except for self-contained systems or where the receiver is an integral part of the condenser or condensing unit.

(4) The outlet of a liquid receiver.

(5) The inlets and outlets of condensers where more than one condenser is used in parallel in the systems.

Exception: Systems that have a refrigerant pumpout function capable of storing the refrigerant charge, or are equipped with the provisions for pumpout of the refrigerant or self-contained systems. [ASHRAE 15:9.12.5]

1110.3 Support. Stop valves installed in copper refrigerant lines of ¾ of an inch (20 mm) or less outside diameter shall be supported independently of the tubing or piping.

1110.4 Access. Stop valves required by Section 1110.0 shall be readily accessible from the refrigeration machinery room floor or a level platform.

1110.5 Identification. Stop valves shall be identified by tagging in accordance with the reference standard for identification. A valve chart shall be mounted under glass at an approved location near the principal entrance to a refrigeration machinery room.

1111.0 Pressure-Limiting Devices.

1111.1 Where Required. Pressure-limiting devices complying with Section 1111.2 through Section 1111.4 shall be provided for compressors on all systems operating above atmospheric pressure.

Exception: Pressure limiting devices are not required for listed factory-sealed systems containing less than 22 pounds (9.9 kg) of Group A1 refrigerant. [ASHRAE 15:9.9.1]

1111.2 Setting. Pressure limiting devices shall be set in accordance with one the following:

(1) For positive displacement compressors:

  (a) When systems are protected by a highside pressure relief device, the compressor's pressure limiting device shall be set not more than 90 percent of the operating pressure for the highside pressure relief device.

  (b) When systems are not protected by a highside pressure relief device, the compressor's pressure limiting device shall be set not more than the system's highside design pressure.

(2) For nonpositive displacement compressors:

  (a) When systems are protected by a highside pressure relief device, the compressor's pressure limiting device shall be set not more than 90 percent of the operating pressure for the highside pressure relief device.

  (b) When systems are protected by a lowside pressure relief device that is only subject to lowside pressure, and is provided with a permanent relief path between the systems' highside and lowside, without intervening valves, the compressor's pressure limiting device shall be set not more than the systems' highside design pressure. [ASHRAE 15:9.9.2]

1111.3 Location. Stop valves shall not be installed between the pressure imposing element and pressure limiting devices serving compressors. [ASHRAE 15:9.9.3]

1111.4 Emergency Stop. Activation of a pressure limiting device shall stop the action of the pressure imposing element. [ASHRAE 15:9.9.4]

1112.0 Pressure-Relief Devices.

1112.1 General. Refrigeration systems shall be protected by a pressure-relief device or other approved means to safely relieve pressure due to fire or abnormal conditions. [ASHRAE 15:9.4.1]

1112.2 Positive Displacement Compressor. A positive displacement compressor with a stop valve in the discharge connection shall be equipped with a pressure-relief device that is sized, and with a pressure setting, in accordance with the compressor manufacturer to prevent rupture of the compressor or to prevent the pressure from increasing to more than 10 percent above the maximum allowable working pressure of components located in the discharge line between the compressor and the stop valve or in accordance with Section 1113.5, whichever is larger. The pressure-relief device shall discharge into the low-pressure side of the system or in accordance with Section 1112.10.

Exception: Hermetic refrigerant motor-compressors that are listed and have a displacement not more than 50 cubic feet per minute (1.42 m^3/min).

The relief device(s) shall be sized based on compressor flow at the following conditions:

(1) For compressors in single-stage systems and high-stage compressors of other systems, the flow shall be calculated based on 50°F (10°C) saturated suction temperature at the compressor suction.

(2) For low-stage or booster compressors in compound systems, the compressors that are capable of running only where discharging to the suction of a high-stage compressor, the flow shall be calculated based on the saturated suction temperature equal to the design operating intermediate temperature.

(3) For low-stage compressors in cascade systems, the compressors that are located in the lower-temperature stage(s) of cascade systems, the flow shall be calculated based on the suction pressure being equal to the pressure setpoint of the pressure-relieving devices that protect the lowside of the stage against overpressure.

Exceptions: For Section 1112.2(1), Section 1112.2(2), and Section 1112.2(3), the discharge capacity of the relief device shall be permitted to be the minimum regulated flow rate of the compressor where the following conditions are met:

(1) The compressor is equipped with capacity regulation.

(2) Capacity regulation actuates to a flow at not less than 90 percent of the pressure-relief device setting.

(3) A pressure-limiting device is installed and set in accordance with the requirements of Section 1111.0. [ASHRAE 15:9.8]

1112.3 Liquid-Containing Portions of Systems. Liquid-containing portions of systems, including piping, that is isolated from pressure-relief devices required elsewhere, and that develops pressures exceeding their working design pressures due to temperature rise, shall be protected by the installation of pressure-relief devices.

1112.4 Evaporators. Heat exchanger coils located downstream, or upstream within 18 inches (457 mm), of a heating source and capable of being isolated shall be fitted with a pressure-relief device that discharges to another part of the system in accordance with Section 1112.5 through Section 1112.5.2 or outside any enclosed space in accordance with Section 1112.10. The pressure relief device shall be connected at the highest possible location of the heat exchanger or piping between the heat exchanger and its manual isolation valves.

Exceptions:

(1) Relief valves shall not be required on heat exchanger coils that have a design pressure more than 110 percent of refrigerant saturation pressure when exposed to the maximum heating source temperature.

(2) A relief valve shall not be required on self-contained or unit systems where the volume of the lowside of the system, which is shut off by valves, is more than the specific volume of the refrigerant at critical conditions of temperature and pressure, as determined in accordance with Equation 1112.4.

$$V_1 / [W_1 - (V_2 - V_1) / V_{gt}] \qquad \text{(Equation 1112.4)}$$

Shall be more than V_{gc}

Where:

V_1 = Lowside volume, cubic foot (m^3).

V_2 = Total volume of system, cubic foot (m^3).

W_1 = Total weight of refrigerant in system, pounds (kg).

V_{gt} = Specific volume of refrigerant vapor at 110°F (43°C), cubic feet per pound (m^3/kg).

V_{gc} = Specific volume at critical temperature and pressure, cubic feet per pound (m^3/kg). [ASHRAE 15:9.4.4]

1112.5 Hydrostatic Expansion. Pressure rise resulting from hydrostatic expansion due to temperature rise of liquid refrigerant trapped in or between closed valves shall be addressed in accordance with Section 1112.5.1 and Section 1112.5.2. [ASHRAE 15:9.4.3]

1112.5.1 Hydrostatic Expansion During Normal Operation. Where trapping of liquid with subsequent hydrostatic expansion is capable of occurring automatically during normal operation or during standby, shipping, or power failure, engineering controls shall be used that are capable of preventing the pressure from exceeding the design pressure. Acceptable engineering controls include but are not limited to the following:

(1) Pressure relief device to relieve hydrostatic pressure to another part of the system.

(2) Reseating pressure relief valve to relieve the hydrostatic pressure to an approved treatment system. [ASHRAE 15:9.4.3.1]

1112.5.2 Hydrostatic Expansion During Maintenance. Where trapping of liquid with subsequent hydrostatic expansion is capable of occurring only during maintenance—i.e., when personnel are performing maintenance tasks—either engineering or administrative controls shall be used to relieve or prevent the hydrostatic overpressure. [ASHRAE 15:9.4.3.2]

1112.6 Actuation. Pressure-relief devices shall be direct-pressure actuated or pilot operated. Pilot-operated pressure-relief valves shall be self-actuated, and the main valve shall open automatically at the set pressure and, where an essential part of the pilot fails, shall discharge its full rated capacity. [ASHRAE 15:9.4.5]

» **1112.7 Stop Valves Prohibited.** Stop valves shall not be located between a pressure-relief device and parts of the system protected thereby. A three-way valve, used in conjunction with the dual relief valve in accordance with Section 1113.6, shall not be considered a stop valve. [ASHRAE 15:9.4.6]

» **1112.8 Location.** Pressure-relief devices shall be connected directly to the pressure vessel or other parts of the system protected thereby. These devices shall be connected above the liquid refrigerant level and installed so that they are accessible for inspection and repair, and so that they are not capable of being readily rendered inoperative.

Exception: Where fusible plugs are used on the highside, they shall be located above or below the liquid refrigerant level. [ASHRAE 15:9.4.8]

» **1112.9 Materials.** The seats and discs of pressure-relief devices shall be constructed of compatible material to resist refrigerant corrosion or other chemical action caused by the refrigerant. Seats or discs of cast iron shall not be used. Seats and discs shall be limited in distortion, by pressure or other cause, to a set pressure change of not more than 5 percent in a span of five years. [ASHRAE 15:9.4.9]

» **1112.10 Pressure-Relief Device Settings.** Pressure-relief valves shall start to function at a pressure not exceeding the design pressure of the parts of the system protected.

Exception: Relief valves that discharge into other parts of the system shall comply with Section 1112.10.1. [ASHRAE 15:9.5.1]

» **1112.10.1 Rupture Member Setting.** Rupture members used in lieu of, or in series with, a relief valve shall have a nominal rated rupture pressure not exceeding the design pressure of the parts of the system protected. The conditions of application shall comply with ASME BPVC Section VIII. The size of rupture members installed ahead of relief valves shall not be less than the relief-valve inlet. [ASHRAE 15:9.5.2]

» **1112.11 Discharge from Pressure-Relief Devices.** Pressure-relief systems designed for vapor shall comply with Section 1112.11.1 through Section 1112.11.4.1.

» **1112.11.1 Discharging Location Interior to Building.** Pressure-relief devices, including fusible plugs, serving refrigeration systems shall be permitted to discharge to the interior of a building where in accordance with the following:

(1) The system contains less than 110 pounds (49.9 kg) of a Group A1 refrigerant.

(2) The system contains less than 6.6 pounds (2.99 kg) of a Group A2, B1 or B2 refrigerant.

(3) The system does not contain any quantity of a Group A3 or B3 refrigerant.

(4) The system is not required to be installed in a machinery room in accordance with Section 1106.0.

(5) The refrigerant concentration limits in Section 1104.0 are not exceeded. Refrigeration systems that do not comply with the above requirements shall comply with the requirements of Section 1112.11.2 through Section 1112.11.4. [ASHRAE 15:9.7.8.1]

1112.11.2 Discharging Location Exterior to Building. Pressure-relief devices designed to discharge external to the refrigeration system shall be arranged to discharge outside of a building and shall be in accordance with the following: «

(1) The point of vent discharge shall be located not less than 15 feet (4572 mm) above the adjoining ground level.

Exception: Outdoor systems containing Group A1 refrigerant shall be permitted to discharge at any elevation where the point of discharge is located in an access-controlled area accessible to authorized personnel only.

(2) The point of vent discharge shall be located not less than 20 feet (6096 mm) from windows, building ventilation openings, pedestrian walkways, or building exits.

(3) For heavier-than-air refrigerants, the point of vent discharge shall be located not less than 20 feet (6096 mm) horizontally from below-grade walkways, entrances, pits or ramps where a release of the entire system charge into such a space would yield a concentration of refrigerant in excess of the RCL. The direct discharge of a relief vent into enclosed outdoor spaces, such as a courtyard with walls on all sides, shall not be permitted where a release of the entire system charge into such a space would yield a concentration of refrigerant in excess of the RCL. The volume for the refrigerant concentration calculation shall be determined using the gross area of the space and a height of 8.2 feet (2499 mm), regardless of the actual height of the enclosed space.

(4) The termination point of a vent discharge line shall be made in a manner that prevents discharged refrigerant from spraying directly onto personnel that are capable of being in the vicinity.

(5) The termination point of vent discharge line shall be made in a manner that prevents foreign material or debris from entering the discharge piping.

(6) Relief vent lines that terminate vertically upward and are subject to moisture entry shall be provided with a drip pocket having a length of not less than 24 inches (610 mm) and having the size of the vent discharge pipe. The drip pocket shall be installed to extend below the first change in vent pipe direction and shall be fitted with a valve or drain plug to permit removal of accumulated moisture. [ASHRAE 15:9.7.8.2]

1112.11.3 Internal Relief. Pressure-relief valves designed to discharge from a higher-pressure vessel into «

a lower pressure vessel internal to the system shall comply with the following:

(1) The pressure-relief valve that protects the higher-pressure vessel shall be selected to deliver capacity in accordance with Section 1113.5 without exceeding the maximum allowable working pressure of the higher-pressure vessel accounting for the change in mass flow capacity due to the elevated backpressure.

(2) The capacity of the pressure-relief valve protecting the part of the system receiving a discharge from a pressure-relief valve protecting a higher-pressure vessel shall be not less than the sum of the capacity required in Section 1113.5 plus the mass flow capacity of the pressure-relief valve discharging into that part of the system.

(3) The design pressure of the body of the relief valve used on the higher-pressure vessel shall be rated for operation at the design pressure of the higher-pressure vessel in both pressure-containing areas of the valve. [ASHRAE 15:9.7.8.3]

1112.11.4 Discharge Location, Special Requirements. Additional requirements for relief device discharge location and allowances shall apply for specific refrigerants in accordance with Section 1112.11.4.1. [ASHRAE 15:9.7.8.4]

1112.11.4.1 Water (R-718). Where water is the refrigerant, discharge to a floor drain shall be permitted where the following conditions are met:

(1) The pressure-relief device set pressure shall not exceed 15 psig (103 kPa).

(2) The floor drain shall be sized to handle the flow rate from a single broken tube in a refrigerant-containing heat exchanger.

(3) The Authority Having Jurisdiction finds it acceptable that the working fluid, corrosion inhibitor, and other additives used in this type of refrigeration system are permitted to infrequently be discharged to the sewer system, or a catch tank that is sized to handle the expected discharge shall be installed and equipped with a normally closed drain valve and an overflow line to drain. [ASHRAE 15:9.7.8.4.1]

1112.12 Discharge Piping. The piping used for pressure-relief device discharge shall be in accordance with Section 1112.12.1 through Section 1112.12.5. [ASHRAE 15:9.7.9]

1112.12.1 Piping Connection. Piping connected to the discharge side of a fusible plug or rupture member shall have provisions to prevent plugging of the pipe upon operation of a fusible plug or rupture member. [ASHRAE 15:9.7.9.1]

1112.12.2 Pipe Size. The size of the discharge pipe from the pressure-relief device or fusible plug shall be not less than the outlet size of the pressure-relief device or fusible plug. [ASHRAE 15:9.7.9.2]

1112.12.3 Maximum Length. The maximum length of the discharge piping installed on the outlet of pressure-relief devices and fusible plugs discharging to the atmosphere shall be determined in accordance with Section 1112.12.4 and Section 1112.12.5. See Table 1112.12.3 for the allowable flow capacity of various equivalent lengths of single discharge piping vents for conventional pressure-relief valves. [ASHRAE 15:9.7.9.3]

TABLE 1112.12.3
ATMOSPHERIC PRESSURE AT NOMINAL INSTALLATION ELEVATION (P_a)
[ASHRAE 15: TABLE 9.7.9.3.2]

ELEVATION ABOVE SEA LEVEL, FEET	POUNDS PER SQUARE INCH, ABSOLUTE (P_a)
0	14.7
500	14.4
1000	14.2
1500	13.9
2000	13.7
2500	13.4
3000	13.2
3500	12.9
4000	12.7
4500	12.5
5000	12.2
6000	11.8
7000	11.3
8000	10.9
9000	10.5
1000	10.1

For SI units: 1 foot = 304.8 mm, 1 pound-force per square inch = 6.8947 kPa

1112.12.4 Design Back Pressure. The design back pressure due to flow in the discharge piping at the outlet of pressure-relief devices and fusible plugs, discharging to atmosphere, shall be limited by the allowable equivalent length of piping determined in accordance with Equation 1112.12.4(1).

[Equation 1112.12.4(1)]

$$L = \frac{0.2146 \cdot d^5 (P_0^2 - P_2^2)}{f \cdot C_r^2} - \frac{d \cdot \ln(\frac{P_0}{P_2})}{6 \cdot f}$$

Where:

L = Equivalent length of discharge piping, feet.

Cr = Rated capacity as stamped on the relief device in pounds per minute (lb/min), or in SCFM multiplied by 0.0764, or as calculated in Section 1112.14 for a rupture member or fusible plug, or as adjusted for reduced capacity due to piping in accordance with the manufacturer of the device, or as adjusted for reduced capacity due to piping as estimated by an approved method.

f = Moody friction factor in fully turbulent flow.

d = Inside diameter of pipe or tube, inches.

ln = Natural logarithm.

P_2 = Absolute pressure at outlet of discharge piping, psia.

P_0 = Allowed back pressure (absolute) at the outlet of pressure relief device, (psia).

For SI units: 1 foot = 304.8 mm, 1 pound-force per square inch = 6.8947 kPa, 1 pound per minute = 0.00756 kg/s

Unless the maximum allowable back pressure (P_0) is specified by the relief valve manufacturer, the following maximum allowable back pressure values shall be used for P_0, where P is the set pressure and P_a is atmospheric pressure at the nominal elevation of the installation (see Table 1112.12.3):

For conventional relief valves: 15 percent of set pressure:

$$P_0 = (0.15 \cdot P) + P_a \quad \text{[Equation 1112.12.4(2)]}$$

For balanced relief valves: 25 percent of set pressure:

$$P_0 = (0.25 \cdot P) + P_a \quad \text{[Equation 1112.12.4(3)]}$$

For rupture disks alone: fusible plugs, and pilot operated relief devices, 50 percent of set pressure:

$$P_0 = (0.50 \cdot P) + P_a \quad \text{[Equation 1112.12.4(4)]}$$

For fusible plugs, P shall be the saturated absolute pressure for the stamped temperature melting point of the fusible plug or the critical pressure of the refrigerant used, whichever is smaller. [ASHRAE 15:9.7.9.3.1, 9.7.9.3.2]

» **1112.12.5 Simultaneous Operation.** Where outlets of two or more relief devices or fusible plugs, which are expected to operate simultaneously, connect to a common discharge pipe, the common pipe shall be sized large enough to prevent the outlet pressure at each relief device from exceeding the maximum allowable outlet pressure in accordance with Section 1112.12.4. [ASHRAE 15:9.7.9.3.3]

» **1112.13 Rating of Pressure-Relief Device.** The rated discharge capacity of a pressure-relief device expressed in pounds of air per minute (kg/s), shall be determined in accordance with ASME BPVC Section VIII. Pipe and fittings between the pressure-relief valve and the parts of the system it protects shall have not less than the area of the pressure-relief valve inlet area. [ASHRAE 15:9.7.6]

» **1112.14 Rating of Rupture Members and Fusible Plugs.** The rated discharge capacity of a rupture member or fusible plug discharging to atmosphere under critical flow conditions, in pounds of air per minute (kg/s), shall be determined in accordance with the following formulas:

$$C = 0.64 P_1 d^2 \quad \text{[Equation 1112.14(1)]}$$

$$d = 1.25 \sqrt{C/P_1} \quad \text{[Equation 1112.14(2)]}$$

Where:

C = Rated discharge capacity of air, pounds per minute.

d = Smallest internal diameter of the inlet pipe, retaining flanges, fusible plug, or rupture member; inches.

For rupture members:

[Equation 1112.14(3)]

P_1 = (rated pressure in psig x 1.1) + 14.7

For fusible plugs:

P_1 = Absolute saturation pressure, corresponding to the stamped temperature melting point of the fusible plug or the critical pressure of the refrigerant used, whichever is smaller, pound-force per square inch atmosphere, psia. [ASHRAE 15:9.7.7]

For SI units: 1 inch = 25.4 mm, 1 pound-force per square inch = 6.8947 kPa, 1 pound per minute = 0.00756 kg/s

1113.0 Overpressure Protection.

1113.1 General. Pressure vessels shall be provided with overpressure protection in accordance with ASME BPVC Section VIII. Pressure vessels containing liquid refrigerant that are capable of being isolated by stop valves from other parts of the refrigerating system shall be provided with overpressure protection. Pressure-relief devices or fuse plugs shall be sized in accordance with Section 1113.5. [ASHRAE 15:9.7.1, 9.7.2]

1113.2 Type of Protection. Pressure vessels with an internal gross volume of 3 cubic feet (0.1 m^3) or less shall use one or more pressure-relief devices or a fusible plug. Pressure vessels of more than 3 cubic feet (0.1 m^3) but less than 10 cubic feet (0.28 m^3) internal gross volume shall use one or more pressure-relief devices; fusible plugs shall not be used. [ASHRAE 15:9.7.2.1, 9.7.2.2]

1113.3 Discharging into Lowside of System. For pressure-relief valves discharging into the lowside of the system, a single relief valve (not rupture member) of the required relieving capacity shall not be used on vessels of 10 cubic feet (0.28 m^3) or more internal gross volume except under the conditions permitted in Section 1112.10.1. [ASHRAE 15:9.7.3]

1113.4 Parallel Pressure-Relief Devices. Two or more pressure-relief devices in parallel to obtain the required capacity shall be considered as one pressure-relief device. The discharge capacity shall be the sum of the capacities required for each pressure vessel being protected.

1113.5 Discharge Capacity. The minimum required discharge capacity of the pressure-relief device or fusible plug for a pressure vessel shall be determined in accordance with Equation 1113.5:

$$C = fDL \quad \text{(Equation 1113.5)}$$

Where:

C = Minimum required discharge capacity of the relief device expressed as mass flow of air, pounds per minute (kg/s).

D = Outside diameter of vessel, feet (m).

L = Length of vessel, feet (m).

f = Factor dependent upon type of refrigerant from Table 1113.5.

Where combustible materials are used within 20 ft (6096 mm) of a pressure vessel, the value of f shall be multiply by 2.5. Equation 1113.5 is based on fire conditions, other heat sources shall be calculated separately. Where one pressure-relief device or fusible plug is used to protect more than one pressure vessel, the required capacity shall be the sum of the capacity required for every pressure vessel. [ASHRAE 15:9.7.5]

TABLE 1113.5
RELIEF DEVICES CAPACITY FACTOR*
[ASHRAE 15:TABLE 9.7.5]

REFRIGERANT	VALUE OF *f*
Where used on the lowside of a limited-charge cascade system:	
R-23, R-170, R-744, R-1150, R-508A, R-508B	1
R-13, R-13B1, R-503	2
R-14	2.5
Other applications:	
R-718	0.2
R-717	0.5
R-11, R-32, R-113, R-123, R-142b, R-152a, R-290, R-600, R-600a, R-764	1
R-12, R-22, R-114, R-124, R-134a, R-401A, R-401B, R-401C, R- 405A, R-406A, R-407C, R-407D, R-407E, R-409A, R-409B, R-411A, R-411B, R-411C, R-412A, R-414A, R-414B, R-500, R-1270	1.6
R-143a, R-402B, R-403A, R-407A, R-408A, R-413A	2
R-115, R-402A, R-403B, R-404A, R-407B, R-410A, R-410B, R- 502, R-507A, R-509A	2.5

* In accordance with Section 1102.2, ammonia refrigeration systems are not regulated by this chapter. R-717 (ammonia) is included in this table because the table is extracted from ASHRAE 15 and is not capable of being modified.

1113.6 Three-Way Valve. Pressure vessels of 10 cubic feet (0.28 m^3) or more internal gross volume shall use one or more rupture member(s) or dual pressure-relief valves where discharging to the atmosphere. Dual pressure-relief valves shall be installed with a three-way valve to allow testing or repair. Where dual relief valves are used, the valve shall comply with Section 1113.5.

Exception: A single relief valve shall be permitted on pressure vessels of 10 cubic feet (0.28 m^3) or more internal gross volume where in accordance with the following conditions:

(1) The relief valves are located on the lowside of the system.

(2) The vessel is provided with shutoff valves designed to allow pumpdown of the refrigerant charge of the pressure vessel.

(3) Other pressure vessels in the system are separately protected in accordance with Section 1113.1. [ASHRAE 15:9.7.2.3]

1114.0 Special Discharge Requirements.

1114.1 General. Systems containing other than Group A1 or B1 refrigerants shall discharge to atmosphere through an approved flaring device.

Exceptions:

(1) Where the Authority Having Jurisdiction determines upon review of a rational engineering analysis that fire, health, or environmental hazards will not result from the proposed atmospheric release.

(2) Lithium bromide absorption system using water as the refrigerant.

1114.2 Design Requirements. Flaring devices shall be designed to incinerate the entire discharge. The products of refrigerant incineration shall not pose health or environmental hazards. Incineration shall be automatic upon initiation of discharge, shall be designed to prevent blow-back, and shall not expose structures or materials to the threat of fire. Standby fuel, such as LP-Gas, and standby power shall have the capacity to operate for one and a half times the required time for complete incineration of the charge.

1114.3 Testing. Flaring systems shall be tested to demonstrate their safety and effectiveness. A report from an approved agency shall be submitted detailing the emission products from the system as installed.

1115.0 Labeling and Identification.

1115.1 General. In addition to labels required elsewhere in this chapter, a refrigeration system shall be provided with identification labels in accordance with Section 1115.2 and Section 1115.3.

1115.2 Volume and Type. A condenser, receiver, absorber, accumulator and similar equipment having an internal volume of more than 3 cubic feet (0.1 m^3) and containing refrigerant shall be equipped with a permanent label setting forth the type of refrigerant in such vessel.

1115.3 Permanent Sign. In a refrigeration machinery room and for a direct refrigerating system of more than 10 horsepower (7.5 kW), there shall be a permanent sign at an approved location giving the following information:

(1) Name of contractor installing the equipment.

(2) Name and number designation of refrigerant in system.

(3) Pounds of refrigerant in system.

1115.4 Marking of Pressure-Relief Devices. Pressure-relief valves for refrigerant-containing components shall be set and sealed by the manufacturer or an assembler in accordance with ASME BPVC Section VIII. Pressure-relief valves shall be marked by the manufacturer or assembler with the data required in accordance with ASME BPVC Section VIII.

Exception: Relief valves for systems with design pressures of 15 pounds-force per square inch gauge (psig) (103 kPa) or less shall be marked by the manufacturer with the pressure-setting capacity. [ASHRAE 15:9.6.1]

1115.4.1 Rupture Members. Rupture members for refrigerant pressure vessels shall be marked with the data required in accordance with ASME BPVC Section VIII. [ASHRAE 15:9.6.2]

1115.4.2 Fusible Plugs. Fusible plugs shall be marked with the melting temperatures in °F (°C). [ASHRAE 15:9.6.3]

1116.0 Testing of Refrigeration Equipment.

1116.1 Factory Tests. Refrigerant-containing parts of unit systems shall be tested and proved tight by the manufacturer at not less than the design pressure for which they are rated. Pressure vessels shall be tested in accordance with Section 1117.0. [ASHRAE 15:9.14.1]

1116.1.1 Testing Procedure. Tests shall be performed with dry nitrogen or another nonflammable, nonreactive, dried gas. Oxygen, air, or mixtures containing them shall not be used. The means used to build up the test pressure shall have a pressure-limiting device or a pressure-reducing device and a gage on the outlet side. The pressure-relief device shall be set above the test pressure but low enough to prevent permanent deformation of the system's components.

Exceptions:

(1) Mixtures of dry nitrogen, inert gases, nonflammable refrigerants permitted for factory tests.

(2) Mixtures of dry nitrogen, inert gases, or a combination of them with flammable refrigerants in concentrations not exceeding the lesser of a refrigerant weight fraction (mass fraction) of 5 percent or 25 percent of the LFL shall be permitted for factory tests.

(3) Compressed air without added refrigerant shall be permitted for factory tests provided the system is subsequently evacuated to less than 0.039 inch of mercury (0.132 kPa) before charging with refrigerant. The required evacuation level is atmospheric pressure for systems using R-718 (water) or R-744 (carbon dioxide) as the refrigerant. [ASHRAE 15:9.14.1.1]

1116.1.2 Applied Pressure. The test pressure applied to the highside of each factory-assembled refrigerating system shall be not less than the design pressure of the highside. The test pressure applied to the lowside of a factory assembled refrigerating system shall be not less than the design pressure of the lowside.

1116.1.3 Design Pressure of 15 psig or Less. Units with a design pressure of 15 psig (103 kPa) or less shall be tested at a pressure not less than 1.33 times the design pressure, and shall be proved leak-tight at not less than the lowside design pressure. [ASHRAE 15:9.14.3]

1116.2 Field Tests. Refrigerant-containing parts of a system that is field-erected shall be tested and proved tight after complete installation and before the operation. The high and low sides of each system shall be tested and proved tight at not less than the lower of the pressure in Table 1116.2 or the setting of the pressure-relief device.

Exceptions:

(1) Compressors, condensers, evaporators, coded pressure vessels, safety devices, pressure gauges, control mechanisms, and systems that are factory tested.

(2) Refrigeration systems containing Group R-22, not exceeding 5 tons of refrigeration capacity (18 kW), and field-piped using approved, factory-charged line sets shall be permitted to be proved tight by observing retention of pressure on a set of charging gauges and soaping connections while the system is operating.

TABLE 1116.2
FIELD LEAK TEST PRESSURES (psig)*

REFRIGERANT NUMBER	HIGHSIDE WATER COOLED	HIGHSIDE AIR COOLED	LOWSIDE
11	15	35	15
12	140	220	140
22	230	360	230
113	15	15	15
114	40	80	40
115	275	340	275
123	15	30	15
134a	150	250	150
152a	130	220	130
500	165	265	165
502	250	385	250
744*	—	—	—

For SI units: 1 pound-force per square inch gauge = 6.8947 kPa

* Special design required; test pressures typically exceed 1000 psig (6895 kPa).

1116.3 Test Gases. Tests shall be performed with dry nitrogen or other nonflammable, nonreactive, dried gas. Oxygen, air, or mixtures containing them shall not be used. The means used to build up the test pressure shall have either a pressure-limiting device or a pressure-reducing device and a gauge on the outlet side. The pressure-relief device shall be set above the test pressure but low enough to prevent permanent deformation of the system's components.

Exceptions:

(1) Mixtures of dry nitrogen, inert gases, or a combination of them with nonflammable refrigerant in concentrations of a refrigerant weight fraction (mass fraction) not exceeding 5 percent shall be permitted for tests.

(2) Mixtures of dry nitrogen, inert gases, or a combination of them with flammable refrigerants in concentrations not exceeding the lower of a refrigerant weight fraction (mass fraction) of 5 percent or 25 percent of the LFL shall be permitted for tests.

(3) Compressed air without added refrigerants shall be permitted for tests, provided the system is subsequently evacuated to less than 1000 microns (0.1333 kPa) before charging with refrigerant. The required evacuation level is atmospheric pressure for systems using R-718 (water) or R-744 (carbon dioxide) as the refrigerant.

(4) Systems erected on the premises using Group A1 refrigerant and with copper tubing not exceeding 0.62 of an inch (15.7 mm) outside diameter shall be tested by means of the refrigerant charged into the system at the saturated vapor pressure of the refrigerant at not less than 68°F (20°C). [ASHRAE 15:10.1.2]

1116.4 Declaration. A dated declaration of test shall be provided for systems containing more than 55 pounds (24.9 kg) of refrigerant. The declaration shall give the name of the refrigerant and the field test pressure applied to the highside and the lowside of the system. The declaration of test shall be signed by the installer and, where an inspector is present at the tests, the inspector shall also sign the declaration. Where requested, copies of this declaration shall be furnished to the Authority Having Jurisdiction. [ASHRAE 15:10.2]

1116.5 Brine Systems. Brine-containing portions of a system shall be tested at one and a half times the design pressure of the system using brine as the test fluid.

1117.0 Refrigerant-Containing Pressure Vessels.

1117.1 Inside Dimensions 6 inches or Less. Pressure vessels having inside dimensions of 6 inches (152 mm) or less shall comply with the following:

(1) Be listed individually or as part of an assembly.

(2) Marked directly on the vessel or on a nameplate attached to the vessel in accordance with ASME BPVC Section VIII.

(3) Where requested by the Authority Having Jurisdiction, the manufacturer shall provide documentation to confirm that the vessel design, fabrication, and testing requirements are in accordance with ASME BPVC Section VIII.

Exception: Vessels having an internal or external design pressure of 15 psig (103 kPa) or less.

Pressure vessels having inside dimensions of 6 inches (152 mm) or less shall be protected by either a pressure-relief device or a fusible plug. [ASHRAE 15:9.3.1.1]

1117.1.1 Pressure-Relief Device. Where a pressure-relief device is used to protect a pressure vessel having an inside dimension of 6 inches (152 mm) or less, the ultimate strength of the pressure vessel so protected shall withstand a pressure of not less than 3.0 times the design pressure. [ASHRAE 15:9.3.1.2]

1117.1.2 Fusible Plug. Where a fusible plug is used to protect a pressure vessel having an inside diameter of 6 inches (152 mm) or less, the ultimate strength of the pressure vessel so protected shall withstand a pressure 2.5 times the saturation pressure of the refrigerant used at the temperature stamped on the fusible plug or 2.5 times the critical pressure of the refrigerant used, whichever is less. [ASHRAE 15:9.3.1.3]

1117.2 Inside Dimensions More than 6 inches. Pressure vessels having an inside diameter exceeding 6 inches (152 mm) and having an internal or external design pressure of more than 15 psig (103 kPa) shall be directly marked, or marked on a nameplate in accordance with ASME BPVC Section VIII. [ASHRAE 15:9.3.2]

1117.3 Pressure Vessels for 15 psig or Less. Pressure vessels having an internal or external design pressure of 15 psig (103 kPa) or less shall have an ultimate strength to withstand not less than 3.0 times the design pressure and shall be tested with a pneumatic test pressure of not less than 1.25 times the design pressure or a hydrostatic test pressure of not less than 1.5 times the design pressure. [ASHRAE 15:9.3.3]

1118.0 Maintenance and Operation.

1118.1 General. Refrigeration systems shall be operated and maintained as required by the fire code.

Part II – Cooling Towers.

1119.0 General.

1119.1 Applicability. Cooling towers, evaporative condensers, and fluid coolers shall be readily accessible. Where located on roofs, such equipment having combustible exterior surfaces shall be protected with an approved automatic fire-extinguishing system.

1120.0 Support and Anchorage.

1120.1 General. Cooling towers, evaporative condensers, and fluid coolers shall be supported on noncombustible grillage designed in accordance with the building code. Seismic restraints shall be as required by the building code.

1121.0 Drainage.

1121.1 General. Drains, overflows, and blow-down provisions shall have an indirect connection to an approved disposal location. Discharge of chemical waste shall be as approved by the regulatory authority.

1122.0 Chemical Treatment Systems.

1122.1 General. Chemical treatment systems shall comply with the fire code. Where chemicals used present a contact hazard to personnel, approved emergency eye-wash and shower facilities shall be installed.

1122.2 Automated Control of Cycles of Concentration. Cooling towers, evaporative condensers, and fluid coolers shall include controls that automate system bleed based on conductivity, fraction of metered makeup volume, metered bleed volume, recirculating pump run time, or bleed time.

1123.0 Location.

1123.1 General. Cooling towers, evaporative condensers, and fluid coolers shall be located such that their plumes cannot enter occupied spaces. Plume discharges shall be not less than 25 feet (7620 mm) away from a ventilation inlet to a building. Location on the property shall be as required for buildings by the building code.

1124.0 Electrical.

1124.1 General. Electrical systems shall be in accordance with the electrical code. Equipment shall be provided with a vibration switch to shut off fans operating with excessive vibration. In climates commonly subject to electrical storms, lightning protection shall be provided on roof-mounted equipment.

1125.0 Refrigerants and Hazardous Fluids.

1125.1 General. Equipment containing refrigerants as a part of a closed-cycle refrigeration system shall comply with Part I of this chapter. Equipment containing other fluids that are flammable, combustible, or hazardous shall be in accordance with this code and the fire code.

1126.0 Drift Eliminators.

1126.1 General. Cooling towers, evaporative condensers, and fluid coolers shall be equipped with drift eliminators that have a drift rate of not more than 0.005 percent of the circulated water flow rate in accordance with the equipment manufacturer's instructions.

TABLE 1102.3
REFRIGERANT GROUPS, PROPERTIES, AND ALLOWABLE QUANTITIES[8]
[ASHRAE 34: TABLE 4-1, TABLE 4-2]

REFRIGERANT	CHEMICAL FORMULA	CHEMICAL NAME[1] (COMPOSITION FOR BLENDS)	SAFETY GROUP[7]	OEL[2] (ppm)	POUNDS PER 1000 CUBIC FEET OF SPACE
R-11	CCl_3F	Trichlorofluoromethane	A1	C1000	0.39
R-12	CCl_2F_2	Dichlorodifluoromethane	A1	1000	5.6
R-12B1	$CBrClF_2$	Bromochlorodifluoromethane	—	—	—
R-13	$CClF_3$	Chlorotrifluoromethane	A1	1000	—
R-13B1	$CBrF_3$	Bromotrifluoromethane	A1	1000	—
R-14	CF_4	Tetrafluoromethane (carbon tetrafluoride)	A1	1000	25
R-21	$CHCl_2F$	Dichlorofluoromethane	B1	—	—
R-22	$CHClF_2$	Chlorodifluoromethane	A1	1000	13
R-23	CHF_3	Trifluoromethane	A1	1000	7.3
R-30	CH_2Cl_2	Dichloromethane (methylene chloride)	B1	—	—
R-31	CH_2ClF	Chlorofluoromethane	—	—	—
R-32	CH_2F_2	Difluoromethane (methylene fluoride)	A2L	1000	4.8
R-40	CH_3Cl	Chloromethane (methyl chloride)	B2	—	—
R-41	CH_3F	Fluoromethane (methyl fluoride)	—	—	—
R-50	CH_4	Methane	A3	1000	—
R-113	CCl_2FCClF_2	1, 1, 2-trichloro-1, 2, 2 – trifluoroethane	A1	1000	1.2
R-114	$CClF_2CClF_2$	1, 2-dichloro-1, 1, 2, 2 tetrafluoroethane	A1	1000	8.7
R-115	$CClF_2CF_3$	Chloropentafluoroethane	A1	1000	47
R-116	CF_3CF_3	Hexafluoroethane	A1	1000	34
R-123	$CHCl_2CF_3$	2, 2-dichloro-1, 1, 1, - trifluoroethane	B1	50	3.5
R-124	$CHClFCF_3$	2-chloro-1, 1, 1, 2 - tetrafluoroethane	A1	1000	3.5
R-125	CHF_2CF_3	Pentafluoroethane	A1	1000	23
R-134a	CH_2FCF_3	1, 1, 1, 2-tetrafluoroethane	A1	1000	13
R-141b	CH_3CCl_2F	1, 1-dichloro-1-fluoroethane	—	500	0.78
R-142b	CH_3CClF_2	1-chloro-1, 1-difluoroethane	A2	1000	5.1
R-143a	CH_3CF_3	1, 1, 1-trifluoroethane	A2L	1000	4.5
R-152a	CH_3CHF_2	1, 1-difluoroethane	A2	1000	2.0
R-170	CH_3CH_3	Ethane	A3	1000	0.54
R-E170	CH_3OCH_3	Methoxymethane (Dimethyl ether)	A3	1000	1.0
R-218	CF_3CF2CF_3	Octafluoropropane	A1	1000	43
R-227ea	CF_3CHFCF_3	1, 1, 1, 2, 3, 3, 3- heptafluoropropane	A1	1000	36
R-236fa	$CF_3CH_2CF_3$	1, 1, 1, 3, 3, 3-hexafluoropropane	A1	1000	21
R-245fa	$CHF_2CH_2CF_3$	1, 1, 1, 3, 3-pentafluoropropane	B1	300	12
R-290	$CH_3CH_2CH_3$	Propane	A3	1000	0.56
R-C318	$-(CF_2)_4-$	Octafluorocyclobutane	A1	1000	41
R-400	zeotrope	R-12/114 (50.0/50.0)	A1	1000	10
R-400	zeotrope	R-12/114 (60.0/40.0)	A1	1000	11
R-401A	zeotrope	R-22/152a/124 (53.0/13.0/34.0)	A1	1000	6.6
R-401B	zeotrope	R-22/152a/124 (61.0/11.0/28.0)	A1	1000	7.2
R-401C	zeotrope	R-22/152a/124 (33.0/15.0/52.0)	A1	1000	5.2
R-402A	zeotrope	R-125/290/22 (60.0/2.0/38.0)	A1	1000	17

TABLE 1102.3 (continued)
REFRIGERANT GROUPS, PROPERTIES, AND ALLOWABLE QUANTITIES[8]
[ASHRAE 34: TABLE 4-1, TABLE 4-2]

REFRIGERANT	CHEMICAL FORMULA	CHEMICAL NAME[1] (COMPOSITION FOR BLENDS)	SAFETY GROUP[7]	OEL[2] (ppm)	POUNDS PER 1000 CUBIC FEET OF SPACE
R-402B	zeotrope	R-125/290/22 (38.0/2.0/60.0)	A1	1000	15
R-403A	zeotrope	R-290/22/218 (5.0/75.0/20.0)	A2	1000	7.6
R-403B	zeotrope	R-290/22/218 (5.0/56.0/39.0)	A1	1000	18
R-404A	zeotrope	R-125/143a/134a (44.0/52.0/4.0)	A1	1000	31
R-405A	zeotrope	R-22/152a/142b/C318 (45.0/7.0/5.5/42.5)	—	1000	16
R-406A	zeotrope	R-22/600a/142b (55.0/4.0/41.0)	A2	1000	4.7
R-407A	zeotrope	R-32/125/134a (20.0/40.0/40.0)	A1	1000	19
R-407B	zeotrope	R-32/125/134a (10.0/70.0/20.0)	A1	1000	21
R-407C	zeotrope	R-32/125/134a (23.0/25.0/52.0)	A1	1000	18
R-407D	zeotrope	R-32/125/134a (15.0/15.0/70.0)	A1	1000	16
R-407E	zeotrope	R-32/125/134a (25.0/15.0/60.0)	A1	1000	17
R-407F	zeotrope	R-32/125/134a (30.0/30.0/40.0)	A1	1000	20
R-407G	zeotrope	R-32/125/134a (2.5/2.5/95.0)	A1	1000	13
R-408A	zeotrope	R-125/143a/22 (7.0/46.0/47.0)	A1	1000	21
R-409A	zeotrope	R-22/124/142b (60.0/25.0/15.0)	A1	1000	7.1
R-409B	zeotrope	R-22/124/142b (65.0/25.0/10.0)	A1	1000	7.3
R-410A	zeotrope	R-32/125 (50.0/50.0)	A1	1000	26
R-410B	zeotrope	R-32/125 (45.0/55.0)	A1	—	27
R-411A[6]	zeotrope	R-1270/22/152a (1.5/87.5/11.0)	A2	990	2.9
R-411B[6]	zeotrope	R-1270/22/152a (3.0/94.0/3.0)	A2	980	2.8
R-412A	zeotrope	R-22/218/142b (70.0/5.0/25.0)	A2	1000	5.1
R-413A	zeotrope	R-218/134a/600a (9.0/88.0/3.0)	A2	1000	5.8
R-414A	zeotrope	R-22/124/600a/142b (51.0/28.5/4.0/16.5)	A1	1000	6.4
R-414B	zeotrope	R-22/124/600a/142b (50.0/39.0/1.5/9.5)	A1	1000	6.0
R-415A	zeotrope	R-22/152a (82.0/18.0)	A2	1000	2.9
R-415B	zeotrope	R-22/152a (25.0/75.0)	A2	1000	2.1
R-416A[6]	zeotrope	R-134a/124/600 (59.0/39.5/1.5)	A1	1000	3.9
R-417A[6]	zeotrope	R-125/134a/600 (46.6/50.0/3.4)	A1	1000	3.5
R-417B	zeotrope	R-125/134a/600 (79.0/18.3/2.7)	A1	1000	4.3
R-417C	zeotrope	R-125/134a/600 (19.5/78.8/1.7)	A1	1000	5.4
R-418A	zeotrope	R-290/22/152a (1.5/96.0/2.5)	A2	1000	4.8
R-419A	zeotrope	R-125/134a/E170 (77.0/19.0/4.0)	A2	1000	4.2
R-419B	zeotrope	R-125/134a/E170 (48.5/48.0/3.5)	A2	1000	4.6
R-420A	zeotrope	R-134a/142b (88.0/12.0)	A1	1000	12
R-421A	zeotrope	R-125/134a (58.0/42.0)	A1	1000	17
R-421B	zeotrope	R-125/134a (85.0/15.0)	A1	1000	21
R-422A	zeotrope	R-125/134a/600a (85.1/11.5/3.4)	A1	1000	18
R-422B	zeotrope	R-125/134a/600a (55.0/42.0/3.0)	A1	1000	16
R-422C	zeotrope	R-125/134a/600a (82.0/15.0/3.0)	A1	1000	18
R-422D	zeotrope	R-125/134a/600a (65.1/31.5/3.4)	A1	1000	16
R-422E	zeotrope	R-125/134a/600a (58.0/39.3/2.7)	A1	1000	16
R-423A	zeotrope	R-134a/227ea (52.5/47.5)	A1	1000	19

ASHRAE
ASHRAE
ASHRAE
ASHRAE
ASHRAE

TABLE 1102.3 (continued)
REFRIGERANT GROUPS, PROPERTIES, AND ALLOWABLE QUANTITIES[8]
[ASHRAE 34: TABLE 4-1, TABLE 4-2]

ASHRAE ASHRAE ASHRAE ASHRAE ASHRAE

REFRIGERANT	CHEMICAL FORMULA	CHEMICAL NAME[1] (COMPOSITION FOR BLENDS)	SAFETY GROUP[7]	OEL[2] (ppm)	POUNDS PER 1000 CUBIC FEET OF SPACE
R-424A[6]	zeotrope	R-125/134a/600a/600/601a (50.5/47.0/0.9/1.0/0.6)	A1	970	6.2
R-425A	zeotrope	R-32/134a/227ea (18.5/69.5/12.0)	A1	1000	16
R-426A[6]	zeotrope	R-125/134a/600/601a (5.1/93.0/1.3/0.6)	A1	990	5.2
R427A	zeotrope	R-32/125/143a/134a (15.0/25.0/10.0/50.0)	A1	1000	18
R428A	zeotrope	R-125/143a/290/600a (77.5/20.0/0.6/1.9)	A1	1000	23
R-429A	zeotrope	R-E170/152a/600a (60.0/10.0/30.0)	A3	1000	0.81
R-430A	zeotrope	R-152a/600a (76.0/24.0)	A3	1000	1.3
R-431A	zeotrope	R-290/152a (71.0/29.0)	A3	1000	0.69
R-432A	zeotrope	R-1270/E170 (80.0/20.0)	A3	700	0.13
R-433A	zeotrope	R-1270/290 (30.0/70.0)	A3	880	0.34
R-433B	zeotrope	R-1270/290 (5.0/95.0)	A3	950	0.51
R-433C	zeotrope	R-1270/290 (25.0/75.0)	A3	790	0.41
R-434A	zeotrope	R-125/143a/134a/600a (63.2/18.0/16.0/2.8)	A1	1000	20
R-435A	zeotrope	R-E170/152a (80.0/20.0)	A3	1000	1.1
R-436A	zeotrope	R-290/600a (56.0/44.0)	A3	1000	0.50
R-436B	zeotrope	R-290/600a (52.0/48.0)	A3	1000	0.51
R-437A	zeotrope	R-125/134a/600/601 (19.5/78.5/1.4/0.6)	A1	990	5.0
R-438A	zeotrope	R-32/125/134a/600/601a (8.5/45.0/44.2/1.7/0.6)	A1	990	4.9
R-439A	zeotrope	R-32/125/600a (50.0/47.0/3.0)	A2	990	4.7
R-440A	zeotrope	R-290/134a/152a (0.6/1.6/97.8)	A2	1000	1.9
R-441A	zeotrope	R-170/290/600a/600 (3.1/54.8/6.0/36.1)	A3	1000	0.39
R-442A	zeotrope	R-32/125/134a/152a/227ea (31.0/31.0/30.0/3.0/5.0)	A1	1000	21
R-443A	zeotrope	R-1270/290/600a (55.0/40.0/5.0)	A3	580	0.19
R-444A	zeotrope	R-32/152a/1234ze(E) (12.0/5.0/83.0)	A2L	850	5.1
R-444B	zeotrope	R-32/152a/1234ze(E) (41.5/10.0/48.5)	A2L	890	4.3
R-445A	zeotrope	R-744/134a/1234ze (E) (6.0/9.0/85.0)	A2L	930	4.2
R-446A	zeotrope	R-32/1234ze(E)/600 (68.0/29.0/3.0)	A2L	960	2.5
R-447A	zeotrope	R-32/125/1234ze(E) (68.0/3.5/28.5)	A2L	900	2.6
R-447B	zeotrope	R-32/125/1234ze(E) (68.0/8.0/24.0)	A2L	970	23
R-448A	zeotrope	R-32/125/1234yf/134a/1234ze(E) (26.0/26.0/20.0/21.0/7.0)	A1	890	24
R-449A	zeotrope	R-32 /125 /1234yf /134a (24.3/24.7/25.3/25.7)	A1	830	23
R-449B	zeotrope	R-32/125/1234yf/134a (25.2/24.3/23.2/27.3)	A1	850	23
R-449C	zeotrope	R-32/125/1234yf/134a (20.0/20.0/31.0/29.0)	A1	800	23
R-450A	zeotrope	R-134a/1234ze(E) (42.0/58.0)	A1	880	20
R-451A	zeotrope	R-1234yf/134a (89.8/10.2)	A2L	520	5.3
R-451B	zeotrope	R-1234yf/134a (88.8/11.2)	A2L	530	5.3
R-452A	zeotrope	R-32/125/1234yf (11.0/59.0/30.0)	A1	780	27

TABLE 1102.3 (continued)
REFRIGERANT GROUPS, PROPERTIES, AND ALLOWABLE QUANTITIES[8]
[ASHRAE 34: TABLE 4-1, TABLE 4-2]

REFRIGERANT	CHEMICAL FORMULA	CHEMICAL NAME[1] (COMPOSITION FOR BLENDS)	SAFETY GROUP[7]	OEL[2] (ppm)	POUNDS PER 1000 CUBIC FEET OF SPACE
R-452B	zeotrope	R-32/125/1234yf (67.0/7.0/26.0)	A2L	870	23
R-452C	zeotrope	R-32/125/1234yf (12.5/61.0/26.5)	A1	800	27
R-453A	zeotrope	R-32/125/134a/227ea/600/601a (20.0/20.0/53.8/5.0/0.6/0.6)	A1	1000	7.8
R-454A	zeotrope	R-32/1234yf (35.0/65.0)	A2L	690	28
R-454B	zeotrope	R-32/1234yf (68.9/31.1)	A2L	850	22
R-454C	zeotrope	R-32/1234yf (21.5/78.5)	A2L	620	29
R-455A	zeotrope	R-744/32/1234yf (3.0/21.5/75.5)	A2L	650	23
R-456A	zeotrope	R-32/134a/1234ze(E) (6.0/45.0/49.0)	A1	900	20
R-457A	zeotrope	R-32/1234yf/152a (18.0/70.0/12.0)	A2L	650	25
R-458A	zeotrope	R-32/125/134a/227ea/236fa (20.5/4.0/61.4/13.5/0.6)	A1	1000	18
R-500	azeotrope[3]	R-12/152a (73.8/26.2)	A1	1000	7.6
R-501	azeotrope[3]	R-22/12 (75.0/25.0)[4]	A1	1000	13
R-502	azeotrope[3]	R-22/115 (48.8/51.2)	A1	1000	21
R-503	azeotrope[3]	R-23/13 (40.1/59.9)	—	1000	—
R-504	azeotrope[3]	R-32/115 (48.2/51.8)	—	1000	28
R-505	azeotrope[3]	R-12/31 (78.0/22.0)[4]	—	—	—
R-506	azeotrope[3]	R-31/114 (55.1/44.9)	—	—	—
R-507A[5]	azeotrope[3]	R-125/143a (50.0/50.0)	A1	1000	32
R-508A[5]	azeotrope[3]	R-23/116 (39.0/61.0)	A1	1000	14
R-508B	azeotrope[3]	R-23/116 (46.0/54.0)	A1	1000	13
R-509A[5]	azeotrope[3]	R-22/218 (44.0/56.0)	A1	1000	24
R-510A	azeotrope[3]	R-E170/600a (88.0/12.0)	A3	1000	0.87
R-511A	azeotrope[3]	R-290/E170 (95.0/5.0)	A3	1000	0.59
R-512A	azeotrope[3]	R-134a/152a (5.0/95.0)	A2	1000	1.9
R-513A	azeotrope[3]	R-1234yf/134a (56.0/44.0)	A1	650	20
R-513B	azeotrope[3]	R-1234yf/134a (58.5/41.5)	A1	640	21
R-514A	azeotrope[3]	R-1336mzz(Z)/1130 (E) (74.7/25.3)	B1	320	0.86
R-515A	azeotrope[3]	R-1234ze(E)/227ea (88.0/12.0)	A1	810	19
R-600	$CH_3CH_2CH_2CH_3$	Butane	A3	1000	0.15
R-600a	$CH(CH_3)_2CH_3$	2-methylpropane (isobutene)	A3	1000	0.59
R-601	$CH_3CH_2CH_2CH_2\,CH_3$	Pentane	A3	600	0.18
R-601a	$(CH_3)_2CHCH_2CH_3$	2-methylbutane (isopentane)	A3	600	0.18
R-610	$CH_3CH_2OCH_2CH_3$	Ethoxyethane (ethyl ether)	—	400	—
R-611	$HCOOCH_3$	Methyl formate	B2	100	—
R-620	—	(Reserved for future assignment)	—	—	—
R-630	CH_3NH_2	Methanamine (methyl amine)	—	—	—
R-631	$CH_3CH_2(NH_2)$	Ethanamine (ethyl amine)	—	—	—
R-702	H_2	Hydrogen	A3	—	—
R-704	He	Helium	A1	—	—
R-717	NH_3	Ammonia	B2L	25	0.014
R-718	H_2O	Water	A1	—	—
R-720	Ne	Neon	A1	—	—
R-728	N_2	Nitrogen	A1	—	—
R-732	O_2	Oxygen	—	—	—
R-740	Ar	Argon	A1	—	—
R-744	CO_2	Carbon dioxide	A1	5000	3.4

TABLE 1102.3 (continued)
REFRIGERANT GROUPS, PROPERTIES, AND ALLOWABLE QUANTITIES[8]
[ASHRAE 34: TABLE 4-1, TABLE 4-2]

ASHRAE ASHRAE ASHRAE ASHRAE ASHRAE

REFRIGERANT	CHEMICAL FORMULA	CHEMICAL NAME[1] (COMPOSITION FOR BLENDS)	SAFETY GROUP[7]	OEL[2] (ppm)	POUNDS PER 1000 CUBIC FEET OF SPACE
R-744A	N_2O	Nitrous oxide	—	—	—
R-764	SO_2	Sulfur dioxide	B1	—	—
R-1130(E)	CHCl=CHCl	Trans-1,2-dichloroethene	B1	200	0.25
R-1150	$CH_2=CH_2$	Ethene (ethylene)	A3	200	—
R-1233zd(E)	$CF_3CH=CHCl$	Trans-1-chloro-3,3,3-trifluoro-1- propene	A1	800	5.3
R-1234yf	$CF_3CF=CH_2$	2, 3, 3, 3-tetrafluoro-1-propene	A2L	500	4.7
R-1234ze(E)	$CF_3CH=CHF$	Trans-1,3,3,3- tetrafluoro-1-propene	A2L	800	4.7
R-1270	$CH_3CH=CH_2$	Propene (propylene)	A3	500	0.11
1336mzz(Z)	$CF_3CHCHCF_3$	Cis-1,1,1,4,4,4-hexaflouro-2-butene	A1	500	5.4

For SI units: 1 pound = 0.453 kg, 1 cubic foot = 0.0283 m³

Notes:

1 The preferred chemical name is followed by the popular name in parenthesis.

ASHRAE ASHRAE 2 The OELs are 8-hour TWAs; a "C" designation denotes a ceiling limit.

3 Azeotropic refrigerants exhibit some segregation of components at conditions of temperature and pressure other than those at which they were formulated. The extent of segregation depends on the particular azeotrope and hardware system configuration.

4 The exact composition of this azeotrope is in question and additional experimental studies are needed.

ASHRAE ASHRAE 5 R-507, R-508, and R-509 are allowed alternative designations for R-507A, R-508A, and R-509A due to a change in designations after assignment of R-500 through R-509. Corresponding changes were not made for R-500 through R-506.

6 The RCL values for these refrigerant blends are approximated in the absence of adequate data for a component comprising less than 4 percent m/m of the blend and expected to have a small influence in an acute, accidental release.

7 Refrigerant flammability classification of Class 2L shall comply with the requirements for flammability classification of Class 2.

8 In accordance with Section 1102.2, ammonia refrigeration systems are not regulated by this chapter. R-717 (ammonia) is included in this table because the table is extracted from ASHRAE 34 and is not capable of being modified.

TABLE 1104.1
PERMISSIBLE REFRIGERATION SYSTEMS[1]

	OCCUPANCY GROUP[3]	HIGH-PROBABILITY SYSTEM	LOW PROBABILITY SYSTEM	MACHINERY ROOM
	A-1	Group A1 only	Any	Any
	A-2	Group A1 only	Any	Any
	A-3	Group A1 only	Any	Any
	A-4	Group A1 only	Any	Any
	B	Group A1[2] only	Any	Any
	E	Group A1 only	Any	Any
	F-1	Group A1[2] only	Any	Any
	F-2	Any[2]	Any	Any
	H-1	Any	Any	Any
	H-2	Any	Any	Any
	H-3	Any	Any	Any
	H-4	Group A1 only	Any	Any
	H-5	Group A1 only	Any	Any
	I-1	None	Any	Any
	I-2	Group A1 only	Any	Any
[OSHPD 1, 1R, 2, 3, 4 & 5]	*I-2.1*	*Group A1 only*	*Any*	*Any*
	I-3	None	Any	Any
	I-4	Group A1 only	Any	Any
	M	Group A1[2] only	Any	Any
	R-1	Group A1 only	Any	Any
	R-2	Group A1 only	Any	Any
	R-3	Group A1 only	Any	Any
	R-4	Group A1 only	Any	Any
	S-1	Group A1[2] only	Any	Any
	S-2	Any[2]	Any	Any
	U	Any	Any	Any

Notes:

[1] See Section 1104.0.

[2] A refrigerant shall be permitted to be used within a high-probability system where the room or space is in accordance with Section 1104.4.

[3] Occupancy classifications are defined in the building code.

CALIFORNIA MECHANICAL CODE – MATRIX ADOPTION TABLE
CHAPTER 12 – HYDRONICS

(Matrix Adoption Tables are non-regulatory, intended only as an aid to the code user. See Chapter 1 for state agency authority and building applications.)

Adopting Agency	BSC	BSC-CG	SFM	HCD			DSA			OSHPD						BSCC	DPH	AGR	DWR	CEC	CA	SL	SLC
				1	2	1-AC	AC	SS	SS/CC	1	1R	2	3	4	5								
Adopt Entire Chapter	X			X	X			X	X				X										
Adopt Entire Chapter as amended (amended sections listed below)										X	X	X		X	X								
Adopt only those sections that are listed below																							
Chapter/Section																							
1205.2, *Note*				X	X																		
1210.2										X	X	X		X	X								
1211.6, *Note*				X	X																		
1217.5.2				X	X																		
1217.5.4, *[HCD 1 & HCD 2]*				X	X																		
1220.2.5, *[HCD 1 & HCD 2]*				X	X																		

This state agency does not adopt sections identified with the following symbol: †

The Office of the State Fire Marshal's adoption of this chapter or individual sections is applicable to structures regulated by other state agencies pursuant to Section 1.11.0.

CHAPTER 12
HYDRONICS

1201.0 General.

1201.1 Applicability. This chapter shall apply to hydronic piping systems that are part of heating, cooling, ventilation, refrigeration, and air conditioning systems. Such piping systems include steam, hot water, chilled water, steam condensate, condenser water, and ground source heat pump systems. The regulations of this chapter shall govern the construction, location, and installation of hydronic piping systems.

1201.2 Insulation. Surfaces within reach of building occupants shall not exceed 140°F (60°C). Where sleeves are installed, the insulation shall continue full size through them.

Coverings and insulation used for piping shall be of material approved for the operating temperature of the system and the installation environment. Where installed in a plenum, the insulation, jackets, and lap-seal adhesives, including pipe coverings and linings, shall have a flame-spread index not to exceed 25 and a smoke-developed index not to exceed 50 where tested in accordance with ASTM E84 or UL 723.

1201.3 Water Hammer. The piping system shall be designed to prevent water hammer.

1201.4 Terminal Units. Terminal units, valves, and flow control devices shall be installed in accordance with the manufacturer's installation instructions.

1201.5 Return-Water Low-Temperature Protection. Where a minimum return-water temperature to the heat source is specified by the manufacturer, the heating system shall be designed and installed to meet or exceed the minimum return-water temperature during the normal operation of the heat source.

1202.0 Protection of Potable Water Supply.

1202.1 Prohibited Sources. Hydronic systems or parts thereof shall be constructed in such a manner that polluted, contaminated water or substances shall not enter a portion of the potable water system either during normal use or where the system is subject to pressure that exceeds the operating pressure in the potable water system. Piping, components and devices in contact with the potable water shall be approved for such use and where an additive is used it shall not affect the performance of the system.

1202.2 Chemical Injection. Where systems include an additive, chemical injection or provisions for such injection, the potable water supply shall be protected by a reduced-pressure principle backflow prevention assembly listed or labeled in accordance with ASSE 1013. Such additive or chemical shall be compatible with system components.

1202.3 Compatibility. Fluids used in hydronic systems shall be compatible with all components that will contact the fluid. Where a heat exchanger is installed with a dual purpose water heater, such application shall comply with the requirements for a single wall heat exchanger in Section 1218.1.

1203.0 Capacity of Heat Source.

1203.1 Heat Source. The heat source shall be sized to the design load.

1203.2 Dual Purpose Water Heater. Water heaters utilized for combined space-heating and water-heating applications shall be listed or labeled in accordance with the standards referenced in Table 1203.2, and shall be installed in accordance with the manufacturer's installation instructions. The total heating capacity of a dual purpose water heater shall be based on the sum of the potable hot water requirements and the space heating design requirements corrected for hot water first-hour draw recovery.

TABLE 1203.2
WATER HEATERS

TYPE	STANDARDS
Gas, 75 000 Btu/h or less	CSA Z21.10.1
Gas, Above 75 000 Btu/h	CSA Z21.10.3
Electric, Space Heating	UL 834
Solid Fuel	UL 2523

For SI units: 1000 British thermal units per hour = 0.293kW

1203.3 Tankless Water Heater. Tankless water heaters shall be rated by the manufacturer for space- heating applications, and the output performance shall be determined by the temperature rise and flow rate of water through the unit. The ratings shall be expressed by the water temperature rise at a given flow rate. Manufacturers flow rates shall not be exceeded.

1204.0 Identification of a Potable and Nonpotable Water System.

1204.1 General. In buildings where potable water and nonpotable water systems are installed, each system shall be clearly identified in accordance with Section 1204.2 through Section 1204.5.

1204.2 Color and Information. Each system shall be identified with a colored pipe or band and coded with paint, wraps, and materials compatible with the piping.

1204.3 Potable Water. Potable water systems shall be identified with a green background with white lettering. The minimum size of letters and length of the color field shall be in accordance with Table 1204.3.

1204.4 Nonpotable Water. Nonpotable water systems shall have a yellow background with black uppercase lettering, with the words "CAUTION: NONPOTABLE WATER, DO NOT DRINK." Each nonpotable system shall be identified to designate the liquid being conveyed, and the direction of normal flow shall be clearly shown. The minimum size of the letters and length of the color field shall comply with Table 1204.3.

TABLE 1204.3
MINIMUM LENGTH OF COLOR FIELD AND SIZE OF LETTERS

OUTSIDE DIAMETER OF PIPE OR COVERING (inches)	MINIMUM LENGTH OF COLOR FIELD (inches)	MINIMUM SIZE OF LETTERS (inches)
1/2 to 1 1/4	8	1/2
1 1/2 to 2	8	3/4
2 1/2 to 6	12	1 1/4
8 to 10	24	2 1/2
over 10	32	3 1/2

For SI units: 1 inch = 25.4 mm

1204.5 Location of Piping Identification. The background color and required information shall be indicated every 20 feet (6096 mm) but not less than once per room, and shall be visible from the floor level.

1204.6 Flow Directions. Flow directions shall be indicated on the system.

1205.0 Installation, Testing, and Inspection.

1205.1 Operating Instructions. Operating and maintenance information shall be provided to the building owner.

1205.2 Pressure Testing. System piping and components shall be tested with a pressure of not less than one and one-half times the operating pressure but not less than 100 psi (689 kPa). Piping shall be tested with water or air except that plastic pipe shall not be tested with air. Test pressures shall be held for a period of not less than 30 minutes with no perceptible drop in pressure. These tests shall be made in the presence of the Authority Having Jurisdiction.

Exception: For PEX, PP-R, PP-RCT, PEX-AL-PEX, PE-RT, and PE-AL-PE piping systems, testing with air shall be permitted where authorized by the manufacturer's instructions for the PEX, PP-R, PP-RCT, PEX-AL-PEX, PE-RT, and PE-AL-PE pipe and fittings products, and air testing is not prohibited by applicable codes, laws, or regulations outside this code.

Note: [HCD 1 & HCD 2] *PEX-AL-PEX is not adopted in the California Plumbing Code for use in potable water supply and distribution systems.*

1205.3 Flushing. Heating and cooling sources, system piping and tubing shall be flushed after installation with water or a cleaning solution. Cleaning and flushing of the heating and cooling sources shall comply with the manufacturer's instructions. The cleaning solution shall be compatible with all system components and shall be used in accordance with the manufacturer's instructions.

1206.0 Pressure and Safety Devices.

1206.1 General. Each closed hydronic system shall be protected against pressures exceeding design limitations with not less than one pressure relief valve. Each closed section of the system containing a heat source shall have a relief valve located so that the heat source is not capable of being isolated from a relief device. Pressure relief valves shall be installed in accordance with their listing and the manufacturer's installation instructions.

1206.2 Discharge Piping. The discharge piping serving a temperature relief valve, pressure relief valve, or combination of both shall have no valves, obstructions, or means of isolation and be provided with the following:

(1) Equal to the size of the valve outlet and shall discharge full size to the flood level of the area receiving the discharge and pointing down.

(2) Materials shall be rated at not less than the operating temperature of the system and approved for such use.

(3) Discharge pipe shall discharge independently by gravity through an air gap into the drainage system or outside of the building with the end of the pipe not exceeding 2 feet (610 mm) and not less than 6 inches (152 mm) above the ground and pointing downwards.

(4) Discharge in such a manner that does not cause personal injury or structural damage.

(5) No part of such discharge pipe shall be trapped or subject to freezing.

(6) The terminal end of the pipe shall not be threaded.

(7) Discharge from a relief valve into a water heater pan shall be prohibited.

1207.0 Heating Appliances and Equipment.

1207.1 General. Heating appliances, equipment, safety and operational controls shall be listed for its intended use in a hydronic heating system and installed in accordance with the manufacturer's installation instructions.

1207.2 Boilers. Boilers and their control systems shall comply with Section 1002.0.

1207.2.1 Condensing Boilers. A condensing boiler, in which the heat exchanger and venting system are designed to operate with condensing flue gases, shall be permitted to be connected directly to the panel heating system without a protective mixing device.

1207.2.2 Noncondensing Boilers. Where the heat exchanger and venting system are not designed to operate with condensed flue gases, the boiler shall be permitted to connect directly to the panel heating system where protected from flue gas condensation. The operating temperature of the boiler shall be more than the fluid temperature in accordance with the manufacturer's instructions.

1207.3 Dual-Purpose Water Heaters. Water heaters used for combined space- and water-heating applications shall be in accordance with the standards referenced in Table 1203.2, and shall be installed in accordance with the manufacturer's installation instructions. Water used as the heat transfer fluid in the hydronic heating system shall be isolated from the potable water supply and distribution in accordance with Section 312.1, Section 1202.0, and Section 1218.0.

1207.3.1 Temperature Limitations. Where a combined space- and water-heating application requires water for space heating at temperatures exceeding 140°F (60°C), a thermostatic mixing valve that is in accordance with ASSE 1017 shall be installed to temper the water supplied to the potable water distribution system to a temperature of 140°F (60°C) or less.

1207.4 Solar Heat Collector Systems. Solar water heating systems used in hydronic panel radiant heating systems shall be installed in accordance with the Uniform Solar Energy Code and Hydronics Code (USEHC).

1208.0 Circulators and Pumps.

1208.1 General. Circulators and pumps shall be selected for their intended use based on the heat transfer fluid, intended operating temperature range and pressure. Circulators and pumps shall be installed to allow for service and maintenance. The manufacturer's installation instructions shall be followed for correct orientation and installation. Motor Operated pumps rated 600V or less shall be listed and labeled in accordance with UL 778.

1208.2 Mounting. The circulator or pump shall be installed in such a way that strain from the piping is not transferred to the circulator or pump housing. The circulator or pump shall be permitted to be directly connected to the piping, provided the piping is supported on each side of the circulator or pump. Where the installation of a circulator or pump will cause strain on the piping, the circulator or pump shall be installed on a mounting bracket or base plate. Where means for controlling vibration of a circulator or pump is required, an approved means for support and restraint shall be provided.

1208.3 Sizing. The selection and sizing of a circulator or pump shall be based on all of the following:

(1) Loop or system head pressure, feet of head (m)

(2) Capacity, gallons per minute (L/s)

(3) Maximum and minimum temperature, °F (°C)

(4) Maximum working pressure, pounds per square inch (kPa)

(5) Fluid type

1209.0 Expansion Tanks.

1209.1 General. An expansion tank shall be installed in each closed hydronic system to control system pressure due to thermal expansion. Expansion tanks shall be of the closed or open type. Tanks shall be rated for the pressure of the system.

1209.2 Installation. Expansion tanks shall be accessible for maintenance and shall be installed in accordance with the manufacturer's installation instructions. Each tank shall be equipped with a shutoff device that will remain open during operation of the heating system. Valve handles shall be locked open or removed to prevent from being inadvertently shut off. Provisions shall be made for draining the tank without emptying the system. Expansion tanks shall be securely fastened to the structure. Supports shall be capable of carrying twice the weight of the tank filled with water without placing a strain on connecting piping. Hot-water-heating systems incorporating hot water tanks or fluid relief columns shall be installed to prevent freezing under normal operating conditions.

1209.3 Open-Type Expansion Tanks. Open type expansion tanks shall be located not less than 3 feet (914 mm) above the highest point of the system. An overflow with a diameter of not less than one-half the size of the supply or not less than 1 inch (25 mm) in diameter shall be installed at the top of the tank. The overflow shall discharge through an air gap into the drainage system.

1209.4 Closed-Type Tanks. Closed-type expansion tanks shall be designed for a hydrostatic test pressure of two and one-half times the allowable working pressure of the system. Expansion tanks for systems designed to operate at more than 30 pounds-force per square inch (psi) (207 kPa) shall comply with ASME BPVC Section VIII.

1209.5 Sizing. Expansion tanks shall be sized to accept the full expansion volume of the fluid in the system. The minimum capacity of a closed-type expansion tank shall be sized in accordance with Section 1004.4.

1210.0 Materials.

1210.1 Piping, Tubing, and Fittings. Hydronic pipe and tubing shall comply with the applicable standards referenced in Table 1210.1 and shall be approved for use based on the intended purpose. Materials shall be rated for the operating temperature and pressure of the system and shall be compatible with the type of heat transfer fluid. Pipe fittings and valves shall be approved for the specific installation with the piping, materials to be installed and shall comply with the applicable standards referenced in Table 1210.1. Where required, exterior piping shall be protected against freezing, UV radiation, corrosion and degradation. Embedded pipe or tubing shall comply with Section 1221.2.

1210.2 Expansion and Contraction. Pipe and tubing shall be so installed that it will not be subject to undue strains or stresses, and provisions shall be made for expansion, contraction, and structural settlement. ***[OSHPD 1, 1R, 2, 4 & 5]*** *Pipe connections less than 2½ inches (64 mm) to heating coils, cooling coils, humidifiers, and similar equipment shall have flexible connectors or three (3) 90-degree offsets in close proximity of the connection.*

1210.3 Hangers and Supports. Pipe and tubing shall be supported in accordance with Section 313.3. Equipment that is part of the piping system shall be provided with additional support in accordance with this code and manufacturer's installation instructions.

1210.4 Oxygen Diffusion Corrosion. PEX and PE-RT tubing in closed hydronic systems shall contain an oxygen barrier.

Exception: Closed hydronic systems without ferrous components in contact with the hydronic fluid.

1211.0 Joints and Connections.

1211.1 General. Joints and connections shall be of an approved type. Joints shall be gas and watertight and designed for the pressure of the hydronic system. Changes in direction shall be made by the use of fittings or with pipe bends. Pipe bends shall have a radius of not less than six times the outside diameter of the tubing or shall be in accordance with the manufacturer's installation instructions. Joints between pipe and fittings shall be installed in accordance with the manufacturer's installation instructions.

1211.2 Chlorinated Polyvinyl Chloride (CPVC) Pipe. Joints between chlorinated polyvinyl chloride (CPVC) pipe or fittings shall be installed in accordance with one of the following methods:

TABLE 1210.1
MATERIALS FOR HYDRONIC SYSTEM PIPING, TUBING, AND FITTINGS

MATERIAL	STANDARDS	
	PIPING/TUBING	FITTINGS
Copper/Copper Alloy	ASTM B42, ASTM B43, ASTM B75, ASTM B88, ASTM B135, ASTM B251[2], ASTM B302, ASTM B447	ASME B16.15, ASME B16.18, ASME B16.22, ASME B16.23, ASME B16.24, ASME B16.26, ASME B16.29, ASME B16.51, ASSE 1061
Ductile Iron	AWWA C115/A21.15, AWWA C151/A21.51	AWWA C110/A21.10[1], AWWA C153/A21.53
Steel	ASTM A53, ASTM A106, ASTM A254	ASME B16.5, ASME B16.9, ASME B16.11, ASTM A420
Gray Iron	—	ASTM A126
Malleable Iron	—	ASME B16.3
Chlorinated Polyvinyl Chloride (CPVC)	ASTM D2846, ASTM F441, ASTM F442, CSA B137.6	ASSE 1061, ASTM D2846, ASTM F437, ASTM F438, ASTM F439, ASTM F1970, CSA B137.6
Polyethylene (PE)	ASTM D1693, ASTM D2513, ASTM D2683, ASTM D2737, ASTM D3035, ASTM D3350, ASTM F714, AWWA C901, CSA B137.1, NSF 358-1	ASTM D2609, ASTM D2683, ASTM D3261, ASTM F1055, CSA B137.1, NSF 358-1
Cross-Linked Polyethylene (PEX)	ASTM F876, CSA B137.5, NSF 358-3	ASSE 1061, ASTM F877, ASTM F1807, ASTM F1960, ASTM F1961, ASTM F2080, ASTM F2098, ASTM F2159, ASTM F2735, CSA B137.5, NSF 358-3
Polypropylene (PP)	ASTM F2389, NSF 358-2	ASTM F2389, NSF 358-2
Polyvinyl Chloride (PVC)	ASTM D1785, ASTM D2241, CSA B137.3	ASTM D2464, ASTM D2466, ASTM D2467, ASTM F1970, CSA B137.2, CSA B137.3
Raised Temperature Polyethylene (PE-RT)	ASTM F2623, ASTM F2769, CSA B137.18	ASSE 1061, ASTM F1807, ASTM F2159, ASTM F2735, ASTM F2769, ASTM D3261, ASTM F1055, CSA B137.18
Cross-Linked Polyethylene/ Aluminum/Cross-Linked Polyethylene (PEX-AL-PEX)	ASTM F1281, ASTM F2262, CSA B137.10	ASTM F1281, ASTM F1974, ASTM F2434, CSA B137.10
Polyethylene/Aluminum/Polyethylene (PE-AL-PE)	ASTM F1282, CSA B137.9	ASTM F1282, ASTM F1974, CSA B137.9
Chlorinated Polyvinyl Chloride/Aluminum/ Chlorinated Polyvinyl Chloride (CPVC/AL/CPVC)	ASTM F2855	ASTM D2846

Note:

[1] Ductile and gray iron.

[2] Only type K, L, or M tubing allowed to be installed.

(1) Mechanical joints shall include flanged, grooved, and push fit fittings. Removable and non-removable push fit fittings with an elastomeric o-ring that employ quick assembly push fit connectors shall be in accordance with ASSE 1061.

(2) Solvent cement joints for CPVC pipe and fittings shall be clean from dirt and moisture. Solvent cements in accordance with ASTM F493, requiring the use of a primer shall be orange in color. The primer shall be colored and be in accordance with ASTM F656. Listed solvent cement in accordance with ASTM F493 that does not require the use of primers, yellow or red in color, shall be permitted for pipe and fittings manufactured in accordance with ASTM D2846, $^{1}/_{2}$ of an inch (15 mm) through 2 inches (50 mm) in diameter or ASTM F442, $^{1}/_{2}$ of an inch (15 mm) through 3 inches (80 mm) in diameter. Apply primer where required inside the fitting and to the depth of the fitting on pipe. Apply liberal coat of cement to the outside surface of pipe to depth of fitting and inside of fitting. Place pipe inside fitting to forcefully bottom the pipe in the socket and hold together until joint is set.

(3) Threaded joints for CPVC pipe shall be made with pipe threads in accordance with ASME B1.20.1. A minimum of Schedule 80 shall be permitted to be threaded, and the pressure rating shall be reduced by 50 percent. The use of

molded fittings shall not result in a 50 percent reduction in the pressure rating of the pipe provided that the molded fittings shall be fabricated so that the wall thickness of the material is maintained at the threads. Thread sealant compound that is compatible with the pipe and fitting, insoluble in water, and nontoxic shall be applied to male threads. Caution shall be used during assembly to prevent over tightening of the CPVC components once the thread sealant has been applied. Female CPVC threaded fittings shall be used with plastic male threads only.

1211.3 CPVC/AL/CPVC Plastic Pipe and Joints. Joints between chlorinated polyvinyl chloride/aluminum/ chlorinated polyvinyl chloride (CPVC/AL/CPVC) pipe or fittings shall be installed in accordance with one of the following methods:

(1) Mechanical joints shall include flanged and grooved.

(2) Solvent cement joints for CPVC/AL/CPVC pipe and fittings shall be clean from dirt and moisture. Solvent cements in accordance with ASTM F493, requiring the use of a primer shall be orange in color. The primer shall be colored and be in accordance with ASTM F656. Listed solvent cement in accordance with ASTM F493 that does not require the use of primers, yellow in color, shall be permitted for pipe and fittings manufactured in accordance with ASTM D2846, ½ of an inch (15 mm) through 2 inches (50 mm) in diameter, ½ of an inch (15 mm) through 3 inches (80 mm) in diameter. Apply primer where required inside the fitting and to the depth of the fitting on pipe. Apply liberal coat of cement to the outside surface of pipe to depth of fitting and inside of fitting. Place pipe inside fitting to forcefully bottom the pipe in the socket and hold together until joint is set.

›› **1211.4 Copper or Copper Alloy Pipe and Tubing.** Joints between copper pipe, tubing, or fittings shall be installed in accordance with one of the following methods:

(1) Brazed joints between copper or copper alloy pipe, tubing, or fittings shall be made with brazing alloys having a liquid temperature above 1000°F (538°C). The joint surfaces to be brazed shall be cleaned bright by either manual or mechanical means. Tubing shall be cut square and reamed to full inside diameter. Brazing flux shall be applied to the joint surfaces where required by manufacturer's recommendation. Brazing filler metal in accordance with AWS A5.8 shall be applied at the point where the pipe or tubing enters the socket of the fitting.

(2) Flared joints for soft copper or copper alloy tubing shall be made with fittings that are in accordance with the applicable standards referenced in Table 1210.1. Pipe or tubing shall be cut square using an appropriate tubing cutter. The tubing shall be reamed to full inside diameter, resized to round, and expanded with a proper flaring tool.

(3) Mechanically formed tee fittings shall have extracted collars that shall be formed in a continuous operation consisting of drilling a pilot hole and drawing out the pipe or tube surface to form a collar having a height not less than three times the thickness of the branch tube wall. The branch pipe or tube shall be notched to conform to the inner curve of the run pipe or tube and shall have two dimple depth stops to ensure that penetration of the branch pipe or tube into the collar is of a depth for brazing and that the branch pipe or tube does not obstruct the flow in the main line pipe or tube. Dimple depth stops shall be in line with the run of the pipe or tube. The second dimple shall be ¼ of an inch (6.4 mm) above the first and shall serve as a visual point of inspection. Fittings and joints shall be made by brazing. Soldered joints shall not be permitted.

(4) Pressed fittings for copper or copper alloy pipe or tubing shall have an elastomeric o-ring that forms the joint. The pipe or tubing shall be fully inserted into the fitting, and the pipe or tubing marked at the shoulder of the fitting. Pipe or tubing shall be cut square, chamfered, and reamed to full inside diameter. The fitting alignment shall be checked against the mark on the pipe or tubing to ensure the pipe or tubing is inserted into the fitting. The joint shall be pressed using the tool recommended by the manufacturer.

(5) Removable and nonremovable push fit fittings for copper or copper alloy tubing or pipe that employ quick assembly push fit connectors shall be in accordance with ASSE 1061. Push fit fittings for copper pipe or tubing shall have an approved elastomeric o-ring that forms the joint. Pipe or tubing shall be cut square, chamfered, and reamed to full inside diameter. The tubing shall be fully inserted into the fitting, and the tubing marked at the shoulder of the fitting. The fitting alignment shall be checked against the mark on the tubing to ensure the tubing is inserted into the fitting and gripping mechanism has engaged on the pipe.

(6) Soldered joints between copper or copper alloy pipe, tubing, or fittings shall be made in accordance with ASTM B828. Pipe or tubing shall be cut square and reamed to the full inside diameter including the removal of burrs on the outside of the pipe or tubing. Surfaces to be joined shall be cleaned bright by manual or mechanical means. Flux shall be applied to pipe or tubing and fittings and shall be in accordance with ASTM B813, and shall become noncorrosive and nontoxic after soldering. Insert pipe or tubing into the base of the fitting and remove excess flux. Pipe or tubing and fitting shall be supported to ensure a uniform capillary space around the joint. Solder in accordance with ASTM B32 shall be applied to the joint surfaces until capillary action draws the molten solder into the cup. Joint surfaces shall not be disturbed until cool, and any remaining flux residue shall be cleaned.

(7) Threaded joints for copper or copper alloy pipe shall be made with pipe threads in accordance with ASME B1.20.1. Thread sealant tape or compound shall be applied only on male threads, and such material shall be of approved types, insoluble in water, and nontoxic.

1211.5 Cross-Linked Polyethylene (PEX) Pipe. Joints between cross-linked polyethylene (PEX) pipe or fittings shall be installed with fittings for PEX tubing that comply with the applicable standards referenced in Table 1210.1. PEX tubing labeled in accordance with ASTM F876 shall be marked with the applicable standard designation for the fittings specified for use with the tubing. Mechanical joints shall be installed in accordance with the manufacturer's installation instructions. ‹‹

1211.6 Cross-Linked Polyethylene/Aluminum/Cross-Linked Polyethylene (PEX-AL-PEX) Pipe. Joints between cross-linked polyethylene/aluminum/cross-linked polyethylene (PEX-AL-PEX) pipe or fittings shall be installed in accordance with one of the following methods:

(1) Mechanical joints between PEX-AL-PEX pipe or fittings shall include mechanical and compression type fittings and insert fittings with a crimping ring. Insert fittings utilizing a crimping ring shall be in accordance with ASTM F1974 or ASTM F2434. Crimp joints for crimp insert fittings shall be joined to PEX-AL-PEX pipe by the compression of a crimp ring around the outer circumference of the pipe, forcing the pipe material into annular spaces formed by ribs on the fitting.

(2) Compression joints shall include compression insert fittings and shall be joined to PEX-AL-PEX pipe through the compression of a split ring or compression nut around the outer circumference of the pipe, forcing the pipe material into the annular space formed by the ribs on the fitting.

Note: [HCD 1 & HCD 2] *PEX-AL-PEX is not adopted in the California Plumbing Code for use in potable water supply and distribution systems.*

1211.7 Ductile Iron Pipe. Joints between ductile iron pipe or fittings shall be installed in accordance with one of the following methods:

(1) Mechanical joints for ductile iron pipe or fittings shall consist of a bell that is cast integrally with the pipe or fitting and provided with an exterior flange having bolt holes and a socket with annular recesses for the sealing gasket and the plain end of the pipe or fitting. The elastomeric gasket shall comply with AWWA C111. Lubricant recommended for the application by the pipe manufacturer shall be applied to the gasket and plain end of the pipe.

(2) Push-on joints for ductile iron pipe or fittings shall consist of a single elastomeric gasket that shall be assembled by positioning the elastomeric gasket in an annular recess in the pipe or fitting socket and forcing the plain end of the pipe or fitting into the socket. The plain end shall compress the elastomeric gasket to form a positive seal and shall be designed so that the elastomeric gasket shall be locked in place against displacement. The elastomeric gasket shall comply with AWWA C111. Lubricant recommended for the application by the pipe manufacturer shall be applied to the gasket and plain end of the pipe.

1211.8 Polyethylene (PE) Plastic Pipe/Tubing. Joints between polyethylene (PE) plastic pipe, tubing, or fittings shall be installed in accordance with one of the following:

(1) Butt-fusion joints shall be installed in accordance with ASTM F2620 and shall be made by heating the squared ends of two pipes, pipe and fitting, or two fittings by holding ends against a heated element. The heated element shall be removed where the proper melt is obtained, and joined ends shall be placed together with applied force.

(2) Electro-fusion joints shall be heated internally by a conductor at the interface of the joint. Align and restrain fitting to pipe to prevent movement and apply electric current to the fitting. Turn off the current when the proper time has elapse to heat the joint. The joint shall fuse together and remain undisturbed until cool.

(3) Socket-fusion joints shall be installed in accordance ASTM F2620 and shall be made by simultaneously heating the outside surface of a pipe end and the inside of a fitting socket. Where the proper melt is obtained, the pipe and fitting shall be joined by inserting one into the other with applied force. The joint shall fuse together and remain undisturbed until cool.

(4) Mechanical joints between PE pipe, tubing, or fittings shall include insert and mechanical compression fittings that provide a pressure seal resistance to pullout. Joints for insert fittings shall be made by cutting the pipe square, using a cutter designed for plastic piping, and removal of sharp edges. Two stainless steel clamps shall be placed over the end of the pipe. Fittings shall be checked for proper size based on the diameter of the pipe. The end of pipe shall be placed over the barbed insert fitting, making contact with the fitting shoulder. Clamps shall be positioned equal to 180 degrees (3.14 rad) apart and shall be tightened to provide a leak tight joint. Compression type couplings and fittings shall be permitted for use in joining PE piping and tubing. Stiffeners that extend beyond the clamp or nut shall be prohibited. Bends shall be not less than 30 pipe diameters or the coil radius where bending with the coil. Bends shall not be permitted closer than 10 pipe diameters of a fitting or valve. Mechanical joints shall be designed for their intended use.

1211.9 Polyethylene/Aluminum/Polyethylene (PE-AL-PE). Joints between polyethylene/aluminum/ polyethylene (PE-AL-PE) pipe or fittings shall be installed in accordance with one of the following methods:

(1) Mechanical joints for PE-AL-PE pipe, tubing, or fittings shall be either of the metal insert fittings with a split ring and compression nut or metal insert fittings with copper crimp rings. Metal insert fittings shall comply with ASTM F1974. Crimp insert fittings shall be joined to the pipe by placing the copper crimp ring around the outer circumference of the pipe, forcing the pipe material into the space formed by the ribs on the fitting until the pipe contacts the shoulder of the fitting. The crimp ring shall then be positioned on the pipe so the edge of the crimp ring is $^1/_8$ of an inch (3.2 mm) to $^1/_4$ of an inch (6.4 mm) from the end of the pipe. The jaws of the crimping tool shall be centered over the crimp ring and tool perpendicular to the barb. The jaws shall be closed around the crimp ring and shall not be crimped more than once.

(2) Compression joints for PE-AL-PE pipe, tubing, or fittings shall be joined through the compression of a split ring, by a compression nut around the circumference of the pipe. The compression nut and split ring shall be placed around the pipe. The ribbed end of the fitting shall be inserted onto the pipe until the pipe contacts the shoulder of the fitting. Position and compress the split ring by tightening the compression nut onto the insert fitting.

1211.10 Polyethylene of Raised Temperature (PE-RT). Joints between polyethylene of raised temperature (PE-RT) tubing or fittings shall be installed with fittings for PE-RT tubing that comply with the applicable standards referenced in Table 1210.1. Metal insert fittings, metal compression fittings, and plastic fittings shall be manufactured to and marked in accordance with the standards for fittings in Table 1210.1.

1211.11 Polypropylene (PP) Pipe. Joints between polypropylene pipe or fittings shall be installed in accordance with one of the following methods:

(1) Heat-fusion joints for polypropylene (PP) pipe shall be installed with socket-type heat-fused polypropylene fittings, butt-fusion polypropylene fittings or pipe, or electrofusion polypropylene fittings. Joint surfaces shall be clean and free from moisture. The joint shall be undisturbed until cool. Joints shall be made in accordance with ASTM F2389 or CSA B137.11.

(2) Mechanical and compression sleeve joints shall be installed in accordance with the manufacturer's installation instructions. Polypropylene pipe shall not be threaded. Polypropylene transition fittings for connection to other piping materials shall only be threaded by the use of copper alloy or stainless steel inserts molded in the fitting.

1211.12 Polyvinyl Chloride (PVC) Pipe. Joints between polyvinyl chloride pipe or fittings shall be installed in accordance with one of the following methods:

(1) Mechanical joints shall be designed to provide a permanent seal and shall be of the mechanical or push-on joint. The mechanical joint shall include a pipe spigot that has a wall thickness to withstand without deformation or collapse; the compressive force exerted where the fitting is tightened. The push-on joint shall have a minimum wall thickness of the bell at any point between the ring and the pipe barrel. The elastomeric gasket shall comply with ASTM D3139, and be of such size and shape as to provide a compressive force against the spigot and socket after assembly to provide a positive seal.

(2) Solvent cement joints for PVC pipe or fittings shall be clean from dirt and moisture. Pipe shall be cut square and pipe shall be deburred. Where surfaces to be joined are cleaned and free of dirt, moisture, oil, and other foreign material, apply primer purple in color in accordance with ASTM F656. Primer shall be applied until the surface of the pipe and fitting is softened. Solvent cements in accordance with ASTM D2564 shall be applied to all joint surfaces. Joints shall be made while both the inside socket surface and outside surface of pipe are wet with solvent cement. Hold joint in place and undisturbed for 1 minute after assembly.

(3) Threads shall comply with ASME B1.20.1. A minimum of Schedule 80 shall be permitted to be threaded; however, the pressure rating shall be reduced by 50 percent. The use of molded fittings shall not result in a 50 percent reduction in the pressure rating of the pipe provided that the molded fittings shall be fabricated so that the wall thickness of the material is maintained at the threads. Thread sealant compound that is compatible with the pipe and fitting, insoluble in water, and nontoxic shall be applied to male threads. Caution shall be used during assembly to prevent over tightening of the PVC components once the thread sealant has been applied. Female PVC threaded fittings shall be used with plastic male threads only.

1211.13 Steel Pipe and Tubing. Joints between steel pipe, tubing, or fittings shall be installed in accordance with one of the following methods:

(1) Mechanical joints shall be made with an approved and listed elastomeric gasket.

(2) Threaded joints shall be made with pipe threads that are in accordance with ASME B1.20.1. Thread sealant tape or compound shall be applied only on male threads, and such material shall be of approved types, insoluble in water, and nontoxic.

(3) Welded joints shall be made by electrical arc or oxygen/acetylene method. Joint surfaces shall be cleaned by an approved procedure. Joints shall be welded by an approved filler metal.

(4) Pressed joints shall have an elastomeric o-ring that forms the connection. The pipe or tubing shall be fully inserted into the fitting, and the pipe or tubing marked at the shoulder of the fittings. Pipe or tubing shall be cut square, chamfered, and reamed to full inside diameter. The fitting alignment shall be checked against the mark on the pipe or tubing to ensure the pipe or tubing is fully inserted into the fitting. The joint shall be pressed using the tool recommended by the manufacturer.

1211.14 Joints Between Various Materials. Joints between various materials shall be installed in accordance with the manufacturer's installation instructions and shall comply with Section 1211.14.1 and Section 1211.14.2.

1211.14.1 Copper or Copper Alloy Pipe or Tubing to Threaded Pipe Joints. Joints from copper or copper alloy pipe or tubing to threaded pipe shall be made by the use of copper alloy adapter, copper alloy nipple [minimum 6 inches (152 mm)], dielectric fitting, or dielectric union in accordance with ASSE 1079. The joint between the copper or copper alloy pipe or tubing and the fitting shall be a soldered, brazed, flared, or pressed joint and the connection between the threaded pipe and the fitting shall be made with a standard pipe size threaded joint.

1211.14.2 Plastic Pipe to Other Materials. Where connecting plastic pipe to other types of piping, approved types of adapter or transition fittings designed for the specific transition intended shall be used.

1212.0 Valves.

1212.1 General. Valves shall be rated for the operating temperature and pressure of the system. Valves shall be compatible with the type of heat transfer medium and piping material.

1212.2 Where Required. Valves shall be installed in hydronic piping systems in accordance with Section 1212.3 through Section 1212.11.

1212.3 Heat Exchanger. Isolation valves shall be installed on the supply and return side of the heat exchanger.

1212.4 Pressure Vessels. Isolation valves shall be installed on connections to pressure vessels.

1212.5 Pressure Reducing Valves. Isolation valves shall be installed on both sides of a pressure reducing valve.

1212.6 Equipment, Components, and Appliances. Serviceable equipment, components, and appliances within the system shall have isolation valves installed upstream and downstream of such devices.

1212.7 Expansion Tank. Isolation valves shall be installed at connections to non-diaphragm-type expansion tanks.

1212.8 Flow Balancing Valves. Where flow balancing valves are installed, such valves shall be capable of increasing or decreasing the amount of flow by means of adjustment.

1212.9 Mixing or Temperature Control Valves. Where mixing or temperature control valves are installed, such valves shall be capable of obtaining the design water temperature and design flow requirements.

1212.10 Thermosiphoning. An approved type check valve shall be installed on liquid heat transfer piping to control thermosiphoning of heated liquids.

1212.11 Air Removal Device or Air Vents. Isolation valves shall be installed where air removal devices or automatic air vents are utilized to permit cleaning, inspection, or repair without shutting the system down.

1213.0 System Controls.

1213.1 Water Temperature Controls. A heat source or system of commonly connected heat sources shall be protected by a water-temperature-activated operating control to stop heat output of the heat source where the system water reaches a pre-set operating temperature.

» **1213.2 Operating Steam Controls.** A steam heat source or system of commonly connected steam heat sources shall be protected by a pressure-actuated control to shut off the fuel supply where the system pressure reaches a pre-set operating pressure.

» **1213.2.1 Water-Level Controls.** A primary water-level control shall be installed on a steam heat source to control the water level in the heat source. The control shall be installed in accordance with the manufacturer's installation instructions.

» **1213.3 Occupied Spaces.** A temperature-sensing device shall be installed in the occupied space to regulate the operation of the hydronic system.

1214.0 Pressure and Flow Controls.

1214.1 Balancing. A means for balancing distribution loops, heat emitting devices, and multiple boiler installations shall be provided in accordance with the manufacturer's instructions. A means for balancing and flow control shall include the piping design, pumping equipment, or balancing devices.

1214.2 Low-Water Control. Direct-fired heat sources within a closed heating system shall have a low-water fuel cut-off device, except as specified in Section 1214.3. Where a low-water control is integral with the heat source as part of the appliance's integrated control and is listed for such use, a separate low-water control shall not be required. An external cut-off device shall be installed in accordance with the heat-source manufacturer's installation instructions. No valve shall be located between the external low-water fuel cut-off and the heat-source unit. Where a pumped condensate return is installed, a second low-water cut-off shall be provided.

1214.3 Flow-Sensing Devices. A direct-fired heat source, requiring forced circulation to prevent overheating, shall have a flow-sensing device installed with the appliance, or such device shall be integral with the appliance. A low-water fuel cut-off device shall not be required.

1214.4 Automatic Makeup Fluid. Where an automatic makeup fluid supply fill device is used to maintain the water content of the heat-source unit, or any closed loop in the system, the makeup supply shall be located at the expansion tank connection or other approved location.

A pressure-reducing valve shall be installed on a makeup water feed line. The pressure of the feed line shall be set in accordance with the design of the system, and connections to potable water shall be in accordance with Section 1202.0 to prevent contamination due to backflow.

1214.5 Differential Pressure Regulation. Provisions shall be made to control zone flows in a multi-zone hydronic system where the closing of some or all of the two-way zone valves causes excess flow through the open zones or deadheading of a fixed-speed pump.

1214.5.1 Differential Pressure Bypass Valve. Where a differential pressure bypass valve is used for the purpose specified in Section 1214.5, it shall be installed and adjusted to provide bypass of the distribution system when most or all of the zones are closed.

1214.6 Air-Removal Device. Provision shall be made for the removal of air in the heat-distribution piping system. The air-removal device shall be located in the area of the hydronic piping system where air is likely to accumulate. Air-removal devices shall be installed to facilitate their removal for examination, repair, or replacement.

Exception: Drainback type solar thermal systems shall not require an air-removal device.

1214.7 Air-Separation Device. To assist with the removal of entrained air, an air-separation device shall be installed in hydronic systems. The device shall be located in accordance with the manufacturer's installation instructions or at the point of no mechanically-induced pressure change within the distribution system.

1214.8 Secondary Loops. Secondary loops that are isolated from the primary heat-distribution loop by a heat exchanger are closed-loop hydronic systems and shall have an expansion tank in accordance with Section 1209.0, an air-removal device in accordance with Section 1214.6, and an air-separation device in accordance with Section 1214.7.

1215.0 Hydronic Space Heating.

1215.1 General. Based on the system design, the heat-distribution units shall be selected in accordance with the manufacturer's specifications.

1215.2 Installation. Heat-distribution units shall be installed in accordance with the manufacturer's installation instructions and this code.

1215.3 Freeze Protection. Hydronic systems and components shall be designed, installed, and protected from freezing.

1215.4 Balancing. System loops shall be installed so that the design flow rates are achieved within the system.

1215.5 Heat Transfer Fluid. The ignitable flash point of heat transfer fluid in a hydronic piping system shall be a minimum of 50°F (28°C) above the maximum system operating temperature. The heat transfer fluid shall be compatible with the makeup fluid supplied to the system.

1216.0 Steam Systems.

1216.1 Steam Traps. For other than one-pipe steam systems, each heat-distribution unit shall be supplied with a steam trap that is listed for the application.

1216.2 Sloping for Two-Pipe System. Two-pipe steam system piping and heat-distribution units shall be sloped down at not less than $^1/_8$ inch per foot (10.4 mm/m) in the direction of the steam flow.

1216.3 Sloping for One-Pipe System. One-pipe steam system piping and heat-distribution units shall be sloped down at not less than $^1/_8$ inch per foot (10.4 mm/m) towards the steam boiler, without trapping.

1216.4 Automatic Air Vents. Steam automatic air vents shall be installed to eliminate air pressure in heat-distribution units on gravity steam piping systems. Air vents shall not be used on a vacuum system.

1216.5 Condensate Flow. System piping shall be installed to allow condensate to flow to the condensate receiver or steam boiler either by gravity or pump-assisted.

1216.6 Steam-Distribution Piping. Where multi-row elements are installed in an enclosure, they shall be top fed and piped in parallel down to the steam trap. A single steam trap for each row of heating elements shall be installed. Where the size of the return header is increased by a minimum of one pipe size, a single steam trap shall be permitted to be installed for multiple rows. Where multiple steam unit heaters are installed, an individual steam trap for each unit shall be installed.

1217.0 Radiant Heating and Cooling.

1217.1 Installation. Radiant heating and cooling systems shall be installed in accordance with the system design.

1217.1.1 Manifolds. Manifolds shall be equipped with isolation valves on the supply and return lines. Manifolds shall be capable of withstanding the pressure and temperature of the system. The material of the manifold shall be compatible with the system fluid and shall be installed in accordance with the manufacturer's installation instructions.

1217.2 Radiant Under-Floor Heating. Floor finished surface temperatures shall not exceed the following temperatures for space heating applications:

(1) 85°F (29°C) in general occupied applications.

(2) 90°F (32°C) in bathrooms, foyers, distribution areas such as hallways and indoor swimming pools.

(3) 88°F (31°C) in industrial spaces

(4) 95°F (35°C) in radiant panel perimeter areas, i.e., up to 2.5 feet (762 mm) from outside walls.

The radiant heating system temperature shall not exceed the maximum temperature rating of the materials used in its construction.

1217.3 Radiant Cooling Systems. Radiant cooling systems shall be designed to minimize the potential for condensation. To prevent condensation on any cooled radiant surface, the supply water temperature for a radiant cooling system shall be not less than 3°F (2°C) above the anticipated space dewpoint temperature, or in accordance with the manufacturer's recommendation. Chilled water piping, valves, and fittings shall be insulated and vapor sealed to prevent surface condensation.

1217.4 Tube Placement. Hydronic radiant system tubing shall be installed in accordance with the manufacturer's installation instructions and with the tube layout and spacing in accordance with the system design. Except for distribution mains, tube spacing and the individual loop lengths shall be installed with a variance of not more than ±10 percent from the design. The maximum loop length of continuous tubing from a supply-and-return manifold shall not exceed the lengths specified by the manufacturer or, in the absence of manufacturer's specifications, the lengths specified in Table 1217.4. Actual loop lengths shall be determined by spacing, flow rate, and pressure drop requirements as specified in the system design.

For the purpose of system balancing, each individual loop shall have a tag securely affixed to the manifold to indicate the length of the loop and the room(s) and area(s) served.

TABLE 1217.4
MAXIMUM LENGTH OF CONTINUOUS TUBING FROM A SUPPLY-AND-RETURN MANIFOLD ARRANGEMENT

NOMINAL TUBE SIZE (inches)	MAXIMUM LOOP LENGTH (feet)
$^1/_4$	125
$^5/_{16}$	200
$^3/_8$	250
$^1/_2$	300
$^5/_8$	400
$^3/_4$	500
1	750

For SI units: 1 inch = 25.4 mm, 1 foot = 304.8 mm

1217.5 Poured Floor Structural Concrete Slab Systems (Thermal Mass). Where tubing is embedded in a structural concrete slab such tubes shall not be larger in outside dimension than one-third of the overall thickness of the slab and shall be spaced not less than three diameters on center except within 10 feet (3048 mm) of the distribution manifold. The top of the tubing shall be embedded in the slab not less than 2 inches (51 mm) below the surface.

1217.5.1 Slab Penetration Tube and Joint Protection. Where embedded in or installed under a concrete slab, tubing shall be protected from damage at penetrations of the slab with protective sleeving approved by the tubing manufacturer. The space between the tubing and sleeve shall be sealed with an approved sealant compatible with the tubing. The tubing at the location of an expansion joint in a concrete slab shall be encased in protective pipe sleeving that covers the tubing not less than 12 inches (305 mm) on either side of the expansion joint or the tubing shall be installed below the slab.

1217.5.2 Insulation. Where a poured concrete radiant floor system is installed in contact with the soil, insulation approved for such an application with a minimum R-value of 5 shall be placed between the soil and the concrete; extend to the outside edges of the concrete; and be placed on all slab edges.

[HCD 1 & HCD 2] *See California Energy Code Section 110.8(g) for insulation requirements for heated slab floors.*

1217.5.3 Joist Systems and Subfloors. Where tubing is installed below a subfloor, the tube spacing shall be in accordance with the system design and joist space limitations.

Where tubing is installed above or in the subfloor, the tube spacing shall not exceed 12 inches (305 mm) center-to-center for living areas.

Where tubing is installed in the joist cavity, the cavity shall be insulated with not less than R-12 material below the heated space.

An air space of not less than 1 inch (25.4 mm) and not more than 3 inches (76 mm) shall be maintained between the top of the insulation and the underside of the floor unless a conductive plate is installed in accordance with manufacturer's instructions.

Where tubing is installed in panels above or in the subfloor and not embedded in concrete, the floor assembly shall be insulated with not less than R-5 material below the tubing when installed over habitable space.

1217.5.4 Wall and Ceiling Panels. Where piping is installed in the stud wall cavity or the ceiling joist cavity, the cavity shall be insulated with not less than R-12 material. The insulation shall be installed in such a manner as to prevent heating or cooling loss from the space intended to be controlled.

An air space of not less than 1 inch (25.4 mm) and not more than 3 inches (76 mm) shall be maintained between the insulation and the interior surface of the panel unless a conductive plate is installed.

[HCD 1 & HCD 2] *See California Energy Code Sections 150.0(j)2 and 120.3(c) for pipe insulation requirements based on pipe diameter.*

1217.6 Radiant Heating and Cooling Panels. Radiant heating and cooling panels shall be installed in accordance with the manufacturer's installation instructions and shall be listed for the application.

1217.6.1 Electric Heating Panel Systems. Clearances for electric heating panels or between outlets, junction boxes, mounting luminaries, ventilating, or other openings shall comply with *the California Electrical Code.*

1217.6.2 Radiant Wall and Ceiling Panels. Radiant panels attached to wood, steel, masonry, or concrete framing members shall be fastened by means of anchors, bolts, or approved screws of sufficient size and anchorage to support the loads applied. Panels shall be installed with corrosion-resistant fasteners. Piping systems shall be designed for thermal expansion to prevent the load being transmitted to the panel.

1218.0 Heat Exchangers.

1218.1 General. Systems utilizing heat exchangers shall protect the potable water system from being contaminated by the heat transfer medium. Systems that incorporate a single-wall heat exchanger to separate potable water from the heat-transfer fluid shall meet the following requirements:

(1) Heat transfer medium is either potable water or contains fluids recognized as safe by the Food and Drug Administration (FDA) as food grade.

(2) A tag or label shall be securely affixed to the heat source with the word, "CAUTION" and the following statements:

   (a) The heat transfer medium shall be water or other nontoxic fluid recognized as safe by the FDA.

   (b) The maximum operating pressure of the heat exchanger shall not exceed the maximum operating pressure of the potable water supply.

(3) The word "CAUTION" and the statements listed above shall have an uppercase height of not less than 0.120 of an inch (3.048 mm). The vertical spacing between lines of type shall be not less than 0.046 of an inch (1.168 mm). Lowercase letters shall be not less than compatible with the uppercase letter size specification.

Systems that do not comply with the requirements for a single-wall heat exchanger shall install a double wall heat exchanger. Double-wall heat exchangers shall separate the potable water from the heat transfer medium by providing a space between the two walls that are vented to the atmosphere.

1219.0 Indirect-Fired Domestic Hot-Water Storage Tanks.

1219.1 General. Domestic hot-water heat exchangers, whether internal or external to the heating appliance, shall be permitted to be used to heat water in domestic hot-water storage tanks. Tanks used to store hot water shall be listed for the intended use and constructed in accordance with nationally recognized standards. A pressure- and temperature-relief valve with a set pressure not exceeding 150 percent of the maximum operating pressure of the system, and at a temperature of 210°F (99°C), shall be installed on the storage tank.

Where the normal operating temperature of the boiler or dual-purpose water heater that provides heat input for domestic hot water exceeds 140°F (60°C), a thermostatically controlled mixing valve as specified in Section 1207.3.1 shall be installed to limit the water supplied to the potable hot water system to a temperature of 140°F (60°C) or less. The potability of the water shall be maintained throughout the system.

1220.0 Auxiliary Systems.

1220.1 Use of Chemical Additives and Corrosive Fluids. Where auxiliary systems contain chemical additives, corrosive fluids, or both not intended or designed for use in the primary system, a double wall heat exchanger shall be used in accordance with Section 1218.1. The chemical additives in the auxiliary systems shall be compatible with auxiliary system components and accepted for use by the heat exchanger manufacturer.

1220.2 Snow and Ice Melt Controls. An automatic thermostatically operating control device that controls the supply hydronic solution temperature to the snow and ice melt area shall be installed in the system. Snow and ice melt systems shall be protected from freezing with a mixture of propylene glycol or ethylene glycol, and water or other approved fluid. Automotive antifreeze shall not be used.

1220.2.1 Tube Placement. Snow and ice melt tubing shall be installed in accordance with the manufacturer's installation instructions and with the tube layout and spacing in accordance with the system design. Except for distribution mains, tube and the individual loop lengths shall be installed with a variance of not more than ±10 percent from the design.

The maximum loop length of continuous tubing from a supply-and-return manifold arrangement shall not exceed the lengths specified by the manufacturer or, in the absence of manufacturer's specifications, the lengths specified in Table 1220.2.1. Actual loop lengths shall be determined by spacing, flow rate, and pressure drop in accordance with the system design.

TABLE 1220.2.1
LOOP LENGTHS FOR SNOW AND ICE MELT SYSTEMS[1,2]

NOMINAL TUBE SIZE (inches)	MAXIMUM ACTIVE LOOP LENGTH (feet)	TOTAL LOOP LENGTH (feet)
PE-RT and PEX Tubing		
1/2	115	140
5/8	225	250
3/4	300	325
1	450	475
Copper Tubing[3]		
1/2	–	140
3/4	–	280

For SI units: 1 inch = 25.4 mm, 1 foot = 304.8 mm

Notes:

1 The total PE-RT and PEX loop lengths consist of two separate sections, the active loop, and the leader length. The active loop is installed within the heated slab. The leader length is the total distance to and from the manifold and heated slab, including any vertical distances.

2 The manifolds shall be installed as close to the snow melt area as possible.

3 In concrete use minimum Type L copper water tubing. In bituminous pavement use a Type K copper water tubing.

1220.2.2 Poured Concrete Slab Systems (Thermal Mass). Where tubes are embedded in a concrete slab, such tubes shall not be larger in outside dimension than one-third of the overall thickness of the slab and shall be spaced not less than three diameters on center. The top of the tubing shall be embedded in the slab not less than 2 inches (51 mm) below the surface.

1220.2.3 Slab Penetration Tube and Joint Protection. Where embedded in or installed under a concrete slab, tubing shall be protected from damage at penetrations of the slab with protective sleeving approved by the tubing manufacturer. The space between the tubing and sleeve shall be sealed with an approved sealant compatible with the tubing. The tubing at the location of an expansion joint in a concrete slab shall be encased in a protective pipe sleeve that covers the tubing not less than 12 inches (305 mm) on either side of the joint or the tubing shall be installed below the slab.

1220.2.4 Concrete Slab Preparation. A solid foundation shall be prepared before the tubing is installed. Compaction shall be used for slabs, sidewalks, and driveways.

1220.2.5 Insulation. Where a poured concrete snow melt system is installed in contact with the soil, insulation that has a R-5 value shall be placed between the concrete and the compacted grade; extend as close as practical to the outside edges of the concrete; and be placed on vertical slab edges that are in contact with plants or landscaping.

[HCD 1 & HCD 2] *See California Energy Code Section 110.8(g) for insulation requirements for heated slab floors.*

1220.2.6 Testing and Flushing. Testing and flushing of auxiliary systems shall be in accordance with Section 1205.0.

1220.3 Hydronic Makeup Air Units. Hydronic makeup air units that are affected by freezing shall be protected against freezing by a hydronic solution.

1221.0 Piping Installation.

1221.1 General. Piping, fittings, and connections shall be installed in accordance with the conditions of their approval and manufacturer's installation instructions.

1221.2 Embedded Piping Materials and Joints. Piping embedded in concrete shall be steel pipe, Type L copper tubing or plastic pipe or tubing rated at not less than 100 psi at 180°F (689 kPa at 82°C). Joints of pipe or tubing that are embedded in a portion of the building, such as concrete or plaster shall be installed in accordance with the requirements of Section 1221.2.1 through Section 1221.2.3.

1221.2.1 Steel Pipe. Steel pipe shall be welded by electrical arc or oxygen/acetylene method.

1221.2.2 Copper Tubing. Copper tubing shall be joined by brazing with filler metals having a melting point not less than 1000°F (538°C).

1221.2.3 Plastics. Plastic pipe and tubing shall be installed in continuous lengths or shall be joined by heat fusion method.

1221.3 Pressure Testing. Piping to be embedded in concrete shall be pressure-tested in accordance with Section 1205.2 prior to pouring concrete. During the pour, the pipe system shall maintain the test pressure of not less than one and one-half times the hydronic system operating pressure and not less than 100 psi (689 kPa). During freezing or the possibility of freezing conditions, testing shall be done with air where permitted by the manufacturer.

1221.4 System Drainage. Hydronic piping systems shall be installed to permit the system to be drained. The system shall drain by indirect waste in accordance with Section 1001.4. Embedded piping underground or under floors is not required to be designed for draining the system.

1221.5 Condensate Drainage. Condensate drains from dehumidifying coils shall be constructed and sloped for condensate removal. Such drains shall be installed in accordance with Section 310.0.

1221.6 Clearance to Combustibles. Hydronic piping where the exterior temperature exceeds 250°F (121°C) shall have a clearance of not less 1 inch (25.4 mm) to combustible materials.

CALIFORNIA MECHANICAL CODE – MATRIX ADOPTION TABLE
CHAPTER 13 – FUEL GAS PIPING

(Matrix Adoption Tables are non-regulatory, intended only as an aid to the code user. See Chapter 1 for state agency authority and building applications.)

Adopting Agency	BSC	BSC-CG	SFM	HCD			DSA			OSHPD						BSCC	DPH	AGR	DWR	CEC	CA	SL	SLC
				1	2	1-AC	AC	SS	SS/CC	1	1R	2	3	4	5								
Adopt Entire Chapter	X		X	X	X			X	X	X	X	X	X	X	X								
Adopt Entire Chapter as amended (amended sections listed below)																							
Adopt only those sections that are listed below																							
Chapter/Section																							

This state agency does not adopt sections identified with the following symbol: †

The Office of the State Fire Marshal's adoption of this chapter or individual sections is applicable to structures regulated by other state agencies pursuant to Section 1.11.0.

CHAPTER 13
FUEL GAS PIPING

1301.0 Scope of Gas Piping.

1301.1 Applicability. The regulations of this chapter shall govern the installation of fuel gas piping in or in connection with a building, structure or within the property lines of premises up to 5 pounds-force per square inch (psi) (34 kPa) for natural gas and 10 psi (69 kPa) for undiluted propane, other than service pipe. Fuel oil piping systems shall be installed in accordance with NFPA 31.

1302.0 Coverage of Piping System.

1302.1 General. Coverage of piping systems shall extend from the point of delivery to the appliance connections. For other than undiluted liquefied petroleum gas (LP-Gas) systems, the point of delivery shall be the outlet of the service meter assembly or the outlet of the service regulator or service shutoff valve where no meter is provided. For undiluted LP-Gas systems, the point of delivery shall be considered to be the outlet of the final pressure regulator, exclusive of line gas regulators where no meter is installed. Where a meter is installed, the point of delivery shall be the outlet of the meter. [NFPA 54:1.1.1.1(A)]

1302.2 Piping System Requirements. Requirements for piping systems shall include design, materials, components, fabrication, assembly, installation, testing, inspection, operation, and maintenance. [NFPA 54:1.1.1.1(C)]

1302.3 Applications. This code shall not apply to the following items (reference standards for some of which appear in Chapter 17):

(1) Portable LP-Gas appliances and equipment of all types that are not connected to a fixed fuel piping system.

(2) Installation of appliances such as brooders, dehydrators, dryers, and irrigation equipment used for agricultural purposes.

(3) Raw material (feedstock) applications except for piping to special atmosphere generators.

(4) Oxygen-fuel gas cutting and welding systems.

(5) Industrial gas applications using such gases as acetylene and acetylenic compounds, hydrogen, ammonia, carbon monoxide, oxygen, and nitrogen.

(6) Petroleum refineries, pipeline compressor or pumping stations, loading terminals, compounding plants, refinery tank farms, and natural gas processing plants.

(7) Large integrated chemical plants or portions of such plants where flammable or combustible liquids or gases are produced by chemical reactions or used in chemical reactions.

(8) LP-Gas installations at utility gas plants.

(9) Liquefied natural gas (LNG) installations.

(10) Fuel gas piping in electric utility power plants.

(11) Proprietary items of equipment, apparatus, or instruments such as gas-generating sets, compressors, and calorimeters.

(12) LP-Gas equipment for vaporization, gas mixing, and gas manufacturing.

(13) LP-Gas piping for buildings under construction or renovations that is not to become part of the permanent building piping system—that is, temporary fixed piping for building heat.

(14) Installation of LP-Gas systems for railroad switch heating.

(15) Installation of LP-Gas and compressed natural gas (CNG) systems on vehicles.

(16) Gas piping, meters, gas-pressure regulators, and other appurtenances used by the serving gas supplier in distribution of gas, other than undiluted LP-Gas.

(17) Building design and construction, except as specified herein.

(18) Fuel gas systems on recreational vehicles manufactured in accordance with NFPA 1192.

(19) Fuel gas systems using hydrogen as a fuel.

(20) Construction of appliances. [NFPA 54:1.1.1.2]

1303.0 Inspection.

1303.1 Inspection Notification. Upon completion of the installation, alteration, or repair of gas piping, and prior to the use thereof, the Authority Having Jurisdiction shall be notified that such gas piping is ready for inspection.

1303.2 Excavation. Excavations required for the installation of underground piping shall be kept open until such time as the piping has been inspected and approved. Where such piping is covered or concealed before such approval, it shall be exposed upon the direction of the Authority Having Jurisdiction.

1303.3 Type of Inspections. The Authority Having Jurisdiction shall make the following inspections and either shall approve that portion of the work as completed or shall notify the permit holder wherein the same fails to be in accordance with this code.

1303.3.1 Rough Piping Inspection. This inspection shall be made after gas piping authorized by the permit has been installed before such piping has been covered or concealed, or before fixture or appliance has been attached thereto. This inspection shall include a determination that the gas piping size, material, and installation meet the requirements of this code.

1303.3.2 Final Piping Inspection. This inspection shall be made after piping authorized by the permit has been installed and after portions thereof that are to be covered or concealed are so concealed and before fixture, appliance, or shutoff valve has been attached thereto. This inspection shall comply with Section 1313.1. Test gauges used in conducting tests shall be in accordance with Section 1303.3.3 through Section 1303.3.3.4.

1303.3.3 Test Gauges. Tests required by this code, which are performed utilizing dial gauges, shall be limited to gauges having the following pressure graduations or increments.

1303.3.3.1 Pressure Tests (10 psi or less). Required pressure tests of 10 psi (69 kPa) or less shall be performed with gauges of 0.10 psi (0.69 kPa) increments or less.

1303.3.3.2 Pressure Tests (greater than 10 psi to 100 psi). Required pressure tests exceeding 10 psi (69 kPa) but less than or equal to 100 psi (689 kPa) shall be performed with gauges of 1 psi (7 kPa) increments or less.

1303.3.3.3 Pressure Tests (exceeding 100 psi). Required pressure tests exceeding 100 psi (689 kPa) shall be performed with gauges of 2 percent increments or less of the required test pressure.

1303.3.3.4 Pressure Range. Test gauges shall have a pressure range not exceeding twice the test pressure applied.

1303.4 Inspection Waived. In cases where the work authorized by the permit consists of a minor installation of additional piping to piping already connected to a gas meter, the foregoing inspections shall be permitted to be waived at the discretion of the Authority Having Jurisdiction. In this event, the Authority Having Jurisdiction shall make such inspection as deemed advisable in order to be assured that the work has been performed in accordance with the intent of this code.

1304.0 Certificate of Inspection.

1304.1 Issuance. Where upon final piping inspection, the installation is found to be in accordance with the provisions of this code, a certificate of inspection shall be permitted to be issued by the Authority Having Jurisdiction.

1304.2 Gas Supplier. A copy of the certificate of such final piping inspection shall be issued to the serving gas supplier supplying gas to the premises.

1304.3 Unlawful. It shall be unlawful for a serving gas supplier or person furnishing gas, to turn on or cause to be turned on, fuel gas or a gas meter or meters until such certificate of final inspection, as herein provided, has been issued.

1305.0 Authority to Render Gas Service.

1305.1 Authorized Personnel. It shall be unlawful for a person, firm, or corporation, excepting an authorized agent or employee of a person, firm, or corporation engaged in the business of furnishing or supplying gas and whose service pipes supply or connect with the particular premises, to turn on or reconnect gas service in or on a premises where gas service is, at the time, not being rendered.

1305.2 Outlets. It shall be unlawful to turn on or connect gas in or on the premises unless outlets are securely connected to gas appliances or capped or plugged with screw joint fittings.

1306.0 Authority to Disconnect.

1306.1 Disconnection. The Authority Having Jurisdiction or the serving gas supplier is hereby authorized to disconnect gas piping or appliance or both that shall be found not to be in accordance with the requirements of this code or that are found defective and in such condition as to endanger life or property.

1306.2 Notice. Where such disconnection has been made, a notice shall be attached to such gas piping or appliance or both that shall state the same has been disconnected, together with the reasons thereof.

1306.3 Capped Outlets. It shall be unlawful to remove or disconnect gas piping or gas appliance without capping or plugging with a screw joint fitting, the outlet from which said pipe or appliance was removed. Outlets to which gas appliances are not connected shall be left capped and gastight on a piping system that has been installed, altered, or repaired.

Exception: Where an approved listed quick-disconnect device is used.

1307.0 Temporary Use of Gas.

1307.1 General. Where temporary use of gas is desired, and the Authority Having Jurisdiction deems the use necessary, a permit shall be permitted to be issued for such use for a period of time not to exceed that designated by the Authority Having Jurisdiction, provided that such gas piping system otherwise is in accordance with to the requirements of this code regarding material, sizing, and safety.

1308.0 Gas Piping System Design, Materials, and Components.

1308.1 Installation of Piping System. Where required by the Authority Having Jurisdiction, a piping sketch or plan shall be prepared before proceeding with the installation. The plan shall show the proposed location of piping, the size of different branches, the various load demands, and the location of the point of delivery. [NFPA 54:5.1.1]

1308.1.1 Addition to Existing System. When additional appliances are being connected to a gas piping system, the existing piping shall be checked to determine whether it has adequate capacity. If inadequate, the existing system shall be enlarged as required, or separate gas piping of adequate capacity shall be provided. [NFPA 54:5.1.2 – 5.1.2.2]

1308.2 Provision for Location of Point of Delivery. The location of the point of delivery shall be acceptable to the serving gas supplier. [NFPA 54:5.2]

1308.3 Interconnections Between Gas Piping Systems. Where two or more meters, or two or more service regulators where meters are not provided, are located on the same premises and supply separate users, the gas piping systems shall not be interconnected on the outlet side of the meters or service regulators. [NFPA 54:5.3.1]

1308.3.1 Interconnections for Standby Fuels. Where a supplementary gas for standby use is connected downstream from a meter or a service regulator where a meter is not provided, equipment to prevent backflow shall be installed. A three-way valve installed to admit the standby supply and at the same time shut off the regular supply shall be permitted to be used for this purpose. [NFPA 54:5.3.2 – 5.3.2.2]

1308.4 Sizing of Gas Piping Systems. Gas piping systems shall be of such size and so installed as to provide a supply of gas to meet the maximum demand and supply gas to each appliance inlet at not less than the minimum supply pressure required by the appliance. [NFPA 54:5.4.1]

1308.4.1 Maximum Gas Demand. The volumetric flow rate of gas to be provided shall be the sum of the maximum inputs of the appliances served. The volumetric flow rate of gas to be provided shall be adjusted for altitude where the installation is above 2 000 feet (610 m). [NFPA 54:5.4.2.1 – 5.4.2.2] Where the input rating is not indicated, the gas supplier, appliance manufacturer, or a qualified agency shall be contacted or the rating from Table 1308.4.1 shall be used for estimating the volumetric flow rate of gas to be supplied.

The total connected hourly load shall be used as the basis for pipe sizing, assuming all appliances are operating at full capacity simultaneously.

Exception: Sizing shall be permitted to be based upon established load diversity factors. [NFPA 54:5.4.2.3]

1308.4.2 Sizing Methods. Gas piping shall be sized in accordance with one of the following:

(1) Pipe sizing tables or sizing equations in this chapter.

(2) Other approved engineering methods acceptable to the Authority Having Jurisdiction.

(3) Sizing tables included in a listed piping system manufacturer's instructions. [NFPA 54:5.4.3]

1308.4.3 Allowable Pressure Drop. The design pressure loss in any piping system under maximum probable flow conditions, from the point of delivery to the inlet connection of the appliance, shall be such that the supply pressure at the appliance is greater than or equal to the minimum pressure required by the appliance. [NFPA 54:5.4.4]

1308.5 Acceptable Piping Materials and Joining Methods. Materials used for piping systems shall either comply with the requirements of this chapter or be acceptable to the Authority Having Jurisdiction. [NFPA 54:5.6.1.1]

1308.5.1 Materials. Pipe, fittings, valves, or other materials shall not be used again unless they are free of foreign materials and have been ascertained to be adequate for the service intended. [NFPA 54:5.6.1.2]

1308.5.1.1 Other Materials. Material not covered by the standards specifications listed herein shall meet the following criteria:

(1) Be investigated and tested to determine that it is safe and suitable for the proposed service.

(2) Be recommended for that service by the manufacturer.

(3) Be acceptable to the Authority Having Jurisdiction. [NFPA 54:5.6.1.3]

TABLE 1308.4.1
APPROXIMATE GAS INPUT FOR TYPICAL APPLIANCES
[NFPA 54: TABLE A.5.4.2.1]

APPLIANCE	INPUT (Btu/h approx.)
Space Heating Units	
Warm air furnace	
Single family	100 000
Multifamily, per unit	60 000
Hydronic boiler	
Single family	100 000
Multifamily, per unit	60 000
Space and Water Heating Units	
Hydronic boiler	
Single family	120 000
Multifamily, per unit	75 000
Water Heating Appliances	
Water heater, automatic storage 30 to 40 gallon tank	35 000
Water heater, automatic storage 50 gallon tank	50 000
Water heater, automatic instantaneous	
Capacity at 2 gallons per minute	142 800
Capacity at 4 gallons per minute	285 000
Capacity at 6 gallons per minute	428 400
Water heater, domestic, circulating or side-arm	35 000
Cooking Appliances	
Range, freestanding, domestic	65 000
Built-in oven or broiler unit, domestic	25 000
Built-in top unit, domestic	40 000
Other Appliances	
Refrigerator	3000
Clothes dryer, Type 1 (domestic)	35 000
Gas fireplace direct vent	40 000
Gas log	80 000
Barbecue	40 000
Gaslight	2500

For SI units: 1000 British thermal units per hour = 0.293 kW

1308.5.2 Metallic Pipe. Cast-iron pipe shall not be used. [NFPA 54:5.6.2.1]

1308.5.2.1 Steel and Wrought-Iron. Steel and wrought-iron pipe shall be not less than standard weight (Schedule 40) and shall comply with one of the following standards:

(1) ASME B36.10

(2) ASTM A53

(3) ASTM A106 [NFPA 54:5.6.2.2]

1308.5.2.2 Copper and Copper Alloy. Copper and copper alloy pipe shall not be used where the gas contains more than an average of 0.3 grains of hydrogen sulfide per 100 standard cubic feet (scf) of gas (0.7 mg/100 L).

Threaded copper, copper alloy, or aluminum alloy pipe shall not be used with gases corrosive to such material.

1308.5.2.3 Aluminum Alloy. Aluminum alloy pipe shall comply with ASTM B241 (except that the use of alloy 5456 is prohibited), and shall be marked at each end of each length indicating compliance. Aluminum alloy pipe shall be coated to protect against external corrosion where it is in contact with masonry, plaster, or insulation or is subject to repeated wettings by such liquids as water, detergents, or sewage. [NFPA 54:5.6.2.5]

Aluminum alloy pipe shall not be used in exterior locations or underground. [NFPA 54:5.6.2.6]

1308.5.3 Metallic Tubing. Seamless copper, aluminum alloy, or steel tubing shall not be used with gases corrosive to such material. [NFPA 54:5.6.3]

1308.5.3.1 Steel. Steel tubing shall comply with ASTM A254. [NFPA 54:5.6.3.1]

1308.5.3.2 Copper and Copper Alloy. Copper and copper alloy tubing shall not be used where the gas contains more than an average of 0.3 grains of hydrogen sulfide per 100 scf of gas (0.7 mg/100 L). Copper tubing shall comply with standard Type K or L of ASTM B88 or ASTM B280.

1308.5.3.3 Aluminum Alloy. Aluminum alloy tubing shall comply with ASTM B210 or ASTM B241. Aluminum alloy tubing shall be coated to protect against external corrosion where it is in contact with masonry, plaster, insulation, or is subject to repeated wettings by such liquids as water, detergent, or sewage. Aluminum alloy tubing shall not be used in exterior locations or underground. [NFPA 54:5.6.3.3]

1308.5.3.4 Corrugated Stainless Steel. Corrugated stainless steel tubing shall be listed in accordance with CSA LC-1. [NFPA 54:5.6.3.4]

1308.5.4 Plastic Pipe, Tubing, and Fittings. Polyethylene plastic pipe, tubing, and fittings used to supply fuel gas shall conform to ASTM D2513. Pipe to be used shall be marked "gas" and "ASTM D2513." [NFPA 54:5.6.4.1.1] Polyvinyl chloride (PVC) and chlorinated polyvinyl chloride (CPVC) plastic pipe, tubing, and fittings shall not be used to supply fuel gas. [NFPA 54:5.6.4.1.3]

1308.5.4.1 Regulator Vent Piping. Plastic pipe and fittings used to connect regulator vents to remote vent terminations shall be PVC in accordance with UL 651. PVC vent piping shall not be installed indoors. [NFPA 54:5.6.4.2]

1308.5.4.2 Anodeless Risers. Anodeless risers shall comply with Section 1308.5.4.2.1 through Section 1308.5.4.2.3. [NFPA 54:5.6.4.3]

1308.5.4.2.1 Factory-Assembled Anodeless Risers. Factory-assembled anodeless risers shall be recommended by the manufacturer for the gas used and shall be leak-tested by the manufacturer in accordance with written procedures. [NFPA 54:5.6.4.3(1)]

1308.5.4.2.2 Service Head Adapters and Field-Assembled Anodeless Risers. Service head adapters and field assembled anodeless risers incorporating service head adapters shall be recommended by the manufacturer for the gas used and shall be design-certified to be in accordance with the requirements of Category I of ASTM D2513. The manufacturer shall provide the user qualified installation instructions. [NFPA 54:5.6.4.3(2)]

1308.5.4.2.3 Undiluted Liquefied Petroleum Gas Piping. The use of plastic pipe, tubing, and fittings in undiluted LP-Gas piping systems shall be in accordance with NFPA 58. [NFPA 54:5.6.4.3(3)]

1308.5.5 Workmanship and Defects. Gas pipe, tubing, and fittings shall be clear and free from cutting burrs and defects in structure or threading, and shall be thoroughly brushed and chip and scale blown. Defects in pipe, tubing, and fittings shall not be repaired. Defective pipe, tubing, and fittings shall be replaced. [NFPA 54:5.6.5]

1308.5.6 Protective Coating. Where in contact with material or atmosphere exerting a corrosive action, metallic piping and fittings coated with a corrosion-resistant material shall be used. External or internal coatings or linings used on piping or components shall not be considered as adding strength. [NFPA 54:5.6.6]

1308.5.7 Metallic Pipe Threads. Metallic pipe and fitting threads shall be taper pipe threads and shall comply with ASME B1.20.1. [NFPA 54:5.6.7.1]

1308.5.7.1 Damaged Threads. Pipe with threads that are stripped, chipped, corroded, or otherwise damaged shall not be used. Where a weld opens during the operation of cutting or threading, that portion of the pipe shall not be used. [NFPA 54:5.6.7.2]

1308.5.7.2 Number of Threads. Field threading of metallic pipe shall be in accordance with Table 1308.5.7.2. [NFPA 54:5.6.7.3]

TABLE 1308.5.7.2
SPECIFICATIONS FOR THREADING METALLIC PIPE
[NFPA 54: TABLE 5.6.7.3]

IRON PIPE SIZE (inches)	APPROXIMATE LENGTH OF THREADED PORTION (inches)	APPROXIMATE NUMBER OF THREADS TO BE CUT
½	¾	10
¾	¾	10
1	⅞	10
1¼	1	11
1½	1	11
2	1	11
2½	1½	12
3	1½	12
4	1⅝	13

For SI units: 1 inch = 25.4 mm

1308.5.7.3 Thread Joint Compounds. Thread joint compounds shall be resistant to the action of LP-Gas or to any other chemical constituents of the gases to be conducted through the piping. [NFPA 54:5.6.7.4]

1308.5.8 Metallic Piping Joints and Fittings. The type of piping joint used shall be suitable for the pressure and temperature conditions and shall be selected giving consideration to joint tightness and mechanical strength under the service conditions. The joint shall be able to sustain the maximum end force due to the internal pressure and any additional forces due to temperature expansion or contraction, vibration, fatigue, or the weight of the pipe and its contents. [NFPA 54:5.6.8]

1308.5.8.1 Pipe Joints. Pipe joints shall be threaded, flanged, brazed, welded, or press-connect fittings made in accordance with CSA LC-4. Where nonferrous pipe is brazed, the brazing materials shall have a melting point in excess of 1000°F (538°C). Brazing alloys shall not contain more than 0.05 percent phosphorus.

1308.5.8.2 Tubing Joints. Tubing joints shall either be made with approved gas tubing fittings, be brazed with a material having a melting point in excess of 1000°F (538°C), or made by press-connect fittings in accordance with CSA LC-4. Brazing alloys shall not contain more than 0.05 percent phosphorus. [NFPA 54:5.6.8.2]

1308.5.8.3 Flared Joints. Flared joints shall be used in systems constructed from nonferrous pipe and tubing where experience or tests have demonstrated that the joint is approved for the conditions and where provisions are made in the design to prevent separation of the joints. [NFPA 54:5.6.8.3]

1308.5.8.4 Metallic Pipe Fittings (Including Valves, Strainers, Filters). Metallic pipe fittings shall comply with the following:

(1) Threaded fittings in sizes exceeding 4 inches (100 mm) shall not be used unless acceptable to the Authority Having Jurisdiction.

(2) Fittings used with steel or wrought-iron pipe shall be steel, copper alloy, bronze, malleable iron, or cast-iron.

(3) Fittings used with copper or copper alloy pipe shall be copper or copper alloy.

(4) Fittings used with aluminum alloy pipe shall be of aluminum alloy.

(5) Cast-iron fittings shall comply with the following:

   (a) Flanges shall be permitted.

   (b) Bushings shall not be used.

   (c) Fittings shall not be used in systems containing flammable gas-air mixtures.

   (d) Fittings in sizes 4 inches (100 mm) and larger shall not be used indoors unless approved by the Authority Having Jurisdiction.

   (e) Fittings in sizes 6 inches (150 mm) and larger shall not be used unless approved by the Authority Having Jurisdiction.

(6) Aluminum alloy fitting threads shall not form the joint seal.

(7) Zinc-aluminum alloy fittings shall not be used in systems containing flammable gas-air mixtures.

(8) Special fittings such as couplings; proprietary-type joints; saddle tees; gland-type compression fittings; and flared, flareless, or compression-type tubing fittings shall be as follows:

   (a) Used within the fitting manufacturer's pressure-temperature recommendations.

   (b) Used within the service conditions anticipated with respect to vibration, fatigue, thermal expansion, or contraction.

   (c) Installed or braced to prevent separation of the joint by gas pressure or external physical damage.

   (d) Acceptable to the Authority Having Jurisdiction.

1308.5.9 Plastic Piping, Joints, and Fittings. Plastic pipe, tubing, and fittings shall be installed in accordance with the manufacturer's installation instructions. Section 1308.5.9.1 through Section 1308.5.9.4 shall be observed where making such joints. [NFPA 54:5.6.9]

1308.5.9.1 Joint Design. The joint shall be designed and installed so that the longitudinal pull-out resistance of the joint will be at least equal to the tensile strength of the plastic piping material. [NFPA 54:5.6.9(1)]

1308.5.9.2 Heat-Fusion Joint. Heat-fusion joints shall be made in accordance with qualified procedures that have been established and proven by test to produce gastight joints as strong as the pipe or tubing being joined. Joints shall be made with the joining method recommended by the pipe manufacturer. Heat-fusion fittings shall be marked "ASTM D2513." [NFPA 54:5.6.9(2)]

1308.5.9.3 Compression-Type Mechanical Joints. Where compression-type mechanical joints are used, the gasket material in the fitting shall be compatible with the plastic piping and with the gas distributed by the system. An internal tubular rigid stiffener shall be used in conjunction with the fitting. The stiffener shall be flush with the end of the pipe or tubing and shall extend not less than the outside end of the compression fitting where installed. The stiffener shall be free of rough or sharp edges and shall not be a forced fit in the plastic. Split tubular stiffeners shall not be used. [NFPA 54:5.6.9(3)]

1308.5.9.4 Liquefied Petroleum Gas Piping Systems. Plastic piping joints and fittings for use in LP-Gas piping systems shall be in accordance with NFPA 58. [NFPA 54:5.6.9(4)]

1308.5.10 Flange Specification. Cast iron flanges shall be in accordance with ASME B16.1. [NFPA 54:5.6.10.1.1]

1308.5.10.1 Steel Flanges. Steel flanges shall be in accordance with the following:

(1) ASME B16.5 or

(2) ASME B16.47. [NFPA 54:5.6.10.1.2]

1308.5.10.2 Non-Ferrous Flanges. Non-ferrous flanges shall be in accordance with ASME B16.24. [NFPA 54:5.6.10.1.3]

1308.5.10.3 Ductile Iron Flanges. Ductile iron flanges shall be in accordance with ASME B16.42. [NFPA 54:5.6.10.1.4]

1308.5.10.4 Dissimilar Flange Connections. Raised-face flanges shall not be joined to flat-faced cast iron, ductile iron or nonferrous material flanges. [NFPA 54:5.6.10.2]

1308.5.10.5 Flange Facings. Standard facings shall be permitted for use under this code. Where 150 psi (1034 kPa) steel flanges are bolted to Class 125 cast-iron flanges, the raised face on the steel flange shall be removed. [NFPA 54:5.6.10.3]

1308.5.10.6 Lapped Flanges. Lapped flanges shall be used only aboveground or in exposed locations accessible for inspection. [NFPA 54:5.6.10.4]

1308.5.11 Flange Gaskets. The material for gaskets shall be capable of withstanding the design temperature and pressure of the piping system and the chemical constituents of the gas being conducted without change to its chemical and physical properties. The effects of fire exposure to the joint shall be considered in choosing the material. [NFPA 54:5.6.11]

1308.5.11.1 Flange Gasket Materials. Acceptable materials shall include the following:

(a) Metal (plain or corrugated)

(b) Composition

(c) Aluminum "O" rings

(d) Spiral-wound metal gaskets

(e) Rubber-faced phenolic

(f) Elastomeric [NFPA 54:5.6.11.1]

1308.5.11.2 Metallic Flange Gaskets. Metallic flange gaskets shall be in accordance with ASME B16.20. [NFPA 54:5.6.11.2.1]

1308.5.11.3 Non-Metallic Flange Gaskets. Non-metallic flange gaskets shall be in accordance with ASME B16.21. [NFPA 54:5.6.11.2.2]

1308.5.11.4 Full-Face Flange Gasket. Full-face flange gaskets shall be used with all non-steel flanges. [NFPA 54:5.6.11.3]

1308.5.11.5 Separated Flanges. When a flange joint is separated, the gasket shall be replaced. [NFPA 54:5.6.11.4]

1308.6 Gas Meters. Gas meters shall be selected for the maximum expected pressure and permissible pressure drop. [NFPA 54:5.7.1]

1308.6.1 Location. Gas meters shall be located in ventilated spaces readily accessible for examination, reading, replacement, or necessary maintenance. [NFPA 54:5.7.2.1]

1308.6.1.1 Subject to Damage. Gas meters shall not be placed where they will be subjected to damage, such as adjacent to a driveway, under a fire escape, in public passages, halls, or where they will be subject to excessive corrosion or vibration. [NFPA 54:5.7.2.2]

1308.6.1.2 Extreme Temperatures. Gas meters shall not be located where they will be subjected to extreme temperatures or sudden extreme changes in temperature or in areas where they are subjected to temperatures beyond those recommended by the manufacturer. [NFPA 54:5.7.2.3]

1308.6.2 Supports. Gas meters shall be supported or connected to rigid piping so as not to exert a strain on the meters. Where flexible connectors are used to connect a gas meter to downstream piping at mobile homes in mobile home parks, the meter shall be supported by a post or bracket placed in a firm footing or by other means providing equivalent support. [NFPA 54:5.7.3]

1308.6.3 Meter Protection. Meters shall be protected against overpressure, backpressure, and vacuum. [NFPA 54:5.7.4]

1308.6.4 Identification. Gas piping at multiple meter installations shall be marked by a metal tag or other permanent means designating the building or the part of the building being supplied and attached by the installing agency. [NFPA 54:5.7.5]

1308.7 Gas Pressure Regulators. A line pressure regulator or gas appliance pressure regulator, as applicable, shall be installed where the gas supply pressure is higher than that at which the branch supply line or appliances are designed to operate or vary beyond design pressure limits. [NFPA 54:5.8.1]

1308.7.1 Listing. Line pressure regulators shall be listed in accordance with CSA Z21.80. [NFPA 54:5.8.2]

1308.7.2 Location. The gas pressure regulator shall be accessible for servicing. [NFPA 54:5.8.3]

1308.7.3 Regulator Protection. Pressure regulators shall be protected against physical damage. [NFPA 54:5.8.4]

1308.7.4 Venting of Line Pressure Regulators. Line pressure regulators shall comply with all of the following:

(1) An independent vent to the exterior of the building, sized in accordance with the regulator manufac-

turer's instructions, shall be provided where the location of a regulator is such that a ruptured diaphragm will cause a hazard. Where more than one regulator is at a location, each regulator shall have a separate vent to the outdoors or, if approved by the Authority Having Jurisdiction, the vent lines shall be permitted to be manifolded in accordance with accepted engineering practices to minimize backpressure in the event of diaphragm failure. Materials for vent piping shall be in accordance with Section 1308.5 through Section 1308.5.11.5.

Exception: A regulator and vent limiting means combination listed as complying with CSA Z21.80 shall be permitted to be used without a vent to the outdoors.

(2) The vent shall be designed to prevent the entry of water, insects, or other foreign materials that could cause blockage.

(3) The regulator vent shall terminate at least 3 feet (914 mm) from a source of ignition.

(4) At locations where regulators might be submerged during floods, a special antiflood-type breather vent fitting shall be installed, or the vent line shall be extended above the height of the expected flood waters.

(5) A regulator shall not be vented to the appliance flue or exhaust system. [NFPA 54:5.8.5.1]

1308.7.5 Venting of Gas Appliance Pressure Regulators. For venting of gas appliance pressure regulators, see Section 902.15. [NFPA 54:5.8.5.2]

1308.7.6 Discharge of Vents. The discharge of vents shall be in accordance with the following:

(1) The discharge stacks, vents, or outlet parts of all pressure-relieving and pressure-limiting devices shall be located so that gas is safely discharged to the outdoors.

(2) Discharge stacks or vents shall be designed to prevent the entry of water, insects, or other foreign material that could cause blockage. The discharge stack or vent line shall be at least the same size as the outlet of the pressure-relieving device. [NFPA 54:5.9.8.1, 5.9.8.2]

1308.7.7 Bypass Piping. Valved and regulated bypasses shall be permitted to be placed around gas line pressure regulators where continuity of service is imperative. [NFPA 54:5.8.6]

1308.7.8 Identification. Line pressure regulators at multiple regulator installations shall be marked by a metal tag or other permanent means designating the building or the part of the building being supplied. [NFPA 54:5.8.7]

1308.8 Overpressure Protection. Where the serving gas supplier delivers gas at a pressure greater than 2 psi for piping systems serving appliances designed to operate at a gas pressure of 14 inches water column or less, overpressure protection devices shall be installed. Piping systems serving equipment designed to operate at inlet pressures greater than 14 inches water column (3.5 kPa) shall be equipped with overpressure protection devices as required by the appliance manufacturer's installation instructions. [NFPA 54:5.9.1]

1308.9 Pressure Limitation Requirements. Where piping systems serving appliances designed to operate with a gas supply pressure of 14 inches water column or less are required to be equipped with overpressure protection by Section 1308.8, each overpressure protection device shall be adjusted to limit the gas pressure to each connected appliance to 2 psi or less upon a failure of the line pressure regulator. [NFPA 54:5.9.2.1]

1308.9.1 Overpressure Protection Required. Where piping systems serving appliances designed to operate with a gas supply pressure greater than 14 inches water column are required to be equipped with overpressure protection by Section 1308.8, each overpressure protection device shall be adjusted to limit the gas pressure to each connected appliance as required by the appliance manufacturer's installation instructions. [NFPA 54:5.9.2.2]

1308.9.2 Overpressure Protection Devices. Each overpressure protection device installed to meet the requirements of this section shall be capable of limiting the pressure to its connected appliance(s) as required by this section independently of any other pressure control equipment in the piping system. [NFPA 54:5.9.2.3]

1308.9.3 Detection of Failure. Each gas piping system for which an overpressure protection device is required by this section shall be designed and installed so that a failure of the primary pressure control device(s) is detectable. [NFPA 54:5.9.2.4]

1308.9.4 Flow Capacity. If a pressure relief valve is used to meet the requirements of this section, it shall have a flow capacity such that the pressure in the protected system is maintained at or below the limits specified in Section 1308.9 under the following conditions:

(1) The line pressure regulator for which the relief valve is providing overpressure protection has failed wide open.

(2) The gas pressure at the inlet of the line pressure regulator for which the relief valve is providing overpressure protection is not less than the regulator's normal operating inlet pressure. [NFPA 54:5.9.2.5]

1308.10 Backpressure Protection. Protective devices shall be installed as close to the equipment as practical where the design of the equipment connected is such that air, oxygen, or standby gases are capable of being forced into the gas supply system. Gas and air combustion mixers incorporating double diaphragm "zero" or "atmosphere" governors or regulators shall require no further protection unless connected directly to compressed air or oxygen at pressures of 5 psi (34 kPa) or more. [NFPA 54:5.10.1]

1308.10.1 Protective Devices. Protective devices shall include, but not be limited to the following:

(1) Check valves.

(2) Three-way valves (of the type that completely closes one side before starting to open the other side).

(3) Reverse flow indicators controlling positive shutoff valves.

(4) Normally closed air-actuated positive shutoff pressure regulators. [NFPA 54:5.10.2]

1308.11 Low-Pressure Protection. A protective device shall be installed between the meter and the appliance or equipment where the operation of the appliance or equipment is such that it is capable of producing a vacuum or a dangerous reduction in gas pressure at the meter. Such protective devices include, but are not limited to, mechanical, diaphragm-operated, or electrically operated low-pressure shutoff valves. [NFPA 54:5.11]

1308.12 Shutoff Valves. Shutoff valves shall be approved and shall be selected giving consideration to pressure drop, service involved, emergency use, and reliability of operation. Shutoff valves of size 1 inch (25 mm) National Pipe Thread and smaller shall be listed. [NFPA 54:5.12]

1308.13 Expansion and Flexibility. Piping systems shall be designed to prevent failure from thermal expansion or contraction. [NFPA 54:5.14.1]

1308.13.1 Special Local Conditions. Where local conditions include earthquake, tornado, unstable ground, or flood hazards, special consideration shall be given to increased strength and flexibility of piping supports and connections. [NFPA 54:5.14.2]

1309.0 Excess Flow Valve.

1309.1 General. Where automatic excess flow valves are installed, they shall be listed to CSA Z21.93 and shall be sized and installed in accordance with the manufacturer's instructions. [NFPA 54:5.13]

1310.0 Gas Piping Installation.

1310.1 Piping Underground. Underground gas piping shall be installed with approved clearance from any other underground structure to avoid contact therewith, to allow maintenance, and to protect against damage from proximity to other structures. In addition, underground plastic piping shall be installed with approved clearance or shall be insulated from any source of heat so as to prevent the heat from impairing the serviceability of the pipe. [NFPA 54:7.1.1]

1310.1.1 Cover Requirements. Underground piping systems shall be installed with a minimum of 12 inches (305 mm) of cover. The minimum cover shall be increased to 18 inches (457 mm) if external damage to the pipe or tubing from external forces is likely to result. Where a minimum of 12 inches (305 mm) of cover cannot be provided, the pipe shall be installed in conduit or bridged (shielded). [NFPA 54:7.1.2.1]

1310.1.2 Trenches. The trench shall be graded so that the pipe has a firm, substantially continuous bearing on the bottom of the trench. [NFPA 54:7.1.2.2]

1310.1.2.1 Backfilling. Where flooding of the trench is done to consolidate the backfill, care shall be exercised to see that the pipe is not floated from its firm bearing on the trench bottom. [NFPA 54:7.1.2.3]

1310.1.3 Protection Against Corrosion. Steel pipe and steel tubing installed underground shall be installed in accordance with Section 1310.1.3.1 through Section 1310.1.3.9. [NFPA 54:7.1.3]

1310.1.3.1 Zinc Coating. Zinc coating (galvanizing) shall not be deemed adequate protection for underground gas piping. [NFPA 54:7.1.3.1]

1310.1.3.2 Underground Piping. Underground piping shall comply with one or more of the following unless approved technical justification is provided to demonstrate that protection is unnecessary:

(1) The piping shall be made of corrosion-resistant material that is suitable for the environment in which it will be installed.

(2) Pipe shall have a factory-applied, electrically insulating coating. Fittings and joints between sections of coated pipe shall be coated in accordance with the coating manufacturer's instructions.

(3) The piping shall have a cathodic protection system installed, and the system shall be maintained in accordance with Section 1310.1.3.3 or Section 1310.1.3.6. [NFPA 54:7.1.3.2]

1310.1.3.3 Cathodic Protection. Cathodic protection systems shall be monitored by testing and the results shall be documented. The test results shall demonstrate one of the following:

(1) A pipe-to-soil voltage of −0.85 volts or more negative is produced, with reference to a saturated copper-copper sulfate half cell.

(2) A pipe-to-soil voltage of −0.78 volts or more negative is produced, with reference to a saturated KCl calomel half cell.

(3) A pipe-to-soil voltage of −0.80 volts or more negative is produced, with reference to a silver-silver chloride half cell.

(4) Compliance with a method described in Appendix D of Title 49 of the Code of Federal Regulations, Part 192. [NFPA 54:7.1.3.3]

1310.1.3.4 Sacrificial Anodes. Sacrificial anodes shall be tested in accordance with the following:

(1) Upon installation of the cathodic protection system, except where prohibited by climatic conditions, in which case the testing shall be performed not later than 180 days after the installation of the system.

(2) 12 to 18 months after the initial test.

(3) Upon successful verification testing in accordance with Section 1310.1.3.4(1) and Section

1310.1.3.4(2), periodic follow-up testing shall be performed at intervals not to exceed 36 months. [NFPA 54:7.1.3.4]

1310.1.3.5 System Failing Tests. Systems failing a test shall be repaired not more than 180 days after the date of the failed testing. The testing schedule shall be restarted as required in Section 1310.1.3.4(1) and Section 1310.1.3.4(2), and the results shall comply with Section 1310.1.3.3. [NFPA 54:7.1.3.5]

1310.1.3.6 Impressed Current Cathodic Protection. Impressed current cathodic protection systems shall be inspected and tested in accordance with the following schedule:

(1) The impressed current rectifier voltage output shall be checked at intervals not exceeding two months.

(2) The pipe-to-soil voltage shall be tested at least annually. [NFPA 54:7.1.3.6]

1310.1.3.7 Documentation. Documentation of the results of the two most recent tests shall be retained. [NFPA 54:7.1.3.7]

1310.1.3.8 Dissimilar Metals. Where dissimilar metals are joined underground, an insulating coupling or fitting shall be used. [NFPA 54:7.1.3.8]

1310.1.3.9 Steel Risers. Steel risers, other than anodeless risers, connected to plastic piping shall be cathodically protected by means of a welded anode. [NFPA 54:7.1.3.9]

1310.1.4 Protection Against Freezing. Where the formation of hydrates or ice is known to occur, piping shall be protected against freezing. [NFPA 54:7.1.4]

1310.1.5 Piping through Foundation Wall. Underground piping, where installed through the outer foundation or basement wall of a building shall be encased in a protective sleeve or protected by an approved device or method. The space between the gas piping and the sleeve and between the sleeve and the wall shall be sealed to prevent entry of gas and water. [NFPA 54:7.1.5]

1310.1.6 Piping Underground Beneath Buildings. Where gas piping is installed underground beneath buildings, the piping shall be either of the following:

(1) Encased in an approved conduit designed to withstand the imposed loads and installed in accordance with Section 1310.1.6.1 or Section 1310.1.6.2.

(2) A piping/encasement system listed for installation beneath buildings. [NFPA 54:7.1.6]

1310.1.6.1 Conduit with One End Terminating Outdoors. The conduit shall extend into an accessible portion of the building and, at the point where the conduit terminates in the building, the space between the conduit and the gas piping shall be sealed to prevent the possible entrance of any gas leakage. Where the end sealing is of a type that retains the full pressure of the pipe, the conduit shall be designed for the same pressure as the pipe. The conduit shall extend at least 4 inches (102 mm) outside the building, be vented outdoors above finished ground level, and be installed so as to prevent the entrance of water and insects. [NFPA 54:7.1.6.1]

1310.1.6.2 Conduit with Both Ends Terminating Indoors. Where the conduit originates and terminates within the same building, the conduit shall originate and terminate in an accessible portion of the building and shall not be sealed. [NFPA 54:7.1.6.2]

1310.1.7 Plastic Piping. Plastic piping shall be installed outdoors, underground only.

Exceptions:

(1) Plastic piping shall be permitted to terminate aboveground where an anodeless riser is used.

(2) Plastic piping shall be permitted to terminate with a wall head adapter aboveground in buildings, including basements, where the plastic piping is inserted in a piping material permitted for use in buildings. [NFPA 54:7.1.7.1]

1310.1.7.1 Connections Between Metallic and Plastic Piping. Connections made between metallic and plastic piping shall be made with fittings conforming to one of the following:

(1) ASTM D2513 Category I transition fittings

(2) ASTM F1973

(3) ASTM F2509 [NFPA 54:7.1.7.2]

1310.1.7.2 Tracer Wire. An electrically continuous corrosion-resistant tracer shall be buried with the plastic pipe to facilitate locating. The tracer shall be one of the following:

(1) A product specifically designed for that purpose.

(2) Insulated copper conductor not less than 14 AWG.

Where tracer wire is used, access shall be provided from aboveground or one end of the tracer wire or tape shall be brought aboveground at a building wall or riser. [NFPA 54:7.1.7.3 – 7.1.7.3.2]

1310.2 Installation of Piping. Piping installed aboveground shall be securely supported and located where it will be protected from physical damage. Where passing through an exterior wall, the piping shall also be protected from corrosion by coating or wrapping with an inert material approved for such applications. The piping shall be sealed around its circumference at the point of the exterior penetration to prevent the entry of water, insects, and rodents. Where piping is encased in a protective pipe sleeve, the annular spaces between the gas piping and the sleeve and between the sleeve and the wall opening shall be sealed. [NFPA 54:7.2.1]

1310.2.1 Building Structure. The installation of gas piping shall not cause structural stresses within building components to exceed allowable design limits. Approval shall be obtained before any beams or joists are cut or notched. [NFPA 54:7.2.2 – 7.2.2.2]

1310.2.2 Gas Piping to be Sloped. Piping for other than dry gas conditions shall be sloped not less than ¼ inch in 15 feet (1.4 mm/m) to prevent traps. [NFPA 54:7.2.3]

1310.2.2.1 Ceiling Locations. Gas piping shall be permitted to be installed in accessible spaces between a fixed ceiling and a dropped ceiling, whether or not such spaces are used as a plenum. Valves shall not be located in such spaces.

Exception: Appliance or equipment shutoff valves required by this code shall be permitted to be installed in accessible spaces containing vented appliances.

1310.2.3 Prohibited Locations. Gas piping inside any building shall not be installed in or through a clothes chute, chimney or gas vent, dumbwaiter, elevator shaft, or air duct, other than combustion air ducts. [NFPA 54:7.2.4]

Exception: Ducts used to provide ventilation air in accordance with Section 701.0 or to above-ceiling spaces in accordance with Section 1310.2.2.1.

1310.2.4 Hangers, Supports, and Anchors. Piping shall be supported with metal pipe hooks, metal pipe straps, metal bands, metal brackets, metal hangers, or building structural components, approved for the size of piping, of adequate strength and quality, and located at intervals so as to prevent or damp out excessive vibration. Piping shall be anchored to prevent undue strains on connected appliances and equipment and shall not be supported by other piping. Pipe hangers and supports shall conform to the requirements of MSS SP-58. [NFPA 54:7.2.5.1]

1310.2.4.1 Spacing. Spacing of supports in gas piping installations shall not be greater than shown in Table 1310.2.4.1. Spacing of supports of CSST shall be in accordance with the CSST manufacturer's instructions. [NFPA 54:7.2.5.2]

1310.2.4.2 Expansion and Contraction. Supports, hangers, and anchors shall be installed so as not to interfere with the free expansion and contraction of the piping between anchors. All parts of the supporting system shall be designed and installed so they are not disengaged by movement of the supported piping. [NFPA 54:7.2.5.3]

1310.2.4.3 Piping on Roof Tops. Gas piping installed on the roof surfaces shall be elevated above the roof surface and shall be supported in accordance with Table 1310.2.4.1. [NFPA 54:7.2.5.4]

1310.2.5 Removal of Piping. Where piping containing gas is to be removed, the line shall be first disconnected from sources of gas and then thoroughly purged with air, water, or inert gas before cutting or welding is done.

TABLE 1310.2.4.1
SUPPORT OF PIPING
[NFPA 54: TABLE 7.2.5.2]

STEEL PIPE, NOMINAL SIZE OF PIPE(inches)	SPACING OF SUPPORTS (feet)	NOMINAL SIZE OF TUBING SMOOTH-WALL (inches O.D.)	SPACING OF SUPPORTS (feet)
½	6	½	4
¾ or 1	8	⅝ or ¾	6
1¼ or larger (horizontal)	10	⅞ or 1 (horizontal)	8
1¼ or larger (vertical)	Every floor level	1 or larger (vertical)	Every floor level

For SI units: 1 inch = 25 mm, 1 foot = 304.8 mm

1310.3 Concealed Piping in Buildings. Gas piping in concealed locations shall be installed in accordance with this section. [NFPA 54:7.3.1]

1310.3.1 Connections. Where gas piping is to be concealed, connections shall be of the following type:

(1) Pipe fittings, such as elbows, tees, couplings, and right/left nipple/couplings.

(2) Joining tubing by brazing (see Section 1308.5.8.2).

(3) Fittings listed for use in concealed spaces or that have been demonstrated to sustain, without leakage, forces due to temperature expansion or contraction, vibration, or fatigue based on their geographic location, application, or operation.

(4) Where necessary to insert fittings in gas pipe that has been installed in a concealed location, the pipe shall be reconnected by welding, flanges, or the use of a right/left nipple/coupling.

1310.3.2 Piping in Partitions. Concealed gas piping shall not be located in solid partitions. [NFPA 54:7.3.3]

1310.3.3 Tubing in Partitions. This provision shall not apply to tubing that pierces walls, floors, or partitions. Tubing installed vertically and horizontally inside hollow walls or partitions without protection along its entire concealed length shall meet the following requirements:

(1) A steel striker barrier not less than 0.0508 of an inch (1.3 mm) thick, or equivalent, is installed between the tubing and the finished wall and extends at least 4 inches (102 mm) beyond concealed penetrations of plates, firestops, wall studs, and so on.

(2) The tubing is installed in single runs and is not rigidly secured. [NFPA 54:7.3.4]

1310.3.4 Piping in Floors. In industrial occupancies, gas piping in solid floors such as concrete shall be laid in channels in the floor and covered to permit access to the piping with a minimum of damage to the building. Where piping in floor channels could be exposed to excessive moisture or corrosive substances, the piping shall be protected in an approved manner. [NFPA 54:7.3.5.1]

Exception: In other than industrial occupancies and where approved by the Authority Having Jurisdiction, gas piping embedded in concrete floor slabs constructed with portland cement shall be surrounded with a minimum of 1½ inches (38 mm) of concrete and shall not be in physical contact with other metallic structures such as reinforcing rods or electrically neutral conductors. All piping, fittings, and risers shall be protected against corrosion in accordance with Section 1308.5.6. Piping shall not be embedded in concrete slabs containing quick-set additives or cinder aggregate. [NFPA 54:7.3.5.2]

1310.4 Piping in Vertical Chases. Where gas piping exceeding 5 psi (34 kPa) is located within vertical chases in accordance with Section 1310.5(2), the requirements of Section 1310.4.1 through Section 1310.4.3 shall apply. [NFPA 54:7.4]

1310.4.1 Pressure Reduction. Where pressure reduction is required in branch connections for compliance with Section 1310.5, such reduction shall take place either inside the chase or immediately adjacent to the outside wall of the chase. Regulator venting and downstream overpressure protection shall comply with Section 1308.7.4 and Section 1308.8 through Section 1308.9.4. The regulator shall be accessible for service and repair, and vented in accordance with one of the following:

(1) Where the fuel gas is lighter than air, regulators equipped with a vent limiting means shall be permitted to be vented into the chase. Regulators not equipped with a vent limiting means shall be permitted to be vented either directly to the outdoors or to a point within the top 1 foot (305 mm) of the chase.

(2) Where the fuel gas is heavier than air, the regulator vent shall be vented only directly to the outdoors. [NFPA 54:7.4.1]

1310.4.2 Construction. Chase construction shall comply with local building codes with respect to fire resistance and protection of horizontal and vertical openings. [NFPA 54:7.4.2]

1310.4.3 Ventilation. A chase shall be ventilated to the outdoors and only at the top. The opening(s) shall have a minimum free area [in square inches (square meters)] equal to the product of one-half of the maximum pressure in the piping [in pounds per square inch (kilopascals)] times the largest nominal diameter of that piping [in inches (millimeters)], or the cross-sectional area of the chase, whichever is smaller. Where more than one fuel gas piping system is present, the free area for each system shall be calculated and the largest area used. [NFPA 54:7.4.3]

1310.5 Maximum Design Operating Pressure. The maximum design operating pressure for piping systems located inside buildings shall not exceed 5 psi (34 kPa) unless one or more of the following conditions are met:

(1) The piping system is welded.

(2) The piping is located in a ventilated chase or otherwise enclosed for protection against accidental gas accumulation.

(3) The piping is located inside buildings or separate areas of buildings used exclusively for one of the following:

(a) Industrial processing or heating

(b) Research

(c) Warehousing

(d) Boiler or mechanical rooms

(4) The piping is a temporary installation for buildings under construction.

(5) The piping serves appliances or equipment used for agricultural purposes.

(6) The piping system is an LP-Gas piping system with a design operating pressure greater than 20 psi (138 kPa) and complies with NFPA 58. [NFPA 54:5.5.1]

1310.5.1 LP-Gas Systems. LP-Gas systems designed to operate below -5°F (-21°C) or with butane or a propane-butane mix shall be designed to either accommodate liquid LP-Gas or to prevent LP-Gas vapor from condensing back into a liquid. [NFPA 54:5.5.2]

1310.6 Appliance Overpressure Protection. The maximum operating pressure for piping systems serving appliances designed to operate at 14 inches water column (3.5 kPa) inlet pressure or less shall be 2 pounds-force per square inch gauge (psig) (14 kPa) unless an over pressure protection device designed to limit pressure at the appliance to 2 psig (14 kPa) upon failure of the line gas pressure regulator is installed.

1310.7 Gas Pipe Turns. Changes in direction of gas pipe shall be made by the use of fittings, factory bends, or field bends. [NFPA 54:7.5]

1310.7.1 Metallic Pipe. Metallic pipe bends shall comply with the following:

(1) Bends shall be made only with bending tools and procedures intended for that purpose.

(2) All bends shall be smooth and free from buckling, cracks, or other evidence of mechanical damage.

(3) The longitudinal weld of the pipe shall be near the neutral axis of the bend.

(4) Pipe shall not be bent through an arc of more than 90 degrees.

(5) The inside radius of a bend shall be not less than six times the outside diameter of the pipe. [NFPA 54:7.5.1]

1310.7.2 Plastic Pipe. Plastic pipe bends shall comply with the following:

(1) The pipe shall not be damaged, and the internal diameter of the pipe shall not be effectively reduced.

(2) Joints shall not be located in pipe bends.

(3) The radius of the inner curve of such bends shall not be less than 25 times the inside diameter of the pipe.

(4) Where the piping manufacturer specifies the use of special bending tools or procedures, such tools or procedures shall be used. [NFPA 54:7.5.2]

1310.7.3 Elbows. Factory-made welding elbows or transverse segments cut therefrom shall have an arc length measured along the crotch of at least 1 inch (25 mm) for pipe sizes 2 inches (50 mm) and larger. [NFPA 54:7.5.3]

1310.8 Drips and Sediment Traps. For other than dry gas conditions, a drip shall be provided at any point in the line of pipe where condensate could collect. Where required by the Authority Having Jurisdiction or the serving gas supplier, a drip shall also be provided at the outlet of the meter. This drip shall be installed so as to constitute a trap wherein an accumulation of condensate shuts off the flow of gas before it runs back into the meter. [NFPA 54:7.6.1]

1310.8.1 Location of Drips. All drips shall be installed only in such locations that they are readily accessible to permit cleaning or emptying. A drip shall not be located where the condensate is likely to freeze. [NFPA 54:7.6.2]

1310.8.2 Sediment Traps. The installation of sediment traps shall be in accordance with Section 1312.9. [NFPA 54:7.6.3]

1310.9 Outlets. Outlets shall be located and installed in accordance with the following requirements:

(1) The outlet fittings or piping shall be securely fastened in place.

(2) Outlets shall not be located behind doors.

(3) Outlets shall be located far enough from floors, walls, patios, slabs, and ceilings to permit the use of wrenches without straining, bending, or damaging the piping.

(4) The unthreaded portion of gas piping outlets shall extend not less than 1 inch (25.4 mm) through finished ceilings or indoor or outdoor walls.

(5) The unthreaded portion of gas piping outlets shall extend not less than 2 inches (51 mm) above the surface of floors or outdoor patios or slabs.

(6) The provisions of Section 1310.9(4) and Section 1310.9(5) shall not apply to listed quick-disconnect devices of the flush-mounted type or listed gas convenience outlets. Such devices shall be installed in accordance with the manufacturer's installation instructions. [NFPA 54:7.7.1.1 – 7.7.1.6]

1310.9.1 Cap Outlets. Each outlet, including a valve, shall be closed gastight with a threaded plug or cap immediately after installation and shall be left closed until the appliance or equipment is connected thereto. When an appliance or equipment is disconnected from an outlet and the outlet is not to be used again immediately, it shall be capped or plugged gastight.

Exceptions:

(1) Laboratory appliances installed in accordance with Section 1312.3.1 shall be permitted.

(2) The use of a listed quick-disconnect device with integral shutoff or listed gas convenience outlet shall be permitted. [NFPA 54:7.7.2.1]

1310.9.1.1 Appliance Shutoff Valves. Appliance shutoff valves installed in fireplaces shall be removed and the piping capped gastight where the fireplace is used for solid-fuel burning. [NFPA 54:7.7.2.2]

1310.10 Branch Pipe Connection. When a branch outlet is placed on a main supply line before it is known what size pipe will be connected to it, the outlet shall be of the same size as the line that supplies it. [NFPA 54:7.8]

1310.11 Manual Gas Shutoff Valves. An accessible gas shutoff valve shall be provided upstream of each gas pressure regulator. Where two gas pressure regulators are installed in series in a single gas line, a manual valve shall not be required at the second regulator. [NFPA 54:7.9.1]

1310.11.1 Valves Controlling Multiple Systems. Main gas shutoff valves controlling several gas piping systems shall be readily accessible for operation and installed so as to be protected from physical damage. They shall be marked with a metal tag or other permanent means attached by the installing agency so that the gas piping systems supplied through them can be readily identified. [NFPA 54:7.9.2.1]

1310.11.1.1 Shutoff Valves for Multiple House Lines. In multiple-tenant buildings supplied through a master meter, through one service regulator where a meter is not provided, or where meters or service regulators are not readily accessible from the appliance or equipment location, an individual shutoff valve for each apartment or tenant line shall be provided at a convenient point of general accessibility. In a common system serving a number of individual buildings, shutoff valves shall be installed at each building. [NFPA 54:7.9.2.2]

1310.11.2 Emergency Shutoff Valves. An exterior shutoff valve to permit turning off the gas supply to each building in an emergency shall be provided. The emergency shutoff valves shall be plainly marked as such and their locations posted as required by the Authority Having Jurisdiction. [NFPA 54:7.9.2.3]

1310.11.3 Shutoff Valve for Laboratories. Each laboratory space containing two or more gas outlets installed on tables, benches, or in hoods in educational, research, commercial and industrial occupancies shall have a single shutoff valve through which all such gas outlets are supplied. The shutoff valve shall be accessible, located within the laboratory or adjacent to the laboratory's egress door, and identified. [NFPA 54:7.9.2.4]

1310.12 Prohibited Devices. No device shall be placed inside the gas piping or fittings that reduces the cross-sectional area or otherwise obstructs the free flow of gas, except where proper allowance in the piping system design has been made for such a device and where approved by the Authority Having Jurisdiction. [NFPA 54:7.10]

1310.13 Systems Containing Gas-Air Mixtures Outside the Flammable Range. Where gas-air mixing machines are employed to produce mixtures above or below the flammable range, they shall be provided with stops to prevent adjustment of the mixture to within or approaching the flammable range. [NFPA 54:7.11]

1310.14 Systems Containing Flammable Gas-Air Mixtures. Systems containing flammable gas-air mixtures shall be in accordance with Section 1310.14.1 through Section 1310.14.6.

1310.14.1 Required Components. A central premix system with a flammable mixture in the blower or compressor shall consist of the following components:

(1) Gas-mixing machine in the form of an automatic gas-air proportioning device combined with a downstream blower or compressor.

(2) Flammable mixture piping, minimum Schedule 40.

(3) Automatic firecheck(s).

(4) Safety blowout(s) or backfire preventers for systems utilizing flammable mixture lines above 2½ inches (64 mm) nominal pipe size or the equivalent. [NFPA 54:7.12.1]

1310.14.2 Optional Components. The following components shall also be permitted to be utilized in any type of central premix system:

(1) Flowmeter(s)

(2) Flame arrester(s) [NFPA 54:7.12.2]

1310.14.3 Additional Requirements. Gas-mixing machines shall have nonsparking blowers and shall be constructed so that a flashback does not rupture machine casings. [NFPA 54:7.12.3]

1310.14.4 Special Requirements for Mixing Blowers. A mixing blower system shall be limited to applications with minimum practical lengths of mixture piping, limited to a maximum mixture pressure of 10 inches water column (2.5 kPa) and limited to gases containing no more than 10 percent hydrogen.

The blower shall be equipped with a gas-control valve at its air entrance arranged so that gas is admitted to the airstream, entering the blower in proper proportions for correct combustion by the type of burners employed, the said gas-control valve being of either the zero governor or mechanical ratio valve type that controls the gas and air adjustment simultaneously. No valves or other obstructions shall be installed between the blower discharge and the burner or burners. [NFPA 54:7.12.4]

1310.14.5 Installation of Gas-Mixing Machines. Installation of gas-mixing machines shall comply with the following:

(1) The gas-mixing machine shall be located in a well-ventilated area or in a detached building or cutoff room provided with room construction and explosion vents in accordance with sound engineering principles. Such rooms or below-grade installations shall have adequate positive ventilation.

(2) Where gas-mixing machines are installed in well-ventilated areas, the type of electrical equipment shall be in accordance with *the California Electrical Code*, for general service conditions unless other hazards in the area prevail. Where gas-mixing machines are installed in small detached buildings or cutoff rooms, the electrical equipment and wiring shall be installed in accordance with *the California Electrical Code* for hazardous locations (Articles 500 and 501, Class I, Division 2).

(3) Air intakes for gas-mixing machines using compressors or blowers shall be taken from outdoors whenever practical.

(4) Controls for gas-mixing machines shall include interlocks and a safety shutoff valve of the manual reset type in the gas supply connection to each machine arranged to automatically shut off the gas supply in the event of high or low gas pressure. Except for open-burner installations only, the controls shall be interlocked so that the blower or compressor stops operating following a gas supply failure. Where a system employs pressurized air, means shall be provided to shut off the gas supply in the event of air failure.

(5) Centrifugal gas-mixing machines in parallel shall be reviewed by the user and equipment manufacturer before installation, and means or plans for minimizing the effects of downstream pulsation and equipment overload shall be prepared and utilized as needed. [NFPA 54:7.12.5.1 – 7.12.5.5]

1310.14.6 Use of Automatic Firechecks, Safety Blowouts, or Backfire Preventers. Automatic firechecks and safety blowouts or backfire preventers shall be provided in piping systems distributing flammable air-gas mixtures from gas-mixing machines to protect the piping and the machines in the event of flashback, in accordance with the following:

(1) Approved automatic firechecks shall be installed upstream as close as practical to the burner inlets following the firecheck manufacturer's instructions.

(2) A separate manually operated gas valve shall be provided at each automatic firecheck for shutting off the flow of the gas-air mixture through the firecheck after a flashback has occurred. The valve shall be located upstream as close as practical to the inlet of the automatic firecheck.

Caution: These valves shall not be reopened after a flashback has occurred until the firecheck has cooled sufficiently to prevent re-ignition of the flammable mixture and has been reset properly.

(3) A safety blowout or backfiring preventer shall be provided in the mixture line near the outlet of each gas-mixing machine where the size of the piping is

larger than 2½ inches (65 mm) NPS, or equivalent, to protect the mixing equipment in the event of an explosion passing through an automatic firecheck. The manufacturer's instructions shall be followed when installing these devices, particularly after a disc has burst. The discharge from the safety blowout or backfire preventer shall be located or shielded so that particles from the ruptured disc cannot be directed towards personnel. Wherever there are interconnected installations of gas-mixing machines with safety blowouts or backfire preventers, provision shall be made to keep the mixture from other machines from reaching any ruptured disc opening. Check valves shall not be used for this purpose.

(4) Large-capacity premix systems provided with explosion heads (rupture discs) to relieve excessive pressure in pipelines shall be located at and vented to a safe outdoor location. Provisions shall be provided for automatically shutting off the supply of the gas-air mixture in the event of rupture. [NFPA 54:7.12.6]

1311.0 Electrical Bonding and Grounding.

1311.1 Pipe and Tubing Other than CSST. Each above-ground portion of a gas piping system other than CSST that is likely to become energized shall be electrically continuous and bonded to an effective ground-fault current path. Gas piping, other than CSST, shall be considered to be bonded when it is connected to appliances that are connected to the appliance grounding conductor of the circuit supplying that appliance. [NFPA 54:7.13.1]

1311.2 Bonding of CSST Gas Piping. CSST gas piping systems, and gas piping systems containing one or more segments of CSST, shall be bonded to the electrical service grounding electrode system or, where provided, lightning protection grounding electrode system. [NFPA 54:7.13.2]

1311.2.1 Bonding Jumper Connection. The bonding jumper shall connect to a metallic pipe, pipe fitting, or CSST fitting. [NFPA 54:7.13.2.1]

1311.2.2 Bonding Jumper Size. The bonding jumper shall not be smaller than 6 AWG copper wire or equivalent. [NFPA 54:7.13.2.2]

1311.2.3 Bonding Jumper Length. The length of the jumper between the connection to the gas piping system and the grounding electrode system shall not exceed 75 feet (22 860 mm). Any additional electrodes shall be bonded to the electrical service grounding electrode system or, where provided, lightning protection grounding electrode system. [NFPA 54:7.13.2.3]

1311.2.4 Bonding Connections. Bonding connections shall be in accordance with *the California Electrical Code.* [NFPA 54:7.13.2.4]

1311.2.5 Devices Used for Bonding. Devices used for the bonding connection shall be listed for the application in accordance with UL 467. [NFPA 54:7.13.2.5]

1311.3 Grounding Conductor of Electrode. Gas piping shall not be used as a grounding conductor or electrode. [NFPA 54:7.13.3]

1311.4 Lightning Protection System. Where a lightning protection system is installed, the bonding of the gas piping shall be in accordance with NFPA 780. [NFPA 54:7.13.4]

1311.5 Electrical Circuits. Electrical circuits shall not utilize gas piping or components as conductors.

Exception: Low-voltage (50V or less) control circuits, ignition circuits, and electronic flame detection device circuits shall be permitted to make use of piping or components as a part of an electric circuit. [NFPA 54:7.14]

1311.6 Electrical Connections. All electrical connections between wiring and electrically operated control devices in a piping system shall conform to the requirements of *the California Electrical Code.* [NFPA 54:7.15.1]

1311.6.1 Safety Control. Any essential safety control depending on electric current as the operating medium shall be of a type that shuts off (fail safe) the flow of gas in the event of current failure. [NFPA 54:7.15.2]

1312.0 Appliance Connections to Building Piping.

1312.1 Connecting Appliances and Equipment. Appliances and equipment shall be connected to the building piping in compliance with Section 1312.6 through Section 1312.8 by one of the following:

(1) Rigid metallic pipe and fittings.

(2) Semirigid metallic tubing and metallic fittings. Aluminum alloy tubing shall not be used in exterior locations.

(3) A listed connector in compliance with CSA Z21.24. The connector shall be used in accordance with the manufacturer's installation instructions and shall be in the same room as the appliance. Only one connector shall be used per appliance.

(4) A listed connector in compliance with CSA Z21.75. Only one connector shall be used per appliance.

(5) CSST where installed in accordance with the manufacturer's installation instructions.

(6) Listed nonmetallic gas hose connectors in accordance with Section 1312.3.

(7) Unlisted gas hose connectors for use in laboratories and educational facilities in accordance with Section 1312.4. [NFPA 54:9.6.1]

1312.1.1 Commercial Cooking Appliances. Connectors used with commercial cooking appliances that are moved for cleaning and sanitation purposes shall be installed in accordance with the connector manufacturer's installation instructions. Such connectors shall be listed in accordance with CSA Z21.69. [NFPA 54:9.6.1.3]

1312.1.2 Restraining Device. Movement of appliances with casters shall be limited by a restraining device installed in accordance with the connector and appliance manufacturer's installation instructions. [NFPA 54:9.6.1.4]

1312.2 Suspended Low-Intensity Infrared Tube Heaters. Suspended low-intensity infrared tube heaters shall be connected to the building piping system with a connector listed for the application in accordance with CSA Z21.24 as follows:

(1) The connector shall be installed in accordance with the tube heater installation instructions, and shall be in the same room as the appliance.

(2) Only one connector shall be used per appliance. [NFPA 54:9.6.1.5]

1312.3 Use of Gas Hose Connectors. Listed gas hose connectors shall be used in accordance with the manufacturer's installation instructions and in accordance with Section 1312.3.1 or Section 1312.3.2. [NFPA 54:9.6.2]

1312.3.1 Indoor. Indoor gas hose connectors shall be used only to connect laboratory, shop, and ironing appliances requiring mobility during operation and installed in accordance with the following:

(1) An appliance shutoff valve shall be installed where the connector is attached to the building piping.

(2) The connector shall be of minimum length and shall not exceed 6 feet (1829 mm).

(3) The connector shall not be concealed and shall not extend from one room to another or pass through wall partitions, ceilings, or floors. [NFPA 54:9.6.2(1)]

1312.3.2 Outdoor. Where outdoor gas hose connectors are used to connect portable outdoor appliances, the connector shall be listed in accordance with CSA Z21.54 and installed in accordance with the following:

(1) An appliance shutoff valve, a listed quick-disconnect device, or a listed gas convenience outlet shall be installed where the connector is attached to the supply piping and in such a manner so as to prevent the accumulation of water or foreign matter.

(2) This connection shall be made only in the outdoor area where the appliance is to be used. [NFPA 54:9.6.2(2)]

(3) The connector length shall not exceed 15 feet (4572 mm).

1312.4 Injection (Bunsen) Burners. Injection (Bunsen) burners used in laboratories and educational facilities shall be permitted to be connected to the gas supply by an unlisted hose. [NFPA 54:9.6.3]

1312.5 Connection of Portable and Mobile Industrial Appliances. Where portable industrial appliances or appliances requiring mobility or subject to vibration are connected to the building gas piping system by the use of a flexible hose, the hose shall be suitable and safe for the conditions under which it can be used. [NFPA 54:9.6.4.1]

1312.5.1 Swivel Joints or Couplings. Where industrial appliances requiring mobility are connected to the rigid piping by the use of swivel joints or couplings, the swivel joints or couplings shall be suitable for the service required and only the minimum number required shall be installed. [NFPA 54:9.6.4.2]

1312.5.2 Metal Flexible Connectors. Where industrial appliances subject to vibration are connected to the building piping system by the use of all metal flexible connectors, the connectors shall be suitable for the service required. [NFPA 54:9.6.4.3]

1312.5.3 Flexible Connectors. Where flexible connections are used, they shall be of the minimum practical length and shall not extend from one room to another or pass through any walls, partitions, ceilings, or floors. Flexible connections shall not be used in any concealed location. They shall be protected against physical or thermal damage and shall be provided with gas shutoff valves in readily accessible locations in rigid piping upstream from the flexible connections. [NFPA 54:9.6.4.4]

1312.6 Appliance Shutoff Valves and Connections. Each appliance connected to a piping system shall have an accessible, approved manual shutoff valve with a nondisplaceable valve member, or a listed gas convenience outlet. Appliance shutoff valves and convenience outlets shall serve a single appliance only. The shutoff valve shall be located within 6 feet (1829 mm) of the appliance it serves. Where a connector is used, the valve shall be installed upstream of the connector. A union or flanged connection shall be provided downstream from the valve to permit removal of appliance controls. Shutoff valves serving decorative appliances shall be permitted to be installed in fireplaces if listed for such use. [NFPA 54:9.6.5, 9.6.5.1(A)(B)]

Exceptions:

(1) Shutoff valves shall be permitted to be accessibly located inside or under an appliance where such appliance is removed without removal of the shutoff valve.

(2) Shutoff valves shall be permitted to be accessibly located inside wall heaters and wall furnaces listed for recessed installation where necessary maintenance is performed without removal of the shutoff valve.

1312.7 Quick-Disconnect Devices. Quick-disconnect devices used to connect appliances to the building piping shall be listed to CSA Z21.41. Where installed indoors, an approved manual shutoff valve with a nondisplaceable valve member shall be installed upstream of the quick-disconnect device. [NFPA 54:9.6.6 – 9.6.6.2]

1312.8 Gas Convenience Outlets. Appliances shall be permitted to be connected to the building piping by means of a listed gas convenience outlet, in conjunction with a listed appliance connector, installed in accordance with the manufacturer's installation instructions.

Gas convenience outlets shall be listed in accordance with CSA Z21.90 and installed in accordance with the manufacturer's installation instructions. [NFPA 54:9.6.7]

1312.9 Sediment Trap. Where a sediment trap is not incorporated as a part of the appliance, a sediment trap shall be installed downstream of the appliance shutoff valve as close to the inlet of the appliance as practical, but before the flex connector, where used at the time of appliance installation. The sediment trap shall be either a tee fitting with a capped nipple in the bottom outlet, as illustrated in Figure 1312.9 or other device recognized as an effective sediment trap. Illuminating appliances, ranges, clothes dryers, decorative appliances for installation in vented fireplaces, gas fireplaces, and outdoor grills shall not be required to be so equipped.

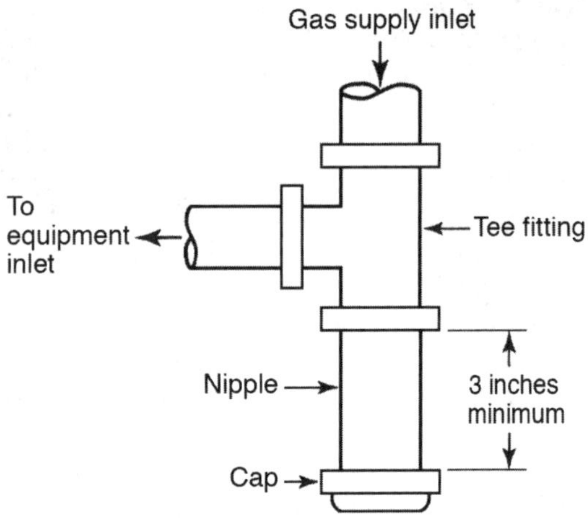

For SI units: 1 inch = 25.4 mm

FIGURE 1312.9
METHOD OF INSTALLING A TEE FITTING SEDIMENT TRAP
[NFPA 54: FIGURE 9.6.8]

1312.10 Installation of Piping. Piping shall be installed in a manner not to interfere with inspection, maintenance, or servicing of the appliances. [NFPA 54:9.6.9]

1312.11 Liquefied Petroleum Gas Facilities and Piping. Liquefied petroleum gas facilities shall comply with NFPA 58.

1313.0 Pressure Testing and Inspection.

1313.1 Piping Installations. Prior to acceptance and initial operation, all piping installations shall be visually inspected and pressure tested to determine that the materials, design, fabrication, and installation practices comply with the requirements of this code. [NFPA 54:8.1.1.1]

1313.1.1 Inspection Requirements. Inspection shall consist of visual examination, during or after manufacture, fabrication, assembly, or pressure tests. [NFPA 54:8.1.1.2]

1313.1.2 Repairs and Additions. Where repairs or additions are made following the pressure test, the affected piping shall be tested. Minor repairs and additions are not required to be pressure tested, provided that the work is inspected and connections are tested with a noncorrosive leak-detecting fluid or other leak-detecting methods approved by the Authority Having Jurisdiction. [NFPA 54:8.1.1.3]

1313.1.3 New Branches. Where new branches are installed to new appliance(s), only the newly installed branch(es) shall be required to be pressure tested. Connections between the new piping and the existing piping shall be tested with a noncorrosive leak-detecting fluid or approved leak-detecting methods. [NFPA 54:8.1.1.4]

1313.1.4 Piping System. A piping system shall be tested as a complete unit or in sections. Under no circumstances shall a valve in a line be used as a bulkhead between gas in one section of the piping system and test medium in an adjacent section, unless a double block and bleed valve system is installed. A valve shall not be subjected to the test pressure unless it can be determined that the valve, including the valve-closing mechanism, is designed to safely withstand the pressure. [NFPA 54:8.1.1.5]

1313.1.5 Regulators and Valves. Regulator and valve assemblies fabricated independently of the piping system in which they are to be installed shall be permitted to be tested with inert gas or air at the time of fabrication. [NFPA 54:8.1.1.6]

1313.1.6 Test Medium. The test medium shall be air, nitrogen, carbon dioxide, or an inert gas. OXYGEN SHALL NEVER BE USED. [NFPA 54:8.1.2]

1313.2 Test Preparation. Test preparation shall comply with Section 1313.2.1 through Section 1313.2.6.

1313.2.1 Pipe Joints. Pipe joints, including welds, shall be left exposed for examination during the test.

Exception: Covered or concealed pipe end joints that have been previously tested in accordance with this code. [NFPA 54:8.1.3.1]

1313.2.2 Expansion Joints. Expansion joints shall be provided with temporary restraints, if required, for the additional thrust load under test. [NFPA 54:8.1.3.2]

1313.2.3 Appliances and Equipment. Appliances and equipment that are not to be included in the test shall be either disconnected from the piping or isolated by blanks, blind flanges, or caps. Flanged joints at which blinds are inserted to blank off other equipment during the test shall not be required to be tested. [NFPA 54:8.1.3.3]

1313.2.4 Designed for (less than) Operating Pressures. Where the piping system is connected to appliances or equipment designed for operating pressures of less than the test pressure, such appliances or equipment shall be isolated from the piping system by disconnecting them and capping the outlets. [NFPA 54:8.1.3.4]

1313.2.5 Designed for (equal to or more than) Operating Pressures. Where the piping system is connected to appliances or equipment designed for operating pressures equal to or greater than the test pressure, such appliances or equipment shall be isolated from the piping system by closing the individual appliance or equipment shutoff valve(s). [NFPA 54:8.1.3.5]

1313.2.6 Safety. All testing of piping systems shall be performed in a manner that protects the safety of employees and the public during the test. [NFPA 54:8.1.3.6]

1313.3 Test Pressure. This inspection shall include an air, CO_2, or nitrogen pressure test, at which time the gas piping shall stand a pressure of not less than 10 psi (69 kPa) gauge pressure. Test pressures shall be held for a length of time satisfactory to the Authority Having Jurisdiction but in no case less than 15 minutes with no perceptible drop in pressure. For welded piping, and for piping carrying gas at pressures in excess of 14 inches water column (3.5 kPa) pressure, the test pressure shall be not less than 60 psi (414 kPa) and shall be continued for a length of time satisfactory to the Authority Having Jurisdiction, but in no case for less than 30 minutes. For CSST carrying gas at pressures in excess of 14 inches water column (3.5 kPa) pressure, the test pressure shall be 30 psi (207 kPa) for 30 minutes. These tests shall be made using air, CO_2, or nitrogen pressure and shall be made in the presence of the Authority Having Jurisdiction. Necessary apparatus for conducting tests shall be furnished by the permit holder. Test gauges used in conducting test shall be in accordance with Section 1303.3.3.1 through Section 1303.3.3.4.

1313.4 Detection of Leaks and Defects. The piping system shall withstand the test pressure specified without showing any evidence of leakage or other defects. Any reduction of test pressures as indicated by pressure gauges shall be deemed to indicate the presence of a leak unless such reduction can be readily attributed to some other cause. [NFPA 54:8.1.5.1]

1313.4.1 Detecting Leaks. The leakage shall be located by means of an approved gas detector, a noncorrosive leak detection fluid, or other approved leak detection methods. [NFPA 54:8.1.5.2]

1313.4.2 Repair or Replace. Where leakage or other defects are located, the affected portion of the piping system shall be repaired or replaced and retested. [NFPA 54:8.1.5.3]

1313.5 Piping System Leak Test. Leak checks using fuel gas shall be permitted in piping systems that have been pressure-tested in accordance with Section 1313.0. [NFPA 54:8.2.1]

1313.5.1 Turning Gas On. During the process of turning gas on into a system of new gas piping, the entire system shall be inspected to determine that there are no open fittings or ends and that all valves at unused outlets are closed and plugged or capped. [NFPA 54:8.2.2]

1313.5.2 Leak Check. Immediately after the gas is turned on into a new system or into a system that has been initially restored after an interruption of service, the piping system shall be checked for leakage. Where leakage is indicated, the gas supply shall be shut off until the necessary repairs have been made. [NFPA 54:8.2.3]

1313.5.3 Placing Appliances and Equipment in Operation. Appliances and equipment shall not be placed in operation until after the piping system has been checked for leakage in accordance with Section 1313.5.2; the piping system is purged in accordance with Section 1313.6, and connections to the appliance are checked for leakage. [NFPA 54:8.2.4]

1313.6 Purging Requirements. The purging of piping shall be in accordance with Section 1313.6.1 through Section 1313.6.3. [NFPA 54:8.3]

1313.6.1 Piping Systems Required to be Purged Outdoors. The purging of piping systems shall be in accordance with Section 1313.6.1.1 through Section 1313.6.1.4 where the piping system meets either of the following:

(1) The design operating gas pressure is greater than 2 psig (14 kPag).

(2) The piping being purged contains one or more sections of pipe or tubing meeting the size and length criteria of Table 1313.6.1. [NFPA 54:8.3.1]

TABLE 1313.6.1
SIZE AND LENGTH OF PIPING
[NFPA 54: TABLE 8.3.1]*

NOMINAL PIPING SIZE(inches)	LENGTH OF PIPING(feet)
≥ 2½ < 3	> 50
≥ 3 < 4	> 30
≥ 4 < 6	> 15
≥ 6 < 8	> 10
≥ 8	Any length

For SI units: 1 inch = 25 mm; 1 foot = 304.8 mm

* CSST EHD size of 62 is equivalent to nominal 2 inches (50 mm) pipe or tubing size.

1313.6.1.1 Removal from Service. Where existing gas piping is opened, the section that is opened shall be isolated from the gas supply and the line pressure vented in accordance with Section 1313.6.1.3. Where gas piping meeting the criteria of Table 1313.6.1 is removed from service, the residual fuel gas in the piping shall be displaced with an inert gas. [NFPA 54:8.3.1.1]

1313.6.1.2 Placing in Operation. Where gas piping containing air and meeting the criteria of Table 1313.6.1 is placed in operation, the air in the piping shall first be displaced with an inert gas. The inert gas shall then be displaced with fuel gas in accordance with Section 1313.6.1.3. [NFPA 54:8.3.1.2]

1313.6.1.3 Outdoor Discharge of Purged Gases. The open end of a piping system being pressure vented or purged shall discharge directly to an outdoor location. Purging operations shall comply with all of the following requirements:

(1) The point of discharge shall be controlled with a shutoff valve.

(2) The point of discharge shall be located at least 10 feet (3048 mm) from sources of ignition, at least 10 feet (3048 mm) from building openings, and at least 25 feet (7620 mm) from mechanical air intake openings.

(3) During discharge, the open point of discharge shall be continuously attended and monitored with a combustible gas indicator that complies with Section 1313.6.1.4.

(4) Purging operations introducing fuel gas shall be stopped when 90 percent fuel gas by volume is detected within the pipe.

(5) Persons not involved in the purging operations shall be evacuated from all areas within 10 feet (3048 mm) of the point of discharge. [NFPA 54:8.3.1.3]

1313.6.1.4 Combustible Gas Indicator. Combustible gas indicators shall be listed and calibrated in accordance with the manufacturer's instructions. Combustible gas indicators shall numerically display a volume scale from 0 percent to 100 percent in 1 percent or smaller increments. [NFPA 54:8.3.1.4]

1313.6.2 Piping Systems Allowed to be Purged Indoors or Outdoors. The purging of piping systems shall be in accordance with the provisions of Section 1313.6.2.1 where the piping system meets both of the following:

(1) The design operating pressure is 2 psig (14 kPag) or less.

(2) The piping being purged is constructed entirely from pipe or tubing not meeting the size and length criteria of Table 1313.6.1. [NFPA 54:8.3.2]

1313.6.2.1 Purging Procedure. The piping system shall be purged in accordance with one or more of the following:

(1) The piping shall be purged with fuel gas and shall discharge to the outdoors.

(2) The piping shall be purged with fuel gas and shall discharge to the indoors or outdoors through an appliance burner not located in a combustion chamber. Such burner shall be provided with a continuous source of ignition.

(3) The piping shall be purged with fuel gas and shall discharge to the indoors or outdoors through a burner that has a continuous source of ignition and that is designed for such purpose.

(4) The piping shall be purged with fuel gas that is discharged to the indoors or outdoors, and the point of discharge shall be monitored with a listed combustible gas detector in accordance with Section 1313.6.2.2. Purging shall be stopped when fuel gas is detected.

(5) The piping shall be purged by the gas supplier in accordance with written procedures. [NFPA 54:8.3.2.1]

1313.6.2.2 Combustible Gas Detector. Combustible gas detectors shall be listed and calibrated or tested in accordance with the manufacturer's instructions. Combustible gas detectors shall be capable of indicating the presence of fuel gas. [NFPA 54:8.3.2.2]

1313.6.3 Purging Appliances and Equipment. After the piping system has been placed in operation, appliances and equipment shall be purged before being placed into operation. [NFPA 54:8.3.3]

1314.0 Required Gas Supply.

1314.1 General. The following regulations, shall comply with this section and Section 1315.0, shall be the standard for the installation of gas piping. Natural gas regulations and tables are based on the use of gas having a specific gravity of 0.60 and for undiluted liquefied petroleum gas having a specific gravity of 1.50.

Where gas of a different specific gravity is to be delivered, the serving gas supplier shall be permitted to be contacted for specific gravity conversion factors to use in sizing piping systems from the pipe sizing tables in this chapter.

1314.2 Volume. The hourly volume of gas required at each piping outlet shall be taken as not less than the maximum hourly rating as specified by the manufacturer of the appliance or appliances to be connected to each such outlet.

1314.3 Gas Appliances. Where the gas appliances to be installed have not been definitely specified, Table 1308.4.1 shall be permitted to be used as a reference to estimate requirements of typical appliances. To obtain the cubic feet per hour (m^3/h) of gas required, divide the input of the appliances by the average Btu (kW•h) heating value per cubic foot (m^3) of the gas. The average Btu (kW•h) per cubic foot (m^3) of the gas in the area of the installation shall be permitted to be obtained from the serving gas supplier.

1314.4 Size of Piping Outlets. The size of the supply piping outlet for a gas appliance shall be not less than ½ of an inch (15 mm).

The size of a piping outlet for a mobile home shall be not less than ¾ of an inch (20 mm).

1315.0 Required Gas Piping Size.

1315.1 Pipe Sizing Methods. Where the pipe size is to be determined using a method in Section 1315.1.1 through Section 1315.1.3, the diameter of each pipe segment shall be obtained from the pipe sizing tables in Section 1315.2 or from the sizing equations in Section 1315.3. [NFPA 54:6.1]

1315.1.1 Longest Length Method. The pipe size of each section of gas piping shall be determined using the longest length of piping from the point of delivery to the most remote outlet and the load of the section. [NFPA 54:6.1.1]

1315.1.2 Branch Length Method. Pipe shall be sized as follows:

(1) The pipe size of each section of the longest pipe run from the point of delivery to the most remote outlet shall be determined using the longest run of piping and the load of the section.

(2) The pipe size of each section of branch piping not previously sized shall be determined using the length

of piping from the point of delivery to the most remote outlet in each branch and the load of the section. [NFPA 54:6.1.2]

1315.1.3 Hybrid Pressure. The pipe size for each section of higher pressure gas piping shall be determined using the longest length of piping from the point of delivery to the most remote line pressure regulator. The pipe size from the line pressure regulator to each outlet shall be determined using the length of piping from the regulator to the most remote outlet served by the regulator. [NFPA 54:6.1.3]

1315.2 Tables for Sizing Gas Piping Systems. Table 1315.2(1) through Table 1315.2(36) shall be used to size gas piping in conjunction with one of the methods described in Section 1315.1.1 through Section 1315.1.3. [NFPA 54:6.2]

1315.3 Sizing Equations. The inside diameter of smooth wall pipe or tubing shall be determined by Equation 1315.3(1), Equation 1315.3(2), Table 1315.3, and using the equivalent pipe length determined by Section 1315.1.1 through Section 1315.1.3. [NFPA 54:6.4]

EQUATION 1315.3(1)
LOW-PRESSURE GAS FORMULA (LESS THAN 1.5 psi)
[NFPA 54:6.4.1]

$$D = \frac{Q^{0.381}}{19.17\left(\frac{\Delta H}{Cr \times L}\right)^{0.206}}$$

Where:

D = inside diameter of pipe, inches

Q = input rate appliance(s), cubic feet per hour at 60°F and 30 inch mercury column

L = equivalent length of pipe, feet

ΔH = pressure drop, in. water column

Cr = in accordance with Table 1315.3

EQUATION 1315.3(2)
HIGH-PRESSURE GAS FORMULA (1.5 psi AND ABOVE)
[NFPA 54:6.4.2]

$$D = \frac{Q^{0.381}}{18.93\left[\frac{(P_1^2 - P_2^2) \cdot Y}{Cr \times L}\right]^{0.206}}$$

Where:

D = inside diameter of pipe, inches

Q = input rate of appliance(s), cubic feet per hour at 60°F and 30 inch mercury column

P_1 = upstream pressure, psia (P_1 + 14.7)

P_2 = downstream pressure, psia (P_2 + 14.7)

L = equivalent length of pipe, feet

Cr = in accordance with Table 1315.3

Y = in accordance with Table 1315.3

For SI units: 1 cubic foot = 0.0283 m^3, 1000 British thermal units per hour = 0.293 kW, 1 inch = 25 mm, 1 foot = 304.8 mm, 1 pound-force per square inch = 6.8947 kPa, °C = (°F-32)/1.8, 1 inch mercury column = 3.39 kPa, 1 inch water column = 0.249 kPa

TABLE 1315.3
***Cr* AND *Y* FOR NATURAL GAS AND UNDILUTED PROPANE AT STANDARD CONDITIONS**
[NFPA 54: TABLE 6.4.2]

GAS	FORMULA FACTORS	
	Cr	*Y*
Natural Gas	0.6094	0.9992
Undiluted Propane	1.2462	0.9910

1315.4 Sizing of Piping Sections. To determine the size of each section of pipe in a system within the range of Table 1315.2(1) through Table 1315.2(36), proceed as follows:

(1) Measure the length of the pipe from the gas meter location to the most remote outlet on the system.

(2) Select the length in feet column and row showing the distance, or the next longer distance where the table does not give the exact length.

(3) Starting at the most remote outlet, find in the just selected the gas demand for that outlet. Where the exact figure of demand is not shown, choose the next larger figure in the row.

(4) At the top of the column in the table will be found the correct size of pipe.

(5) Using this same row, proceed in a similar manner for each section of pipe serving this outlet. For each section of pipe, determine the total gas demand supplied by that section. Where gas piping sections serve both heating and cooling appliances and the installation prevents both units from operating simultaneously, the larger of the two demand loads needs to be used in sizing these sections.

(6) Size each section of branch piping not previously sized by measuring the distance from the gas meter location to the most remote outlet in that branch and follow the procedures of steps 2, 3, 4, and 5 above. Size branch piping in the order of their distance from the meter location, beginning with the most distant outlet not previously sized.

1315.5 Engineering Methods. For conditions other than those covered by Section 1315.1, such as longer runs or greater gas demands, the size of each gas piping system shall be determined by standard engineering methods acceptable to the Authority Having Jurisdiction, and each such system shall be so designed that the total pressure drop between the meter or other point of supply and an outlet where full demand is being supplied to outlets, shall be in accordance with the requirements of Section 1308.4.

1315.6 Variable Gas Pressures. Where the supply gas pressure exceeds 5 psi (34 kPa) for natural gas and 10 psi (69 kPa) for undiluted propane or is less than 6 inches (1.5 kPa) of water column, or where diversity demand factors are used, the design, pipe, sizing, materials, location, and use of such systems first shall be approved by the Authority Having Jurisdiction. Piping systems designed for pressures exceeding the serving gas supplier's standard delivery pressure shall have prior verification from the gas supplier of the availability of the design pressure.

»

FIGURE 1315.1.1
EXAMPLE ILLUSTRATING USE OF TABLE 1308.4.1 AND TABLE 1315.2(1)

Problem: Determine the required pipe size of each section and outlet of the piping system shown in Figure 1315.1.1. Gas to be used has a specific gravity of 0.60 and 1100 British thermal units (Btu) per cubic foot (0.0114 kW•h/L), delivered at 8 inch water column (1.9 kPa) pressure.

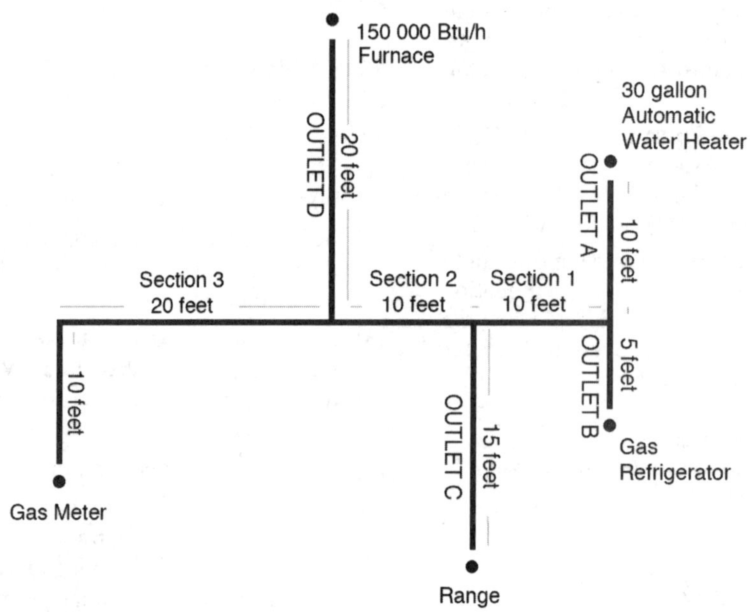

For SI units: 1 foot = 304.8 mm, 1 gallon = 3.785 L, 1000 British thermal units per hour = 0.293 kW, 1 cubic foot per hour = 0.0283 m^3/h

Solution:

(1) Maximum gas demand of Outlet A –
32 cubic feet per hour (0.91 m^3/h) (from Table 1308.4.1).
Maximum gas demand of Outlet B –
3 cubic feet per hour (0.08 m^3/h) (from Table 1308.4.1).
Maximum gas demand of Outlet C –
59 cubic feet per hour (1.67 m^3/h) (from Table 1308.4.1).
Maximum gas demand of Outlet D –
136 cubic feet per hour (3.85 m^3/h) [150 000 Btu/hour (44 kW)] divided by 1100 Btu per cubic foot (0.0114 kW•h/L)

(2) The length of pipe from the gas meter to the most remote outlet (Outlet A) is 60 feet (18 288 mm).

(3) Using the length in feet column row marked 60 feet (18 288 mm) in Table 1315.2(1):
Outlet A, supplying 32 cubic feet per hour (0.91 m^3/h), requires ½ of an inch (15 mm) pipe.
Section 1, supplying Outlets A and B, or 35 cubic feet per hour (0.99 m^3/h) requires ½ of an inch (15 mm) pipe.
Section 2, supplying Outlets A, B, and C, or 94 cubic feet per hour (2.66 m^3/h) requires ¾ of an inch (20 mm) pipe.
Section 3, supplying Outlets A, B, C, and D, or 230 cubic feet per hour (6.51 m^3/h), requires 1 inch (25 mm) pipe.

(4) Using the column marked 60 feet (18 288 mm) in Table 1315.2(1) [no column for actual length of 55 feet (16 764 mm)]:
Outlet B supplying 3 cubic feet per hour (0.08 m^3/h), requires ½ of an inch (15 mm) pipe.
Outlet C, supplying 59 cubic feet per hour (1.67 m^3/h), requires ½ of an inch (15 mm) pipe.
Using the column marked 60 feet (18 288 mm) in Table 1315.2(1):
Outlet D, supplying 136 cubic feet per hour (3.85 m^3/h), requires ¾ of an inch (20 mm) pipe.

TABLE 1315.2(1) SCHEDULE 40 METALLIC PIPE [NFPA 54: TABLE 6.2(b)][1,2]

											GAS:	NATURAL		
											INLET PRESSURE:	LESS THAN 2 psi		
											PRESSURE DROP:	0.5 in. w.c.		
											SPECIFIC GRAVITY:	0.60		
	PIPE SIZE (inch)													
NOMINAL:	½	¾	1	1¼	1½	2	2½	3	4	5	6	8	10	12
ACTUAL ID:	0.622	0.824	1.049	1.380	1.610	2.067	2.469	3.068	4.026	5.047	6.065	7.981	10.020	11.938
LENGTH (feet)	CAPACITY IN CUBIC FEET OF GAS PER HOUR													
10	172	360	678	1390	2090	4020	6400	11 300	23 100	41 800	67 600	139 000	252 000	399 000
20	118	247	466	957	1430	2760	4400	7780	15 900	28 700	46 500	95 500	173 000	275 000
30	95	199	374	768	1150	2220	3530	6250	12 700	23 000	37 300	76 700	139 000	220 000
40	81	170	320	657	985	1900	3020	5350	10 900	19 700	31 900	65 600	119 000	189 000
50	72	151	284	583	873	1680	2680	4740	9660	17 500	28 300	58 200	106 000	167 000
60	65	137	257	528	791	1520	2430	4290	8760	15 800	25 600	52 700	95 700	152 000
70	60	126	237	486	728	1400	2230	3950	8050	14 600	23 600	48 500	88 100	139 000
80	56	117	220	452	677	1300	2080	3670	7490	13 600	22 000	45 100	81 900	130 000
90	52	110	207	424	635	1220	1950	3450	7030	12 700	20 600	42 300	76 900	122 000
100	50	104	195	400	600	1160	1840	3260	6640	12 000	19 500	40 000	72 600	115 000
125	44	92	173	355	532	1020	1630	2890	5890	10 600	17 200	35 400	64 300	102 000
150	40	83	157	322	482	928	1480	2610	5330	9650	15 600	32 100	58 300	92 300
175	37	77	144	296	443	854	1360	2410	4910	8880	14 400	29 500	53 600	84 900
200	34	71	134	275	412	794	1270	2240	4560	8260	13 400	27 500	49 900	79 000
250	30	63	119	244	366	704	1120	1980	4050	7320	11 900	24 300	44 200	70 000
300	27	57	108	221	331	638	1020	1800	3670	6630	10 700	22 100	40 100	63 400
350	25	53	99	203	305	587	935	1650	3370	6100	9880	20 300	36 900	58 400
400	23	49	92	189	283	546	870	1540	3140	5680	9190	18 900	34 300	54 300
450	22	46	86	177	266	512	816	1440	2940	5330	8620	17 700	32 200	50 900
500	21	43	82	168	251	484	771	1360	2780	5030	8150	16 700	30 400	48 100
550	20	41	78	159	239	459	732	1290	2640	4780	7740	15 900	28 900	45 700
600	19	39	74	152	228	438	699	1240	2520	4560	7380	15 200	27 500	43 600
650	18	38	71	145	218	420	669	1180	2410	4360	7070	14 500	26 400	41 800
700	17	36	68	140	209	403	643	1140	2320	4190	6790	14 000	25 300	40 100
750	17	35	66	135	202	389	619	1090	2230	4040	6540	13 400	24 400	38 600
800	16	34	63	130	195	375	598	1060	2160	3900	6320	13 000	23 600	37 300
850	16	33	61	126	189	363	579	1020	2090	3780	6110	12 600	22 800	36 100
900	15	32	59	122	183	352	561	992	2020	3660	5930	12 200	22 100	35 000
950	15	31	58	118	178	342	545	963	1960	3550	5760	11 800	21 500	34 000
1000	14	30	56	115	173	333	530	937	1910	3460	5600	11 500	20 900	33 100
1100	14	28	53	109	164	316	503	890	1810	3280	5320	10 900	19 800	31 400
1200	13	27	51	104	156	301	480	849	1730	3130	5070	10 400	18 900	30 000
1300	12	26	49	100	150	289	460	813	1660	3000	4860	9980	18 100	28 700
1400	12	25	47	96	144	277	442	781	1590	2880	4670	9590	17 400	27 600
1500	11	24	45	93	139	267	426	752	1530	2780	4500	9240	16 800	26 600
1600	11	23	44	89	134	258	411	727	1480	2680	4340	8920	16 200	25 600
1700	11	22	42	86	130	250	398	703	1430	2590	4200	8630	15 700	24 800
1800	10	22	41	84	126	242	386	682	1390	2520	4070	8370	15 200	24 100
1900	10	21	40	81	122	235	375	662	1350	2440	3960	8130	14 800	23 400
2000	NA	20	39	79	119	229	364	644	1310	2380	3850	7910	14 400	22 700

For SI units: 1 inch = 25 mm, 1 foot = 304.8 mm, 1 cubic foot per hour = 0.0283 m^3/h, 1 pound-force per square inch = 6.8947 kPa, 1 inch water column = 0.249 kPa

Notes:

1 Table entries are rounded to 3 significant digits.

2 NA means a flow of less than 10 ft^3/h (0.283 m^3/h).

TABLE 1315.2(2)
SCHEDULE 40 METALLIC PIPE [NFPA 54: TABLE 6.2(c)]*

							GAS:	NATURAL
							INLET PRESSURE:	LESS THAN 2 psi
							PRESSURE DROP:	3.0 in. w.c.
							SPECIFIC GRAVITY:	0.60
INTENDED USE: INITIAL SUPPLY PRESSURE OF 8.0 IN. W.C. OR GREATER								
	PIPE SIZE (inch)							
NOMINAL:	½	¾	1	1¼	1½	2	2½	3	4
ACTUAL ID:	0.622	0.824	1.049	1.380	1.610	2.067	2.469	3.068	4.026
LENGTH (feet)	CAPACITY IN CUBIC FEET OF GAS PER HOUR								
10	454	949	1790	3670	5500	10 600	16 900	29 800	60 800
20	312	652	1230	2520	3780	7280	11 600	20 500	41 800
30	250	524	986	2030	3030	5840	9310	16 500	33 600
40	214	448	844	1730	2600	5000	7970	14 100	28 700
50	190	397	748	1540	2300	4430	7060	12 500	25 500
60	172	360	678	1390	2090	4020	6400	11 300	23 100
70	158	331	624	1280	1920	3690	5890	10 400	21 200
80	147	308	580	1190	1790	3440	5480	9690	19 800
90	138	289	544	1120	1670	3230	5140	9090	18 500
100	131	273	514	1060	1580	3050	4860	8580	17 500
125	116	242	456	936	1400	2700	4300	7610	15 500
150	105	219	413	848	1270	2450	3900	6890	14 100
175	96	202	380	780	1170	2250	3590	6340	12 900
200	90	188	353	726	1090	2090	3340	5900	12 000
250	80	166	313	643	964	1860	2960	5230	10 700
300	72	151	284	583	873	1680	2680	4740	9660
350	66	139	261	536	803	1550	2470	4360	8890
400	62	129	243	499	747	1440	2290	4050	8270
450	58	121	228	468	701	1350	2150	3800	7760
500	55	114	215	442	662	1280	2030	3590	7330
550	52	109	204	420	629	1210	1930	3410	6960
600	50	104	195	400	600	1160	1840	3260	6640
650	47	99	187	384	575	1110	1760	3120	6360
700	46	95	179	368	552	1060	1690	3000	6110
750	44	92	173	355	532	1020	1630	2890	5890
800	42	89	167	343	514	989	1580	2790	5680
850	41	86	162	332	497	957	1530	2700	5500
900	40	83	157	322	482	928	1480	2610	5330
950	39	81	152	312	468	901	1440	2540	5180
1000	38	79	148	304	455	877	1400	2470	5040
1100	36	75	141	289	432	833	1330	2350	4780
1200	34	71	134	275	412	794	1270	2240	4560
1300	33	68	128	264	395	761	1210	2140	4370
1400	31	65	123	253	379	731	1160	2060	4200
1500	30	63	119	244	366	704	1120	1980	4050
1600	29	61	115	236	353	680	1080	1920	3910
1700	28	59	111	228	342	658	1050	1850	3780
1800	27	57	108	221	331	638	1020	1800	3670
1900	27	56	105	215	322	619	987	1750	3560
2000	26	54	102	209	313	602	960	1700	3460

For SI units: 1 inch = 25 mm, 1 foot = 304.8 mm, 1 cubic foot per hour = 0.0283 m^3/h, 1 pound-force per square inch = 6.8947 kPa, 1 inch water column = 0.249 kPa

* Table entries are rounded to 3 significant digits.

›› **TABLE 1315.2(3)**
SCHEDULE 40 METALLIC PIPE [NFPA 54: TABLE 6.2(d)]*

					GAS:	NATURAL		
					INLET PRESSURE:	LESS THAN 2 psi		
					PRESSURE DROP:	6.0 in. w.c.		
					SPECIFIC GRAVITY:	0.60		
INTENDED USE: INITIAL SUPPLY PRESSURE OF 11.0 IN. W.C. OR GREATER									
	PIPE SIZE (inch)								
NOMINAL:	½	¾	1	1¼	1½	2	2½	3	4
ACTUAL ID:	0.622	0.824	1.049	1.38	1.61	2.067	2.469	3.068	4.026
LENGTH (feet)	CAPACITY IN CUBIC FEET OF GAS PER HOUR								
10	660	1380	2600	5340	8000	15 400	24 600	43 400	88 500
20	454	949	1790	3670	5500	10 600	16 900	29 800	60 800
30	364	762	1440	2950	4410	8500	13 600	24 000	48 900
40	312	652	1230	2520	3780	7280	11 600	20 500	41 800
50	276	578	1090	2240	3350	6450	10 300	18 200	37 100
60	250	524	986	2030	3030	5840	9310	16 500	33 600
70	230	482	907	1860	2790	5380	8570	15 100	30 900
80	214	448	844	1730	2600	5000	7970	14 100	28 700
90	201	420	792	1630	2440	4690	7480	13 200	27 000
100	190	397	748	1540	2300	4430	7060	12 500	25 500
125	168	352	663	1360	2040	3930	6260	11 100	22 600
150	153	319	601	1230	1850	3560	5670	10 000	20 500
175	140	293	553	1140	1700	3270	5220	9230	18 800
200	131	273	514	1056	1580	3050	4860	8580	17 500
250	116	242	456	936	1400	2700	4300	7610	15 500
300	105	219	413	848	1270	2450	3900	6890	14 100
350	96	202	380	780	1170	2250	3590	6340	12 900
400	90	188	353	726	1090	2090	3340	5900	12 000
450	84	176	332	681	1020	1960	3130	5540	11 300
500	80	166	313	643	964	1860	2960	5230	10 700
550	76	158	297	611	915	1760	2810	4970	10 100
600	72	151	284	583	873	1680	2680	4740	9660
650	69	144	272	558	836	1610	2570	4540	9250
700	66	139	261	536	803	1550	2470	4360	8890
750	64	134	252	516	774	1490	2380	4200	8560
800	62	129	243	499	747	1440	2290	4050	8270
850	60	125	235	483	723	1390	2220	3920	8000
900	58	121	228	468	701	1350	2150	3800	7760
950	56	118	221	454	681	1310	2090	3690	7540
1000	55	114	215	442	662	1280	2030	3590	7330
1100	52	109	204	420	629	1210	1930	3410	6960
1200	50	104	195	400	600	1160	1840	3260	6640
1300	47	99	187	384	575	1110	1760	3120	6360
1400	46	95	179	368	552	1060	1690	3000	6110
1500	44	92	173	355	532	1020	1630	2890	5890
1600	42	89	167	343	514	989	1580	2790	5680
1700	41	86	162	332	497	957	1530	2700	5500
1800	40	83	157	322	482	928	1480	2610	5330
1900	39	81	152	312	468	901	1440	2540	5180
2000	38	79	148	304	455	877	1400	2470	5040

For SI units: 1 inch = 25 mm, 1 foot = 304.8 mm, 1 cubic foot per hour = 0.0283 m^3/h, 1 pound-force per square inch = 6.8947 kPa, 1 inch water column = 0.249 kPa

* Table entries are rounded to 3 significant digits.

TABLE 1315.2(4)
SCHEDULE 40 METALLIC PIPE [NFPA 54: TABLE 6.2(e)]*

							GAS:	**NATURAL**
							INLET PRESSURE:	**2.0 psi**
							PRESSURE DROP:	**1.0 psi**
							SPECIFIC GRAVITY:	**0.60**
	PIPE SIZE (inch)							
NOMINAL:	½	¾	1	1¼	1½	2	2½	3	4
ACTUAL ID:	0.622	0.824	1.049	1.380	1.610	2.067	2.469	3.068	4.026
LENGTH (feet)	**CAPACITY IN CUBIC FEET OF GAS PER HOUR**								
10	1510	3040	5560	11 400	17 100	32 900	52 500	92 800	189 000
20	1070	2150	3930	8070	12 100	23 300	37 100	65 600	134 000
30	869	1760	3210	6590	9880	19 000	30 300	53 600	109 000
40	753	1520	2780	5710	8550	16 500	26 300	46 400	94 700
50	673	1360	2490	5110	7650	14 700	23 500	41 500	84 700
60	615	1240	2270	4660	6980	13 500	21 400	37 900	77 300
70	569	1150	2100	4320	6470	12 500	19 900	35 100	71 600
80	532	1080	1970	4040	6050	11 700	18 600	32 800	67 000
90	502	1010	1850	3810	5700	11 000	17 500	30 900	63 100
100	462	934	1710	3510	5260	10 100	16 100	28 500	58 200
125	414	836	1530	3140	4700	9060	14 400	25 500	52 100
150	372	751	1370	2820	4220	8130	13 000	22 900	46 700
175	344	695	1270	2601	3910	7530	12 000	21 200	43 300
200	318	642	1170	2410	3610	6960	11 100	19 600	40 000
250	279	583	1040	2140	3210	6180	9850	17 400	35 500
300	253	528	945	1940	2910	5600	8920	15 800	32 200
350	232	486	869	1790	2670	5150	8210	14 500	29 600
400	216	452	809	1660	2490	4790	7640	13 500	27 500
450	203	424	759	1560	2330	4500	7170	12 700	25 800
500	192	401	717	1470	2210	4250	6770	12 000	24 400
550	182	381	681	1400	2090	4030	6430	11 400	23 200
600	174	363	650	1330	2000	3850	6130	10 800	22 100
650	166	348	622	1280	1910	3680	5870	10 400	21 200
700	160	334	598	1230	1840	3540	5640	9970	20 300
750	154	322	576	1180	1770	3410	5440	9610	19 600
800	149	311	556	1140	1710	3290	5250	9280	18 900
850	144	301	538	1100	1650	3190	5080	8980	18 300
900	139	292	522	1070	1600	3090	4930	8710	17 800
950	135	283	507	1040	1560	3000	4780	8460	17 200
1000	132	275	493	1010	1520	2920	4650	8220	16 800
1100	125	262	468	960	1440	2770	4420	7810	15 900
1200	119	250	446	917	1370	2640	4220	7450	15 200
1300	114	239	427	878	1320	2530	4040	7140	14 600
1400	110	230	411	843	1260	2430	3880	6860	14 000
1500	106	221	396	812	1220	2340	3740	6600	13 500
1600	102	214	382	784	1180	2260	3610	6380	13 000
1700	99	207	370	759	1140	2190	3490	6170	12 600
1800	96	200	358	736	1100	2120	3390	5980	12 200
1900	93	195	348	715	1070	2060	3290	5810	11 900
2000	91	189	339	695	1040	2010	3200	5650	11 500

For SI units: 1 inch = 25 mm, 1 foot = 304.8 mm, 1 cubic foot per hour = 0.0283 m^3/h, 1 pound-force per square inch = 6.8947 kPa

* Table entries are rounded to 3 significant digits.

TABLE 1315.2(5) SCHEDULE 40 METALLIC PIPE [NFPA 54: TABLE 6.2(f)]*

						GAS:	NATURAL		
						INLET PRESSURE:	3.0 psi		
						PRESSURE DROP:	2.0 psi		
						SPECIFIC GRAVITY:	0.60		
	PIPE SIZE (inch)								
NOMINAL:	½	¾	1	1¼	1½	2	2½	3	4
ACTUAL ID:	0.622	0.824	1.049	1.380	1.610	2.067	2.469	3.068	4.026
LENGTH (feet)	CAPACITY IN CUBIC FEET OF GAS PER HOUR								
10	2350	4920	9270	19 000	28 500	54 900	87 500	155 000	316 000
20	1620	3380	6370	13 100	19 600	37 700	60 100	106 000	217 000
30	1300	2720	5110	10 500	15 700	30 300	48 300	85 400	174 000
40	1110	2320	4380	8990	13 500	25 900	41 300	73 100	149 000
50	985	2060	3880	7970	11 900	23 000	36 600	64 800	132 000
60	892	1870	3520	7220	10 800	20 800	33 200	58 700	120 000
70	821	1720	3230	6640	9950	19 200	30 500	54 000	110 000
80	764	1600	3010	6180	9260	17 800	28 400	50 200	102 000
90	717	1500	2820	5800	8680	16 700	26 700	47 100	96 100
100	677	1420	2670	5470	8200	15 800	25 200	44 500	90 800
125	600	1250	2360	4850	7270	14 000	22 300	39 500	80 500
150	544	1140	2140	4400	6590	12 700	20 200	35 700	72 900
175	500	1050	1970	4040	6060	11 700	18 600	32 900	67 100
200	465	973	1830	3760	5640	10 900	17 300	30 600	62 400
250	412	862	1620	3330	5000	9620	15 300	27 100	55 300
300	374	781	1470	3020	4530	8720	13 900	24 600	50 100
350	344	719	1350	2780	4170	8020	12 800	22 600	46 100
400	320	669	1260	2590	3870	7460	11 900	21 000	42 900
450	300	627	1180	2430	3640	7000	11 200	19 700	40 200
500	283	593	1120	2290	3430	6610	10 500	18 600	38 000
550	269	563	1060	2180	3260	6280	10 000	17 700	36 100
600	257	537	1010	2080	3110	5990	9550	16 900	34 400
650	246	514	969	1990	2980	5740	9150	16 200	33 000
700	236	494	931	1910	2860	5510	8790	15 500	31 700
750	228	476	897	1840	2760	5310	8470	15 000	30 500
800	220	460	866	1780	2660	5130	8180	14 500	29 500
850	213	445	838	1720	2580	4960	7910	14 000	28 500
900	206	431	812	1670	2500	4810	7670	13 600	27 700
950	200	419	789	1620	2430	4670	7450	13 200	26 900
1000	195	407	767	1580	2360	4550	7240	12 800	26 100
1100	185	387	729	1500	2240	4320	6890	12 200	24 800
1200	177	369	695	1430	2140	4120	6570	11 600	23 700
1300	169	353	666	1370	2050	3940	6290	11 100	22 700
1400	162	340	640	1310	1970	3790	6040	10 700	21 800
1500	156	327	616	1270	1900	3650	5820	10 300	21 000
1600	151	316	595	1220	1830	3530	5620	10 000	20 300
1700	146	306	576	1180	1770	3410	5440	9610	19 600
1800	142	296	558	1150	1720	3310	5270	9320	19 000
1900	138	288	542	1110	1670	3210	5120	9050	18 400
2000	134	280	527	1080	1620	3120	4980	8800	18 000

For SI units: 1 inch = 25 mm, 1 foot = 304.8 mm, 1 cubic foot per hour = 0.0283 m^3/h, 1 pound-force per square inch = 6.8947 kPa

* Table entries are rounded to 3 significant digits.

TABLE 1315.2(6)
SCHEDULE 40 METALLIC PIPE [NFPA 54: TABLE 6.2(g)]*

					GAS:	NATURAL			
					INLET PRESSURE:	5.0 psi			
					PRESSURE DROP:	3.5 psi			
					SPECIFIC GRAVITY:	0.60			
	PIPE SIZE (inch)								
NOMINAL:	½	¾	1	1¼	1½	2	2½	3	4
ACTUAL ID:	0.622	0.824	1.049	1.380	1.610	2.067	2.469	3.068	4.026
LENGTH (feet)	CAPACITY IN CUBIC FEET OF GAS PER HOUR								
10	3190	6430	11 800	24 200	36 200	69 700	111 000	196 000	401 000
20	2250	4550	8320	17 100	25 600	49 300	78 600	139 000	283 000
30	1840	3720	6790	14 000	20 900	40 300	64 200	113 000	231 000
40	1590	3220	5880	12 100	18 100	34 900	55 600	98 200	200 000
50	1430	2880	5260	10 800	16 200	31 200	49 700	87 900	179 000
60	1300	2630	4800	9860	14 800	28 500	45 400	80 200	164 000
70	1200	2430	4450	9130	13 700	26 400	42 000	74 300	151 000
80	1150	2330	4260	8540	12 800	24 700	39 300	69 500	142 000
90	1060	2150	3920	8050	12 100	23 200	37 000	65 500	134 000
100	979	1980	3620	7430	11 100	21 400	34 200	60 400	123 000
125	876	1770	3240	6640	9950	19 200	30 600	54 000	110 000
150	786	1590	2910	5960	8940	17 200	27 400	48 500	98 900
175	728	1470	2690	5520	8270	15 900	25 400	44 900	91 600
200	673	1360	2490	5100	7650	14 700	23 500	41 500	84 700
250	558	1170	2200	4510	6760	13 000	20 800	36 700	74 900
300	506	1060	1990	4090	6130	11 800	18 800	33 300	67 800
350	465	973	1830	3760	5640	10 900	17 300	30 600	62 400
400	433	905	1710	3500	5250	10 100	16 100	28 500	58 100
450	406	849	1600	3290	4920	9480	15 100	26 700	54 500
500	384	802	1510	3100	4650	8950	14 300	25 200	51 500
550	364	762	1440	2950	4420	8500	13 600	24 000	48 900
600	348	727	1370	2810	4210	8110	12 900	22 900	46 600
650	333	696	1310	2690	4030	7770	12 400	21 900	44 600
700	320	669	1260	2590	3880	7460	11 900	21 000	42 900
750	308	644	1210	2490	3730	7190	11 500	20 300	41 300
800	298	622	1170	2410	3610	6940	11 100	19 600	39 900
850	288	602	1130	2330	3490	6720	10 700	18 900	38 600
900	279	584	1100	2260	3380	6520	10 400	18 400	37 400
950	271	567	1070	2190	3290	6330	10 100	17 800	36 400
1000	264	551	1040	2130	3200	6150	9810	17 300	35 400
1100	250	524	987	2030	3030	5840	9320	16 500	33 600
1200	239	500	941	1930	2900	5580	8890	15 700	32 000
1300	229	478	901	1850	2770	5340	8510	15 000	30 700
1400	220	460	866	1780	2660	5130	8180	14 500	29 500
1500	212	443	834	1710	2570	4940	7880	13 900	28 400
1600	205	428	806	1650	2480	4770	7610	13 400	27 400
1700	198	414	780	1600	2400	4620	7360	13 000	26 500
1800	192	401	756	1550	2330	4480	7140	12 600	25 700
1900	186	390	734	1510	2260	4350	6930	12 300	25 000
2000	181	379	714	1470	2200	4230	6740	11 900	24 300

For SI units: 1 inch = 25 mm, 1 foot = 304.8 mm, 1 cubic foot per hour = 0.0283 m^3/h, 1 pound-force per square inch = 6.8947 kPa

* Table entries are rounded to 3 significant digits.

TABLE 1315.2(7) SEMI-RIGID COPPER TUBING [NFPA 54: TABLE 6.2(h)][1, 2]

						GAS:	NATURAL		
						INLET PRESSURE:	LESS THAN 2 psi		
						PRESSURE DROP:	0.3 in. w.c.		
						SPECIFIC GRAVITY:	0.60		
	TUBE SIZE (inch)								
NOMINAL: K & L:	¼	⅜	½	⅝	¾	1	1¼	1½	2
NOMINAL: ACR:	⅜	½	⅝	¾	⅞	1⅛	1⅜	–	–
OUTSIDE:	0.375	0.500	0.625	0.750	0.875	1.125	1.375	1.625	2.125
INSIDE:[3]	0.305	0.402	0.527	0.652	0.745	0.995	1.245	1.481	1.959
LENGTH (feet)	CAPACITY IN CUBIC FEET OF GAS PER HOUR								
10	20	42	85	148	210	448	806	1270	2650
20	14	29	58	102	144	308	554	873	1820
30	11	23	47	82	116	247	445	701	1460
40	10	20	40	70	99	211	381	600	1250
50	NA	17	35	62	88	187	337	532	1110
60	NA	16	32	56	79	170	306	482	1000
70	NA	14	29	52	73	156	281	443	924
80	NA	13	27	48	68	145	262	413	859
90	NA	13	26	45	64	136	245	387	806
100	NA	12	24	43	60	129	232	366	761
125	NA	11	22	38	53	114	206	324	675
150	NA	10	20	34	48	103	186	294	612
175	NA	NA	18	31	45	95	171	270	563
200	NA	NA	17	29	41	89	159	251	523
250	NA	NA	15	26	37	78	141	223	464
300	NA	NA	13	23	33	71	128	202	420
350	NA	NA	12	22	31	65	118	186	387
400	NA	NA	11	20	28	61	110	173	360
450	NA	NA	11	19	27	57	103	162	338
500	NA	NA	10	18	25	54	97	153	319
550	NA	NA	NA	17	24	51	92	145	303
600	NA	NA	NA	16	23	49	88	139	289
650	NA	NA	NA	15	22	47	84	133	277
700	NA	NA	NA	15	21	45	81	128	266
750	NA	NA	NA	14	20	43	78	123	256
800	NA	NA	NA	14	20	42	75	119	247
850	NA	NA	NA	13	19	40	73	115	239
900	NA	NA	NA	13	18	39	71	111	232
950	NA	NA	NA	13	18	38	69	108	225
1000	NA	NA	NA	12	17	37	67	105	219
1100	NA	NA	NA	12	16	35	63	100	208
1200	NA	NA	NA	11	16	34	60	95	199
1300	NA	NA	NA	11	15	32	58	91	190
1400	NA	NA	NA	10	14	31	56	88	183
1500	NA	NA	NA	NA	14	30	54	84	176
1600	NA	NA	NA	NA	13	29	52	82	170
1700	NA	NA	NA	NA	13	28	50	79	164
1800	NA	NA	NA	NA	13	27	49	77	159
1900	NA	NA	NA	NA	12	26	47	74	155
2000	NA	NA	NA	NA	12	25	46	72	151

For SI units: 1 inch = 25 mm, 1 foot = 304.8 mm, 1 cubic foot per hour = 0.0283 m^3/h, 1 pound-force per square inch = 6.8947 kPa, 1 inch water column = 0.249 kPa

Notes:

1 Table entries are rounded to 3 significant digits.

2 NA means a flow of less than 10 ft^3/h (0.283 m^3/h).

3 Table capacities are based on Type K copper tubing inside diameter (shown), which has the smallest inside diameter of the copper tubing products.

» **TABLE 1315.2(8)**
SEMI-RIGID COPPER TUBING [NFPA 54: TABLE 6.2(i)][1, 2]

					GAS:	NATURAL		
					INLET PRESSURE:	LESS THAN 2 psi		
					PRESSURE DROP:	0.5 in. w.c.		
					SPECIFIC GRAVITY:	0.60		
	TUBE SIZE (inch)							
NOMINAL: K & L:	¼	⅜	½	⅝	¾	1	1¼	1½	2
NOMINAL: ACR:	⅜	½	⅝	¾	⅞	1⅛	1⅜	–	–
OUTSIDE:	0.375	0.500	0.625	0.750	0.875	1.125	1.375	1.625	2.125
INSIDE:[3]	0.305	0.402	0.527	0.652	0.745	0.995	1.245	1.481	1.959
LENGTH (feet)	CAPACITY IN CUBIC FEET OF GAS PER HOUR								
10	27	55	111	195	276	590	1060	1680	3490
20	18	38	77	134	190	406	730	1150	2400
30	15	30	61	107	152	326	586	925	1930
40	13	26	53	92	131	279	502	791	1650
50	11	23	47	82	116	247	445	701	1460
60	10	21	42	74	105	224	403	635	1320
70	NA	19	39	68	96	206	371	585	1220
80	NA	18	36	63	90	192	345	544	1130
90	NA	17	34	59	84	180	324	510	1060
100	NA	16	32	56	79	170	306	482	1000
125	NA	14	28	50	70	151	271	427	890
150	NA	13	26	45	64	136	245	387	806
175	NA	12	24	41	59	125	226	356	742
200	NA	11	22	39	55	117	210	331	690
250	NA	NA	20	34	48	103	186	294	612
300	NA	NA	18	31	44	94	169	266	554
350	NA	NA	16	28	40	86	155	245	510
400	NA	NA	15	26	38	80	144	228	474
450	NA	NA	14	25	35	75	135	214	445
500	NA	NA	13	23	33	71	128	202	420
550	NA	NA	13	22	32	68	122	192	399
600	NA	NA	12	21	30	64	116	183	381
650	NA	NA	12	20	29	62	111	175	365
700	NA	NA	11	20	28	59	107	168	350
750	NA	NA	11	19	27	57	103	162	338
800	NA	NA	10	18	26	55	99	156	326
850	NA	NA	10	18	25	53	96	151	315
900	NA	NA	NA	17	24	52	93	147	306
950	NA	NA	NA	17	24	50	90	143	297
1000	NA	NA	NA	16	23	49	88	139	289
1100	NA	NA	NA	15	22	46	84	132	274
1200	NA	NA	NA	15	21	44	80	126	262
1300	NA	NA	NA	14	20	42	76	120	251
1400	NA	NA	NA	13	19	41	73	116	241
1500	NA	NA	NA	13	18	39	71	111	232
1600	NA	NA	NA	13	18	38	68	108	224
1700	NA	NA	NA	12	17	37	66	104	217
1800	NA	NA	NA	12	17	36	64	101	210
1900	NA	NA	NA	11	16	35	62	98	204
2000	NA	NA	NA	11	16	34	60	95	199

For SI units: 1 inch = 25 mm, 1 foot = 304.8 mm, 1 cubic foot per hour = 0.0283 m^3/h, 1 pound-force per square inch = 6.8947 kPa, 1 inch water column = 0.249 kPa

Notes:

[1] Table entries are rounded to 3 significant digits.

[2] NA means a flow of less than 10 ft^3/h (0.283 m^3/h).

[3] Table capacities are based on Type K copper tubing inside diameter (shown), which has the smallest inside diameter of the copper tubing products.

TABLE 1315.2(9) SEMI-RIGID COPPER TUBING [NFPA 54: TABLE 6.2(j)][1, 2]

						GAS:		NATURAL	
						INLET PRESSURE:		LESS THAN 2 psi	
						PRESSURE DROP:		1.0 in. w.c.	
						SPECIFIC GRAVITY:		0.60	
INTENDED USE: TUBE SIZING BETWEEN HOUSE LINE REGULATOR AND THE APPLIANCE									
	TUBE SIZE (inch)								
NOMINAL: K & L:	¼	⅜	½	⅝	¾	1	1¼	1½	2
NOMINAL: ACR:	⅜	½	⅝	¾	⅞	1⅛	1⅜	–	–
OUTSIDE:	0.375	0.500	0.625	0.750	0.875	1.125	1.375	1.625	2.125
INSIDE:[3]	0.305	0.402	0.527	0.652	0.745	0.995	1.245	1.481	1.959
LENGTH (feet)	CAPACITY IN CUBIC FEET OF GAS PER HOUR								
10	39	80	162	283	402	859	1550	2440	5080
20	27	55	111	195	276	590	1060	1680	3490
30	21	44	89	156	222	474	853	1350	2800
40	18	38	77	134	190	406	730	1150	2400
50	16	33	68	119	168	359	647	1020	2130
60	15	30	61	107	152	326	586	925	1930
70	13	28	57	99	140	300	539	851	1770
80	13	26	53	92	131	279	502	791	1650
90	12	24	49	86	122	262	471	742	1550
100	11	23	47	82	116	247	445	701	1460
125	NA	20	41	72	103	219	394	622	1290
150	NA	18	37	65	93	198	357	563	1170
175	NA	17	34	60	85	183	329	518	1080
200	NA	16	32	56	79	170	306	482	1000
250	NA	14	28	50	70	151	271	427	890
300	NA	13	26	45	64	136	245	387	806
350	NA	12	24	41	59	125	226	356	742
400	NA	11	22	39	55	117	210	331	690
450	NA	10	21	36	51	110	197	311	647
500	NA	NA	20	34	48	103	186	294	612
550	NA	NA	19	32	46	98	177	279	581
600	NA	NA	18	31	44	94	169	266	554
650	NA	NA	17	30	42	90	162	255	531
700	NA	NA	16	28	40	86	155	245	510
750	NA	NA	16	27	39	83	150	236	491
800	NA	NA	15	26	38	80	144	228	474
850	NA	NA	15	26	36	78	140	220	459
900	NA	NA	14	25	35	75	135	214	445
950	NA	NA	14	24	34	73	132	207	432
1000	NA	NA	13	23	33	71	128	202	420
1100	NA	NA	13	22	32	68	122	192	399
1200	NA	NA	12	21	30	64	116	183	381
1300	NA	NA	12	20	29	62	111	175	365
1400	NA	NA	11	20	28	59	107	168	350
1500	NA	NA	11	19	27	57	103	162	338
1600	NA	NA	10	18	26	55	99	156	326
1700	NA	NA	10	18	25	53	96	151	315
1800	NA	NA	NA	17	24	52	93	147	306
1900	NA	NA	NA	17	24	50	90	143	297
2000	NA	NA	NA	16	23	49	88	139	289

For SI units: 1 inch = 25 mm, 1 foot = 304.8 mm, 1 cubic foot per hour = 0.0283 m^3/h, 1 pound-force per square inch = 6.8947 kPa, 1 inch water column = 0.249 kPa

Notes:

1 Table entries are rounded to 3 significant digits.

2 NA means a flow of less than 10 ft^3/h (0.283 m^3/h).

3 Table capacities are based on Type K copper tubing inside diameter (shown), which has the smallest inside diameter of the copper tubing products.

TABLE 1315.2(10)
SEMI-RIGID COPPER TUBING [NFPA 54: TABLE 6.2(k)][2]

							GAS:	NATURAL
							INLET PRESSURE:	LESS THAN 2 psi
							PRESSURE DROP:	17.0 in. w.c.
							SPECIFIC GRAVITY:	0.60
	TUBE SIZE (inch)							
NOMINAL: K & L:	¼	⅜	½	⅝	¾	1	1¼	1½	2
NOMINAL: ACR:	⅜	½	⅝	¾	⅞	1⅛	1⅜	–	–
OUTSIDE:	0.375	0.500	0.625	0.750	0.875	1.125	1.375	1.625	2.125
INSIDE:[1]	0.305	0.402	0.527	0.652	0.745	0.995	1.245	1.481	1.959
LENGTH (feet)	CAPACITY IN CUBIC FEET OF GAS PER HOUR								
10	190	391	796	1390	1970	4220	7590	12 000	24 900
20	130	269	547	956	1360	2900	5220	8230	17 100
30	105	216	439	768	1090	2330	4190	6610	13 800
40	90	185	376	657	932	1990	3590	5650	11 800
50	79	164	333	582	826	1770	3180	5010	10 400
60	72	148	302	528	749	1600	2880	4540	9460
70	66	137	278	486	689	1470	2650	4180	8700
80	62	127	258	452	641	1370	2460	3890	8090
90	58	119	243	424	601	1280	2310	3650	7590
100	55	113	229	400	568	1210	2180	3440	7170
125	48	100	203	355	503	1080	1940	3050	6360
150	44	90	184	321	456	974	1750	2770	5760
175	40	83	169	296	420	896	1610	2540	5300
200	38	77	157	275	390	834	1500	2370	4930
250	33	69	140	244	346	739	1330	2100	4370
300	30	62	126	221	313	670	1210	1900	3960
350	28	57	116	203	288	616	1110	1750	3640
400	26	53	108	189	268	573	1030	1630	3390
450	24	50	102	177	252	538	968	1530	3180
500	23	47	96	168	238	508	914	1440	3000
550	22	45	91	159	226	482	868	1370	2850
600	21	43	87	152	215	460	829	1310	2720
650	20	41	83	145	206	441	793	1250	2610
700	19	39	80	140	198	423	762	1200	2500
750	18	38	77	135	191	408	734	1160	2410
800	18	37	74	130	184	394	709	1120	2330
850	17	35	72	126	178	381	686	1080	2250
900	17	34	70	122	173	370	665	1050	2180
950	16	33	68	118	168	359	646	1020	2120
1000	16	32	66	115	163	349	628	991	2060
1100	15	31	63	109	155	332	597	941	1960
1200	14	29	60	104	148	316	569	898	1870
1300	14	28	57	100	142	303	545	860	1790
1400	13	27	55	96	136	291	524	826	1720
1500	13	26	53	93	131	280	505	796	1660
1600	12	25	51	89	127	271	487	768	1600
1700	12	24	49	86	123	262	472	744	1550
1800	11	24	48	84	119	254	457	721	1500
1900	11	23	47	81	115	247	444	700	1460
2000	11	22	45	79	112	240	432	681	1420

For SI units: 1 inch = 25 mm, 1 foot = 304.8 mm, 1 cubic foot per hour = 0.0283 m^3/h, 1 pound-force per square inch = 6.8947 kPa, 1 inch water column = 0.249 kPa

Notes:

[1] Table capacities are based on Type K copper tubing inside diameter (shown), which has the smallest inside diameter of the copper tubing products.

[2] Table entries are rounded to 3 significant digits.

» **TABLE 1315.2(11)**
SEMI-RIGID COPPER TUBING [NFPA 54: TABLE 6.2(I)][2]

							GAS:	NATURAL	
							INLET PRESSURE:	2.0 psi	
							PRESSURE DROP:	1.0 psi	
							SPECIFIC GRAVITY:	0.60	
	TUBE SIZE (inch)								
NOMINAL: K & L:	¼	⅜	½	⅝	¾	1	1¼	1½	2
NOMINAL: ACR:	⅜	½	⅝	¾	⅞	1⅛	1⅜	–	–
OUTSIDE:	0.375	0.500	0.625	0.750	0.875	1.125	1.375	1.625	2.125
INSIDE:[1]	0.305	0.402	0.527	0.652	0.745	0.995	1.245	1.481	1.959
LENGTH (feet)	CAPACITY IN CUBIC FEET OF GAS PER HOUR								
10	245	506	1030	1800	2550	5450	9820	15 500	32 200
20	169	348	708	1240	1760	3750	6750	10 600	22 200
30	135	279	568	993	1410	3010	5420	8550	17 800
40	116	239	486	850	1210	2580	4640	7310	15 200
50	103	212	431	754	1070	2280	4110	6480	13 500
60	93	192	391	683	969	2070	3730	5870	12 200
70	86	177	359	628	891	1900	3430	5400	11 300
80	80	164	334	584	829	1770	3190	5030	10 500
90	75	154	314	548	778	1660	2990	4720	9820
100	71	146	296	518	735	1570	2830	4450	9280
125	63	129	263	459	651	1390	2500	3950	8220
150	57	117	238	416	590	1260	2270	3580	7450
175	52	108	219	383	543	1160	2090	3290	6850
200	49	100	204	356	505	1080	1940	3060	6380
250	43	89	181	315	448	956	1720	2710	5650
300	39	80	164	286	406	866	1560	2460	5120
350	36	74	150	263	373	797	1430	2260	4710
400	33	69	140	245	347	741	1330	2100	4380
450	31	65	131	230	326	696	1250	1970	4110
500	30	61	124	217	308	657	1180	1870	3880
550	28	58	118	206	292	624	1120	1770	3690
600	27	55	112	196	279	595	1070	1690	3520
650	26	53	108	188	267	570	1030	1620	3370
700	25	51	103	181	256	548	986	1550	3240
750	24	49	100	174	247	528	950	1500	3120
800	23	47	96	168	239	510	917	1450	3010
850	22	46	93	163	231	493	888	1400	2920
900	22	44	90	158	224	478	861	1360	2830
950	21	43	88	153	217	464	836	1320	2740
1000	20	42	85	149	211	452	813	1280	2670
1100	19	40	81	142	201	429	772	1220	2540
1200	18	38	77	135	192	409	737	1160	2420
1300	18	36	74	129	183	392	705	1110	2320
1400	17	35	71	124	176	376	678	1070	2230
1500	16	34	68	120	170	363	653	1030	2140
1600	16	33	66	116	164	350	630	994	2070
1700	15	31	64	112	159	339	610	962	2000
1800	15	30	62	108	154	329	592	933	1940
1900	14	30	60	105	149	319	575	906	1890
2000	14	29	59	102	145	310	559	881	1830

For SI units: 1 inch = 25 mm, 1 foot = 304.8 mm, 1 cubic foot per hour = 0.0283 m^3/h, 1 pound-force per square inch = 6.8947 kPa

Notes:

[1] Table capacities are based on Type K copper tubing inside diameter (shown), which has the smallest inside diameter of the copper tubing products.

[2] Table entries are rounded to 3 significant digits.

TABLE 1315.2(12)
SEMI-RIGID COPPER TUBING [NFPA 54: TABLE 6.2(m)][3]

						GAS:	NATURAL		
						INLET PRESSURE:	2.0 psi		
						PRESSURE DROP:	1.5 psi		
						SPECIFIC GRAVITY:	0.60		
INTENDED USE: PIPE SIZING BETWEEN POINT OF DELIVERY AND THE HOUSE LINE REGULATOR. TOTAL LOAD SUPPLIED BY A SINGLE HOUSE LINE REGULATOR NOT EXCEEDING 150 CUBIC FEET PER HOUR[2].									
	TUBE SIZE (inch)								
NOMINAL: K & L:	1/4	3/8	1/2	5/8	3/4	1	1 1/4	1 1/2	2
NOMINAL: ACR:	3/8	1/2	5/8	3/4	7/8	1 1/8	1 3/8	–	–
OUTSIDE:	0.375	0.500	0.625	0.750	0.875	1.125	1.375	1.625	2.125
INSIDE:[1]	0.305	0.402	0.527	0.652	0.745	0.995	1.245	1.481	1.959
LENGTH (feet)	CAPACITY IN CUBIC FEET OF GAS PER HOUR								
10	303	625	1270	2220	3150	6740	12 100	19 100	39 800
20	208	430	874	1530	2170	4630	8330	13 100	27 400
30	167	345	702	1230	1740	3720	6690	10 600	22 000
40	143	295	601	1050	1490	3180	5730	9030	18 800
50	127	262	532	931	1320	2820	5080	8000	16 700
60	115	237	482	843	1200	2560	4600	7250	15 100
70	106	218	444	776	1100	2350	4230	6670	13 900
80	98	203	413	722	1020	2190	3940	6210	12 900
90	92	190	387	677	961	2050	3690	5820	12 100
100	87	180	366	640	907	1940	3490	5500	11 500
125	77	159	324	567	804	1720	3090	4880	10 200
150	70	144	294	514	729	1560	2800	4420	9200
175	64	133	270	472	670	1430	2580	4060	8460
200	60	124	252	440	624	1330	2400	3780	7870
250	53	110	223	390	553	1180	2130	3350	6980
300	48	99	202	353	501	1070	1930	3040	6320
350	44	91	186	325	461	984	1770	2790	5820
400	41	85	173	302	429	916	1650	2600	5410
450	39	80	162	283	402	859	1550	2440	5080
500	36	75	153	268	380	811	1460	2300	4800
550	35	72	146	254	361	771	1390	2190	4560
600	33	68	139	243	344	735	1320	2090	4350
650	32	65	133	232	330	704	1270	2000	4160
700	30	63	128	223	317	676	1220	1920	4000
750	29	60	123	215	305	652	1170	1850	3850
800	28	58	119	208	295	629	1130	1790	3720
850	27	57	115	201	285	609	1100	1730	3600
900	27	55	111	195	276	590	1060	1680	3490
950	26	53	108	189	268	573	1030	1630	3390
1000	25	52	105	184	261	558	1000	1580	3300
1100	24	49	100	175	248	530	954	1500	3130
1200	23	47	95	167	237	505	910	1430	2990
1300	22	45	91	160	227	484	871	1370	2860
1400	21	43	88	153	218	465	837	1320	2750
1500	20	42	85	148	210	448	806	1270	2650
1600	19	40	82	143	202	432	779	1230	2560
1700	19	39	79	138	196	419	753	1190	2470
1800	18	38	77	134	190	406	731	1150	2400
1900	18	37	74	130	184	394	709	1120	2330
2000	17	36	72	126	179	383	690	1090	2270

For SI units: 1 inch = 25 mm, 1 foot = 304.8 mm, 1 cubic foot per hour = 0.0283 m^3/h, 1 pound-force per square inch = 6.8947 kPa

Notes:

[1] Table capacities are based on Type K copper tubing inside diameter (shown), which has the smallest inside diameter of the copper tubing products.

[2] Where this table is used to size the tubing upstream of a line pressure regulator, the pipe or tubing downstream of the line pressure regulator shall be sized using a pressure drop no greater than 1 inch water column (0.249 kPa).

[3] Table entries are rounded to 3 significant digits.

TABLE 1315.2(13)
SEMI-RIGID COPPER TUBING [NFPA 54: TABLE 6.2(n)][2]

							GAS:	NATURAL	
							INLET PRESSURE:	5.0 psi	
							PRESSURE DROP:	3.5 psi	
							SPECIFIC GRAVITY:	0.60	
	TUBE SIZE (inch)								
NOMINAL: K & L:	¼	⅜	½	⅝	¾	1	1¼	1½	2
NOMINAL: ACR:	⅜	½	⅝	¾	⅞	1⅛	1⅜	–	–
OUTSIDE:	0.375	0.500	0.625	0.750	0.875	1.125	1.375	1.625	2.125
INSIDE:[1]	0.305	0.402	0.527	0.652	0.745	0.995	1.245	1.481	1.959
LENGTH (feet)	CAPACITY IN CUBIC FEET OF GAS PER HOUR								
10	511	1050	2140	3750	5320	11 400	20 400	32 200	67 100
20	351	724	1470	2580	3650	7800	14 000	22 200	46 100
30	282	582	1180	2070	2930	6270	11 300	17 800	37 000
40	241	498	1010	1770	2510	5360	9660	15 200	31 700
50	214	441	898	1570	2230	4750	8560	13 500	28 100
60	194	400	813	1420	2020	4310	7750	12 200	25 500
70	178	368	748	1310	1860	3960	7130	11 200	23 400
80	166	342	696	1220	1730	3690	6640	10 500	21 800
90	156	321	653	1140	1620	3460	6230	9820	20 400
100	147	303	617	1080	1530	3270	5880	9270	19 300
125	130	269	547	955	1360	2900	5210	8220	17 100
150	118	243	495	866	1230	2620	4720	7450	15 500
175	109	224	456	796	1130	2410	4350	6850	14 300
200	101	208	424	741	1050	2250	4040	6370	13 300
250	90	185	376	657	932	1990	3580	5650	11 800
300	81	167	340	595	844	1800	3250	5120	10 700
350	75	154	313	547	777	1660	2990	4710	9810
400	69	143	291	509	722	1540	2780	4380	9120
450	65	134	273	478	678	1450	2610	4110	8560
500	62	127	258	451	640	1370	2460	3880	8090
550	58	121	245	429	608	1300	2340	3690	7680
600	56	115	234	409	580	1240	2230	3520	7330
650	53	110	224	392	556	1190	2140	3370	7020
700	51	106	215	376	534	1140	2050	3240	6740
750	49	102	207	362	514	1100	1980	3120	6490
800	48	98	200	350	497	1060	1910	3010	6270
850	46	95	194	339	481	1030	1850	2910	6070
900	45	92	188	328	466	1000	1790	2820	5880
950	43	90	182	319	452	967	1740	2740	5710
1000	42	87	177	310	440	940	1690	2670	5560
1100	40	83	169	295	418	893	1610	2530	5280
1200	38	79	161	281	399	852	1530	2420	5040
1300	37	76	154	269	382	816	1470	2320	4820
1400	35	73	148	259	367	784	1410	2220	4630
1500	34	70	143	249	353	755	1360	2140	4460
1600	33	68	138	241	341	729	1310	2070	4310
1700	32	65	133	233	330	705	1270	2000	4170
1800	31	63	129	226	320	684	1230	1940	4040
1900	30	62	125	219	311	664	1200	1890	3930
2000	29	60	122	213	302	646	1160	1830	3820

For SI units: 1 inch = 25 mm, 1 foot = 304.8 mm, 1 cubic foot per hour = 0.0283 m^3/h, 1 pound-force per square inch = 6.8947 kPa

Notes:

[1] Table capacities are based on Type K copper tubing inside diameter (shown), which has the smallest inside diameter of the copper tubing products.

[2] Table entries are rounded to 3 significant digits.

TABLE 1315.2(14) CORRUGATED STAINLESS STEEL TUBING (CSST) [NFPA 54: TABLE 6.2(o)][1, 2]

											GAS:	**NATURAL**		
											INLET PRESSURE:	**LESS THAN 2 psi**		
											PRESSURE DROP:	**0.5 in. w.c.**		
											SPECIFIC GRAVITY:	**0.60**		
	TUBE SIZE (EHD)[3]													
FLOW DESIGNATION:	**13**	**15**	**18**	**19**	**23**	**25**	**30**	**31**	**37**	**39**	**46**	**48**	**60**	**62**
LENGTH (feet)	**CAPACITY IN CUBIC FEET OF GAS PER HOUR**													
5	46	63	115	134	225	270	471	546	895	1037	1790	2070	3660	4140
10	32	44	82	95	161	192	330	383	639	746	1260	1470	2600	2930
15	25	35	66	77	132	157	267	310	524	615	1030	1200	2140	2400
20	22	31	58	67	116	137	231	269	456	536	888	1050	1850	2080
25	19	27	52	60	104	122	206	240	409	482	793	936	1660	1860
30	18	25	47	55	96	112	188	218	374	442	723	856	1520	1700
40	15	21	41	47	83	97	162	188	325	386	625	742	1320	1470
50	13	19	37	42	75	87	144	168	292	347	559	665	1180	1320
60	12	17	34	38	68	80	131	153	267	318	509	608	1080	1200
70	11	16	31	36	63	74	121	141	248	295	471	563	1000	1110
80	10	15	29	33	60	69	113	132	232	277	440	527	940	1040
90	10	14	28	32	57	65	107	125	219	262	415	498	887	983
100	9	13	26	30	54	62	101	118	208	249	393	472	843	933
150	7	10	20	23	42	48	78	91	171	205	320	387	691	762
200	6	9	18	21	38	44	71	82	148	179	277	336	600	661
250	5	8	16	19	34	39	63	74	133	161	247	301	538	591
300	5	7	15	17	32	36	57	67	95	148	226	275	492	540

For SI units: 1 inch = 25 mm, 1 foot = 304.8 mm, 1 cubic foot per hour = 0.0283 m^3/h, 1 pound-force per square inch = 6.8947 kPa, 1 inch water column = 0.249 kPa

Notes:

1 Table entries are rounded to 3 significant digits.

2 Table includes losses for four 90 degree (1.57 rad) bends and two end fittings. Tubing runs with larger numbers of bends, fittings, or both shall be increased by an equivalent length of tubing to the following equation: $L = 1.3\ n$, where L is additional length (ft) of tubing and n is the number of additional fittings, bends, or both.

3 EHD = Equivalent Hydraulic Diameter, which is a measure of the relative hydraulic efficiency between different tubing sizes. The greater the value of EHD, the greater the gas capacity of the tubing.

TABLE 1315.2(15)
CORRUGATED STAINLESS STEEL TUBING (CSST) [NFPA 54: TABLE 6.2(p)][1, 2]

								GAS:		NATURAL			
								INLET PRESSURE:		LESS THAN 2 psi			
								PRESSURE DROP:		3.0 in. w.c.			
								SPECIFIC GRAVITY:		0.60			
INTENDED USE: INITIAL SUPPLY PRESSURE OF 8.0 INCH WATER COLUMN OR GREATER													
	TUBE SIZE (EHD)[3]												
FLOW DESIGNATION:	13	15	18	19	23	25	30	31	37	46	48	60	62
LENGTH (feet)	CAPACITY IN CUBIC FEET OF GAS PER HOUR												
5	120	160	277	327	529	649	1180	1370	2140	4430	5010	8800	10 100
10	83	112	197	231	380	462	828	958	1530	3200	3560	6270	7160
15	67	90	161	189	313	379	673	778	1250	2540	2910	5140	5850
20	57	78	140	164	273	329	580	672	1090	2200	2530	4460	5070
25	51	69	125	147	245	295	518	599	978	1960	2270	4000	4540
30	46	63	115	134	225	270	471	546	895	1790	2070	3660	4140
40	39	54	100	116	196	234	407	471	778	1550	1800	3180	3590
50	35	48	89	104	176	210	363	421	698	1380	1610	2850	3210
60	32	44	82	95	161	192	330	383	639	1260	1470	2600	2930
70	29	41	76	88	150	178	306	355	593	1170	1360	2420	2720
80	27	38	71	82	141	167	285	331	555	1090	1280	2260	2540
90	26	36	67	77	133	157	268	311	524	1030	1200	2140	2400
100	24	34	63	73	126	149	254	295	498	974	1140	2030	2280
150	19	27	52	60	104	122	206	240	409	793	936	1660	1860
200	17	23	45	52	91	106	178	207	355	686	812	1440	1610
250	15	21	40	46	82	95	159	184	319	613	728	1290	1440
300	13	19	37	42	75	87	144	168	234	559	665	1180	1320

For SI units: 1 foot = 304.8 mm, 1 cubic foot per hour = 0.0283 m^3/h, 1 pound-force per square inch = 6.8947 kPa, 1 inch water column = 0.249 kPa

Notes:

[1] Table entries are rounded to 3 significant digits.

[2] Table includes losses for four 90 degree (1.57 rad) bends and two end fittings. Tubing runs with larger numbers of bends, fittings, or both shall be increased by an equivalent length of tubing to the following equation: $L = 1.3\ n$, where L is additional length (ft) of tubing and n is the number of additional fittings, bends, or both.

[3] EHD = Equivalent Hydraulic Diameter, which is a measure of the relative hydraulic efficiency between different tubing sizes. The greater the value of EHD, the greater the gas capacity of the tubing.

TABLE 1315.2(16) CORRUGATED STAINLESS STEEL TUBING (CSST) [NFPA 54: TABLE 6.2(q)][1, 2]

								GAS:	NATURAL				
								INLET PRESSURE:	LESS THAN 2 psi				
								PRESSURE DROP:	6.0 in. w.c.				
								SPECIFIC GRAVITY:	0.60				
INTENDED USE: INITIAL SUPPLY PRESSURE OF 11.0 INCH WATER COLUMN OR GREATER													
	TUBE SIZE (EHD)[3]												
FLOW DESIGNATION:	13	15	18	19	23	25	30	31	37	46	48	60	62
LENGTH (feet)	CAPACITY IN CUBIC FEET OF GAS PER HOUR												
5	173	229	389	461	737	911	1690	1950	3000	6280	7050	12 400	14 260
10	120	160	277	327	529	649	1180	1370	2140	4430	5010	8800	10 100
15	96	130	227	267	436	532	960	1110	1760	3610	4100	7210	8260
20	83	112	197	231	380	462	828	958	1530	3120	3560	6270	7160
25	74	99	176	207	342	414	739	855	1370	2790	3190	5620	6400
30	67	90	161	189	313	379	673	778	1250	2540	2910	5140	5850
40	57	78	140	164	273	329	580	672	1090	2200	2530	4460	5070
50	51	69	125	147	245	295	518	599	978	1960	2270	4000	4540
60	46	63	115	134	225	270	471	546	895	1790	2070	3660	4140
70	42	58	106	124	209	250	435	505	830	1660	1920	3390	3840
80	39	54	100	116	196	234	407	471	778	1550	1800	3180	3590
90	37	51	94	109	185	221	383	444	735	1460	1700	3000	3390
100	35	48	89	104	176	210	363	421	698	1380	1610	2850	3210
150	28	39	73	85	145	172	294	342	573	1130	1320	2340	2630
200	24	34	63	73	126	149	254	295	498	974	1140	2030	2280
250	21	30	57	66	114	134	226	263	447	870	1020	1820	2040
300	19	27	52	60	104	122	206	240	409	793	936	1660	1860

For SI units: 1 foot = 304.8 mm, 1 cubic foot per hour = 0.0283 m^3/h, 1 pound-force per square inch = 6.8947 kPa, 1 inch water column = 0.249 kPa

Notes:

1 Table entries are rounded to 3 significant digits.

2 Table includes losses for four 90 degree (1.57 rad) bends and two end fittings. Tubing runs with larger numbers of bends, fittings, or both shall be increased by an equivalent length of tubing to the following equation: $L = 1.3\ n$, where L is additional length (ft) of tubing and n is the number of additional fittings, bends, or both.

3 EHD = Equivalent Hydraulic Diameter, which is a measure of the relative hydraulic efficiency between different tubing sizes. The greater the value of EHD, the greater the gas capacity of the tubing.

TABLE 1315.2(17) CORRUGATED STAINLESS STEEL TUBING (CSST) [NFPA 54: TABLE 6.2(r)][1, 2, 3, 4]

											GAS:	NATURAL		
											INLET PRESSURE:	2.0 psi		
											PRESSURE DROP:	1.0 psi		
											SPECIFIC GRAVITY:	0.60		
	TUBE SIZE (EHD)[5]													
FLOW DESIGNATION:	13	15	18	19	23	25	30	31	37	39	46	48	60	62
LENGTH (feet)	CAPACITY IN CUBIC FEET OF GAS PER HOUR													
10	270	353	587	700	1100	1370	2590	2990	4510	5037	9600	10 700	18 600	21 600
25	166	220	374	444	709	876	1620	1870	2890	3258	6040	6780	11 900	13 700
30	151	200	342	405	650	801	1480	1700	2640	2987	5510	6200	10 900	12 500
40	129	172	297	351	567	696	1270	1470	2300	2605	4760	5380	9440	10 900
50	115	154	266	314	510	624	1140	1310	2060	2343	4260	4820	8470	9720
75	93	124	218	257	420	512	922	1070	1690	1932	3470	3950	6940	7940
80	89	120	211	249	407	496	892	1030	1640	1874	3360	3820	6730	7690
100	79	107	189	222	366	445	795	920	1470	1685	3000	3420	6030	6880
150	64	87	155	182	302	364	646	748	1210	1389	2440	2800	4940	5620
200	55	75	135	157	263	317	557	645	1050	1212	2110	2430	4290	4870
250	49	67	121	141	236	284	497	576	941	1090	1890	2180	3850	4360
300	44	61	110	129	217	260	453	525	862	999	1720	1990	3520	3980
400	38	52	96	111	189	225	390	453	749	871	1490	1730	3060	3450
500	34	46	86	100	170	202	348	404	552	783	1330	1550	2740	3090

For SI units: 1 foot = 304.8 mm, 1 cubic foot per hour = 0.0283 m^3/h, 1 pound-force per square inch = 6.8947 kPa

Notes:

1 Table does not include effect of pressure drop across the line regulator. Where regulator loss exceeds 0.75 psi (5.17 kPa), DO NOT USE THIS TABLE. Consult with regulator manufacturer for pressure drops and capacity factors. Pressure drops across a regulator are capable of varying with flow rate.

2 CAUTION: Capacities shown in table are capable of exceeding maximum capacity for a selected regulator. Consult with regulator or tubing manufacturer for guidance.

3 Table includes losses for four 90 degree (1.57 rad) bends and two end fittings. Tubing runs with larger numbers of bends, fittings, or both shall be increased by an equivalent length of tubing according to the following equation: $L = 1.3\ n$, where L is additional length (ft) of tubing and n is the number of additional fittings, bends, or both.

4 Table entries are rounded to 3 significant digits.

5 EHD = Equivalent Hydraulic Diameter, which is a measure of the relative hydraulic efficiency between different tubing sizes. The greater the value of EHD, the greater the gas capacity of the tubing.

TABLE 1315.2(18)
CORRUGATED STAINLESS STEEL TUBING (CSST) [NFPA 54: TABLE 6.2(s)][1, 2, 3, 4]

											GAS:	NATURAL		
											INLET PRESSURE:	5.0 psi		
											PRESSURE DROP:	3.5 psi		
											SPECIFIC GRAVITY:	0.60		
	TUBE SIZE (EHD)[5]													
FLOW DESIGNATION:	**13**	**15**	**18**	**19**	**23**	**25**	**30**	**31**	**37**	**39**	**46**	**48**	**60**	**62**
LENGTH (feet)	CAPACITY IN CUBIC FEET OF GAS PER HOUR													
10	523	674	1080	1300	2000	2530	4920	5660	8300	9140	18 100	19 800	34 400	40 400
25	322	420	691	827	1290	1620	3080	3540	5310	5911	11 400	12 600	22 000	25 600
30	292	382	632	755	1180	1480	2800	3230	4860	5420	10 400	11 500	20 100	23 400
40	251	329	549	654	1030	1280	2420	2790	4230	4727	8970	10 000	17 400	20 200
50	223	293	492	586	926	1150	2160	2490	3790	4251	8020	8930	15 600	18 100
75	180	238	403	479	763	944	1750	2020	3110	3506	6530	7320	12 800	14 800
80	174	230	391	463	740	915	1690	1960	3020	3400	6320	7090	12 400	14 300
100	154	205	350	415	665	820	1510	1740	2710	3057	5650	6350	11 100	12 800
150	124	166	287	339	548	672	1230	1420	2220	2521	4600	5200	9130	10 500
200	107	143	249	294	478	584	1060	1220	1930	2199	3980	4510	7930	9090
250	95	128	223	263	430	524	945	1090	1730	1977	3550	4040	7110	8140
300	86	116	204	240	394	479	860	995	1590	1813	3240	3690	6500	7430
400	74	100	177	208	343	416	742	858	1380	1581	2800	3210	5650	6440
500	66	89	159	186	309	373	662	766	1040	1422	2500	2870	5060	5760

For SI units: 1 foot = 304.8 mm, 1 cubic foot per hour = 0.0283 m^3/h, 1 pound-force per square inch = 6.8947 kPa

Notes:

[1] Table does not include effect of pressure drop across the line regulator. Where regulator loss exceeds 1 psi (7 kPa), DO NOT USE THIS TABLE. Consult with regulator manufacturer for pressure drops and capacity factors. Pressure drops across regulator are capable of varying with the flow rate.

[2] CAUTION: Capacities shown in table are capable of exceeding the maximum capacity of selected regulator. Consult tubing manufacturer for guidance.

[3] Table includes losses for four 90 degree (1.57 rad) bends and two end fittings. Tubing runs with larger numbers of bends, fittings, or both shall be increased by an equivalent length of tubing to the following equation: $L = 1.3\ n$, where L is additional length (feet) of tubing and n is the number of additional fittings, bends, or both.

[4] Table entries are rounded to 3 significant digits.

[5] EHD = Equivalent Hydraulic Diameter, which is a measure of the relative hydraulic efficiency between different tubing sizes. The greater the value of EHD, the greater the gas capacity of the tubing.

›› TABLE 1315.2(19)
POLYETHYLENE PLASTIC PIPE [NFPA 54: TABLE 6.2(t)]*

						GAS:	NATURAL
						INLET PRESSURE:	LESS THAN 2 psi	
						PRESSURE DROP:	0.3 in. w.c.	
						SPECIFIC GRAVITY:	0.60	
	PIPE SIZE (inch)							
NOMINAL OD:	½	¾	1	1 ¼	1 ½	2	3	4
DESIGNATION:	SDR 9.3	SDR 11	SDR 11	SDR 10	SDR 11	SDR 11	SDR 11	SDR 11
ACTUAL ID:	0.660	0.860	1.077	1.328	1.554	1.943	2.864	3.682
LENGTH (feet)	CAPACITY IN CUBIC FEET OF GAS PER HOUR							
10	153	305	551	955	1440	2590	7170	13 900
20	105	210	379	656	991	1780	4920	9520
30	84	169	304	527	796	1430	3950	7640
40	72	144	260	451	681	1220	3380	6540
50	64	128	231	400	604	1080	3000	5800
60	58	116	209	362	547	983	2720	5250
70	53	107	192	333	503	904	2500	4830
80	50	99	179	310	468	841	2330	4500
90	46	93	168	291	439	789	2180	4220
100	44	88	159	275	415	745	2060	3990
125	39	78	141	243	368	661	1830	3530
150	35	71	127	221	333	598	1660	3200
175	32	65	117	203	306	551	1520	2940
200	30	60	109	189	285	512	1420	2740
250	27	54	97	167	253	454	1260	2430
300	24	48	88	152	229	411	1140	2200
350	22	45	81	139	211	378	1050	2020
400	21	42	75	130	196	352	974	1880
450	19	39	70	122	184	330	914	1770
500	18	37	66	115	174	312	863	1670

For SI units: 1 inch = 25 mm, 1 foot = 304.8 mm, 1 cubic foot per hour = 0.0283 m^3/h, 1 pound-force per square inch = 6.8947 kPa, 1 inch water column = 0.249 kPa

* Table entries are rounded to 3 significant digits.

» **TABLE 1315.2(20)**
POLYETHYLENE PLASTIC PIPE [NFPA 54: TABLE 6.2(u)]*

					GAS:	NATURAL		
					INLET PRESSURE:	LESS THAN 2 psi		
					PRESSURE DROP:	0.5 in. w.c.		
					SPECIFIC GRAVITY:	0.60		
	PIPE SIZE (inch)							
NOMINAL OD:	½	¾	1	1 ¼	1 ½	2	3	4
DESIGNATION:	SDR 9.3	SDR 11	SDR 11	SDR 10	SDR 11	SDR 11	SDR 11	SDR 11
ACTUAL ID:	0.660	0.860	1.077	1.328	1.554	1.943	2.864	3.682
LENGTH (feet)	CAPACITY IN CUBIC FEET OF GAS PER HOUR							
10	201	403	726	1260	1900	3410	9450	18 260
20	138	277	499	865	1310	2350	6490	12 550
30	111	222	401	695	1050	1880	5210	10 080
40	95	190	343	594	898	1610	4460	8630
50	84	169	304	527	796	1430	3950	7640
60	76	153	276	477	721	1300	3580	6930
70	70	140	254	439	663	1190	3300	6370
80	65	131	236	409	617	1110	3070	5930
90	61	123	221	383	579	1040	2880	5560
100	58	116	209	362	547	983	2720	5250
125	51	103	185	321	485	871	2410	4660
150	46	93	168	291	439	789	2180	4220
175	43	86	154	268	404	726	2010	3880
200	40	80	144	249	376	675	1870	3610
250	35	71	127	221	333	598	1660	3200
300	32	64	115	200	302	542	1500	2900
350	29	59	106	184	278	499	1380	2670
400	27	55	99	171	258	464	1280	2480
450	26	51	93	160	242	435	1200	2330
500	24	48	88	152	229	411	1140	2200

For SI units: 1 inch = 25 mm, 1 foot = 304.8 mm, 1 cubic foot per hour = 0.0283 m^3/h, 1 pound-force per square inch = 6.8947 kPa, 1 inch water column = 0.249 kPa
* Table entries are rounded to 3 significant digits.

TABLE 1315.2(21)
POLYETHYLENE PLASTIC PIPE [NFPA 54: TABLE 6.2(v)]*

					GAS:	NATURAL	
					INLET PRESSURE:	2.0 psi	
					PRESSURE DROP:	1.0 psi	
					SPECIFIC GRAVITY:	0.60	
	PIPE SIZE (inch)						
NOMINAL OD:	½	¾	1	1 ¼	1 ½	2	3	4
DESIGNATION:	SDR 9.3	SDR 11	SDR 11	SDR 10	SDR 11	SDR 11	SDR 11	SDR 11
ACTUAL ID:	0.660	0.860	1.077	1.328	1.554	1.943	2.864	3.682
LENGTH (feet)	CAPACITY IN CUBIC FEET OF GAS PER HOUR							
10	1860	3720	6710	11 600	17 600	31 600	87 300	169 000
20	1280	2560	4610	7990	12 100	21 700	60 000	116 000
30	1030	2050	3710	6420	9690	17 400	48 200	93 200
40	878	1760	3170	5490	8300	14 900	41 200	79 700
50	778	1560	2810	4870	7350	13 200	36 600	70 700
60	705	1410	2550	4410	6660	12 000	33 100	64 000
70	649	1300	2340	4060	6130	11 000	30 500	58 900
80	603	1210	2180	3780	5700	10 200	28 300	54 800
90	566	1130	2050	3540	5350	9610	26 600	51 400
100	535	1070	1930	3350	5050	9080	25 100	48 600
125	474	949	1710	2970	4480	8050	22 300	43 000
150	429	860	1550	2690	4060	7290	20 200	39 000
175	395	791	1430	2470	3730	6710	18 600	35 900
200	368	736	1330	2300	3470	6240	17 300	33 400
250	326	652	1180	2040	3080	5530	15 300	29 600
300	295	591	1070	1850	2790	5010	13 900	26 800
350	272	544	981	1700	2570	4610	12 800	24 700
400	253	506	913	1580	2390	4290	11 900	22 900
450	237	475	856	1480	2240	4020	11 100	21 500
500	224	448	809	1400	2120	3800	10 500	20 300
550	213	426	768	1330	2010	3610	9990	19 300
600	203	406	733	1270	1920	3440	9530	18 400
650	194	389	702	1220	1840	3300	9130	17 600
700	187	374	674	1170	1760	3170	8770	16 900
750	180	360	649	1130	1700	3050	8450	16 300
800	174	348	627	1090	1640	2950	8160	15 800
850	168	336	607	1050	1590	2850	7890	15 300
900	163	326	588	1020	1540	2770	7650	14 800
950	158	317	572	990	1500	2690	7430	14 400
1000	154	308	556	963	1450	2610	7230	14 000
1100	146	293	528	915	1380	2480	6870	13 300
1200	139	279	504	873	1320	2370	6550	12 700
1300	134	267	482	836	1260	2270	6270	12 100
1400	128	257	463	803	1210	2180	6030	11 600
1500	124	247	446	773	1170	2100	5810	11 200
1600	119	239	431	747	1130	2030	5610	10 800
1700	115	231	417	723	1090	1960	5430	10 500
1800	112	224	404	701	1060	1900	5260	10 200
1900	109	218	393	680	1030	1850	5110	9900
2000	106	212	382	662	1000	1800	4970	9600

For SI units: 1 inch = 25 mm, 1 foot = 304.8 mm, 1 cubic foot per hour = 0.0283 m^3/h, 1 pound-force per square inch = 6.8947 kPa

* Table entries are rounded to 3 significant digits.

TABLE 1315.2(22)
POLYETHYLENE PLASTIC TUBING [NFPA 54: TABLE 6.2(w)][2, 3]

	GAS:	NATURAL
	INLET PRESSURE:	LESS THAN 2.0 psi
	PRESSURE DROP:	0.3 in. w.c.
	SPECIFIC GRAVITY:	0.60
	PLASTIC TUBING SIZE (CTS)[1] (inch)	
NOMINAL OD:	½	1
DESIGNATION:	SDR 7	SDR 11
ACTUAL ID:	0.445	0.927
LENGTH (feet)	CAPACITY IN CUBIC FEET OF GAS PER HOUR	
10	54	372
20	37	256
30	30	205
40	26	176
50	23	156
60	21	141
70	19	130
80	18	121
90	17	113
100	16	107
125	14	95
150	13	86
175	12	79
200	11	74
225	10	69
250	NA	65
275	NA	62
300	NA	59
350	NA	54
400	NA	51
450	NA	47
500	NA	45

For SI units: 1 inch = 25 mm, 1 foot = 304.8 mm, 1 cubic foot per hour = $0.0283 m^3/h$, 1 pound-force per square inch = 6.8947 kPa, 1 inch water column = 0.249 kPa

Notes:

[1] CTS = Copper tube size.

[2] Table entries are rounded to 3 significant digits.

[3] NA means a flow of less than 10 ft^3/h (0.283 m^3/h).

TABLE 1315.2(23)
POLYETHYLENE PLASTIC TUBING [NFPA 54: TABLE 6.2(x)][2, 3]

	GAS:	NATURAL
	INLET PRESSURE:	LESS THAN 2.0 psi
	PRESSURE DROP:	0.5 in. w.c.
	SPECIFIC GRAVITY:	0.60
	PLASTIC TUBING SIZE (CTS)[1] (inch)	
NOMINAL OD:	½	1
DESIGNATION:	SDR 7	SDR 11
ACTUAL ID:	0.445	0.927
LENGTH (feet)	CAPACITY IN CUBIC FEET OF GAS PER HOUR	
10	72	490
20	49	337
30	39	271
40	34	232
50	30	205
60	27	186
70	25	171
80	23	159
90	22	149
100	21	141
125	18	125
150	17	113
175	15	104
200	14	97
225	13	91
250	12	86
275	11	82
300	11	78
350	10	72
400	NA	67
450	NA	63
500	NA	59

For SI units: 1 inch = 25 mm, 1 foot = 304.8 mm, 1 cubic foot per hour = $0.0283 m^3/h$, 1 pound-force per square inch = 6.8947 kPa, 1 inch water column = 0.249 kPa

Notes:

[1] CTS = Copper tube size.

[2] Table entries are rounded to 3 significant digits.

[3] NA means a flow of less than 10 ft^3/h (0.283 m^3/h).

» **TABLE 1315.2(24)**
SCHEDULE 40 METALLIC PIPE [NFPA 54: TABLE 6.3(a)]*

							GAS:	UNDILUTED PROPANE
							INLET PRESSURE:	10.0 psi
							PRESSURE DROP:	1.0 psi
							SPECIFIC GRAVITY:	1.50
INTENDED USE: PIPE SIZING BETWEEN FIRST STAGE (HIGH PRESSURE) REGULATOR AND SECOND STAGE (LOW PRESSURE) REGULATOR									
	PIPE SIZE (inch)								
NOMINAL INSIDE:	½	¾	1	1¼	1½	2	2½	3	4
ACTUAL:	0.622	0.824	1.049	1.380	1.610	2.067	2.469	3.068	4.026
LENGTH (feet)	**CAPACITY IN THOUSANDS OF BTU PER HOUR**								
10	3320	6950	13 100	26 900	40 300	77 600	124 000	219 000	446 000
20	2280	4780	9000	18 500	27 700	53 300	85 000	150 000	306 000
30	1830	3840	7220	14 800	22 200	42 800	68 200	121 000	246 000
40	1570	3280	6180	12 700	19 000	36 600	58 400	103 000	211 000
50	1390	2910	5480	11 300	16 900	32 500	51 700	91 500	187 000
60	1260	2640	4970	10 200	15 300	29 400	46 900	82 900	169 000
70	1160	2430	4570	9380	14 100	27 100	43 100	76 300	156 000
80	1080	2260	4250	8730	13 100	25 200	40 100	70 900	145 000
90	1010	2120	3990	8190	12 300	23 600	37 700	66 600	136 000
100	956	2000	3770	7730	11 600	22 300	35 600	62 900	128 000
125	848	1770	3340	6850	10 300	19 800	31 500	55 700	114 000
150	768	1610	3020	6210	9300	17 900	28 600	50 500	103 000
175	706	1480	2780	5710	8560	16 500	26 300	46 500	94 700
200	657	1370	2590	5320	7960	15 300	24 400	43 200	88 100
250	582	1220	2290	4710	7060	13 600	21 700	38 300	78 100
300	528	1100	2080	4270	6400	12 300	19 600	34 700	70 800
350	486	1020	1910	3930	5880	11 300	18 100	31 900	65 100
400	452	945	1780	3650	5470	10 500	16 800	29 700	60 600
450	424	886	1670	3430	5140	9890	15 800	27 900	56 800
500	400	837	1580	3240	4850	9340	14 900	26 300	53 700
550	380	795	1500	3070	4610	8870	14 100	25 000	51 000
600	363	759	1430	2930	4400	8460	13 500	23 900	48 600
650	347	726	1370	2810	4210	8110	12 900	22 800	46 600
700	334	698	1310	2700	4040	7790	12 400	21 900	44 800
750	321	672	1270	2600	3900	7500	12 000	21 100	43 100
800	310	649	1220	2510	3760	7240	11 500	20 400	41 600
850	300	628	1180	2430	3640	7010	11 200	19 800	40 300
900	291	609	1150	2360	3530	6800	10 800	19 200	39 100
950	283	592	1110	2290	3430	6600	10 500	18 600	37 900
1000	275	575	1080	2230	3330	6420	10 200	18 100	36 900
1100	261	546	1030	2110	3170	6100	9720	17 200	35 000
1200	249	521	982	2020	3020	5820	9270	16 400	33 400
1300	239	499	940	1930	2890	5570	8880	15 700	32 000
1400	229	480	903	1850	2780	5350	8530	15 100	30 800
1500	221	462	870	1790	2680	5160	8220	14 500	29 600
1600	213	446	840	1730	2590	4980	7940	14 000	28 600
1700	206	432	813	1670	2500	4820	7680	13 600	27 700
1800	200	419	789	1620	2430	4670	7450	13 200	26 900
1900	194	407	766	1570	2360	4540	7230	12 800	26 100
2000	189	395	745	1530	2290	4410	7030	12 400	25 400

For SI units: 1 inch = 25 mm, 1 foot = 304.8 mm, 1000 British thermal units per hour = 0.293 kW, 1 pound-force per square inch = 6.8947 kPa

* Table entries are rounded to 3 significant digits.

TABLE 1315.2(25)
SCHEDULE 40 METALLIC PIPE [NFPA 54: TABLE 6.3(b)]*

							GAS:	UNDILUTED PROPANE	
							INLET PRESSURE:	10.0 psi	
							PRESSURE DROP:	3.0 psi	
							SPECIFIC GRAVITY:	1.50	
INTENDED USE: PIPE SIZING BETWEEN FIRST STAGE (HIGH PRESSURE) REGULATOR AND SECOND STAGE (LOW PRESSURE) REGULATOR									
	PIPE SIZE (inch)								
NOMINAL INSIDE:	½	¾	1	1¼	1½	2	2½	3	4
ACTUAL:	0.622	0.824	1.049	1.380	1.610	2.067	2.469	3.068	4.026
LENGTH (feet)	CAPACITY IN THOUSANDS OF BTU PER HOUR								
10	5890	12 300	23 200	47 600	71 300	137 000	219 000	387 000	789 000
20	4050	8460	15 900	32 700	49 000	94 400	150 000	266 000	543 000
30	3250	6790	12 800	26 300	39 400	75 800	121 000	214 000	436 000
40	2780	5810	11 000	22 500	33 700	64 900	103 000	183 000	373 000
50	2460	5150	9710	19 900	29 900	57 500	91 600	162 000	330 000
60	2230	4670	8790	18 100	27 100	52 100	83 000	147 000	299 000
70	2050	4300	8090	16 600	24 900	47 900	76 400	135 000	275 000
80	1910	4000	7530	15 500	23 200	44 600	71 100	126 000	256 000
90	1790	3750	7060	14 500	21 700	41 800	66 700	118 000	240 000
100	1690	3540	6670	13 700	20 500	39 500	63 000	111 000	227 000
125	1500	3140	5910	12 100	18 200	35 000	55 800	98 700	201 000
150	1360	2840	5360	11 000	16 500	31 700	50 600	89 400	182 000
175	1250	2620	4930	10 100	15 200	29 200	46 500	82 300	167 800
200	1160	2430	4580	9410	14 100	27 200	43 300	76 500	156 100
250	1030	2160	4060	8340	12 500	24 100	38 400	67 800	138 400
300	935	1950	3680	7560	11 300	21 800	34 800	61 500	125 400
350	860	1800	3390	6950	10 400	20 100	32 000	56 500	115 300
400	800	1670	3150	6470	9690	18 700	29 800	52 600	107 300
450	751	1570	2960	6070	9090	17 500	27 900	49 400	100 700
500	709	1480	2790	5730	8590	16 500	26 400	46 600	95 100
550	673	1410	2650	5450	8160	15 700	25 000	44 300	90 300
600	642	1340	2530	5200	7780	15 000	23 900	42 200	86 200
650	615	1290	2420	4980	7450	14 400	22 900	40 500	82 500
700	591	1240	2330	4780	7160	13 800	22 000	38 900	79 300
750	569	1190	2240	4600	6900	13 300	21 200	37 400	76 400
800	550	1150	2170	4450	6660	12 800	20 500	36 200	73 700
850	532	1110	2100	4300	6450	12 400	19 800	35 000	71 400
900	516	1080	2030	4170	6250	12 000	19 200	33 900	69 200
950	501	1050	1970	4050	6070	11 700	18 600	32 900	67 200
1000	487	1020	1920	3940	5900	11 400	18 100	32 000	65 400
1100	463	968	1820	3740	5610	10 800	17 200	30 400	62 100
1200	442	923	1740	3570	5350	10 300	16 400	29 000	59 200
1300	423	884	1670	3420	5120	9870	15 700	27 800	56 700
1400	406	849	1600	3280	4920	9480	15 100	26 700	54 500
1500	391	818	1540	3160	4740	9130	14 600	25 700	52 500
1600	378	790	1490	3060	4580	8820	14 100	24 800	50 700
1700	366	765	1440	2960	4430	8530	13 600	24 000	49 000
1800	355	741	1400	2870	4300	8270	13 200	23 300	47 600
1900	344	720	1360	2780	4170	8040	12 800	22 600	46 200
2000	335	700	1320	2710	4060	7820	12 500	22 000	44 900

For SI units: 1 inch = 25 mm, 1 foot = 304.8 mm, 1000 British thermal units per hour = 0.293 kW, 1 pound-force per square inch = 6.8947 kPa

* Table entries are rounded to 3 significant digits.

TABLE 1315.2(26)
SCHEDULE 40 METALLIC PIPE [NFPA 54: TABLE 6.3(c)]*

						GAS:	UNDILUTED PROPANE		
						INLET PRESSURE:	2.0 psi		
						PRESSURE DROP:	1.0 psi		
						SPECIFIC GRAVITY:	1.50		
INTENDED USE: PIPE SIZING BETWEEN 2 PSI SERVICE AND LINE PRESSURE REGULATOR									
	PIPE SIZE (inch)								
NOMINAL:	½	¾	1	1¼	1½	2	2½	3	4
ACTUAL ID:	0.622	0.824	1.049	1.380	1.610	2.067	2.469	3.068	4.026
LENGTH (feet)	CAPACITY IN THOUSANDS OF BTU PER HOUR								
10	2680	5590	10 500	21 600	32 400	62 400	99 500	176 000	359 000
20	1840	3850	7240	14 900	22 300	42 900	68 400	121 000	247 000
30	1480	3090	5820	11 900	17 900	34 500	54 900	97 100	198 000
40	1260	2640	4980	10 200	15 300	29 500	47 000	83 100	170 000
50	1120	2340	4410	9060	13 600	26 100	41 700	73 700	150 000
60	1010	2120	4000	8210	12 300	23 700	37 700	66 700	136 000
70	934	1950	3680	7550	11 300	21 800	34 700	61 400	125 000
80	869	1820	3420	7020	10 500	20 300	32 300	57 100	116 000
90	815	1700	3210	6590	9880	19 000	30 300	53 600	109 000
100	770	1610	3030	6230	9330	18 000	28 600	50 600	103 000
125	682	1430	2690	5520	8270	15 900	25 400	44 900	91 500
150	618	1290	2440	5000	7490	14 400	23 000	40 700	82 900
175	569	1190	2240	4600	6890	13 300	21 200	37 400	76 300
200	529	1110	2080	4280	6410	12 300	19 700	34 800	71 000
250	469	981	1850	3790	5680	10 900	17 400	30 800	62 900
300	425	889	1670	3440	5150	9920	15 800	27 900	57 000
350	391	817	1540	3160	4740	9120	14 500	25 700	52 400
400	364	760	1430	2940	4410	8490	13 500	23 900	48 800
450	341	714	1340	2760	4130	7960	12 700	22 400	45 800
500	322	674	1270	2610	3910	7520	12 000	21 200	43 200
550	306	640	1210	2480	3710	7140	11 400	20 100	41 100
600	292	611	1150	2360	3540	6820	10 900	19 200	39 200
650	280	585	1100	2260	3390	6530	10 400	18 400	37 500
700	269	562	1060	2170	3260	6270	9990	17 700	36 000
750	259	541	1020	2090	3140	6040	9630	17 000	34 700
800	250	523	985	2020	3030	5830	9300	16 400	33 500
850	242	506	953	1960	2930	5640	9000	15 900	32 400
900	235	490	924	1900	2840	5470	8720	15 400	31 500
950	228	476	897	1840	2760	5310	8470	15 000	30 500
1000	222	463	873	1790	2680	5170	8240	14 600	29 700
1100	210	440	829	1700	2550	4910	7830	13 800	28 200
1200	201	420	791	1620	2430	4680	7470	13 200	26 900
1300	192	402	757	1550	2330	4490	7150	12 600	25 800
1400	185	386	727	1490	2240	4310	6870	12 100	24 800
1500	178	372	701	1440	2160	4150	6620	11 700	23 900
1600	172	359	677	1390	2080	4010	6390	11 300	23 000
1700	166	348	655	1340	2010	3880	6180	10 900	22 300
1800	161	337	635	1300	1950	3760	6000	10 600	21 600
1900	157	327	617	1270	1900	3650	5820	10 300	21 000
2000	152	318	600	1230	1840	3550	5660	10 000	20 400

For SI units: 1 inch = 25 mm, 1 foot = 304.8 mm, 1000 British thermal units per hour = 0.293 kW, 1 pound-force per square inch = 6.8947 kPa

* Table entries are rounded to 3 significant digits.

TABLE 1315.2(27)
SCHEDULE 40 METALLIC PIPE [NFPA 54: TABLE 6.3(d)]*

						GAS:	UNDILUTED PROPANE		
						INLET PRESSURE:	11.0 in. w.c.		
						PRESSURE DROP:	0.5 in. w.c.		
						SPECIFIC GRAVITY:	1.50		
INTENDED USE: PIPE SIZING BETWEEN SINGLE OR SECOND STAGE (LOW PRESSURE) REGULATOR AND APPLIANCE									
	PIPE SIZE (inch)								
NOMINAL INSIDE:	½	¾	1	1¼	1½	2	2½	3	4
ACTUAL:	0.622	0.824	1.049	1.380	1.610	2.067	2.469	3.068	4.026
LENGTH (feet)	CAPACITY IN THOUSANDS OF BTU PER HOUR								
10	291	608	1150	2350	3520	6790	10 800	19 100	39 000
20	200	418	787	1620	2420	4660	7430	13 100	26 800
30	160	336	632	1300	1940	3750	5970	10 600	21 500
40	137	287	541	1110	1660	3210	5110	9030	18 400
50	122	255	480	985	1480	2840	4530	8000	16 300
60	110	231	434	892	1340	2570	4100	7250	14 800
80	101	212	400	821	1230	2370	3770	6670	13 600
100	94	197	372	763	1140	2200	3510	6210	12 700
125	89	185	349	716	1070	2070	3290	5820	11 900
150	84	175	330	677	1010	1950	3110	5500	11 200
175	74	155	292	600	899	1730	2760	4880	9950
200	67	140	265	543	814	1570	2500	4420	9010
250	62	129	243	500	749	1440	2300	4060	8290
300	58	120	227	465	697	1340	2140	3780	7710
350	51	107	201	412	618	1190	1900	3350	6840
400	46	97	182	373	560	1080	1720	3040	6190
450	42	89	167	344	515	991	1580	2790	5700
500	40	83	156	320	479	922	1470	2600	5300
550	37	78	146	300	449	865	1380	2440	4970
600	35	73	138	283	424	817	1300	2300	4700
650	33	70	131	269	403	776	1240	2190	4460
700	32	66	125	257	385	741	1180	2090	4260
750	30	64	120	246	368	709	1130	2000	4080
800	29	61	115	236	354	681	1090	1920	3920
850	28	59	111	227	341	656	1050	1850	3770
900	27	57	107	220	329	634	1010	1790	3640
950	26	55	104	213	319	613	978	1730	3530
1000	25	53	100	206	309	595	948	1680	3420
1100	25	52	97	200	300	578	921	1630	3320
1200	24	50	95	195	292	562	895	1580	3230
1300	23	48	90	185	277	534	850	1500	3070
1400	22	46	86	176	264	509	811	1430	2930
1500	21	44	82	169	253	487	777	1370	2800
1600	20	42	79	162	243	468	746	1320	2690
1700	19	40	76	156	234	451	719	1270	2590
1800	19	39	74	151	226	436	694	1230	2500
1900	18	38	71	146	219	422	672	1190	2420
2000	18	37	69	142	212	409	652	1150	2350

For SI units: 1 inch = 25 mm, 1 foot = 304.8 mm, 1000 British thermal units per hour = 0.293 kW, 1 inch water column = 0.249 kPa

* Table entries are rounded to 3 significant digits.

TABLE 1315.2(28) SEMI-RIGID COPPER TUBING [NFPA 54: TABLE 6.3(e)][2]

							GAS:	UNDILUTED PROPANE	
							INLET PRESSURE:	10.0 psi	
							PRESSURE DROP:	1.0 psi	
							SPECIFIC GRAVITY:	1.50	
INTENDED USE: TUBE SIZING BETWEEN FIRST STAGE (HIGH PRESSURE) REGULATOR AND SECOND STAGE (LOW PRESSURE) REGULATOR									
	TUBE SIZE (inch)								
NOMINAL: K & L:	¼	⅜	½	⅝	¾	1	1¼	1½	2
NOMINAL: ACR:	⅜	½	⅝	¾	⅞	1⅛	1⅜	–	–
OUTSIDE:	0.375	0.500	0.625	0.750	0.875	1.125	1.375	1.625	2.125
INSIDE:[1]	0.305	0.402	0.527	0.652	0.745	0.995	1.245	1.481	1.959
LENGTH (feet)	CAPACITY IN THOUSANDS OF BTU PER HOUR								
10	513	1060	2150	3760	5330	11 400	20 500	32 300	67 400
20	352	727	1480	2580	3670	7830	14 100	22 200	46 300
30	283	584	1190	2080	2940	6290	11 300	17 900	37 200
40	242	500	1020	1780	2520	5380	9690	15 300	31 800
50	215	443	901	1570	2230	4770	8590	13 500	28 200
60	194	401	816	1430	2020	4320	7780	12 300	25 600
70	179	369	751	1310	1860	3980	7160	11 300	23 500
80	166	343	699	1220	1730	3700	6660	10 500	21 900
90	156	322	655	1150	1630	3470	6250	9850	20 500
100	147	304	619	1080	1540	3280	5900	9310	19 400
125	131	270	549	959	1360	2910	5230	8250	17 200
150	118	244	497	869	1230	2630	4740	7470	15 600
175	109	225	457	799	1130	2420	4360	6880	14 300
200	101	209	426	744	1060	2250	4060	6400	13 300
250	90	185	377	659	935	2000	3600	5670	11 800
300	81	168	342	597	847	1810	3260	5140	10 700
350	75	155	314	549	779	1660	3000	4730	9840
400	70	144	292	511	725	1550	2790	4400	9160
450	65	135	274	480	680	1450	2620	4130	8590
500	62	127	259	453	643	1370	2470	3900	8120
550	59	121	246	430	610	1300	2350	3700	7710
600	56	115	235	410	582	1240	2240	3530	7350
650	54	111	225	393	558	1190	2140	3380	7040
700	51	106	216	378	536	1140	2060	3250	6770
750	50	102	208	364	516	1100	1980	3130	6520
800	48	99	201	351	498	1060	1920	3020	6290
850	46	96	195	340	482	1030	1850	2920	6090
900	45	93	189	330	468	1000	1800	2840	5910
950	44	90	183	320	454	970	1750	2750	5730
1000	42	88	178	311	442	944	1700	2680	5580
1100	40	83	169	296	420	896	1610	2540	5300
1200	38	79	161	282	400	855	1540	2430	5050
1300	37	76	155	270	383	819	1470	2320	4840
1400	35	73	148	260	368	787	1420	2230	4650
1500	34	70	143	250	355	758	1360	2150	4480
1600	33	68	138	241	343	732	1320	2080	4330
1700	32	66	134	234	331	708	1270	2010	4190
1800	31	64	130	227	321	687	1240	1950	4060
1900	30	62	126	220	312	667	1200	1890	3940
2000	29	60	122	214	304	648	1170	1840	3830

For SI units: 1 inch = 25 mm, 1 foot = 304.8 mm, 1000 British thermal units per hour = 0.293 kW, 1 pound-force per square inch = 6.8947 kPa

Notes:

[1] Table capacities are based on Type K copper tubing inside diameter (shown), which has the smallest inside diameter of the copper tubing products.

[2] Table entries are rounded to 3 significant digits.

TABLE 1315.2(29)
SEMI-RIGID COPPER TUBING [NFPA 54: TABLE 6.3(f)][2, 3]

						GAS:	**UNDILUTED PROPANE**		
						INLET PRESSURE:	**11.0 in. w.c.**		
						PRESSURE DROP:	**0.5 in. w.c.**		
						SPECIFIC GRAVITY:	**1.50**		
INTENDED USE: TUBE SIZING BETWEEN SINGLE OR SECOND STAGE (LOW PRESSURE) REGULATOR AND APPLIANCE									
	TUBE SIZE (inch)								
NOMINAL: K & L:	¼	⅜	½	⅝	¾	1	1¼	1½	2
NOMINAL: ACR:	⅜	½	⅝	¾	⅞	1⅛	1⅜	–	–
OUTSIDE:	0.375	0.500	0.625	0.750	0.875	1.125	1.375	1.625	2.125
INSIDE:[1]	0.305	0.402	0.527	0.652	0.745	0.995	1.245	1.481	1.959
LENGTH (feet)	**CAPACITY IN THOUSANDS OF BTU PER HOUR**								
10	45	93	188	329	467	997	1800	2830	5890
20	31	64	129	226	321	685	1230	1950	4050
30	25	51	104	182	258	550	991	1560	3250
40	21	44	89	155	220	471	848	1340	2780
50	19	39	79	138	195	417	752	1180	2470
60	17	35	71	125	177	378	681	1070	2240
70	16	32	66	115	163	348	626	988	2060
80	15	30	61	107	152	324	583	919	1910
90	14	28	57	100	142	304	547	862	1800
100	13	27	54	95	134	287	517	814	1700
125	11	24	48	84	119	254	458	722	1500
150	10	21	44	76	108	230	415	654	1360
175	NA	20	40	70	99	212	382	602	1250
200	NA	18	37	65	92	197	355	560	1170
250	NA	16	33	58	82	175	315	496	1030
300	NA	15	30	52	74	158	285	449	936
350	NA	14	28	48	68	146	262	414	861
400	NA	13	26	45	63	136	244	385	801
450	NA	12	24	42	60	127	229	361	752
500	NA	11	23	40	56	120	216	341	710
550	NA	11	22	38	53	114	205	324	674
600	NA	10	21	36	51	109	196	309	643
650	NA	NA	20	34	49	104	188	296	616
700	NA	NA	19	33	47	100	180	284	592
750	NA	NA	18	32	45	96	174	274	570
800	NA	NA	18	31	44	93	168	264	551
850	NA	NA	17	30	42	90	162	256	533
900	NA	NA	17	29	41	87	157	248	517
950	NA	NA	16	28	40	85	153	241	502
1000	NA	NA	16	27	39	83	149	234	488
1100	NA	NA	15	26	37	78	141	223	464
1200	NA	NA	14	25	35	75	135	212	442
1300	NA	NA	14	24	34	72	129	203	423
1400	NA	NA	13	23	32	69	124	195	407
1500	NA	NA	13	22	31	66	119	188	392
1600	NA	NA	12	21	30	64	115	182	378
1700	NA	NA	12	20	29	62	112	176	366
1800	NA	NA	11	20	28	60	108	170	355
1900	NA	NA	11	19	27	58	105	166	345
2000	NA	NA	11	19	27	57	102	161	335

For SI units: 1 inch = 25 mm, 1 foot = 304.8 mm, 1000 British thermal units per hour = 0.293 kW, 1 inch water column = 0.249 kPa

Notes:

1 Table capacities are based on Type K copper tubing inside diameter (shown), which has the smallest inside diameter of the copper tubing products.

2 Table entries are rounded to 3 significant digits.

3 NA means a flow of less than 10 000 Btu/h (2.93 kW).

» **TABLE 1315.2(30)**
SEMI-RIGID COPPER TUBING [NFPA 54: TABLE 6.3(g)][2]

						GAS:	UNDILUTED PROPANE		
						INLET PRESSURE:	2.0 psi		
						PRESSURE DROP:	1.0 psi		
						SPECIFIC GRAVITY:	1.50		
INTENDED USE: TUBE SIZING BETWEEN 2 PSIG SERVICE AND LINE PRESSURE REGULATOR									
	TUBE SIZE (inch)								
NOMINAL: K & L:	¼	⅜	½	⅝	¾	1	1¼	1½	2
NOMINAL: ACR:	⅜	½	⅝	¾	⅞	1⅛	1⅜	–	–
OUTSIDE:	0.375	0.500	0.625	0.750	0.875	1.125	1.375	1.625	2.125
INSIDE:[1]	0.305	0.402	0.527	0.652	0.745	0.995	1.245	1.481	1.959
LENGTH (feet)	CAPACITY IN THOUSANDS OF BTU PER HOUR								
10	413	852	1730	3030	4300	9170	16 500	26 000	54 200
20	284	585	1190	2080	2950	6310	11 400	17 900	37 300
30	228	470	956	1670	2370	5060	9120	14 400	29 900
40	195	402	818	1430	2030	4330	7800	12 300	25 600
50	173	356	725	1270	1800	3840	6920	10 900	22 700
60	157	323	657	1150	1630	3480	6270	9880	20 600
70	144	297	605	1060	1500	3200	5760	9090	18 900
80	134	276	562	983	1390	2980	5360	8450	17 600
90	126	259	528	922	1310	2790	5030	7930	16 500
100	119	245	498	871	1240	2640	4750	7490	15 600
125	105	217	442	772	1100	2340	4210	6640	13 800
150	95	197	400	700	992	2120	3820	6020	12 500
175	88	181	368	644	913	1950	3510	5540	11 500
200	82	168	343	599	849	1810	3270	5150	10 700
250	72	149	304	531	753	1610	2900	4560	9510
300	66	135	275	481	682	1460	2620	4140	8610
350	60	124	253	442	628	1340	2410	3800	7920
400	56	116	235	411	584	1250	2250	3540	7370
450	53	109	221	386	548	1170	2110	3320	6920
500	50	103	209	365	517	1110	1990	3140	6530
550	47	97	198	346	491	1050	1890	2980	6210
600	45	93	189	330	469	1000	1800	2840	5920
650	43	89	181	316	449	959	1730	2720	5670
700	41	86	174	304	431	921	1660	2620	5450
750	40	82	168	293	415	888	1600	2520	5250
800	39	80	162	283	401	857	1540	2430	5070
850	37	77	157	274	388	829	1490	2350	4900
900	36	75	152	265	376	804	1450	2280	4750
950	35	72	147	258	366	781	1410	2220	4620
1000	34	71	143	251	356	760	1370	2160	4490
1100	32	67	136	238	338	721	1300	2050	4270
1200	31	64	130	227	322	688	1240	1950	4070
1300	30	61	124	217	309	659	1190	1870	3900
1400	28	59	120	209	296	633	1140	1800	3740
1500	27	57	115	201	286	610	1100	1730	3610
1600	26	55	111	194	276	589	1060	1670	3480
1700	26	53	108	188	267	570	1030	1620	3370
1800	25	51	104	182	259	553	1000	1570	3270
1900	24	50	101	177	251	537	966	1520	3170
2000	23	48	99	172	244	522	940	1480	3090

For SI units: 1 inch = 25 mm, 1 foot = 304.8 mm, 1000 British thermal units per hour = 0.293 kW, 1 pound-force per square inch = 6.8947 kPa

Notes:

1 Table capacities are based on Type K copper tubing inside diameter (shown), which has the smallest inside diameter of the copper tubing products.

2 Table entries are rounded to 3 significant digits.

TABLE 1315.2(31)
CORRUGATED STAINLESS STEEL TUBING (CSST) [NFPA 54: TABLE 6.3(h)][1, 2]

									GAS:	**UNDILUTED PROPANE**			
									INLET PRESSURE:	**11.0 in. w.c.**			
									PRESSURE DROP:	**0.5 in. w.c.**			
									SPECIFIC GRAVITY:	**1.50**			
INTENDED USE: CSST SIZING BETWEEN SINGLE OR SECOND STAGE (LOW PRESSURE) REGULATOR AND APPLIANCE SHUTOFF VALVE													

	TUBE SIZE (EHD)[3]													
FLOW DESIGNATION:	**13**	**15**	**18**	**19**	**23**	**25**	**30**	**31**	**37**	**39**	**46**	**48**	**60**	**62**
LENGTH (feet)	**CAPACITY IN THOUSANDS OF BTU PER HOUR**													
5	72	99	181	211	355	426	744	863	1420	1638	2830	3270	5780	6550
10	50	69	129	150	254	303	521	605	971	1179	1990	2320	4110	4640
15	39	55	104	121	208	248	422	490	775	972	1620	1900	3370	3790
20	34	49	91	106	183	216	365	425	661	847	1400	1650	2930	3290
25	30	42	82	94	164	192	325	379	583	762	1250	1480	2630	2940
30	28	39	74	87	151	177	297	344	528	698	1140	1350	2400	2680
40	23	33	64	74	131	153	256	297	449	610	988	1170	2090	2330
50	20	30	58	66	118	137	227	265	397	548	884	1050	1870	2080
60	19	26	53	60	107	126	207	241	359	502	805	961	1710	1900
70	17	25	49	57	99	117	191	222	330	466	745	890	1590	1760
80	15	23	45	52	94	109	178	208	307	438	696	833	1490	1650
90	15	22	44	50	90	102	169	197	286	414	656	787	1400	1550
100	14	20	41	47	85	98	159	186	270	393	621	746	1330	1480
150	11	15	31	36	66	75	123	143	217	324	506	611	1090	1210
200	9	14	28	33	60	69	112	129	183	283	438	531	948	1050
250	8	12	25	30	53	61	99	117	163	254	390	476	850	934
300	8	11	23	26	50	57	90	107	147	234	357	434	777	854

For SI units: 1 foot = 304.8 mm, 1000 British thermal units per hour = 0.293 kW, 1 inch water column = 0.249 kPa

Notes:

1 Table includes losses for four 90 degree (1.57 rad) bends and two end fittings. Tubing runs with larger numbers of bends, fittings, or both shall be increased by an equivalent length of tubing to the following equation: $L = 1.3\ n$, where L is additional length (ft) of tubing and n is the number of additional fittings, bends, or both.

2 Table entries are rounded to 3 significant digits.

3 EHD = Equivalent Hydraulic Diameter, which is a measure of the relative hydraulic efficiency between different tubing sizes. The greater the value of EHD, the greater the gas capacity of the tubing.

TABLE 1315.2(32)
CORRUGATED STAINLESS STEEL TUBING (CSST) [NFPA 54: TABLE 6.3(i)][1, 2, 3, 4]

								GAS:	UNDILUTED PROPANE					
								INLET PRESSURE:	2.0 psi					
								PRESSURE DROP:	1.0 psi					
								SPECIFIC GRAVITY:	1.50					
INTENDED USE: CSST SIZING BETWEEN 2 PSI SERVICE AND LINE PRESSURE REGULATOR														
	TUBE SIZE (EHD)[5]													
FLOW DESIGNATION:	13	15	18	19	23	25	30	31	37	39	46	48	60	62
LENGTH (feet)	CAPACITY IN THOUSANDS OF BTU PER HOUR													
10	426	558	927	1110	1740	2170	4100	4720	7130	7958	15 200	16 800	29 400	34 200
25	262	347	591	701	1120	1380	2560	2950	4560	5147	9550	10 700	18 800	21 700
30	238	316	540	640	1030	1270	2330	2690	4180	4719	8710	9790	17 200	19 800
40	203	271	469	554	896	1100	2010	2320	3630	4116	7530	8500	14 900	17 200
50	181	243	420	496	806	986	1790	2070	3260	3702	6730	7610	13 400	15 400
75	147	196	344	406	663	809	1460	1690	2680	3053	5480	6230	11 000	12 600
80	140	189	333	393	643	768	1410	1630	2590	2961	5300	6040	10 600	12 200
100	124	169	298	350	578	703	1260	1450	2330	2662	4740	5410	9530	10 900
150	101	137	245	287	477	575	1020	1180	1910	2195	3860	4430	7810	8890
200	86	118	213	248	415	501	880	1020	1660	1915	3340	3840	6780	7710
250	77	105	191	222	373	448	785	910	1490	1722	2980	3440	6080	6900
300	69	96	173	203	343	411	716	829	1360	1578	2720	3150	5560	6300
400	60	82	151	175	298	355	616	716	1160	1376	2350	2730	4830	5460
500	53	72	135	158	268	319	550	638	1030	1237	2100	2450	4330	4880

For SI units: 1 foot = 304.8 mm, 1000 British thermal units per hour = 0.293 kW, 1 pound-force per square inch = 6.8947 kPa

Notes:

1 Table does not include effect of pressure drop across the line regulator. Where regulator loss exceeds 0.5 psi (3.4 kPa) [based on 13 inch water column (3.2 kPa) outlet pressure], DO NOT use THIS TABLE. Consult with regulator manufacturer for pressure drops and capacity factors. Pressure drops across a regulator are capable of varying with flow rate.

2 CAUTION: Capacities shown in table are capable of exceeding the maximum capacity for a selected regulator. Consult with regulator or tubing manufacturer for guidance.

3 Table includes losses for four 90 degree (1.57 rad) bends and two end fittings. Tubing runs with larger numbers of bends, fittings, or both shall be increased by an equivalent length of tubing to the following equation: $L = 1.3\ n$, where L is additional length (ft) of tubing and n is the number of additional fittings, bends, or both.

4 Table entries are rounded to 3 significant digits.

5 EHD = Equivalent Hydraulic Diameter, which is a measure of the relative hydraulic efficiency between different tubing sizes. The greater the value of EHD, the greater the gas capacity of the tubing.

TABLE 1315.2(33) CORRUGATED STAINLESS STEEL TUBING (CSST) [NFPA 54: TABLE 6.3(j)][1, 2, 3, 4]

									GAS:		UNDILUTED PROPANE			
									INLET PRESSURE:		5.0 psi			
									PRESSURE DROP:		3.5 psi			
									SPECIFIC GRAVITY:		1.50			
	TUBE SIZE (EHD)[5]													
FLOW DESIGNATION:	13	15	18	19	23	25	30	31	37	39	46	48	60	62
LENGTH (feet)	CAPACITY IN THOUSANDS OF BTU PER HOUR													
10	826	1070	1710	2060	3150	4000	7830	8950	13 100	14 441	28 600	31 200	54 400	63 800
25	509	664	1090	1310	2040	2550	4860	5600	8400	9339	18 000	19 900	34 700	40 400
30	461	603	999	1190	1870	2340	4430	5100	7680	8564	16 400	18 200	31 700	36 900
40	396	520	867	1030	1630	2030	3820	4400	6680	7469	14 200	15 800	27 600	32 000
50	352	463	777	926	1460	1820	3410	3930	5990	6717	12 700	14 100	24 700	28 600
75	284	376	637	757	1210	1490	2770	3190	4920	5539	10 300	11 600	20 300	23 400
80	275	363	618	731	1170	1450	2680	3090	4770	5372	9990	11 200	19 600	22 700
100	243	324	553	656	1050	1300	2390	2760	4280	4830	8930	10 000	17 600	20 300
150	196	262	453	535	866	1060	1940	2240	3510	3983	7270	8210	14 400	16 600
200	169	226	393	464	755	923	1680	1930	3050	3474	6290	7130	12 500	14 400
250	150	202	352	415	679	828	1490	1730	2740	3124	5620	6390	11 200	12 900
300	136	183	322	379	622	757	1360	1570	2510	2865	5120	5840	10 300	11 700
400	117	158	279	328	542	657	1170	1360	2180	2498	4430	5070	8920	10 200
500	104	140	251	294	488	589	1050	1210	1950	2247	3960	4540	8000	9110

For SI units: 1 foot = 304.8 mm, 1000 British thermal units per hour = 0.293 kW, 1 pound-force per square inch = 6.8947 kPa

Notes:

1 Table does not include effect of pressure drop across the line regulator. Where regulator loss exceeds 0.5 psi (3.4 kPa) [based on 13 inch water column (3.2 kPa) outlet pressure], DO NOT USE THIS TABLE. Consult with regulator manufacturer for pressure drops and capacity factors. Pressure drops across a regulator are capable of varying with flow rate.

2 CAUTION: Capacities shown in table are capable of exceeding the maximum capacity for a selected regulator. Consult with regulator or tubing manufacturer for guidance.

3 Table includes losses for four 90 degree (1.57 rad) bends and two end fittings. Tubing runs with larger numbers of bends, fittings, or both shall be increased by an equivalent length of tubing to the following equation: $L = 1.3\ n$, where L is additional length (ft) of tubing and n is the number of additional fittings, bends, or both.

4 Table entries are rounded to 3 significant digits.

5 EHD = Equivalent Hydraulic Diameter, which is a measure of the relative hydraulic efficiency between different tubing sizes. The greater the value of EHD, the greater the gas capacity of the tubing.

TABLE 1315.2(34)
POLYETHYLENE PLASTIC PIPE [NFPA 54: TABLE 6.3(k)]*

					GAS:	UNDILUTED PROPANE	
					INLET PRESSURE:	11.0 in. w.c.	
					PRESSURE DROP:	0.5 in. w.c.	
					SPECIFIC GRAVITY:	1.50	
INTENDED USE: PE SIZING BETWEEN INTEGRAL SECOND-STAGE REGULATOR AT TANK OR SECOND-STAGE (LOW PRESSURE) REGULATOR AND BUILDING							
	PIPE SIZE (inch)							
NOMINAL OD:	½	¾	1	1¼	1½	2	3	4
DESIGNATION:	SDR 9.3	SDR 11	SDR 11	SDR 10	SDR 11	SDR 11	SDR 11	SDR 11
ACTUAL ID:	0.660	0.860	1.077	1.328	1.554	1.943	2.864	3.682
LENGTH (feet)	CAPACITY IN THOUSANDS OF BTU PER HOUR							
10	340	680	1230	2130	3210	5770	16 000	30 900
20	233	468	844	1460	2210	3970	11 000	21 200
30	187	375	677	1170	1770	3180	8810	17 000
40	160	321	580	1000	1520	2730	7540	14 600
50	142	285	514	890	1340	2420	6680	12 900
60	129	258	466	807	1220	2190	6050	11 700
70	119	237	428	742	1120	2010	5570	10 800
80	110	221	398	690	1040	1870	5180	10 000
90	103	207	374	648	978	1760	4860	9400
100	98	196	353	612	924	1660	4590	8900
125	87	173	313	542	819	1470	4070	7900
150	78	157	284	491	742	1330	3690	7130
175	72	145	261	452	683	1230	3390	6560
200	67	135	243	420	635	1140	3160	6100
250	60	119	215	373	563	1010	2800	5410
300	54	108	195	338	510	916	2530	4900
350	50	99	179	311	469	843	2330	4510
400	46	92	167	289	436	784	2170	4190
450	43	87	157	271	409	736	2040	3930
500	41	82	148	256	387	695	1920	3720

For SI units: 1 inch = 25 mm, 1 foot = 304.8 mm, 1000 British thermal units per hour = 0.293 kW, 1 inch water column = 0.249 kPa

* Table entries are rounded to 3 significant digits.

TABLE 1315.2(35) POLYETHYLENE PLASTIC PIPE [NFPA 54: TABLE 6.3(I)]*

					GAS:		UNDILUTED PROPANE	
					INLET PRESSURE:		2.0 psi	
					PRESSURE DROP:		1.0 psi	
					SPECIFIC GRAVITY:		1.50	
INTENDED USE: PE PIPE SIZING BETWEEN 2 PSI SERVICE REGULATOR AND LINE PRESSURE REGULATOR								
	PIPE SIZE (inch)							
NOMINAL OD:	½	¾	1	1¼	1½	2	3	4
DESIGNATION:	SDR 9.3	SDR 11	SDR 11	SDR 10	SDR 11	SDR 11	SDR 11	SDR 11
ACTUAL ID:	0.660	0.860	1.077	1.328	1.554	1.943	2.864	3.682
LENGTH (feet)	CAPACITY IN THOUSANDS OF BTU PER HOUR							
10	3130	6260	11 300	19 600	29 500	53 100	147 000	284 000
20	2150	4300	7760	13 400	20 300	36 500	101 000	195 000
30	1730	3450	6230	10 800	16 300	29 300	81 100	157 000
40	1480	2960	5330	9240	14 000	25 100	69 400	134 100
50	1310	2620	4730	8190	12 400	22 200	61 500	119 000
60	1190	2370	4280	7420	11 200	20 100	55 700	108 000
70	1090	2180	3940	6830	10 300	18 500	51 300	99 100
80	1010	2030	3670	6350	9590	17 200	47 700	92 200
90	952	1910	3440	5960	9000	16 200	44 700	86 500
100	899	1800	3250	5630	8500	15 300	42 300	81 700
125	797	1600	2880	4990	7530	13 500	37 500	72 400
150	722	1450	2610	4520	6830	12 300	33 900	65 600
175	664	1330	2400	4160	6280	11 300	31 200	60 300
200	618	1240	2230	3870	5840	10 500	29 000	56 100
250	548	1100	1980	3430	5180	9300	25 700	49 800
300	496	994	1790	3110	4690	8430	23 300	45 100
350	457	914	1650	2860	4320	7760	21 500	41 500
400	425	851	1530	2660	4020	7220	12 000	38 600
450	399	798	1440	2500	3770	6770	18 700	36 200
500	377	754	1360	2360	3560	6390	17 700	34 200
550	358	716	1290	2240	3380	6070	16 800	32 500
600	341	683	1230	2140	3220	5790	16 000	31 000
650	327	654	1180	2040	3090	5550	15 400	29 700
700	314	628	1130	1960	2970	5330	14 700	28 500
750	302	605	1090	1890	2860	5140	14 200	27 500
800	292	585	1050	1830	2760	4960	13 700	26 500
850	283	566	1020	1770	2670	4800	13 300	25 700
900	274	549	990	1710	2590	4650	12 900	24 900
950	266	533	961	1670	2520	4520	12 500	24 200
1000	259	518	935	1620	2450	4400	12 200	23 500
1100	246	492	888	1540	2320	4170	11 500	22 300
1200	234	470	847	1470	2220	3980	11 000	21 300
1300	225	450	811	1410	2120	3810	10 600	20 400
1400	216	432	779	1350	2040	3660	10 100	19 600
1500	208	416	751	1300	1960	3530	9760	18 900
1600	201	402	725	1260	1900	3410	9430	18 200
1700	194	389	702	1220	1840	3300	9130	17 600
1800	188	377	680	1180	1780	3200	8850	17 100
1900	183	366	661	1140	1730	3110	8590	16 600
2000	178	356	643	1110	1680	3020	8360	16 200

For SI units: 1 inch = 25 mm, 1 foot = 304.8 mm, 1000 British thermal units per hour = 0.293 kW, 1 pound-force per square inch = 6.8947 kPa

* Table entries are rounded to 3 significant digits.

»

TABLE 1315.2(36)
POLYETHYLENE PLASTIC TUBING [NFPA 54: TABLE 6.3(m)][2]

	GAS:	UNDILUTED PROPANE
	INLET PRESSURE:	11.0 in. w.c.
	PRESSURE DROP:	0.5 in. w.c.
	SPECIFIC GRAVITY:	1.50
INTENDED USE: PE TUBE SIZING BETWEEN INTEGRAL SECOND-STAGE REGULATOR AT TANK OR SECOND-STAGE (LOW PRESSURE) REGULATOR AND BUILDING		
	PLASTIC TUBING SIZE (CTS)[1] (inch)	
NOMINAL OD:	½	1
DESIGNATION:	SDR 7	SDR 11
ACTUAL ID:	0.445	0.927
LENGTH (feet)	CAPACITY IN THOUSANDS OF BTU PER HOUR	
10	121	828
20	83	569
30	67	457
40	57	391
50	51	347
60	46	314
70	42	289
80	39	269
90	37	252
100	35	238
125	31	211
150	28	191
175	26	176
200	24	164
225	22	154
250	21	145
275	20	138
300	19	132
350	18	121
400	16	113
450	15	106
500	15	100

For SI units: 1 inch = 25 mm, 1 foot = 304.8 mm, 1000 British thermal units per hour = 0.293 kW, 1 inch water column = 0.249 kPa

Notes:

[1] CTS = Copper tube size.

[2] Table entries are rounded to 3 significant digits.

CALIFORNIA MECHANICAL CODE – MATRIX ADOPTION TABLE
CHAPTER 14 – PROCESS PIPING

(Matrix Adoption Tables are non-regulatory, intended only as an aid to the code user. See Chapter 1 for state agency authority and building applications.)

Adopting Agency	BSC	BSC-CG	SFM	HCD			DSA			OSHPD						BSCC	DPH	AGR	DWR	CEC	CA	SL	SLC
				1	2	1-AC	AC	SS	SS/CC	1	1R	2	3	4	5								
Adopt Entire Chapter	X		X					X	X	X	X	X	X	X	X								
Adopt Entire Chapter as amended (amended sections listed below)																							
Adopt only those sections that are listed below																							
Chapter/Section																							

This state agency does not adopt sections identified with the following symbol: †

The Office of the State Fire Marshal's adoption of this chapter or individual sections is applicable to structures regulated by other state agencies pursuant to Section 1.11.0.

CHAPTER 14
PROCESS PIPING

1401.0 General.

1401.1 Applicability. Except as otherwise addressed in this code, this chapter shall govern the installation of process piping in or in conjunction with a building or structure located upon the premises.

1402.0 Permit.

1402.1 General. It shall be unlawful to install, alter, or repair or cause to be installed, altered, or repaired process material piping without first obtaining a permit.

Permits for process piping shall show the total number of outlets to be provided for on each system and such other information as required by the Authority Having Jurisdiction.

Fees for process piping permits are included in Table 104.5.

1403.0 Plans Required.

1403.1 General. Plans, engineering calculations, diagrams, and other data shall be submitted in one or more sets with each application for a permit. The Authority Having Jurisdiction shall be permitted to require plans, computations, and specifications to be prepared and designed by a registered design professional.

Where plans or other data are submitted for review, a plan review fee shall be paid, as provided in Section 104.3.2.

1404.0 Workmanship.

1404.1 General. Process piping shall not be strained or bent, nor shall tanks, vessels, vats, appliances, or cabinets be supported by or develop strain or stress on the piping.

1405.0 Inspections.

1405.1 General. Upon completion of the installation, alteration, or repair of process piping, and prior to the use thereof, the Authority Having Jurisdiction shall be notified that such piping is ready for inspection.

Excavations required for the installation of underground piping shall be kept open until such time as the piping has been inspected and approved. Where such piping is covered or concealed before such approval, it shall be exposed upon the direction of the Authority Having Jurisdiction.

1405.2 Required Inspections. The Authority Having Jurisdiction shall make the following inspections and shall either approve that portion of the work as completed or shall notify the permit holder wherein the same fails to be in accordance with this code.

1405.2.1 Rough Piping Inspection. This inspection shall be made after process piping authorized by the permit has been installed and before piping has been covered or concealed. This inspection shall include a determination that the piping size, material, and installation are in accordance with the requirements of this code.

1405.2.2 Final Piping Inspection. This inspection shall be made after piping authorized by the permit has been installed and after portions thereof that are to be covered or concealed are so concealed. This inspection shall include a pressure test, at which time the piping shall stand a pressure of not less than one-and-one-half times the maximum designed operating pressure where hydraulic testing is conducted or 110 percent where testing is conducted pneumatically. Test pressures shall be held for a length of time satisfactory to the Authority Having Jurisdiction, but in no case for less than 30 minutes with no perceptible drop in pressure. HPM drain, waste, and vent piping shall be tested in accordance with the plumbing code. Tests shall be made in the presence of the Authority Having Jurisdiction. Necessary apparatus for conducting tests shall be furnished by the permit holder.

1405.3 Other Inspections. In addition to the inspections required by this section, the Authority Having Jurisdiction shall be permitted to require a special inspector, as specified in the building code, during installation of piping systems. In cases where the work authorized was installed in accordance with plans and specifications prepared by a registered design professional, the Authority Having Jurisdiction shall be permitted to require a final signed report stating that the work was installed in accordance with approved plans and specifications and the applicable provisions of this chapter.

1406.0 Pipe, Tubing, and Fittings.

1406.1 General. Process pipe, tubing, and fittings shall comply with the applicable standards in Table 1701.1 and shall be installed in accordance with the manufacturer's installation instructions. Materials shall be rated for the operating temperatures and pressures of the system, and shall be compatible with the type of liquid.

1406.2 Hazardous Process Piping (HPP). HPP supply piping or tubing in service corridors shall be exposed to view. HPP piping shall be identified in accordance with nationally recognized standards to indicate the material being transported. Liquid HPP piping shall have an approved means for directing spilled materials to an approved containment or drainage system.

Liquid HPP waste or drainage systems shall be installed in accordance with the plumbing code.

1406.2.1 Installation in Exit Corridors and Above Other Occupancies. Hazardous process supply pipe shall not be located within exit corridors, within a portion of a means of egress required to be enclosed in fire-resistive construction, or in concealed spaces in or above areas not classified as Group H Occupancies, except as permitted by this subsection.

Hazardous production material piping and tubing shall be permitted to be installed within the space defined by the walls of exit corridors and the floor or roof above, or in concealed spaces above other occupancies in accordance with Section 1406.2.1.1 through Section 1406.2.1.6.

1406.2.1.1 Automatic Sprinklers. Automatic sprinklers shall be installed within the space unless the space is less than 6 inches (152 mm) in the least dimension.

1406.2.1.2 Ventilation. Ventilation at not less than 6 air changes per hour (ACH) shall be provided. The space shall not be used to convey air from other areas.

1406.2.1.3 Receptor. Where the piping or tubing is used to transport HPP liquids, a receptor shall be installed below such piping or tubing. The receptor shall be designed to collect discharge or leakage and drain it to an approved location. The 1 hour enclosure shall not be used as part of the receptor.

1406.2.1.4 Separation. HPP supply piping and tubing and HPP nonmetallic waste lines shall be separated from the exit corridor and from an occupancy other than a semi-conductor fabrication facility classified as a Group H Occupancy by construction, as required for walls or partitions that have a fire-protection rating of not less than 1 hour. Where gypsum wallboard is used, joints on the piping side of the enclosure need not be taped, provided the joints occur over framing members. Access openings into the enclosure shall be protected by approved fire assemblies.

1406.2.1.5 Emergency Shutoff Valves. Readily accessible manual or automatic remotely activated fail-safe emergency shutoff valves shall be installed on piping and tubing other than waste lines at the following locations:

(1) At branch connections into the fabrication area.

(2) At entries into exit corridors. Excess flow valves shall be installed as required by the fire code.

1406.2.1.6 Electrical Wiring. Electrical wiring and equipment located in the piping space shall be approved for Class I, Division 2, Hazardous Locations.

Exception: Occasional transverse crossing of the corridors by supply piping that is enclosed within the corridor need not comply with Section 1406.2.1.1 through Section 1406.2.1.6.

1406.3 Special Requirements for HPP Gases. In addition to other requirements of this section, HPP gases shall comply with this subsection and the fire code.

1406.3.1 Special Provisions. Where HPP supply gas is carried in pressurized piping, a fail-safe system shall shut off flow due to a rupture in the piping. Where the piping originates from outside the building, the valve shall be located outside the building as close to the bulk source as practical.

1406.3.2 Piping and Tubing Installation. Piping and tubing shall be installed in accordance with approved standards. Supply piping for hazardous production materials having a health hazard ranking of 3 or 4 shall have welded connections throughout, unless an exhausted enclosure is provided.

Exception: Material that is incompatible with ferrous piping shall be permitted to be installed in nonmetallic piping with approved connections.

1406.3.3 Gas-Detection System. Where hazardous production material gas is used or dispensed and the physiological warning properties for the gas are at a higher level than the accepted permissible exposure limit for the gas, a continuous gas-monitoring system shall be provided to detect the presence of a short-term hazard condition. Where dispensing occurs and flammable gases or vapors are capable of being present in quantities in excess of 20 percent of the lower explosive limit, a continuous gas-monitoring system shall be provided. The monitoring system shall be connected to the emergency control station.

CALIFORNIA MECHANICAL CODE – MATRIX ADOPTION TABLE
CHAPTER 15 – SOLAR ENERGY SYSTEMS

(Matrix Adoption Tables are non-regulatory, intended only as an aid to the code user. See Chapter 1 for state agency authority and building applications.)

Adopting Agency	BSC	BSC-CG	SFM	HCD			DSA			OSHPD						BSCC	DPH	AGR	DWR	CEC	CA	SL	SLC
				1	2	1-AC	AC	SS	SS/CC	1	1R	2	3	4	5								
Adopt Entire Chapter								X	X														
Adopt Entire Chapter as amended (amended sections listed below)				X	X																		
Adopt only those sections that are listed below																							
Chapter/Section																							
1502.0				X	X																		

This state agency does not adopt sections identified with the following symbol: †

The Office of the State Fire Marshal's adoption of this chapter or individual sections is applicable to structures regulated by other state agencies pursuant to Section 1.11.0.

CHAPTER 15
SOLAR ENERGY SYSTEMS

1501.0 General.

1501.1 Applicability. See Section 1203.0 and the Uniform Solar Energy and Hydronics Code (USEHC), published by the International Association of Plumbing and Mechanical Officials. The Uniform Solar Energy and Hydronics Code (USEHC) provides requirements that shall be permitted to be adopted as part of the code by the Authority Having Jurisdiction.

1502.0 General. [HCD 1 & HCD 2] *For applications listed in Section 1.8.2 regulated by the Department of Housing and Community Development, see Health and Safety Code Section 17959.1, Civil Code Sections 714 and 801.5, and Government Code Section 65850.5.*

CALIFORNIA MECHANICAL CODE – MATRIX ADOPTION TABLE
CHAPTER 16 – STATIONARY POWER PLANTS

(Matrix Adoption Tables are non-regulatory, intended only as an aid to the code user. See Chapter 1 for state agency authority and building applications.)

Adopting Agency	BSC	BSC-CG	SFM	HCD			DSA			OSHPD						BSCC	DPH	AGR	DWR	CEC	CA	SL	SLC
				1	2	1-AC	AC	SS	SS/CC	1	1R	2	3	4	5								
Adopt Entire Chapter	X		X	X	X			X	X														
Adopt Entire Chapter as amended (amended sections listed below)																							
Adopt only those sections that are listed below																							
Chapter/Section																							

This state agency does not adopt sections identified with the following symbol: †

The Office of the State Fire Marshal's adoption of this chapter or individual sections is applicable to structures regulated by other state agencies pursuant to Section 1.11.0.

CHAPTER 16

STATIONARY POWER PLANTS

1601.0 Stationary Fuel Cell Power Plants.

1601.1 General. Fuel cell power plants with a power output of less than 50 kW shall be listed and installed in accordance with the manufacturer's instructions. Fuel cell power plants with a power output of greater than 50 kW shall be installed in accordance with NFPA 853. [NFPA 54:10.30] Stationary fuel cell power plants shall be tested in accordance with CSA FC-1.

1602.0 Stationary Gas Engines and Generators.

1602.1 General. The installation of gas engines shall conform to NFPA 37. [NFPA 54:10.23]

1602.2 Connection to the Gas Supply Piping. Stationary gas engines shall not be rigidly connected to the gas supply piping. [NFPA 54:10.23.1]

1602.3 Stationary Engine Generators. Stationary engine generators shall be tested in accordance with UL 2200, and shall be installed in accordance with NFPA 37 and the manufacturer's installation instructions.

CALIFORNIA MECHANICAL CODE – MATRIX ADOPTION TABLE
CHAPTER 17 – REFERENCED STANDARDS

(Matrix Adoption Tables are non-regulatory, intended only as an aid to the code user. See Chapter 1 for state agency authority and building applications.)

Adopting Agency	BSC	BSC-CG	SFM	HCD			DSA			OSHPD						BSCC	DPH	AGR	DWR	CEC	CA	SL	SLC
				1	2	1-AC	AC	SS	SS/CC	1	1R	2	3	4	5								
Adopt Entire Chapter	X			X	X																		
Adopt Entire Chapter as amended (amended sections listed below)			X					X	X	X	X	X	X	X	X								
Adopt only those sections that are listed below																							
Chapter/Section																							
Table 1701.1	X		X				X	X	X	X	X	X	X	X									
Table 1701.2								X	X														

This state agency does not adopt sections identified with the following symbol: †

The Office of the State Fire Marshal's adoption of this chapter or individual sections is applicable to structures regulated by other state agencies pursuant to Section 1.11.0.

CHAPTER 17
REFERENCED STANDARDS

1701.0 General.

1701.1 Standards. The standards listed in Table 1701.1 are referenced in various sections of this code and shall be considered part of the requirements of this document. The standards are listed herein by the standard number and effective date, the title, application and the section(s) of this code that reference the standard. The application of the referenced standard(s) shall be as specified in Section 302.1.2.

The promulgating agency acronym referred to in Table 1701.1 are defined in a list found at the end of the table.

TABLE 1701.1
REFERENCED STANDARDS

STANDARD NUMBER	STANDARD TITLE	APPLICATION	REFERENCED SECTION
AABC-2016	Total System Balance, 7th Edition	Balancing	314.1(1)
ACCA Manual B-2009	Balancing and Testing Air and Hydronic Systems	Balancing	314.1(1)
ACCA Manual D-2016	Residential Duct Systems	Ducts, Balancing	601.2
ACCA 4 QM-2013	Maintenance of Residential HVAC Systems	HVAC Systems	102.3.2
AHRI 700-2016a	Specifications for Refrigerants	Refrigerants	1104.7, 1104.7.3
AMCA 540-2013	Louvers Impacted by Wind Borne Debris	Louvers	315.1.2
AMCA 550-2015	Test Method for High Velocity Wind Driven Rain Resistance Louvers	Louvers	315.1.1
ASHRAE 15-2016	Safety Standard for Refrigeration Systems	Refrigeration Systems	1102.1, 1106.1, Table 1113.5
ASHRAE 34-2016	Designation and Safety Classification of Refrigerants	Refrigeration Classifications	1102.3, 1103.1, Table 1102.3, Table 1106.2.5.2
ASHRAE 62.1-2016	Ventilation for Acceptable Indoor Air Quality	Indoor Air Quality Ventilation	402.4.1
ASHRAE 62.2-2016	Ventilation and Acceptable Indoor Air Quality in Residential Buildings	Ventilation	402.1.2
ASHRAE 111-2008	Measurement, Testing, Adjusting, and Balancing of Building HVAC Systems	Balancing	314.1(3)
ASHRAE 129-1997 (R2002)	Measuring Air-Change Effectiveness	Air Change Effectiveness	Table 403.2.2
ASHRAE 154-2016	Ventilation for Commercial Cooking Operations	Commercial Kitchens	510.5.6
ASHRAE 170-2013	Ventilation of Health Care Facilities	Ventilation	402.1.3
ASHRAE/ACCA 180-2012	Inspection and Maintenance of Commercial Building HVAC Systems	Maintenance	102.3.1, 1013.3
ASHRAE Handbook-2013	Fundamentals	Climatic Conditions	Figure 803.1.2(6)
ASME B1.20.1-2013	Pipe Threads, General Purpose (Inch)	Joints	1211.2(3), 1211.4(7), 1211.12(3), 1211.13(2), 1308.5.7
ASME B16.1-2015	Gray Iron Pipe Flanges and Flanged Fittings: Classes 25, 125, and 250	Fittings	1308.5.10
ASME B16.3-2011	Malleable Iron Threaded Fittings: Classes 150 and 300	Fittings	Table 1210.1
ASME B16.5-2013	Pipe Flanges and Flanged Fittings: NPS ½ through NPS 24 Metric/Inch	Fittings	Table 1210.1, 1308.5.10.1
ASME B16.9-2012	Factory-Made Wrought Buttwelding Fittings	Fittings	Table 1210.1
ASME B16.11-2011	Forged Fittings, Socket-Welding and Threaded	Fittings	Table 1210.1
ASME B16.15-2013	Cast Copper Alloy Threaded Fittings: Classes 125 and 250	Fittings	Table 1210.1
ASME B16.18-2012	Cast Copper Alloy Solder Joint Pressure Fittings	Fittings	Table 1210.1

TABLE 1701.1 (continued)
REFERENCED STANDARDS

STANDARD NUMBER	STANDARD TITLE	APPLICATION	REFERENCED SECTION
ASME B16.20-2012	Metallic Gaskets for Pipe Flanges: Ring-Joint, Spiral-Wound, and Jacketed	Joints	1308.5.11.2
ASME B16.21-2011	Nonmetallic Flat Gaskets for Pipe Flanges	Fuel Gas Piping	1308.5.11.3
ASME B16.22-2013	Wrought Copper and Copper Alloy Solder-Joint Pressure Fittings	Fittings	Table 1210.1
ASME B16.23-2011	Cast Copper Alloy Solder Joint Drainage Fittings: DWV	Fittings	Table 1210.1
ASME B16.24-2011	Cast Copper Alloy Pipe Flanges and Flanged Fittings: Classes 150, 300, 600, 900, 1500, and 2500	Fittings	Table 1210.1, 1308.5.10.2
ASME B16.26-2013	Cast Copper Alloy Fittings for Flared Copper Tubes	Fittings	Table 1210.1
ASME B16.29-2012	Wrought Copper and Wrought Copper Alloy Solder-Joint Drainage Fittings – DWV	Fittings	Table 1210.1
ASME B16.42-2011	Ductile Iron Pipe Flanges and Flanged Fittings	Fuel Gas Piping	1308.5.10.3
ASME B16.47-2011	Large Diameter Steel Flanges	Fuel Gas Piping	1308.5.10.1
ASME B16.51-2013	Copper and Copper Alloy Press-Connect Pressure Fittings	Fittings	Table 1210.1
ASME B31.5-2016	Refrigeration Piping and Heat Transfer Components	Refrigeration Piping	1109.1.1, 1109.1.3
ASME B36.10M-2015	Welded and Seamless Wrought Steel Pipe	Fuel Gas Piping	1308.5.2.1(1)
ASME BPVC Section I-2015	Rules for Construction of Power Boilers	Boilers	1002.1(1), Table 1003.2.1
ASME BPVC Section IV-2015	Rules for Construction of Heating Boilers	Boilers	1002.1(2)
ASME BPVC Section VIII-2015	Rules for Construction of Pressure Vessels Division 1	Pressure Vessels	1002.1, 1004.3, 1112.10.1, 1112.13, 1113.1, 1115.4, 1115.4.1, 1117.1(2), 1117.1(3), 1117.2, 1209.4
ASME CSD-1-2015	Controls and Safety Devices for Automatically Fired Boilers	Boilers	1003.2.1
ASSE 1013-2011	Reduced Pressure Principle Backflow Preventers and Reduced Pressure Principle Fire Protection Backflow Preventers	Backflow Protection	1202.2
ASSE 1017-2009	Temperature Actuated Mixing Valves for Hot Water Distribution Systems	Valves	1207.3.1
ASSE 1061-2015	Push-Fit Fittings	Fittings	1211.2(1), 1211.4(5), Table 1210.1
ASTM A53/A53M-2012	Pipe, Steel, Black and Hot-Dipped, Zinc-Coated, Welded and Seamless	Piping	1308.5.2.1(2), Table 1210.1
ASTM A106/A106M-2015	Seamless Carbon Steel Pipe for High-Temperature Service	Piping	1308.5.2.1(3), Table 1210.1
ASTM A126-2004 (R2014)	Gray Iron Castings for Valves, Flanges, and Pipe Fittings	Piping	Table 1210.1
ASTM A254/A254M-2012	Copper-Brazed Steel Tubing	Piping	1308.5.3.1, Table 1210.1
ASTM A420/A420M-2016	Piping Fittings of Wrought Carbon Steel and Alloy Steel for Low-Temperature Service	Fittings	Table 1210.1
ASTM B32-2008 (R2014)	Solder Metal	Joints	1211.4(6)
ASTM B42-2015a	Seamless Copper Pipe, Standard Sizes	Piping	Table 1210.1
ASTM B43-2015	Seamless Red Brass Pipe, Standard Sizes	Piping	Table 1210.1

TABLE 1701.1 (continued)
REFERENCED STANDARDS

STANDARD NUMBER	STANDARD TITLE	APPLICATION	REFERENCED SECTION
ASTM B75/B75M-2011	Seamless Copper Tube	Piping	Table 1210.1
ASTM B88-2016	Seamless Copper Water Tube	Piping	1308.5.3.2, Table 1210.1
ASTM B135-2010	Seamless Brass Tube	Piping	Table 1210.1
ASTM B210-2012	Aluminum and Aluminum-Alloy Drawn Seamless Tubes	Piping	1308.5.3.3
ASTM B241/B241M-2016	Aluminum and Aluminum-Alloy Seamless Pipe and Seamless Extruded Tube	Piping	1308.5.2.3, 1308.5.3.3
ASTM B251-2010	General Requirements for Wrought Seamless Copper and Copper-Alloy Tube	Piping	Table 1210.1
ASTM B280-2016	Seamless Copper Tube for Air Conditioning and Refrigeration Field Service	Piping	1109.1.2, 1308.5.3.2
ASTM B302-2012	Threadless Copper Pipe, Standard Sizes	Piping	Table 1210.1
ASTM B447-2012a	Welded Copper Tube	Piping	Table 1210.1
ASTM B813-2016	Liquid and Paste Fluxes for Soldering of Copper and Copper Alloy Tube	Joints	1211.4(6)
ASTM B828-2016	Making Capillary Joints by Soldering of Copper and Copper Alloy Tube and Fittings	Joints	1211.4(6)
ASTM B1003-2016	Seamless Copper Tube for Linesets	Piping	1109.1.2
ASTM C315-2007 (R2016)	Clay Flue Liners and Chimney Pots	Venting Systems	803.1.11.1
ASTM C411-2011	Hot-Surface Performance of High-Temperature Thermal Insulation	Duct Coverings and Linings	604.1.2
ASTM D1693-2015	Environmental Stress-Cracking of Ethylene Plastics	Piping	Table 1210.1
ASTM D1785-2015	Poly (Vinyl Chloride) (PVC) Plastic Pipe, Schedules 40, 80, and 120	Piping	Table 1210.1
ASTM D2241-2015	Poly (Vinyl Chloride) (PVC) Pressure-Rated Pipe (SDR Series)	Piping	Table 1210.1
ASTM D2464-2015	Threaded Poly (Vinyl Chloride) (PVC) Plastic Pipe Fittings, Schedule 80	Fittings	Table 1210.1
ASTM D2466-2015	Poly (Vinyl Chloride) (PVC) Plastic Pipe Fittings, Schedule 40	Fittings	Table 1210.1
ASTM D2467-2015	Poly (Vinyl Chloride) (PVC) Plastic Pipe Fittings, Schedule 80	Fittings	Table 1210.1
ASTM D2513-2014[e1]	Polyethylene (PE) Gas Pressure Pipe, Tubing, and Fittings	Piping	1308.5.4, 1308.5.4.2.2, 1308.5.9.2, 1310.1.7.1(1), Table 1210.1
ASTM D2564-2012	Solvent Cements for Poly (Vinyl Chloride) (PVC) Plastic Piping Systems	Joints	1211.12(2)
ASTM D2609-2015	Plastic Insert Fittings for Polyethylene (PE) Plastic Pipe	Fittings	Table 1210.1
ASTM D2683-2014	Socket-Type Polyethylene Fittings for Outside Diameter-Controlled Polyethylene Pipe and Tubing	Fittings	Table 1210.1
ASTM D2737-2012a	Polyethylene (PE) Plastic Tubing	Piping, Plastic	Table 1210.1
ASTM D2846/D2846M-2014	Chlorinated Poly (Vinyl Chloride) (CPVC) Plastic Hot- and Cold-Water Distribution Systems	Piping	1211.2(2), 1211.3(2), Table 1210.1
ASTM D3035-2015	Polyethylene (PE) Plastic Pipe (DR-PR) Based on Controlled Outside Diameter	Piping	Table 1210.1
ASTM D3139-1998 (R2011)	Joints for Plastic Pressure Pipes Using Flexible Elastomeric Seals	Joints	1211.12(1)

TABLE 1701.1 (continued)
REFERENCED STANDARDS

STANDARD NUMBER	STANDARD TITLE	APPLICATION	REFERENCED SECTION
ASTM D3261-2016	Butt Heat Fusion Polyethylene (PE) Plastic Fittings for Polyethylene (PE) Plastic Pipe and Tubing	Fittings	Table 1210.1
ASTM D3350-2014	Polyethylene Plastics Pipe and Fittings Materials	Piping, Fittings	Table 1210.1
ASTM E84-2016	Surface Burning Characteristics of Building Materials	Miscellaneous	508.3.4, 602.2, 604.1.2, 1201.2
ASTM E814-2013a	Fire Tests of Penetration Firestop Systems	Miscellaneous	507.4.4, 507.4.5
ASTM E2231-2015	Specimen Preparation and Mounting of Pipe and Duct Insulation Materials to Assess Surface Burning Characteristics	Insulation of Ducts	604.1.2
ASTM E2336-2016	Fire Resistive Grease Duct Enclosure Systems	Grease Ducts	507.4.2.2, 507.4.5
ASTM F437-2015	Threaded Chlorinated Poly (Vinyl Chloride) (CPVC) Plastic Pipe Fittings, Schedule 80	Fittings	Table 1210.1
ASTM F438-2015	Socket-Type Chlorinated Poly (Vinyl Chloride) (CPVC) Plastic Pipe Fittings, Schedule 40	Fittings	Table 1210.1
ASTM F439-2013	Chlorinated Poly (Vinyl Chloride) (CPVC) Plastic Pipe Fittings, Schedule 80	Fittings	Table 1210.1, 1211.2(2)
ASTM F441/F441M-2015	Chlorinated Poly (Vinyl Chloride) (CPVC) Plastic Pipe, Schedules 40 and 80	Piping	Table 1210.1
ASTM F442/442M-2013^{e1}	Chlorinated Poly (Vinyl Chloride) (CPVC) Plastic Pipe (SDR-PR)	Piping	Table 1210.1, 1211.2(2)
ASTM F493-2014	Solvent Cements for Chlorinated Poly (Vinyl Chloride) (CPVC) Plastic Pipe and Fittings	Joints	1211.2(2), 1211.3(2)
ASTM F656-2015	Primers for Use in Solvent Cement Joints of Poly (Vinyl Chloride) (PVC) Plastic Pipe and Fittings	Joints	1211.2(2), 1211.3(2), 1211.12(2)
ASTM F714-2013	Polyethylene (PE) Plastic Pipe (DR-PR) Based on Outside Diameter	Piping, Plastic	Table 1210.1
ASTM F876-2015a	Crosslinked Polyethylene (PEX) Tubing	Piping	1211.5, Table 1210.1
ASTM F877-2011a	Crosslinked Polyethylene (PEX) Hot- and Cold-Water Distribution Systems	Piping	Table 1210.1
ASTM F1055-2016a	Electrofusion Type Polyethylene Fittings for Outside Diameter Controlled Polyethylene and Crosslinked Polyethylene (PEX) Pipe and Tubing	Fittings	Table 1210.1
ASTM F1281-2011	Crosslinked Polyethylene/Aluminum/Crosslinked Polyethylene (PEX-AL-PEX) Pressure Pipe	Piping	Table 1210.1
ASTM F1282-2010	Polyethylene/Aluminum/Polyethylene (PE-AL-PE) Composite Pressure Pipe	Piping	Table 1210.1
ASTM F1807-2015	Metal Insert Fittings Utilizing a Copper Crimp Ring for SDR9 Cross-linked Polyethylene (PEX) Tubing and SDR9 Polyethylene of Raised Temperature (PE-RT) Tubing	Fittings	Table 1210.1
ASTM F1960-2015	Cold Expansion Fittings with PEX Reinforcing Rings for Use with Cross-linked Polyethylene (PEX) Tubing	Fittings	Table 1210.1
ASTM F1961-2009	Metal Mechanical Cold Flare Compression Fittings with Disc Spring for Crosslinked Polyethylene (PEX) Tubing	Fittings	Table 1210.1

TABLE 1701.1 (continued)
REFERENCED STANDARDS

STANDARD NUMBER	STANDARD TITLE	APPLICATION	REFERENCED SECTION
ASTM F1970-2012^{e1}	Special Engineered Fittings, Appurtenances or Valves for Use in Poly (Vinyl Chloride) (PVC) or Chlorinated Poly (Vinyl Chloride) (CPVC) Systems	Piping	Table 1210.1
ASTM F1973-2013^{e1}	Factory Assembled Anodeless Risers and Transition Fittings in Polyethylene (PE) and Polyamide 11 (PA11) and Polyamide 12 (PA12) Fuel Gas Distribution Systems	Fuel Gas	1310.1.7.1(2)
ASTM F1974-2009 (R2015)	Metal Insert Fittings for Polyethylene/Aluminum/Polyethylene and Crosslinked Polyethylene/Aluminum/ Crosslinked Polyethylene Composite Pressure Pipe	Fittings	1211.6(1), 1211.9(1), Table 1210.1
ASTM F2080-2016	Cold-Expansion Fittings with Metal Compression-Sleeves for Crosslinked Polyethylene (PEX) Pipe and SDR9 Polyethylene of Raised Temperature (PE-RT) Pipe	Fittings	Table 1210.1
ASTM F2098-2015	Stainless Steel Clamps for Securing SDR9 Cross-linked Polyethylene (PEX) Tubing to Metal Insert and Plastic Insert Fittings	Joints	Table 1210.1
ASTM F2159-2014	Plastic Insert Fittings Utilizing a Copper Crimp Ring for SDR9 Cross-linked Polyethylene (PEX) Tubing and SDR9 Polyethylene of Raised Temperature (PE-RT) Tubing	Fittings	Table 1210.1
ASTM F2262-2009	Crosslinked Polyethylene/Aluminum/Crosslinked Polyethylene Tubing OD Controlled SDR9	Piping, Plastic	Table 1210.1
ASTM F2389-2015	Pressure-Rated Polypropylene (PP) Piping Systems	Piping	1211.11(1), Table 1210.1
ASTM F2434-2014	Metal Insert Fittings Utilizing a Copper Crimp Ring for SDR9 Cross-linked Polyethylene (PEX) Tubing and SDR9 Cross-linked Polyethylene/Aluminum/Cross-linked Polyethylene (PEX-AL-PEX) Tubing	Fittings	1211.6(1), Table 1210.1
ASTM F2509-2015	Field-Assembled Anodeless Riser Kits for Use on Outside Diameter Controlled Polyethylene and Polyamide-11 (PA11) Gas Distribution Pipe and Tubing	Fuel Gas	1310.1.7.1(3)
ASTM F2620-2013	Standard Practice for Heat Fusion Joining of Polyethylene Pipe and Fittings	Joints	1211.8(1), 1211.8(3)
ASTM F2623-2014	Polyethylene of Raised Temperature (PE-RT) SDR9 Tubing	Piping	Table 1210.1
ASTM F2735-2009 (R2016)	Plastic Insert Fittings for SDR9 Cross-linked Polyethylene (PEX) and Polyethylene of Raised Temperature (PE-RT) Tubing	Fittings	Table 1210.1
ASTM F2769-2016	Polyethylene of Raised Temperature (PE-RT) Plastic Hot and Cold-Water Tubing and Distribution Systems	Piping, Fittings	Table 1210.1
ASTM F2855-2012	Specification for Chlorinated Poly (Vinyl Chloride)/Aluminum/Chlorinated Poly (Vinyl Chloride) (CPVC-AL-CPVC) Composite Pressure Tubing	Piping, Plastic	Table 1210.1
AWS A5.8-2011	Filler Metals for Brazing and Braze Welding	Joints	1211.4(1)
AWWA C110-2012	Ductile-Iron and Gray-Iron Fittings	Fittings	Table 1210.1

TABLE 1701.1 (continued)
REFERENCED STANDARDS

STANDARD NUMBER	STANDARD TITLE	APPLICATION	REFERENCED SECTION
AWWA C111-2012	Rubber-Gasket Joints for Ductile-Iron Pressure Pipe and Fittings	Joints	1211.7(1)
AWWA C115-2011	Flanged Ductile-Iron Pipe with Ductile-Iron or Gray-Iron Treaded Flanges	Piping	Table 1210.1
AWWA C151-2009	Ductile-Iron Pipe, Centrifugally Cast	Piping, Ferrous	Table 1210.1
AWWA C153-2011	Ductile-Iron Compact Fittings	Fittings	Table 1210.1
AWWA C901-2008	Polyethylene (PE) Pressure Pipe and Tubing, ½ in. (13 mm) Through 3 in. (76 mm) for Water Service	Piping, Plastic	Table 1210.1
CSA B137.1-2013	Polyethylene (PE) Pipe, Tubing, and Fittings for Cold-Water Pressure Services	Piping	Table 1210.1
CSA B137.2-2013	Polyvinylchloride (PVC) Injection-Moulded Gasketed Fittings for Pressure Applications	Piping, Plastic	Table 1210.1
CSA B137.3-2013	Rigid Polyvinylchloride (PVC) Pipe and Fittings for Pressure Applications	Piping, Plastic	Table 1210.1
CSA B137.5-2013	Crosslinked Polyethylene (PEX) Tubing Systems for Pressure Applications	Piping	Table 1210.1
CSA B137.6-2013	Chlorinated Polyvinylchloride (CPVC) Pipe, Tubing, and Fittings for Hot-and Cold-Water Distribution Systems	Piping, Plastic	Table 1210.1
CSA B137.9-2013	Polyethylene/Aluminum/Polyethylene (PE-AL-PE) Composite Pressure-Pipe Systems	Piping	Table 1210.1
CSA B137.10-2013	Crosslinked Polyethylene/Aluminum/Crosslinked Polyethylene (PEX-AL-PEX) Composite Pressure-Pipe Systems	Piping	Table 1210.1
CSA B137.11-2013	Polypropylene (PP-R) Pipe and Fittings for Pressure Applications	Piping	1211.11(1)
CSA B137.18-2013	Polyethylene of Raised Temperature Resistance (PE-RT) Tubing Systems for Pressure Applications	Piping, Plastic	Table 1210.1
CSA FC 1-2014	Fuel Cell Technologies-Part 3-100: Stationary Fuel Cell Power Systems-Safety	Fuel Cell Power Plants	1601.1
CSA LC 1-2016	Fuel Gas Piping Systems Using Corrugated Stainless Steel Tubing (same as CSA 6.26)	Fuel Gas	1308.5.3.4
CSA LC 4a-2013	Press-Connect Metallic Fittings For Use in Fuel Gas Distribution Systems (same as CSA 6.32a)	Fuel Gas	1308.5.8.1, 1308.5.8.2
CSA Z21.8-1994 (R2012)	Installation of Domestic Gas Conversion Burners	Fuel Gas, Appliances	909.1
CSA Z21.10.1-2014	Gas Water Heaters-Volume I, Storage Water Heaters with Input Ratings of 75,000 Btu Per Hour or Less (same as CSA 4.1)	Fuel Gas, Appliances	Table 1203.2
CSA Z21.10.3-2015	Gas-Fired Water Heaters, Volume III, Storage Water Heaters with Input Ratings Above 75,000 Btu Per Hour, Circulating and Instantaneous (same as CSA 4.3)	Fuel Gas, Appliances	Table 1203.2
CSA Z21.24-2015	Connectors for Gas Appliances (same as CSA 6.10)	Fuel Gas	1312.1(3), 1312.2
CSA Z21.41-2014	Quick Disconnect Devices for Use with Gas Fuel Appliances (same as CSA 6.9)	Fuel Gas	1312.7

**TABLE 1701.1 (continued)
REFERENCED STANDARDS**

STANDARD NUMBER	STANDARD TITLE	APPLICATION	REFERENCED SECTION
CSA Z21.54-2014	Gas Hose Connectors for Portable Outdoor Gas-Fired Appliances (same as CSA 8.4)	Fuel Gas	1312.3.2
CSA Z21.69-2015	Connectors for Moveable Gas Appliances (same as CSA 6.16)	Fuel Gas	1312.1.1
CSA Z21.75-2016	Connectors for Outdoor Gas Appliances and Manufactured Homes (same as CSA 6.27)	Appliances	1312.1(4)
CSA Z21.80a-2012	Line Pressure Regulators (same as CSA 6.22a)	Fuel Gas	1308.7.1, 1308.7.4(1)
CSA Z21.90-2015	Gas Convenience Outlets and Optional Enclosures (same as CSA 6.24)	Gas Outlets	1312.8
CSA Z21.93-2013	Excess Flow Valves for Natural and LP Gas with Pressures up to 5 psig (same as CSA 6.30)	Fuel Gas	1309.1
CSA Z83.4-2015	Non-Recirculating Direct Gas-Fired Industrial Air Heaters (same as CSA 3.7)	Air Heaters, Non-Recirculating, Non-Recirculating Heaters	914.1
CSA Z83.18-2015	Recirculating Direct Gas-Fired Industrial Air Heaters	Air Heaters, Industrial Heaters, Recirculating	915.1
IIAR 2-2014	Standard for the Safe Design of Closed-Circuit Ammonia Refrigeration Systems	Ammonia Refrigeration Systems	1102.2
IIAR 3-2012	Ammonia Refrigeration Valves	Ammonia Refrigeration Systems	1102.2
IIAR 4-2015	Installation of Closed-Circuit Ammonia Refrigeration Systems	Ammonia Refrigeration	1102.2
IIAR 5-2013	Start-Up and Commissioning of Closed Circuit Ammonia Refrigeration Systems	Ammonia Refrigeration Systems	1102.2
MSS SP-58-2009	Pipe Hangers and Supports-Materials, Design, Manufacture, Selection, Application, and Installation	Miscellaneous	1310.2.4
NEBB-2015	Procedural Standard for Testing, Adjusting and Balancing of Environmental Systems, 8th Edition	Balancing	314.1(4)
NFPA 2-2016	Hydrogen Technologies Code	Gaseous Hydrogen Systems	937.1
NFPA 10-2013	Portable Fire Extinguishers	Fire Extinguishing	513.12, 517.7.4
NFPA 12-2015	Carbon Dioxide Extinguishing Systems	Fire Extinguishing	513.2.3(1)
NFPA 13-2016	Installation of Sprinkler Systems	Miscellaneous	513.2.3(2), 517.7.6
NFPA 17-2017	Dry Chemical Extinguishing Systems	Fire Extinguishing	513.2.3(3), 513.3.2
NFPA 17A-2017	Wet Chemical Extinguishing Systems	Fire Extinguishing	513.2.3(4), 513.3.2
NFPA 30A-2015	Motor Fuel Dispensing Facilities and Repair Garages	Miscellaneous	303.11.1
NFPA 31-2016	Installation of Oil-Burning Equipment	Fuel Gas, Appliances	301.5, 1002.2.2, 1301.1
NFPA 37-2015	Installation and Use of Stationary Combustion Engines and Gas Turbines	Generators	1602.1, 1602.3
NFPA 51-2013	Design and Installation of Oxygen-Fuel Gas Systems for Welding, Cutting, and Allied Processes	Fuel Gas Systems	902.7
NFPA 52-2016	Vehicular Natural Gas Fuel Systems Code	CNG Vehicle Fuel Systems	938.1
NFPA 54/Z223.1-2015	National Fuel Gas Code	Fuel Gas	516.2.1
NFPA 58-2017	Liquefied Petroleum Gas Code	Fuel Gas	303.7, 516.2.1, 1308.5.4.2.3, 1308.5.9.4, 1310.5(6), 1312.11

TABLE 1701.1 (continued)
REFERENCED STANDARDS

STANDARD NUMBER	STANDARD TITLE	APPLICATION	REFERENCED SECTION
NFPA 68-2013	Explosion Protection by Deflagration Venting	Product Conveying Ducts	505.3.3
NFPA 69-2014	Explosion Prevention Systems	Explosion Prevention	505.3.1
NFPA 70-2017	National Electrical Code	Miscellaneous	301.4, 511.1.6, 512.2.5, 516.2.7, 516.2.9(4), 602.2.1, 905.8.2, 1104.4(6), 1107.1.7, 1107.1.8, 1217.6.1, 1310.14.5(2), 1311.2.4, 1311.6
NFPA 80-2016	Fire Doors and Other Opening Protectives	Fire Doors	510.7.7
NFPA 82-2014	Incinerators and Waste and Linen Handling Systems and Equipment	Incinerator Chutes	802.2.8, Table 802.4, 925.1
NFPA 85-2015	Boiler and Combustion Systems Hazards Code	Appliances	1002.1(3), 1011.1, Table 1003.2.1
NFPA 86-2015	Ovens and Furnaces	Product Conveying Ducts	505.3.2
NFPA 88A-2015	Parking Structures	Miscellaneous	303.11
NFPA 90A-2015	Installation of Air-Conditioning and Ventilating Systems	HVAC	904.7
NFPA 90B-2015	Installation of Warm Air Heating and Air-Conditioning Systems	HVAC	904.7
NFPA 96-2014	Ventilation Control and Fire Protection of Commercial Cooking Operations	Commercial Cooking	922.4
NFPA 211-2016	Chimneys, Fireplaces, Vents, and Solid Fuel-Burning Appliances	Fuel Gas Appliances	517.7, 517.7.1, 801.2, 801.3, 802.5.2, 802.5.3, 802.5.7.1, 802.5.7.3, 902.10
NFPA 221-2015	High Challenge Fire Walls, Fire Walls, and Fire Barrier Walls	Building Fire Walls, Fire Barrier	506.3
NFPA 262-2015	Standard Method of Test for Flame Travel and Smoke of Wires and Cables for Use in Air-Handling Spaces	Certification	602.2.1
NFPA 409-2016	Aircraft Hangars	Miscellaneous	303.12
NFPA 654-2017	Prevention of Fire and Dust Explosions from the Manufacturing, Processing, and Handling of Combustible Particulate Solids	Dust Explosion Prevention, Gutters	506.6, Table 505.9
NFPA 780-2017	Installation of Lightning Protection Systems	Fuel Gas	1311.4
NFPA 853-2015	Installation of Stationary Fuel Cell Power Systems	Fuel Cell Power Plants, Fuel Cells	1601.1
NFPA 1192-2015	Recreational Vehicles	Fuel Gas Piping	1302.3
NSF 358-1-2014	Polyethylene Pipe and Fittings for Water-Based Ground-Source "Geothermal" Heat Pump Systems	Piping, Plastic	Table 1210.1
NSF 358-2-2012	Polypropylene Pipe and Fittings for Water-Based Ground-Source "Geothermal" Heat Pump Systems	Piping, Plastic	Table 1210.1
NSF 358-3-2016	Cross-linked polyethylene (PEX) pipe and fittings for water-based ground-source (geothermal) heat pump systems	Piping, Plastic	Table 1210.1
SMACNA-2002	HVAC Systems Testing, Adjusting and Balancing, 3rd Edition	Balancing	314.1(5)
SMACNA-2006	HVAC Duct Construction Standards Metal and Flexible, 3rd Edition	Ducts, Metal and Flexible	504.4.5, 506.2, 602.3, 603.3, 603.4, 603.5, 603.8, 603.10, 603.12, 604.1

**TABLE 1701.1 (continued)
REFERENCED STANDARDS**

STANDARD NUMBER	STANDARD TITLE	APPLICATION	REFERENCED SECTION
SMACNA-2012	HVAC Air Duct Leakage Test Manual, 2nd Edition	Ducts	603.10.1
SMACNA-2015	Phenolic Duct Construction Standards	Duct Systems	602.4
UL 17-2008	Vent or Chimney Connector Dampers for Oil-Fired Appliances (with revisions through September 25, 2013)	Vent Dampers	802.15.1
UL 103-2010	Factory-Built Chimneys for Residential Type and Building Heating Appliances (with revisions through July 27, 2012)	Fuel Gas, Appliances	802.5.1.1, 802.5.1.2
UL 127-2011	Factory-Built Fireplaces (with revisions through July 27. 2016)	Fireplaces	802.5.1.1, 913.1, 913.1.1
UL 181-2013	Factory-Made Air Ducts and Air Connectors	Air Connectors, Air Ducts	602.6, 603.4, 603.5, 603.8, 604.1.1
UL 181A-2013	Closure Systems for Use with Rigid Air Ducts	Air Ducts	603.10, Table 603.10
UL 181B-2013	Closure Systems for Use with Flexible Air Ducts and Air Connectors	Air Connectors, Air Ducts	603.10, Table 603.10
UL 197-2010	Commercial Electric Cooking Appliances (with revisions through September 17, 2014)	Appliances, Commercial Cooking, Electric Appliances	921.1
UL 207-2009	Refrigerant-Containing Components and Accessories, Nonelectrical (with revisions through June 27, 2014)	Refrigeration Components	1109.2
UL 268A-2008	Smoke Detectors for Duct Application (with revisions through August 12, 2016)	Smoke Detectors	608.1
UL 295-2007	Commercial-Industrial Gas Burners (with revisions through January 30, 2015)	Gas Burners	910.2
UL 296-2003	Oil Burners (with revisions through June 11, 2015)	Fuel Gas, Appliances	910.1
UL 300-2005	Fire Testing of Fire Extinguishing Systems for Protection of Commercial Cooking Equipment (with revisions through December 16, 2014)	Certification	513.2.2, 513.2.5, 517.3.1.1(6)
UL 378-2006	Draft Equipment (with revisions through September 17, 2013)	Fuel Gas, Appliances	802.15.1
UL 391-2010	Solid-Fuel and Combination-Fuel Central and Supplementary Furnaces (with revisions through June 12, 2014)	Furnaces, Solid Fuel	904.11
UL 412-2011	Refrigeration Unit Coolers (with revisions through September 17, 2013)	Refrigeration	934.2
UL 427-2011	Refrigerating Units (with revisions through February 28, 2014)	Refrigeration Systems	934.3
UL 441-2016	Gas Vents (with revisions through July 27, 2016)	Fuel Gas	802.1
UL 467-2013	Grounding and Bonding Equipment	Grounding and Bonding	1311.2.5
UL 471-2010	Commercial Refrigerators and Freezers (with revisions December 8, 2016)	Freezers, Refrigerators	934.1
UL 499-2014	Electric Heating Appliances	Kilns	931.4
UL 555-2006	Fire Dampers (with revisions through October 21, 2016)	Dampers	605.2
UL 555C-2014	Ceiling Dampers	Dampers	605.3
UL 555S-2014	Smoke Dampers (with revisions through October 27, 2016)	Dampers	605.1

TABLE 1701.1 (continued)
REFERENCED STANDARDS

STANDARD NUMBER	STANDARD TITLE	APPLICATION	REFERENCED SECTION
UL 641-2010	Type L Low-Temperature Venting Systems (with revisions through June 12, 2013)	Equipment	802.1
UL 651-2011	Schedule 40, 80, Type EB and A Rigid PVC Conduit and Fittings (with revisions through June 15, 2016)	Piping, Plastic	1308.5.4.1
UL 705-2004	Power Ventilators (with revisions through December 17, 2013)	Power Ventilators	504.4.2.3
UL 710-2012	Exhaust Hoods for Commercial Cooking Equipment (with revisions through November 5, 2013)	Exhaust Hoods, Hoods	507.3.1, 508.2, 508.2.1
UL 710B-2011	Recirculating Systems (with revisions through August 14, 2014)	Exhaust Hoods	508.1, 513.2.2, 516.2.2, 516.2.3
UL 710C-2006	Ultraviolet Radiation Systems For Use In The Ventilation Control of Commercial Cooking Operations	Exhaust Hoods, Hoods	508.2.1
UL 723-2008	Test for Surface Burning Characteristics of Building Materials (with revisions through August 12, 2013)	Miscellaneous	508.3.4, 602.2, 604.1.2, 1201.2
UL 726-1995	Oil-Fired Boiler Assemblies (with revisions through October 9, 2013)	Fuel Gas, Appliances	1002.2.1
UL 727-2006	Oil-Fired Central Furnaces (with revisions through October 9, 2013)	Fuel Gas, Appliances	904.12
UL 729-2003	Oil-Fired Floor Furnaces (with revisions through November 22, 2016)	Furnaces, Floor, Oil Fired Furnaces	906.13
UL 730-2003	Oil-Fired Wall Furnaces (with revisions through November 22, 2016)	Furnaces, Wall, Oil Fired	907.4
UL 731-1995	Oil-Fired Unit Heaters (with revisions through November 22, 2016)	Heaters, Oil Fired	917.6
UL 737-2011	Fireplace Stoves (with revisions through August 19, 2015)	Fireplace Stoves	913.2
UL 762-2013	Power Roof Ventilators for Restaurant Exhaust Appliances	Ventilators	511.1
UL 778-2016	Motor Operater Water Pumps (with revisions through February 22, 2017)	Pumps	1208.1
UL 791-2006	Residential Incinerators (with revisions through November 7, 2014)	Incinerators	925.3
UL 795-2016	Commercial-Industrial Gas Heating Equipment	Heating Equipment, Gas Fired	904.13
UL 834-2004	Heating, Water Supply, and Power Boilers - Electric (with revisions through December 9, 2013)	Appliances	1002.3, Table 1203.2
UL 858-2014	Household Electric Ranges (with revisions through April 6, 2016)	Electric Ranges, Ranges	920.1
UL 867-2011	Electrostatic Air Cleaners (with revisions through September 16, 2016)	Filters	936.1
UL 896-1993	Oil-Burning Stoves (with revisions through November 22, 2016)	Stoves, Oil Fired	921.3
UL 900-2015	Air Filter Units	Air Filters, Filters	311.2
UL 907-2016	Fireplace Accessories	Fireplace Accessories	913.3
UL 921-2016	Commercial Dishwashers	Appliances	519.1
UL 923-2013	Microwave Cooking Appliances (with revisions through November 18, 2015)	Microwaves	920.3.2(3)
UL 959-2010	Medium Heat Appliance Factory-Built Chimneys (with revisions through June 12, 2014)	Fuel Gas, Appliances	802.5.1.2

TABLE 1701.1 (continued)
REFERENCED STANDARDS

STANDARD NUMBER	STANDARD TITLE	APPLICATION	REFERENCED SECTION
UL 1046-2010	Grease Filters for Exhaust Ducts (with revisions through January 13, 2012)	Filters-Grease	509.1, 509.1.1, 518.1(4)
UL 1240-2005	Electric Commercial Clothes-Drying Equipment (with revisions through October 17, 2012)	Clothes Dryers, Commercial	908.1
UL 1479-2015	Fire Tests of Penetration Firestops	Miscellaneous	507.4.4, 507.4.5
UL 1482-2011	Solid-Fuel Type Room Heaters (with revisions through August 19, 2015)	Room Heaters, Solid Fuel Heaters	802.5.1.1, 916.3
UL 1777-2015	Chimney Liners	Chimneys, Liners	803.1.11.2
UL 1812-2013	Ducted Heat Recovery Ventilators (with revisions through April 17, 2014)	Heat Recovery Ventilators	504.5
UL 1815-2012	Nonducted Heat Recovery Ventilators (with revisions through April 17, 2014)	Heat Recovery Ventilators	504.5
UL 1820-2004	Fire Test of Pneumatic Tubing for Flame and Smoke Characteristics (with revisions through May 10, 2013)	Surface Burning Test, Pneumatic Tubing	602.2.3
UL 1887-2004	Fire Test of Plastic Sprinkler Pipe for Visible Flame and Smoke Characteristics (with revisions through May 3, 2013)	Surface Burning Test, Fire Sprinkler Pipe	602.2.2
UL 1963-2011	Refrigerant Recovery/Recycling Equipment (with revisions through October 11, 2013)	Refrigerant Equipment	1101.2
UL 1978-2010	Grease Ducts (with revisions through September 19, 2013)	Ducts, Grease	507.3.1, 510.4.1, 510.5.2, 510.5.3
UL 1995-2015	Heating and Cooling Equipment	HVAC, Electric	903.1, 904.14
UL 1996-2009	Electric Duct Heaters (with revisions through June 13, 2014)	Duct Heaters	905.8
UL 2021-2015	Fixed and Location-Dedicated Electric Room Heaters	Product, Heating, Electric	916.1
UL 2043-2013	Fire Test for Heat and Visible Smoke Release For Discrete Products and Their Accessories Installed in Air-Handling Spaces	Surface Burning Test, Discrete Products	602.2.4
UL 2158-2015	Electric Clothes Dryers	Clothes Dryers, Electric	908.1
UL 2158A-2013	Clothes Dryer Transition Duct	Clothes Dryer Exhaust	504.4
UL 2162-2014	Commercial Wood-Fired Baking Ovens-Refractory Type	Baking Ovens	921.2
UL 2200-2012	Stationary Engine Generator Assemblies (with revisions through July 29, 2015)	Assemblies	1602.3
UL 2221-2010	Tests of Fire Resistive Grease Duct Enclosure Assemblies	Duct Wrap, Grease Duct Enclosure	507.4.4
UL 2518-2005	Outline of Investigation for Air Dispersion Systems Materials	Duct Systems	603.13
UL 2523-2009	Solid Fuel-Fired Hydronic Heating Appliances, Water Heaters, and Boilers (with revisions through February 8, 2013)	Appliances	1002.4, Table 1203.2
UL 2790-2010	Commercial Incinerators (with revisions through October 8, 2014)	Incinerators	925.2
UL 60335-2-40-2012	Household and Similar Electrical Appliances, Part 2-40: Particular Requirements for Electrical Heat Pumps, Air-Conditioners and Dehumidifiers	Appliances	903.1

1701.2 Standards, Publications, Practices, and Guides. The standards, publications, practices and guides listed in Table 1701.2 are not referenced in other sections of this code. The application of the referenced standards, publications, practices and guides shall be as specified in Section 302.1.2. The promulgating agency acronyms are found at the end of the table.

TABLE 1701.2
STANDARDS, PUBLICATIONS, PRACTICES, AND GUIDES

DOCUMENT NUMBER	DOCUMENT TITLE	APPLICATION
ACCA Manual J- 2016	Residential Load Calculations	Ducts
ACCA Manual N-2012	Commercial Load Calculations	Ducts
ACCA Manual Q-1990	Low Pressure, Low Velocity Duct System Design	Ducts
AHRI 1200-2013	Performance Rating of Commercial Refrigerated Display Merchandisers and Storage Cabinets	Commercial Refrigerated Display Merchandisers and Storage Cabinets
AHRI 1230-2010a	Performance Rating of Variable Refrigerant Flow (VRF) Multi-Split Air-Conditioning and Heat Pump Equipment	Refrigerants
AMCA 500-D-2012	Laboratory Methods of Testing Dampers for Rating	Dampers
ASCE 25-2016	Earthquake-Actuated Automatic Gas Shutoff Devices	Fuel Gas
ASHRAE 52.2-2012	General Ventilation Air-Cleaning Devices for Removal Efficiency by Particle Size	Cleaning Devices
ASHRAE 55-2013	Thermal Environmental Conditions for Human Occupancy	Miscellaneous
ASHRAE 90.1-2016	Energy Standard for Buildings Except Low-Rise Residential Buildings	Energy
ASHRAE 90.2-2007	Energy-Efficient Design of Low-Rise Residential Buildings	Energy, Dwellings
ASHRAE 127-2012	Method of Testing for Rating Computer and Data Processing Room Unitary Air Conditioners	Air Conditioners
ASHRAE/ACCA 183-2007 (R2014)	Peak Cooling and Heating Load Calculations in Buildings Except Low-Rise Residential Buildings	Cooling and Heating Load, Miscellaneous
ASHRAE Handbook-2012	HVAC Systems and Equipment	Design
ASME A13.1-2007 (R2013)	Scheme for the Identification of Piping Systems	Piping
ASME A112.18.6/CSA B125.6-2009 (R2014)	Flexible Water Connectors	Piping
ASME B1.20.3-1976 (R2013)	Dryseal Pipe Threads (Inch)	Joints
ASME B16.33-2012	Manually Operated Metallic Gas Valves for Use in Gas Piping Systems up to 175 psi (Sizes NPS ½ through NPS 2)	Valves
ASME B16.50-2013	Wrought Copper and Copper Alloy Braze-Joint Pressure Fittings	Fittings
ASTM A312/A312M-2016a	Seamless, Welded, and Heavily Cold Worked Austenitic Stainless Steel Pipes	Piping, Ferrous
ASTM A568/A568M-2015	Steel, Sheet, Carbon, Structural, and High-Strength, Low-Alloy, Hot-Rolled and Cold-Rolled, General Requirements for	Piping
ASTM A653/A653M-2015^{e1}	Steel Sheet, Zinc-Coated (Galvanized) or Zinc-Iron Alloy-Coated (Galvannealed) by the Hot-Dip Process	Piping, Ferrous
ASTM A733-2015	Welded and Seamless Carbon Steel and Austenitic Stainless Steel Pipe Nipples	Piping, Ferrous
ASTM B370-2012	Standard Specification for Copper Sheet and Strip for Building Construction	Miscellaneous
ASTM B687-1999 (R2016)	Brass, Copper, and Chromium-Plated Pipe Nipples	Piping, Copper Alloy
ASTM C518-2015	Steady-State Thermal Transmission Properties by Means of the Heat Flow Meter Apparatus	Certification: Calibration, Error Analysis, Heat Flow Meter Apparatus, Heat Flux, Instrument Verification, Thermal Conductivity, Thermal Resistance, Thermal Testing, Tunnel Test
ASTM D93-2016a	Flash Point by Pensky-Martens Closed Cup Tester	Certification
ASTM D396-2016	Fuel Oils	Boiler
ASTM D2517-2006 (R2011)	Reinforced Epoxy Resin Gas Pressure Pipe and Fittings	Piping, Plastic
ASTM E96/E96M-2016	Water Vapor Transmission of Materials	Miscellaneous
ASTM E136-2016a	Behavior of Materials in a Vertical Tube Furnace at 750°C	Furnace

TABLE 1701.2 (continued)
STANDARDS, PUBLICATIONS, PRACTICES, AND GUIDES

DOCUMENT NUMBER	DOCUMENT TITLE	APPLICATION
ASTM F1476-2007 (R2013)	Performance of Gasketed Mechanical Couplings for Use in Piping Applications	Joints
ASTM F2158-2008 (R2016)	Residential Central-Vacuum Tube and Fittings	Central Vacuum, Fittings, PVC, Tube
AWS B2.4-2012	Welding Procedure and Performance Qualification for Thermoplastics	Joints, Certification
CSA C448-2016	Design and Installation of Ground Source Heat Pump Systems for Commercial and Residential Buildings	Miscellaneous
CSA Z21.1-2016	Household Cooking Gas Appliances	Household Cooking Appliances, Gas Fired
CSA Z21.5.1-2016	Gas Clothes Dryers- Volume I, Type 1 Clothes Dryers (same as CSA 7.1)	Fuel Gas, Appliances
CSA Z21.5.2-2016	Gas Clothes Dryers-Volume II, Type 2 Clothes Dryers (same as CSA 7.2)	Fuel Gas, Appliances
CSA Z21.11.2-2013	Gas-Fired Room Heaters, Volume II, Unvented Room Heaters	Room Heaters, Unvented Heaters
CSA Z21.12b-1994 (R2010)	Draft Hoods	Fuel Gas, Appliances
CSA Z21.13-2014	Gas-Fired Low Pressure Steam and Hot Water Boilers (same as CSA 4.9)	Fuel Gas, Appliances
CSA Z21.15b-2013 (R2014)	Manually Operated Gas Valves for Appliances, Appliance Connector Valves and Hose End Valves (same as CSA 9.1b)	Fuel Gas
CSA Z21.17a-2008	Domestic Gas Conversion Burners (same as CSA 2.7a)	Conversion Burner Installation, Gas Burners
CSA Z21.18b-2012 (R2012)	Gas Appliance Pressure Regulators (same as CSA 6.3b)	Appliance Regulators, Gas Refrigerators, Pressure Regulators
CSA Z21.19-2014	Refrigerators Using Gas Fuel (same as CSA 1.4)	Gas Refrigerators, Refrigerators
CSA Z21.20a-2008	Automatic Gas Ignition Systems and Components	Automatic Ignition, Gas Ignition, Ignition
CSA Z21.21-2015	Automatic Valves for Gas Appliances (same as CSA 6.5)	Appliance Valves, Automatic Gas Valves, Valves
CSA Z21.22-2015	Relief Valves for Hot Water Supply Systems (same as CSA 4.4)	Valves
CSA Z21.40.1a-1997/CGA 2.91a-M97 (R2012)	Gas-Fired, Heat Activated Air-Conditioning and Heat Pump Appliances	Gas Fired Air Conditioning, Gas Fired Heat Pump
CSA Z21.42-2013	Gas-Fired Illuminating Appliances	Illuminating Appliances
CSA Z21.47-2012	Gas-Fired Central Furnaces (same as CSA 2.3)	Fuel Gas, Appliances
CSA Z21.50-2014	Vented Gas Fireplaces (same as CSA 2.22)	Appliances, Decorative Appliances
CSA Z21.56-2014	Gas-Fired Pool Heaters (same as CSA 4.7)	Fuel Gas, Swimming Pools, Spas, and Hot Tubs
CSA Z21.58-2015	Outdoor Cooking Gas Appliances (same as CSA 1.6)	Cooking Appliances
CSA Z21.60-2012	Decorative Gas Appliances for Installation in Solid-Fuel Burning Fireplaces (same as CSA 2.26)	Decorative Appliances, Decorative Fireplace Appliances
CSA Z21.61-1983 (R2004)	Gas-Fired Toilets	Toilets
CSA Z21.66-2015	Automatic Damper Devices for Use with Gas-Fired Appliances (same as CSA 6.14)	Vent Dampers, Automatic Dampers
CSA Z21.71a-2005 (R2007)	Automatic Intermittent Ignition Systems for Field Installation	Automatic Pilot Ignition, Pilot Ignition
CSA Z21.86-2008	Vented Gas-Fired Space Heating Appliances (same as CSA 2.32)	Fuel Gas Appliances
CSA Z21.88-2014	Vented Gas Fireplace Heaters (same as CSA 2.33)	Fireplace Heaters
CSA Z83.8-2016	Gas Unit Heaters, Gas Packaged Heaters, Gas Utility Heaters, and Gas-Fired Duct Furnaces (same as CSA 2.6)	Fuel Gas Appliances
CSA Z83.11-2016	Gas Food Service Equipment (same as CSA 1.8)	Fuel Gas, Appliances

TABLE 1701.2 (continued)
STANDARDS, PUBLICATIONS, PRACTICES, AND GUIDES

DOCUMENT NUMBER	DOCUMENT TITLE	APPLICATION
CSA Z83.19a-2011	Gas-Fired High-Intensity Infrared Heaters (same as CSA 2.35a)	High Intensity Heaters, Infrared Heaters
CSA Z83.20b-2011	Gas-Fired Low-Intensity Infrared Heaters (same as CSA 2.34b)	Infrared Heaters, Low Intensity Heaters
IAPMO IGC 293-2012	Tubing and Fittings for Special Hydronic Radiant Drywall Panels	HVAC, Fittings, Tubing
IAPMO PS 117-2012a[e1]	Press and Nail Connections	Fittings
IAPMO PS 120-2004	Flashing and Stand Combination for Air Conditioning Units (Residential or Commercial Unit Curb)	Air Conditioning Flashing Stand
ISO 13256-1-1998 (R2012)	Water Source Heat Pumps-Testing and Rating for Performance-Water-to-Air and Brine-to-Air Heat Pumps	Water-Source Heat Pumps
ISO 13256-2-1998	Water Source Heat Pumps-Testing and Rating for Performance-Water-to-Water and Brine-to-Water Heat Pumps	Water-Source Heat Pumps
MSS SP-67-2011	Butterfly Valves	Valves
MSS SP-80-2013	Bronze Gate, Globe, Angle, and Check Valves	Valves
MSS SP-104-2012	Wrought Copper Solder-Joint Pressure Fittings	Fittings
MSS SP-106-2012	Cast Copper Alloy Flanges and Flanged Fittings: Class 125, 150, and 300	Fittings
MSS SP-109-2012	Weld-Fabricated Copper Solder-Joint Pressure Fittings	Fittings
NFPA 30-2015	Flammable and Combustible Liquids Code	Combustible Liquids, Flammable Liquids
NFPA 91-2015	Exhaust Systems for Air Conveying of Vapors, Gases, Mists, and Particulate Solids	Product Conveying Ducts
NFPA 259-2013	Potential Heat of Building Materials	Testing, Building Materials
NFPA 274-2013	Test Method to Evaluate Fire Performance Characteristics of Pipe Insulation	Pipe Insulation
NFPA 501A-2017	Fire Safety Criteria for Manufactured Home Installations, Sites, and Communities	Miscellaneous
NFPA 5000-2015	Building Construction and Safety Code	Miscellaneous
SAE J512-1997	Automotive Tube Fittings	Fittings
SMACNA-2003	Fibrous Glass Duct Construction Standard, 7th Edition	Fiberglass Ducts
SMACNA-2008	IAQ Guidelines for Occupied Buildings Under Construction, 2nd Edition	Ventilation
UL 21-2014	LP-Gas Hose (with revisions through September 29, 2015)	Hose, LPG Hose
UL 33-2010	Heat Responsive Links for Fire-Protection Service (with revisions through April 14, 2015)	Fusible Links
UL 51-2013	Power-Operated Pumps and Bypass Valves for Anhydrous Ammonia, LP-Gas, and Propylene (with revisions through May 18, 2015)	Ammonia Pumps, LPG Pumps, Pumps
UL 80-2007	Steel Tanks for Oil-Burner Fuels and Other Combustible Liquids (with revisions through January 16, 2014)	Fuel Gas
UL 125-2014	Flow Control Valves for Anhydrous Ammonia and LP-Gas (with revisions through February 18, 2015)	Fuel Gas
UL 132-2015	Safety Relief Valves for Anhydrous Ammonia and LP-Gas (with revisions through February 17, 2016)	Fuel Gas
UL 144-2012	LP-Gas Regulators (with revisions through November 05, 2014)	Fuel Gas
UL 174-2004	Household Electric Storage Tank Water Heaters (with revisions through April 10, 2015)	Appliances
UL 180-2012	Liquid-Level Gauges for Oil Burner Fuels and other combustible liquids	Gauges, Level Gauges
UL 252-2010	Compressed Gas Regulators (with revisions through January 28, 2015)	Fuel Gas
UL 343-2008	Pumps for Oil-Burning Appliances (with revisions through June 12, 2013)	Fuel Gas, Appliances
UL 353-1994	Limit Controls (with revisions through November 8, 2011)	Controls
UL 404-2010	Gauges, Indicating Pressure, for Compressed Gas Service (with revisions through February 11, 2015)	Fuel Gas
UL 429-2013	Electrically Operated Valves	Valves
UL 443-2006	Steel Auxiliary Tanks for Oil-Burner Fuel (with revisions through March 8, 2013)	Fuel Gas

TABLE 1701.2 (continued)
STANDARDS, PUBLICATIONS, PRACTICES, AND GUIDES

DOCUMENT NUMBER	DOCUMENT TITLE	APPLICATION
UL 525-2008	Flame Arresters (with revisions through August 10, 2012)	Flame Arrestors
UL 565-2013	Liquid-Level Gauges for Anhydrous Ammonia and LP-Gas	Fuel Gas
UL 569-2013	Pigtails and Flexible Hose Connectors for LP-Gas	Fuel Gas
UL 732-1995	Oil-Fired Storage Tank Water Heaters (with revisions through October 9, 2013)	Fuel Gas, Appliances
UL 733-1993	Oil-Fired Air Heaters and Direct-Fired Heaters (with revisions through October 9, 2013)	Water Heaters, Direct Fired, Oil Fired
UL 842-2015	Valves for Flammable Fluids	Valves
UL 984-1996	Hermetic Refrigerant Motor-Compressors (with revisions through September 23, 2005)	Compressors, Refrigeration
UL 1453-2016	Electric Booster and Commercial Storage Tank Water Heaters	Appliances
UL 1746-2007	External Corrosion Protection Systems for Steel Underground Storage Tanks (with revisions through December 19, 2014)	Tanks

ABBREVIATIONS IN TABLE 1701.1 AND TABLE 1701.2

AABC	Associated Air Balance Council, 1518 K Street NW, Suite 503, Washington, DC 20005.
ACCA	Air Conditioning Contractors of America Association, Inc., 2800 S Shirlington Road, Suite 300, Arlington, VA 22206.
AHRI	Air-Conditioning, Heating, and Refrigeration Institute, 2111 Wilson Boulevard, Suite 500, Arlington, VA 22201.
AMCA	Air Movement and Control Association, 30 West University Drive, Arlington Heights, IL 60004-1806.
ANSI	American National Standards Institute, Inc., 25 W. 43rd Street, 4th Floor, New York, NY 10036.
ASCE	American Society of Civil Engineers, 1801 Alexander Bell Drive, Reston, VA 20191-4400.
ASHRAE	American Society of Heating, Refrigerating, and Air Conditioning Engineers, Inc., 1791 Tullie Circle, NE, Atlanta, GA 30329-2305.
ASME	American Society of Mechanical Engineering, Two Park Avenue, New York, NY 10016-5990.
ASSE	American Society of Sanitary Engineering, 18927 Hickory Creek Drive, Suite 220, Mokena, IL 60448.
ASTM	ASTM International, 100 Barr Harbor Drive, West Conshohocken, PA 19428-2959.
AWS	American Welding Society, 8669 NW 36 Street, #130, Miami, FL 33166-6672.
AWWA	American Water Works Association, 6666 W. Quincy Avenue, Denver, CO 80235.
CGA	Canadian Gas Association, 243 Consumers Road, Suite 1200, North York, Ontario, Canada M2J 5E3.
CSA	Canadian Standards Association, 178 Rexdale Boulevard, Toronto, Ontario, Canada M9W 1R3.
e1	An editorial change since the last revision or reapproval.
IAPMO	International Association of Plumbing and Mechanical Officials, 4755 E. Philadelphia Street, Ontario, CA 91761.
IIAR	International Institute of Ammonia Refrigeration, 1001 N. Fairfax Street, Suite 503, Alexandria, VA 22314.
ISO	International Organization for Standardization, 1 ch. de la Voie-Creuse, Casa Postale 56, CH-1211 Geneva 20, Switzerland.
MSS	Manufacturers Standardization Society of the Valve and Fittings Industry, 127 Park Street NE, Vienna, VA 22180.
NEBB	NEBB, 8575 Grovemont Circle, Gaithersburg, MD 20877.
NFPA	National Fire Protection Association, 1 Batterymarch Park, Quincy, MA 02169-7471.
NSF	NSF International, 789 N. Dixboro Road, Ann Arbor, MI 48105.
SAE	Society of Automotive Engineers, 400 Commonwealth Drive, Warrendale, PA 15096.
SMACNA	Sheet Metal and Air Conditioning Contractors National Association, 4201 Lafayette Center Drive, Chantilly, VA 20151-1219.
UL	Underwriters Laboratories, Inc., 333 Pfingsten Road, Northbrook, IL 60062.

APPENDICES

The appendices are intended to supplement the provisions of the installation requirements of this code. The definitions in Chapter 2 are also applicable to the appendices.

TABLE OF CONTENTS

CALIFORNIA MECHANICAL CODE – MATRIX ADOPTION TABLE
APPENDIX A – RESIDENTIAL PLAN EXAMINER REVIEW FORM FOR HVAC SYSTEM DESIGN

(Matrix Adoption Tables are non-regulatory, intended only as an aid to the code user. See Chapter 1 for state agency authority and building applications.)

Adopting Agency	BSC	BSC-CG	SFM	HCD			DSA			OSHPD						BSCC	DPH	AGR	DWR	CEC	CA	SL	SLC
				1	2	1-AC	AC	SS	SS/CC	1	1R	2	3	4	5								
Adopt Entire Chapter																							
Adopt Entire Chapter as amended (amended sections listed below)																	X						
Adopt only those sections that are listed below																							
Chapter/Section																							

This state agency does not adopt sections identified with the following symbol: †

The Office of the State Fire Marshal's adoption of this chapter or individual sections is applicable to structures regulated by other state agencies pursuant to Section 1.11.0.

APPENDIX A

RESIDENTIAL PLAN EXAMINER REVIEW FORM FOR HVAC SYSTEM DESIGN (Loads, Equipment, Ducts) [ACCA]

The provisions contained in this appendix are not mandatory unless specifically adopted by a state agency, or referenced in the adopting ordinance.

The following Residential Plans Examiner Review Form for HVAC System Design (Loads, Equipment, Ducts), Form RPER1, is included here for the convenience of the users of the Uniform Mechanical Code.

ACCA® Air Conditioning Contractors of America

Residential Plans Examiner Review Form for HVAC System Design (Loads, Equipment, Ducts)

Form RPER 1
15 Mar 09

County, Town, Municipality, Jurisdiction
Header Information

Contractor ____________________

Mechanical License # ____________________

Building Plan # ____________________

REQUIRED ATTACHMENTS	ATTACHED	
Manual J1 Form (and supporting worksheets):	Yes ☐	No ☐
or MJ1AE Form* (and supporting worksheets):	Yes ☐	No ☐
OEM performance data (heating, cooling, blower):	Yes ☐	No ☐
Manual D Friction Rate Worksheet:	Yes ☐	No ☐
Duct distribution system sketch:	Yes ☐	No ☐

Home Address (Street or Lot#, Block, Subdivision) ____________________

HVAC LOAD CALCULATION (See Section 1105.1)

Design Conditions

Winter Design Conditions

- Outdoor temperature ______ °F
- Indoor temperature ______ °F
- Total heat loss ______ Btu

Summer Design Conditions

- Outdoor temperature ______ °F
- Indoor temperature ______ °F
- Grains difference ______ Δ Gr @ ______ % Rh
- Sensible heat gain ______ Btu
- Latent heat gain ______ Btu
- Total heat gain ______ Btu

Building Construction Information

Building

- Orientation (Front door faces) ______
  North, East, West, South, Northeast, Northwest, Southeast, Southwest
- Number of bedrooms ______
- Conditioned floor area ______ Sq Ft
- Number of occupants ______

Windows

- Eave overhang depth ______ Ft
- Internal shade ______
  Blinds, drapes, etc
- Number of skylights ______

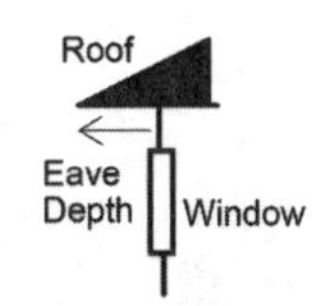

HVAC EQUIPMENT SELECTION

Heating Equipment Data

- Equipment type ______
  Furnace, Heat pump, Boiler, etc.
- Model ______
- Heating output capacity ______ Btu
  Heat pumps - capacity at winter design outdoor conditions
- Auxiliary heat output capacity ______ Btu

Cooling Equipment Data

- Equipment type ______
  Air Conditioner, Heat pump, etc
- Model ______
- Sensible cooling capacity ______ Btu
- Latent cooling capacity ______ Btu
- Total cooling capacity ______ Btu

Blower Data

- Heating CFM ______ CFM
- Cooling CFM ______ CFM
- Static pressure ______ IWC
  Fan's rated external static pressure for design airflow

HVAC DUCT DISTRIBUTION SYSTEM DESIGN (See Section 601.2)

- Design airflow ______ CFM
- External Static Pressure (ESP) ______ IWC
- Component Pressure Losses (CPL) ______ IWC
- **Available Static Pressure (ASP)** ______ IWC
  ASP = ESP - CPL

- Longest supply duct: ______ Ft
- Longest return duct: ______ Ft
- **Total Effective Length (TEL)** ______ Ft
- **Friction Rate:** ______ IWC
  Friction Rate = (ASP × 100) ÷ TEL

Duct Materials Used (circle)

Trunk Duct: Duct board, Flex, Sheet metal, Lined sheet metal, Other (specify) ______

Branch Duct: Duct board, Flex, Sheet metal, Lined sheet metal, Other (specify) ______

I declare the load calculation, equipment selection, and duct system design were rigorously performed based on the building plan listed above. I understand the claims made on these forms will be subject to review and verification.

Contractor's Printed Name ____________________ Date __________

Contractor's Signature ____________________

Reserved for County, Town, Municipality, or Authority having jurisdiction use.

* Home qualifies for MJ1AE Form based on Abridged Edition Checklist.

CALIFORNIA MECHANICAL CODE – MATRIX ADOPTION TABLE
APPENDIX B – PROCEDURES TO BE FOLLOWED TO PLACE GAS EQUIPMENT IN OPERATION

(Matrix Adoption Tables are non-regulatory, intended only as an aid to the code user. See Chapter 1 for state agency authority and building applications.)

Adopting Agency	BSC	BSC-CG	SFM	HCD			DSA			OSHPD						BSCC	DPH	AGR	DWR	CEC	CA	SL	SLC
				1	2	1-AC	AC	SS	SS/CC	1	1R	2	3	4	5								
Adopt Entire Chapter	X									X	X	X	X	X	X								
Adopt Entire Chapter as amended (amended sections listed below)																							
Adopt only those sections that are listed below																							
Chapter/Section																							

This state agency does not adopt sections identified with the following symbol: †

The Office of the State Fire Marshal's adoption of this chapter or individual sections is applicable to structures regulated by other state agencies pursuant to Section 1.11.0.

APPENDIX B

PROCEDURES TO BE FOLLOWED TO PLACE GAS EQUIPMENT IN OPERATION

The provisions contained in this appendix are not mandatory unless specifically adopted by a state agency, or referenced in the adopting ordinance.

B 101.0 Adjusting the Burner Input.

B 101.1 Adjusting Input. The input rate of the burner shall be adjusted to the value in accordance with the appliance manufacturer's instructions. Firing at a rate in excess of the nameplate rating shall be prohibited. The input rate shall be adjusted by changing the size of a fixed orifice, changing the adjustment of an adjustable orifice, or readjusting the appliance's gas pressure regulator outlet pressure (where a regulator is provided in the appliance). [NFPA 54:11.1.1]

B 101.2 High Altitude. Gas input ratings of appliances shall be used for elevations up to 2000 feet (610 m). The input rating of appliances operating at elevations above 2000 feet (610 m), shall be reduced in accordance with one of the following methods:

(1) At the rate of 4 percent for each 1000 feet (305 m) above sea level before selecting appropriately sized appliances.

(2) As permitted by the Authority Having Jurisdiction.

(3) In accordance with the manufacturer's installation instructions. [NFPA 54:11.1.2]

B 102.0 Primary Air Adjustment.

B 102.1 General. The primary air for injection (Bunsen)-type burners shall be adjusted for flame characteristics in accordance with the appliance manufacturer's instructions. After setting the primary air, the adjustment means shall be secured in position. [NFPA 54:11.2]

B 103.0 Safety Shutoff Devices.

B 103.1 General. Where a safety shutoff device is provided, it shall be checked for operation and adjustment in accordance with the appliance manufacturer's instructions. Where the device does not function properly to turn off the gas supply in the event of pilot outage or other improper operation, it shall be serviced or replaced with a new device. [NFPA 54:11.3]

B 104.0 Automatic Ignition.

B 104.1 General. Appliances supplied with means for automatic ignition shall be checked for operation within the parameters provided by the manufacturer. Any adjustments made shall be in accordance with the manufacturer's instructions. [NFPA 54:11.4]

B 105.0 Protective Devices.

B 105.1 General. Where required by the manufacturer's instructions, protective devices furnished with the appliance such as a limit control, fan control to blower, temperature-and pressure-relief valve, low-water cutoff device, or manual operating features, shall be checked for operation within the parameters provided by the manufacturer. Adjustments made shall be in accordance with the manufacturer's instructions. [NFPA 54:11.5]

B 106.0 Checking the Draft.

B 106.1 General. Draft hood-equipped appliances shall be checked to verify that there is no draft hood spillage after 5 minutes of main burner operation. [NFPA 54:11.6]

B 107.0 Operating Instructions.

B 107.1 General. Operating instructions shall be furnished and shall be left in a prominent position near the appliance for the use of the consumer. [NFPA 54:11.7]

CALIFORNIA MECHANICAL CODE – MATRIX ADOPTION TABLE
APPENDIX C – INSTALLATION AND TESTING OF OIL (LIQUID) FUEL-FIRED EQUIPMENT

(Matrix Adoption Tables are non-regulatory, intended only as an aid to the code user. See Chapter 1 for state agency authority and building applications.)

Adopting Agency	BSC	BSC-CG	SFM	HCD			DSA			OSHPD						BSCC	DPH	AGR	DWR	CEC	CA	SL	SLC
				1	2	1-AC	AC	SS	SS/CC	1	1R	2	3	4	5								
Adopt Entire Chapter	X									X	X	X	X	X	X								
Adopt Entire Chapter as amended (amended sections listed below)																							
Adopt only those sections that are listed below																							
Chapter/Section																							

This state agency does not adopt sections identified with the following symbol: †

The Office of the State Fire Marshal's adoption of this chapter or individual sections is applicable to structures regulated by other state agencies pursuant to Section 1.11.0.

APPENDIX C
INSTALLATION AND TESTING OF OIL (LIQUID) FUEL-FIRED EQUIPMENT

The provisions contained in this appendix are not mandatory unless specifically adopted by a state agency, or referenced in the adopting ordinance.

C 101.0 General.

C 101.1 Applicability. Appendix C governs the installation, testing, or repair of oil or liquid fuel burners, oil or liquid fuel-burning systems, oil or liquid fuel-burning equipment, and the oil or liquid fuel piping systems used in connection with buildings or structures and equipment within the property lines of the premises.

C 102.0 Definitions.

Anti-Flooding Device. A primary safety control that causes the flow of oil or fuel to be shut off after a rise in oil or fuel level, or after receiving excess oil or fuel, and that operates before the hazardous discharge of oil or fuel can occur.

Burner, Automatically Ignited. A burner equipped so that main burner fuel may be turned on and ignited automatically.

Burner, Manually Ignited. A burner equipped, so that main burner fuel is turned on only by hand and ignited under supervision.

Burner, Mechanical Draft Type. A burner that includes a power-driven fan, blower, or other mechanism as the primary means for supplying the air for combustion.

Burner, Natural Draft Type. A burner that depends primarily on the natural draft created in the chimney or venting system to induce air required for combustion into the burner.

Constant Level Valve. A device for maintaining within a reservoir a constant level of oil or fuel for delivery to a burner.

Control Limit. An automatic safety control that is responsive to changes in fluid flow or level, pressure, or temperature and that is normally set beyond the operating range for limiting the operation of the controlled equipment by shutting off the energy supply.

Control Safety. Automatic interlock controls, including relays, switches, and other auxiliary equipment used in conjunction with them, to form a safety control system that is intended to prevent unsafe operation of the controlled equipment.

Draft Booster. A power-operated fan, blower, or other device installed in the chimney connector to increase the natural draft developed in the connected chimney.

Draft Regulator, Barometric. A device built into a fuel-burning appliance or made part of a chimney connector or vent connector that functions to reduce excessive draft through an appliance to a desired value by admitting ambient air into the appliance chimney, chimney connector, vent, or vent connector.

Fuel. Natural, manufactured, or liquefied petroleum gas, or a mixture of these gases; all grades of fuel oil, wood, or any other combustible or flammable material or any mixture of combustible or flammable materials.

Fuel Burner. A device used to convey the appropriate fuel into the combustion chamber zone in close proximity to its primary and secondary air supply to permit a stable controlled heat release compatible with the burner design, listing, and applicable approvals in a boiler, furnace, device or appliance. It includes but is not limited to burning oil or liquid fuel.

Fuel Burner System. The fuel burner and a conveyance system or piping system for the purpose of introducing the appropriate fuel into the combustion chamber zone.

Fuel-Burning Equipment/Appliance. An oil or fuel burner of any type including all oil or liquid fuel burners, oil or liquid fuel-fired units, dual, or multi-fuel burners and heating and cooking appliances with their fuel burner system and with their tank or fuel storage system, piping system, vent connectors, vent flues, fans, blowers, valves, control devices, combustion air, wiring, controls, and related devices including all accessories and appurtenances for safe and proper operation of the appliance.

Fuel Oil. Hydrocarbon oil as specified by ASTM D396, or the Canadian Government Specification Board, 3-GP-28, and having a flashpoint of not less than 100°F (38°C).

Fuel-Piping System. Method of conveying liquid, vapor, steam, gases, or slurry from one point to another, including accessories, appurtenances, and equipment necessary for its proper operation.

Indirect-Fired Appliance. An oil or fuel-burning appliance in which products of combustion (flue gasses) are not mixed in the appliance with the air or other medium being heated.

Labeled. Having attached a label, symbol, or other identifying mark of an organization acceptable to the Authority Having Jurisdiction and concerned with product evaluation that maintains periodic inspection of production of labeled equipment or materials and by whose labeling the manufacturer indicates compliance with appropriate standards or performance in a specified manner.

Premixing. A power burner in which all or nearly all of the air for combustion is mixed with the gas as primary air.

Pump, Oil or Fuel Transfer. An oil or fuel pump, automatically or manually operated, that transfers oil or fuel through continuous piping from a supply tank to an oil or fuel-burning appliance or to an auxiliary tank, and that is not designed to stop pumping automatically in case of total breakage of the oil or fuel supply line between the pump and the appliance.

Tank, Auxiliary. A tank having a capacity of not over 60 gallons (227 L) listed for installation in the supply piping between a burner and its main fuel supply tank. It shall be permitted to be included as an integral part of an automatic pump or a transfer pump, or it shall be permitted to be a separate tank.

Tank, Gravity. A supply tank from which the oil or fuel is delivered directly to the burner by gravity.

Tank, Integral. A tank that is furnished by the manufacturer as an integral part of an oil or fuel-burning appliance.

Tank Storage. A separate tank that is not connected to the oil or fuel-burning appliance.

Tank Supply. A separate tank connected directly or by a pump to the oil or fuel-burning appliance.

Tank, Vacuum or Barometric. A tank not exceeding 5 gallons (19 L) capacity that maintains a definite level of oil or fuel in a sump or similar receptacle by barometric feed. Fuel is delivered from the sump to the burner by gravity.

Valve, Manual Oil, Gas, or Fuel Shutoff. A manually operated valve in a fuel line for the purpose of turning on or completely shutting off the fuel supply to the burner.

Valve, Oil, Gas, or Fuel Control. An automatically or manually operated device consisting essentially of a fuel valve for controlling the fuel supply to a burner.

C 103.0 Standards and Accepted Practices.

C 103.1 General. The installation, testing, and repair of oil or liquid fuel-burning equipment systems shall be in accordance with Section C 103.0, the standards listed in Chapter 17, and other information outlined in this code such as, but not limited to, combustion air, flue and breeching requirements, room clearances and dimensions, and control requirements.

C 104.0 Approval of Equipment.

C 104.1 General. Oil or liquid fuel-burning equipment shall be approved.

C 105.0 Placing Equipment in Operation.

C 105.1 General. After completion of installations, the installer shall test safety and operating controls and venting before placing the burner in service. The correct input of liquid fuel shall be determined, and the fuel-to-air ratio shall be set. Each oil or liquid fuel burner shall be adjusted to its input according to the manufacturer's instructions. Overrating the burners or the appliance is prohibited. The input range shall be appropriate to the appliance:

(1) For conversion burners installed in hot water (liquid) boilers or warm air furnaces, the rate of flow of the oil or liquid fuel in British thermal units per hour (Btu/h) (kW) shall be adjusted to within plus or minus 5 percent of the design load, not to exceed the design rate of the appliance.

(2) For conversion burners installed in steam boilers, the oil or liquid fuel hourly input demand shall be adjusted to be in accordance with the steam load requirements. The oil or liquid fuel input demand necessitated by an oversized boiler shall be established and added to total input demand.

C 106.0 Pilot Operation.

C 106.1 General. Igniter or pilot flames shall be effective to ignite the oil or liquid fuel at the main burner or burners and shall be adequately protected from drafts. Pilot flames shall not become extinguished during the pilot cycle where the main burner or burners are turned on or off in a normal manner either manually or by automatic controls.

C 107.0 Burner Operation.

C 107.1 General. In making tests to determine compliance with the requirements of Section C 107.1, the following care shall be exercised to prevent the accumulation of unburned liquid fuel in the appliance that will result in an explosion or fire:

(1) The flames from the burner shall ignite freely the liquid fuel where operating at the lowest firing position.

(2) Burner flames shall not flash back where the liquid fuel is turned on or off by an automatic control mechanism.

(3) Main burner flames shall ignite freely from the pilot where the pilot flame is reduced to a minimum point that will actuate the pilot safety device.

(4) Where ignition is made in a normal manner, the flame will not flash outside the appliance.

(5) Burners shall not expel liquid fuel through air openings where operating at prevailing pressure.

(6) Burners shall have a liquid-fuel air mixture to ensure smooth ignition of the main burner.

C 108.0 Method of Test.

C 108.1 General Test Methods.

(1) The flue gas, venting, safety and operating controls of the appliance shall be checked by the installer to ensure their proper and safe operation.

(2) Method of test – atmospheric, induced draft or fan assisted types.

   (a) The appliance shall be allowed to operate until the stack temperature becomes stabilized, after which a sample of the undiluted flue products shall be taken from the appliance flue outlet. The sample taken shall be analyzed for carbon monoxide, carbon dioxide, and oxygen. Stack temperature shall be noted.

   Appliance designs incorporating induced-draft assemblies require a flue gas sample to be taken ahead of the draft regulator or induced draft fan.

(3) Performance standards for atmospheric type shall be provided in accordance with the following:

   (a) Shall be not less than 75 percent efficiency as determined by flue gas analysis method at the appliance flue outlet.

   (b) Carbon monoxide concentration in flue gas shall not exceed 0.04 percent.

   (c) Stack temperature shall not exceed 700°F (371°C) plus ambient.

(d) Carbon dioxide concentration shall be between 8 percent and 13 percent.

(e) Oxygen concentration shall be between 4 percent and 10 percent.

(f) Smoke test shall not exceed number 2 for light oils or number 4 for oils heavier than number 4.

(g) Draft shall comply with the burner manufacturer's instructions.

(4) Performance standards for induced-draft or fan-assisted types shall comply with the following:

(a) Shall be not less than 75 percent efficiency, as determined by flue gas analysis method at appliance flue outlet.

(b) Carbon monoxide concentration in flue gas not exceeding 0.04 percent.

(c) Stack temperature shall not exceed 700°F (371°C) plus ambient.

(d) Carbon dioxide concentration shall be between 8 percent and 13 percent.

(e) Oxygen concentration shall be between 4 percent and 10 percent.

(f) Smoke test shall not exceed number 2 for light oils or number 4 for oils heavier than number 4.

(g) Draft shall comply with the burner manufacturer's instructions.

Induced-draft and fan-assisted types of appliances require a sample be taken after the induced-draft fan that will cause oxygen figures in excess of the limits stated. In such cases, safe liquid fuel combustion ratios shall be maintained and be consistent with approvals and listings of the appliance.

(5) Method of test – power type.

(a) The appliance shall be allowed to operate until the stack temperature becomes stabilized; after that, a sample of the undiluted flue products shall be taken from the appliance flue outlet. The sample shall be analyzed for carbon monoxide, carbon dioxide, and oxygen. Stack temperature shall be noted.

(6) Performance standards for power type.

(a) Shall be not less than 80 percent efficiency as determined by flue gas analysis method at the appliance flue outlet.

(b) Carbon monoxide concentration in the flue gas shall not exceed 0.04 percent.

(c) Stack temperature shall not exceed 700°F (371°C) plus ambient.

(d) Carbon dioxide concentration shall be between 8 percent and 13 percent.

(e) Oxygen concentration shall be between 4 percent and 10 percent.

(f) Smoke test shall not exceed number 2 for light oils or number 4 for oils heavier than number 4.

(g) Draft shall comply with the burner manufacturer's instructions.

(7) After completion of the test of newly installed oil or liquid fuel burner equipment as provided in this section, the installer shall file with the Authority Having Jurisdiction complete records of the test on a form approved by the Authority Having Jurisdiction. The tag stating the date of the test and the name of the installer shall be attached to the appliance at the main valve.

(8) Listing and approval.

(a) The concentration of oxygen in the undiluted flue products of oil or liquid fuel burners shall in no case be less than 3 percent nor exceed 10 percent and shall be in accordance with performance standards and shall be consistent with the listing and approval of the equipment.

(b) The allowable limit of carbon monoxide shall not exceed 0.04 percent.

(c) The flue gas temperature of an oil appliance, as taken on the appliance side of the draft regulator, shall not exceed applicable performance standards and shall be consistent with the listing and approvals of the equipment.

(9) The oxygen figures shall not apply where there is an approved oxygen trim system on the burner that is designed for that use, including a low oxygen interlock where approved by the Authority Having Jurisdiction.

(10) Supervision shall be as follows:

(a) Supervised startup shall be required to verify the safe operation of an oil or liquid fuel burner and to provide documentation that operation is consistent with this code, listing, and approval. Supervised startup shall be required for liquid fuel burners in Section C 109.1(2), Section C 109.1(3), and Section C 109.1(4). Supervised startup requires that the liquid-fuel burner shall be tested in the presence of the mechanical official in a manner set forth by the Authority Having Jurisdiction before the installation is approved. Testing shall include safety and operating controls, input, flue gas analysis, and venting. Flue gas shall be tested at high, medium, and low fires. Provisions shall be made in the system to allow a firing test in warm weather. After completion of the test of newly installed oil or liquid fuel burner equipment, as provided in this section, the installer shall file with the Authority Having Jurisdiction complete records of the test on a form approved by the Authority Having Jurisdiction. The tag stating the date of the test and the name of the tester shall be attached to the appliance at the main valve.

(b) Oil and liquid fuel burners of 1 000 000 Btu/h (293 kW) input or more require a supervised startup in accordance with Section C 108.1(10)(a).

(c) Installation of oxygen trim systems, modulating dampers, or other draft control or combustion devices require a supervised startup in accordance with Section C 108.1(10)(a).

(d) Direct-fired heaters shall require a supervised startup in accordance with Section C 108.1(10)(a).

(11) The complete control diagram of the installation and operating instructions shall be supplied and posted by the installer of the appliance.

C 109.0 Special Requirements Based on Btu/h Input.

C 109.1 General.

(1) Zero to 400 000 Btu/h (0 kW to 117 kW) per burner.

(a) One approved manual shutoff valve lever handle.

(b) One approved fuel oil filter, installed on the supply piping.

(c) Approved automatic safety shutoff valve to provide 100 percent shutoff of all oil.

(d) A flame safeguard control capable of providing 100 percent shutoff in the event of flame failure. Flame failure response timing shall not exceed the control manufacturer's instructions.

(e) Two controls, one operating and one high limit, activated by temperature or pressure, as appropriate.

(f) Burners relying on mechanical means to provide air for combustion shall have actual proof-of-air interlock device.

(g) Installations with dampered combustion air openings shall prove damper open position before trial for burner ignition.

(h) Vent dampers and flue dampers shall be properly interlocked to prevent burner ignition unless safely open.

(2) Four hundred thousand and one to 999 999 Btu/h (117.2 kW to 292.9 kW) per burner.

(a) One approved manual shutoff valve lever handle.

(b) One approved fuel-oil filter, installed on the supply piping.

(c) Two safety shutoff valves in series, maximum five seconds closing time.

(d) One electronic flame safeguard pilot control providing a separately supervised and proven pilot, 100 percent shutoff manual reset. Flame failure response time shall not exceed the control manufacturer's instructions.

Direct-spark ignition shall be allowed where approved by the Authority Having Jurisdiction and where used on number 2 or lighter oil.

(e) Two controls, one operating and one high limit, activated by temperature or pressure, as appropriate.

(f) Burners relying on mechanical means to provide air for combustion shall have actual proof-of-air interlock device.

(g) Power burners shall include proven prepurge of not less than 60 seconds at high-fire damper settings. This prepurge shall occur before every burner cycle, regardless of reason.

(h) Installations with dampered combustion air openings shall prove damper open position before trial for burner ignition.

(i) Vent dampers and flue dampers shall be interlocked to prevent burner ignition unless safely open.

(j) One high oil or liquid fuel-pressure interlock, reset from flame safeguard or manually.

(k) Where hot water or steam, one low water cutoff.

(l) An atomizing medium proving switch.

(m) A low oil temperature switch for oil or liquid fuel requiring preheating.

(n) A high oil temperature interlock for oil or liquid fuel requiring preheating.

(o) The burner oil pump shall automatically not operate or rotate while the alternate fuel is firing.

(p) A pressure-relief valve shall be provided between safety shutoff valves and between pump and safety valves where an integral valve is used with a pump.

(q) A separate relief device is required on each transfer pump.

(3) One million to 2 499 999 Btu/h (293 kW to 732 kW) input per burner.

(a) One approved manual shutoff valve lever handle.

(b) One approved fuel-oil filter, installed on the supply piping.

(c) Two safety shutoff valves in series with a combined flame failure response and valve closing time not to exceed 5 seconds with strainer directly before the valves.

(d) Programmed electronic flame safeguard including proven low-fire start, manual reset lockout, 100 percent shutoff (both pilot and main burner), and a separately supervised and proven pilot.

Flame-sensing systems utilizing a UV scanner shall prove pilot and interrupt ignition spark prior to main burner valves being energized.

(e) Two controls, one operating and one high limit, activated by temperature or pressure.

(f) Burners relying on mechanical means to provide air for combustion shall have actual proof-of-air interlock device.

(g) Power burners shall include proven prepurge of not less than 60 seconds at high-fire damper settings. This prepurge shall occur before every burner cycle, regardless of reason.

(h) Installations with dampered combustion air openings shall prove damper open position before trial for burner ignition.

(i) Vent dampers and flue dampers shall be interlocked to prevent burner ignition unless safely open.

(j) One high oil or liquid fuel-pressure interlock, reset from flame safeguard or manually.

(k) Where hot water or steam, two low water cutoffs.

(l) An atomizing medium proving switch.

(m) A low oil temperature switch for oil or liquid fuel requiring preheating.

(n) A high oil temperature interlock for oil or liquid fuel requiring preheating.

(o) The burner oil pump shall automatically not operate or rotate while the alternate fuel is firing.

(p) A pressure-relief valve shall be provided between safety shutoff valves and between pump and safety valves where an integral valve is used with a pump.

(q) A separate relief device is required on each transfer pump.

(r) One low oil or liquid fuel-pressure interlock, reset from flame safeguard or manually.

(s) Burners with automatic controls, prepurge, proof-of-closure, modulation, or postpurge shall not use relays external to the flame safeguard to accomplish these functions.

(4) Two million five hundred thousand to 12 499 999 Btu/h (733 kW to 3663.3 kW) per burner.

(a) One approved manual shutoff valve lever handle.

(b) One approved fuel-oil filter, installed on the supply piping.

(c) Two safety shutoff valves in series, with a combined flame failure response and valve closing time not to exceed 5 seconds with strainer directly before the valves.

(d) Programmed electronic flame safeguard including proven low-fire start, manual reset lockout, 100 percent shutoff (both pilot and main burner), and a separately supervised and proven pilot.

Flame-sensing systems utilizing a UV scanner shall prove pilot and interrupt ignition spark prior to main burner valves being energized.

(e) Two controls, one operating and one high limit, activated by temperature or pressure.

(f) Burners relying on mechanical means to provide air for combustion shall have actual proof-of-air interlock device.

(g) Power burners shall include proven prepurge of not less than 60 seconds at high-fire damper settings. This prepurge shall occur before every burner cycle, regardless of reason.

(h) Installations with dampered combustion air openings shall prove damper open position before trial for burner ignition.

(i) Vent dampers and flue dampers shall be interlocked to prevent burner ignition unless safely open.

(j) One high oil or liquid fuel-pressure interlock, reset from flame safeguard or manually.

(k) Where hot water or steam, two low water cutoffs.

(l) An atomizing medium proving switch.

(m) A low oil temperature switch for oil or liquid fuel requiring preheating.

(n) A high oil temperature interlock for oil or liquid fuel requiring preheating.

(o) A separate firing rate control valve.

(p) The burner oil pump shall automatically not operate or rotate while the alternate fuel is firing.

(q) A pressure-relief valve shall be provided between safety shutoff valves and between pump and safety valves where an integral valve is used with a pump.

(r) A separate relief device is required on each transfer pump.

(s) One low oil or liquid fuel-pressure interlock reset from flame safeguard or manually.

(t) Burners with automatic controls, prepurge, proof-of-closure, modulation, or postpurge shall not use relays external to the flame safeguard to accomplish these functions.

(5) More than 12 500 000 Btu/h (3663.4 kW) per burner inputs. These burners shall comply with the requirements of the appropriate standards listed in Chapter 17 and the following:

(a) One approved manual shutoff valve lever handle.

(b) One approved fuel-oil filter, installed on the supply piping.

(c) Two safety shutoff valves in series, one with proof of closure, with a combined flame failure response and valve closing time not to exceed 2 seconds with strainer directly before the valves.

(d) Programmed electronic flame safeguard including proven low-fire start, manual reset lockout, 100 percent shutoff (both pilot and main burner), and a separately supervised and proven pilot.

Flame-sensing systems utilizing a UV scanner shall prove pilot and interrupt ignition spark prior to main burner valves being energized.

(e) Two controls, one operating and one high limit, activated by temperature or pressure.

(f) Burners relying on mechanical means to provide air for combustion shall have actual proof-of-air interlock device.

(g) Power burners must include proven prepurge of not less than 60 seconds at high-fire damper settings. This prepurge shall occur before every burner cycle, regardless of reason.

(h) Installations with dampered combustion air openings shall prove damper open position before trial for burner ignition.

(i) Vent dampers and flue dampers shall be interlocked to prevent burner ignition unless safely open.

(j) One high oil or liquid fuel-pressure interlock; reset from flame safeguard or manually.

(k) A manual firing cock.

(l) Where hot water or steam, two low water cut-offs.

(m) An atomizing medium proving switch.

(n) A low oil temperature switch for oil or liquid fuel requiring preheating.

(o) A high oil temperature interlock for oil or liquid fuel requiring pre-heating.

(p) A separate firing rate control valve.

(q) The burner oil pump shall automatically not operate or rotate while the alternate fuel is firing.

(r) A pressure-relief valve shall be provided between safety shutoff valves and between pump and safety valves where an integral valve is used with a pump.

(s) A separate relief device is required on each transfer pump.

(t) One low oil or liquid fuel-pressure interlock, reset from flame safeguard or manually.

(u) Burners with automatic controls, prepurge, proof-of-closure, modulation, or postpurge shall not use relays external to the flame safeguard to accomplish these functions.

(6) Shutoff Valve.

(a) Oil or liquid fuel burner installations shall include a non-electric shutoff valve that is held open by a fusible link designed to close at 165°F (74°C), installed near the burner in the same room as the burner. This shall prevent the flow of oil or liquid fuel to the burner through the supply pipe. A check valve is required in the return line if the tank is higher than the burner.

CALIFORNIA MECHANICAL CODE – MATRIX ADOPTION TABLE

APPENDIX D – FUEL SUPPLY: MANUFACTURED/MOBILE HOME PARKS AND RECREATIONAL VEHICLE PARKS

(Matrix Adoption Tables are non-regulatory, intended only as an aid to the code user. See Chapter 1 for state agency authority and building applications.)

Adopting Agency	BSC	BSC-CG	SFM	HCD			DSA			OSHPD						BSCC	DPH	AGR	DWR	CEC	CA	SL	SLC
				1	2	1-AC	AC	SS	SS/CC	1	1R	2	3	4	5								
Adopt Entire Chapter								X	X														
Adopt Entire Chapter as amended (amended sections listed below)																							
Adopt only those sections that are listed below																							
Chapter/Section																							

This state agency does not adopt sections identified with the following symbol: †

The Office of the State Fire Marshal's adoption of this chapter or individual sections is applicable to structures regulated by other state agencies pursuant to Section 1.11.0.

APPENDIX D

FUEL SUPPLY: MANUFACTURED/MOBILE HOME PARKS AND RECREATIONAL VEHICLE PARKS

The provisions contained in this appendix are not mandatory unless specifically adopted by a state agency, or referenced in the adopting ordinance.

D 101.0 Fuel Gas Piping Systems.

D 101.1 General. Fuel gas piping systems serving manufactured homes, accessory buildings, or structures and communities shall be designed and constructed in accordance with the applicable provisions of NFPA 54 and NFPA 58. NFPA 31 shall apply to oil fuel-burning systems and shall comply with the criteria of the Authority Having Jurisdiction. [NFPA 501A:4.1.1]

D 101.2 Gas Supply Connections. Gas supply connections at sites, where provided from an underground gas supply piping system, shall be located and arranged to permit attachment to a manufactured home (M/H) occupying the site. For the installation of liquefied petroleum gas (LPG) storage systems, the applicable provisions of NFPA 58 shall be followed. [NFPA 501A:4.1.2]

D 101.3 Location of Gas Supply Connection. The gas supply to the M/H shall be located within 4 feet (1219 mm) of the M/H stand.

Exception: Gas supply connections for manufactured homes located on all-weather wood, concrete, concrete block foundation systems or on foundations constructed in accordance with the local building code or, in the absence of a local code, with a recognized model building code. [NFPA 501A:4.1.3]

D 101.4 Recreational Vehicle Park Fuel-Gas Equipment and Installations. Fuel gas equipment and installations shall comply with this appendix, except as otherwise permitted or required by this code.

D 102.0 Single and Multiple Manufactured Home Site Fuel Supply Systems.

D 102.1 Underground Installation. Underground gas piping system installations shall comply with any applicable building code and Section D 102.1.1 and Section D 102.1.2. [NFPA 501A:4.2.1]

D 102.1.1 Open-Ended Gastight Conduit. Underground gas piping shall not be installed beneath that portion of a M/H site reserved for the location of a manufactured home or M/H accessory building or structure unless installed in the open-ended gastight conduit of Section D 102.1.2. [NFPA 501A:4.2.1.1]

D 102.1.2 Requirements. The open-ended gastight conduit shall comply with the following:

(1) The conduit shall be not less than Schedule 40 pipe that is approved for underground installation beneath buildings.

(2) The interior diameter of the conduit shall be not less than ½ of an inch (15 mm) larger than the outside diameter of the gas piping.

(3) The conduit shall extend to a point not less than 4 inches (102 mm) beyond the outside wall of the M/H, accessory building, or structure, and the outer ends shall not be sealed.

(4) Where the conduit terminates within a M/H, accessory building, or structure, it shall be accessible, and the space between the conduit and the gas piping shall be sealed to prevent leakage of gas into the building. [NFPA 501A:4.2.1.2]

D 103.0 Manufactured Home Site Gas Shutoff Valve.

D 103.1 General. Each M/H site shall have a listed gas shutoff valve installed upstream of the M/H site gas outlet. The gas shutoff valve shall be located on the outlet riser at a height of not less than 6 inches (152 mm) above grade. A gas shutoff valve shall not be located under a M/H. The outlet shall be equipped with a cap or plug to prevent discharge of gas where the M/H site outlet is not connected to a M/H.

Exception: Gas shutoff valves for manufactured homes located on foundations constructed in accordance with the local building code or, in the absence of a local code, with a recognized model building code. [NFPA 501A:4.2.2]

D 104.0 Gas Meters.

D 104.1 Support of Meters. Where installed, gas meters shall be supported by a post or bracket placed on a firm footing or other means providing equivalent support and shall not depend on the gas outlet riser for support. [NFPA 501A:4.2.3.1]

D 104.2 Location of Meters. Each gas meter shall be installed in an accessible location and shall be provided with unions or other fittings so that the meter is removed easily and replaced in an upright position. Meters shall not be installed in unventilated or inaccessible locations or closer than 3 feet (914 mm) to sources of ignition. [NFPA 501A:4.2.3.2]

D 104.3 Meter Shutoff Valve or Cock. Gas meter installations shall be provided with shutoff valves or cocks located adjacent to and on the inlet side of the meters. In the case of a single meter installation utilizing an LP-Gas container, the container service valve shall be permitted to be used in lieu of the shutoff valve or cock. Gas meter installations shall be provided with test tees located adjacent to and on the outlet side of the meters. [NFPA 501A:4.2.4]

D 105.0 Cathodic Protection Requirements.

D 105.1 General. Cathodic protection shall be installed for corrosion control of buried or submerged metallic gas piping in accordance with the following requirements:

(1) Where amphoteric metals are included in a buried or submerged pipeline containing a metal of different anodic potential the following protection shall be provided:

(a) The buried or submerged pipeline shall be cathodically protected at a negative (cathodic) voltage of 0.85 volt, measured between the structure surface and a saturated copper-copper sulfate half cell contacting the electrolyte.

(b) The amphoteric metals shall be electrically isolated from the remainder of the pipeline with insulating flanges, or equivalent, and cathodically protected.

(2) The amount of cathodic protection shall be such that the protective coating and the pipe are not damaged.

D 106.0 Manufactured Home Community LPG Supply Systems.

D 106.1 General. Where 10 or more customers are served by one LPG supply system, the installation of the gas supply system shall be in accordance with 49 CFR 192. Other types of liquefied petroleum gas supply systems and the storage and handling of LPG shall be in accordance with NFPA 58 (see Section D 113.0). [NFPA 501A:4.3.2]

D 107.0 Required Gas Supply.

D 107.1 General. The minimum hourly volume of gas required at each M/H site outlet or a section of the M/H community gas piping system shall be calculated as shown in Table D 107.1. [NFPA 501A:4.3.4.1]

TABLE D 107.1
DEMAND FACTORS FOR USE IN CALCULATING GAS PIPING SYSTEMS IN M/H COMMUNITIES*
[NFPA 501A: TABLE 4.3.4.1]

NUMBER OF M/H SITES	BRITISH THERMAL UNITS PER HOUR PER M/H SITE
1	125 000
2	117 000
3	104 000
4	96 000
5	92 000
6	87 000
7	83 000
8	81 000
9	79 000
10	77 000
11–20	66 000
21–30	62 000
31–40	58 000
41–60	55 000
Over 60	50 000

For SI units: 1000 British thermal units per hour = 0.293 kW
* In extreme climate areas, additional capacities shall be considered.

D 108.0 Gas Pipe Sizing and Pressure.

D 108.1 Size. The size of each section of a gas piping system shall be determined in accordance with NFPA 54, or by other standard engineering methods acceptable to the Authority Having Jurisdiction. [NFPA 501A:4.3.5.1]

D 108.2 Pressure. Where connected appliances are operated at their rated capacity, the gas supply pressure shall be not less than 7 inches of water column (1.7 kPa). The gas supply pressure shall not exceed 14 inches of water column (3.5 kPa). [NFPA 501A:4.3.5.2]

D 109.0 Gas Piping Materials.

D 109.1 Metal. Metal gas pipe shall be standard-weight wrought iron or steel (galvanized or black), yellow brass containing not more than 75 percent copper, or internally tinned or treated copper of iron pipe size. Galvanizing shall not be considered protection against corrosion.

Seamless copper or steel tubing shall be permitted to be used with gases not corrosive to such material. Steel tubing shall comply with ASTM A254. Copper tubing shall comply with ASTM B88 (Type K or Type L) or ASTM B280. Copper tubing (unless tin-lined) shall not be used if the gas contains more than an average of 0.3 grains of hydrogen sulfide per 100 standard cubic feet (0.7 mg/100 L) of gas. [NFPA 501A:4.3.6.1 – 4.3.6.1.6]

D 109.2 Protection Coatings for Metal Gas Piping. Buried or submerged metallic gas piping shall be protected from corrosion by approved coatings or wrapping materials. Gas pipe protective coatings shall be approved types, machine applied, and shall comply with recognized standards. Field wrapping shall provide equivalent protection and is restricted to those short sections and fittings that are stripped for threading or welding. Risers shall be coated or wrapped to a point not less than 6 inches (152 mm) aboveground. [NFPA 501A:4.3.6.2]

D 109.3 Plastic. Plastic piping shall only be used underground and shall meet the requirements of ASTM D2513 or ASTM D2517, as well as the design pressure and design limitations of 49 CFR (Section 192.123), and shall otherwise conform to the installation requirements thereof. [NFPA 501A:4.3.6.3]

D 110.0 Gas Piping Installations.

D 110.1 Minimum Burial Below Ground Level and Clearances. Gas piping installed belowground level shall have an earth cover of not less than 18 inches (457 mm) and shall be installed with not less than 12 inches (305 mm) of clearance from other underground utility systems. [NFPA 501A:4.3.7.1]

D 110.2 Metallic Gas Piping. Metallic gas piping systems shall be installed in accordance with approved construction documents, including provisions for cathodic protection. Each cathodic protection system shall be designed and installed in accordance with the provisions of 49 CFR 192.

D 110.2.1 Cathodic Protection. Where the cathodic protection system is designed to protect the gas piping system, the gas piping system shall be electrically isolated from other underground metallic systems or installations. Where the gas piping system is cathodically protected against corrosion, a dielectric fitting shall be used in the M/H gas connection to insulate the M/H from the underground gas piping system. [NFPA 501A:4.3.7.2.3, 4.3.7.2.4]

D 110.2.2 Underground Metallic Systems. Where a cathodic protection system is designed to provide underground metallic systems and installations with protection against corrosion, such systems and installations shall be electrically bonded together and protected as a whole. [NFPA 501A:4.3.7.2.5]

D 110.3 Plastic Gas Piping. Plastic gas piping shall be used underground and shall be installed with an electrically conductive wire for locating the pipe. The wire used to locate the plastic pipe shall be copper, not less than No. 18 AWG, with insulation approved for direct burial. Portions of a plastic gas piping system consisting of metallic pipe shall be cathodically protected against corrosion. [NFPA 501A:4.3.7.3]

D 110.4 Gas Piping System Shutoff Valve. An accessible and identifiable shutoff valve controlling the flow of gas to the entire M/H community gas piping system shall be installed in a location approved by the Authority Having Jurisdiction and near the point of connection to the service piping or to the supply connection of an LPG container. [NFPA 501A:4.3.7.4]

D 111.0 Liquefied Petroleum Gas Appliances.

D 111.1 General. LP-Gas equipment shall be installed in accordance with the applicable provisions of NFPA 58. [NFPA 501A:4.3.8]

D 112.0 Oil Supply.

D 112.1 General. The following three methods of supplying oil to an individual manufactured home site shall be permitted:

(1) Supply from an outside underground tank (see Section D 113.6).

(2) Supply from a centralized oil distribution system designed and installed in accordance with accepted engineering practices and in compliance with NFPA 31.

(3) Supply from an outside aboveground tank (see Section D 113.6). [NFPA 501A:4.3.9]

D 112.2 Minimum Oil Supply Tank Size. Oil supply tanks shall have a minimum capacity equal to 20 percent of the average annual oil consumption. [NFPA 501A:4.3.10]

D 112.3 Oil Supply Connections. Oil supply connections at manufactured home sites, where provided from a centralized oil distribution system, shall be located and arranged to permit attachment to a manufactured home utilizing the stand. [NFPA 501A:4.3.11.1] The installation of such facilities shall comply with the following requirements:

(1) The main distribution pipeline shall be permitted to be connected to a tank or tanks having an aggregate capacity not exceeding 20 000 gallons (75 708 L) at a point below the liquid level.

(2) Where this piping is so connected, a readily accessible internal or external shutoff valve shall be installed in the piping as close as practicable to the tank.

(3) If external and aboveground, the shutoff valve and its tank connections shall be made of steel.

(4) Connections between the tank(s) and the main pipeline shall be made with double swing joints or flexible connectors, or shall otherwise be arranged to permit the tank(s) to settle without damaging the system.

(5) If located aboveground, the connections specified in Section D 112.3(4) shall be located within the diked area.

(6) A readily accessible and identified manual shutoff valve shall be installed either inside or outside of the structure in each branch supply pipeline that enters a building, mobile home, travel trailer, or other structure. If outside, the valve shall be protected from weather and damage. If inside, the valve shall be located directly adjacent to the point at which the supply line enters the structure.

(7) A device shall be provided in the supply line at or ahead of the point where it enters the interior of the structure that will automatically shut off the oil supply, if the supply line between this device and the appliance is broken. This device shall be located on the appliance side of the manual shutoff valve required in Section D 112.3(6) and shall be solidly supported and protected from damage.

(8) Means shall be provided to limit the oil pressure at the appliance inlet to a maximum gauge pressure of 3 pound-force per square inch gauge (psig) (21 kPa). If a pressure-reducing valve is used, it shall be a type approved for the service.

(9) A device shall be provided that will automatically shut off the oil supply to the appliance if the oil pressure at the appliance inlet exceeds a gauge pressure of 8 psig (55 kPa). The device shall not be required under either of the following conditions:

   (a) Where the distribution system is supplied from a gravity tank and the maximum hydrostatic head of oil in the tank is such that the oil pressure at the appliance inlet will not exceed a gauge pressure of 8 psig (55 kPa).

   (b) Where a means is provided to automatically shut off the oil supply if the pressure-regulating device provided in accordance with Section D 112.3(8) fails to regulate the pressure as required.

(10) Only appliances equipped with primary safety controls specifically listed for the appliance shall be connected to a centralized oil distribution system. [NFPA 31:9.2.10–9.2.15]

D 113.0 Fuel Supply Systems Installation.

D 113.1 Flexible Gas Connector. Except for manufactured homes located on an all-weather wood, concrete, or concrete block foundation system or on a foundation constructed in accordance with the local building code or, in the absence of a local code, with a recognized model building code, each gas supply connector shall be listed for outside manufactured home use, shall be not more than 6 feet (1829 mm) in length, and shall have a capacity rating to supply the connected load. [NFPA 501A:4.4.1]

D 113.2 Use of Approved Pipe and Fittings of Extension. Where it is necessary to extend the M/H inlet to permit connection of the 6 foot (1829 mm) listed connector to the site gas outlet, the extension shall be of approved materials of the same size as the M/H inlet and shall be supported at not more than 4 foot (1219 mm) intervals to the M/H. [NFPA 501A:4.4.2]

D 113.3 Mechanical Protection. Gas outlet risers, regulators, meters, valves, or other exposed equipment shall be protected against accidental damage. [NFPA 501A:4.4.3]

D 113.4 Special Rules on Atmospherically Controlled Regulators. Atmospherically controlled regulators shall be installed in such a manner that moisture cannot enter the regulator vent and accumulate above the diaphragm. Where the regulator vent is obstructed due to snow and icing conditions; shields, hoods, or other approved devices shall be provided to guard against closing of the vent opening. [NFPA 501A:4.4.4]

D 113.5 Fuel Gas Piping Test. The M/H fuel gas piping system shall be tested with air before it is connected to the gas supply. The M/H gas piping system shall be subjected to a pressure test with appliance shutoff valves in their closed positions. [NFPA 501A:4.4.5]

D 113.5.1 Procedures. The fuel gas piping test shall consist of air pressure of not less than 10 inches water column or more than 14 inches water column (2.5 kPa to 3.5 kPa). The fuel gas piping system shall be isolated from the air pressure source and shall maintain this pressure for not less than 10 minutes without perceptible leakage. Upon satisfactory completion of the fuel gas piping test, the appliance valves shall be opened, and the gas appliance connectors shall be tested with soapy water or bubble solution while under the pressure remaining in the piping system. Solutions used for testing for leakage shall not contain corrosive chemicals. Pressure shall be measured with either a manometer, slope gauge, or gauge that is calibrated in either water inch (mm) or psi (kPa), with increments of either $\frac{1}{10}$ of an inch (2.5 mm) or $\frac{1}{10}$ psi (0.7 kPa gauge), as applicable. Upon satisfactory completion of the fuel gas piping test, the M/H gas supply connector shall be installed, and the connections shall be tested with soapy water or bubble solution. [NFPA 501A:4.4.5.1]

D 113.5.2 Warning. The following warning shall be supplied to the installer:

WARNING

Do not overpressurize the fuel gas piping system. Damage to valves, regulators, and appliances is capable of occurring due to pressurization beyond the maximums specified. [NFPA 501A:4.4.5.2]

D 113.5.3 Vents. Gas appliance vents shall be visually inspected to ensure that they have not been dislodged in transit and are connected securely to the appliance. [NFPA 501A:4.4.5.3]

D 113.6 Oil Tanks. Not more than one 660 gallon (2498 L) tank or two tanks with aggregate capacity of 660 gallons (2498 L) or less shall be connected to one oil-burning appliance. Two supply tanks, where used, shall be cross-connected and provided with a single fill and single vent in accordance with NFPA 31, and shall be on a common slab and rigidly secured one to the other. Tanks having a capacity of 660 gallons (2498 L) or less shall be securely supported by rigid, noncombustible supports to prevent settling, sliding, or lifting. [NFPA 501A:4.4.6]

D 113.6.1 Installation. Oil supply tanks shall be installed in accordance with the applicable provisions of NFPA 31. [NFPA 501A:4.4.6.1]

D 113.6.2 Capacity. A tank with a capacity not larger than 60 gallons (227 L) shall be permitted to be a DOT-5 shipping container (drum) and so marked, or a tank constructed in accordance with the provisions of UL 80. Tanks other than DOT-5 shipping containers having a capacity of not more than 660 gallons (2498 L) shall be constructed in accordance with the provisions of UL 80. Pressure tanks shall be constructed in accordance with Section VIII of the ASME Boiler and Pressure Vessel Code. [NFPA 501A:4.4.6.2]

D 113.6.3 Location. Tanks, as described in Section D 113.6 and Section D 113.6.2, that are adjacent to buildings shall be located not less than 10 feet (3048 mm) from a property line that is permitted to be built upon. [NFPA 501A:4.4.6.3]

D 113.6.4 Vent. Tanks with a capacity not larger than 660 gallons (2498 L) shall be equipped with an open vent not smaller than $1\frac{1}{2}$ inch (40 mm) iron pipe size; tanks with a 500 gallon (1892 L) or less capacity shall have a vent of $1\frac{1}{4}$ inch (32 mm) iron pipe size. [NFPA 501A:4.4.6.4]

D 113.6.5 Liquid Level. Tanks shall be provided with a means of determining the liquid level. [NFPA 501A:4.4.6.5]

D 113.6.6 Fill Opening. The fill opening shall be a size and in a location that permits filling without spillage. [NFPA 501A:4.4.6.6]

D 114.0 Manufactured Home Accessory Building Fuel Supply Systems.

D 114.1 General. Fuel gas supply systems installed in a M/H accessory building or structure shall be in accordance with the applicable provisions of NFPA 54 and NFPA 58. Fuel oil supply systems shall comply with the applicable provisions of NFPA 31. [NFPA 501A:4.5]

D 115.0 Community Building Fuel Supply Systems in Manufactured Home Communities.

D 115.1 Fuel Gas Piping and Equipment Installations. Fuel gas piping and equipment installed within a permanent building in a M/H community shall be in accordance with nationally recognized appliance and fuel gas piping codes and standards adopted by the Authority Having Jurisdiction. Where the state or other political subdivision does not assume jurisdiction, such fuel gas piping and equipment installations shall be designed and installed in accordance with the applicable provisions of NFPA 54 or NFPA 58. [NFPA 501A:4.6.1]

D 115.2 Oil Supply Systems in M/H Communities. Oil-burning equipment and installations within a M/H community shall be designed and constructed in accordance with the applicable codes and standards adopted by the Authority Having Jurisdiction. Where the state or other political subdivision does not assume jurisdiction, such installations shall be designed and constructed in accordance with the applicable provisions of NFPA 31. [NFPA 501A:4.6.2]

D 115.3 Oil-Burning Equipment and Installation. Oil-burning equipment and installations within a building constructed in a M/H community in accordance with the local building code or a nationally recognized building code shall be in accordance with nationally recognized codes and standards adopted by the Authority Having Jurisdiction. Where the state or other political subdivision does not assume jurisdiction, such oil-burning equipment and installation shall be designed and installed in accordance with NFPA 31. [NFPA 501A:4.6.3]

D 115.4 Inspection and Tests. Inspections and tests for fuel gas piping shall be made in accordance with Chapter 1 and Chapter 13 of this code.

CALIFORNIA MECHANICAL CODE – MATRIX ADOPTION TABLE
APPENDIX E – SUSTAINABLE PRACTICES

(Matrix Adoption Tables are non-regulatory, intended only as an aid to the code user. See Chapter 1 for state agency authority and building applications.)

Adopting Agency	BSC	BSC-CG	SFM	HCD			DSA			OSHPD						BSCC	DPH	AGR	DWR	CEC	CA	SL	SLC
				1	2	1-AC	AC	SS	SS/CC	1	1R	2	3	4	5								
Adopt Entire Chapter																							
Adopt Entire Chapter as amended (amended sections listed below)																							
Adopt only those sections that are listed below																							
Chapter/Section																							

This state agency does not adopt sections identified with the following symbol: †

The Office of the State Fire Marshal's adoption of this chapter or individual sections is applicable to structures regulated by other state agencies pursuant to Section 1.11.0.

APPENDIX E

SUSTAINABLE PRACTICES

The provisions contained in this appendix are not mandatory unless specifically adopted by a state agency, or referenced in the adopting ordinance.

E 101.0 General.

E 101.1 Applicability. The purpose of this appendix is to provide a comprehensive set of technically sound provisions that encourage sustainable practices and works towards enhancing the design and construction of mechanical systems that result in a positive long-term environmental impact. This appendix is not intended to circumvent the health, safety, and general welfare requirements of this code.

E 101.2 Definition of Terms. For the purposes of this code, the definitions shall apply to this appendix.

No attempt is made to define ordinary words, which are used in accordance with their established dictionary meanings, except where a word has been used loosely, and it is necessary to define its meaning as used in this appendix to avoid misunderstanding.

The definitions of terms are arranged alphabetically according to the first word of the term.

E 201.0 Definitions.

E 201.1 Cycles of Concentration for Cooling Towers. Cycles of concentration equals the specific conductance of the water in the cooling tower basin divided by the combined flow-weighted average specific conductance of the makeup water(s) to the cooling tower.

E 201.2 Duct Wall Penetrations. Includes pipe, tubing, rods, and wire. Screws and other fasteners are not considered to be ductwork penetrations.

E 201.3 Energy Star. A joint program of the U.S. Environmental Protection Agency and the U.S. Department of Energy. Energy Star is a voluntary program designed to identify and promote energy-efficient products and practices.

E 201.4 Geothermal. Renewable energy generated by deep-earth.

E 201.5 Heating Seasonal Performance Factor (HSPF). The total heating output of a heat pump during its normal annual usage period for heating in British thermal units (Btu) (kW•h) divided by the total electric energy input during the same period. [ASHRAE 90.1:3.2]

E 201.6 Integrated Energy Efficiency Ratio (IEER). A single-number figure of merit expressing cooling part-load EER efficiency for commercial unitary air-conditioning and heat pump equipment on the basis of weighted operation at various load capacities for the equipment. [ASHRAE 90.1:3.2]

E 201.7 Integrated Part-Load Value (IPLV). A single-number figure of merit based on part-load EER, COP, or kW/ton expressing part-load efficiency for air-conditioning and heat pump equipment on the basis of weighted operation at various load capacities for the equipment. [ASHRAE 90.1:3.2]

E 201.8 Joint, Transverse. Connections of two duct sections oriented perpendicular to airflow.

E 201.9 Maintenance. The upkeep of property or equipment by the owner of the property in accordance with the requirements of this appendix.

E 201.10 Minimum Efficiency Reporting Value (MERV). Filter minimum efficiency reporting value, in accordance with ASHRAE 52.2.

E 201.11 Multi-Occupant Spaces. Indoor spaces used for presentations and training, including classrooms and conference rooms.

E 201.12 Recirculation System. A system of hot water supply and return piping with shutoff valves, balancing valves, circulating pumps, and a method of controlling the circulating system.

E 201.13 Seam, Longitudinal. Joints oriented in the direction of airflow.

E 201.14 Seasonal Energy Efficiency Ratio (SEER). The total cooling output of an air conditioner during its normal annual usage period for cooling in Btu (kW•h) divided by the total electric energy input during the same period in Btu (kW•h). [ASHRAE 90.1:3.2]

E 301.0 General Regulations.

E 301.1 Installation. Mechanical systems covered by this appendix shall be installed in accordance with this code, other applicable codes, and the manufacturer's installation and operating instructions.

E 301.2 Qualifications. Where permits are required, the Authority Having Jurisdiction shall have the authority to require contractors, installers, or service technicians to demonstrate competency. Where determined by the Authority Having Jurisdiction, the contractor, installer or service technician shall be licensed to perform such work.

E 302.0 Disposal of Liquid Waste.

E 302.1 Disposal. It shall be unlawful for a person to cause, suffer, or permit the disposal of liquid wastes, in a place or manner, except through and by means of an approved drainage system, installed and maintained in accordance with the provisions of the plumbing code.

E 302.2 Connections to Plumbing System Required. Equipment and appliances, used to receive or discharge liquid wastes or sewage, shall be connected to the drainage system of the building or premises in accordance with the requirements of the plumbing code and this appendix.

E 303.0 Abandonment.

E 303.1 General. An abandoned system or part thereof covered under the scope of this appendix shall be disconnected from remaining systems, drained, plugged, and capped in an approved manner.

E 401.0 Water Conservation and Efficiency.

E 401.1 General. The provisions of this section establish the means of conserving potable and nonpotable water used in and around a building.

E 402.0 Meters.

E 402.1 Required. A water meter shall be required for buildings connected to a public water system, including municipally supplied reclaimed (recycled) water. In other than single-family houses, multi-family structures not exceeding three stories above grade, and modular houses, a separate meter or submeter shall be installed in the following locations:

(1) The makeup water supply to cooling towers, evaporative condensers, and fluid coolers.

(2) The makeup water supply to one or more boilers collectively exceeding 1 000 000 British thermal units per hour (Btu/h) (293 kW).

(3) The water supply to a water-using process where the consumption exceeds 1000 gallons per day (gal/d) (0.0438 L/s), except for manufacturing processes.

(4) The makeup water supply to an evaporative cooler having an air flow exceeding 30 000 cubic feet per minute (ft^3/min) (14.1584 m^3/s).

E 402.2 Consumption Data. A means of communicating water consumption data from submeters to the water consumer shall be provided.

E 402.3 Access. Meters and submeters shall be accessible.

E 403.0 HVAC Water Use.

E 403.1 Once-Through Cooling. Once-through cooling using potable water is prohibited.

E 403.2 Cooling Towers and Evaporative Coolers. Cooling towers and evaporative coolers shall be equipped with makeup water and blow down meters, conductivity controllers, and overflow alarms. Cooling towers shall be equipped with efficiency drift eliminators that achieve drift reduction to 0.002 percent of the circulated water volume for counterflow towers and 0.005 percent for cross-flow towers.

E 403.3 Cooling Tower Makeup Water. Not less than 5 cycles of concentration is required for air-conditioning cooling tower makeup water having a total hardness of less than 11 grains per gallon (gr/gal) (188 mg/L) expressed as calcium carbonate. Not less than 3.5 cycles of concentration is required for air-conditioning cooling tower makeup water having a total hardness equal to or exceeding 11 gr/gal (188 mg/L) expressed as calcium carbonate.

Exception: Air-conditioning cooling tower makeup water having discharge conductivity range not less than 7 gr/gal (120 mg/L) to 9 gr/gal (154 mg/L) of silica measured as silicon dioxide.

E 403.4 Evaporative Cooler Water Use. Evaporative cooling systems shall use 3.5 gallons (13.2 L) or less of water per ton-hour (kW•h) of cooling where system controls are set to maximum water use. Water use expressed in maximum water use per ton-hour (kW•h) of cooling, shall be marked on the device and included in the product user manual, product information literature, and manufacturer's installation instructions. Water use information shall be readily available at the time of code compliance inspection.

E 403.4.1 Overflow Alarm. Cooling systems shall be equipped with an overflow alarm to alert building owners, tenants, or maintenance personnel where the water refill valve continues to allow water to flow into the reservoir where the reservoir is full. The alarm shall have a sound pressure level rating of not less than 85 dBa measured at a distance of 10 feet (3048 mm).

E 403.4.2 Automatic Pump Shut-Off. Cooling systems shall automatically cease pumping water to the evaporation pads where airflow across evaporation pads ceases.

E 403.4.3 Cooler Reservoir Discharge. A water quality management system (either timer or water quality sensor) shall be provided. Where timers are used, the time interval between discharge of reservoir water shall be set to 6 or more hours of cooler operation. Where water quality sensors are used, the discharge of reservoir water shall be set for 800 ppm or more of total dissolved solids (TDS). Continuous discharge or continuous bleed systems shall not be installed.

E 403.4.4 Discharge Water Reuse. Discharge water shall be reused where applications exist on site. Where a nonpotable water source system exists on site, evaporative cooler discharge water shall be collected and discharged to the collection system.

Exception: Where the reservoir water affects the quality of the nonpotable water supply making the nonpotable water unusable for its intended purposes.

E 403.4.5 Discharge Water to Drain. Where discharge water is not recovered for reuse, the sump overflow line shall not be directly connected to a drain. Where the discharge water is discharged into a sanitary drain, an air gap of not less than 6 inches (152 mm) shall be provided between the termination of the discharge line and the drain opening. The discharge line shall terminate in a location that is visible to the building owner, tenants, or maintenance personnel.

E 403.5 Use of Reclaimed (Recycled) and On-Site Treated Nonpotable Water for Cooling. Where approved for use by the water or wastewater utility and the Authority Having Jurisdiction, reclaimed (recycled), or on-site treated nonpotable water shall be permitted to be used for industrial and commercial cooling or air-conditioning.

E 403.5.1 Drift Eliminator. A drift eliminator shall be utilized in a cooling system, utilizing alternate sources of water, where the aerosolized water is capable of coming in contact with employees or members of the public.

E 403.5.2 Disinfection. A biocide shall be used to treat the cooling system recirculation water where the recycled water is capable of coming in contact with employees or members of the public.

E 501.0 Heating, Ventilation, and Air-Conditioning Systems and Equipment – Energy Efficiency.

E 501.1 Scope. The provisions of this section shall establish the means of enhancing energy efficiency associated with mechanical systems in a building.

E 502.0 Heating, Ventilation, and Air-Conditioning Low-Rise Residential Buildings.

E 502.1 General. The heating, ventilating, air-conditioning, for single-family houses, multi-family structures not exceeding three stories above grade, and modular houses shall be in accordance with Section E 502.2 through Section E 502.12. The heating, ventilation, and air-conditioning system of other buildings shall be in accordance with Section E 503.0.

E 502.2 Heating, Ventilating, and Air-Conditioning Systems and Equipment. This section shall regulate only equipment using single-phase electric power, air conditioners, and heat pumps with rated cooling capacities less than 65 000 British thermal units per hour (Btu/h) (19 kW), warm air furnaces with rated heating capacities less than 225 000 Btu/h (66 kW), boilers less than 300 000 Btu/h (88 kW) input, and heating-only heat pumps with rated heating capacities less than 65 000 Btu/h (19 kW). [ASHRAE 90.2:6.2]

E 502.2.1 Non-Residential Type Systems and Equipment. Heating, ventilating, and air-conditioning systems and equipment that do not fall under the requirements of Section E 502.0 shall be in accordance with the applicable requirements of Section E 503.0.

E 502.3 Balancing. The air distribution system design, including outlet grilles, shall provide a means for balancing the air distribution system unless the design procedure provides a system intended to operate within plus or minus 10 percent of design air quantities. [ASHRAE 90.2:6.3]

E 502.3.1 Balancing Dampers. Balancing dampers shall be installed in branch ducts, and the axis of the damper shall be installed parallel to the direction of airflow in the main duct.

E 502.4 Ducts. Ducts shall be sized, installed, and tested in accordance with Section E 502.4.1 through Section E 502.4.4.

E 502.4.1 Insulation for Ducts. Portions of the air distribution system installed in or on buildings for heating and cooling shall be R-8. Where the mean outdoor dew-point temperature in a month exceeds 60°F (16°C), vapor retarders shall be installed on conditioned-air supply ducts. Vapor retarders shall have a water vapor permeance not exceeding 0.5 perm [2.87 E-11 kg/(Pa•s•m^2)] where tested in accordance with Procedure A in ASTM E96.

Insulation shall not be required where the ducts are within the conditioned space. [ASHRAE 90.2:6.4]

E 502.4.2 Ducts and Register Penetrations. Joints, seams, and penetrations of duct systems shall be made airtight by means of mastics, gasketing, or other means in accordance with this code. Register penetrations shall be sealed to the wall or floor assemblies. Where HVAC duct penetrates a conditioned space, the duct penetration shall be sealed to the wall or floor assembly to prevent leakage into an unconditioned space.

E 502.4.3 Duct Leakage Test. For systems with a duct or air handler outside of the conditioned space, a duct leakage test shall be performed in accordance with Section E 502.4.3.1.

E 502.4.3.1 Duct Leakage Verification Test. Ductwork shall be tested to the maximum permitted leakage in 1 cubic foot per minute (ft^3/min) per 100 square feet [0.0001 (m^3/s)/m^2] of duct surface area in accordance with SMACNA HVAC Air Duct Leakage Test Manual. Register penetrations shall be sealed during the test. The test shall be conducted with a pressure differential of 0.1 inch water gauge (0.02 kPa) across the tested system.

E 502.4.4 Duct Sizing. Duct systems shall be sized in accordance with ACCA Manual D or other methods approved by the Authority Having Jurisdiction with the velocity in the main duct not to exceed 1000 feet per minute (ft/min) (5.08 m/s) and the velocity in the secondary branch duct not to exceed 600 ft/min (3.048 m/s).

E 502.5 Insulation for Piping. HVAC system piping installed to serve buildings and within buildings shall be thermally insulated in accordance with Table E 502.5. [ASHRAE 90.2:6.5]

E 502.6 Ventilation and Combustion Air. The building shall be designed to have the capability to provide the ventilation air specified in Table E 502.6. Mechanical ventilation shall be calculated in accordance with Equation E 502.6. [ASHRAE 90.2:6.6.1]

(Equation E 502.6)

$$\textit{Mechanical Ventilation} = [(0.35 - \textit{Summer}) \times \textit{Volume}] / 60$$

Where:

Mechanical Ventilation = required mechanical ventilation rate to supplement summer infiltration, cfm (m^3/s)

Summer = summer design infiltration rate, ACH

Volume = volume of conditioned space, ft^3 (m^3)

E 502.6.1 Combustion Air. Combustion air for fossil fuel heating equipment shall comply with this code or with one of the following:

(1) Natural gas and propane heating equipment, NFPA 54

(2) Oil heating equipment, NFPA 31

(3) Solid fuel burning equipment, NFPA 211 [ASHRAE 90.2:6.6.2]

E 502.7 Electric Heating Systems. Electric heating systems shall be installed in accordance with the following requirements. [ASHRAE 90.2:6.7]

E 502.7.1 Wall, Floor, or Ceiling Electric-Resistance Heating. Where wall, floor, or ceiling electric-resistance heating units are used, the structure shall be zoned and heaters installed in each zone in accordance with the heat loss of that zone. Where living and sleeping zones are separate, the number of zones shall be not less than two. Where two or more heaters are installed in one room, they shall be controlled by one thermostat. [ASHRAE 90.2:6.7.1]

E 502.7.2 Electric Central Warm Air Heating. Where electric central warm air heating is to be installed, an electric heat pump or an off-peak electric heating system with thermal storage shall be used.

Exceptions:

(1) Electric resistance furnaces where the ducts are located inside the conditioned space, and not less than two zones are provided where the living and sleeping zones are separate.

(2) Packaged air-conditioning units with supplemental electric heat. [ASHRAE 90.2:6.7.2]

E 502.8 Bath Ceiling Units. Bath ceiling units providing a combination of heat, light, or ventilation shall be provided with controls permitting separate operation of the heating function. [ASHRAE 90.2:6.8]

TABLE E 502.5
MINIMUM PIPE INSULATION THICKNESS[1, 5]
[ASHRAE 90.2: TABLE 6.5]

	INSULATION CONDUCTIVITY		NOMINAL PIPE DIAMETER (inches)				
FLUID DESIGN OPERATING TEMPERATURE RANGE (°F)	Btu•inch/(h•ft^2•°F)	MEAN RATING TEMPERATURE(°F)	<1	1 TO 1¼	1½ TO 3½	4 TO 6	EQUAL TO OR GREATER THAN 8
HEATING SYSTEMS (STEAM, STEAM CONDENSATE, AND HOT WATER)[2, 3]							
201–250	0.27–0.30	150	1.5	1.5	2.0	2.0	2.0
141–200	0.25–0.29	125	1.0	1.0	1.0	1.5	1.5
105–140	0.22–0.28	100	0.5	0.5	1.0	1.0	1.0
COOLING SYSTEMS (CHILLED WATER, BRINE, AND REFRIGERANT)[4]							
40–55	0.22–0.28	100	0.5	0.5	1.0	1.0	1.0
Below 40	0.22–0.28	100	0.5	1.0	1.0	1.0	1.5

For SI Units: °C= (°F-32)/1.8, 1 British thermal unit inch per hour square foot degree Fahrenheit = [0.1 W/(m•K)], 1 inch = 25 mm

Notes:

1 For insulation outside the stated conductivity range, the minimum thickness (T) shall be determined as follows:

$T = r\{(1 + t/r)^{K/k} - 1\}$

Where:

T = minimum insulation thickness (inches).

r = actual outside radius of pipe (inches) (mm).

t = insulation thickness listed in this table for applicable fluid temperature and pipe size.

K = conductivity of alternate material at mean rating temperature indicated for the applicable fluid temperature [Btu•in/(h•ft^2•°F)] [W/(m•K)].

k = the upper value of the conductivity range listed in this table for the applicable fluid temperature.

2 These thicknesses are based on energy efficiency considerations only. Additional insulation is sometimes required relative to safety issues/surface temperature.

3 Piping insulation is not required between the control valve and coil on run-outs where the control valve is located within 4 feet (1219 mm) of the coil and the pipe size is 1 inch (25 mm) or less.

4 These thicknesses are based on energy efficiency considerations only. Issues such as water vapor permeability or surface condensation sometimes require vapor retarders, additional insulation or both.

5 For piping exposed to outdoor air, increase insulation thickness by ½ of an inch (12.7 mm). The outdoor air is defined as any portion of insulation that is exposed to outdoor air. For example, attic spaces and crawlspaces are considered exposed to outdoor air.

TABLE E 502.6
VENTILATION AIR
[ASHRAE 90.2: TABLE 6.6.1]

CATEGORY	MINIMUM REQUIREMENT	CONDITIONS
Mechanical ventilation[1]	50 ft^3/min outdoor air	Where summer design infiltration rate calculated in accordance with reference standard (a) or (b) is less than 0.35 ACH[2].
Kitchen exhaust	100 ft^3/min intermittent	All conditions
Bath exhaust	intermittent	All conditions

For SI units: 1 cubic foot per minute = 0.00047 m^3/s

Notes:

1 Calculate in accordance with Equation E 502.6.

2 Reference standards:

(a) ACCA Manual J

(b) ASHRAE GRP-158

E 502.9 HVAC Equipment, Rated Combinations. HVAC system equipment and system components shall be furnished with the input(s), the output(s), and the value of the appropriate performance descriptor of HVAC products in accordance with federal law or in accordance with Table E 502.9, as applicable. These shall be based on newly produced equipment or components. Manufacturer's instructions shall be furnished with and attached to the equipment. The manufacturer of electric-resistance heating equipment shall furnish full-load energy input over the range of voltages at which the equipment is intended to operate. [ASHRAE 90.2:6.9]

E 502.10 Controls. Each system or each zone within a system shall be provided with not less than one thermostat capable of being set from 55°F (13°C) to 85°F (29°C) and capable of operating the system's heating and cooling. The thermostat or control system, or both, shall have an adjustable deadband, the range of which includes a setting of 10°F (6°C) between heating and cooling where automatic changeover is provided. Wall-mounted temperature controls shall be mounted on an inside wall. [ASHRAE 90.2:6.10.1]

E 502.10.1 Initial Control Setting. The control shall initially be set for a maximum heating temperature of 70°F (21°C) and a cooling temperature of not less than 78°F (26°C).

E 502.10.2 Ventilation Control. Each mechanical ventilation system (supply, exhaust, or both) shall be equipped with a readily accessible switch or other means for shutoff. Manual or automatic dampers installed for the purpose of isolating outside air intakes and exhausts from the air distribution system shall be designed for tight shutoff. [ASHRAE 90.2:6.10.2]

E 502.10.3 Humidity Control. Where additional energy-consuming equipment is provided for adding moisture to maintain specific selected relative humidities in spaces or zones, a humidistat shall be provided. This device shall be capable of being set to prevent energy from being used to produce relative humidity within the space above 30 percent. [ASHRAE 90.2:6.10.3.1]

E 502.10.3.1 Cooling. Where additional energy-consuming equipment is provided for reducing humidity, it shall be equipped with controls capable of being set to prevent energy from being used to produce a relative humidity within the space below 50 percent during periods of human occupancy and below 60 percent during unoccupied periods. [ASHRAE 90.2:6.10.3.2]

E 502.10.4 Freeze Protection Systems. Freeze protection systems, such as heat tracing of outdoor piping and heat exchangers, including self-regulating heat tracing, shall include automatic controls capable of and configured to shut off the systems where outdoor air temperatures are above 40°F (4°C) or where the conditions of the protected fluid will prevent freezing. Snow- and ice-melting systems shall include automatic controls capable of and configured to shut off the systems where the pavement temperature is above 50°F (10°C) and no precipitation is falling and an automatic or manual control that will allow shutoff where the outdoor temperature is above 40°F (4°C) so that the potential for snow or ice accumulation is negligible. [ASHRAE 90.1:6.4.3.7]

E 502.10.5 Other Controls. Where setback, zoned, humidity and cooling controls and equipment are provided, they shall be designed and installed in accordance with Section E 502.10 through Section E 502.10.3.1. [ASHRAE 90.2:6.10.3.3]

E 502.11 Whole House Fans. Whole house exhaust fans shall have insulated louvers or covers which close where the fan is off. Covers or louvers shall have an insulation value of not less than R-4.2, and shall be installed in accordance with the manufacturer's installation instructions. The attic openings shall be sufficient to accommodate the ventilation capacity of the whole house fan. The operation of the whole house fan shall be considered in determining the adequacy of providing combustion air in accordance with this code.

E 502.12 Dampers. Dampers shall be installed to close off outdoor air inlets and exhaust outlets where the ventilation system is not operating.

E 503.0 Heating, Ventilation, and Air-Conditioning – Other than Low-Rise Residential Buildings.

E 503.1 General. The heating, ventilation, and air-conditioning in buildings, other than single-family houses, multi-family structures of not more than three stories above grade, and modular houses, shall be in accordance with Section E 503.0.

TABLE E 502.9
MINIMUM REQUIREMENTS FOR NON-FEDERALLY COVERED HVAC EQUIPMENT
[ASHRAE 90.2: TABLE 6.9]

EQUIPMENT TYPE	SUBCATEGORY OR RATING CONDITION	MINIMUM EFFICIENCY	TEST PROCEDURE
Groundwater source heat pump*	Cooling Mode	11.0 EER at 70°F Ent. Water	ARI 325
		11.5 EER at 50°F Ent. Water	
	Heating Mode	3.4 COP at 70°F Ent. Water	
		3.0 COP at 50°F Ent. Water	
Unitary A/C	Water cooled split system	9.3 EER at 85°F Ent. Water	ARI 210/240
		8.3 IPLV at 75°F Ent. Water	
	Evaporatively cooled split system	9.3 EER at 95°F Out. Amb.	
		8.5 IPLV at 80°F Out. Amb.	

For SI units: °C = (°F-32)/1.8
* Performance for electrically powered equipment with capacity less than 65 000 Btu/h (19 kW) where rated in accordance with ARI 325.

E 503.1.1 New Buildings. Mechanical equipment and systems serving the heating, cooling, ventilating, or refrigeration needs of new buildings shall be in accordance with the requirements of this section as described in Section E 503.2. [ASHRAE 90.1:6.1.1.1]

E 503.1.2 Additions to Existing Buildings. Mechanical equipment and systems serving the heating, cooling, ventilating, or refrigeration needs of additions to existing buildings shall be in accordance with the requirements of this section as described in Section E 503.2.

Exception: Where HVACR to an addition is provided by existing HVACR systems and equipment, such existing systems and equipment shall not be required to be in accordance with this appendix. A new system or equipment installed shall be in accordance with specific requirements applicable to those systems and equipment. [ASHRAE 90.1:6.1.1.2]

E 503.1.3 Alterations to Heating, Ventilating, Air-Conditioning, and Refrigeration in Existing Buildings. New HVACR equipment as a direct replacement of existing HVACR equipment shall be in accordance with the following sections as applicable for the equipment being replaced:

(1) Section E 503.3
(2) Section E 503.4
(3) Section E 503.4.6
(4) Section E 503.4.6.2
(5) Section E 503.4.6.3
(6) Section E 503.4.6.4
(7) Section E 503.4.6.8
(8) Section E 503.4.6.9
(9) Section E 503.4.6.11
(10) Section E 503.5.1
(11) Section E 503.5.3
(12) Section E 503.5.6.1.2
(13) Section E 503.5.6.2
(14) Section E 503.5.6.5
(15) Section E 503.5.7
(16) Section E 503.5.7.2
(17) Section E 503.5.8.1. [ASHRAE 90.1:6.1.1.3.1]

E 503.1.3.1 New Cooling Systems. New cooling systems installed to serve previously uncooled spaces shall be in accordance with this section as described in Section E 503.2. [ASHRAE 90.1:6.1.1.3.2]

E 503.1.3.2 Existing Cooling Systems. Alterations to existing cooling systems shall not decrease economizer capability unless the system is in accordance with Section E 503.5 through Section E 503.5.4.1. [ASHRAE 90.1:6.1.1.3.3]

E 503.1.3.3 Ductwork. New and replacement ductwork shall comply with Section E 503.4.7.1 through Section E 503.4.7.2.1. [ASHRAE 90.1:6.1.1.3.4]

E 503.1.3.4 Piping. New and replacement piping shall comply with Section E 503.4.7.1.

Exceptions:

(1) For equipment that is being modified or repaired but not replaced, provided that such modifications or repairs will not result in an increase in the annual energy consumption of the equipment using the same energy type.
(2) Where a replacement or alteration of equipment requires extensive revisions to other systems, equipment, or elements of a building, and such replaced or altered equipment is a like-for-like replacement.
(3) For a refrigerant change of existing equipment.
(4) For the relocation of existing equipment.
(5) For ducts and piping where there is insufficient space or access to comply with these requirements. [ASHRAE 90.1:6.1.1.3.5]

E 503.2 Compliance Path(s). Section E 503.0 shall be achieved in accordance with the requirements of Section E 503.1.1 through Section E 503.1.3.4, Section E 503.6, Section E 503.7, and one of the following:

(1) Section E 503.3 and Section E 503.3.1
(2) Section E 503.4
(3) Section E 503.4 and Section E 503.8 [ASHRAE 90.1:6.2.1]

E 503.2.1 Projects Using Energy Cost Budget Method. Projects using the energy cost budget method in accordance with ASHRAE 90.1 shall comply with Section E 503.4, the mandatory provisions of this section, as a portion of that compliance path. [ASHRAE 90.1:6.2.2]

E 503.3 Simplified Approach Option for HVAC Systems. The simplified approach shall be an optional path for compliance where the following conditions are met:

(1) The building is not more than two stories in height.
(2) Gross floor area is less than 25 000 square feet (2322.6 m^2).
(3) The HVAC system in the building is in accordance with the requirements listed in Section E 503.3.1. [ASHRAE 90.1:6.3.1]

E 503.3.1 Criteria. The HVAC system shall comply with the following criteria:

(1) The system serves a single HVAC zone.
(2) The equipment shall comply with the variable flow requirements of Section E 503.5.6.2.
(3) Cooling (where any) shall be provided by a unitary packaged or split-system air conditioner that is either air-cooled or evaporatively cooled, with efficiency that is in accordance with the requirements shown in Table E 503.7.1(1), Table E 503.7.1(2), or Table E 503.7.1(4) for the applicable equipment category.
(4) The system shall have an air economizer in accordance with Section E 503.5 and Section E 503.4.6.13.

(5) Heating (where any) shall be provided by a unitary packaged or split-system heat pump that is in accordance with the applicable efficiency requirements shown in Table E 503.7.1(2) or Table E 503.7.1(4), a fuel-fired furnace that is in accordance with the applicable efficiency requirements shown in Table E 503.7.1(5), an electric resistance heater, or a baseboard system connected to a boiler that is in accordance with the applicable efficiency requirements shown in Table E 503.7.1(6).

(6) The system shall comply with the exhaust air energy recovery requirements in accordance with Section E 503.5.10.

(7) The system shall be controlled by a manual changeover or dual setpoint thermostat.

(8) Where a heat pump equipped with auxiliary internal electric resistance heaters is installed, controls shall be provided that prevent supplemental heater operation where the heating load is capable of being met by the heat pump alone during both steady-state operation and setback recovery. Supplemental heater operation shall be permitted during outdoor coil defrost cycles. The heat pump shall be controlled in accordance with one of the following:

(a) A digital or electronic thermostat designed for heat pump use that energizes auxiliary heat where the heat pump has insufficient capacity to maintain setpoint or to warm up the space at a sufficient rate.

(b) A multistage space thermostat and an outdoor air thermostat wired to energize auxiliary heat on the last stage of the space thermostat and where outdoor air temperature is less than 40°F (4°C).

Exceptions: Heat Pumps that comply with the following:

(1) Have a minimum efficiency regulated by NAECA.

(2) In accordance with the requirements shown in Table E 503.7.1(2).

(3) Include all usage of internal electric resistance heating.

(9) The system controls shall not permit reheat or other form of simultaneous heating and cooling for humidity control.

(10) Systems serving spaces other than hotel or motel guest rooms, and other than those requiring continuous operation, which have both a cooling or heating capacity more than 15 000 Btu/h (4.4 kW) and a supply fan motor power more than 0.75 horsepower (hp) (0.56 kW), shall be provided with a time clock that is in accordance with the following:

(a) Can start and stop the system under different schedules for seven different day-types per week.

(b) Is capable of retaining programming and time setting during a loss of power for a period of not less than 10 hours.

(c) Includes an accessible manual override that allows temporary operation of the system for up to 2 hours.

(d) Is capable of and configured with temperature setback down to 55°F (13°C) during off hours.

(e) Is capable of and configured with temperature setup to 90°F (32°C) during off hours.

(11) Except for piping within manufacturer's units, HVAC piping shall be insulated in accordance with Table E 503.7.3(1) and Table E 503.7.3(2). Insulation exposed to weather shall be suitable for outdoor service (e.g., protected by aluminum, sheet metal, painted canvas, or plastic cover). Cellular foam insulation shall be protected as above or painted with a coating that is water retardant and provides shielding from solar radiation.

(12) Ductwork and plenums shall be insulated in accordance with Table E 503.7.2 and shall be sealed in accordance with Section E 503.4.7.2.

(13) Construction documents shall require a ducted system to be air balanced in accordance with industry-accepted procedures.

(14) Outdoor air intake and exhaust systems shall comply with Section E 503.4.6.4 through Section E 503.4.6.5.

(15) Where separate heating and cooling equipment serves the same temperature zone, thermostats shall be interlocked to prevent simultaneous heating and cooling.

(16) Systems with a design supply air capacity more than 10 000 ft³/min (4.7195 m³/s) shall have optimum start controls.

(17) The system shall comply with the demand control ventilation requirements of Section E 503.4.6.9 and the ventilation design requirements of Section E 503.5.6.6.

(18) The system shall comply with the door switch requirements of Section E 503.5.14. [ASHRAE 90.1:6.3.2]

E 503.3.2 Climate Zone Determination. Climate zones identified in this appendix shall be determined in accordance with ASHRAE 90.1. For locations in the United States and its territories, the assigned climate zone and, where required, the assigned climate zone letter shall be in accordance with ASHRAE 169.

Exception: Where recorded historical climatic data are available for a construction site, it is permitted to be used to determine compliance where approved by the Authority Having Jurisdiction. [ASHRAE 90.1:5.1.4.1]

E 503.4 Mandatory Provisions. Equipment shown in Table E 503.7.1(1) through Table E 503.7.1(16) shall have a minimum performance at the specified rating conditions where tested in accordance with the specified test procedure. Where multiple rating conditions or performance requirements are provided, the equipment shall satisfy the stated requirements unless otherwise exempted by footnotes in the table. Equipment covered under the Federal Energy Policy Act of 1992 (EPACT) shall have no minimum efficiency requirements for operation at minimum capacity or other than standard rating conditions. Equipment used to provide service water heating functions as part of a combination system shall satisfy the stated requirements for the appropriate space heating or cooling category.

Tables are as follows:

(1) Table E 503.7.1(1), "Electrically Operated Unitary Air Conditioners and Condensing Units-Minimum Efficiency Requirements"

(2) Table E 503.7.1(2), "Electrically Operated Unitary and Applied Heat Pumps-Minimum Efficiency Requirements"

(3) Table E 503.7.1(3), "Water-Chilling Packages-Efficiency Requirements" (See Section E 503.4.1 for water-cooled centrifugal water-chilling packages that are designed to operate at nonstandard conditions.)

(4) Table E 503.7.1(4), "Electrically Operated Packaged Terminal Air Conditioners, Packaged Terminal Heat Pumps, Single-Package Vertical Air Conditioners, Single-Package Vertical Heat Pumps, Room Air Conditioners, and Room Air Conditioner Heat Pumps-Minimum Efficiency Requirements"

(5) Table E 503.7.1(5), "Warm-Air Furnaces and Combination Warm-Air Furnaces/Air-Conditioning Units, Warm-Air Duct Furnaces, and Unit Heaters-Minimum Efficiency Requirements" Heating, Ventilating, and Air Conditioning

(6) Table E 503.7.1(6), "Gas- and Oil-Fired Boilers-Minimum Efficiency Requirements"

(7) Table E 503.7.1(7), "Performance Requirements for Heat-Rejection Equipment"

(8) Table E 503.7.1(8), "Heat Transfer Equipment"

(9) Table E 503.7.1(9), "Electrically Operated Variable-Refrigerant-Flow Air Conditioners- Minimum Efficiency Requirements"

(10) Table E 503.7.1(10), "Electrically Operated Variable-Refrigerant-Flow and Applied Heat Pumps-Minimum Efficiency Requirements"

(11) Table E 503.7.1(11), "Air Conditioners and Condensing Units Serving Computer Rooms"

(12) Table E 503.7.1(12), "Commercial Refrigerators and Freezers-Minimum Efficiency Requirements"

(13) Table E 503.7.1(13), "Commercial Refrigeration-Minimum Efficiency Requirements"

(14) Table E 503.7.1(14), "Vapor-Compression-Based Indoor Pool Dehumidifiers-Minimum Efficiency Requirements"

(15) Table E 503.7.1(15), "Electrically Operated DX-DOAS Units, Single-Package and Remote Condenser, without Energy Recovery-Minimum Efficiency Requirements"

(16) Table E 503.7.1(16), "Electrically Operated DX-DOAS Units, Single-Package and Remote Condenser, with Energy Recovery-Minimum Efficiency Requirements" [ASHRAE 90.1:6.4.1.1]

E 503.4.1 Water-Cooled Centrifugal Chilling Packages. Equipment not designed for operation in accordance with AHRI 550/590 test conditions of 44°F (7°C) leaving chilled fluid temperature and 2.4 gallons per minute per ton (gpm/ton) (0.00015 L/s/kg) evaporator fluid flow and 85°F (29°C) entering condenser-fluid temperature with 3.0 gpm/ton (0.00018 L/s/kg) condenser-fluid flow shall have maximum full-load kW/ton (FL) and part-load rating requirements adjusted in accordance with Equation E 503.4.1(1) through Equation E 503.4.1(3):

$$FL_{adj} = FL/K_{adj} \quad \text{[Equation E 503.4.1(1)]}$$

$$PLV_{adj} = IPLV/K_{adj} \quad \text{[Equation E 503.4.1(2)]}$$

$$K_{adj} = A \times B \quad \text{[Equation E 503.4.1(3)]}$$

Where:

FL = full-load kW/ton value from Table E 503.7.1(3)

FL_{adj} = maximum full-load kW/ton rating, adjusted for nonstandard conditions

$IPLV$ = $IPLV$ value from Table E 503.7.1(3)

$IPLV_{adj}$ = maximum NPLV rating, adjusted for nonstandard conditions

A = $0.00000014592 \times (LIFT)^4 - 0.0000346496 \times (LIFT)^3 + 0.00314196 \times (LIFT)^2 - 0.147199 \times (LIFT) + 3.9302$

B = $0.0015 \times LvgEvap + 0.934$

$LIFT$ = $LvgCond - LvgEvap$

$LvgCond$ = Full-load condenser leaving fluid temperature (°F)

$LvgEvap$ = Full-load evaporator leaving temperature (°F)

The FL_{adj} and PLV_{adj} values shall only be applicable for centrifugal chillers in accordance with the following full-load design ranges:

(1) Minimum Evaporator Leaving Temperature: 36°F (2°C)

(2) Maximum Condenser Leaving Temperature: 115°F (46°C)

(3) *LIFT* is not less than 20°F (-6°C) and not more than 80°F (27°C)

Manufacturers shall calculate the FL_{adj} and PLV_{adj} before determining whether to label the chiller in accordance with Section E 503.4.4. Chillers that are in accordance with ASHRAE 90.1 shall be labeled on chillers in accordance with the scope of ASHRAE 90.1.

Centrifugal chillers designed to operate outside of these ranges shall not be covered under this appendix.

Example: Path A, 600 ton (600 000 kg) centrifugal chiller Table E 503.7.1(3) efficiencies.

F = 0.560 kW/ton

$IPLV$ = 0.500 kW/ton

$LvgCond$ = 91.16°F

$LvgEvap$ = 42°F

$LIFT$ = 91.16°F – 42°F = 49.16°F

K_{adj} = $A \times B$

A = 0.00000014592 x $(49.16)^4$ - 0.0000346496 x $(49.16)^3$ + 0.00314196 x $(49.16)^2$ - 0.147199 x (49.16) + 3.9302 = 1.0228

B = 0.0015 x 42 + 0.934 = 0.9970

FL_{ajd} = 0.560/(1.0228 x 0.9970) = 0.549 kW/ton

PLV_{adj} = 0.500/(1.0228 x 0.9970) = 0.490 kW/ton [ASHRAE 90.1:6.4.1.2.1]

For SI units: 1 metric ton = 1000 kg, 1000 British thermal units per hour = 0.293 kW, 1 gallon per minute = 0.06 L/s, °C = (°F-32)/1.8

E 503.4.1.1 Positive Displacement (air- and water-cooled) Chilling Packages. Equipment with an evaporator leaving fluid temperature more than 32°F (0°C) and water-cooled positive displacement chilling packages with a condenser leaving fluid temperature less than 115°F (46°C) shall be in accordance with Table E 503.7.1(3) where tested or certified with water at standard rating conditions, in accordance with the referenced test procedure. [ASHRAE 90.1:6.4.1.2.2]

E 503.4.2 Equipment not Listed. Equipment not listed in the tables referenced in Section E 503.4 and Section E 503.4.1 shall be permitted to be used. [ASHRAE 90.1:6.4.1.3]

E 503.4.3 Verification of Equipment Efficiencies. Equipment efficiency information supplied by manufacturers shall be verified in accordance with one of the following:

(1) Equipment covered under EPACT shall be in accordance with U.S. Department of Energy certification requirements.

(2) Where a certification program exists for a covered product, and it includes provisions for verification and challenge of equipment efficiency ratings, then the product shall be listed in the certification program.

(3) Where a certification program exists for a covered product, and it includes provisions for verification and challenge of equipment efficiency ratings, but the product is not listed in the existing certification program, the ratings shall be verified by an independent laboratory test report.

(4) Where no certification program exists for a covered product, the equipment efficiency ratings shall be supported by data furnished by the manufacturer.

(5) Where components such as indoor or outdoor coils from different manufacturers are used, the system designer shall specify component efficiencies whose combined efficiency is in accordance with the minimum equipment efficiency requirements in Section E 503.4 through Section E 503.4.4.1.

(6) Requirements for plate-type liquid-to-liquid heat exchangers are listed in Table E 503.7.1(8). [ASHRAE 90.1:6.4.1.4]

E 503.4.4 Labeling. Mechanical equipment that is not covered by the U.S. National Appliance Energy Conservation Act (NAECA) of 1987 shall carry a permanent label installed by the manufacturer stating that the equipment is in accordance with the requirements of ASHRAE 90.1. [ASHRAE 90.1:6.4.1.5.1]

E 503.4.4.1 Packaged Terminal Air Conditioners. Nonstandard-size packaged terminal air conditioners and heat pumps with existing sleeves having an external wall opening of less than 16 inches (406 mm) high or less than 42 inches (1067 mm) wide and having a cross-sectional area less than 670 square inches (0.432 m^2) shall be factory labeled in accordance with the following:

"Manufactured for nonstandard-size applications only: not to be installed in new construction projects." [ASHRAE 90.1:6.4.1.5.2]

E 503.4.5 Load Calculations. Heating and cooling system design loads for the purpose of sizing systems and equipment shall be determined in accordance with ASHRAE/ACCA 183. [ASHRAE 90.1:6.4.2.1]

E 503.4.5.1 Pump Head. Pump differential pressure (head) for the purpose of sizing pumps shall be determined in accordance with generally accepted engineering standards and handbooks acceptable to the Authority Having Jurisdiction. The pressure drop through each device and pipe segment in the critical circuit at design conditions shall be calculated. [ASHRAE 90.1:6.4.2.2]

E 503.4.6 Zone Thermostatic Controls. The supply of heating and cooling energy to each zone shall be individually controlled by thermostatic controls responding to temperature within the zone. For the purposes of Section E 503.4.6, a dwelling unit shall be permitted to be considered a single zone. ASHRAE ASHRAE

Exceptions: Independent perimeter systems that are designed to offset only building envelope loads shall be permitted to serve one or more zones also served by an interior system provided:

(1) the perimeter system includes not less than one thermostatic control zone for each building exposure having walls facing only one orientation for 50 contiguous feet (15 240 mm) or more and

(2) the perimeter system heating and cooling supply is controlled by thermostatic controls located within the zones served by the system.

Exterior walls and semiexterior walls are considered to have different orientations where the exposures they face differ by more than 45 degrees (0.79 rad). [ASHRAE 90.1:6.4.3.1.1]

E 503.4.6.1 Dead Band. Where used to control both heating and cooling, zone thermostatic controls shall be capable of and configured to provide a temperature range or dead band of not less than 5°F (3°C) within which the supply of heating and cooling energy to the zone is shut off or reduced to a minimum. ASHRAE ASHRAE

Exceptions:

(1) Thermostats that require manual changeover between heating and cooling modes.

(2) Special occupancy or special applications where wide temperature ranges are not acceptable (such as retirement homes, process applications, museums, some areas of hospitals) and are approved by the Authority Having Jurisdiction. [ASHRAE 90.1:6.4.3.1.2]

E 503.4.6.2 Setpoint Overlap Restriction. Where heating and cooling to a zone are controlled by separate zone thermostatic controls located within the zone, means (such as limit switches, mechanical stops, or, for DDC systems, software programming) shall be provided to prevent the heating setpoint from exceeding the cooling setpoint minus any applicable proportional band. [ASHRAE 90.1:6.4.3.2]

E 503.4.6.3 Off-Hour Controls. HVAC systems shall have the off-hour controls required by Section E 503.4.6.3.1 through Section E 503.4.6.3.4.

Exceptions:

(1) HVAC systems intended to operate continuously.

(2) HVAC systems having a design heating capacity and cooling capacity less than 15 000 Btu/h (4.4 kW) that are equipped with readily accessible manual ON/OFF controls. [ASHRAE 90.1:6.4.3.3]

E 503.4.6.3.1 Automatic Shutdown. HVAC systems shall be equipped with not less than one of the following:

(1) Controls that can start and stop the system under different time schedules for seven different day-types per week, are capable of retaining programming and time setting during loss of power for a period of not less than 10 hours, and include an accessible manual override, or equivalent function, that allows temporary operation of the system for up to 2 hours.

(2) An occupant sensor that is capable of shutting the system off where no occupant is sensed for a period of up to 30 minutes.

(3) A manually operated timer capable of being adjusted to operate the system for up to 2 hours.

(4) An interlock to a security system that shuts the system off where the security system is activated.

Exception: Residential occupancies shall be permitted to use controls that can start and stop the system under two different time schedules per week. [ASHRAE 90.1:6.4.3.3.1]

E 503.4.6.3.2 Setback Controls. Heating systems shall be equipped with controls capable of and configured to automatically restart and temporarily operate the system as required to maintain zone temperatures above an adjustable heating setpoint of not less than 10°F (6°C) below the occupied heating setpoint. Cooling systems shall be equipped with controls capable of and configured to automatically restart and temporarily operate the mechanical cooling system as required to maintain zone temperatures below an adjustable cooling setpoint of not less than 5°F (3°C) above the occupied cooling setpoint or to prevent high space humidity levels.

Exception: Radiant heating systems capable of and configured with a setback heating setpoint at not less than 4°F (2°C) below the occupied heating setpoint. [ASHRAE 90.1:6.4.3.3.2]

E 503.4.6.3.3 Optimum Start Controls. Individual heating and cooling systems with setback controls and DDC shall have optimum start controls. The control algorithm shall, as a minimum, be a function of the difference between space temperature and occupied setpoint, the outdoor temperature, and the amount of time prior to scheduled occupancy. Mass radiant floor slab systems shall incorporate floor temperature into the optimum start algorithm. [ASHRAE 90.1:6.4.3.3.3]

E 503.4.6.3.4 Zone Isolation. HVAC systems serving zones that are intended to operate or be occupied nonsimultaneously shall be divided into isolation areas. Zones shall be permitted to be grouped into a single isolation area provided it does not exceed 25 000 square feet (2322.6 m^2) of conditioned floor area and does not include more than one floor. Each isolation area shall be equipped with isolation devices capable of and configured to automatically shut off the supply of conditioned air and outdoor air to and exhaust air from the area. Each isolation area shall be controlled independently by a device meeting the requirements of Section E 503.4.6.3.1. For central systems and plants, controls and devices shall be provided to allow stable system and equipment operation for any length of time while serving only the smallest isolation area served by the system or plant.

Exceptions: Isolation devices and controls are not required for the following:

(1) Exhaust air and outdoor air connections to isolation zones where the fan system to which they connect is not more than 5000 ft^3/min (2.3597 m^3/s).

(2) Exhaust airflow from a single isolation zone of less than 10 percent of the design airflow of the exhaust system to which it connects.

(3) Zones intended to operate continuously or intended to be inoperative only when all other zones are inoperative. [ASHRAE 90.1:6.4.3.3.4]

E 503.4.6.4 Ventilation System Controls. Stair and elevator shaft vents shall be equipped with motorized dampers that are capable of and configured to automatically close during normal building operation and are interlocked to open as required by fire and smoke detection systems. [ASHRAE 90.1:6.4.3.4.1]

E 503.4.6.4.1 Shutoff Damper Controls. Outdoor air intake and exhaust systems shall be equipped with motorized dampers that will automatically shut when the systems or spaces served are not in use. Ventilation outdoor air and exhaust or relief dampers shall be capable of and configured to automatically shut off during preoccupancy building warm-up, cooldown, and setback, except when ventilation reduces energy costs or when ventilation shall be supplied to comply with the code requirements.

Exceptions:

(1) Backdraft gravity (nonmotorized) dampers shall be permitted for exhaust and relief in buildings less than three stories in height and for ventilation air intakes and exhaust and relief dampers in buildings of any height located in Climate Zones 0, 1, 2 and 3. Back-draft dampers for ventilation air intakes shall be protected from direct exposure to wind.

(2) Back-draft gravity (nonmotorized) dampers shall be permitted in systems with a design outdoor air intake or exhaust capacity of 300 ft^3/min (0.142 m^3/s) or less.

(3) Dampers shall not be required in ventilation or exhaust systems serving unconditioned spaces.

(4) Dampers shall not be required in exhaust systems serving Type 1 kitchen exhaust hoods. [ASHRAE 90.1:6.4.3.4.2]

E 503.4.6.4.2 Dampers Leakage. Where outdoor air supply, and exhaust or relief dampers are required in Section E 503.4.6.4, they shall have a maximum leakage rate in accordance with Table E 503.4.6.4.2 where tested in accordance with AMCA 500D. [ASHRAE 90.1:6.4.3.4.3]

E 503.4.6.4.3 Ventilation Fan Controls. Fans with motors more than 0.75 hp (0.56 kW) shall have automatic controls in accordance with Section E 503.4.6.3.1 that are capable of and configured to shut off fans when not required.

Exception: HVAC systems intended to operate continuously. [ASHRAE 90.1:6.4.3.4.4]

E 503.4.6.5 Enclosed Parking Garage Ventilation. Enclosed parking garage ventilation systems shall automatically detect contaminant levels and stage fans or modulate fan airflow rates to 50 percent or less of design capacity, provided acceptable contaminant levels are maintained.

Exceptions:

(1) Garages not more than 30 000 square feet (2787.09 m^2) with ventilation systems that do not utilize mechanical cooling or mechanical heating.

(2) Garages that have a garage area to ventilation system motor nameplate hp ratio that exceeds 1500 square feet per horsepower (ft^2/hp) (186.8 m^2/kW) and do not utilize mechanical cooling or heating.

(3) Where not permitted by the Authority Having Jurisdiction. [ASHRAE 90.1:6.4.3.4.5]

TABLE E 503.4.6.4.2
MAXIMUM DAMPER LEAKAGE
(cubic foot per minute per square foot) at 1.0 in. w.g
[ASHRAE 90.1: TABLE 6.4.3.4.3]

	VENTILATION AIR INTAKE		EXHAUST/RELIEF	
CLIMATE ZONE	NONMOTORIZED*	MOTORIZED	NONMOTORIZED*	MOTORIZED
0, 1, 2	–	–	–	–
any height	20	4	20	4
3	–	–	–	–
any height	20	10	20	10
4, 5b, 5c	–	–	–	–
less than 3 stories	not allowed	10	20	10
3 or more stories	not allowed	10	not allowed	10
5a, 6, 7, 8	–	–	–	–
less than 3 stories	not allowed	4	20	4
3 or more stories	not allowed	4	not allowed	4

For SI units: 1 cubic foot per minute = 0.00047 m^3/s, 1 square foot = 0.0929 m^2, 1 inch water gauge = 0.249 kPa

* Dampers smaller than 24 inches (610 mm) in either dimension shall be permitted to have leakage of 40 ft^3/min per square foot [0.203 (m^3/s)/m^2].

E 503.4.6.6 Heat Pump Auxiliary Heat Control. Heat pumps equipped with internal electric resistance heaters shall have controls that prevent supplemental heater operation where the heating load is capable of being met by the heat pump alone during both steady-state operation and setback recovery. Supplemental heater operation shall be permitted during outdoor coil defrost cycles.

Exception: Heat pumps whose minimum efficiency is regulated by U.S. National Appliance Energy Conservation Act (NAECA) and whose ratings are in accordance with the requirements shown in Table E 503.7.1(2) and includes the use of an internal electric resistance heating. [ASHRAE 90.1:6.4.3.5]

E 503.4.6.7 Humidification and Dehumidification. Humidity control shall prevent the use of fossil fuel or electricity to produce relative humidity (RH) more than 30 percent in the warmest zone served by the humidification system and to reduce the RH valve to less than 60 percent in the coldest zone served by the dehumidification system. Where a zone is served by a system or systems with both humidification and dehumidification capability, means (such as limit switches, mechanical stops, or, for DDC systems, software programming) shall be provided capable of preventing simultaneous operation of humidification and dehumidification equipment.

Exceptions:

(1) Zones served by desiccant systems used with direct evaporative cooling in series.

(2) Systems serving zones where specific humidity levels are required, such as museums and hospitals, and approved by the Authority Having Jurisdiction or required by accreditation standards and humidity controls are configured to maintain a deadband of not less than 10 percent RH where no active humidification or dehumidification takes place.

(3) Systems serving zones where humidity levels are required to be maintained with precision of not more than ± 5 percent RH to comply with applicable codes or accreditation standards or as approved by the Authority Having Jurisdiction. [ASHRAE 90.1:6.4.3.6]

E 503.4.6.8 Freeze Protection and Snow or Ice Melting Systems. Freeze protection systems, such as heat tracing of outdoor piping and heat exchangers, including self-regulating heat tracing, shall include automatic controls capable of and configured to shut off the systems when outdoor air temperatures are more than 40°F (4°C) or when the conditions of the protected fluid will prevent freezing. Snow and ice melting systems shall include automatic controls capable of and configured to shut off the systems when the pavement temperature is more than 50°F (10°C) and no precipitation is falling, and an automatic or manual control that will allow shutoff when the outdoor temperature is more than 40°F (4°C) so that the potential for snow or ice accumulation is negligible. [ASHRAE 90.1:6.4.3.7]

E 503.4.6.9 Ventilation Controls for High-Occupancy Areas. Demand control ventilation (DCV) shall be required for spaces that are more than 500 square feet (46.45 m^2) and with a design occupancy for ventilation of not less than 25 people per 1000 square feet (92.9 m^2) of floor area and served by systems with one or more of the following:

(1) Air- economizer.

(2) Automatic modulating control of outdoor air damper.

(3) Design outdoor airflow more than 3000 ft^3/min (1.4158 m^3/s).

Exceptions:

(1) Systems with exhaust air energy recovery in accordance with Section E 503.5.10.

(2) Multiple-zone systems without DDC of individual zones communicating with a central control panel.

(3) Systems with a design outdoor airflow less than 750 ft^3/min (0.3540 m^3/s).

(4) Spaces where more than 75 percent of the space design outdoor airflow is required for makeup air that is exhausted from the space or transfer air that is required for makeup air that is exhausted from other spaces.

(5) Spaces with one of the following occupancy categories in accordance with Chapter 4 or ASHRAE 62.1: correctional cells, daycare sickrooms, science labs, barbers, beauty and nail salons, and bowling alley seating. [ASHRAE 90.1:6.4.3.8]

E 503.4.6.10 Outdoor Heating. Radiant heat systems shall be used to provide heat outdoors. Outdoor radiant heating systems shall be provided with controls that sense the presence of occupants or other device that automatically shuts down the system where no occupants are in the heating area.

E 503.4.6.11 Heated or Cooled Vestibules. Heating for vestibules and for air curtains with integral heating shall include automatic controls capable of and configured to shut off the heating system when outdoor air temperatures are more than 45°F (7.2°C) Vestibule heating and cooling systems shall be controlled by a thermostat in the vestibule capable of and configured to limit heating to a maximum of 60°F (15.5°C) and cooling to a minimum of 85°F (29.4°C).

Exception: Heating or cooling provided by site-recovered energy or by transfer air that would otherwise be exhausted. [ASHRAE 90.1:6.4.3.9]

E 503.4.6.12 Direct Digital Control (DDC) Requirements. Direct digital control shall be required in accordance with Section E 503.4.6.12.1 through Section E 503.4.6.12.3. [ASHRAE 90.1:6.4.3.10]

E 503.4.6.12.1 DDC Applications. DDC shall be provided in the applications and qualifications in accordance with Table E 503.4.6.12.1.

Exception: DDC is not required for systems using the simplified approach to compliance in accordance with Section E 503.3. [ASHRAE 90.1:6.4.3.10.1]

E 503.4.6.12.2 DDC Controls. Where DDC is required by Section E 503.4.6.12.1, the DDC system shall be capable of and configured with all of the following, as required, to provide the control logic required in Section E 503.5:

(1) Monitoring zone and system demand for fan pressure, pump pressure, heating, and cooling.

(2) Transferring zone and system demand information from zones to air distribution system controllers and from air distribution systems to heating and cooling plant controllers.

(3) Automatically detecting those zones and systems that are capable of excessively driving the reset logic and generate an alarm or other indication to the system operator.

(4) Readily allowing operator removal of zone(s) from the reset algorithm. [ASHRAE 90.1:6.4.3.10.2]

E 503.4.6.12.3 DDC Display. Where DDC is required in accordance with Section E 503.4.6.12.1 for new buildings, the DDC system shall be capable of trending and graphically displaying input and output points. [ASHRAE 90.1:6.4.3.10.3]

E 503.4.6.13 Economizer Fault Detection Diagnostics (FDD). Air-cooled direct-expansion cooling units listed in Tables E 503.7.1(1) and E 503.7.1(2), where an air economizer is installed in accordance with Section E 503.5, shall include a fault detection and diagnostics (FDD) system complying with the following:

(1) The following temperature sensors shall be permanently installed to monitor system operation:

   (a) Outdoor air

   (b) Supply air

   (c) Return air, where required for economizer control

(2) The system shall have the capability of displaying the value of each sensor.

(3) The FDD system or unit controls shall be capable of and configured to provide system status by indicating the following:

   (a) Free cooling available

   (b) Economizer enabled

   (c) Compressor enabled

   (d) Heating enabled

   (e) Mixed-air low-limit cycle active

(4) The FDD system or unit controls shall have provisions to manually initiate each operating mode so that the operation of compressors, economizers, fans, and the heating system can be independently tested and verified.

TABLE E 503.4.6.12.1
DDC APPLICATIONS AND QUALIFICATIONS
[ASHRAE 90.1:6.4.3.10.1]

BUILDING STATUS	APPLICATION	QUALIFICATIONS
New building	Air-handling system and all zones served by the system	Individual systems supplying more than three zones and with fan system bhp of 10 hp or more
New building	Chilled-water plant and all coils and terminal units served by the system	Individual plants supplying more than three zones and with design cooling capacity of 300 000 Btu/h or more
New building	Hot-water plant and all coils and terminal units served by the system	Individual plants supplying more than three zones and with design heating capacity of 300 000 Btu/h or more
Alteration or addition	Zone terminal unit such as VAV box	Where existing zones served by the same air-handling, chilled-water, or hot-water system have DDC
Alteration or addition	Air-handling system or fan coil	Where existing air-handling system(s) and fan-coil (s) served by the same chilled- or hot-water plant have DDC
Alteration or addition	New air-handling system and all new zones served by the system	Individual systems with fan system bhp of 10 hp or more and supplying more than three zones and more than 75 percent of zones are new
Alteration or addition	New or upgraded chilled-water plant	Where all chillers are new and plant design cooling capacity is 300 000 Btu/h or more
Alteration or addition	New or upgraded hot-water plant	Where all boilers are new and plant design heating capacity is 300 000 Btu/h or more

For SI units: 1000 British thermal units = 0.293 kW, 1 horsepower = 0.746 kW

(5) The FDD system shall be capable of and configured to detect the following faults:
   (a) Air temperature sensor failure/fault
   (b) Not economizing when the unit should be economizing
   (c) Economizing when the unit should not be economizing
   (d) Damper not modulating
   (e) Excess outdoor air

(6) The FDD system shall be capable of and configured to report faults to a fault management application or DDC system accessible by operating or service personnel, or annunciated locally on zone thermostats. [ASHRAE 90.1: 6.4.3.12]

E 503.4.7 HVAC System Construction and Insulation. HVAC Ducts shall be constructed in accordance with provisions contained in the SMACNA HVAC Duct Construction Standard. HVAC system construction and insulation shall comply with Section E 503.4.7.1 and Section E 503.4.7.2.

E 503.4.7.1 Insulation. Insulation required by this section shall be installed in accordance with industry-accepted standards. These requirements shall not apply to HVAC equipment. Insulation shall be protected from damage, including that due to sunlight, moisture, equipment maintenance, and wind, but not limited to the following:

(1) Insulation exposed to weather shall be suitable for outdoor service (e.g., protected by aluminum, sheet metal, painted canvas, or plastic cover). Cellular foam insulation shall be protected as above or painted with a coating that is water retardant and provides shielding from solar radiation that is capable of causing degradation of the material.

(2) Insulation covering chilled-water piping, refrigerant suction piping, or cooling ducts located outside the conditioned space shall include a vapor retardant located outside the insulation (unless the insulation is inherently vapor retardant), penetrations and joints of which shall be sealed. [ASHRAE 90.1:6.4.4.1.1]

E 503.4.7.1.1 Duct and Plenum Insulation. Supply and return ducts and plenums installed as part of an HVAC air distribution system shall be thermally insulated in accordance with Table E 503.7.2.

Exceptions:

(1) Factory-installed plenums, casings, or ductwork furnished as a part of HVAC equipment tested and rated in accordance with Section E 503.4 through Section E 503.4.4.1.

(2) Ducts or plenums located in heated spaces, semi-heated spaces, or cooled spaces.

(3) For runouts less than 10 feet (3048 mm) in length to air terminals or air outlets, the rated R-value of insulation shall not be required to exceed R-3.5.

(4) Backs of air outlets and outlet plenums exposed to unconditioned or indirectly conditioned spaces with face areas exceeding 5 square feet (0.5 m^2) shall not be required to exceed R-2; those not exceeding 5 square feet (0.5 m^2) shall not be required to be insulated. [ASHRAE 90.1:6.4.4.1.2]

E 503.4.7.1.2 Piping Insulation. Piping shall be thermally insulated in accordance with Table E 503.7.3(1) and Table E 503.7.3(2).

Exceptions:

(1) Factory-installed piping within HVAC equipment tested and rated in accordance with Section E 503.4 through Section E 503.4.4.1.

(2) Piping that conveys fluids having a design operating temperature range between 60°F (16°C) and 105°F (41°C), inclusive.

(3) Piping that conveys fluids that have not been heated or cooled through the use of fossil fuels or electricity (such as roof and condensate drains, domestic cold water supply, or natural gas piping).

(4) Where heat gain or heat loss will not increase energy usage (such as liquid refrigerant piping).

(5) For piping 1 inch (25.4 mm) or less, insulation shall not be required for strainers, control valves, and balancing valves. [ASHRAE 90.1:6.4.4.1.3]

E 503.4.7.1.3 Sensible Heating Panel. Thermally ineffective panel surfaces of sensible heating panels, including U-bends and headers, shall be insulated with not less than R-3.5. Adjacent building envelope insulation shall be applied to this insulation value. [ASHRAE 90.1:6.4.4.1.4]

E 503.4.7.1.4 Radiant Floor Heating. The bottom surfaces of floor structures incorporating radiant heating shall be insulated not less than R-3.5. Adjacent building envelope insulation shall be applied to this insulated value.

Exception: Heated slab-on-grade floors incorporating radiant heating shall be in accordance with ASHRAE 90.1. [ASHRAE 90.1:6.4.4.1.5]

E 503.4.7.2 Ductwork and Plenum Leakage. Transverse joints, longitudinal seams, and duct wall penetrations shall be sealed. Pressure-sensitive tape shall not be used as the primary sealant, unless it has been certified to comply with UL 181A or UL 181B by an independent testing laboratory and the tape is used in accordance with that certification. All other connections shall be considered transverse joints, including but not limited to spin-ins, taps, other branch connections, access door frames and jambs, and duct connections to equipment.

Exceptions:

(1) Rods that penetrate the duct wall that shall be permitted to move in order to function properly (control rod for volume damper) shall not be sealed in a fashion that prevents them from working properly.

(2) Spiral lock seams in a round or flat oval duct.

E 503.4.7.2.1 Duct Leakage Tests. Ductwork shall be leak-tested in accordance with the SMACNA HVAC Air Duct Leakage Test Manual. Representative sections totaling not less than 20 percent of the total installed duct area shall be tested. Where the tested 20 percent fail to comply with the requirements of this section, then 40 percent of the total installed duct area shall be tested. Where the tested 40 percent fail to comply with the requirements of this section, then 100 percent of the total installed duct area shall be tested. Sections shall be selected by the building owner or designated representative of the building owner. Positive pressure leakage testing shall be permitted for negative pressure ductwork. The permitted duct leakage shall be not more than the following:

$$L_{max} = C_L P^{0.65} \quad \text{(Equation E 503.4.7.2.1)}$$

Where:

L_{max} = maximum permitted leakage, (ft³/min)/100 square feet [0.0001 (m³/s)/m²] duct surface area.

C_L = Six, duct leakage class, (ft³/min)/100 square feet [0.0001 (m³/s)/m²] duct surface area at 1 inch water column (0.2 kPa).

P = test pressure, which shall be equal to the design duct pressure class rating, inch water column (kPa).

E 503.5 Prescriptive Path, Economizers. Cooling systems shall include either an air economizer or fluid economizer in accordance with Section E 503.5.1 through Section E 503.5.4.1.

Exceptions: Economizers shall not be required for the following systems:

(1) Individual fan-cooling units with a supply capacity less than the minimum listed in Table E 503.5(1).

(2) Chilled-water cooling systems without a fan or that use induced airflow, where the total capacity of these systems is less than 1 000 000 Btu/h (293 kW) in Climate Zones 0, 1B, and 2 through 4; less than 1 400 000 Btu/h (410 kW) in Climate Zones 5 through 8; or any size in Climate Zone 1A.

(3) Systems that include nonparticulate air treatment in accordance with ASHRAE 62.1.

(4) In hospitals and ambulatory surgery centers, where more than 75 percent of the air designed to be supplied by the system is to spaces that are required to be humidified more than 35°F (2°C) dew-point temperature to comply with applicable codes or accreditation standards; in all other buildings, where more than 25 percent of the air designed to be supplied by the system is to spaces that are designed to be humidified more than 35°F (2°C) dew-point temperature to satisfy process needs. This exception shall not apply to computer rooms.

(5) Systems that include a condenser heat recovery system with a minimum capacity in accordance with Section E 503.5.10.1.2.

(6) Systems that serve residential spaces where the system capacity is less than five times the requirement listed in Table E 503.5(1).

(7) Systems that serve spaces whose sensible cooling load at design conditions, excluding transmission and infiltration loads, is less than or equal to transmission and infiltration losses at an outdoor temperature of 60°F (16°C).

(8) Systems expected to operate less than 20 hours per week.

(9) Where the use of outdoor air for cooling will affect supermarket open refrigerated casework systems.

(10) For comfort cooling where the cooling efficiency is not less than the efficiency improvement requirements in accordance with Table E 503.5(2).

(11) Systems primarily serving computer rooms where in accordance with one of the following:

(a) The total design cooling load of all computer rooms in the building is less than 3 000 000 Btu/h (879 kW) and the building in which they are located is not served by a centralized chilled water plant.

(b) The room total design cooling load is less than 600 000 Btu/h (176 kW) and the building in which they are located is served by a centralized chilled water plant.

(c) The local water authority does not permit cooling towers.

(d) Less than 600 000 Btu/h (176 kW) of computer room cooling equipment capacity is being added to an existing building.

(12) Dedicated systems for computer rooms where a minimum of 75 percent of the design load serves one of the following:

(a) Spaces classified as an essential facility.

(b) Spaces having a design of Tier IV in accordance with TIA 942.

(c) Spaces classified as Critical Operations Power Systems (COPS) in accordance with NFPA 70.

(d) Spaces where core clearing and settlement services are performed such that their failure to settle pending financial transactions is capable of systemic risk in accordance with "The Interagency Paper on Sound Practices to Strengthen the Resilience of the US Financial System" (April 7, 2003). [ASHRAE 90.1:6.5.1]

TABLE E 503.5(1)
MINIMUM FAN-COOLING UNIT SIZE WHERE AN ECONOMIZER IS REQUIRED
[ASHRAE 90.1: TABLE 6.5.1-1]

CLIMATE ZONES	COOLING CAPACITY WHERE AN ECONOMIZER IS REQUIRED
0A, 0B, 1A, 1B	No economizer requirement
2A, 2B, 3A, 4A, 5A, 6A, 3B, 3C, 4B, 4C, 5B, 5C, 6B, 7, 8	≥54 000 Btu/h

For SI units: 1000 British thermal units per hour = 0.293 kW

TABLE E 503.5(2)
ELIMINATE REQUIRED ECONOMIZER FOR COMFORT COOLING BY INCREASING COOLING EFFICIENCY
[ASHRAE 90.1: TABLE 6.5.1-2]

CLIMATE ZONES	EFFICIENCY IMPROVEMENT*
2A	17%
2B	21%
3A	27%
3B	32%
3C	65%
4A	42%
4B	49%
4C	64%
5A	49%
5B	59%
5C	74%
6A	56%
6B	65%
7	72%
8	77%

* Where a unit is rated with an IPLV, IEER or SEER, to eliminate the required economizer, the minimum cooling efficiency of the HVAC unit shall be increased by the percentage shown. Where the HVAC unit is rated with a full load metric like EER cooling, these shall be increased by the percentage shown.

E 503.5.1 Air Economizers, Design Capacity. Air economizer systems shall be capable of and configured to modulate outdoor air and return air dampers to provide up to 100 percent of the design supply air quantity as outdoor air for cooling. [ASHRAE 90.1:6.5.1.1.1]

E 503.5.1.1 Control Signal. Economizer controls shall be capable of and configured to sequence the dampers with the mechanical cooling equipment and shall not be controlled by only mixed air temperature.

Exception: The use of mixed air temperature limit control shall be permitted for systems controlled from space temperature (such as single-zone systems). [ASHRAE 90.1:6.5.1.1.2]

E 503.5.1.2 High-Limit Shutoff. Air economizers shall be capable of and configured to automatically reduce outdoor air intake to the design minimum outdoor air quantity where outdoor air intake will no longer reduce cooling energy use. High-limit shutoff control types and associated setpoints for specific climate zones shall be chosen from Table E 503.5.1.2. [ASHRAE 90.1:6.5.1.1.3]

E 503.5.1.3 Dampers. Return air, exhaust or relief, and outdoor air dampers shall comply with Section E 503.4.6.4.2. [ASHRAE 90.1:6.5.1.1.4]

E 503.5.1.4 Relief of Excess Outdoor Air. Systems shall provide a means to relieve excess outdoor air during air economizer operation to prevent overpressurizing the building. The relief air outlet shall be located to avoid recirculation into the building. [ASHRAE 90.1:6.5.1.1.5]

E 503.5.1.5 Sensor Accuracy. Outdoor air, return air, mixed air, and supply air sensors shall be calibrated within the following accuracies:

(1) Dry-bulb and wet-bulb temperatures shall be accurate to ±2°F (1.1°C) over the range of 40°F (4.4°C) to 80°F (27°C).

(2) Enthalpy and the value of a differential enthalpy sensor shall be accurate to ±3 Btu/lb (7 E+03 J/kg) over the range of 20 Btu/lb (4.6 E+04 J/kg) to 36 Btu/lb (8.4 E+04 J/kg).

(3) Relative humidity shall be accurate to ±5 percent over the range of 20 percent to 80 percent relative humidity. [ASHRAE 90.1:6.5.1.1.6]

E 503.5.2 Fluid Economizers, Design Capacity. Fluid economizer systems shall be capable of providing up to 100 percent of the expected system cooling load at outdoor air temperatures of not more than 50°F (10°C) dry bulb or 45°F (7°C) wet bulb.

Exceptions:

(1) Systems primarily serving computer rooms in which 100 percent of the expected system cooling load at the dry bulb and wet bulb temperatures in accordance with Table E 503.5.2 is met with water-cooled fluid economizers.

(2) Systems primarily serving computer rooms in which 100 percent of the expected system cooling load at the dry bulb temperatures listed in Table E 503.5.2 is met with air-cooled fluid economizers.

(3) Systems where dehumidification requirements are not capable of being met using outdoor air temperatures of 50°F (10°C) dry bulb or 45°F (7°C) wet bulb and where 100 percent of the expected system cooling load at 45°F (7°C) dry bulb or 40°F (4°C) wet bulb is met with water-cooled fluid economizers. [ASHRAE 90.1:6.5.1.2.1]

TABLE E 503.5.1.2
HIGH-LIMIT SHUTOFF CONTROL SETTINGS FOR AIR ECONOMIZERS[2]
[ASHRAE 90.1: TABLE 6.5.1.1.3]

CONTROL TYPE	ALLOWED ONLY IN CLIMATE ZONE AT LISTED SETPOINT	REQUIRED HIGH LIMIT (ECONOMIZER OFF WHERE):	
		EQUATION	DESCRIPTION
Fixed dry bulb temperature	0B, 1B, 2B, 3B, 3C, 4B, 4C, 5B, 5C, 6B, 7, 8	$T_{oa} > 75°F$	Outdoor air temperature exceeds 75°F
	5A, 6A	$T_{oa} > 70°F$	Outdoor air temperature exceeds 70°F
	0A, 1A, 2A, 3A, 4A	$T_{oa} > 65°F$	Outdoor air temperature exceeds 65°F
Differential dry bulb temperature	0B, LB, 2B, 3B, 3C, 4B, 4C, 5A, 5B, 5C, 6A, 6B, 7, 8	$T_{oa} > T_{ra}$	Outdoor air temperature exceeds return air temperature
Fixed enthalpy with fixed dry-bulb temperature	All	$h_{oa} > 28$ Btu/lb[1] or $T_{oa} > 75°F$	Outdoor air enthalpy exceeds 28 Btu/lb[1] of dry air[1] or outdoor air temperature exceeds 75°F
Differential enthalpy with fixed dry-bulb temperature	All	$h_{oa} > h_{ra}$ or $T_{oa} > 75°F$	Outdoor air enthalpy exceeds return air enthalpy or outdoor air temperature exceeds 75°F

For SI units: °C = (°F-32)/1.8, 1 British thermal unit per pound = 2326 J/kg

Notes:

[1] At altitudes substantially different than sea level, the fixed enthalpy limit shall be set to the enthalpy value at 75°F (24°C) and 50 percent relative humidity. As an example, at approximately 6000 feet (1829 m) elevation, the fixed enthalpy limit shall be approximately 30.7 Btu/lb (71 408 J/kg).

[2] Devices with selectable rather than adjustable setpoints shall be capable of being set to within 2°F (1°C) and 2 Btu/lb (4649 J/kg) of the setpoint listed.

TABLE E 503.5.2
WATER ECONOMIZER SIZING DRY-BULB AND WET-BULB REQUIREMENTS FOR COMPUTER ROOMS*
[ASHRAE 90.1: TABLE 6.5.1.2.1]

CLIMATE ZONE		WATER COOLED		AIR COOLED
		DRY BULB, °F	WET BULB, °F	DRY BULB, °F
0	A	NR	NR	NR
0	B	NR	NR	NR
1	A	NR	NR	NR
1	B	NR	NR	NR
2	A	40.0	35.0	30.0
2	B	35.0	30.0	30.0
3	A	40.0	35.0	25.0
3	B	30.0	25.0	25.0
3	C	30.0	25.0	30.0
4	A	40.0	35.0	25.0
4	B	30.0	25.0	25.0
4	C	30.0	25.0	25.0
5	A	40.0	35.0	20.0
5	B	30.0	25.0	20.0
5	C	30.0	25.0	25.0
6	A	35.0	30.0	20.0
6	B	30.0	25.0	20.0
7	—	30.0	25.0	20.0
8	—	30.0	25.0	20.0

For SI units: °C = (°F-32)/1.8
NR = Not Required

E 503.5.2.1 Maximum Hydronic Pressure Drop. Precooling coils and fluid-to-water heat exchangers used as part of a fluid economizer system shall either have a water-side pressure drop of less than 15 feet of water (45 kPa), or a secondary loop shall be created so that the coil or heat exchanger pressure drop is not seen by the circulating pumps where the system is in the normal cooling (non-economizer) mode. [ASHRAE 90.1:6.5.1.2.2]

E 503.5.3 Integrated Economizer Control. Economizer systems shall be integrated with the mechanical cooling system and be capable of and configured to provide partial cooling even where additional mechanical cooling is required to be in accordance with the remainder of the cooling load. Controls shall not false load the mechanical cooling systems by limiting or disabling the economizer or by other means, such as hot gas bypass, except at the lowest stage of mechanical cooling.

Units that include an air economizer shall comply with the following:

(1) Unit controls shall have the mechanical cooling capacity control interlocked with the air economizer controls such that the outdoor air damper is at the 100 percent open position when mechanical cooling is on, and the outdoor air damper does not begin to close to prevent coil freezing due to minimum compressor run time until the leaving air temperature is less than 45°F (7°C).

(2) DX units with a rated capacity no less than 65 000 Btu/h (18 kW) that control the capacity of the mechanical cooling directly based on occupied space temperature shall have not less than two stages of mechanical cooling capacity.

(3) Other DX units, including those that control space temperature by modulating the airflow to the space, shall comply with the requirements of Table E 503.5.3. [ASHRAE 90.1:6.5.1.3]

TABLE E 503.5.3
DX COOLING STAGE REQUIREMENTS FOR MODULATING AIRFLOW UNITS
[ASHRAE 90.1:6.5.1.3]

RATING CAPACITY, Btu/h	MINIMUM NUMBER OF MECHANICAL COOLING STAGES	MINIMUM COMPRESSOR DISPLACEMENT*
≥65 000 and <240 000	3	≤35% of full load
≥240 000	4	≤25% full load

For SI units: 1000 British thermal units = 0.293 kW

* For mechanical cooling stage control that does not use variable compressor displacement the percent displacement shall be equivalent to the mechanical cooling capacity reduction evaluated at the full load rating conditions for the compressor.

E 503.5.4 Economizer Heating System Impact. HVAC system design and economizer controls shall be such that economizer operation does not increase the building heating energy use during normal operation.

Exception: Economizers on variable air valve (VAV) systems that cause zone level heating to increase due to a reduction in supply air temperature. [ASHRAE 90.1:6.5.1.4]

E 503.5.4.1 Economizer Humidification System Impact. Systems with hydronic cooling and humidification systems designed to maintain inside humidity at a dew-point temperature more than 35°F (2°C) shall use a fluid economizer where an economizer is required in accordance with Section E 503.5 through Section E 503.5.4.1. [ASHRAE 90.1:6.5.1.5]

E 503.5.5 Simultaneous Heating and Cooling Limitation, Zone Controls. Zone thermostatic controls shall prevent the following:

(1) Reheating.

(2) Recooling.

(3) Mixing or simultaneously supplying air that has been previously mechanically heated and air that has been previously cooled, either by mechanical cooling or by economizer systems.

(4) Other simultaneous operation of heating and cooling systems to the same zone.

Exceptions:

(1) Zones for which the volume of air that is reheated, recooled, or mixed is less than the larger of the following:

(a) Twenty percent of the zone design peak supply for systems with DDC and 30 percent for other systems.

(b) The outdoor airflow rate required to be in accordance with the ventilation requirements of Chapter 4 or ASHRAE 62.1 for the zone.

(c) A higher rate that is capable of demonstrating, to the satisfaction of the Authority Having Jurisdiction, to reduce overall system annual energy usage by offsetting reheat or recool energy losses through a reduction in outdoor air intake for the system.

(d) The airflow rate required to be in accordance with applicable codes or accreditation standards, such as pressure relationships or minimum air change rates.

(2) Zones with DDC that comply with the following:

(a) The airflow rate in dead band between heating and cooling does not exceed the larger of the following:

(1) Twenty percent of the zone design peak supply rate.

(2) The outdoor airflow rate required to be in accordance with the ventilation requirements of Chapter 4 or ASHRAE 62.1 for the zone.

(3) A higher rate that is capable of demonstrating, to the satisfaction of the Authority Having Jurisdiction, to reduce overall system annual energy usage by offsetting reheat or recool energy losses through a reduction in outdoor air intake.

(4) The airflow rate required in accordance with applicable codes or accreditation standards, such as pressure relationships or minimum air change rates.

(b) The airflow rate that is reheated, recooled, or mixed shall be less than 50 percent of the zone design peak supply rate.

(c) The first stage of heating consists of modulating the zone supply air temperature setpoint up to a maximum setpoint while the airflow is maintained at the dead band flow rate.

(d) The second stage of heating consists of modulating the airflow rate from the dead band flow rate up to the heating maximum flow rate.

(3) Laboratory exhaust systems in accordance with Section E 503.5.11.3.

(4) Zones where not less than 75 percent of the energy for reheating or for providing warm air in mixing systems is provided from a site-recovered (including condenser heat) or site-solar energy source. [ASHRAE 90.1:6.5.2.1]

E 503.5.5.1 Supply Air Temperature Reheat Limit. Where reheating is permitted in accordance with this appendix, zones that have both supply and return or exhaust air openings more than 6 feet (1829 mm) above the floor shall not supply heating air more than 20°F (11°C) above the space temperature setpoint.

Exceptions:

(1) Laboratory exhaust systems in accordance with Section E 503.5.11.3.

(2) During preoccupancy building warm-up and setback. [ASHRAE 90.1:6.5.2.1.1]

E 503.5.5.2 Hydronic System Controls. The heating of fluids in hydronic systems that have been previously mechanically cooled and the cooling of fluids that have been previously mechanically heated shall be limited in accordance with Section E 503.5.5.2.1 through Section E 503.5.5.2.3. [ASHRAE 90.1:6.5.2.2]

E 503.5.5.2.1 Three-Pipe System. Hydronic systems that use a common return system for both hot water and chilled water shall not be used. [ASHRAE 90.1:6.5.2.2.1]

E 503.5.5.2.2 Two-Pipe Changeover System. Systems that use a common distribution system to supply both heated and chilled water are acceptable where in accordance with the following:

(1) The system is designed to allow a dead band between changeover from one mode to the other of not less than 15°F (8°C) outdoor air temperature.

(2) The system is designed to operate and is provided with controls that will allow operation in one mode for not less than 4 hours before changing over to the other mode.

(3) Reset controls are provided that allow heating and cooling supply temperatures at the changeover point to be not more than 30°F (17°C) apart. [ASHRAE 90.1:6.5.2.2.2]

E 503.5.5.2.3 Hydronic (Water Loop) Heat Pump Systems. Hydronic heat pumps connected to a common heat pump water loop with central devices for heat rejection (e.g., cooling tower) and heat addition (e.g., boiler) shall have the following:

(1) Controls that are capable of and configured to provide a heat pump water supply temperature dead band of not less than 20°F (11°C) between initiation of heat rejection and heat addition by the central devices (e.g., tower and boiler).

(2) For climate zone 3 through zone 8, where a closed-circuit tower (fluid cooler) is used, either an automatic valve shall be installed to bypass all but a minimal flow of water around the tower (for freeze protection) or low-leakage positive closure dampers shall be provided. Where an open-circuit tower is used directly in the heat pump loop, an automatic valve shall be installed to bypass heat pump water flow around the tower. Where an open-circuit tower is used in conjunction with a separate heat exchanger to isolate the tower from the heat pump loop, then heat loss shall be controlled by shutting down the circulation pump on the cooling tower loop.

Exception: Where a system loop temperature optimization controller is used to determine the most efficient operating temperature based on real-time conditions of demand and capacity, dead bands of less than 20°F (11°C) shall be permitted. [ASHRAE 90.1:6.5.2.2.3]

E 503.5.5.3 Dehumidification. Where humidity controls are provided, such controls shall prevent reheating, mixing of hot and cold airstreams, or other means of simultaneous heating and cooling of the same airstream.

Exceptions:

(1) The system is capable of and configured to reduce supply air volume to 50 percent or less of the design airflow rate or the minimum outdoor air ventilation rate in accordance with ASHRAE 62.1 or other applicable federal, state, or local code or recognized standard, whichever is larger before simultaneous heating and cooling takes place.

(2) The individual fan cooling unit has a design cooling capacity of not more than 65 000 Btu/h (19 kW) and is capable of and configured to unload to 50 percent capacity before simultaneous heating and cooling takes place.

(3) The individual mechanical cooling unit has a design cooling capacity of not more than 40 000 Btu/h (11.7 kW). An individual mechanical cooling unit is a single system composed of a fan or fans and a cooling coil capable of providing mechanical cooling.

(4) Systems serving spaces where specific humidity levels are required to satisfy process needs, such as vivariums, museums, surgical suites, pharmacies, and buildings with refrigerating systems, such as supermarkets, refrigerated warehouses, and ice arenas, and where the building includes site-recovered energy or site solar energy that provide energy equal to 75 percent or more of the annual energy for reheating or for providing warm air in mixing systems. This exception shall not apply to computer rooms.

(5) Not less than 90 percent of the annual energy for reheating or for providing warm air in mixing systems is provided from site-recovered energy (including condenser heat) or site-solar energy.

(6) Systems where the heat added to the airstream is the result of the use of a desiccant system and 75 percent of the heat added by the desiccant system is removed by a heat exchanger, either before or after the desiccant system with energy recovery. [ASHRAE 90.1:6.5.2.3]

E 503.5.5.4 Humidifier Preheat. Humidifiers with preheating jackets mounted in the airstream shall be provided with an automatic valve to shut off preheat where humidification is not required. [ASHRAE 90.1:6.5.2.4.1]

E 503.5.5.4.1 Insulation. Humidification system dispersion tube hot surfaces in the airstreams of ducts or air-handling units shall be insulated with a product with an insulating value of not less than R-0.5.

Exception: Systems where mechanical cooling, including economizer operation, does not occur simultaneously with humidification. [ASHRAE 90.1:6.5.2.4.2]

E 503.5.5.5 Preheat Coils. Preheat coils shall have controls that stop their heat output where mechanical cooling, including economizer operation, is occurring. [ASHRAE 90.1:6.5.2.5]

E 503.5.6 Air System Design and Control. HVAC air system design and control shall be in accordance with the provisions of Section E 503.5.6.1 through Section E 503.5.6.6.

E 503.5.6.1 Fan System Power and Efficiency. Each HVAC system having a total fan system motor nameplate horsepower (kW) exceeding 5 hp (3.7 kW) at fan system design conditions shall not exceed the allowable fan system motor nameplate horsepower (kW) (Option 1) or fan system brake horsepower (kW) (Option 2) as shown in Table E 503.5.6.1(1). This shall include supply fans, return or relief fans, exhaust fans, and fan-powered terminal units associated with systems providing heating or cooling capability that operate at fan system design conditions. Single-zone VAV systems shall comply with the constant-volume fan power limitation.

Exceptions:

(1) Hospital, vivarium, and laboratory systems that utilize flow control devices on exhaust, return, or both to maintain space pressure relationships necessary for occupant health and safety, or environmental control shall be permitted to use variable-volume fan power limitation.

(2) Individual exhaust fans with motor nameplate horsepower of 1 hp (0.7 kW) or less. [ASHRAE 90.1:6.5.3.1.1]

E 503.5.6.1.1 Motor Nameplate Horsepower. For each fan, the selected fan motor shall be not larger than the first available motor size more than the brake horsepower (bhp) (kW). The fan brake horsepower shall be indicated on the design documents to allow for compliance verification by the Authority Having Jurisdiction.

TABLE E 503.5.6.1(1)
FAN POWER LIMITATION*
[ASHRAE 90.1: TABLE 6.5.3.1-1]

	LIMIT	CONSTANT VOLUME	VARIABLE VOLUME
Option 1: Fan system motor nameplate (hp)	Allowable nameplate motor (hp)	$hp \leq CFM_S \cdot 0.0011$	$hp \leq CFM_S \cdot 0.0015$
Option 2: Fan system (bhp)	Allowable fan system (bhp)	$bhp \leq CFM_S \cdot 0.00094 + A$	$bhp \leq CFM_S \cdot 0.0013 + A$

For SI units: 1 horsepower = 0.746 kW, 1 cubic foot per minute = 0.00047 m³/s

* Where:

CFM_S = the maximum design supply airflow rate to conditioned spaces served by the system in cubic feet per minute (m³/s)

hp = the maximum combined motor nameplate horsepower (kW)

bhp = the maximum combined fan brake horsepower (kW)

A = sum of (PD x CFM_D/4131)

PD = each applicable pressure drop adjustment from Table E 503.5.6.1(2) in inch water column (kPa)

CFM_D = the design airflow through each applicable device from Table E 503.5.6.1(2) in cubic feet per minute (m³/s)

Exceptions:

(1) For fans less than 6 bhp (4.5 kW), where the first available motor larger than the bhp (kW) has a nameplate rating within 50 percent of the bhp (kW), the next larger nameplate motor size shall be selected.

(2) For fans 6 bhp (4.5 kW) and larger, where the first available motor larger than the bhp (kW) has a nameplate rating within 30 percent of the bhp (kW), the next larger nameplate motor size shall be selected.

(3) Systems that are in accordance with Section E 503.5.6.1, Option 1.

(4) Fans with motor nameplate horsepower of less than 1 hp (0.7 kW). [ASHRAE 90.1:6.5.3.1.2]

E 503.5.6.1.2 Fan Efficiency. Fans shall have a fan efficiency grade (FEG) of 67 or more, based on manufacturers' certified data in accordance with AMCA 205. The total efficiency of the fan at the design point of operation shall be within 15 percentage points of the maximum total efficiency of the fan.

Exceptions:

(1) Individual fans with a motor nameplate horsepower of 5 hp (3.7 kW) or less that are not part of a group operated as the functional equivalent of a single fan.

(2) Multiple fans in series or parallel (e.g., fan arrays) that have a combined motor nameplate horsepower of 5 hp (3.7 kW) or less and are operated as the functional equivalent of a single fan.

(3) Fans that are part of equipment listed under Section E 503.4.

(4) Fans included in equipment bearing a third party-certified seal for air or energy performance of the equipment package.

(5) Powered wall/roof ventilators (PRV).

(6) Fans outside the scope of AMCA 205.

(7) Fans that are intended to only operate during emergency conditions. [ASHRAE 90.1:6.5.3.1.3]

E 503.5.6.2 Fan Airflow Control. Cooling systems listed in Table E 503.5.6.2 shall be designed to vary the indoor fan airflow as a function of load and shall be in accordance with the following:

(1) DX and chilled-water cooling units that control the capacity of the mechanical cooling directly based on space temperature shall have a minimum of two stages of fan control. Low or minimum speed shall not exceed 66 percent of full speed. At low or minimum speed, the fan system shall draw not more than 40 percent of the fan

TABLE E 503.5.6.1(2)
FAN POWER LIMITATION PRESSURE DROP ADJUSTMENT
[ASHRAE 90.1: TABLE 6.5.3.1-2]

DEVICE	ADJUSTMENT
CREDITS	
Return or exhaust systems required by code or accreditation standards to be fully ducted, or systems required to maintain air pressure differentials between adjacent rooms	0.5 in. w.c. (2.15 in w.c. for laboratory and vivarium systems)
Return, exhaust, or both airflow control devices	0.5 in. w.c.
Exhaust filters, scrubbers, or other exhaust treatment	The pressure drop of device calculated at fan system design condition
Particulate Filtration Credit: MERV 9 through 12	0.5 in. w.c.
Particulate Filtration Credit: MERV 13 through 15	0.9 in. w.c.
Particulate Filtration Credit: MERV 16 and greater, and electronically enhanced filters	Pressure drop calculated at 2x clean filter pressure drop at fan system design condition
Carbon and other gas-phase air cleaners	Clean filter pressure drop at fan system design condition
Biosafety cabinet	Pressure drop of device at fan system design condition
Energy recovery device, other than coil runaround loop	For each airstream [(2.2 x enthalpy recovery ratio) - 0.5] in w.c.
Coil runaround loop	0.6 in. w.c. for each airstream
Evaporative humidifier or cooler in series with another cooling coil	Pressure drop of device at fan system design condition
Sound attenuation section (fans serving spaces with design background noise goals below NC35)	0.15 in. w.c.
Exhaust system serving fume hoods	0.35 in. w.c.
Laboratory and vivarium exhaust systems in high-rise buildings	0.25 in. w.c. per 100 feet of vertical duct exceeding 75 ft
DEDUCTIONS	
Systems without central cooling device	–0.6 in. wc
Systems without central heating device	–0.3 in. wc
Systems with central electric resistance heat	–0.2 in. wc

For SI units: 1 inch water column = 0.249 kPa, 1 foot = 304.8 mm power at full fan speed. Low or minimum speed shall be used during periods of low cooling load and ventilation-only operation.

(2) Other units, including DX cooling units and chilled water units that control the space temperature by modulating the airflow to the space, shall have modulating fan control. Minimum speed shall not exceed 50 percent of full speed. At minimum speed, the fan system shall draw not more than 30 percent of the power at full fan speed. Low or minimum speed shall be used during periods of low cooling load and ventilation-only operation.

(3) Units that include an air-side economizer to comply with Section E 503.5 through Section E 503.5.4.1 shall have not less than of two speeds of fan control during economizer operation.

Exceptions:

(1) Modulating fan control shall not be required for chilled-water and evaporative cooling units with less than1 hp (0.7 kW) fan motors where the units are not used to provide ventilation air and the indoor fan cycles with the load.

(2) Where the volume of outdoor air required to comply with the ventilation requirements of Chapter 4 or ASHRAE 62.1 at low speed exceeds the air that would be delivered at the speed defined in Section E 503.5.6.2(1), or Section E 503.5.6.2(2), then the minimum speed shall be selected to provide the required ventilation air. [ASHRAE 90.1:6.5.3.2.1]

TABLE E 503.5.6.2
FAN AIRFLOW CONTROL
[ASHRAE 90.1: TABLE 6.5.3.2.1]

COOLING SYSTEM TYPE	FAN MOTOR SIZE, (hp)	MECHANICAL COOLING CAPACITY, (Btu/h)
DX cooling	Any	≥65 000
Chilled-water and evaporative cooling	≥1/4	Any

For SI units: 1000 British thermal units per hour = 0.293 kW, 1 horsepower = 0.746 kW, 1 cubic foot per minute = 0.00047 m^3/s

E 503.5.6.2.1 VAV Static Pressure Sensor Location. Static pressure sensors used to control VAV fans shall be located such that the controller setpoint is not more than 1.2 inches water column (0.30 kPa). Where this results in the sensor being located downstream of major duct splits, sensors shall be installed in each major branch to ensure that static pressure is maintained in each.

Exception: Systems that are in accordance with Section E 503.5.6.2.2. [ASHRAE 90.1:6.5.3.2.2]

E 503.5.6.2.2 VAV Setpoint Reset. For multiple-zone VAV systems having a total fan system motor nameplate horsepower exceeding 5 hp (3.7 kW) with DDC of individual zones reporting to the central control panel, static pressure setpoint shall be reset based on the zone requiring the most pressure, such as the setpoint is reset lower until one zone damper is nearly wide open. Controls shall provide the following:

(1) Monitor zone damper positions or other indicator of need for static pressure.

(2) Automatically detect those zones that are capable of excessively driving the reset logic and generate an alarm to the system operator.

(3) Readily allow operator removal of zones from the reset algorithm. [ASHRAE 90.1:6.5.3.2.3]

E 503.5.6.3 Multiple-Zone VAV System Ventilation Optimization Control. Multiple-zone VAV systems with DDC individual zone boxes reporting to a central control panel shall include a means to automatically reduce outdoor air intake flow below design rates in response to changes in system ventilation efficiency in accordance with ASHRAE 62.1.

Exceptions:

(1) VAV systems with zonal transfer fans that recirculate air from other zones without directly mixing it with outdoor air, dual-duct dual-fan VAV systems, and VAV systems with fan-powered terminal units.

(2) Systems where total design exhaust airflow is more than 70 percent of total design outdoor air intake flow requirements. [ASHRAE 90.1:6.5.3.3]

E 503.5.6.4 Supply Air Temperature Reset Controls. Multiple zone HVAC systems shall include controls that automatically reset the supply air temperature in response to representative building loads, or to outdoor air temperature. The controls shall reset the supply air temperature to not less than 25 percent of the difference between the design supply air temperature and the design room air temperature. Controls that adjust the reset based on zone humidity shall be permitted. Zones that are expected to experience relatively constant loads, such as electronic equipment rooms, shall be designed for the fully reset supply temperature.

Exceptions:

(1) Climate zones 0A, 1A, 2A, and 3A.

(2) Systems that prevent reheating, recooling, or mixing of heated and cooled supply air.

(3) Systems where not less than 75 percent of the energy for reheating, on an annual basis, is from site recovered or site solar energy sources. [ASHRAE 90.1:6.5.3.5]

E 503.5.6.5 Fractional Horsepower Fan Motors. Motors for fans that are 1/12 hp (62.1 W) or more and less than 1 hp (0.7 kW) shall be electronically-commutated motors or shall have a motor efficiency of not less than 70 percent where rated in accordance with DOE 10 CFR 431. These motors shall also have the means to adjust motor speed for either balancing or remote control. Belt-driven fans shall be permitted to use sheave adjustments for airflow balancing in lieu of a varying motor speed.

Exceptions:

(1) Motors in the airstream within fan coils and terminal units that operate when providing heating to the space served.

(2) Motors installed in space conditioning equipment certified in accordance with Section E 503.4 through Section E 503.4.4.1.

(3) Motors shown in Table E 503.5.6.5(1) or Table E 503.5.6.5(2). [ASHRAE 90.1:6.5.3.6]

E 503.5.6.6 Ventilation Design. The required minimum outdoor air rate is the larger of the minimum outdoor air rate or the minimum exhaust air rate required by ASHRAE 62.1, ASHRAE 170, or applicable codes or accreditation standards. Outdoor air ventilation systems shall comply with one of the following:

(1) Design minimum system outdoor air provided shall not exceed 135 percent of the required minimum outdoor air rate.

TABLE E 503.5.6.5(1)
MINIMUM AVERAGE FULL-LOAD EFFICIENCY FOR POLYPHASE SMALL ELECTRIC MOTORS*
[ASHRAE 90.1: TABLE 10.8-3]

FULL-LOAD EFFICIENCY, %			
NUMBER OF POLES	OPEN MOTORS		
	2	4	6
SYNCHRONOUS SPEED (RPM)	3600	1800	1200
MOTOR HORSEPOWER	EFFICIENCY, %		
0.25	65.6	69.5	67.5
0.33	69.5	73.4	71.4
0.50	73.4	78.2	75.3
0.75	76.8	81.1	81.7
1	77.0	83.5	82.5
1.5	84.0	86.5	83.8
2	85.5	86.5	N/A
3	85.5	86.9	N/A

* Average full-load efficiencies shall be established in accordance with 10 CFR 431.

TABLE E 503.5.6.5(2)
MINIMUM AVERAGE FULL-LOAD EFFICIENCY FOR CAPACITOR-START CAPACITOR-RUN AND CAPACITOR-START INDUCTION-RUN SMALL ELECTRIC MOTORS*
[ASHRAE 90.1: TABLE 10.8-4]

FULL-LOAD EFFICIENCY, %			
NUMBER OF POLES	OPEN MOTORS		
	2	4	6
SYNCHRONOUS SPEED (RPM)	3600	1800	1200
MOTOR HORSEPOWER	EFFICIENCY, %		
0.25	66.6	68.5	62.2
0.33	70.5	72.4	66.6
0.50	72.4	76.2	76.2
0.75	76.2	81.8	80.2
1	80.4	82.6	81.1
1.5	81.5	83.8	N/A
2	82.9	84.5	N/A
3	84.1	N/A	N/A

* Average full-load efficiencies shall be established in accordance with 10 CFR 431.

(2) Dampers, ductwork, and controls shall be provided that allow the system to supply no more than the required minimum outdoor air rate with a single setpoint adjustment.

(3) The system includes exhaust air energy recovery complying with Section E 503.5.10. [ASHRAE 90.1:6.5.3.7]

E 503.5.7 Hydronic System Design and Control. Boiler systems with design input of 1 000 000 Btu/h (293 kW) or more shall comply with the turndown ratio in accordance with Table E 503.5.7.

The system turndown requirement shall use multiple single-input boilers, one or more modulating boilers, or a combination of single-input and modulating boilers.

Boilers shall comply with the minimum efficiency requirements in Table E 503.7.1(6). [ASHRAE 90.1:6.5.4.1]

TABLE E 503.5.7
BOILER TURNDOWN
[ASHRAE 90.1: TABLE 6.5.4.1]

BOILER SYSTEM DESIGN INPUT, Btu/h	MINIMUM TURNDOWN RATIO
≥1 000 000 and ≤5 000 000	3 to 1
>5 000 000 and ≤10 000 000	4 to 1
>10 000 000	5 to 1

For SI units: 1000 British thermal units per hour = 0.293 kW

ASHRAE ASHRAE ASHRAE

E 503.5.7.1 Hydronic Variable Flow Systems. Chilled- and hot-water distribution systems that include three or more control valves designed to modulate or step open and close as a function of load shall be designed for variable fluid flow and shall be capable of and configured to reduce pump flow rates to not more than the larger of 25 percent of the design flow rate or the minimum flow required by the heating/cooling equipment manufacturer for the proper operation of equipment. Individual or parallel pumps serving variable-flow heating-water or chilled-water systems, where the nameplate horsepower of the motor or combined parallel motors is not less than the power shown in Table E 503.5.7.1, shall have controls or devices that will result in pump motor demand of not more than 30 percent of design wattage at 50 percent of design water flow. The controls or devices shall be controlled as a function of desired flow or to maintain a minimum required differential pressure. Differential pressure shall be measured at or near the most remote heat exchanger or the heat exchanger requiring the greatest differential pressure. The differential pressure setpoint shall not exceed 110 percent of that required to achieve design flow through the heat exchanger. Where differential pressure control is used to comply with this section, and DDC systems are used, the setpoint shall be reset downward based on valve positions until one valve is nearly wide open.

Exceptions:

(1) Differential pressure set-point reset is not required where valve position is used to comply with Section E 503.5.7.3.

(2) Variable-pump flow control is not required on heating-water pumps where more than 50 percent of annual heat is generated by an electric boiler.

(3) Variable flow is not required for primary pumps in a primary/secondary system.

(4) Variable flow is not required for a coil pump provided for freeze protection.

(5) Variable flow is not required for heat recovery coil runaround loops. [ASHRAE 90.1:6.5.4.2]

TABLE E 503.5.7.1
PUMP FLOW CONTROL REQUIREMENTS
[ASHRAE 90.1: TABLE 6.5.4.2]

CHILLED WATER PUMPS IN THESE CLIMATE ZONES	HEATING WATER PUMPS IN THESE CLIMATE ZONES	MOTOR NAMEPLATE HORSEPOWER
0A, 0B, 1A, 1B, 2B	NR	≥2 hp
2A, 3B	NR	≥3 hp
3A, 3C, 4A, 4B	7, 8	≥5 hp
4C, 5A, 5B, 5C, 6A, 6B	3C, 5A, 5C, 6A, 6B	≥7.5 hp
—	4A, 4C, 5B	≥10 hp
7, 8	4B	≥15 hp
—	2A, 2B, 3A, 3B	≥25 hp
—	1B	≥100 hp
—	0A, 0B, 1A	≥200 hp

For SI units: 1 horsepower = 0.746 kW

E 503.5.7.2 Chiller and Boiler Isolation. Where a chilled-water plant includes more than one chiller, provisions shall be made so that the fluid flow through the chiller is automatically shut off where the chiller is shut down. Chillers piped in series for the purpose of increased temperature differential, shall be considered as one chiller. Where constant-speed chilled-water or condenser water pumps are used to serve multiple chillers, the number of pumps shall be not less than the number of chillers and staged on and off with the chillers. [ASHRAE 90.1:6.5.4.3.1]

E 503.5.7.2.1 Boiler Isolation. Where a boiler plant includes more than one boiler, provisions shall be made so that the flow through the boiler is automatically shut off where the boiler is shut down. Where constant-speed hot-water pumps are used to serve multiple boilers, the number of pumps shall be not less than the number of boilers and staged on and off with the boilers. [ASHRAE 90.1:6.5.4.3.2]

E 503.5.7.3 Chilled- and Hot-Water Temperature Reset Controls. Chilled- and hotwater systems with a design capacity exceeding 300 000 Btu/h (88 kW) supplying chilled or heated water (or both) to comfort conditioning systems shall include controls that automatically reset supply water temperatures by representative building loads (including return water temperature) or by outdoor air temperature. Where DDC is used to control valves, the set point shall be reset based on valve positions until one valve is nearly wide open or setpoint limits of the system equipment or application have been reached.

Exceptions:

(1) Where chilled-water supply is already cold, such as chilled water supplied from a district cooling or thermal energy storage system, such that blending would be required to achieve the reset chilled-water supply temperature.

(2) Where a specific temperature is required for a process.

(3) Water temperature reset is not required where valve position is used to comply with Section E 503.5.7. [ASHRAE 90.1:6.5.4.4]

E 503.5.7.4 Hydronic (Water Loop) Heat Pump and Water-Cooled Unitary Air Conditioners. Hydronic heat pumps and water-cooled unitary air-conditioners shall have a two-position automatic valve interlocked to shut off water flow when the compressor is off.

Exception: Units employing water economizers. [ASHRAE 90.1:6.5.4.5.1]

E 503.5.7.4.1 Controls. Hydronic heat pumps and water-cooled unitary air-conditioners having a total pump system power exceeding 5 hp (3.7 kW) shall have controls, devices, or both (such as variable speed control) that will result in pump motor demand of not more than 30 percent of design wattage at 50 percent of design water flow. [ASHRAE 90.1:6.5.4.5.2]

E 503.5.7.5 Pipe Sizing. Chilled-water and condenser-water piping shall be designed such that the design flow rate in a pipe segment does not exceed the values listed in Table E 503.5.7.5 for the appropriate total annual hours of operation. Pipe size selections for systems that operate under variable flow conditions, such as modulating two-way control valves at coils, and that contain variable-speed pump motors shall be permitted to be made from the "Variable Flow/Variable Speed" columns. All others shall be made from the "Other" columns.

Exceptions:

(1) Design flow rates exceeding the values in Table E 503.5.7.5 shall be permitted in specific sections of pipe where the pipe is not in the critical circuit at design conditions and is not predicted to be in the critical circuit during 30 percent or more of operating hours.

(2) Piping systems that have not more than the total pressure drop than the same system constructed with standard weight steel pipe with piping and fittings sized in accordance with Table E 503.5.7.5. [ASHRAE 90.1:6.5.4.6]

E 503.5.8 Heat Rejection Equipment. Section E 503.5.8 through Section E 503.5.9 apply to heat rejection equipment used in comfort cooling systems such as air-cooled condensers, dry coolers, open-circuit cooling towers, closed-circuit cooling towers, and evaporative condensers.

Exception: Heat rejection devices whose energy usage is included in the equipment efficiency ratings listed in Table E 503.7.1(1) through Table E 503.7.1(4). [ASHRAE 90.1:6.5.5.1]

E 503.5.8.1 Fan Speed Control. The fan system on a heat-rejection device powered by an individual motor or an array of motors with a connected power, including the motor service factor, totaling 5 hp (3.7 kW) or more shall have controls and/or devices (such as variable-speed control) that shall result in fan motor demand of no more than 30 percent of design wattage at 50 percent of the design airflow

TABLE E 503.5.7.5
PIPING SYSTEM DESIGN MAXIMUM FLOW RATE (gallons per minute)
[ASHRAE 90.1: TABLE 6.5.4.6]

OPERATING HOURS/YEAR	≤2000 HOURS/YEAR		>2000 AND ≤ 4400 HOURS/YEAR		>4400 HOURS/YEAR	
NOMINAL PIPE SIZE, (inches)	OTHER	VARIABLE FLOW/ VARIABLE SPEED	OTHER	VARIABLE FLOW/ VARIABLE SPEED	OTHER	VARIABLE FLOW/ VARIABLE SPEED
2½	120	180	85	130	68	110
3	180	270	140	210	110	170
4	350	530	260	400	210	320
5	410	620	310	470	250	370
6	740	1100	570	860	440	680
8	1200	1800	900	1400	700	1100
10	1800	2700	1300	2000	1000	1600
12	2500	3800	1900	2900	1500	2300
Maximum velocity for pipes over 14-24 inches in size	8.5 ft/s	13.0 ft/s	6.5 ft/s	9.5 ft/s	5.0 ft/s	7.5 ft/s

For SI units: 1 gallon per minute = 0.06 L/s, 1 foot per second = 0.3048 m/s, 1 inch = 25.4 mm and that shall automatically change the fan speed to control the leaving fluid temperature or condensing temperature or pressure of the heat rejection device.

Exceptions:

(1) Condenser fans serving multiple refrigerant circuits or fluid cooling circuits.

(2) Condenser fans serving flooded condensers. [ASHRAE 90.1:6.5.5.2.1]

E 503.5.8.2 Variable-Speed Fan Drives. Multicell heat rejection equipment with variable-speed fan drives shall:

(1) Operate the maximum number of fans allowed that comply with the manufacturer's requirements for all system components.

(2) Control all fans to the same fan speed required for the instantaneous cooling duty, as opposed to staged (on/off) operation. Minimum fan speed shall comply with the minimum allowable speed of the fan drive system per the manufacturer's recommendations. [ASHRAE 90.1:6.5.5.2.2]

E 503.5.9 Limitation on Centrifugal Fan Open-Circuit Cooling Towers. Centrifugal fan open-circuit cooling towers with a combined rated capacity of 1100 gallons per minute (gpm) (69.39 L/s) or greater at 95°F (35°C) condenser water return, 85°F (29°C) condenser water supply, and 75°F (24°C) outdoor air wet-bulb temperature shall comply with the energy efficiency requirement for axial fan open-circuit cooling towers in accordance with Table E 503.7.1(7).

Exception: Centrifugal open-circuit cooling towers that are ducted (inlet or discharge) or require external sound attenuation. [ASHRAE 90.1:6.5.5.3]

E 503.5.9.1 Tower Flow Turndown. Open-circuit cooling towers used on water-cooled chiller systems that are configured with multiple- or variable-speed condenser water pumps shall be designed so that all open-circuit cooling tower cells can be run in parallel with the larger of the following:

(1) The flow that is produced by the smallest pump at its minimum expected flow rate.

(2) Fifty percent of the design flow for the cell. [ASHRAE 90.1:6.5.5.4]

ASHRAE ASHRAE **E 503.5.10 Exhaust Air Energy Recovery.** Each fan system shall have an energy recovery system where the design supply fan airflow rate exceeds the value listed in Table E 503.5.10(1) and Table E 503.5.10(2), based on the climate zone and percentage of outdoor air at design airflow conditions. Table E 503.5.10(1) shall be used for all ventilation systems that operate less than 8000 hours per year and Table E 503.5.10(2) shall be used for all ventilation systems that operate 8000 or more hours per year.

Energy recovery systems required by this section shall result in an enthalpy recovery ratio of not less than 50 percent. A fifty percent enthalpy recovery ratio shall mean a change in the enthalpy of the outdoor air supply equal to 50 percent of the difference between the outdoor air and entering exhaust air enthalpies at design conditions. Provision shall be provided to bypass or control the energy recovery system to permit air economizer operation in accordance with Section E 503.5.1.

Exceptions:

(1) Laboratory systems that are in accordance with Section E 503.5.11.3.

(2) Systems serving spaces that are not cooled and that are heated to less than 60°F (16°C).

(3) Where more than 60 percent of the outdoor air heating energy is provided from site-recovered energy or site-solar energy.

(4) Heating energy recovery in Climate Zones 0, 1, and 2.

(5) Cooling energy recovery in climate zones 3C, 4C, 5B, 5C, 6B, 7, and 8.

(6) Where the sum of the airflow rates exhausted and relieved within 20 feet (6096 mm) of each other is less than 75 percent of the design outdoor airflow rate, excluding exhaust air that is;

(a) used for another energy recovery system,

(b) not allowed by ASHRAE 170 for use in energy recovery systems with leakage potential, or

(c) of Class 4 as defined in ASHRAE 62.1.

(7) Systems requiring dehumidification that employ energy recovery in series with the cooling coil.

(8) Systems expected to operate less than 20 hours per week at the outdoor air percentage in accordance with Table E 503.5.10(1). [ASHRAE 90.1:6.5.6.1]

E 503.5.10.1 Heat Recovery for Service Water Heating. Heat recovery shall comply with Section E 503.5.10.1.1 and Section E 503.5.10.1.2.

E 503.5.10.1.1 Condenser Heat Recovery Systems. Condenser heat recovery systems shall be installed for the heating or preheating of service hot water where the following conditions exist:

(1) The facility operates 24 hours a day.

(2) The total installed heat rejection capacity of the water-cooled system is more than 6 000 000 Btu/h (1757 kW) of heat rejection.

(3) The design service water heating load is more than 1 000 000 Btu/h (293 kW). [ASHRAE 90.1:6.5.6.2.1]

E 503.5.10.1.2 Capacity. The required heat recovery system shall have the capacity to provide the smaller of:

**TABLE E 503.5.10(1)
EXHAUST AIR ENERGY RECOVERY REQUIREMENTS FOR VENTILATION
SYSTEMS OPERATING LESS THAN 8000 HOURS PER YEAR*
[ASHRAE 90.1: TABLE 6.5.6.1-1]**

CLIMATE ZONE	PERCENT OUTDOOR AIR AT FULL DESIGN AIRFLOW RATE							
	≥10% and <20%	≥20% and <30%	≥30% and <40%	≥40% and <50%	≥50% and <60%	≥60% and <70%	≥70% and <80%	≥80%
	DESIGN SUPPLY FAN AIRFLOW RATE (cubic feet per minute)							
3B, 3C, 4B, 4C, 5B	NR	NR	NR	NR	NR	NR	NR	NR
0B, 1B, 2B, 5C	NR	NR	NR	NR	≥26 000	≥12 000	≥5000	≥4000
6B	≥28 000	≥26 500	≥11 000	≥5500	≥4500	≥3500	≥2500	≥1500
0A, 1A, 2A, 3A, 4A, 5A, 6A	≥26 000	≥16 000	≥5500	≥4500	≥3500	≥2000	≥1000	≥120
7, 8	≥4500	≥4000	≥2500	≥1000	≥140	≥120	≥100	≥80

For SI units: 1 cubic foot per minute = 0.00047 m^3/s
* NR = Not Required

**TABLE E 503.5.10(2)
EXHAUST AIR ENERGY RECOVERY REQUIREMENTS FOR VENTILATION
SYSTEMS OPERATING NOT LESS THAN 8000 HOURS PER YEAR*
[ASHRAE 90.1: TABLE 6.5.6.1-2]**

CLIMATE ZONE	PERCENT OUTDOOR AIR AT FULL DESIGN AIRFLOW RATE							
	≥10% and <20%	≥20% and <30%	≥30% and <40%	≥40% and <50%	≥50% and <60%	≥60% and <70%	≥70% and <80%	≥80%
	DESIGN SUPPLY FAN AIRFLOW RATE (cubic feet per minute)							
3C	NR	NR	NR	NR	NR	NR	NR	NR
0B, 1B, 2B, 3B, 4C, 5C	NR	≥19 500	≥9000	≥5000	≥4000	≥3000	≥1500	≥120
0A, 1A, 2A, 3A, 4B, 5B	≥2500	≥2000	≥1000	≥500	≥140	≥120	≥100	≥80
4A, 5A, 6A, 6B, 7, 8	≥200	≥130	≥100	≥80	≥70	≥60	≥50	≥40

For SI units: 1 cubic foot per minute = 0.00047 m^3/s
* NR = Not Required

(1) Sixty percent of the peak heat rejection load at design conditions.

(2) Preheat of the peak service hot water draw to 85°F (29°C).

Exceptions:

(1) Facilities that employ condenser heat recovery for space heating with a heat recovery design of more than 30 percent of the peak water-cooled condenser load at design conditions.

(2) Facilities that provide 60 percent of their service water heating from site-solar, site-recovered energy, or from other sources. [ASHRAE 90.1:6.5.6.2.2]

E 503.5.11 Exhaust Systems. Exhaust systems shall comply with Section E 503.5.11.1 through Section E 503.5.11.3.

E 503.5.11.1 Kitchen Exhaust Systems. Replacement air introduced directly into the hood cavity of kitchen exhaust hoods shall not exceed 10 percent of the hood exhaust airflow rate. [ASHRAE 90.1:6.5.7.2.1]

E 503.5.11.1.1 Conditioned Supply Air. Conditioned supply air delivered to a space with a kitchen hood shall not exceed the greater of the following:

(1) The supply flow required to be in accordance with the space heating or cooling load.

(2) The hood exhaust flow minus the available transfer air from adjacent spaces. Available transfer air is that portion of outdoor ventilation air not required to satisfy other exhaust needs, such as restrooms, and not required to maintain pressurization of adjacent spaces. [ASHRAE 90.1:6.5.7.1.2]

E 503.5.11.2 Exhaust Flow Rate. Where a kitchen or dining facility has a total kitchen hood exhaust airflow rate exceeding 5000 ft^3/min (2.3597 m^3/s), each hood shall have an exhaust rate in accordance with Table E 503.5.11.2. Where a single hood, or hood section, is installed over appliances with different duty ratings, the maximum allowable flow rate for the hood or hood section shall not exceed the values in Table E 503.5.11.2 for the highest

TABLE E 503.5.11.2
MAXIMUM NET EXHAUST FLOW RATE, CFM PER LINEAR FOOT OF HOOD LENGTH
[ASHRAE 90.1: TABLE 6.5.7.2.2]

TYPE OF HOOD	LIGHT DUTY EQUIPMENT	MEDIUM DUTY EQUIPMENT	HEAVY DUTY EQUIPMENT	EXTRA HEAVY DUTY EQUIPMENT
Wall-mounted canopy	140	210	280	385
Single island	280	350	420	490
Double island (per side)	175	210	280	385
Eyebrow	175	175	Not allowed	Not allowed
Backshelf/ Pass-over	210	210	280	Not allowed

For SI units: 1 foot = 304.8 mm, 1 cubic foot per minute = 0.00047 m^3/s

appliance duty rating under the hood or hood section. Refer to ASHRAE 154 for definitions of hood type, appliance duty, and net exhaust flow rate.

Exception: Seventy-five percent or more of the total replacement air is transfer air that would otherwise be exhausted. [ASHRAE 90.1:6.5.7.2.2]

E 503.5.11.2.1 Kitchen or Dining Facility. Where a kitchen or dining facility has a total kitchen hood exhaust airflow rate more than 5000 ft^3/min (2.3597 m^3/s), then one of the following shall be provided:

ASHRAE ASHRAE (1) Fifty percent or more of all replacement air is transfer air that would otherwise be exhausted.

ASHRAE ASHRAE (2) Demand ventilation systems on 75 percent or more of the exhaust air. Such systems shall be capable of and configured to provide 50 percent or more reduction in exhaust and replacement air system airflow rates, including controls necessary to modulate airflow in response to appliance operation and to maintain full capture and containment of smoke, effluent, and combustion products during cooking and idle.

ASHRAE ASHRAE (3) Listed energy recovery devices that result in a sensible energy recovery ratio of 40 percent or more on 50 percent or more of the total exhaust airflow. A 40 percent sensible energy recovery ratio shall mean a change in the dry-bulb temperature of the outdoor air supply equal to 40 percent of the difference between the outdoor air and entering exhaust air dry-bulb temperatures at design conditions. [ASHRAE 90.1:6.5.7.2.3]

E 503.5.11.2.2 Performance Testing. An approved field test method shall be used to evaluate design air flow rates and demonstrate proper capture and containment performance of installed commercial kitchen exhaust systems. Where demand ventilation systems are utilized to be in accordance with Section E 503.5.11.2.1, additional performance testing shall be provided to demonstrate proper capture and containment at minimum airflow. [ASHRAE 90.1:6.5.7.2.4]

E 503.5.11.3 Laboratory Exhaust Systems. Buildings with laboratory exhaust systems having a total exhaust rate of more than 5000 ft^3/min (2.3597 m^3/s) shall include not less than one of the following features:

(1) VAV laboratory exhaust and room supply systems capable of and configured to reduce exhaust and makeup airflow rates, incorporate a heat recovery system to precondition makeup air from laboratory exhaust, or both and shall be in accordance with the following: ASHRAE ASHRAE ASHRAE ASHRAE

$$A+B\bullet(E/M) \geq 50\% \qquad \text{(Equation E 503.5.11.3)}$$

Where:

A = Percentage that the exhaust and makeup airflow rates are capable of being reduced from design conditions.

B = Sensible energy recovery ratio. ASHRAE ASHRAE

E = Exhaust airflow rate through the heat recovery device at design conditions.

M = Makeup airflow rate of the system at design conditions.

(2) VAV laboratory exhaust and room supply systems that are required to have minimum circulation rates to be in accordance with the codes or accreditation standards shall be capable of and configured to reduce zone exhaust and makeup airflow rates to the regulated minimum circulation values, or the minimum required to maintain pressurization relationship requirements. Systems serving nonregulated zones shall be capable of and configured to reduce exhaust and makeup airflow rates to 50 percent of the zone design values, or the minimum required to maintain pressurization relationship requirements. ASHRAE ASHRAE ASHRAE ASHRAE ASHRAE ASHRAE

(3) Direct makeup (auxiliary) air supply of 75 percent or more of the exhaust airflow rate, heated not more than 2°F (1°C) below room setpoint, cooled to not less than 3°F (2°C) above room setpoint, no humidification added, and no simultaneous heating and cooling are used for dehumidification control. [ASHRAE 90.1:6.5.7.3]

E 503.5.12 Radiant Heating Systems. Radiant heating shall be used when heating is required for unenclosed spaces.

Exception: Loading docks equipped with air curtains. [ASHRAE 90.1:6.5.8.1]

E 503.5.12.1 Heating Enclosed Spaces. Radiant heating systems that are used as primary or supplemental enclosed space heating shall be in accordance with this appendix, including, but not limited to, the following:

(1) Radiant hydronic ceiling or floor panels (used for heating or cooling).

(2) Combination or hybrid systems incorporating radiant heating (or cooling) panels.

(3) Radiant heating (or cooling) panels used in conjunction with other systems such as VAV or thermal storage systems. [ASHRAE 90.1:6.5.8.2]

E 503.5.13 Hot Gas Bypass Limitation. Cooling systems shall not use hot gas bypass or other evaporator pressure control systems unless the system is designed with multiple steps of unloading or continuous capacity modulation. The capacity of the hot gas bypass shall be limited as indicated in Table E 503.5.13 for VAV units and single-zone VAV units. Hot-gas bypass shall not be used on constant-volume units. [ASHRAE 90.1:6.5.9]

TABLE E 503.5.13
HOT GAS BYPASS LIMITATION
[ASHRAE 90.1: TABLE 6.5.9]

RATED CAPACITY	MAXIMUM HOT GAS BYPASS (percent of total capacity)
≤240 000 Btu/h	15%
>240 000 Btu/h	10%

For SI units: 1000 British thermal units per hour = 0.293 kW

E 503.5.14 Door Switches. Conditioned spaces with doors, including doors with more than one-half glass, opening to the outdoors shall be provided with controls that when any such door is open, the following shall occur:

(1) Disable mechanical heating or reset the heating setpoint to 55°F (13°C) or lower within five minutes of the door opening.

(2) Disable mechanical cooling or reset the cooling setpoint to 90°F (32°C) or more within five minutes of the door opening. Mechanical cooling shall be permitted to remain enabled where outdoor air temperature is less than the space temperature.

Exceptions:

(1) Building entries with automatic closing devices.

(2) Any space without a thermostat.

(3) Alterations to existing buildings.

(4) Loading docks. [ASHRAE 90.1:6.5.10]

E 503.6 Submittals. The Authority Having Jurisdiction shall require submittal of compliance documentation and supplemental information in accordance with Section E 503.6.1 through Section E 503.6.3.

E 503.6.1 Construction Details. Compliance documents shall show the pertinent data and features of the building, equipment, and systems in sufficient detail to permit a determination of compliance by the building official and to indicate compliance with the requirements of this appendix. [ASHRAE 90.1:4.2.2.1]

E 503.6.2 Supplemental Information. Supplemental information necessary to verify compliance with this appendix, such as calculations, worksheets, compliance forms, vendor literature, or other data, shall be made available where required by the Authority Having Jurisdiction. [ASHRAE 90.1:4.2.2.2]

E 503.6.3 Manuals. Operating and maintenance information shall be provided to the building owner. This information shall include, but not be limited to, the information specified in Section E 503.6.3.1, Section E 503.6.3.2, and Section E 503.6.5.2. [ASHRAE 90.1:4.2.2.3]

E 503.6.3.1 Required Information. Construction documents shall require that an operating manual and maintenance manual be provided to the building owner. The manuals shall include, at a minimum, the following:

(1) Submittal data stating equipment rating and selected options for each piece of equipment requiring maintenance.

(2) Operation manuals and maintenance manuals for each piece of equipment requiring maintenance. Required routine maintenance actions shall be clearly identified.

(3) Names and addresses of not less than one qualified service agency.

(4) A complete narrative of how each system is intended to operate.

The Authority Having Jurisdiction shall only check to ensure that the construction documents required are provided to the owner, and shall not expect copies of any of the materials. [ASHRAE 90.1:8.7.2]

E 503.6.3.2 Lighting Manuals. Construction documents shall require for all lighting equipment and lighting controls that an operating and maintenance manual be provided to the building owner or the designated representative of the building owner within 90 days after the date of system acceptance. These manuals shall include, at a minimum, the following:

(1) Submittal data indicating all selected options for each piece of lighting equipment, including but not limited to lamps, ballasts, drivers, and lighting controls.

(2) Operation and maintenance manuals for each piece of lighting equipment and lighting controls with routine maintenance clearly identified including, as a minimum, a recommended relamping or cleaning program and a schedule for inspecting and recalibrating all lighting controls.

(3) A complete narrative of how each lighting control system is intended to operate including recommended settings. [ASHRAE 90.1:9.7.2.2]

E 503.6.4 Labeling of Material and Equipment. Materials and equipment shall be labeled in a manner that will allow for determination of their compliance with the applicable provisions of this appendix. [ASHRAE 90.1:4.2.3]

E 503.6.5 Completion Requirements. Section E 503.6.5.1 through Section E 503.6.5.4.1 are mandatory provisions and are necessary to comply with this appendix. [ASHRAE 90.1:6.7.2]

E 503.6.5.1 Drawings. Construction documents shall require that, within 90 days after the date of system acceptance, record drawings of the actual installation be provided to the building owner or the designated representative of the building owner. Record drawings shall include, as a minimum, the location and performance data on each piece of equipment, general configuration of duct and pipe distribution system including sizes, and the terminal air or water design flow rates. [ASHRAE 90.1:6.7.2.1]

E 503.6.5.2 Manuals. Construction documents shall require that an operating manual and a maintenance manual be provided to the building owner or the designated representative of the building owner within 90 days after the date of system acceptance. These manuals shall be in accordance with industry-accepted standards and shall include, at a minimum, the following:

(1) Submittal data stating equipment size and selected options for each piece of equipment requiring maintenance.

(2) Operation manuals and maintenance manuals for each piece of equipment and system requiring maintenance, except equipment not furnished as part of the project. Required routine maintenance actions shall be clearly identified.

(3) Names and addresses of not less than one service agency.

(4) HVAC controls system maintenance and calibration information, including wiring diagrams, schematics, and control sequence descriptions. Desired or field-determined setpoints shall be permanently recorded on control drawings at control devices or, for digital control systems, in programming comments.

(5) A complete narrative of how each system is intended to operate, including suggested setpoints. [ASHRAE 90.1:6.7.2.2]

E 503.6.5.3 System Balancing. Construction documents shall require that HVAC systems be balanced in accordance with generally accepted engineering standards. Construction documents shall require that a written balance report be provided to the building owner or the designated representative of the building owner for HVAC systems serving zones with a total conditioned area exceeding 5000 square feet (464.52 m^2). [ASHRAE 90.1:6.7.2.3.1]

E 503.6.5.3.1 Air System Balancing. Air systems shall be balanced in a manner to first minimize throttling losses. Then, for fans with fan system power greater than 1 hp (0.7 kW), fan speed shall be adjusted to meet design flow conditions. [ASHRAE 90.1:6.7.2.3.2]

E 503.6.5.3.2 Hydronic System Balancing. Hydronic systems shall be proportionately balanced in a manner to first minimize throttling losses; then the pump impeller shall be trimmed or pump speed shall be adjusted to meet design flow conditions.

Exceptions: Impellers need not be trimmed nor pump speed adjusted.

(1) For pumps with pump motors of 10 hp (7.5 kW) or less.

(2) Where throttling results is not greater than 5 percent of the nameplate horsepower draw, or 3 hp (2.2 kW), whichever is greater, above that required where the impeller was trimmed. [ASHRAE 90.1:6.7.2.3.3]

E 503.6.5.4 System Commissioning. HVAC control systems shall be tested to ensure that control elements are calibrated, adjusted, and in proper working condition. For projects larger than 50 000 square feet (4645.15 m^2) conditioned area, except warehouses and semiheated spaces, detailed instructions for commissioning HVAC systems shall be provided by the designer in plans and specifications. [ASHRAE 90.1:6.7.2.4]

E 503.6.5.4.1 Minimum Level of Commission. Commissioning shall be performed for HVAC systems in accordance with Level 1, Basic Commissioning of the SMACNA HVAC Systems Commissioning Manual. (See Section E 801.0 for additional information on HVAC system commissioning)

E 503.7 Minimum Equipment Efficiency Tables. The minimum efficiency requirements for equipment shall comply with Section E 503.7.1; duct insulation shall comply with Section E 503.7.2, and pipe insulation shall comply with Section E 503.7.3.

E 503.7.1 Minimum Efficiency Requirement Listed Equipment – Standard Rating and Operating Conditions. The minimum efficiency requirements for equipment shall comply with Table E 503.7.1(1) through Table E 503.7.1(16).

E 503.7.2 Duct Insulation Tables. Duct insulation shall comply with Table E 503.7.2.

E 503.7.3 Pipe Insulation Tables. Pipe insulation shall comply with Table E 503.7.3(1) through Table E 503.7.3(2).

E 503.8 Alternative Compliance Path. HVAC systems serving heating, cooling, or ventilation needs of a computer room shall be in accordance with Section E 503.1, Section E 503.4, Section E 503.8.1 or Section E 503.8.2, Section E 503.8.3, Section E 502.7 through Section E 502.7.2, and Section E 503.7. [ASHRAE 90.1:6.6.1]

E 503.8.1 Computer Room (PUE_1). The computer room PUE_1 shall be not more than the values listed in Table E 503.8.1. Hourly simulation of the proposed design, for purposes of calculating PUE_1, shall be in accordance with ASHRAE 90.1.

Exception: The compliance path shall not be permitted for a proposed computer room design utilizing a combined heat and power system. [ASHRAE 90.1:6.6.1.1]

TABLE E 503.8.1
POWER USAGE EFFECTIVENESS (PUE) MAXIMUM
[ASHRAE 90.1: TABLE 6.6.1]

CLIMATE ZONE	PUE*
0A	1.64
0B	1.62
1A	1.61
2A	1.49
3A	1.41
4A	1.36
5A	1.36
6A	1.34
1B	1.53
2B	1.45
3B	1.42
4B	1.38
5B	1.33
6B	1.33
3C	1.39
4C	1.38
5C	1.36
7	1.32
8	1.30

* PUE_0 and PUE_1 shall not include energy for battery charging.

E 503.8.2 Computer Room (PUE_0). The computer room PUE_0 shall be not more than the values listed in Table E 503.8.1. The PUE_0 shall be the highest value determined at outdoor cooling design temperatures, and shall be limited to systems utilizing electricity for an energy source. The PUE_0 shall be calculated for the following conditions:

(1) One hundred percent design IT equipment energy.

(2) Fifty percent design IT equipment energy. [ASHRAE 90.1:6.6.1.2]

E 503.8.3 Documentation. Documentation on the following components shall be provided, including a breakdown of energy consumption or demand:

(1) IT equipment

(2) Power distribution losses external to the IT equipment

(3) HVAC systems

(4) Lighting [ASHRAE 90.1:6.6.1.3]

E 504.0 Solar Energy Systems.

E 504.1 General. Solar energy systems shall be installed in accordance with the Uniform Solar Energy and Hydronics Code (USEHC).

TABLE E 503.7.1(1)
ELECTRICALLY OPERATED UNITARY AIR CONDITIONERS AND CONDENSING UNITS MINIMUM EFFICIENCY REQUIREMENTS
[ASHRAE 90.1: TABLE 6.8.1-1]

EQUIPMENT TYPE	SIZE CATEGORY	HEATING SECTION TYPE	SUBCATEGORY OR RATING CONDITION	MINIMUM EFFICIENCY	TEST PROCEDURE[1]
Air conditioners, air cooled	<65 000 Btu/h[2]	All	Split system, three phase	13.0 SEER	AHRI 210/240
			Single package, three phase	14 SEER	
Through the wall, air cooled	≤30 000 Btu/h[2]	All	Split system, three phase	12.0 SEER	AHRI 210/240
			Single package, three phase	12.0 SEER	
Small duct, high velocity, air cooled	<65 000 Btu/h[2]	All	Split system, three phase	11.0 SEER	AHRI 210/240
Air conditioners, air cooled	≥65 000 Btu/h and <135 000 Btu/h	Electric resistance (or none)	Split system and single package	11.2 EER 12.9 IEER	AHRI 340/360
		All other		11.0 EER 12.7 IEER	
	≥135 000 Btu/h and <240 000 Btu/h	Electric resistance (or none)		11.0 EER 12.4 IEER	
		All other		10.8 EER 12.2 IEER	
	≥240 000 Btu/h and <760 000 Btu/h	Electric resistance (or none)		10.0 EER 11.6 IEER	
		All other		9.8 EER 11.4 IEER	
	≥760 000 Btu/h	Electric resistance (or none)		9.7 EER 11.2 IEER	
		All other		9.5 EER 11.0 IEER	
Air conditioners, water cooled	<65 000 Btu/h	All	Split system and single package	12.1 EER 12.3 IEER	AHRI 210/240
	≥65 000 Btu/h and <135 000 Btu/h	Electric resistance (or none)		12.1 EER 13.9 IEER	AHRI 340/360
		All other		11.9 EER 13.7 IEER	
	≥135 000 Btu/h and <240 000 Btu/h	Electric resistance (or none)		12.5 EER 13.9 IEER	
		All other		12.3 EER 13.7 IEER	
	≥240 000 Btu/h and <760 000 Btu/h	Electric resistance (or none)		12.4 EER 13.6 IEER	
		All other		12.2 EER 13.4 IEER	
	≥760 000 Btu/h	Electric resistance (or none)		12.2 EER 13.5 IEER	
		All other		12.0 EER 13.3 IEER	

TABLE E 503.7.1(1) (continued)
ELECTRICALLY OPERATED UNITARY AIR CONDITIONERS AND CONDENSING UNITS
MINIMUM EFFICIENCY REQUIREMENTS
[ASHRAE 90.1: TABLE 6.8.1-1]

EQUIPMENT TYPE	SIZE CATEGORY	HEATING SECTION TYPE	SUBCATEGORY OR RATING CONDITION	MINIMUM EFFICIENCY	TEST PROCEDURE[1]
Air conditioners, evaporatively cooled	<65 000 Btu/h[2]	All	Split system and single package	12.1 EER 12.3 IEER	AHRI 210/240
	≥65 000 Btu/h and <135 000 Btu/h	Electric resistance (or none)		12.1 EER 12.3 IEER	AHRI 340/360
		All other		11.9 EER 12.1 IEER	
	≥135 000 Btu/h and <240 000 Btu/h	Electric resistance (or none)		12.0 EER 12.2 IERR	
		All other		11.8 EER 12.0 IEER	
	≥240 000 Btu/h and <760 000 Btu/h	Electric resistance (or none)		11.9 EER 12.1 IEER	
		All other		11.7 EER 11.9 IEER	
	≥760 000 Btu/h	Electric resistance (or none)		11.7 EER 11.9 IEER	
		All other		11.5 EER 11.7 IEER	
Condensing units, air cooled	≥135 000 Btu/h	–	–	10.5 EER 11.8 IEER	AHRI 365
Condensing units, water cooled	≥135 000 Btu/h	–	–	13.5 EER 14.0 IEER	AHRI 365
Condensing units, evaporatively cooled	≥135 000 Btu/h	–	–	13.5 EER 14.0 IEER	AHRI 365

For SI units: 1000 British thermal units per hour = 0.293 kW

Notes:

[1] ASHRAE 90.1 contains a complete specification of the referenced test procedure, including the referenced year version of the test procedure.

[2] Single-phase, air-cooled air conditioners less than 65 000 Btu/h (19 kW) are regulated by the U.S. Department of Energy Code of Federal Regulations 10 CFR 430. SEER values for single-phase products are set by the U.S. Department of Energy.

TABLE E 503.7.1(2)
ELECTRICALLY OPERATED UNITARY AND APPLIED HEAT PUMPS
MINIMUM EFFICIENCY REQUIREMENTS
[ASHRAE 90.1: TABLE 6.8.1-2]

EQUIPMENT TYPE	SIZE CATEGORY	HEATING SECTION TYPE	SUBCATEGORY OR RATING CONDITION	MINIMUM EFFICIENCY	TEST PROCEDURE[1]
Air cooled (cooling mode)	<65 000 Btu/h[2]	All	Split system, three phase	14 SEER	AHRI 210/240
			Single package, three phase	14 SEER	
Through the wall, air cooled (cooling mode)	≤30 000 Btu/h[2]	All	Split system, three phase	12.0 SEER	AHRI 210/240
			Single package, three phase	12.0 SEER	
Small duct, high velocity, air cooled	<65 000 Btu/h[2]	All	Split System, three phase	11.0 SEER	AHRI 210/240
Air cooled (cooling mode)	≥65 000 Btu/h and <135 000 Btu/h	Electric resistance (or none)	Split system and single package	11.0 EER 12.2 IEER	AHRI 340/360
		All other		10.8 EER 12.0 IEER	
	≥135 000 Btu/h and <240 000 Btu/h	Electric resistance (or none)		10.6 EER 11.6 IEER	
		All other		10.4 EER 11.4 IEER	
	≥240 000 Btu/h	Electric resistance (or none)		9.5 EER 10.6 IEER	
		All other		9.3 EER 10.4 IEER	
Water to air, water loop (cooling mode)	<17 000 Btu/h	All	86°F entering water	12.2 EER	ISO 13256-1
	≥17 000 Btu/h and <65 000 Btu/h			13.0 EER	
	≥65 000 Btu/h and <135 000 Btu/h			13.0 EER	
Water to air, groundwater (cooling mode)	<135 000 Btu/h	All	59°F entering water	18.0 EER	ISO 13256-1
Brine to air, ground loop (cooling mode)	<135 000 Btu/h	All	77°F entering water	14.1 EER	ISO 13256-1
Water to water, water loop (cooling mode)	<135 000 Btu/h	All	86°F entering water	10.6 EER	ISO 13256-2
Water to water, groundwater (cooling mode)	<135 000 Btu/h	All	59°F entering water	16.3 EER	ISO 13256-2
Brine to water, ground loop (cooling mode)	<135 000 Btu/h	All	77°F entering water	12.1 EER	ISO 13256-2
Air cooled (heating mode)	<65 000 Btu/h[2] (cooling capacity)	–	Split system, three phase	8.2 HSPF	AHRI 210/240
			Single package, three phase	8.0 HSPF	
Through the wall, air cooled (heating mode)	≤30 000 Btu/h[2] (cooling capacity)	–	Split system, three phase	7.4 HSPF	AHRI 210/240
			Single package, three phase	7.4 HSPF	
Small duct high velocity, air cooled (heating mode)	<65 000 Btu/h[2]	–	Split system, three phase	6.8 HSPF	AHRI 210/240

TABLE E 503.7.1(2) (continued)
ELECTRICALLY OPERATED UNITARY AND APPLIED HEAT PUMPS
MINIMUM EFFICIENCY REQUIREMENTS
[ASHRAE 90.1: TABLE 6.8.1-2]

EQUIPMENT TYPE	SIZE CATEGORY	HEATING SECTION TYPE	SUBCATEGORY OR RATING CONDITION	MINIMUM EFFICIENCY	TEST PROCEDURE[1]
Air cooled (heating mode)	≥65 000 Btu/hc and <135 000 Btu/h (cooling capacity)	–	47°F db/43°F wb outdoor air	3.3 COP_H	AHRI 340/360
			17°F db/15°F wb outdoor air	2.25 COP_H	
	≥135 000 Btu/hc (cooling capacity)		47°F db/43°F wb outdoor air	3.2 COP_H	
			17°F db/15°F wb outdoor air	2.05 COP_H	
Water to air, water loop (heating mode)	<135 000 Btu/h (cooling capacity)	–	68°F entering water	4.3 COP_H	ISO 13256-1
Water to air, groundwater (heating mode)	<135 000 Btu/h (cooling capacity)	–	50°F entering water	3.7 COP_H	ISO 13256-1
Brine to air, ground loop (heating mode)	<135 000 Btu/h (cooling capacity)	–	32°F entering fluid	3.2 COP_H	ISO 13256-1
Water to water, water loop (heating mode)	<135 000 Btu/h (cooling capacity)	–	68°F entering water	3.7 COP_H	ISO 13256-2
Water to water, groundwater (heating mode)	<135 000 Btu/h (cooling capacity)	–	50°F entering water	3.1 COP_H	ISO 13256-2
Brine to water, ground loop (heating mode)	<135 000 Btu/h (cooling capacity)	–	32°F entering fluid	2.5 COP_H	ISO 13256-2

For SI units: 1000 British thermal units per hour = 0.293 kW, °C = (°F-32)/1.8

Notes:

1 ASHRAE 90.1 contains a complete specification of the referenced test procedure, including the referenced year version of the test procedure.

2 Single-phase, air-cooled heat pumps less than 65 000 Btu/h (19 kW) are regulated by the U.S. Department of Energy Code of Federal Regulations 10 CFR 430. SEER and HSPF values for single-phase products are set by the U.S. Department of Energy.

TABLE E 503.7.1(3)
WATER-CHILLING PACKAGES – MINIMUM EFFICIENCY REQUIREMENTS[1, 2, 5]
[ASHRAE 90.1: TABLE 6.8.1-3]

EQUIPMENT TYPE	SIZE CATEGORY	UNITS	PATH A	PATH B	TEST PROCEDURE[3]
Air-cooled chillers	<150 tons	EER (Btu/Wh)	≥10.100 *FL*	≥9.700 *FL*	AHRI 550/590
			≥13.700 *IPLV.IP*	≥15.800 *IPLV.IP*	
	≥150 tons		≥10.100 *FL*	≥9.700 *FL*	
			≥14.000 *IPLV.IP*	≥16.100 *IPLV.IP*	
Air-cooled without condenser, electrically operated	All capacities	EER (Btu/Wh)	Air-cooled chillers without condenser must be rated with matching condensers and comply with air-cooled chiller efficiency requirements		AHRI 550/590
Water-cooled, electrically operated positive displacement	<75 tons	kW/ton	≤0.750 *FL*	≤0.780 *FL*	AHRI 550/590
			≤0.600 *IPLV.IP*	≤0.500 *IPLV.IP*	
	≥75 tons and <150 tons		≤0.720 *FL*	≤0.750 *FL*	
			≤0.560 *IPLV.IP*	≤0.490 *IPLV.IP*	
	≥150 tons and <300 tons		≤0.660 *FL*	≤0.680 *FL*	
			≤0.540 *IPLV.IP*	≤0.440 *IPLV.IP*	
	≥300 tons and <600 tons		≤0.610 *FL*	≤0.625 *FL*	
			≤0.520 *IPLV.IP*	≤0.410 *IPLV.IP*	
	≥600 tons		≤0.560 *FL*	≤0.585 *FL*	
			≤0.500 *IPLV.IP*	≤0.380 *IPLV.IP*	
Water cooled, electrically operated centrifugal	<150 tons	kW/ton	≤0.610 FL	≤0.695 *FL*	AHRI 550/590
			≤0.550 *IPLV.IP*	≤0.440 *IPLV.IP*	
	≥150 tons and <300 tons		≤0.610 *FL*	≤0.635 *FL*	
			≤0.550 *IPLV.IP*	≤0.400 *IPLV.IP*	
	≥300 tons and <400 tons		≤0.560 *FL*	≤0.595 *FL*	
			≤0.520 *IPLV.IP*	≤0.390 *IPLV.IP*	
	≥400 tons and <600 tons		≤0.560 *FL*	≤0.585 *FL*	
			≤0.500 *IPLV.IP*	≤0.380 *IPLV.IP*	
	≥600 tons		≤0.560 *FL*	≤0.585 *FL*	
			≤0.500 *IPLV.IP*	≤0.380 *IPLV.IP*	
Air-cooled absorption, single effect	All capacities	COP (W/W)	≥0.600 *FL*	NA[4]	AHRI 560
Water-cooled absorption, single effect	All capacities	COP (W/W)	≥0.700 *FL*	NA[4]	AHRI 560
Absorption double effect, indirect fired	All capacities	COP (W/W)	≥1.000 *FL*	NA[4]	AHRI 560
			≥1.050 *IPLV.IP*		
Absorption double effect, direct fired	All capacities	COP (W/W)	≥1.000 *FL*	NA[4]	AHRI 560
			≥1.000 *IPLV*		

For SI units: 1 metric ton = 1000 kg, 1000 British thermal units per hour = 0.293 kW

Notes:

1 The requirements for centrifugal chillers shall be adjusted for nonstandard rating conditions per Section E 503.4.1 and are only applicable for the range of conditions listed there. The requirements for air-cooled, water-cooled positive displacement and absorption chillers are at standard rating conditions defined in the reference test procedure.

2 Both the full-load and *IPLV.IP* requirements must be met or exceeded to comply with this appendix. When there is a Path B, compliance can be with either Path A or Path B for any application.

3 ASHRAE 90.1 contains a complete specification of the referenced test procedure, including the referenced year version of the test procedure.

4 NA means the requirements are not applicable for Path B, and only Path A can be used for compliance.

5 *FL* is the full-load performance requirements, and *IPLV.IP* is for the part-load performance requirements.

TABLE E 503.7.1(4)
ELECTRICALLY OPERATED PACKAGED TERMINAL AIR CONDITIONERS, PACKAGED TERMINAL HEAT PUMPS, SINGLE-PACKAGE VERTICAL AIR CONDITIONERS, SINGLE-PACKAGE VERTICAL HEAT PUMPS, ROOM AIR CONDITIONERS, AND ROOM AIR CONDITIONER HEAT PUMPS – MINIMUM EFFICIENCY REQUIREMENTS
[ASHRAE 90.1: TABLE 6.8.1-4]

EQUIPMENT TYPE	SIZE CATEGORY (INPUT)	SUBCATEGORY OR RATING CONDITION	MINIMUM EFFICIENCY	TEST PROCEDURE[1]
PTAC (cooling mode) standard size	All capacities	95°F db outdoor air	13.8 – (0.300 × Cap/1000)[3] (before 1/1/2015)	AHRI 310/ 380
			14.0 – (0.300 × Cap/1000)[3] (as of 1/1/2015)	
PTAC (cooling mode) nonstandard size[1]	All capacities	95°F db outdoor air	10.9 – (0.213 × Cap/1000)[3] EER	AHRI 310/ 380
PTHP (cooling mode) standard size	All capacities	95°F db outdoor air	14.0 – (0.300 × Cap/1000)[3]	AHRI 310/ 380
PTHP (cooling mode) nonstandard size[2]	All capacities	95°F db outdoor air	10.8 – (0.213 × Cap/1000)[3] EER	AHRI 310/ 380
PTHP (heating mode) standard size	All capacities	–	3.7 – (0.052 × Cap/1000)[3] COP_H	AHRI 310/ 380
PTHP (heating mode) nonstandard size[2]	All capacities	–	2.9 – (0.026 × Cap/1000)[3] COP_H	AHRI 310/ 380
SPVAC (cooling mode)	<65,000 Btu/h	95°F db/75°F wb outdoor air	10.0 EER	AHRI 390
	≥65 000 Btu/h and <135 000 Btu/h		10.0 EER	
	≥135 000 Btu/h and <240 000 Btu/h		10.0 EER	
SPVHP (cooling mode)	<65 000 Btu/h	95°F db/75°F wb outdoor air	10.0 EER	AHRI 390
	≥65 000 Btu/h and <135 000 Btu/h		10.0 EER	
	≥135 000 Btu/h and <240 000 Btu/h		10.0 EER	
SPVHP (heating mode)	<65 000 Btu/h	47°F db/43°F wb outdoor air	3.0 COP_H	AHRI 390
	≥65 000 Btu/h and <135 000 Btu/h		3.0 COP_H	
	≥135 000 Btu/h and <240 000 Btu/h		3.0 COP_H	
Room air conditioners with louvered sides	<6000 Btu/h	–	9.7 SEER	AHAM RAC-1
	≥6000 Btu/h and <8000 Btu/h		9.7 SEER	
	≥8000 Btu/h and <14 000 Btu/h		9.8 EER	
	≥14 000 Btu/h and <20 000 Btu/h		9.7 SEER	
	≥20 000 Btu/h		8.5 EER	
SPVAC (cooling mode), nonweatherized space constrained	≤30 000 Btu/h	95°F db/75°F wb outdoor air	9.2 EER	AHRI 390
	>30 000 Btu/h and ≤36 000 Btu/h		9.0 EER	
SPVHP (cooling mode), nonweatherized space constrained	≤30 000 Btu/h	95°F db/75°F wb outdoor air	9.2 EER	AHRI 390
	>30 000 Btu/h and ≤36 000 Btu/h		9.0 EER	
SPVHP (heating mode), nonweatherized space constrained	≤30 000 Btu/h	47°F db/43°F wb outdoor air	3.0 COP_H	AHRI 390
	>30 000 Btu/h and ≤36 000 Btu/h		3.0 COP_H	

TABLE E 503.7.1(4) (continued)
ELECTRICALLY OPERATED PACKAGED TERMINAL AIR CONDITIONERS, PACKAGED TERMINAL HEAT PUMPS, SINGLE-PACKAGE VERTICAL AIR CONDITIONERS, SINGLE-PACKAGE VERTICAL HEAT PUMPS, ROOM AIR CONDITIONERS, AND ROOM AIR CONDITIONER HEAT PUMPS – MINIMUM EFFICIENCY REQUIREMENTS
[ASHRAE 90.1: TABLE 6.8.1-4]

EQUIPMENT TYPE	SIZE CATEGORY (INPUT)	SUBCATEGORY OR RATING CONDITION	MINIMUM EFFICIENCY	TEST PROCEDURE[1]
Room air conditioners without louvered sides	<8000 Btu/h	–	9.0 EER	AHAM RAC-1
	≥8000 Btu/h and <20 000 Btu/h	–	8.5 EER	
	≥20 000 Btu/h	–	8.5 EER	
Room air conditioner heat pumps with louvered sides	<20 000 Btu/h	–	9.0 EER	AHAM RAC-1
	≥20 000 Btu/h		8.5 EER	
Room air conditioner heat pumps without louvered sides	<14 000 Btu/h	–	8.5 EER	AHAM RAC-1
	≥14 000 Btu/h		8.0 EER	
Room air conditioner, casement only	All capacities	–	8.7 EER	AHAM RAC-1
Room air conditioner, casement slider	All capacities	–	9.5 EER	AHAM RAC-1

For SI units: 1000 British thermal units per hour = 0.293 kW, °C = (°F-32)/1.8

Notes:

1 ASHRAE 90.1 contains a complete specification of the referenced test procedure, including the referenced year version of the test procedure.

2 Nonstandard size units must be factory labeled as follows: "MANUFACTURED FOR NONSTANDARD SIZE APPLICATIONS ONLY; NOT TO BE INSTALLED IN NEW STANDARD PROJECTS." Nonstandard size efficiencies apply only to units being installed in existing sleeves having an external wall opening of less than 16 inch (406 mm) high or less than 42 inch (1067 mm) wide and having a cross-sectional area less than 670 square inches (0.432 m^2).

3 "Cap" means the rated cooling capacity of the product in Btu/h (kW). If the unit's capacity is less than 7000 Btu/h (2.05 kW), use 7000 Btu/h (2.05 kW) in the calculation. Where the unit's capacity is more than 15 000 Btu/h (4.4 kW), use 15 000 Btu/h (4.4 kW) in the calculation.

TABLE E 503.7.1(5)
WARM-AIR FURNACES AND COMBINATION WARM-AIR FURNACES/AIR-CONDITIONING UNITS, WARM-AIR DUCT FURNACES, AND UNIT HEATERS – MINIMUM EFFICIENCY REQUIREMENTS
[ASHRAE 90.1: TABLE 6.8.1-5]

EQUIPMENT TYPE	SIZE CATEGORY (INPUT)	SUBCATEGORY OR RATING CONDITION	MINIMUM EFFICIENCY	TEST PROCEDURE[1]
Warm-air furnace, gas fired	<225 000 Btu/h	Maximum capacity[3]	78% *AFUE* or 80% E_t[2, 4]	DOE 10 CFR Part 430 or Section 2.39, Thermal Efficiency, CSA Z21.47
	≥225 000 Btu/h		80% E_t[4]	Section 2.39, Thermal Efficiency, CSA Z21.47
Warm-air furnace, oil fired	<225 000 Btu/h	Maximum capacity[3]	78% *AFUE* or 80% E_t[2, 4]	DOE 10 CFR Part 430 or Section 42, Combustion, UL 727
	≥225 000 Btu/h		81% E_t[4]	Section 42, Combustion, UL 727
Warm-air duct furnaces, gas fired	All capacities	Maximum capacity[3]	80% E_c[5]	Section 2.10, Efficiency, CSA Z83.8
Warm-air unit heaters, gas fired	All capacities	Maximum capacity[3]	80% E_c[5, 6]	Section 2.10, Efficiency, CSA Z83.8
Warm-air unit heaters, oil fired	All capacities	Maximum capacity[3]	80% E_c[5, 6]	Section 40, Combustion, UL 731

For SI units: 1000 British thermal units per hour = 0.293 kW

Notes:

1 ASHRAE 90.1 contains a complete specification of the referenced test procedure, including the referenced year version of the test procedure.

2 Combination units not covered by the U.S. Department of Energy Code of Federal Regulations 10 CFR 430 [three-phase power or cooling capacity greater than or equal to 65 000 Btu/h (19 kW)] may comply with either rating.

3 Compliance of multiple firing rate units shall be at the maximum firing rate.

4 *Et* = thermal efficiency. Units must also include an interrupted or intermittent ignition device (IID), have jacket losses not exceeding 0.75 percent of the input rating, and have either power venting or a flue damper. A vent damper is an acceptable alternative to a flue damper for those furnaces where combustion air is drawn from the conditioned space.

5 *Ec* = combustion efficiency (100 percent less flue losses). See test procedure for detailed discussion.

6 As of August 8, 2008, according to the Energy Policy Act of 2005, units must also include an interrupted or intermittent ignition device (IID) and have either power venting or an automatic flue damper.

TABLE E 503.7.1(6)
GAS- AND OIL-FIRED BOILERS – MINIMUM EFFICIENCY REQUIREMENTS[2, 3]
[ASHRAE 90.1: TABLE 6.8.1-6]

EQUIPMENT TYPE[1]	SUBCATEGORY OR RATING CONDITION	SIZE CATEGORY (INPUT)	MINIMUM EFFICIENCY	EFFICIENCY AS OF 3/2/2020	TEST PROCEDURE
Boilers, hot water	Gas fired	<300 000 Btu/h[6,7]	82% *AFUE*	82% *AFUE*	10 CFR Part 430
		≥300 000 Btu/h and ≤2 500 000 Btu/h[4]	80% E_t	80% E_t	10 CFR Part 431
		>2 500 000 Btu/h[1]	82% E_c	82% E_c	
	Oil fired[5]	<300 000 Btu/h[7]	84% *AFUE*	84% *AFUE*	10 CFR Part 430
		≥300 000 Btu/h and ≤2 500 000 Btu/h[4]	82% E_t	82% E_t	10 CFR Part 431
		>2 500 000 Btu/h[1]	84% E_c	84% E_c	
Boilers, steam	Gas fired	<300 000 Btu/h[7]	80% *AFUE*	80% *AFUE*	10 CFR Part 430
	Gas fired— all, except natural draft	≥300 000 Btu/h and ≤2 500 000 Btu/h[4]	79% E_t	79% E_t	10 CFR Part 430
		>2 500 000 Btu/h[1]	79% E_t	79% E_t	
	Gas fired— natural draft	≥300 000 Btu/h and ≤2 500 000 Btu/h[4]	77% E_t	79% E_t	
		>2 500 000 Btu/h[1]	77% E_t	79% E_t	
	Oil fired[5]	<300 000 Btu/h	82% *AFUE*	82% *AFUE*	10 CFR Part 430
		≥300 000 Btu/h and ≤2 500 000 Btu/h[4]	81% E_t	81% E_t	10 CFR Part 431
		>2 500 000 Btu/h[1]	81% E_t	81% E_t	

For SI units: 1000 British thermal units per hour = 0.293 kW

Notes:

1 These requirements apply to boilers with rated input of 8 000 000 Btu/h (2343 kW) or less that are not packaged boilers and to all packaged boilers. Minimum efficiency requirements for boilers cover all capacities of packaged boilers.

2 *Ec* = combustion efficiency (100 percent less flue losses). See reference document for detailed information.

3 *Et* = thermal efficiency. See reference document for detailed information.

4 Maximum capacity—minimum and maximum ratings as provided for and allowed by the unit's controls.

5 Includes oil-fired (residual).

6 Boilers shall not be equipped with a constant burning pilot light.

7 A boiler not equipped with a tankless domestic water-heating coil shall be equipped with an automatic means for adjusting the temperature of the water such that an incremental change in inferred heat load produces a corresponding incremental change in the temperature of the water supplied.

TABLE E 503.7.1(7)
PERFORMANCE REQUIREMENTS FOR HEAT REJECTION EQUIPMENT
MINIMUM EFFICIENCY REQUIREMENTS
[ASHRAE 90.1: TABLE 6.8.1-7]

EQUIPMENT TYPE	TOTAL SYSTEM HEAT-REJECTION CAPACITY AT RATED CONDITIONS	SUBCATEGORY OR RATING CONDITION[8]	PERFORMANCE REQUIRED[1,2,3,6,7]	TEST PROCEDURE[4,5]
Propeller or axial fan open-circuit cooling towers	All	95°F entering water 85°F leaving water 75°F entering wb	≥40.2 gpm/hp	CTI ATC-105 and CTI STD-201 RS
Centrifugal fan open-circuit cooling towers	All	95°F entering water 85°F leaving water 75°F entering wb	≥20.0 gpm/hp	CTI ATC-105 and CTI STD-201 RS
Propeller or axial fan closed-circuit cooling towers	All	102°F entering water 90°F leaving water 75°F entering wb	≥16.1 gpm/hp	CTI ATC-105S and CTI STD-201 RS
Centrifugal closed- circuit cooling towers	All	102°F entering water 90°F leaving water 75°F entering wb	≥7.0 gpm/hp	CTI ATC-105S and CTI STD-201 RS
Propeller or axial fan evaporative condensers	All	R-507A test fluid 165°F entering gas temperature 105°F condensing temperature 75°F entering wb	≥157 000 Btu/h·hp	CTI ATC-106
Propeller or axial fan evaporative condensers	All	Ammonia test fluid 140°F entering gas temperature 96.3°F condensing temperature 75°F entering wb	≥134 000 Btu/h·hp	CTI ATC-106
Centrifugal fan evaporative condensers	All	R-507A test fluid 165°F entering gas temperature 105°F condensing temperature 75°F entering wb	≥135 000 Btu/h·hp	CTI ATC-106
Centrifugal fan evaporative condensers	All	Ammonia test fluid 140°F entering gas temperature 96.3°F condensing temperature 75°F entering wb	≥110 000 Btu/h·hp	CTI ATC-106
Air cooled condensers	All	125°F condensing temperature 190°F entering gas temperature 15°F subcooling 95°F entering db	≥176 000 Btu/h·hp	AHRI 460

For SI units: °C = (°F-32)/1.8, 1 gallon per minute per horsepower = 0.085 [(L/s)/kW], 1000 British thermal units per hour = 0.293 kW, 1 horsepower = 0.746 kW

Notes:

1 For purposes of this table, open-circuit cooling tower performance is defined as the water flow rating of the tower at the thermal rating condition listed in Table E 503.7.1(7) divided by the fan motor nameplate power.

2 For purposes of this table, closed-circuit cooling tower performance is defined as the process water flow rating of the tower at the thermal rating condition listed in Table E 503.7.1(7) divided by the sum of the fan motor nameplate power and the integral spray pump motor nameplate power.

3 For purposes of this table, air-cooled condenser performance is defined as the heat rejected from the refrigerant divided by the fan motor nameplate power.

4 ASHRAE 90.1 contains a complete specification of the referenced test procedure, including the referenced year version of the test procedure.

5 The efficiencies and test procedures for both open- and closed-circuit cooling towers are not applicable to hybrid cooling towers that contain a combination of separate wet and dry heat exchange sections. The certification requirements do not apply to field-erected cooling towers.

6 All cooling towers shall comply with the minimum efficiency listed in the table for that specific type of tower with the capacity effect of any project-specific accessories and/or options included in the capacity of the cooling tower.

7 For purposes of this table, evaporative condenser performance is defined as the heat rejected at the specified rating condition in the table, divided by the sum of the fan motor nameplate power and the integral spray pump nameplate power.

8 Requirements for evaporative condensers are listed with ammonia (R-717) and R-507A as test fluids in the table. Evaporative condensers intended for use with halocarbon refrigerants other than R-507A must meet the minimum efficiency requirements listed above with R-507A as the test fluid.

TABLE E 503.7.1(8)
HEAT TRANSFER EQUIPMENT – MINIMUM EFFICIENCY REQUIREMENTS
[ASHRAE 90.1: TABLE 6.8.1-8]

EQUIPMENT TYPE	SUBCATEGORY	MINIMUM EFFICIENCY[1]	TEST PROCEDURE[2]
Liquid-to-liquid heat exchangers	Plate type	NR	AHRI 400

Notes:

[1] NR = No Requirement

[2] ASHRAE 90.1 contains a complete specification of the referenced test procedure, including the referenced year version of the test procedure.

TABLE E 503.7.1(9)
ELECTRICALLY OPERATED VARIABLE-REFRIGERANT-FLOW AIR CONDITIONERS – MINIMUM EFFICIENCY REQUIREMENTS
[ASHRAE 90.1: TABLE 6.8.1-9]

EQUIPMENT TYPE	SIZE CATEGORY	HEATING SECTION TYPE	SUBCATEGORY OR RATING CONDITION	MINIMUM EFFICIENCY	TEST PROCEDURE
VRF air conditioners, air cooled	<65 000 Btu/h	All	VRF multisplit system	13.0 SEER	AHRI 1230
	≥65 000 Btu/h and <135 000 Btu/h	Electric resistance (or none)	VRF multisplit system	11.2 EER 13.1 IEER (before 1/1/2017) 15.5 IEER (as of 1/1/2017)	
	≥135 000 Btu/h and <240 000 Btu/h	Electric resistance (or none)	VRF multisplit system	11.0 EER 12.9 IEER (before 1/1/2017) 14.9 IEER (as of 1/1/2017)	
	≥240 000 Btu/h	Electric resistance (or none)	VRF multisplit system	10.0 EER 11.6 IEER (before 1/1/2017) 13.9 IEER (as of 1/1/2017)	

For SI units: 1000 British thermal units per hour = 0.293 kW

TABLE E 503.7.1(10)
ELECTRICALLY OPERATED VARIABLE-REFRIGERANT-FLOW AND APPLIED HEAT PUMPS – MINIMUM EFFICIENCY REQUIREMENTS
[ASHRAE 90.1: TABLE 6.8.1-10]

EQUIPMENT TYPE	SIZE CATEGORY	HEATING SECTION TYPE	SUBCATEGORY OR RATING CONDITION	MINIMUM EFFICIENCY	TEST PROCEDURE
VRF air cooled (cooling mode)	<65 000 Btu/h	All	VRF multisplit system	13.0 SEER	AHRI 1230
	≥65 000 Btu/h and <135 000 Btu/h	Electric resistance (or none)	VRF multisplit system	11.0 EER 12.9 IEER (before 1/1/2017) 14.6 IEER (as of 1/1/2017)	
			VRF multisplit system with heat recovery	10.8 EER 12.7 IEER (before 1/1/2017) 14.4 IEER (as of 1/1/2017)	
	≥135 000 Btu/h and <240 000 Btu/h		VRF multisplit system	10.6 EER 12.3 IEER (before 1/1/2017) 13.9 IEER (as of 1/1/2017)	
			VRF multisplit system with heat recovery	10.4 EER 12.1 IEER (before 1/1/2017) 13.7 IEER (as of 1/1/2017)	
	≥240 000 Btu/h		VRF multisplit system	9.5 EER 11.0 IEER (before 1/1/2017) 12.7 IEER (as of 1/1/2017)	
			VRF multisplit system with heat recovery	9.3 EER 10.8 IEER (before 1/1/2017) 12.5 IEER (as of 1/1/2017)	
VRF water source (cooling mode)	<65 000 Btu/h	All	VRF multisplit systems 86°F entering water	12.0 EER 16.0 IEER (as of 1/1/2018)	AHRI 1230
			VRF multisplit systems with heat recovery 86°F entering water	11.8 EER 15.8 IEER (as of 1/1/2018)	
	≥65 000 Btu/h and <135 000 Btu/h		VRF multisplit system 86°F entering water	12.0 EER 16.0 IEER (as of 1/1/2018)	
			VRF multisplit system with heat recovery 86°F entering water	11.8 EER 15.8 IEER (as of 1/1/2018)	
	≥135 000 Btu/h and <240 000 Btu/h		VRF multisplit system 86°F entering water	10.0 EER 14.0 IEER (as of 1/1/2018)	
			VRF multisplit system with heat recovery 86°F entering water	9.8 EER 13.8 IEER (as of 1/1/2018)	
	≥240 000 Btu/h		VRF multisplit system 86°F entering water	10.0 EER (before 1/1/2018) 12.0 IEER (as of 1/1/2018)	
			VRF multisplit system with heat recovery 86°F entering water	9.8 EER (before 1/1/2018) 11.8 IEER (as of 1/1/2018)	

**TABLE E 503.7.1(10) (continued)
ELECTRICALLY OPERATED VARIABLE-REFRIGERANT-FLOW AND APPLIED HEAT PUMPS – MINIMUM EFFICIENCY REQUIREMENTS
[ASHRAE 90.1: TABLE 6.8.1-10]**

EQUIPMENT TYPE	SIZE CATEGORY	HEATING SECTION TYPE	SUBCATEGORY OR RATING CONDITION	MINIMUM EFFICIENCY	TEST PROCEDURE
VRF groundwater source (cooling mode)	<135 000 Btu/h	All	VRF multisplit system with heat recovery 59°F entering water	16.2 EER	AHRI 1230
			VRF multisplit system with heat recovery 59°F entering water	16.0 EER	
	≥135 000 Btu/h		VRF multisplit system with heat recovery 59°F entering water	13.8 EER	
			VRF multisplit system with heat recovery 59°F entering water	13.6 EER	
VRF ground source (cooling mode)	<135 000 Btu/h	All	VRF multisplit system 77°F entering water	13.4 EER	AHRI 1230
			VRF multisplit system with heat recovery 77°F entering water	13.2 EER	
	≥135 000 Btu/h		VRF multisplit system 77°F entering water	11.0 EER	
			VRF multisplit system with heat recovery 77°F entering water	10.8 EER	
VRF Air cooled (heating mode)	<65 000 Btu/h (cooling capacity)	—	VRF Multi-split system	7.7 HSPF	AHRI 1230
	≥65 000 Btu/h and <135 000 Btu/h	—	VRF Multi-split system 47°F db/43°F wb outdoor air	3.3 COP_H	
			17°F db/15°F wb outdoor air	2.25 COP_H	
	≥135 000 Btu/h (cooling capacity)	—	VRF Multi-split system 47°F db/43°F wb outdoor air	3.2 COP_H	
			17°F db/15°F wb outdoor air	2.05 COP_H	
VRF Water source (heating mode)	<65 000 Btu/h (cooling capacity)	—	*VRF* multisplit system 68°F entering water	4.2 COP_H (before 1/1/2018) 4.3 COP_H (as of 1/1/2018)	AHRI 1230
	≥65 000 Btu/h and <135 000 Btu/h (cooling capacity)	—	*VRF* multisplit system 68°F entering water	4.2 COP_H (before 1/1/2018) 4.3 COP_H (as of 1/1/2018)	
	≥135 000 Btu/h and <240 000 Btu/h (cooling capacity)	—	*VRF* multisplit system 68°F entering water	3.9 COP_H (before 1/1/2018) 4.0 COP_H (as of 1/1/2018)	
	≥240 000 Btu/h (cooling capacity)	—	*VRF* multisplit system 68°F entering water	3.9 COP_H	
VRF Groundwater source (heating mode)	<135 000 Btu/h (cooling capacity)	—	VRF Multi-split system 50°F entering water	3.6 COP_H	AHRI 1230
	≥135 000 Btu/h (cooling capacity)	—	VRF Multi-split system 50°F entering water	3.3 COP_H	
VRF Ground source (heating mode)	<135 000 Btu/h (cooling capacity)	—	VRF Multi-split system 32°F entering water	3.1 COP_H	AHRI 1230
	≥135 000 Btu/h (cooling capacity)	—	VRF Multi-split system 32°F entering water	2.8 COP_H	

For SI units: 1000 British thermal units per hour = 0.293 kW, °C=(°F-32)/1.8

TABLE E 503.7.1(11)
AIR CONDITIONERS AND CONDENSING UNITS SERVING COMPUTER ROOMS
MINIMUM EFFICIENCY REQUIREMENTS
[ASHRAE 90.1: TABLE 6.8.1-11]

EQUIPMENT TYPE	NET SENSIBLE COOLING CAPACITY	STANDARD MODEL	MINIMUM NET SENSIBLE COP_C RETURN AIR DRY-BULB TEMPERATURE/ DEW-POINT TEMPERATURE			TEST PROCEDURE
			CLASS 1 75°F/52°F	CLASS 2 85°F/52°F	CLASS 3 95°F/52°F	
Air cooled	<65 000 Btu/h	Downflow unit		2.30		AHRI 1360
		Upflow unit-ducted		2.10		
		Upflow unit-nonducted	2.09			
		Horizontal-flow unit			2.45	
	≥65 000 Btu/h and <240 000 Btu/h	Downflow unit		2.20		
		Upflow unit-ducted		2.05		
		Upflow unit-nonducted	1.99			
		Horizontal-flow unit			2.35	
	≥240 000 Btu/h	Downflow unit		2.00		
		Upflow unit-ducted		1.85		
		Upflow unit-nonducted	1.79			
		Horizontal-flow unit			2.15	
Water cooled	<65 000 Btu/h	Downflow unit		2.50		AHRI 1360
		Upflow unit-ducted		2.30		
		Upflow unit-nonducted	2.25			
		Horizontal-flow unit			2.70	
	≥65 000 Btu/h and <240 000 Btu/h	Downflow unit		2.40		
		Upflow unit-ducted		2.20		
		Upflow unit-nonducted	2.15			
		Horizontal-flow unit			2.60	
	≥240 000 Btu/h	Downflow unit		2.25		
		Upflow unit-ducted		2.10		
		Upflow unit-nonducted	2.05			
		Horizontal-flow unit			2.45	
Water cooled with fluid economizer	<65 000 Btu/h	Downflow unit		2.45		AHRI 1360
		Upflow unit-ducted		2.25		
		Upflow unit-nonducted	2.20			
		Horizontal-flow unit			2.60	
	≥65 000 Btu/h and <240 000 Btu/h	Downflow unit		2.35		
		Upflow unit-ducted		2.15		
		Upflow unit-nonducted	2.10			
		Horizontal-flow unit			2.55	
	≥240 000 Btu/h	Downflow unit		2.20		
		Upflow unit-ducted		2.05		
		Upflow unit-nonducted	2.00			
		Horizontal-flow unit			2.40	

**TABLE E 503.7.1(11) (continued)
AIR CONDITIONERS AND CONDENSING UNITS SERVING COMPUTER ROOMS
MINIMUM EFFICIENCY REQUIREMENTS
[ASHRAE 90.1: TABLE 6.8.1-11]**

EQUIPMENT TYPE	NET SENSIBLE COOLING CAPACITY	STANDARD MODEL	MINIMUM NET SENSIBLE COP_C			TEST PROCEDURE
			RETURN AIR DRY-BULB TEMPERATURE/ DEW-POINT TEMPERATURE			
			CLASS 1	CLASS 2	CLASS 3	
			75°F/52°F	85°F/52°F	95°F/52°F	
Glycol cooled	<65 000 Btu/h	Downflow unit		2.30		AHRI 1360
		Upflow unit-ducted		2.10		
		Upflow unit-nonducted	2.00			
		Horizontal-flow unit			2.40	
	≥65 000 Btu/h and <240 000 Btu/h	Downflow unit		2.05		
		Upflow unit-ducted		1.85		
		Upflow unit-nonducted	1.85			
		Horizontal-flow unit			2.15	
	≥240 000 Btu/h	Downflow unit		1.95		
		Upflow unit-ducted		1.80		
		Upflow unit-nonducted	1.75			
		Horizontal-flow unit			2.10	
Glycol cooled with fluid economizer	<65 000 Btu/h	Downflow unit		2.25		AHRI 1360
		Upflow unit-ducted		2.10		
		Upflow unit-nonducted	2.00			
		Horizontal-flow unit			2.35	
	≥65 000 Btu/h and <240 000 Btu/h	Downflow unit		1.95		
		Upflow unit-ducted		1.80		
		Upflow unit-nonducted	1.75			
		Horizontal-flow unit			2.10	
	≥240 000 Btu/h	Downflow unit		1.90		
		Upflow unit-ducted		1.80		
		Upflow unit-nonducted	1.70			
		Horizontal-flow unit			2.10	

For SI units: 1000 British thermal units per hour = 0.293 kW, °C=(°F-32)/1.8

**TABLE E 503.7.1(12)
COMMERCIAL REFRIGERATOR AND FREEZERS – MINIMUM EFFICIENCY REQUIREMENTS
[ASHRAE 90.1: TABLE 6.8.1-12]**

EQUIPMENT TYPE	APPLICATION	ENERGY USE LIMITS, KWH/DAY*	TEST PROCEDURE
Refrigerator with solid doors	Holding temperature	$0.10 \times V + 2.04$	AHRI 1200
Refrigerator with transparent doors	Holding temperature	$0.12 \times V + 3.34$	AHRI 1200
Freezers with solid doors	Holding temperature	$0.40 \times V + 1.38$	AHRI 1200
Freezers with transparent doors	Holding temperature	$0.75 \times V + 4.10$	AHRI 1200
Refrigerators/freezers with solid doors	Holding temperature	the greater of $0.12 \times V + 3.34$ or 0.70	AHRI 1200
Commercial refrigerators	Pulldown	$0.126 \times V + 3.51$	AHRI 1200

For SI units: 1000 British thermal units per hour per day = 0.293 kW/day

* V = the chiller or frozen compartment volume (ft^3) as defined in Association of Home Appliance Manufacturers.

TABLE E 503.7.1(13)
COMMERCIAL REFRIGERATION – MINIMUM EFFICIENCY REQUIREMENTS
[ASHRAE 90.1: TABLE 6.8.1-13]

EQUIPMENT TYPE				ENERGY USE LIMITS[2,3] KWH/DAY	TEST PROCEDURE
EQUIPMENT CLASS[1]	FAMILY CODE	OPERATING MODE	RATING TEMPERATURE		
VOP.RC.M	Vertical open	Remote condensing	Medium temperature	0.82 × *TDA* + 4.07	AHRI 1200
SVO.RC.M	Semivertical open	Remote condensing	Medium temperature	0.83 × *TDA* + 3.18	AHRI 1200
HZO.RC.M	Horizontal open	Remote condensing	Medium temperature	0.35 × *TDA* + 2.88	AHRI 1200
VOP.RC.L	Vertical open	Remote condensing	Low temperature	2.27 × *TDA* + 6.85	AHRI 1200
HZO.RC.L	Horizontal open	Remote condensing	Low temperature	0.57 × *TDA* + 6.88	AHRI 1200
VCT.RC.M	Vertical transparent door	Remote condensing	Medium temperature	0.22 × *TDA* + 1.95	AHRI 1200
VCT.RC.L	Vertical transparent door	Remote condensing	Low temperature	0.56 × *TDA* + 2.61	AHRI 1200
SOC.RC.M	Service over counter	Remote condensing	Medium temperature	0.51 × *TDA* + 0.11	AHRI 1200
VOP.SC.M	Vertical open	Self contained	Medium temperature	1.74 × *TDA* + 4.71	AHRI 1200
SVO.SC.M	Semivertical open	Self contained	Medium temperature	1.73 × *TDA* + 4.59	AHRI 1200
HZO.SC.M	Horizontal open	Self contained	Medium temperature	0.77 × *TDA* + 5.55	AHRI 1200
HZO.SC.L	Horizontal open	Self contained	Low temperature	1.92 × *TDA* + 7.08	AHRI 1200
VCT.SC.I	Vertical transparent door	Self contained	Ice cream	0.67 × *TDA* + 3.29	AHRI 1200
VCS.SC.I	Vertical solid door	Self contained	Ice cream	0.38 × *V* + 0.88	AHRI 1200
HCT.SC.I	Horizontal transparent door	Self contained	Ice cream	0.56 × *TDA* + 0.43	AHRI 1200
SVO.RC.L	Semivertical open	Remote condensing	Low temperature	2.27 × *TDA* + 6.85	AHRI 1200
VOP.RC.I	Vertical open	Remote condensing	Ice cream	2.89 × *TDA* + 8.7	AHRI 1200
SVO.RC.I	Semivertical open	Remote condensing	Ice cream	2.89 × *TDA* + 8.7	AHRI 1200
HZO.RC.I	Horizontal open	Remote condensing	Ice cream	0.72 × *TDA* + 8.74	AHRI 1200
VCT.RC.I	Vertical transparent door	Remote condensing	Ice cream	0.66 × *TDA* + 3.05	AHRI 1200
HCT.RC.M	Horizontal transparent door	Remote condensing	Medium temperature	0.16 × *TDA* + 0.13	AHRI 1200
HCT.RC.L	Horizontal transparent door	Remote condensing	Low temperature	0.34 × *TDA* + 0.26	AHRI 1200
HCT.RC.I	Horizontal transparent door	Remote condensing	Ice cream	0.4 × *TDA* + 0.31	AHRI 1200
VCS.RC.M	Vertical solid door	Remote condensing	Medium temperature	0.11 × *V* + 0.26	AHRI 1200
VCS.RC.L	Vertical solid door	Remote condensing	Low temperature	0.23 × *V* + 0.54	AHRI 1200
VCS.RC.I	Vertical solid door	Remote condensing	Ice cream	0.27 × *V* + 0.63	AHRI 1200
HCS.RC.M	Horizontal solid door	Remote condensing	Medium temperature	0.11 × *V* + 0.26	AHRI 1200
HCS.RC.L	Horizontal solid door	Remote condensing	Low temperature	0.23 × *V* + 0.54	AHRI 1200
HCS.RC.I	Horizontal solid door	Remote condensing	Ice cream	0.27 × *V* + 0.63	AHRI 1200
HCS.RC.I	Horizontal solid door	Remote condensing	Ice cream	0.27 × *V* + 0.63	AHRI 1200
SOC.RC.L	Service over counter	Remote condensing	Low temperature	1.08 × *TDA* + 0.22	AHRI 1200
SOC.RC.I	Service over counter	Remote condensing	Ice cream	1.26 × *TDA* + 0.26	AHRI 1200
VOP.SC.L	Vertical open	Self contained	Low temperature	4.37 × *TDA* + 11.82	AHRI 1200
VOP.SC.I	Vertical open	Self contained	Ice cream	5.55 × *TDA* + 15.02	AHRI 1200
SVO.SC.L	Semivertical open	Self contained	Low temperature	4.34 × *TDA* + 11.51	AHRI 1200
SVO.SC.I	Semivertical open	Self contained	Ice cream	5.52 × *TDA* + 14.63	AHRI 1200
HZO.SC.I	Horizontal open	Self contained	Ice cream	2.44 × *TDA* + 9.0	AHRI 1200
SOC.SC.I	Service over counter	Self contained	Ice cream	1.76 × *TDA* + 0.36	AHRI 1200
HCS.SC.I	Horizontal solid door	Self contained	Ice cream	0.38 × *V* + 0.88	AHRI 1200

For SI units: 1000 British thermal units per hour per day = 0.293 kW/day, °C = (°F-32)/1.8

Notes:

[1] Equipment class designations consist of a combination [in sequential order separated by periods (AAA).(BB).(C)] of the following:

(a) (AAA)—An equipment family code (VOP = vertical open, SVO = semivertical open, HZO = horizontal open, VCT = vertical transparent doors, VCS = vertical solid doors, HCT = horizontal transparent doors, HCS = horizontal solid doors, and SOC = service over counter).

(b) (BB)—An operating mode code (RC = remote condensing and SC = self contained).

(c) (C)—A rating temperature code (M = medium temperature [38°F], L = low temperature [0°F], or I = ice cream temperature [15°F]). For example, "VOP.RC.M" refers to the "vertical open, remote condensing, medium temperature" equipment class.

[2] *V* is the volume of the case (ft) as measured in accordance with AHRI 1200.

[3] *TDA* is the total display area of the case (ft) as measured in accordance with AHRI 1200.

TABLE E 503.7.1(14)
VAPOR COMPRESSION BASED INDOOR POOL DEHUMIDIFIERS – MINIMUM EFFICIENCY REQUIREMENTS
[ASHRAE 90.1: TABLE 6.8.1-14]

EQUIPMENT TYPE	SUBCATEGORY OR RATING CONDITION	MINIMUM EFFICIENCY	TEST PROCEDURE
Single package indoor* (with or without economizer)	Rating Conditions: A, B, or C	3.5 MRE	AHRI 910
Single package indoor water-cooled (with or without economizer)		3.5 MRE	
Single package indoor air-cooled (with or without economizer)		3.5 MRE	
Split system indoor air-cooled (with or without economizer)		3.5 MRE	

* Units without air-cooled condenser

TABLE E 503.7.1(15)
ELECTRICALLY OPERATED DX-DOAS UNITS, SINGLE-PACKAGE AND REMOTE CONDENSER, WITHOUT ENERGY RECOVERY – MINIMUM EFFICIENCY REQUIREMENTS
[ASHRAE 90.1: TABLE 6.8.1-15]

EQUIPMENT TYPE	SUBCATEGORY OR CONDITION	MINIMUM EFFICIENCY	TEST PROCEDURE
Air cooled (dehumidification mode)	–	4.0 *ISMRE*	AHRI 920
Air source heat pumps (dehumidification mode)	–	4.0 *ISMRE*	AHRI 920
Water cooled (dehumidification mode)	Cooling tower condenser water	4.9 *ISMRE*	AHRI 920
	Chilled Water	6.0 *ISMRE*	
Air source heat pump (heating mode)	–	2.7 *ISCOP*	AHRI 920
Water source heat pump (dehumidification mode)	Ground source, closed loop	4.8 *ISMRE*	AHRI 920
	Ground-water source	5.0 *ISMRE*	
	Water source	4.0 *ISMRE*	
Water source heat pump (heating mode)	Ground source, closed loop	2.0 *ISCOP*	AHRI 920
	Ground-water source	3.2 *ISCOP*	
	Water source	3.5 *ISCOP*	

TABLE E 503.7.1(16)
ELECTRICALLY OPERATED DX-DOAS UNITS, SINGLE PACKAGE AND REMOTE CONDENSER, WITH ENERGY RECOVERY – MINIMUM EFFICIENCY REQUIREMENTS
[ASHRAE 90.1: TABLE 6.8.1-16]

EQUIPMENT TYPE	SUBCATEGORY OR RATING CONDITION	MINIMUM EFFICIENCY	TEST PROCEDURE
Air cooled (dehumidification mode)	–	5.2 *ISMRE*	AHRI 920
Air source heat pumps (dehumidification mode)	–	5.2 *ISMRE*	AHRI 920
Water cooled (dehumidification mode)	Cooling tower condenser water	5.3 *ISMRE*	AHRI 920
	Chilled Water	6.6 *ISMRE*	
Air source heat pump (heating mode)	–	3.3 *ISCOP*	AHRI 920
Water source heat pump (dehumidification mode)	Ground source, closed loop	5.2 *ISMRE*	AHRI 920
	Ground-water source	5.8 *ISMRE*	
	Water source	4.8 *ISMRE*	
Water source heat pump (heating mode)	Ground source, closed loop	3.8 *ISCOP*	AHRI 920
	Ground-water source	4.0 *ISCOP*	
	Water source	4.8 *ISCOP*	

TABLE E 503.7.2
MINIMUM DUCT INSULATION R-VALUE[1]
[ASHRAE 90.1: TABLE 6.8.2]

CLIMATE ZONE	DUCT LOCATION		
	EXTERIOR[2]	UNCONDITIONED SPACE AND BURIED DUCTS	INDIRECTLY CONDITIONED SPACE[3, 4]
SUPPLY AND RETURN DUCTS FOR HEATING AND COOLING			
0 to 4	R-8	R-6	R-1.9
5 to 8	R-12	R-6	R-1.9
SUPPLY AND RETURN DUCTS FOR HEATING ONLY			
0 to 1	none	none	none
2 to 4	R-6	R-6	R-1.9
5 to 8	R-12	R-6	R-1.9
SUPPLY AND RETURN DUCTS FOR COOLING ONLY			
0 to 6	R-8	R-6	R-1.9
7 to 8	R-1.9	R-1.9	R-1.9

Notes:

1. Insulation R-values, measured in [°F•h•ft²/(Btu•in)] [(m•K)/W], are for the insulation as installed and do not include film resistance. The required minimum thicknesses do not consider water vapor transmission and possible surface condensation. Where portions of the building envelope are used as a plenum enclosure, building envelope insulation shall be as required by the most restrictive condition of Section E 503.4.7.1 or ASHRAE 90.1, depending on whether the plenum is located in the roof, wall, or floor. Insulation resistance measured on a horizontal plane in accordance with ASTM C518 at a mean temperature of 75°F (24°C) at the installed thickness.
2. Includes attics above insulated ceilings, parking garages and crawl spaces.
3. Includes return air plenums, with or without exposed roofs above.
4. Return ducts in this duct location do not require insulation.

TABLE E 503.7.3(1)
MINIMUM PIPE INSULATION THICKNESS FOR HEATING AND HOT WATER SYSTEMS[1, 2, 3, 4, 5]
(STEAM, STEAM CONDENSATE, HOT WATER HEATING, AND DOMESTIC WATER SYSTEMS)
[ASHRAE 90.1: TABLE 6.8.3-1]

FLUID OPERATING TEMPERATURE RANGE (F°) AND USAGE	INSULATION CONDUCTIVITY		NOMINAL PIPE SIZE OR TUBE SIZE (inches)				
	CONDUCTIVITY Btu•inch/(h•ft²•°F)	MEAN RATING TEMPERATURE °F	<1	1 to <1½	1½ to <4	4 to <8	≥8
			INSULATION THICKNESS (inches)				
>350	0.32 - 0.34	250	4.5	5.0	5.0	5.0	5.0
251 - 350	0.29 - 0.32	200	3.0	4.0	4.5	4.5	4.5
201 - 250	0.27 - 0.30	150	2.5	2.5	2.5	3.0	3.0
141 - 200	0.25 - 0.29	125	1.5	1.5	2.0	2.0	2.0
105 - 140	0.22 - 0.28	100	1.0	1.0	1.5	1.5	1.5

For SI units: °C=(°F-32)/1.8, 1 inch = 25 mm, 1 British thermal unit inch per hour square foot degree Fahrenheit = [0.1 W/(m•K)]

Notes:

1 For insulation outside the stated conductivity range, the minimum thickness (T) shall be determined as follows:

$$T = r\{(1 + t/r)^{K/k} - 1\}$$

Where:

T = minimum insulation thickness (inches).
r = actual outside radius of pipe (inches).
t = insulation thickness listed in this table for applicable fluid temperature and pipe size.
K = conductivity of alternate material at mean rating temperature indicated for the applicable fluid temperature [Btu•in/(h•ft²•°F)] [W/(m•K)].
k = the upper value of the conductivity range listed in this table for the applicable fluid temperature.

2 These thicknesses shall be based on energy efficiency considerations only. Additional insulation shall be permitted to required relative to safety issues/surface temperature.

3 Piping 1½ inches (40 mm) or less and located in partitions within conditioned spaces, reduction of insulation thickness by 1 inch (25.4 mm) shall be permitted before thickness adjustment required in footnote 1, but not a thickness less than 1 inch (25.4 mm).

4 For direct-buried heating and hot water system piping, reduction of insulation thickness by 1½ inch (40 mm) shall be permitted before thickness adjustment required in footnote 1, but not a thickness less than 1 inch (25.4 mm).

5 Table E 503.7.3(1) is based on steel pipe. Non-metallic pipes, less than schedule 80 thickness shall use the table values. For other non-metallic pipes having a thermal resistance more than that of steel pipe, reduced insulation thicknesses shall be permitted where documentation is provided showing that the pipe with the proposed insulation has less heat transfer per foot (mm) than a steel pipe of the same size with the insulation thickness shown in Table E 503.7.3(1).

TABLE E 503.7.3(2)
MINIMUM PIPE INSULATION THICKNESS FOR COOLING SYSTEMS (CHILLED WATER, BRINE, AND REFRIGERANT)[1,2,3,4]
[ASHRAE 90.1: TABLE 6.8.3-2]

INSULATION CONDUCTIVITY			NOMINAL PIPE SIZE OR TUBE SIZE (inches)				
FLUID OPERATING TEMPERATURE RANGE (°F) AND USAGE	CONDUCTIVITY Btu•inch/(h•f^2•°F)	MEAN RATING TEMPERATURE °F	<1	1 to <1½	1½ to <4	4 to <8	>8
			INSULATION THICKNESS (inches)				
40°F - 60°F	021 - 0.27	75	0.5	0.5	1.0	1.0	1.0
<40°F	0.20 - 0.26	50	0.5	1.0	1.0	1.0	1.5

For SI units: °C = (°F-32)/1.8, 1 inch = 25 mm, 1 British thermal unit inch per hour square foot degree Fahrenheit = [0.1 W/(m•k)]

Notes:

1 For insulation outside the stated conductivity range, the minimum thickness (T) shall be determined as follows:

$$T = r\{(1 + t/r)^{K/k} - 1\}$$

Where:

T = minimum insulation thickness (inches).

r = actual outside radius of pipe (inches).

t = insulation thickness listed in this table for applicable fluid temperature and pipe size.

K = conductivity of alternate material at mean rating temperature indicated for the applicable fluid temperature [Btu•inch/(h•ft^2•°F)] [W/(m•K)].

k = the upper value of the conductivity range listed in this table for the applicable fluid temperature.

2 These thicknesses shall be based on energy efficiency considerations only. Issues such as water, vapor permeability, or surface condensation require vapor retarders or additional insulation.

3 Insulation shall not be required for direct-buried cooling system piping.

4 Table E 503.7.3(2) is based on steel pipe. Non-metallic pipes less than schedule 80 thickness shall use the table values. For other non-metallic pipes having thermal resistance more than that of steel pipe, reduced insulation thicknesses shall be permitted where documentation is provided showing that the pipe with the proposed insulation has less heat transfer per foot (mm) than a steel pipe of the same size with the insulation thickness shown in Table E 503.7.3(2).

E 505.0 Geothermal Systems.

E 505.1 Applicability. Geothermal systems that use the earth or body of water as a heat source or sink for heating or cooling shall be in accordance with Section E 505.1.1 through Section E 509.2.

E 505.1.1 Design, Installation, and Testing. Geothermal systems shall be designed by a registered design professional. The geothermal system design, installation, and testing shall be in accordance with CSA C448.

E 505.1.2 Heat Pump Approval. Water source heat pumps used in conjunction with geothermal heat exchangers shall be listed and labeled for use in such systems and shall be designed for the minimum and maximum design water temperature.

E 505.2 Ground Source Heat Pump-Loop Systems. Ground source heat pump ground-loop piping and tubing material for water-based systems shall comply with the standards cited in this appendix.

E 505.3 Material Rating. Piping shall be rated for the operating temperature and pressure of the ground source heat pump-loop system. Fittings shall be rated for the temperature and pressure applications and recommended by the manufacturer for installation with the piping material installed. Where used underground, materials shall be approved for burial.

E 505.4 Used Materials. The installation of used pipe, fittings, valves, and other materials shall not be permitted.

E 505.5 Piping and Tubing Materials Standards. Ground source heat pump ground-loop pipe and tubing shall comply with the standards listed in Table E 505.5.

TABLE E 505.5
PLASTIC GROUND SOURCE LOOP PIPING

MATERIAL	STANDARD
Chlorinated polyvinyl chloride (CPVC)	ASTM D2846; ASTM F441; ASTM F442
Cross-linked polyethylene (PEX)	ASTM F876; ASTM F877; CSA B137.5; NSF 358-3
Polyethylene/aluminum/polyethylene (PE-AL-PE) pressure pipe	ASTM F1282; CSA B137.9
High Density Polyethylene (HDPE)	ASTM D2737; ASTM D3035; ASTM F714; AWWA C901; CSA B137.1; CSA C448; NSF 358-1
Polypropylene (PP)	ASTM F2389; CSA B137.11; NSF 358-2
Polyvinyl chloride (PVC)	ASTM D1785; ASTM D2241
Polyethylene Raised Temperature (PE-RT)	ASTM F2623; ASTM F2769

E 505.6 Fittings. Fittings for ground source heat pump systems shall be approved for installation with the piping materials to be installed, and shall comply with the standards listed in Table E 505.6.

TABLE E 505.6
GROUND SOURCE LOOP PIPE FITTINGS

MATERIAL	STANDARD
Chlorinated polyvinyl chloride (CPVC)	ASTM D2846; ASTM F437; ASTM F438; ASTM F439; ASTM F1970; CSA B137.6
Cross-linked polyethylene (PEX)	ASTM F877; ASTM F1807; ASTM F1960; ASTMF2080; ASTM F2159; ASTM F2434; CSA B137.5; NSF 358-3
Polyethylene/aluminum/polyethylene (PE-AL-PE)	ASTM F1282; ASTM F2434; CSA B137.9
High Density Polyethylene (HDPE)	ASTM D2683; ASTM D3261; ASTM F1055; CSA B137.1; CSA C448; NSF 358-1
Polypropylene (PP)	ASTM F2389; CSA B137.11, NSF 358-2
Polyvinyl chloride (PVC)	ASTM D2464; ASTM D2466; ASTM D2467; ASTM F1970; CSA B137.2; CSA B137.3
Polyethylene Raised Temperature (PE-RT)	ASTM D3261; ASTM F1807; ASTM F2159; ASTM F2769; CSA B137.1

E 506.0 Joints and Connections.

E 506.1 Approval. Joints and connections shall be of an approved type. Joints and connections shall be tight for the pressure of the ground source-loop system. Joints and fittings used underground shall be approved for buried applications.

E 506.2 Joints Between Various Materials. Joints between various piping materials shall be made with approved transition fittings.

E 506.3 Preparation of Pipe Ends. Pipe shall be cut square, reamed, and free of burrs and obstructions. Pipe ends shall have full-bore openings and shall not be undercut. CPVC, PE, and PVC pipe shall be chamfered.

E 506.4 Joint Preparation and Installation. Where required by Section E 506.5 through Section E 506.12.2, the preparation and installation of mechanical and thermoplastic-welded joints shall be in accordance with Section E 506.4 and Section E 506.5.

E 506.5 Mechanical Joints. Mechanical joints shall be installed in accordance with the manufacturer's installation instructions.

E 506.6 Thermoplastic Welded Joints. Joint surfaces for thermoplastic welded joints shall be cleaned by an approved procedure. Joints shall be welded in accordance with the manufacturer's installation instructions.

E 506.7 CPVC Plastic Pipe. Joints between CPVC plastic piping and fittings shall comply with Section E 506.7.1 and Section E 506.7.2.

E 506.7.1 Threaded Joints. Threads shall comply with ASME B1.20.1. Schedule 80 or heavier plastic pipe shall be threaded with dies specifically designed for plastic pipe. Thread lubricant, pipe-joint compound or tape shall be applied on the male threads only and shall be approved for application on the piping material.

E 506.7.2 Solvent Cement. Solvent cement joints for CPVC pipe and fittings shall be clean from dirt and moisture. Solvent cements in accordance with ASTM F493, requiring the use of a primer, shall be orange in color. The primer shall be colored and be in accordance with ASTM F656. Listed solvent cement in accordance with ASTM F493 that does not require the use of primers, yellow or red in color shall be permitted for pipe and fittings manufactured in accordance with ASTM D2846, ½ of an inch (15 mm) through 2 inches (50 mm) in diameter.

E 506.8 Cross-Linked Polyethylene (PEX) Plastic Tubing. Compression or plastic to metal transition joints between cross-linked polyethylene plastic tubing and fittings shall comply with Section E 506.8.1 and Section E 506.8.2. Mechanical joints shall comply with Section E 506.5.

E 506.8.1 Compression-Type Fittings. Where compression-type fittings include inserts and ferrules or o-rings, the fittings shall be installed with the inserts and ferrules or o-rings.

E 506.8.2 Plastic-to-Metal Connections. Soldering on the metal portion of the system shall be performed not less than 18 inches (457 mm) from a plastic-to-metal adapter in the same water line.

E 506.9 Polyethylene Plastic Pipe and Tubing. Joints between polyethylene plastic piping shall comply with Section E 506.9.1 through Section E 506.9.3.

E 506.9.1 Heat-Fusion Joints. Joints shall be of the socket-fusion, saddle-fusion, or butt-fusion type and joined in accordance with ASTM D2657. Joint surfaces shall be clean and free of moisture. Joint surfaces shall be heated to melt temperatures and joined. The joint shall be undisturbed until cool. Fittings shall be manufactured in accordance with ASTM D2683 or ASTM D3261.

E 506.9.2 Electrofusion Joints. Joints shall be of the electrofusion type. Joint surfaces shall be clean and free of moisture, and scoured to expose virgin resin. Joint surfaces shall be heated to melt temperatures for the period of time specified by the manufacturer. The joint shall be undisturbed until cool. Fittings shall be manufactured in accordance with ASTM F1055.

E 506.9.3 Stab-Type Insert Fittings. Joint surfaces shall be clean and free of moisture. Pipe ends shall be chamfered and inserted into the fittings to full depth. Fittings shall be manufactured in accordance with ASTM F1924.

E 506.10 Polypropylene (PP) Plastic. Joints between PP plastic pipe and fittings shall comply with Section E 506.10.1 and Section E 506.10.2.

E 506.10.1 Heat-Fusion Joints. Heat-fusion joints for polypropylene (PP) pipe and tubing joints shall be installed with socket-type heat-fused polypropylene fittings, electrofusion polypropylene fittings, or by butt fusion. Joint surfaces shall be clean and free from moisture. The joint shall be undisturbed until cool. Joints shall be made in accordance with ASTM F2389.

E 506.10.2 Mechanical and Compression Sleeve Joints. Mechanical and compression sleeve joints shall be installed in accordance with the manufacturer's installation instructions.

E 506.11 Raised Temperature Polyethylene (PE-RT) Plastic Tubing. Joints between raised temperature polyethylene tubing and fittings shall comply with Section E 506.11.1 and Section E 506.11.2. Mechanical joints shall comply with Section E 506.5.

E 506.11.1 Compression-Type Fittings. Where compression-type fittings include inserts and ferrules or o-rings, the fittings shall be installed without omitting the inserts and ferrules or o-rings.

E 506.11.2 PE-RT-to-Metal Connections. Solder joints in a metal pipe shall not occur within 18 inches (457 mm) of a transition from such metal pipe to PE-RT pipe.

E 506.12 PVC Plastic Pipe. Joints between PVC plastic pipe and fittings shall comply with Section E 506.12.1 and Section E 506.12.2.

E 506.12.1 Solvent Cement Joints. Solvent cement joints for PVC pipe and fittings shall be clean from dirt and moisture. Purple primer in accordance with ASTM F656 shall be applied until the surface of the pipe and fitting is softened. Solvent cement in accordance with ASTM D2564 shall be applied to joint surfaces.

E 506.12.2 Threaded Joints. Threads shall comply with ASME B1.20.1. Schedule 80 or heavier plastic pipe shall be threaded with dies specifically designed for plastic pipe. Thread lubricant, pipe-joint compound or tape shall be applied on the male threads only and shall be approved for application on the piping material.

E 507.0 Valves.

E 507.1 Where Required. Shutoff valves shall be installed in ground source-loop piping systems in the locations indicated in Section E 507.2 through Section E 507.8.

E 507.2 Heat Exchangers. Shutoff valves shall be installed on the supply and return side of a heat exchanger, except where the heat exchanger is integral with a boiler or is a component of a manufacturer's boiler and heat exchanger packaged unit, and is capable of being isolated from the hydronic system by the supply and return valves.

E 507.3 Central Systems. Shutoff valves shall be installed on the building supply and return of a central utility system.

E 507.4 Pressure Vessels. Shutoff valves shall be installed on the connection to a pressure vessel.

E 507.5 Pressure-Reducing Valves. Shutoff valves shall be installed on both sides of a pressure-reducing valve.

E 507.6 Equipment and Appliances. Shutoff valves shall be installed on connections to mechanical equipment and appliances. This requirement does not apply to components of a ground source loop system such as pumps, air separators, metering devices, and similar equipment.

E 507.7 Expansion Tanks. Shutoff valves shall be installed at connections to nondiaphragm-type expansion tanks.

E 507.8 Reduced Pressure. A pressure relief valve shall be installed on the low-pressure side of a hydronic piping system that has been reduced in pressure. The relief valve shall be set at the maximum pressure of the system design.

E 508.0 Installation.

E 508.1 General. Piping, valves, fittings, and connections shall be installed in accordance with the manufacturer's installation instructions.

E 508.2 Protection of Potable Water. Where ground source heat pump ground loop systems have a connection to a potable water supply, the potable water system shall be protected.

E 508.3 Pipe Penetrations. Openings for pipe penetrations in walls, floors, and ceilings shall be larger than the penetrating pipe. Openings through concrete or masonry building elements shall be sleeved. The annular space surrounding pipe penetrations shall be protected in accordance with the building code.

E 508.4 Clearance from Combustibles. A pipe in a ground source heat pump piping system, having an exterior surface temperature exceeding 250°F (121°C), shall have a clearance of not less than 1 inch (25.4 mm) from combustible materials.

E 508.5 Contact with Building Material. A ground source heat pump ground-loop piping system shall not be in direct contact with building materials that cause the piping or fitting material to degrade or corrode, or that interferes with the operation of the system.

E 508.6 Strains and Stresses. Piping shall be installed so as to prevent detrimental strains and stresses in the pipe. Provisions shall be made to protect piping from damage resulting from expansion, contraction, and structural settlement. Piping shall be installed so as to avoid structural stresses or strains within building components.

E 508.7 Flood Hazard. Piping located in a flood hazard area shall be capable of resisting hydrostatic and hydrodynamic loads and stresses, including the effects of buoyancy, during the occurrence of flooding to the design flood elevation.

E 508.8 Pipe Support. Pipe shall be supported in accordance with Section 313.1.

E 508.9 Velocities. Ground source heat pump ground-loop systems shall be designed so that the flow velocities do not exceed the maximum flow velocity recommended by the pipe and fittings manufacturer. Flow velocities shall be controlled to reduce the possibility of water hammer.

E 508.10 Labeling and Marking. Ground source heat pump ground-loop system piping shall be marked with tape, metal tags, or other methods where it enters a building. The marking shall indicate the following words: "GROUND SOURCE HEAT PUMP-LOOP SYSTEM." The marking shall indicate antifreeze used in the system by name and concentration.

E 508.11 Chemical Compatibility. Antifreeze and other materials used in the system shall be chemically compatible with the pipe, tubing, fittings, and mechanical systems.

E 508.12 Transfer Fluid. The transfer fluid shall be compatible with the makeup water supplied to the system.

E 509.0 Testing.

E 509.1 Ground Source Heat Pump Loop System Testing. Before connection header trenches are backfilled, the assembled loop system shall be pressure tested with water at 100 psi (689 kPa) for 15 minutes with no observed leaks. Flow and pressure loss testing shall be performed, and the actual flow rates and pressure drops shall be compared to the calculated design values. Where actual flow rate or pressure drop values differ from calculated design values by more than 10 percent, the cause shall be identified and corrective action taken.

E 509.2 Pressurizing During Installation. Ground source heat pump ground loop piping to be embedded in concrete shall be pressure tested prior to pouring concrete. During pouring, the pipe shall be maintained at the proposed operating pressure.

E 601.0 Indoor Environment.

E 601.1 Scope. The provisions of this section shall establish the means of reducing the quantity of air contaminants that are odorous, irritating, or harmful to the comfort and well-being of a building's installers, occupants, and neighbors.

E 602.0 Fireplaces.

E 602.1 Requirements. A direct-vent sealed-combustion gas or sealed wood-burning fireplace, or a sealed wood stove shall be installed. The fireplace shall comply with Section E 602.1.1 and Section E 602.1.2.

E 602.1.1 Masonry or Factory-Built Fireplace. Masonry and factory-built fireplaces located in conditioned spaces shall be in accordance with Section E 602.1.1.1 through Section E 602.1.1.3.

E 602.1.1.1 Opening Cover. Closeable metal or glass doors covering the entire opening of the firebox shall be installed.

E 602.1.1.2 Combustion Air Intake. A combustion air intake to draw air from the outside of the building directly into the firebox, which is an area of not less than 6 square inches (0.004 m^2) and is equipped with a readily accessible, operable, and tight-fitting damper or combustion-air control device.

E 602.1.1.3 Accessible Damper Control. The flue damper shall have a readily accessible control.

Exception: Where a gas log, log lighter, or decorative gas appliance is installed in a fireplace, the flue damper shall be blocked open where required by this code or the manufacturer's installation instructions.

E 602.1.2 Prohibited. Continuous burning pilot lights and the use of indoor air for cooling a firebox jacket, where the indoor air is vented to the outside of the building, are prohibited.

E 603.0 Pollutant Control.

E 603.1 Indoor Air Quality During Construction. Indoor air quality of a building shall be maintained in accordance with Section E 603.1.1 through Section E 603.1.3.

E 603.1.1 Temporary Ventilation During Construction. Temporary ventilation during construction shall be provided in accordance with the following:

(1) Ventilation during construction shall be achieved through openings in the building shell using fans to produce not less than three air changes per hour.

(2) During dust-producing operations, the supply and return HVAC system openings shall be protected from dust in accordance with Section E 603.1.3.

(3) Where the building is occupied during demolition or construction, ventilation shall be provided in accordance with the Control Measures of the SMACNA IAQ Guidelines for Occupied Buildings Under Construction.

(4) The permanent HVAC system shall not be used during construction to condition and ventilate the building within the required temperature range for material and equipment installation. Where required, a supplemental HVAC system shall be used during construction, return air shall be equipped with filters with a minimum efficiency reporting value (MERV) of 8, in accordance with ASHRAE 52.2, or an average efficiency of 30 percent in accordance with ASHRAE 52.2. Before occupancy, filters shall be replaced with filters having a MERV 13 rating in accordance with Section E 603.3.

Exception: Embedded hydronics system shall be permitted to be used to condition the building during construction.

E 603.1.2 Indoor Air Quality After Construction. After construction ends and interior finishes are installed, flush-out the building to reduce contaminant concentrations by supplying a total outdoor air volume of 14 000 cubic feet per square foot (ft^3/ft^2) (4267.2 m^3/m^2) of occupiable building area. An internal temperature of not less than 60°F (16°C) and relative humidity not higher than 60 percent shall be maintained during the flush-out process. Occupancy shall begin on condition of 3500 ft^3/ft^2 (1066.8 m^3/m^2) of building area, with the remaining 10 500 ft^3/ft^2 (3200.4 m^3/m^2) being accomplished as soon as possible.

Exception: Other means of reducing the contaminant concentration levels shall be permitted where approved by the Authority Having Jurisdiction.

E 603.1.3 Covering of Duct Openings and Protection of Mechanical Equipment During Construction. At the time of rough installation, or during storage on the construction site and until final startup of the heating and cooling equipment, duct and other related air distribution component openings shall be covered with tape, plastic, sheet metal, or other methods acceptable to the enforcing agency to reduce the amount of dust or debris that collects in the system.

E 603.2 Isolation of Pollutant Sources. Rooms where activities produce hazardous fumes or chemicals, including commercial kitchens, garages, janitorial or laundry rooms, and copy or printing rooms, shall be exhausted and isolated from adjacent spaces in accordance with this code.

E 603.3 Filters. In mechanically ventilated buildings, particle filters, or air-cleaning devices shall be provided to clean outdoor and return air prior to its delivery to occupied spaces. The particle or air cleaner shall have a MERV of 13.

Exception: A filter or air cleaning device with a lower MERV value shall be permitted provided it is the highest value commercially available for the specific equipment that is installed.

E 603.4 Ozone Depletion and Global Warming Reductions. Installations of HVAC and refrigeration shall not contain CFCs and shall be in accordance with this code.

E 604.0 Indoor Moisture Control.

E 604.1 Rainwater Control. Roof drainage systems shall discharge to a place of disposal in accordance with the plumbing code. Storm water shall be directed away from the building.

E 605.0 Indoor Air Quality for Low-Rise Residential.

E 605.1 General. Rooms or occupied spaces within single-family homes and multifamily structures of three stories or less above grade shall be designed to have ventilation (outdoor) air for occupants in accordance with Section E 605.1.1 through Section E 605.1.3.2, or the applicable local code.

E 605.1.1 Natural Ventilation. Naturally ventilated spaces shall be permanently open to and within 20 feet (6096 mm) of operable wall or roof openings to the outdoors, the openable area of which is not less than 5 percent of the conditioned floor area of the naturally ventilated space. Where openings are covered with louvers or otherwise obstructed, openable area shall be based on the free unobstructed area through the opening.

E 605.1.1.1 Access to Operable Openings. The means to open required operable openings shall be readily accessible to building occupants where the space is occupied.

E 605.1.2 Mechanical Ventilation. Each space that is not naturally ventilated in accordance with Section E 605.1.1 shall be ventilated with a mechanical system capable of providing an outdoor air rate not less than 15 ft^3/min (0.007 m^3/s) per person times the expected number of occupants. Mechanical ventilation shall comply with this code.

E 605.1.3 Dwelling-Unit Ventilation. A Mechanical exhaust system, supply system, or combination thereof shall be installed to operate for each dwelling unit to provide continuous dwelling-unit ventilation with outdoor air rate not less than the rate specified in Section E 605.1.3.1. [ASHRAE 62.2:4.1]

E 605.1.3.1 Total Ventilation Rate. The total required ventilation rate (Q_{tot}) shall be as specified in Table E 605.1.3.1 or, alternatively, calculated in accordance with Equation E 605.1.3.1.

TABLE E 605.1.3.1
VENTILATION AIR REQUIREMENTS, (cubic foot per minute)
[ASHRAE 62.2: TABLE 4.1a]

FLOOR AREA (ft^2)	BEDROOMS				
	1	2	3	4	5
<500	30	38	45	53	60
501-1000	45	53	60	68	75
1001-1500	60	68	75	83	90
1501-2000	75	83	90	98	105
2001-2500	90	98	105	113	120
2501-3000	105	113	120	128	135
3001-3500	120	128	135	143	150
3501-4000	135	143	150	158	165
4001-4500	150	158	165	173	180
4501-5000	165	173	180	188	195

For SI units: 1 square foot = 0.0929 m^2, 1 cubic foot per minute = 0.00047 m^3/s

(Equation E 605.1.3.1)

$$Q_{tot} = 0.03A_{floor} + 7.5(N_{br} + 1)$$

Where:

Q_{tot} = total required ventilation rate, cubic feet per minute (ft³/min)

A_{floor} = dwelling unit square foot (ft²)

N_{br} = number of bedrooms; not to be less than one

For SI units: 1 cubic foot per minute = 0.00047 m³/s, 1 square foot = 0.0929 m²

Exceptions: Dwelling-unit mechanical ventilation systems shall not be required where the Authority Having Jurisdiction determines that window operation is a locally permissible method of providing ventilation and provided one or more of the following conditions is met:

(1) The building has no mechanical cooling and is located in zone 1 or 2.

(2) The building is thermally conditioned for human occupancy for less than 876 hours per year. [ASHRAE 62.2:4.1.1]

E 605.1.3.2 Effective Annual Average Infiltration Rate (Q_{inf}). Effective Annual Average Infiltration Rate (Q_{inf}) shall be calculated using Equation E 605.1.3.2:

[Equation 605.1.3.2]

$$Q_{inf}(\text{cfm}) = (NL \text{ x } wsf \text{ x } A_{floor}) / (7.3)^*$$

Where:

NL = normalized leakage

wsf = weather and shielding factor from ASHRAE 62.2.

A_{floor} = floor area of residence, ft² (m²)

* Replace 7.3 with 1.44 for metric units. [ASHRAE 62.2:4.1.2(e)]

E 605.1.3.3 Required Mechanical Ventilation Rate (Q_{fan}). Required Mechanical Ventilation Rate (Q_{fan}) shall be calculated using Equation E 605.1.3.3:

$$Q_{fan} = Q_{tot} - (Q_{inf} \times A_{ext}) \quad \text{[Equation 605.1.3.3]}$$

Where:

Q_{fan} = required mechanical ventilation rate, cfm (L/s)

Q_{tot} = total required ventilation rate, cfm (L/s)

Q_{inf} = may be not greater than $\frac{2}{3} \times Q_{tot}$ (see ASHRAE 62.2 for exceptions for existing buildings)

A_{ext} = 1 for single-family detached homes, or the ratio of exterior envelope surface area that is not attached to garages or other dwelling units to total envelope surface area for single-family attached homes. [ASHRAE 62.2:4.1.2(f)]

E 605.1.3.4 Different Occupant Density. Table E 605.1.3.1 and Equation E 605.1.3.1 assume two persons in a studio or one-bedroom dwelling unit and an additional person for each additional bedroom. Where higher occupant densities are known, the rate shall be increased by 7.5 ft³/min (0.003 m³/s) for each additional person. Where approved by the Authority Having Jurisdiction, lower occupant densities shall be permitted to be used. [ASHRAE 62.2:4.1.3]

E 605.1.4 System Type. The dwelling-unit mechanical ventilation system shall consist of one or more supply or exhaust fans and associated ducts and controls. Local exhaust fans shall be permitted to be part of a mechanical exhaust system. Where local exhaust fans are used to provide dwelling-unit ventilation, the local exhaust airflow shall be permitted to be credited towards the whole dwelling-unit ventilation airflow requirement. Outdoor air ducts connected to the return side of an air handler shall be permitted as supply ventilation where manufacturer's requirements for return air temperature are met. See ASHRAE 62.2 for guidance on selection of methods. [ASHRAE 62.2:4.2]

E 605.1.5 Airflow Measurement. The airflow required by this section shall be the quantity of outdoor ventilation air supply, indoor air, or both exhausted by the mechanical ventilation system as installed and shall be measured according to the ventilation equipment manufacturer's instructions, or by using a flow hood, flow grid, or other airflow measuring device at the mechanical ventilation fan's inlet terminals/grilles, outlet terminals/grilles, or in the connected ventilation ducts. Ventilation airflow of systems with multiple operating modes shall be tested in all modes designed to be in accordance with this section. [ASHRAE 62.2:4.3]

E 605.1.6 Control and Operation. A readily accessible manual ON-OFF control, including but not limited to a fan switch or a dedicated branch-circuit overcurrent device, shall be provided. Controls shall include text or an icon indicating the system's function.

Exception: For multifamily dwelling units, the manual ON-OFF control shall not be required to be readily accessible. [ASHRAE 62.2:4.4]

E 605.1.7 Variable Mechanical Ventilation. Dwelling-unit mechanical ventilation systems designed to provide variable ventilation shall comply with Section E 605.1.7.1 or Section E 605.1.7.2 or Section E 605.1.7.3. Section E 605.1.7.2 and Section E 605.1.7.3 also require compliance with ASHRAE 62.2 and require verification with supporting documentation from the manufacturer, designer, or specifier of the ventilation control system that the system meets the requirements of these sections. Where the dwelling-unit ventilation rate varies based on occupancy, occupancy shall be determined by occupancy sensors or by an occupant-programmable schedule. [ASHRAE 62.2:4.5]

E 605.1.7.1 Short-Term Average Ventilation. To comply with this section, a variable ventilation system shall be installed to provide an average dwelling-unit ventilation rate over any three-hour period that is greater than or equal to Q_{fan} as determined in accordance with Section E 605.1.3.3. [ASHRAE 62.2:4.5.1]

E 605.1.7.2 Scheduled Ventilation. This section shall only be allowed to be used where one or more fixed patterns of designed ventilation are known at the time compliance to Section E 605.0 is being determined. Such patterns include those both clock-driven and driven by typical meteorological data. Compliance with this section shall be in accordance with either Section E 605.1.7.2.1 or Section E 605.1.7.2.2. [ASHRAE 62.2:4.5.2]

E 605.1.7.2.1 Annual Average Schedule. An annual schedule of ventilation complies with this section when the annual average relative exposure during occupied periods is not more than unity as calculated in accordance with ASHRAE 62.2. [ASHRAE 62.2:4.5.2.1]

E 605.1.7.2.2 Block Scheduling. The schedule of ventilation complies with this section when it is broken into blocks of time and each block individually has an average relative exposure during occupied periods that is not more than unity as calculated in ASHRAE 62.2. [ASHRAE 62.2:4.5.2.2]

E 605.1.7.3 Real-Time Control. A real-time ventilation controller complies with this section when it is designed to adjust the ventilation system based on real-time input to the ventilation calculations so that the average relative exposure during occupied periods is not more than unity as calculated in ASHRAE 62.2. The averaging period shall be not less than one day but not more than one year and shall be based on simple, recursive or running average, but not extrapolation. [ASHRAE 62.2:4.5.3]

E 605.1.8 Equivalent Ventilation. A dwelling-unit ventilation system shall be designed and operated in such a way as to provide the same or lower annual exposure as would be provided in accordance with Section E 605.1.3. The calculations shall be based on a single zone with a constant contaminant emission rate. The manufacturer, specifier, or designer of the equivalent ventilation system shall certify that the system is in accordance with this intent and provide supporting documentation. [ASHRAE 62.2:4.6]

E 605.2 Bathroom Exhaust Fans. Except where a whole house energy recovery system is used, a mechanical exhaust fan vented to the outdoors shall be provided in each room containing a bathtub, shower, or tub/shower combination. The ventilation rate shall be not less than 50 ft^3/min (0.02 m^3/s) for intermittent operation and 20 ft^3/min (0.009 m^3/s) for continuous operation. Fans shall comply with the Energy Star Program.

E 605.3 Filters. Heating and air conditioning filters shall have a MERV rating of 6 or higher. The air distribution system shall be designed for the pressure drop across the filter.

E 606.0 Indoor Air Quality for Other than Low-Rise Residential Buildings.

E 606.1 Minimum Indoor Air Quality. The building shall comply with this code and ASHRAE 62.1 for ventilation air supply.

E 607.0 Environmental Comfort.

E 607.1 Thermal Comfort Controls. The mechanical systems and controls of building shall be designed to provide and maintain indoor comfort conditions in accordance with ASHRAE 55.

E 607.2 Heating and Air-Conditioning System Design. Heating and air-conditioning systems shall be sized, designed, and have their equipment selected in accordance with the following:

(1) Heat loss and heat gain are established in accordance with ACCA Manual J, ASHRAE handbooks, or other equivalent methods.

(2) Duct systems shall be sized in accordance with ACCA Manual D, ASHRAE handbooks, or other equivalent methods.

(3) Heating and cooling equipment in accordance with ACCA Manual S or other equivalent methods.

E 608.0 Low VOC Solvent Cement and Primer.

E 608.1 General. Primers and solvent cements used to join plastic pipe, and fittings shall be in accordance with Section E 608.1.1 and Section E 608.1.2.

E 608.1.1 Solvent Cement. Solvent cement, including one-step solvent cement, shall have a volatile organic compound (VOC) content of less than or equal to 65 ounces per gallon (oz/gal) (487 g/L) for CPVC cement, 68 oz/gal (509 g/L) for PVC cement, and 43 oz/gal (322 g/L) for ABS cement, as determined by the South Coast Air Quality Management District's Laboratory Methods of Analysis for Enforcement Samples, Method 316A.

E 608.1.2 Primer. Primer shall have a volatile organic compound (VOC) content of less than or equal to 73 oz/gal (546 g/L), as determined by the South Coast Air Quality Management District's Laboratory Methods of Analysis for Enforcement Samples, Method 316A.

E 701.0 Installer Qualifications.

E 701.1 Scope. The provisions of this section address minimum qualifications of installers of mechanical systems covered within the scope of this appendix.

E 702.0 Qualifications.

E 702.1 General. Where permits are required, the Authority Having Jurisdiction shall have the authority to require contractors, installers, or service technicians to demonstrate competency. Where determined by the Authority Having Jurisdiction, the contractor, installer, or service technician shall be licensed to perform such work.

Part I

E 801.0 Heating, Ventilation, and Air Conditioning Systems Commissioning.

E 801.1 Applicability. The provisions of this section apply to the commissioning of commercial and institutional HVAC systems.

E 802.0 Commissioning.

E 802.1 Commissioning Requirements. HVAC commissioning shall be included in the design and construction processes of the project to verify that the HVAC systems and components meet the owner's project requirements and in accordance with this appendix. Commissioning shall be performed in accordance with this appendix by personnel trained and certified in commissioning by a nationally recognized organization. Commissioning requirements shall include the following:

(1) Owner's project requirements

(2) Basis of design

(3) Commissioning measures shown in the construction documents

(4) Commissioning plan

(5) Functional performance

(6) Testing

(7) Post construction documentation and training

(8) Commissioning report

HVAC systems and components covered by this appendix as well as process equipment and controls, and renewable energy systems shall be included in the scope of the commissioning requirements.

E 802.2 Owner's Project Requirements (OPR). The performance goals and requirements of the HVAC system shall be documented before the design phase of the project begins. This documentation shall include not less than the following:

(1) Environmental and sustainability goals

(2) Energy efficiency goals

(3) Indoor environmental quality requirements

(4) Equipment and systems performance goals

(5) Building occupant and O&M personnel expectations

E 802.3 Basis of Design (BOD). A written explanation of how the design of the HVAC system meets the owner's project requirements shall be completed at the design phase of the building project, and updated as necessary during the design and construction phases. The basis of design document shall cover not less than the following systems:

(1) Heating, ventilation, air conditioning (HVAC) systems and controls

(2) Water heating systems

(3) Renewable energy systems

E 802.4 Commissioning Plan. A commissioning plan shall be completed to document the approach to how the project will be commissioned, and shall be started during the design phase of the building project. The commissioning plan shall include not less than the following:

(1) General project information

(2) Commissioning goals

(3) Systems to be commissioned. Plans to test systems and components shall include not less than the following:

 (a) A detailed explanation of the original design intent.

 (b) Equipment and systems to be tested, including the extent of tests.

 (c) Functions to be tested.

 (d) Conditions under which the test shall be performed.

 (e) Measurable criteria for acceptable performance.

(4) Commissioning team information.

(5) Commissioning process activities, schedules, and responsibilities. Plans for the completion of commissioning requirements listed in Section E 802.5 through Section E 802.7 shall be included.

E 802.5 Functional Performance Testing. Functional performance tests shall demonstrate the correct installation and operation of each component, system, and system-to-system interface in accordance with the approved plans and specifications. Functional performance testing reports shall contain information addressing each of the building components tested, the testing methods utilized, and readings and adjustments made.

E 802.6 Post Construction Documentation and Training. A system manual and systems operations training are required.

E 802.6.1 Systems Manual. Documentation of the operational aspects of the HVAC system shall be completed within the systems manual and delivered to the building owner and facilities operator. The systems manual shall include not less than the following:

(1) Site information, including facility description, history, and current requirements.

(2) Site contact information.

(3) Basic O&M, including general site operating procedures, basic troubleshooting, recommended maintenance requirements, and site events log.

(4) Major systems.

(5) Site equipment inventory and maintenance notes.

(6) Equipment/system warranty documentation and information.

(7) "As-Built" design drawings.

(8) Other resources and documentation.

E 802.6.2 Systems Operations Training. The training of the appropriate maintenance staff for each equipment type or system shall include not less than the following:

(1) System/Equipment overview (what it is, what it does, and what other systems or equipment it interfaces with).

(2) Review of the information in the systems manual.

(3) Review of the record drawings on the system/equipment.

E 802.7 Commissioning Report. A complete report of commissioning process activities undertaken through the design, construction, and post-construction phases of the building project shall be completed and provided to the owner.

Part II

E 803.0 Commissioning Acceptance.

E 803.1 General. Part II of this appendix provides a means of verifying the commissioning requirements of Section E 802.1. The activities specified in Part II of this appendix includes three aspects, as described as follows:

(1) Visual inspection of the equipment and installation.

(2) Review of the certification requirements.

(3) Functional tests of the systems and controls.

E 803.2 Construction Documents. Details of commissioning acceptance requirements shall be incorporated into the construction documents, including information that describes the details of the functional tests to be performed. This information shall be permitted to be integrated into the specifications for testing and air balancing, energy management and control system, equipment startup procedures or commissioning. It is possible that the work will be performed by a combination of the test and balance (TAB) contractor, mechanical/electrical contractor, and the energy management control system (EMCS) contractor, so applicable roles and responsibilities shall be clearly called out.

E 803.2.1 Roles and Responsibilities. The roles and responsibilities of the persons involved in commissioning acceptance are included in Section E 803.2.1.1 through Section E 803.2.1.3.

E 803.2.1.1 Field Technician. The field technician shall be responsible for performing and documenting the results of the acceptance procedures on the certificate of acceptance forms. The field technician shall sign the certificate of acceptance to certify that the information he provides on the certificate of acceptance is true and correct.

E 803.2.1.2 Responsible Person. The responsible person shall be the contractor or registered design professional of record. A certificate of acceptance shall be signed by a responsible person to take responsibility for the scope of work specified by the certificate of acceptance document. The responsible person shall perform the field testing and verification work, and where this is the case, the responsible person shall complete and sign both the field technician's signature block and the responsible person's signature block on the certificate of acceptance form. The responsible person assumes responsibility for the acceptance testing work performed by the field technician agent or employee.

E 803.2.1.3 Certificate of Acceptance. The certificate of acceptance shall be submitted to the Authority Having Jurisdiction in order to receive the final certificate of occupancy. The Authority Having Jurisdiction shall not release a final certificate of occupancy unless the submitted certificate of acceptance demonstrates that the specified systems and equipment have been shown to be performing in accordance with the applicable acceptance requirements. The Authority Having Jurisdiction has the authority to require the field technician and responsible person to demonstrate competence, to its satisfaction. Certificate of acceptance forms are located in Section E 806.0.

E 804.0 Commissioning Tests.

E 804.1 General. Functional tests shall be performed on new equipment and systems installed in either new construction or retrofit applications in accordance with this section. The appropriate certificate of acceptance form along with each specific test shall be completed and submitted to the Authority Having Jurisdiction before a final occupancy permit can be granted.

E 804.2 Tests. Functional testing shall be performed on the devices and systems listed in this section. The functional test results are documented using the applicable certificate of acceptance forms shown in parenthesis and located in Section E 806.0. The functional tests shall be performed in accordance with Section E 805.0 using the following forms:

(1) Minimum ventilation controls for constant and variable air volume systems (Form MECH-2A).

(2) Zone temperature and scheduling controls for constant volume, single-zone, unitary air conditioner and heat pump systems (Form MECH-3A).

(3) Duct leakage on a subset of small single-zone systems depending on the ductwork location (Form MECH-4A).

(4) Air economizer controls for economizers that are not factory installed and tested (Form MECH-5A).

(5) Demand-controlled ventilation control systems (Form MECH-6A).

(6) Supply fan variable flow controls (Form MECH-7A).

(7) Valve leakage for hydronic variable flow systems and isolation valves on chillers and boilers in plants with more than one chiller or boiler being served by the same primary pumps through a common header (Form MECH-8A).

(8) Supply water temperature reset control strategies programmed into the building automation system for water systems (e.g., chilled, hot, or condenser water) (Form MECH-9A).

(9) Hydronic variable flow controls on a water system where the pumps are controlled by variable frequency drives (e.g., chilled and hot water systems; water-loop heat pump systems) (Form MECH-10A).

(10) Automatic demand shed control (Form MECH-11A).

(11) Fault detection and diagnostic for DX units (Form MECH-12A).

(12) Automatic fault detection and diagnostic systems (AFDD) (Form MECH-13A).

(13) Distributed energy storage DEC/DX AC systems (Form MECH-14A).

(14) Thermal energy storage (TES) systems (Form MECH-15A).

E 804.3 Acceptance Process. The functional testing process shall comply with Section E 804.3.1 through Section E 804.3.4.

E 804.3.1 Plan Review. The installing contractor, registered design professional of record, owner's agent, or the person responsible for certification of the acceptance testing on the certificate of acceptance (responsible person) shall review the plans and specifications to ensure that they are in accordance with the acceptance requirements. This is typically done prior to signing a certificate of compliance.

E 804.3.2 Construction Inspection. The installing contractor, registered design professional of record, owner's agent, or the person responsible for certification of the acceptance testing on the certificate of acceptance (responsible person) shall perform a construction inspection prior to testing to ensure that the equipment that is installed is capable of complying with the requirements of this appendix and is calibrated. The installation of associated systems and equipment necessary for proper system operation is required to be completed prior to the testing.

E 804.3.3 Acceptance Testing. One or more field technicians shall perform the acceptance testing; identify performance deficiencies; ensure that they are corrected; and where necessary, repeat the acceptance procedures until the specified systems and equipment are performing in accordance with the acceptance requirements. The field technician who performs the testing shall sign the certificate of acceptance to certify the information has been provided to document the results of the acceptance procedures is true and correct.

The responsible person shall review the test results from the acceptance requirement procedures provided by the field technician and sign the certificate of acceptance to certify compliance with the acceptance requirements. The responsible person shall be permitted to perform the field technician's responsibilities, and shall then sign the field technician declaration on the certificate of acceptance to certify that the information on the form is true and correct.

E 804.3.4 Certificate of Occupancy. The Authority Having Jurisdiction shall not issue the final certificate of occupancy until required certificates of acceptance are submitted. Copies of completed, signed certificates of acceptance are required to be posted, or made available with the permit(s), and shall be made available to the Authority Having Jurisdiction.

E 805.0 HVAC System Tests.

E 805.1 Variable Air Volume Systems (Form MECH-2A). This test ensures that adequate outdoor air ventilation is provided through the variable air volume air handling unit at two representative operating conditions. The test consists of measuring outdoor air values at maximum flow and at or near minimum flow. The test verifies that the minimum volume of outdoor air is introduced to the air handling unit where the system is in occupied mode at these two conditions of supply airflow. This test shall be performed in conjunction with supply fan variable flow controls test procedures to reduce the overall system testing time as both tests use the same two conditions of airflow for their measurements.

E 805.1.1 Test Procedure. The procedure for performing a functional test for variable air volume systems shall be in accordance with Section E 805.1.1.1 and Section E 805.1.1.2.

E 805.1.1.1 Construction Inspection. Prior to functional testing, verify and document that the system controlling outside airflow is calibrated either in the field or factory.

E 805.1.1.2 Functional Testing. The functional testing shall be in accordance with the following steps:

Step 1: Where the system has an outdoor air economizer, force the economizer high limit to disable economizer control (e.g., for a fixed drybulb high limit, lower the setpoint below the current outdoor air temperature).

Step 2: Adjust supply airflow to either the sum of the minimum zone airflows or 30 percent of the total design airflow. Verify and document the following:

(1) Measured outside airflow reading is within 10 percent of the total ventilation air called for in the certificate of compliance.

(2) OSA controls stabilize within 5 minutes.

Step 3: Adjust supply airflow to achieve design airflow. Verify and document the following:

(1) Measured outside airflow reading is within 10 percent of the total ventilation air called for in the certificate of compliance.

(2) OSA controls stabilize within 5 minutes.

Step 4: Restore system to "as-found" operating conditions.

E 805.1.2 Acceptance Criteria. System controlling outdoor air flow shall be calibrated in the field or at the factory.

Measured outdoor airflow reading shall be within 10 percent of the total value found on the certificate of compliance under the following conditions:

(1) Minimum system airflow.

(2) Thirty percent of total design flow design supply airflow.

E 805.2 Constant Volume Systems (Form MECH-2A). The purpose of this test is to ensure that adequate outdoor air ventilation is provided through the constant volume air handling unit to the spaces served under operating conditions. The intent of this test is to verify that the minimum volume of outdoor air is introduced to the air handling unit during typical space occupancy.

E 805.2.1 Test Procedure. The procedure for performing a functional test for constant air volume systems shall be in accordance with Section E 805.2.1.1 and Section E 805.2.1.2.

E 805.2.1.1 Construction Inspection. Prior to functional testing, verify and document the following:

(1) Minimum position is marked on the outside air damper.

(2) The system has means of maintaining the minimum outdoor air damper position.

E 805.2.1.2 Functional Testing. Where the system has an outdoor air economizer, force the economizer to the minimum position and stop outside air damper modulation (e.g., for a fixed drybulb high limit, lower the setpoint below the current outdoor air temperature).

E 805.2.2 Acceptance Criteria. The system has a means of maintaining the minimum outdoor air damper position. The minimum damper position is marked on the outdoor air damper. The measured outside airflow reading shall be within 10 percent of the total ventilation air called for in the certificate of compliance.

E 805.3 Constant Volume, Single-Zone, Unitary Air Conditioner and Heat Pumps Systems Acceptance (Form MECH-3A). The purpose of this test is to verify the individual components of a constant volume, single-zone, unitary air conditioner and heat pump system function correctly; including: thermostat installation and programming, supply fan, heating, cooling, and damper operation.

E 805.3.1 Test Procedure. The procedure for performing a functional test for constant volume, single-zone, unitary air conditioner and heat pump systems shall be in accordance with Section E 805.3.1.1 and Section E 805.3.1.2.

E 805.3.1.1 Construction Inspection. Prior to functional testing, verify and document the following:

(1) Thermostat is located within the space-conditioning zone that is served by the HVAC system.

(2) Thermostat shall be in accordance with temperature adjustment and dead band requirements.

(3) Occupied, unoccupied, and holiday schedules shall be programmed per the facility's schedule.

(4) Preoccupancy purge is programmed.

E 805.3.1.2 Functional Testing. The functional testing shall be in accordance with the following steps:

Step 1: Disable economizer and demand control ventilation systems (where applicable).

Step 2: Simulate a heating demand during the occupied condition. Verify and document the following:

(1) Supply fan operates continually.

(2) The unit provides heating.

(3) No cooling is provided by the unit.

(4) Outside air damper is at minimum position.

Step 3: Simulate operation in the dead band during occupied condition. Verify and document the following:

(1) Supply fan operates continually.

(2) Neither heating nor cooling is provided by the unit.

(3) Outside air damper is at minimum position.

Step 4: Simulate cooling demand during occupied condition. Lock out economizer (where applicable). Verify and document the following:

(1) Supply fan operates continually.

(2) The unit provides cooling.

(3) No heating is provided by the unit.

(4) Outside air damper is at minimum position.

Step 5: Simulate operation in the dead band during unoccupied mode. Verify and document the following:

(1) Supply fan is off.

(2) Outside air damper is fully closed.

(3) Neither heating nor cooling is provided by the unit.

Step 6: Simulate heating demand during unoccupied conditions. Verify and document the following:

(1) Supply fan is on (either continuously or cycling).

(2) Heating is provided by the unit.

(3) No cooling is provided by the unit.

(4) Outside air damper is either closed or at minimum position.

Step 7: Simulate cooling demand during unoccupied condition. Lock out economizer (where applicable). Verify and document the following:

(1) Supply fan is on (either continuously or cycling).

(2) Cooling is provided by the unit.

(3) No heating is provided by the unit.

(4) Outside air damper is either closed or at minimum position.

Step 8: Simulate manual override during unoccupied condition. Verify and document the following:

(1) System operates in "occupied" mode.

(2) System reverts to "unoccupied" mode where manual override time period expires.

Step 9: Restore economizer and demand control ventilation systems (where applicable), and remove system overrides initiated during the test.

E 805.3.2 Acceptance Criteria. Thermostat is located within the space-conditioning zone that is served by the respective HVAC system. The thermostat shall comply with temperature adjustment and dead band requirements. Occupied, unoccupied, and holiday schedules shall be programmed per the facility's schedule. Preoccupancy purge is programmed in accordance with the requirements.

E 805.4 Air Distribution Systems (Form MECH-4A). The purpose of this test is to verify duct work associated with non-exempt constant volume, single-zone, HVAC units (e.g., air conditioners, heat pumps, and furnaces) meet the material, installation, and insulation R-values and leakage requirements outlined in this appendix. This test is required for single-zone units serving less than 5000 square feet (464.52 m^2) of floor area where 25 percent or more of the duct surface area is in one of the following spaces:

(1) Outdoors.

(2) In a space directly under a roof where the U-factor of the roof is greater than the U-factor of the ceiling.

(3) In a space directly under a roof with fixed vents or openings to the outside or unconditioned spaces.

(4) In an unconditioned crawlspace.

(5) In other unconditioned spaces.

This test applies to both new duct systems and to existing duct systems being extended or the space conditioning system is altered by the installation or replacement of space conditioning equipment, including: replacement of the air handler; outdoor condensing unit of a split system air conditioner or heat pump; cooling or heating coil; or the furnace heat exchanger. Existing duct systems do not have to be tested where they are insulated or sealed with asbestos.

E 805.4.1 Test Procedure. The procedure for performing a functional test for air distribution systems shall be in accordance with Section E 805.4.1.1 and Section E 805.4.1.2.

E 805.4.1.1 Construction Inspection. Prior to functional testing, verify and document the following:

(1) Duct connections shall comply with the requirements of this appendix and this code.

(2) Flexible ducts are not compressed.

(3) Ducts are fully accessible for testing.

(4) Joints and seams are properly sealed in accordance with the requirements of this appendix.

(5) Insulation R-Values shall comply with the minimum requirements of this appendix.

E 805.4.1.2 Functional Testing. Perform duct leakage test in accordance with Section E 503.4.7.2.1.

E 805.4.2 Acceptance Criteria. Flexible ducts are not compressed or constricted. Duct connections shall comply with the requirements of this appendix and this code (new ducts only). Joints and seams are properly sealed in accordance with the requirements of this appendix and this code (new ducts only). Duct R-values shall comply with the minimum requirements of this appendix (new ducts only). Insulation is protected from damage and suitable for outdoor usage where applicable (new ducts only). The leakage shall not exceed the rate in accordance with Section E 503.4.7.2.

E 805.5 Air Economizer Controls Acceptance (Form MECH-5A). The purpose of functionally testing an air economizer cycle is to verify that an HVAC system uses outdoor air to satisfy space cooling loads where outdoor air conditions are acceptable. There are two types of economizer controls; stand-alone packages and DDC controls. The stand-alone packages are commonly associated with small unitary rooftop HVAC equipment, and DDC controls are typically associated with built-up or large packaged air handling systems. Test procedures for both economizer control types are provided.

For units with economizers that are factory installed and certified operational by the manufacturer to economizer quality control requirements, the in-field economizer functional tests do not have to be conducted. A copy of the manufacturer's certificate shall be attached to the Form MECH-5A. However, the construction inspection, including compliance with high-temperature lockout temperature setpoint, shall be completed regardless of whether the economizer is field or factory installed.

E 805.5.1 Test Procedure. The procedure for performing a functional test for air economizer controls shall comply with Section E 805.5.1.1 and Section E 805.5.1.2.

E 805.5.1.1 Construction Inspection. Prior to functional testing, verify and document the following:

(1) Economizer lockout setpoint is in accordance with this appendix.

(2) Economizer lockout control sensor is located to prevent false readings.

(3) System is designed to provide up to 100 percent outside air without over-pressurizing the building.

(4) For systems with DDC controls lockout sensor(s) are either factory calibrated or field calibrated.

(5) For systems with non-DDC controls, manufacturer's startup and testing procedures are applied.

E 805.5.1.2 Functional Testing. The functional testing shall be in accordance with the following steps:

Step 1: Disable demand control ventilation systems (where applicable).

Step 2: Enable the economizer, and simulate a cooling demand large enough to drive the economizer fully open. Verify and document the following:

(1) Economizer damper is 100 percent opened and return air damper is 100 percent closed.

(2) Where applicable, verify that the economizer remains 100 percent open where the cooling demand can no longer be met by the economizer alone.

(3) Applicable fans and dampers operate as intended to maintain building pressure.

(4) The unit heating is disabled.

Step 3: Disable the economizer and simulate a cooling demand. Verify and document the following:

(1) Economizer damper shall close to its minimum position.

(2) Applicable fans and dampers shall operate as intended to maintain building pressure.

(3) The unit heating is disabled.

Step 4: Simulate a heating demand, and set the economizer so that it is capable of operating (e.g., actual outdoor air conditions are below lockout setpoint). Verify the economizer is at minimum position.

Step 5: Restore demand control ventilation systems (where applicable) and remove system overrides initiated during the test.

E 805.5.2 Acceptance Criteria. Air economizer controls acceptance criteria shall be as follows:

(1) Where the economizer is factory installed and certified, a valid factory certificate is required for acceptance. No additional equipment tests are necessary.

(2) Air economizer lockout setpoint is in accordance with this appendix. Outside sensor location accurately reads true outdoor air temperature and is not affected by exhaust air or other heat sources.

(3) Sensors are located to achieve the desired control.

(4) During economizer mode, the outdoor air damper shall modulate open to a maximum position and return air damper to 100 percent closed.

(5) The outdoor air damper is 100 percent open before mechanical cooling is enabled and for units 75 000 Btu/h (22 kw) and larger remains at 100 percent open while mechanical cooling is enabled (economizer integration where used for compliance).

(6) Where the economizer is disabled, the outdoor air damper closes to a minimum position; the return damper modulates 100 percent open, and mechanical cooling remains enabled.

E 805.6 Demand-Controlled Ventilation Systems Acceptance (Form MECH-6A). The purpose of this test is to verify that systems required to employ demand-controlled ventilation shall be permitted to vary outside ventilation flow rates based on maintaining interior carbon dioxide (CO_2) concentration setpoints. Demand-controlled ventilation refers to an HVAC system's ability to reduce outdoor air ventilation flow below design values where the space served is at less than design occupancy. Carbon dioxide is a good indicator of occupancy load and is the basis used for modulating ventilation flow rates.

E 805.6.1 Test Procedure. The procedure for performing a functional test for demand-control ventilation (DVC) systems shall be in accordance with Section E 805.6.1.1 and Section E 805.6.1.2.

E 805.6.1.1 Construction Inspection. Prior to functional testing, verify and document the following:

(1) Carbon dioxide control sensor is factory calibrated or field-calibrated in accordance with this appendix.

(2) The sensor is located in the high-density space between 3 feet (914 mm) and 6 feet (1829 mm) above the floor or at the anticipated level of the occupants' heads.

(3) DCV control setpoint is at or below the carbon dioxide concentration permitted by this appendix.

E 805.6.1.2 Functional Testing. The functional testing shall be in accordance with the following steps:

Step 1: Disable economizer controls.

Step 2: Simulate a signal at or slightly above the carbon dioxide concentration setpoint required by this appendix. Verify and document the following:

(1) For single zone units, outdoor air damper modulates open to satisfy the total ventilation air called for in the certificate of compliance.

(2) For multiple zone units, either outdoor air damper or zone damper modulate open to satisfy the zone ventilation requirements.

Step 3: Simulate signal well below the carbon dioxide setpoint. Verify and document the following:

(1) For single zone units, outdoor air damper modulates to the design minimum value.

(2) For multiple zone units, either outdoor air damper or zone damper modulate to satisfy the reduced zone ventilation requirements.

Step 4: Restore economizer controls and remove system overrides initiated during the test.

Step 5: With controls restored, apply carbon dioxide calibration gas at a concentration slightly above the setpoint to the sensor. Verify that the outdoor air damper modulates open to satisfy the total ventilation air called for in the certificate of compliance.

E 805.6.2 Acceptance Criteria. Demand-controlled ventilation systems acceptance criteria shall be as follows:

(1) Each carbon dioxide sensor is factory calibrated (with calibration certificate) or field calibrated.

(2) Each carbon dioxide sensor is wired correctly to the controls to ensure proper control of the outdoor air damper.

(3) Each carbon dioxide sensor is located correctly within the space 1 foot (305 mm) to 6 feet (1829 mm) above the floor.

(4) Interior carbon dioxide concentration setpoint is not more than 600 parts per million (ppm) plus outdoor air carbon dioxide value where dynamically measured or not more than 1000 ppm where no OSA sensor is provided.

(5) A minimum OSA setting is provided where the system is in occupied mode in accordance with this appendix regardless of space carbon dioxide readings.

(6) A maximum OSA damper position for DCV control shall be established in accordance with this appendix, regardless of space carbon dioxide readings.

(7) The outdoor air damper shall modulate open where the carbon dioxide concentration within the space exceeds setpoint.

(8) The outdoor air damper modulates closed (toward minimum position) where the carbon dioxide concentration within the space is below setpoint.

E 805.7 Supply Fan Variable Flow Controls (Form MECH-7A). The purpose of this test is to ensure that the supply fan in a variable air volume application modulates to meet system airflow demand. In most applications, the individual VAV boxes serving each space will modulate the amount of air delivered to the space based on heating and cooling requirements. As a result, the total supply airflow provided by the central air handling unit shall vary to maintain sufficient airflow through each VAV box. Airflow shall be controlled using a variable frequency drive (VFD) to modulate supply fan speed and vary system airflow. The most common strategy for controlling the VFD is to measure and maintain static pressure within the duct.

E 805.7.1 Test Procedure. The procedure for performing a functional test for supply fan variable controls shall be in accordance with Section E 805.7.1.1 and Section E 805.7.1.2.

E 805.7.1.1 Construction Inspection. Prior to functional testing, verify and document the following:

(1) Supply fan controls modulate to increase capacity.

(2) Supply fan maintains discharge static pressure within plus or minus 10 percent of the current operating set point.

(3) Supply fan controls stabilize within a 5 minute period.

E 805.7.1.2 Functional Testing. The functional testing shall be in accordance with the following steps:

Step 1: Simulate demand for design airflow. Verify and document the following:

(1) Supply fan controls modulate to increase capacity.

(2) Supply fan maintains discharge static pressure within plus or minus 10 percent of the current operating set point.

(3) Supply fan controls stabilize within a 5 minute period.

Step 2: Simulate demand for minimum airflow. Verify and document the following:

(1) Supply fan controls modulate to decrease capacity.

(2) Current operating setpoint has decreased (for systems with DDC to the zone level).

(3) Supply fan maintains discharge static pressure within plus or minus 10 percent of the current operating setpoint.

(4) Supply fan controls stabilize within a 5 minute period.

Step 3: Restore system to correct operating conditions.

E 805.7.2 Acceptance Criteria. Supply fan variable flow controls acceptance criteria shall be as follows:

(1) Static pressure sensor(s) is factory calibrated (with calibration certificate) or field calibrated.

(2) For systems without DDC controls to the zone level, the pressure sensor setpoint is less than one-third of the supply fan design static pressure.

(3) For systems with DDC controls with VAV boxes reporting to the central control panel, the pressure setpoint is reset by zone demand (box damper position or a trim and respond algorithm).

At full flow:

(1) Supply fan maintains discharge static pressure within plus or minus 10 percent of the current operating control static pressure setpoint.

(2) Supply fan controls stabilizes within a 5 minute period.

(3) At minimum flow (not less than 30 percent of total design flow).

(4) Supply fan controls modulate to decrease capacity.

(5) Current operating setpoint has decreased (for systems with DDC to the zone level).

(6) Supply fan maintains discharge static pressure within plus or minus 10 percent of the current operating setpoint.

E 805.8 Valve Leakage (Form MECH-8A). The purpose of this test is to ensure that control valves serving variable flow systems are designed to withstand the pump pressure over the full range of operation. Valves with insufficient actuators will lift under certain conditions causing water to leak through and loss of control. This test applies to the variable flow systems, chilled and hot-water variable flow systems, chiller isolation valves, boiler isolation valves, and water-cooled air conditioner and hydronic heat pump systems.

E 805.8.1 Test Procedure. The procedure for performing a functional test for valve leakage shall be in accordance with Section E 805.8.1.1 and Section E 805.8.1.2.

E 805.8.1.1 Construction Inspection. Prior to functional testing, verify and document the valve and piping arrangements were installed in accordance with the design drawings.

E 805.8.1.2 Functional Testing. The functional testing shall be in accordance with the following steps:

Step 1: For each pump serving the distribution system, dead head the pumps using the discharge isolation valves at the pumps. Document the following:

(1) Record the differential pressure across the pumps.

(2) Verify that this is within 5 percent of the submittal data for the pump.

Step 2: Reopen the pump discharge isolation valves. Automatically close valves on the systems being tested. Where three-way valves are present, close off the bypass line. Verify and document the following:

(1) The valves automatically close.

(2) Record the pressure differential across the pump.

(3) Verify that the pressure differential is within 5 percent of the reading from Step 1 for the pump that is operating during the valve test.

Step 3: Restore system to correct operating conditions.

E 805.8.2 Acceptance Criteria. System has no flow where coils are closed and the pump is turned on.

E 805.9 Supply Water Temperature Reset Controls (Form MECH-9A). The purpose of this test is to ensure that both the chilled water and hot water supply temperatures are automatically reset based on either building loads or outdoor air temperature, as indicated in the control sequences. Many HVAC systems are served by central chilled and heating hot water plants. The supply water operating temperatures shall meet peak loads where the system is operating at design conditions. As the loads vary, the supply water temperatures shall be permitted to be adjusted to satisfy the new operating conditions. The chilled water supply temperature shall be permitted to be raised as the cooling load decreases, and heating hot water supply temperature shall be permitted to be lowered as the heating load decreases.

This requirement applies to chilled and hot water systems that are not designed for variable flow, and that have a design capacity greater than or equal to 500 000 Btu/h (147 kW).

E 805.9.1 Test Procedure. The procedure for performing a functional test for supply water temperature reset controls shall be in accordance with Section E 805.9.1.1 and Section E 805.9.1.2.

E 805.9.1.1 Construction Inspection. Prior to functional testing, verify and document the supply water temperature sensors shall be either factory or field calibrated.

E 805.9.1.2 Functional Testing. The functional testing shall be in accordance with the following steps:

Step 1: Change reset control variable to its maximum value. Verify and document the following:

(1) Chilled or hot water temperature setpoint is reset to appropriate value.

(2) Actual supply temperature changes to meet setpoint.

(3) Verify that supply temperature is within 2 percent of the control setpoint.

Step 2: Change reset control variable to its minimum value. Verify and document the following:

(1) Chilled or hot water temperature setpoint is reset to appropriate value.

(2) Actual supply temperature changes to meet setpoint.

(3) Verify that supply temperature is within 2 percent of the control setpoint.

Step 3: Restore reset control variable to automatic control. Verify and document the following:

(1) Chilled or hot water temperature setpoint is reset to appropriate value.

(2) Actual supply temperature changes to meet setpoint.

(3) Verify that supply temperature is within 2 percent of the control setpoint.

E 805.9.2 Acceptance Criteria. The supply water temperature sensors are either factory calibrated (with calibration certificates) or field-calibrated. Sensor performance shall comply with the specifications. The supply water reset is operational.

E 805.10 Hydronic System Variable Flow Controls (Form MECH-10A). The purpose of this test is to ensure that hydronic variable flow chilled water and water-loop heat pump systems with circulating pumps larger than 5 hp (3.7 kW) vary system flow rate by modulating pump speed using a variable frequency drive (VFD) or equivalent. As the loads within the building fluctuate, control valves modulate the amount of water passing through each coil and add or remove the desired amount of energy from the air stream to satisfy the load. In the case of water-loop heat pumps, each two-way control valve associated with a heat pump will be closed where that unit is not operating. As each control valve modulates, the pump variable frequency drive (VFD) responds accordingly to meet system water flow requirements. This is not required on heating hot water systems with variable flow designs or for condensing water serving water cooled chillers.

E 805.10.1 Test Procedure. The procedure for performing a functional test for hydronic system variable flow controls shall be in accordance with Section E 805.10.1.1 and Section E 805.10.1.2.

E 805.10.1.1 Construction Inspection. Prior to functional testing, verify and document the pressure sensors are either factory or field calibrated.

E 805.10.1.2 Functional Testing. The functional testing shall comply with the following steps:

Step 1: Open control valves to increase water flow to not less than 90 percent design flow. Verify and document the following:

(1) Pump speed increases.

(2) System pressure is either within plus or minus 5 percent of current operating setpoint, or the pressure is below the setpoint, and the pumps are operating at 100 percent speed.

(3) System operation shall stabilize within 5 minutes after test procedures are initiated.

Step 2: Modulate control valves to reduce water flow to 50 percent of the design flow or less, but not lower than the pump minimum flow. Verify and document the following:

(1) Pump speed decrease.

(2) Current operating setpoint has decreased (for systems with DDC to the zone level).

(3) Current operating setpoint has not increased (for all other systems).

(4) System pressure is within 5 percent of current operating setpoint.

(5) System operation stabilizes within 5 minutes after test procedures are initiated.

E 805.10.2 Acceptance Criteria. The differential pressure sensor is either factory calibrated (with calibration certificates) or field calibrated. The pressure sensor shall be located at or near the most remote HX or control valve. The setpoint system controls shall stabilize.

E 805.11 Automatic Demand Shed Control (Form MECH-11A). The purpose of this test is to ensure that the central demand shed sequences have been properly programmed into the DDC system.

E 805.11.1 Test Procedure. The procedure for performing a functional test for automatic demand shed controls shall be in accordance with Section E 805.11.1.1 and Section E 805.11.1.2.

E 805.11.1.1 Construction Inspection. Prior to functional testing, verify and document that the EMCS interface enables activation of the central demand shed controls.

E 805.11.1.2 Functional Testing. The functional testing shall comply with the following steps:

Step 1: Engage the global demand shed system. Verify and document the following:

(1) That the cooling setpoint in noncritical spaces increases by the proper amount.

(2) That the cooling setpoint in critical spaces do not change.

Step 2: Disengage the global demand shed system. Verify and document the following:

(1) That the cooling setpoint in noncritical spaces return to their original values.

(2) That the cooling setpoint in critical spaces do not change.

E 805.11.2 Acceptance Criteria. The control system changes the setpoints of noncritical zones on activation of a single central hardware or software point then restores the initial setpoints where the point is released.

E 805.12 Fault Detection and Diagnostics (FDD) for Packaged Direct-Expansion (DX) Units (Form MECH-12A). The purpose of this test is to verify proper fault detection and reporting for automated fault detection and diagnostics systems for packaged units. Automated FDD systems ensure proper equipment operation by identifying and diagnosing common equipment problems such as improper refrigerant charge, low airflow, or faulty economizer operation. Qualifying FDD systems receive a compliance credit where using the performance approach. A system that does not meet the eligibility requirements shall be permitted to be installed, but no compliance credit will be given.

E 805.12.1 Test Procedure. The procedure for performing a functional test for fault detection and diagnostics (FDD) for packaged direct-expansion (DX) units shall be in accordance with Section E 805.12.1.1 and Section E 805.12.1.2.

E 805.12.1.1 Construction Inspection. Prior to functional testing, verify and document that the FDD hardware is installed on equipment by the manufacturer, and that equipment make and model include factory-installed FDD hardware that match the information indicated on copies of the manufacturer's cut sheets and on the plans and specifications.

This procedure applies to fault detection and diagnostics (FDD) system for direct-expansion packaged units containing the following features:

(1) The unit shall include a factory-installed economizer and shall limit the economizer deadband to not more than 2°F (-17°C).

(2) The unit shall include direct-drive actuators on outside air and return air dampers.

(3) The unit shall include an integrated economizer with either differential drybulb or differential enthalpy control.

(4) The unit shall include a low temperature lockout on the compressor to prevent coil freeze-up or comfort problems.

(5) Outside air and return air dampers shall have maximum leakage rates in accordance to this appendix.

(6) The unit shall have an adjustable expansion control device such as a thermostatic expansion valve (TXV).

(7) To improve the ability to troubleshoot charge and compressor operation, a high-pressure refrigerant port will be located on the liquid line. A low-pressure refrigerant port will be located on the suction line.

(8) The following sensors shall be permanently installed to monitor system operation, and the controller shall have the capability of displaying the value of each parameter:

(a) Refrigerant suction pressure

(b) Refrigerant suction temperature

(c) Liquid line pressure

(d) Liquid line temperature

(e) Outside air temperature

(f) Outside air relative humidity

(g) Return air temperature

(h) Return air relative humidity

(i) Supply air temperature

(j) Supply air relative humidity

The controller will provide system status by indicating the following conditions:

(1) Compressor enabled

(2) Economizer enabled

(3) Free cooling available

(4) Mixed air low limit cycle active

(5) Heating enabled

The unit controller shall have the capability to manually initiate each operating mode so that the operation of compressors, economizers, fans, and heating system can be independently tested and verified.

E 805.12.1.2 Functional Testing. The functional testing shall be in accordance with the following steps:

Step 1: Test low airflow condition by replacing the existing filter with a dirty filter or appropriate obstruction.

Step 2: Verify that the fault detection and diagnostics system reports the fault.

Step 3: Verify that the system is able to verify the correct refrigerant charge.

Step 4: Calibrate outside air, return air, and supply air temperature sensors.

E 805.12.2 Acceptance Criteria. The system is able to detect a low airflow condition and report the fault. The system is able to detect where refrigerant charge is low or high and the fault is reported.

E 805.13 Automatic Fault Detection Diagnostics (FDD) for Air Handling Units (AHU) and Zone Terminal Units (Form MECH-13A). The purpose of this test is to verify that the system detects common faults in air handling units and terminal units. FDD systems for air handling units and zone terminal units require DDC controls to the zone level. Successful completion of this test provides a compliance credit where using the performance approach. An FDD system that does not pass this test shall be permitted to be installed, but no compliance credit will be given.

E 805.13.1 Test Procedure. The procedure for performing a functional test for automatic fault detection diagnostics (FDD) for Air Handling Units and Zone Terminal Units shall be in accordance with Section E 805.13.1.1.

E 805.13.1.1 Functional Testing. The functional testing shall be in accordance with Section E 805.13.1.1.1 and Section E 805.13.1.1.2.

E 805.13.1.1.1 Functional Testing for Air Handling Units. The functional testing of AHU with FDD controls shall be in accordance with the following steps:

Step 1: Sensor drift/failure:

(1) Disconnect outside air temperature sensor from unit controller.

(2) Verify that the FDD system reports a fault.

(3) Connect OAT sensor to the unit controller.

(4) Verify that FDD indicates normal system operation.

Step 2: Damper/actuator fault:

(1) From the control system workstation, command the mixing box dampers to full open (100 percent outdoor air).

(2) Disconnect power to the actuator and verify that a fault is reported at the control workstation.

(3) Reconnect power to the actuator and command the mixing box dampers to full open.

(4) Verify that the control system does not report a fault.

(5) From the control system workstation, command the mixing box dampers to a full-closed position (0 percent outdoor air).

(6) Disconnect power to the actuator and verify that a fault is reported at the control workstation.

(7) Reconnect power to the actuator and command the dampers closed.

(8) Verify that the control system does not report a fault during normal operation.

Step 3: Valve/actuator fault:

(1) From the control system workstation, command the heating and cooling coil valves to full open or closed, then disconnect power to the actuator and verify that a fault is reported at the control workstation.

Step 4: Inappropriate simultaneous heating, mechanical cooling, and economizing or all functions:

(1) From the control system workstation, override the heating coil valve and verify that a fault is reported at the control workstation.

(2) From the control system workstation, override the cooling coil valve and verify that a fault is reported at the control workstation.

(3) From the control system workstation, override the mixing box dampers and verify that a fault is reported at the control workstation.

E 805.13.1.1.2 Functional Testing for Zone Terminal Units. The functional testing of one of each type of terminal unit (VAV box) in the project not less than 5 percent of the terminal boxes shall be in accordance with the following steps:

Step 1: Sensor drift/failure:

(1) Disconnect the tubing to the differential pressure sensor of the VAV box.

(2) Verify that control system detects and reports the fault.

(3) Reconnect the sensor and verify proper sensor operation.

(4) Verify that the control system does not report a fault.

Step 2: Damper/actuator fault:

(1) Damper stuck open.

(a) Command the damper to full open (room temperature above setpoint).

(b) Disconnect the actuator to the damper.

(c) Adjust the cooling setpoint so that the room temperature is below the cooling setpoint to command the damper to the minimum position. Verify that the control system reports a fault.

(d) Reconnect the actuator and restore to normal operation.

(2) Damper stuck closed.

(a) Set the damper to the minimum position.

(b) Disconnect the actuator to the damper.

(c) Set the cooling setpoint below the room temperature to simulate a call for cooling. Verify that the control system reports a fault.

(d) Reconnect the actuator and restore to normal operation.

Step 3: Valve/actuator fault (for systems with hydronic reheat):

(1) Command the reheat coil valve to full open.

(2) Disconnect power to the actuator. Set the heating setpoint temperature to be lower than the current space temperature, to command the valve closed. Verify that the fault is reported at the control workstation.

(3) Reconnect the actuator and restore normal operation.

Step 4: Feedback loop tuning fault (unstable airflow):

(1) Set the integral coefficient of the box controller to a value 50 times the current value.

(2) The damper cycles continuously and airflow is unstable. Verify that the control system detects and reports the fault.

(3) Reset the integral coefficient of the controller to the original value to restore normal operation.

Step 5: Disconnected inlet duct:

(1) From the control system workstation, command the damper to full closed; then disconnect power to the actuator, and verify that a fault is reported at the control workstation.

E 805.13.2 Acceptance Criteria. The system is able to detect common faults with air-handling units, such as a sensor failure, a failed damper, an actuator, or an improper operating mode.

The system is able to detect and report common faults with zone terminal units, such as a failed damper, an actuator, or a control tuning issue.

E 805.14 Distributed Energy Storage DX AC System (Form MECH-14A). The purpose of this test is to verify the proper operation of distributed energy storage DX systems. Distributed energy systems (DES) reduce peak demand by operating during off-peak hours and storing cooling, usually in the form of ice. During peak cooling hours the ice is melted to avoid compressor operation. The system typically consists of a water tank containing refrigerant coils that cool the water and convert it to ice. As with a standard direction expansion (DX) air conditioner, the refrigerant is compressed in a compressor and then cooled in an air-cooled condenser. The liquid refrigerant then is directed through the coils in the water tank to make ice or to air handler coils to cool the building. This applies to constant or variable volume, direct expansion (DX) systems with distributed energy storage (DES/DXAC).

E 805.14.1 Test Procedure. The procedure for performing a functional test for distributed energy storage DX AC systems shall be in accordance with Section E 805.14.1.1 through Section E 805.14.1.3.

E 805.14.1.1 Construction Inspection. Prior to functional testing, verify and document the following:

(1) The water tank is filled to the proper level.

(2) The water tank is sitting on a foundation with adequate structural strength.

(3) The water tank is insulated and the top cover is in place.

(4) The DES/DXAC is installed correctly (e.g., refrigerant piping, etc.).

(5) Verify that the correct model number is installed and configured.

E 805.14.1.2 Functional Testing. The functional testing shall be in accordance with the following steps:

Step 1: Simulate cooling load during daytime period (e.g., by setting time schedule to include actual time and placing thermostat cooling setpoint below actual temperature). Verify and document the following:

(1) Supply fan operates continually.

(2) Where the DES/DXAC has cooling capacity, DES/DXAC shall run to meet the cooling demand (in ice melt mode).

(3) Where the DES/DXAC has no ice and there is a call for cooling, the DES/DXAC shall run in direct cooling mode.

Step 2: Simulate no cooling load during daytime condition. Verify and document the following:

(1) Supply fan operates in accordance with the facility thermostat or control system.

(2) The DES/DXAC and the condensing unit do not run.

Step 3: Simulate no cooling load during morning shoulder time period. Verify and document the following:

(1) The DES/DXAC is idle.

Step 4: Simulate a cooling load during morning shoulder time period. Verify and document the following:

(1) The DES/DXAC runs in direct cooling mode.

E 805.14.1.3 Calibrating Controls. Set the proper time and date in accordance with the manufacturer's instructions for approved installers.

E 805.14.2 Acceptance Criteria. Distributed energy storage DXAC system acceptance criteria shall be as follows:

(1) Verify night time ice making operation.

(2) Verify that tank discharges during on-peak cooling periods.

(3) Verify that the compressor does not run and the tank does not discharge where there is no cooling demand during on-peak periods.

(4) Verify that the system does not operate during a morning shoulder period where there is no cooling demand.

(5) Verify that the system operates in direct mode (with compressor running) during the morning shoulder time period.

E 805.15 Thermal Energy Storage (TES) System (Form MECH-15A). The purpose of this test is to verify the proper operation of thermal energy storage (TES) systems. TES systems reduce energy consumption during peak demand periods by shifting energy consumption to nighttime. Operation of the thermal energy storage compressor during the night produces cooling energy which is stored in the form of cooled fluid or ice in tanks. During peak cooling hours the thermal storage is used for cooling to prevent the need for chiller operation. This section is limited to the following types of TES systems:

(1) Chilled water storage

(2) Ice-on-coil

(3) Ice harvester

(4) Brine

(5) Ice-slurry

(6) Eutectic salt

(7) Clathrate hydrate slurry (CHS)

E 805.15.1 Test Procedure. The procedure for performing a functional test for thermal energy storage (TES) system shall be in accordance with Section E 805.15.1.1 and Section E 805.15.1.2.

E 805.15.1.1 Construction Inspection. Prior to functional testing, verify and document the following for the chiller and storage tank:

(1) Chiller:

(a) Brand and Model

(b) Type (centrifugal, reciprocating, other)

(c) Capacity (tons) (SIZE)

(d) Starting efficiency (kW/ton) at beginning of ice production (COMP - kW/TON - START)

(e) Ending efficiency (kW/ton) at end of ice production (COMP - kW/TON/END)

(f) Capacity reduction (percent/°F) (PER – COMP - REDUCT/F)

(g) Verify that the efficiency of the chiller meets or exceeds the requirements of Section E 501.0.

(2) Storage Tank:

(a) Storage type (TES-TYPE)

(b) Number of tanks (SIZE)

(c) Storage capacity per tank (ton-hours) (SIZE)

(d) Storage rate (tons) (COOL – STORE - RATE)

(e) Discharge rate (tons) (COOL – SUPPLY - RATE)

(f) Auxiliary power (watts) (PUMPS + AUX - kW)

(g) Tank area (CTANK – LOSS - COEFF)

(h) Tank insulation (R-Value) (CTANK – LOSS – COEFF)

(3) TES System:

(a) The TES system is one of the above eligible systems.

(b) Initial charge rate of the storage tanks (tons).

(c) Final charge rate of the storage tank (tons).

(d) Initial discharge rate of the storage tanks (tons).

(e) Final discharge rate of the storage tank (tons).

(f) Charge test time (hrs).

(g) Discharge test time (hrs).

(h) Tank storage capacity after charge (ton-hrs).

(i) Tank storage capacity after discharge (ton-hrs).

(j) Tank standby storage losses (UA).

(k) Initial chiller efficiency (kW/ton) during charging.

(l) Final chiller efficiency (kW/ton) during charging.

E 805.15.1.2 Functional Testing. The functional testing shall be in accordance with the following steps:

Step 1: Verify that the TES system and the chilled water plant is controlled and monitored by an energy management system (EMS).

Step 2: Force the time to be between 9:00 p.m. and 9:00 a.m., and simulate a partial or no charge of the tank. Simulate no cooling load by setting the indoor temperature setpoint(s) higher than the ambient temperature.

Where the tank is full or nearly full of ice, it shall be permitted to adjust the control settings for this test. In some cases, the control system will not permit the chiller to start the ice-making process unless a portion of the ice has been melted. The controls designer shall be permitted to use an inventory meter (a 4-20 mA sensor that indicates water level) to determine whether or not ice-making can commence (e.g., not allow ice-making unless the inventory meter signal is less than 17 mA). Where this is the case, this limit can be reset to 20 mA during testing to allow ice making to occur.

Verify that the TES system starts charging (storing energy). This shall be checked by verifying flow and inlet and outlet temperatures of the storage tank, or directly by reading an inventory meter where the system has one.

Step 3: Force the time to be between 6:00 p.m. and 9:00 p.m., and simulate a partial charge on the tank. Simulate a cooling load by setting the indoor temperature setpoint lower than the ambient temperature. Verify that the TES system starts discharging. This shall be checked by observing tank inlet and outlet temperatures and system flow, or directly by reading an inventory meter where the system has one. Where the system has no charge, verify that the system will still attempt to meet the load through storage.

Step 4: Force the time to be between noon and 6:00 p.m., and simulate a cooling load by lowering the indoor air temperature setpoint below the ambient temperature. Verify that the tank starts discharging and the compressor is off.

Step 5: Force the time to be between 9:00 a.m. to noon, and simulate a cooling load by lowering the indoor air temperature setpoint below the ambient temperature. Verify that the tank does not discharge and the cooling load is met by the compressor.

Step 6: Force the time to be between 9:00 p.m. and 9:00 a.m. and simulate a full tank charge. This can be done in a couple of ways:

(1) By changing the inventory sensor limit that indicates tank capacity to the energy management system so that it indicates a full tank.

(2) By resetting the coolant temperature that indicates a full charge to a higher temperature than the current tank leaving temperature. Verify that the tank charging is stopped.

Step 7: Force the time to be between noon and 6:00 p.m. and simulate no cooling load by setting the indoor temperature setpoint above the ambient temperature. Verify that the tank does not discharge and the compressor is off.

E 805.15.2 Acceptance Criteria. Thermal energy storage (TES) system acceptance criteria shall be as follows:

(1) Verify that the system is able to charge the storage tank during off-peak periods where there is no cooling load.

(2) Verify that tank discharges during on-peak cooling periods.

(3) Verify that the compressor does not run and the tank does not discharge where there is no cooling demand during on-peak periods.

(4) Verify that the system does not operate during a morning shoulder period where there is no cooling demand.

(5) Verify that the system operates in direct mode (with compressor running) during the morning shoulder time period.

E 806.0 Certificate of Acceptance Forms.

E 806.1 General. This section includes the certificate of acceptance forms referenced in Section E 804.0 and Section E 805.0.

CERTIFICATE OF ACCEPTANCE	MECH-2A
Outdoor Air Acceptance	**(Page 1 of 3)**
Project Name/Address:	
System Name or Identification/Tag:	System Location or Area Served:

Enforcement Agency:	Permit Number:
Note: Submit one Certificate of Acceptance for each system that must demonstrate compliance.	Enforcement Agency Use: Checked by/Date

FIELD TECHNICIAN'S DECLARATION STATEMENT

- I certify under penalty of perjury the information provided on this form is true and correct.
- I am the person who performed the acceptance requirements verification reported on this Certificate of Acceptance (Field Technician).
- I certify that the construction/installation identified on this form complies with the acceptance requirements indicated in the plans and specifications approved by the enforcement agency, and conforms to the applicable acceptance requirements and procedures specified in Section E 801.0 through Section E 806.0.
- I have confirmed that the Installation Certificate(s) for the construction/installation identified on this form has been completed and is posted or made available with the building permit(s) issued for the building.

Company Name:		
Field Technician's Name:		Field Technician's Signature:
	Date Signed:	Position with Company (Title):

RESPONSIBLE PERSON'S DECLARATION STATEMENT

- I certify under penalty of perjury that I am the Field Technician, or the Field Technician is acting on my behalf as my employee or my agent and I have reviewed the information provided on this form.
- I am a licensed contractor or registered design professional who is eligible per the requirements of the Authority Having Jurisdiction to take responsibility for the scope of work specified on this document and attest to the declarations in this statement (responsible person).
- I certify that the information provided on this form substantiates that the construction/installation identified on this form complies with the acceptance requirements indicated in the plans and specifications approved by the enforcement agency, and conforms to the applicable acceptance requirements and procedures specified in Section E 801.0 through Section E 806.0.
- I have confirmed that the Installation Certificate(s) for the construction/installation identified on this form has been completed and is posted or made available with the permit(s) issued for the building.
- I will ensure that a completed, signed copy of this Certificate of Acceptance shall be posted, or made available with the building permit(s) issued for the building, and made available to the enforcement agency for all applicable inspections. I understand that a signed copy of this Certificate of Acceptance is required to be included with the documentation the builder provides to the building owner at occupancy.

Company Name:		Phone:
Responsible Person's Name:		Responsible Person's Signature:
License:	Date Signed:	Position With Company (Title):

CERTIFICATE OF ACCEPTANCE	MECH-2A
Outdoor Air Acceptance	(Page 2 of 3)
Project Name/Address:	
System Name or Identification/Tag:	System Location or Area Served:

Intent: *Verify measured outside airflow reading is within ± 10% of the total required outside airflow value found in Section E 805.1 through Section E 805.2.2*

Construction Inspection

1. Instrumentation to perform test includes, but not limited to:
   a. Watch.
   b. Calibrated means to measure airflow.
2. Check one of the following:
   - ☐ Variable Air Volume (VAV) - Check as appropriate:
     a. Sensor used to control outdoor air flow must have calibration certificate or be field calibrated.
        - ☐ Calibration certificate (attach calibration certification).
        - ☐ Field calibration (attach results).
   - ☐ Constant Air Volume (CAV) - Check as appropriate:
     - ☐ System is designed to provide a fixed minimum OSA when the unit is on.

Outdoor Air Acceptance		
A. Functional Testing. (Check appropriate column)	**CAV**	**VAV**
a. Verify unit is not in economizer mode during test - check appropriate column.		
Step 1: CAV and VAV testing at full supply airflow.		
a. Adjust supply to achieve design airflow.		
b. Measured outdoor airflow reading (ft^3/min).		
c. Required outdoor airflow (ft^3/min).		
d. Time for outside air damper to stabilize after VAV boxes open (minutes).		
e. Return to initial conditions (check).		
Step 2: VAV testing at reduced supply airflow.		
a. Adjust supply airflow to either the sum of the minimum zone airflows or 30% of the total design airflow.		
b. Measured outdoor airflow reading (ft^3/min).		
c. Required outdoor airflow (ft^3/min).		
d. Time for outside air damper to stabilize after VAV boxes open and minimum air flow achieved (minutes).		
e. Return to initial conditions (check).		
B. Testing Calculations and Results.	**CAV**	**VAV**
Percent OSA at full supply airflow ($\%OA_{FA}$ for Step 1).		
a. $\%OA_{FA}$ = Measured outside air reading /Required outside air (Step 1b / Step 1c)	%	%
b. $90\% \leq \%OA_{FA} \leq 110\%$	Y / N	Y / N
c. Outside air damper position stabilizes within 15 minutes (Step 1d < 15 minutes)	Y / N	Y / N
Percent OSA at reduced supply airflow ($\%OA_{RA}$ for Step 2).		
a. $\%OA_{RA}$ = Measured outside air reading/required outside air (Step 2b / Step 2c).	%	%
b. $90\% \leq \%OA_{RA} \leq 110\%$.		Y / N
c. Outside air damper position stabilizes within 15 minutes (Step 2d < 15 minutes).		Y / N
Note: Shaded boxes do not apply for CAV systems.		

For SI units: 1 cubic foot per minute = 0.00047 m^3/s

CERTIFICATE OF ACCEPTANCE	MECH-2A
Outdoor Air Acceptance	**(Page 3 of 3)**
Project Name/Address:	
System Name or Identification/Tag:	System Location or Area Served:

C.	**PASS/FAIL Evaluation** (check one):
☐	PASS: All Construction Inspection responses are complete and Testing Calculations & Results responses are positive (Y – yes).
☐	FAIL: Any Construction Inspection responses are incomplete *OR* there is one or more negative (N – no) responses in Testing Calculations & Results section. Provide explanation below. Use and attach additional pages if necessary.

CERTIFICATE OF ACCEPTANCE	MECH-3A
Constant Volume Single Zone Unitary Air Conditioner and Heat Pump Systems	**(Page 1 of 3)**
Project Name/Address:	
System Name or Identification/Tag:	System Location or Area Served:

Enforcement Agency:	Permit Number:
Note: Submit one Certificate of Acceptance for each system that must demonstrate compliance.	Enforcement Agency Use: Checked by/Date

FIELD TECHNICIAN'S DECLARATION STATEMENT

- I certify under penalty of perjury the information provided on this form is true and correct.
- I am the person who performed the acceptance requirements verification reported on this Certificate of Acceptance (Field Technician).
- I certify that the construction/installation identified on this form complies with the acceptance requirements indicated in the plans and specifications approved by the enforcement agency, and conforms to the applicable acceptance requirements and procedures specified in Section E 801.0 through Section E 806.0.
- I have confirmed that the Installation Certificate(s) for the construction/installation identified on this form has been completed and is posted or made available with the building permit(s) issued for the building.

Company Name:		
Field Technician's Name:		Field Technician's Signature:
	Date Signed:	Position with Company (Title):

RESPONSIBLE PERSON'S DECLARATION STATEMENT

- I certify under penalty of perjury that I am the Field Technician, or the Field Technician is acting on my behalf as my employee or my agent and I have reviewed the information provided on this form.
- I am a licensed contractor or registered design professional who is eligible per the requirements of the Authority Having Jurisdiction to take responsibility for the scope of work specified on this document and attest to the declarations in this statement (responsible person).
- I certify that the information provided on this form substantiates that the construction/installation identified on this form complies with the acceptance requirements indicated in the plans and specifications approved by the enforcement agency, and conforms to the applicable acceptance requirements and procedures specified in Section E 801.0 through Section E 806.0.
- I have confirmed that the Installation Certificate(s) for the construction/installation identified on this form has been completed and is posted or made available with the permit(s) issued for the building.
- I will ensure that a completed, signed copy of this Certificate of Acceptance shall be posted, or made available with the building permit(s) issued for the building, and made available to the enforcement agency for all applicable inspections. I understand that a signed copy of this Certificate of Acceptance is required to be included with the documentation the builder provides to the building owner at occupancy.

Company Name:		Phone:
Responsible Person's Name:		Responsible Person's Signature:
License:	Date Signed:	Position With Company (Title):

CERTIFICATE OF ACCEPTANCE	MECH-3A
Constant Volume Single Zone Unitary Air Conditioner and Heat Pump Systems	(Page 2 of 3)
Project Name/Address:	
System Name or Identification/Tag:	System Location or Area Served:

Intent: *Verify the individual components of a constant volume, single-zone, unitary air conditioner and heat pump system function correctly, including: thermostat installation and programming, supply fan, heating, cooling, and damper operation.*

Construction Inspection

1. Instrumentation to perform test includes, but not limited to:
   a. None required
2. Installation
   - ☐ Thermostat is located within the space-conditioning zone that is served by the HVAC system.
3. Programming (check all of the following):
   - ☐ Thermostat meets the temperature adjustment and dead band requirements.
   - ☐ Occupied, unoccupied, and holiday schedules have been programmed per the facility's schedule.
   - ☐ Preoccupancy purge has been programmed to meet the requirements of Section E 805.3 through Section E 805.3.2.

A. Functional Testing Requirements. **Operating Modes**

- A: Heating load during occupied condition
- B: No-load during occupied condition
- C: Heating load during unoccupied condition
- D: No-load during unoccupied condition
- E: Manual override
- F: Cooling load during occupied condition
- G: Cooling load during unoccupied condition

	Step 1: Check and verify the following for each simulation mode required.	A	B	C	D	E	F	G
a.	Supply fan operates continually.							
b.	Supply fan turns off.							
c.	Supply fan cycles on and off.							
d.	System reverts to "occupied" mode to satisfy any condition.							
e.	System turns off when manual override time period expires.							
f.	Gas-fired furnace, heat pump, or electric heater stages on.							
g.	Neither heating or cooling is provided by the unit.							
h.	No heating is provided by the unit.							
i.	No cooling is provided by the unit.							
j.	Compressor stages on.							
k.	Outside air damper is open to minimum position.							
l.	Outside air damper closes completely.							
m.	System returned to initial operating conditions after all tests have been completed:	Y/N						

B. Testing Results	A	B	C	D	E	F	G
Indicate if Passed (P), Failed (F), or N/A (X), fill in appropriate letter.							

CERTIFICATE OF ACCEPTANCE	MECH-3A
Constant Volume Single Zone Unitary Air Conditioner and Heat Pump Systems	**(Page 3 of 3)**
Project Name/Address:	
System Name or Identification/Tag:	System Location or Area Served:

C.	PASS/FAIL Evaluation. (check one):
☐	PASS: All **Construction Inspection** responses are complete and **Testing Results** responses are "Pass" (P).
☐	FAIL: Any **Construction Inspection** responses are incomplete OR there is one or more "Fail" (F) responses in **Testing Results** section. Provide explanation below. Use and attach additional pages if necessary.

CERTIFICATE OF ACCEPTANCE	MECH-4A
Air Distribution Systems Acceptance	**(Page 1 of 3)**
Project Name/Address:	
System Name or Identification/Tag:	System Location or Area Served:

Enforcement Agency:	Permit Number:
Note: Submit one Certificate of Acceptance for each system that must demonstrate compliance.	Enforcement Agency Use: Checked by/Date

FIELD TECHNICIAN'S DECLARATION STATEMENT

- I certify under penalty of perjury the information provided on this form is true and correct.
- I am the person who performed the acceptance requirements verification reported on this Certificate of Acceptance (Field Technician).
- I certify that the construction/installation identified on this form complies with the acceptance requirements indicated in the plans and specifications approved by the enforcement agency, and conforms to the applicable acceptance requirements and procedures specified in Section E 801.0 through Section E 806.0.
- I have confirmed that the Installation Certificate(s) for the construction/installation identified on this form has been completed and is posted or made available with the building permit(s) issued for the building.

Company Name:		
Field Technician's Name:		Field Technician's Signature:
	Date Signed:	Position with Company (Title):

RESPONSIBLE PERSON'S DECLARATION STATEMENT

- I certify under penalty of perjury that I am the Field Technician, or the Field Technician is acting on my behalf as my employee or my agent and I have reviewed the information provided on this form.
- I am a licensed contractor or registered design professional who is eligible per the requirements of the Authority Having Jurisdiction to take responsibility for the scope of work specified on this document and attest to the declarations in this statement (responsible person).
- I certify that the information provided on this form substantiates that the construction/installation identified on this form complies with the acceptance requirements indicated in the plans and specifications approved by the enforcement agency, and conforms to the applicable acceptance requirements and procedures specified in Section E 801.0 through Section E 806.0.
- I have confirmed that the Installation Certificate(s) for the construction/installation identified on this form has been completed and is posted or made available with the permit(s) issued for the building.
- I will ensure that a completed, signed copy of this Certificate of Acceptance shall be posted, or made available with the building permit(s) issued for the building, and made available to the enforcement agency for all applicable inspections. I understand that a signed copy of this Certificate of Acceptance is required to be included with the documentation the builder provides to the building owner at occupancy.

Company Name:		Phone:
Responsible Person's Name:		Responsible Person's Signature:
License:	Date Signed:	Position With Company (Title):

CERTIFICATE OF ACCEPTANCE	MECH-4A
Air Distribution Systems Acceptance	(Page 2 of 3)
Project Name/Address:	
System Name or Identification/Tag:	System Location or Area Served:

Intent: *New single zone supply ductwork must be less than 6% leakage rate per Section E 805.4 through Section E 805.4.2, existing single zone ductwork must be less than 15% leakage or other compliance path per Section E 805.4 through Section E 805.4.2.*

Construction Inspection

1. Scope of test – New Buildings – this test required on New Buildings only if all check boxes 1(a) through 1(c) are checked.

   Existing Buildings – this test required if 1(a) through 1(d) are checked.

   Ductwork conforms to the following (note if any of these are not checked, then this test is not required):

   ☐ 1(a) Connected to a constant volume, single zone air conditioners, heat pumps, or furnaces.

   ☐ 1(b) Serves less than 5000 square feet of floor area.

   ☐ 1(c) Has more than 25% duct surface area located in one or more of the following spaces.

   – Outdoors.
   – A space directly under a roof where the U-factor of the roof is greater than U-factor of the ceiling.
   – A space directly under a roof with fixed vents or openings to the outside or unconditioned spaces.
   – An unconditioned crawlspace.
   – Other unconditioned spaces.

   ☐ 1(d) A duct is extended or any of the following replaced: air handler, outdoor condensing unit of a split system, cooling or heating coil, or the furnace heat exchanger.

2. Instrumentation to perform test includes:

   a. Duct Pressure Test.

3. Material and Installation. Complying new duct systems shall have a checked box for all of the following categories (a) through (g):

   a. Choice of drawbands. (check one of the following)

☐	Stainless steel worm-drive hose clamps.
☐	UV-resistant nylon duct ties.

☐	b. Flexible ducts are not constricted in any way.
☐	c. Duct leakage tests performed before access to ductwork and connections are blocked.
☐	d. Joints and seams are not sealed with cloth back rubber adhesive tape unless used in combination with mastic and drawbands.
☐	e. Duct R-values are verified R-8 per Section E 805.4 through Section E 805.4.2.
☐	f. Ductwork located outdoors has insulation that is protected from damage and suitable for outdoor service.
☐	g. A sticker has been affixed to the exterior surface of the air handler access door per Section E 805.4 through Section E 805.4.2.

For SI units: 1 square foot = 0.0929 m^2

CERTIFICATE OF ACCEPTANCE	MECH-4A
Air Distribution Systems Acceptance	**(Page 3 of 3)**
Project Name/Address:	
System Name or Identification/Tag:	System Location or Area Served:

Air Distribution System Leakage Diagnostic.
The installing contractor must pressure test every new HVAC systems that meet the requirements of Section E 805.4 through Section E 805.4.2 and every retrofit to existing HVAC systems that meet the requirements of Section E 805.4 through Section E 805.4.2.

	RATED FAN FLOW (applies to all systems)	**Measured Values**	
1.	Cooling capacity or for heating only units heating capacity.		
	(a) Cooling capacity (for all units but heating only units) in tons.		
	(b) Heating capacity (for heating only units) kBtu/h.		
2.	Fan flow calculation		
	(a) Cooling capacity in tons [__________ (Line # 1a) x 400 ft^3/min/ton].		
	(b) Heating only cap. kBtu/h [__________ (Line # 1b) x 21.7 ft^3/min/kBtu/h].		
3.	Total calculated supply fan flow 2(a) or 2(b) ft^3/min.		

	NEW CONSTRUCTION OR ENTIRE NEW DUCT SYSTEM ALTERATION:		
	Duct pressurization test results (ft^3/min @ 25 Pa).		
4.	Enter tested leakage flow in ft^3/min:		✓ ✓
5.	Pass if leakage percentage ≤6%: [(Line #4) /_______ (Line #3)] x 100	%	☐ Pass ☐ Fail
	ALTERATIONS: Pre-existing duct system with duct alteration and/or HVAC equipment change-out.		
6.	Enter tested leakage flow (cubic feet per minute): Pre-test of existing duct system prior to duct system alteration, equipment change-out, or both.		
7.	Enter tested leakage flow (cubic feet per minute): Final test of new duct system or altered duct system for duct system alteration, equipment change-out, or both.		
	TEST OR VERIFICATION STANDARDS: For altered duct system and/or HVAC equipment change-out use one of the following three tests or verification standards for compliance:		
8.	Pass if leakage percentage <15% [_______ (Line # 7) / _______ (Line # 3)] x 100	%	☐ Pass ☐ Fail
9.	Pass if leakage reduction percentage >60% Leakage reduction = [1 - [_______ (Line#7) / _______ (Line#6)] } x 100	%	☐ Pass ☐ Fail
10.	Pass if all accessible leaks are sealed as confirmed by visual inspection and verification by HERS rater (sampling rate 100%).	%	☐ Pass ☐ Fail
	Pass if One of Lines #8 through #10 pass		☐ Pass ☐ Fail

For SI units: 1000 British thermal units per hour = 0.293 kW, 1 cubic foot per minute = 0.00047 m^3/s, 1 metric ton = 1000 kg

CERTIFICATE OF ACCEPTANCE	MECH-5A
Air Economizer Controls Acceptance	(Page 1 of 3)
Project Name/Address:	
System Name or Identification/Tag:	System Location or Area Served:

Enforcement Agency:	Permit Number:
Note: Submit one Certificate of Acceptance for each system that must demonstrate compliance.	Enforcement Agency Use: Checked by/Date

FIELD TECHNICIAN'S DECLARATION STATEMENT

- I certify under penalty of perjury the information provided on this form is true and correct.
- I am the person who performed the acceptance requirements verification reported on this Certificate of Acceptance (Field Technician).
- I certify that the construction/installation identified on this form complies with the acceptance requirements indicated in the plans and specifications approved by the enforcement agency, and conforms to the applicable acceptance requirements and procedures specified in Section E 801.0 through Section E 806.0.
- I have confirmed that the Installation Certificate(s) for the construction/installation identified on this form has been completed and is posted or made available with the building permit(s) issued for the building.

Company Name:		
Field Technician's Name:		Field Technician's Signature:
	Date Signed:	Position with Company (Title):

RESPONSIBLE PERSON'S DECLARATION STATEMENT

- I certify under penalty of perjury that I am the Field Technician, or the Field Technician is acting on my behalf as my employee or my agent and I have reviewed the information provided on this form.
- I am a licensed contractor or registered design professional who is eligible per the requirements of the Authority Having Jurisdiction to take responsibility for the scope of work specified on this document and attest to the declarations in this statement (responsible person).
- I certify that the information provided on this form substantiates that the construction/installation identified on this form complies with the acceptance requirements indicated in the plans and specifications approved by the enforcement agency, and conforms to the applicable acceptance requirements and procedures specified in Section E 801.0 through Section E 806.0.
- I have confirmed that the Installation Certificate(s) for the construction/installation identified on this form has been completed and is posted or made available with the permit(s) issued for the building.
- I will ensure that a completed, signed copy of this Certificate of Acceptance shall be posted, or made available with the building permit(s) issued for the building, and made available to the enforcement agency for all applicable inspections. I understand that a signed copy of this Certificate of Acceptance is required to be included with the documentation the builder provides to the building owner at occupancy.

Company Name:		Phone:
Responsible Person's Name:		Responsible Person's Signature:
License:	Date Signed:	Position With Company (Title):

CERTIFICATE OF ACCEPTANCE	MECH-5A
Air Economizer Controls Acceptance	(Page 2 of 3)
Project Name/Address:	
System Name or Identification/Tag:	System Location or Area Served:

Intent: *Verify that airside economizers function properly.*

Construction Inspection
1. Instrumentation to perform test includes, but not limited to: a. Handheld temperature probes calibration. Date: (must be within last year). b. Multimeter capable of measuring ohms and milliamps. 2. Test method (check one of the following): ☐ Economizer comes from HVAC system manufacturer installed by and has been factory calibrated and tested. Attach documentation and complete certification statement. No functional testing required. ☐ Economizer field installed and field tested or factory installed and field tested. 3. Installation (check all of the following first level boxes). ☐ Economizer lockout setpoint complies with Section E 805.5 through Section E 805.5.2. ☐ Economizer lockout control sensor is located to prevent false readings. ☐ System is designed to provide up to 100% outside air without over-pressurizing the building. ☐ For systems with DDC controls lockout sensor(s) are either factory calibrated or field calibrated. ☐ For systems with non-DDC controls, manufacturer's startup and testing procedures have been applied.

A. Functional Testing.	
Step 1: Disable demand control ventilation systems (if applicable).	
Step 2: Enable the economizer and simulate a cooling demand large enough to drive the economizer fully open (check and verify the following).	
☐ Economizer damper modulates 100% open.	
☐ Return air damper modulates 100% closed.	
☐ Where applicable, verify that the economizer remains 100% open when the cooling demand can no longer be met by the economizer alone.	
☐ All applicable fans and dampers operate as intended to maintain building pressure.	
☐ The unit heating is disabled.	
Step 3: Simulate a cooling load and disable the economizer (check and verify the following).	
☐ Economizer damper closes to its minimum position.	
☐ All applicable fans and dampers operate as intended to maintain building pressure.	
☐ The unit heating is disabled.	
Step 4: Simulate a heating demand and enable the economizer (check and verify the following).	
☐ Economizer damper closes to its minimum position.	
Step 5: System returned to initial operating conditions.	**Y/N**

B. Testing Results.	**PASS**	**FAIL**
Step 1: Simulate cooling load and enable the economizer (all check boxes are complete).		
Step 2: Simulate cooling load and disable the economizer (all check boxes are complete).		
Step 3: Simulate heating demand and enable the economizer (all check boxes are complete).		

CERTIFICATE OF ACCEPTANCE	MECH-5A
Air Economizer Controls Acceptance	(Page 3 of 3)
Project Name/Address:	
System Name or Identification/Tag:	System Location or Area Served:

C.	PASS/FAIL Evaluation (check one):
☐	PASS: All **Construction Inspection** responses are complete and **Testing Results** responses are "Pass."
☐	FAIL: Any **Construction Inspection** responses are incomplete *OR* there is one or more "Fail" responses in **Testing Results** section. Provide explanation below. Use and attach additional pages if necessary.

CERTIFICATE OF ACCEPTANCE	MECH-6A
Demand Control Ventilation Systems Acceptance	**(Page 1 of 3)**
Project Name/Address:	
System Name or Identification/Tag:	System Location or Area Served:

Enforcement Agency:	Permit Number:
Note: Submit one Certificate of Acceptance for each system that must demonstrate compliance.	Enforcement Agency Use: Checked by/Date

FIELD TECHNICIAN'S DECLARATION STATEMENT

- I certify under penalty of perjury the information provided on this form is true and correct.
- I am the person who performed the acceptance requirements verification reported on this Certificate of Acceptance (Field Technician).
- I certify that the construction/installation identified on this form complies with the acceptance requirements indicated in the plans and specifications approved by the enforcement agency, and conforms to the applicable acceptance requirements and procedures specified in Section E 801.0 through Section E 806.0.
- I have confirmed that the Installation Certificate(s) for the construction/installation identified on this form has been completed and is posted or made available with the building permit(s) issued for the building.

Company Name:		
Field Technician's Name:		Field Technician's Signature:
	Date Signed:	Position with Company (Title):

RESPONSIBLE PERSON'S DECLARATION STATEMENT

- I certify under penalty of perjury that I am the Field Technician, or the Field Technician is acting on my behalf as my employee or my agent and I have reviewed the information provided on this form.
- I am a licensed contractor or registered design professional who is eligible per the requirements of the Authority Having Jurisdiction to take responsibility for the scope of work specified on this document and attest to the declarations in this statement (responsible person).
- I certify that the information provided on this form substantiates that the construction/installation identified on this form complies with the acceptance requirements indicated in the plans and specifications approved by the enforcement agency, and conforms to the applicable acceptance requirements and procedures specified in Section E 801.0 through Section E 806.0.
- I have confirmed that the Installation Certificate(s) for the construction/installation identified on this form has been completed and is posted or made available with the permit(s) issued for the building.
- I will ensure that a completed, signed copy of this Certificate of Acceptance shall be posted, or made available with the building permit(s) issued for the building, and made available to the enforcement agency for all applicable inspections. I understand that a signed copy of this Certificate of Acceptance is required to be included with the documentation the builder provides to the building owner at occupancy.

Company Name:			Phone:
Responsible Person's Name:		Responsible Person's Signature:	
License:	Date Signed:	Position With Company (Title):	

CERTIFICATE OF ACCEPTANCE	MECH-6A
Demand Control Ventilation Systems Acceptance	**(Page 2 of 3)**
Project Name/Address:	
System Name or Identification/Tag:	System Location or Area Served:

Intent: *Verify that systems required to employ demand controlled ventilation can vary outside ventilation flow rates based on maintaining interior carbon dioxide (CO_2) concentration setpoints.*

Construction Inspection

1. Instrumentation to perform test includes, but not limited to:
   a. Calibrated handheld CO_2 analyzer.
   b. Manufacturer's calibration kit.
   c. Calibrated CO_2/air mixtures.
2. Installation.
   ☐ The sensor is located in the high density space between 3 feet and 6 feet above the floor or at the anticipated level of the occupants heads.
3. Documentation of all carbon dioxide control sensors includes (check one of the following):
   a. Calibration method.
      ☐ Factory-calibration certificate (certificate must be attached).
      ☐ Field calibrated.
   b. Sensor accuracy.
      ☐ Certified by manufacturer to be no more than +/- 75 ppm calibration certificate must be attached.

A. Functional Testing.	**Results**
a. Disable economizer controls.	
b. Outside air CO_2 concentration (select one of the following).	
☐ Measured dynamically using CO_2 sensor.	______________ ppm
c. Interior CO_2 concentration setpoint (Outside CO_2 concentration + 600 ppm).	______________ ppm
Step 1: Simulate a signal at or slightly above the CO_2 setpoint or follow manufacturers recommended testing procedures.	
☐ For single zone units, outdoor air damper modulates opens to satisfy the total ventilation air called for in the certificate of compliance.	
☐ For multiple zone units, either outdoor air damper or zone damper modulate open to satisfy the zone ventilation requirements.	
Step 2: Simulate signal well below the CO_2 setpoint or follow manufacturers recommended procedures.	
☐ For single zone units, outdoor air damper modulates to the design minimum value.	
☐ For multiple zone units, either outdoor air damper or zone damper modulate to satisfy the reduced zone ventilation requirements.	
Step 3: System returned to initial operating conditions.	**Y/N**

B. Testing Results.	**PASS / FAIL**	
Step 1: Simulate a high CO_2 load (check box complete).		
Step 2: Simulate a low CO_2 load (check box complete).		

For SI units: 1 inch = 25.4 mm

CERTIFICATE OF ACCEPTANCE	MECH-6A
Demand Control Ventilation Systems Acceptance	**(Page 3 of 3)**
Project Name/Address:	
System Name or Identification/Tag:	System Location or Area Served:

C.	PASS/FAIL Evaluation (check one):
☐	PASS: All **Construction Inspection** responses are complete and **Testing Results** responses are "Pass."
☐	FAIL: Any **Construction Inspection** responses are incomplete *OR* there is one or more "Fail" responses in **Testing Results** section. Provide explanation below. Use and attach additional pages if necessary.

CERTIFICATE OF ACCEPTANCE	MECH-7A
Supply Fan VFD Acceptance	**(Page 1 of 2)**
Project Name/Address:	
System Name or Identification/Tag:	System Location or Area Served:

Enforcement Agency:	Permit Number:
Note: Submit one Certificate of Acceptance for each system that must demonstrate compliance.	Enforcement Agency Use: Checked by/Date

FIELD TECHNICIAN'S DECLARATION STATEMENT

- I certify under penalty of perjury the information provided on this form is true and correct.
- I am the person who performed the acceptance requirements verification reported on this Certificate of Acceptance (Field Technician).
- I certify that the construction/installation identified on this form complies with the acceptance requirements indicated in the plans and specifications approved by the enforcement agency, and conforms to the applicable acceptance requirements and procedures specified in Section E 801.0 through Section E 806.0.
- I have confirmed that the Installation Certificate(s) for the construction/installation identified on this form has been completed and is posted or made available with the building permit(s) issued for the building.

Company Name:		
Field Technician's Name:		Field Technician's Signature:
	Date Signed:	Position with Company (Title):

RESPONSIBLE PERSON'S DECLARATION STATEMENT

- I certify under penalty of perjury that I am the Field Technician, or the Field Technician is acting on my behalf as my employee or my agent and I have reviewed the information provided on this form.
- I am a licensed contractor or registered design professional who is eligible per the requirements of the Authority Having Jurisdiction to take responsibility for the scope of work specified on this document and attest to the declarations in this statement (responsible person).
- I certify that the information provided on this form substantiates that the construction/installation identified on this form complies with the acceptance requirements indicated in the plans and specifications approved by the enforcement agency, and conforms to the applicable acceptance requirements and procedures specified in Section E 801.0 through Section E 806.0.
- I have confirmed that the Installation Certificate(s) for the construction/installation identified on this form has been completed and is posted or made available with the permit(s) issued for the building.
- I will ensure that a completed, signed copy of this Certificate of Acceptance shall be posted, or made available with the building permit(s) issued for the building, and made available to the enforcement agency for all applicable inspections. I understand that a signed copy of this Certificate of Acceptance is required to be included with the documentation the builder provides to the building owner at occupancy.

Company Name:			Phone:
Responsible Person's Name:		Responsible Person's Signature:	
License:	Date Signed:	Position With Company (Title):	

CERTIFICATE OF ACCEPTANCE	MECH-7A
Supply Fan VFD Acceptance	**(Page 2 of 2)**
Project Name/Address:	
System Name or Identification/Tag:	System Location or Area Served:

Intent: *Verify that the supply fan in a variable air volume application modulates to meet system airflow demand.*

Construction Inspection

1 Instrumentation to perform test includes, but not limited to:
   a. Calibrated differential pressure gauge.

2 Installation.
   - ☐ Discharge static pressure sensors are either factory calibrated or field-calibrated.
   - ☐ The static pressure location, setpoint, and reset control meets the requirements of Section E 805.7 through Section E 805.7.2.

3 Documentation of all discharge static pressure sensors including (check one of the following):
   - ☐ Field-calibrated.
   - ☐ Calibration complete, all pressure sensors within 10% of calibrated reference sensor.

A. Functional Testing.	**Results**
Step 1: Drive all VAV boxes to achieve design airflow.	
a. Supply fan controls modulate to increase capacity.	**Y / N**
b. Supply fan maintains discharge static pressure within +/-10% of the current operating setpoint.	**Y / N**
c. Supply fan controls stabilize within a 5 minute period.	**Y / N**
Step 2: Drive all VAV boxes to minimum flow.	
a. Supply fan controls modulate to decrease capacity.	**Y / N**
b. Current operating setpoint has decreased (for systems with DDC to the zone level).	**Y / N**
c. Supply fan maintains discharge static pressure within +/-10% of the current operating setpoint.	**Y / N**
d. Supply fan controls stabilize within a 5 minute period.	**Y / N**
Step 3: System returned to initial operating conditions.	**Y / N**

B. Testing Results.	**PASS / FAIL**	
Step 1: Drive all VAV boxes to achieve design airflow.		
Step 2: Drive all VAV boxes to minimum flow.		

C. PASS / FAIL Evaluation (check one):	
☐	PASS: All Construction Inspection responses are complete and all Testing Results responses are "Pass."
☐	FAIL: Any Construction Inspection responses are incomplete *OR* there is one or more "Fail" responses in Testing Results section. Provide explanation below. Use and attach additional pages if necessary.

CERTIFICATE OF ACCEPTANCE	MECH-8A
Valve Leakage Test	**(Page 1 of 2)**
Project Name/Address:	
System Name or Identification/Tag:	System Location or Area Served:

Enforcement Agency:	Permit Number:
Note: Submit one Certificate of Acceptance for each system that must demonstrate compliance.	Enforcement Agency Use: Checked by/Date

FIELD TECHNICIAN'S DECLARATION STATEMENT

- I certify under penalty of perjury the information provided on this form is true and correct.
- I am the person who performed the acceptance requirements verification reported on this Certificate of Acceptance (Field Technician).
- I certify that the construction/installation identified on this form complies with the acceptance requirements indicated in the plans and specifications approved by the enforcement agency, and conforms to the applicable acceptance requirements and procedures specified in Section E 801.0 through Section E 806.0.
- I have confirmed that the Installation Certificate(s) for the construction/installation identified on this form has been completed and is posted or made available with the building permit(s) issued for the building.

Company Name:		
Field Technician's Name:		Field Technician's Signature:
	Date Signed:	Position with Company (Title):

RESPONSIBLE PERSON'S DECLARATION STATEMENT

- I certify under penalty of perjury that I am the Field Technician, or the Field Technician is acting on my behalf as my employee or my agent and I have reviewed the information provided on this form.
- I am a licensed contractor or registered design professional who is eligible per the requirements of the Authority Having Jurisdiction to take responsibility for the scope of work specified on this document and attest to the declarations in this statement (responsible person).
- I certify that the information provided on this form substantiates that the construction/installation identified on this form complies with the acceptance requirements indicated in the plans and specifications approved by the enforcement agency, and conforms to the applicable acceptance requirements and procedures specified in Section E 801.0 through Section E 806.0.
- I have confirmed that the Installation Certificate(s) for the construction/installation identified on this form has been completed and is posted or made available with the permit(s) issued for the building.
- I will ensure that a completed, signed copy of this Certificate of Acceptance shall be posted, or made available with the building permit(s) issued for the building, and made available to the enforcement agency for all applicable inspections. I understand that a signed copy of this Certificate of Acceptance is required to be included with the documentation the builder provides to the building owner at occupancy.

Company Name:			Phone:
Responsible Person's Name:		Responsible Person's Signature:	
License:	Date Signed:	Position With Company (Title):	

CERTIFICATE OF ACCEPTANCE	MECH-8A
Valve Leakage Test	**(Page 2 of 2)**
Project Name/Address:	
System Name or Identification/Tag:	System Location or Area Served:

Intent: *Ensure that control valves serving variable flow systems are designed to withstand the pump pressure over the full range of operation.*

Construction Inspection

1 Instrumentation to perform test includes, but not limited to:
   a. Calibrated differential pressure gauge.
   b. Pump curve submittals showing the shutoff head.

2 Installation.
   ☐ Valve and piping arrangements were installed per the design drawings.

A. Functional Testing. Pump Tag (Id)		**Results**
Step 1: Determine pump dead head pressure.		
a. Close pump discharge isolation valve.		Y / N
b. Measure and record the differential pump pressure.	Feet Water Column =	
c. Record the shutoff head from the submittal.	Feet Water Column =	
d. The measurement across the pump in step 1b is within 5% of the pump submittal in step 1c.		Y / N
e. Open pump discharge isolation valve.		Y / N
Step 2: Automatically close all valves on the systems being tested. If three-way valves are present, close off the bypass line(s).		
a. The 2-way valves automatically close.		Y / N
b. Measure and record the differential pump pressure in feet of water column.	Feet Water Column =	
c. The measurement across the pump in step 2b is within 5% of the measurement in step 1b.		Y / N
Step 3: System returned to initial operating conditions.	Y / N	

B. Testing Results.	**PASS**	**FAIL**
Step 1: Pressure measurement is within 5% of submittal data for all pumps.	☐	☐
Step 2: Pressure measurements are within 5%.	☐	☐

C. PASS / FAIL Evaluation (check one):	
☐	PASS: All **Construction Inspection** responses are complete and all **Testing Results** responses are "Pass."
☐	FAIL: Any **Construction Inspection** responses are incomplete *OR* there is one or more "Fail" responses in **Testing Results** section. Provide explanation below. Use and attach additional pages if necessary.

For SI units: 1 inch water column = 0.249 kPa

CERTIFICATE OF ACCEPTANCE	MECH-9A
Supply Water Temperature Reset Controls Acceptance	**(Page 1 of 2)**
Project Name/Address:	
System Name or Identification/Tag:	System Location or Area Served:

Enforcement Agency:	Permit Number:
Note: Submit one Certificate of Acceptance for each system that must demonstrate compliance.	Enforcement Agency Use: Checked by/Date

FIELD TECHNICIAN'S DECLARATION STATEMENT

- I certify under penalty of perjury the information provided on this form is true and correct.
- I am the person who performed the acceptance requirements verification reported on this Certificate of Acceptance (Field Technician).
- I certify that the construction/installation identified on this form complies with the acceptance requirements indicated in the plans and specifications approved by the enforcement agency, and conforms to the applicable acceptance requirements and procedures specified in Section E 801.0 through Section E 806.0.
- I have confirmed that the Installation Certificate(s) for the construction/installation identified on this form has been completed and is posted or made available with the building permit(s) issued for the building.

Company Name:		
Field Technician's Name:		Field Technician's Signature:
	Date Signed:	Position with Company (Title):

RESPONSIBLE PERSON'S DECLARATION STATEMENT

- I certify under penalty of perjury that I am the Field Technician, or the Field Technician is acting on my behalf as my employee or my agent and I have reviewed the information provided on this form.
- I am a licensed contractor or registered design professional who is eligible per the requirements of the Authority Having Jurisdiction to take responsibility for the scope of work specified on this document and attest to the declarations in this statement (responsible person).
- I certify that the information provided on this form substantiates that the construction/installation identified on this form complies with the acceptance requirements indicated in the plans and specifications approved by the enforcement agency, and conforms to the applicable acceptance requirements and procedures specified in Section E 801.0 through Section E 806.0.
- I have confirmed that the Installation Certificate(s) for the construction/installation identified on this form has been completed and is posted or made available with the permit(s) issued for the building.
- I will ensure that a completed, signed copy of this Certificate of Acceptance shall be posted, or made available with the building permit(s) issued for the building, and made available to the enforcement agency for all applicable inspections. I understand that a signed copy of this Certificate of Acceptance is required to be included with the documentation the builder provides to the building owner at occupancy.

Company Name:		Phone:
Responsible Person's Name:		Responsible Person's Signature:
License:	Date Signed:	Position With Company (Title):

CERTIFICATE OF ACCEPTANCE	MECH-9A
Supply Water Temperature Reset Controls Acceptance	**(Page 2 of 2)**
Project Name/Address:	
System Name or Identification/Tag:	System Location or Area Served:

Intent: *Ensure that both the chilled water and hot water supply temperatures are automatically reset based on either building loads or outdoor air temperature, as indicated in the control sequences.*

Construction Inspection

1 Instrumentation to perform test includes, but not limited to:
   a. Calibrated reference temperature sensor or drywell bath.

2 Installation
   ☐ Supply water temperature sensors have been either factory or field calibrated.

3 Documentation of hydronic system supply temperature sensors including (check one of the following):
   ☐ Field-calibrated
   ☐ Calibration complete, hydronic system supply temperature sensors within 1% of calibrated reference sensor or drywell bath.

A. Functional Testing.	
Step 1: Test maximum reset value.	
a. Change reset control variable to its maximum value.	**Y / N**
b. Verify that chilled or hot water temperature setpoint is reset to appropriate value.	**Y / N**
c. Verify that actual system temperature changes to within 2% of the new setpoint.	**Y / N**
Step 2: Test minimum reset value.	
a. Change reset control variable to its minimum value.	**Y / N**
b. Verify that chilled or hot water temperature setpoint is reset to appropriate value.	**Y / N**
c. Verify that actual system temperature changes to within 2% of the new setpoint.	**Y / N**
Step 3: Test maximum reset value.	
a. Restore reset control variable to automatic control.	**Y / N**
b. Verify that chilled or hot water temperature setpoint is reset to appropriate value.	**Y / N**
c. Verify that actual supply temperature changes to meet setpoint.	**Y / N**
d. Verify that actual supply temperature changes to within 2% of the new setpoint.	**Y / N**

B. Testing Results.	**PASS / FAIL**	
System passes criteria in 1c, 2c, and 3d.	☐	☐

C. PASS / FAIL Evaluation (check one):	
☐	PASS: All **Construction Inspection** responses are complete and all **Testing Results** responses are "Pass."
☐	FAIL: Any **Construction Inspection** responses are incomplete *OR* there is one or more "Fail" responses in **Testing Results** section. Provide explanation below. Use and attach additional pages if necessary.

CERTIFICATE OF ACCEPTANCE	MECH-10A
Hydronic System Variable Flow Control Acceptance	(Page 1 of 3)
Project Name/Address:	
System Name or Identification/Tag:	System Location or Area Served:

Enforcement Agency:	Permit Number:
Note: Submit one Certificate of Acceptance for each system that must demonstrate compliance.	Enforcement Agency Use: Checked by/Date

FIELD TECHNICIAN'S DECLARATION STATEMENT

- I certify under penalty of perjury the information provided on this form is true and correct.
- I am the person who performed the acceptance requirements verification reported on this Certificate of Acceptance (Field Technician).
- I certify that the construction/installation identified on this form complies with the acceptance requirements indicated in the plans and specifications approved by the enforcement agency, and conforms to the applicable acceptance requirements and procedures specified in Section E 801.0 through Section E 806.0.
- I have confirmed that the Installation Certificate(s) for the construction/installation identified on this form has been completed and is posted or made available with the building permit(s) issued for the building.

Company Name:		
Field Technician's Name:		Field Technician's Signature:
	Date Signed:	Position with Company (Title):

RESPONSIBLE PERSON'S DECLARATION STATEMENT

- I certify under penalty of perjury that I am the Field Technician, or the Field Technician is acting on my behalf as my employee or my agent and I have reviewed the information provided on this form.
- I am a licensed contractor or registered design professional who is eligible per the requirements of the Authority Having Jurisdiction to take responsibility for the scope of work specified on this document and attest to the declarations in this statement (responsible person).
- I certify that the information provided on this form substantiates that the construction/installation identified on this form complies with the acceptance requirements indicated in the plans and specifications approved by the enforcement agency, and conforms to the applicable acceptance requirements and procedures specified in Section E 801.0 through Section E 806.0.
- I have confirmed that the Installation Certificate(s) for the construction/installation identified on this form has been completed and is posted or made available with the permit(s) issued for the building.
- I will ensure that a completed, signed copy of this Certificate of Acceptance shall be posted, or made available with the building permit(s) issued for the building, and made available to the enforcement agency for all applicable inspections. I understand that a signed copy of this Certificate of Acceptance is required to be included with the documentation the builder provides to the building owner at occupancy.

Company Name:		Phone:
Responsible Person's Name:		Responsible Person's Signature:
License:	Date Signed:	Position With Company (Title):

CERTIFICATE OF ACCEPTANCE	MECH-10A
Hydronic System Variable Flow Control Acceptance	**(Page 2 of 3)**
Project Name/Address:	
System Name or Identification/Tag:	System Location or Area Served:

Intent: *Ensure that when loads within the building fluctuate, control valves modulate the amount of water passing through each coil and add or remove the desired amount of energy from the air stream to satisfy the load.*

Construction Inspection

1. Instrumentation to perform test includes, but not limited to:
   a. Calibrated differential pressure gauge.
2. Installation
   - ☐ Pressure sensors are either factory calibrated or field-calibrated.
   - ☐ Pressure sensor location, setpoint, and reset control meets the requirements of Section E 805.8 through Section E 805.8.2.
3. Documentation of all control pressure sensors including (check one of the following):
   a. Factory-calibrated (proof required).
      - ☐ Factory-calibration certificate.

   b. Field-calibrated.
      - ☐ Calibration complete, all pressure sensors within 10% of calibrated reference sensor.

A. Functional Testing.		**Results**
Step 1: Design flow test.		
a. Open control valves to achieve a minimum of 90% of design flow.		**Y / N**
b. Verify that the pump speed increases.		**Y / N**
c. Are the pumps operating at 100% speed?		**Y / N**
d. Record the system pressure as measured at the control sensor.	(Feet Water Column) =	
e. Record the system pressure setpoint.	(Feet Water Column) =	
f. Is the pressure reading 1d within 5% of pressure setpoint 1e?		**Y / N**
g. Did the system operation stabilize within 5 minutes after completion of step 1a?		**Y / N**
Step 2: Low flow test		
a. Close coil control valves to achieve a maximum of 50% of design flow.		**Y / N**
b. Verify that the current operating speed decreases (for systems with DDC to the zone level).		**Y / N**
c. Verify that the current operating speed has not increased (for all other systems that are not DDC).		**Y / N**
d. Record the system pressure as measured at the control sensor.	(Feet Water Column) =	
e. Record the system pressure setpoint.	(Feet Water Column) =	
f. Is the setpoint in 2e is less than the setpoint in 1d?		**Y / N**
g. Is the pressure reading 2d within 5% of pressure setpoint 2e?		**Y / N**
h. Did the system operation stabilize within 5 minutes after completion of step 2a?		**Y / N**
Step 3: System returned to initial operating conditions.		**Y / N**

B. Testing Results	**PASS / FAIL**	
Step 1: Select pass if either 1c or 1f are true.	☐	☐
Step 2: Select pass if 2b, 2e, 2f and 2g are true.	☐	☐

For SI units: 1 inch water column = 0.249 kPa

CERTIFICATE OF ACCEPTANCE	MECH-10A
Hydronic System Variable Flow Control Acceptance	**(Page 3 of 3)**
Project Name/Address:	
System Name or Identification/Tag:	System Location or Area Served:

C. PASS / FAIL Evaluation (check one):	
☐	PASS: All **Construction Inspection** responses are complete and all **Testing Results** responses are "Pass."
☐	FAIL: Any **Construction Inspection** responses are incomplete *OR* there is one or more "Fail" responses in **Testing Results** section. Provide explanation below. Use and attach additional pages if necessary.

CERTIFICATE OF ACCEPTANCE	MECH-11A
Automatic Demand Shed Control Acceptance	**(Page 1 of 2)**
Project Name/Address:	
System Name or Identification/Tag:	System Location or Area Served:

Enforcement Agency:	Permit Number:
Note: Submit one Certificate of Acceptance for each system that must demonstrate compliance.	Enforcement Agency Use: Checked by/Date

FIELD TECHNICIAN'S DECLARATION STATEMENT

- I certify under penalty of perjury the information provided on this form is true and correct.
- I am the person who performed the acceptance requirements verification reported on this Certificate of Acceptance (Field Technician).
- I certify that the construction/installation identified on this form complies with the acceptance requirements indicated in the plans and specifications approved by the enforcement agency, and conforms to the applicable acceptance requirements and procedures specified in Section E 801.0 through Section E 806.0.
- I have confirmed that the Installation Certificate(s) for the construction/installation identified on this form has been completed and is posted or made available with the building permit(s) issued for the building.

Company Name:		
Field Technician's Name:		Field Technician's Signature:
	Date Signed:	Position with Company (Title):

RESPONSIBLE PERSON'S DECLARATION STATEMENT

- I certify under penalty of perjury that I am the Field Technician, or the Field Technician is acting on my behalf as my employee or my agent and I have reviewed the information provided on this form.
- I am a licensed contractor or registered design professional who is eligible per the requirements of the Authority Having Jurisdiction to take responsibility for the scope of work specified on this document and attest to the declarations in this statement (responsible person).
- I certify that the information provided on this form substantiates that the construction/installation identified on this form complies with the acceptance requirements indicated in the plans and specifications approved by the enforcement agency, and conforms to the applicable acceptance requirements and procedures specified in Section E 801.0 through Section E 806.0.
- I have confirmed that the Installation Certificate(s) for the construction/installation identified on this form has been completed and is posted or made available with the permit(s) issued for the building.
- I will ensure that a completed, signed copy of this Certificate of Acceptance shall be posted, or made available with the building permit(s) issued for the building, and made available to the enforcement agency for all applicable inspections. I understand that a signed copy of this Certificate of Acceptance is required to be included with the documentation the builder provides to the building owner at occupancy.

Company Name:		Phone:
Responsible Person's Name:		Responsible Person's Signature:
License:	Date Signed:	Position With Company (Title):

CERTIFICATE OF ACCEPTANCE	MECH-11A
Automatic Demand Shed Control Acceptance	(Page 2 of 2)
Project Name/Address:	
System Name or Identification/Tag:	System Location or Area Served:

Intent: *Ensure that the central demand shed sequences have been properly programmed into the DDC system.*

Construction Inspection
1. Instrumentation to perform test includes, but not limited to: a. None.
2. Installation. ☐ The EMCS front end interface enables activation of the central demand shed controls.

A. Functional Testing. Pump Tag (Id)	
Step 1: Engage the demand shed controls.	
a. Engage the central demand shed control signal.	Y / N
b. Verify that the current operating temperature setpoint in a sample of noncritical spaces increases by the proper amount.	Y / N
c. Verify that the current operating temperature setpoint in a sample of critical spaces does not change.	Y / N
Step 2: Disengage the demand shed controls.	
a. Disengage the central demand shed control signal.	Y / N
b. Verify that the current operating temperature setpoint in the sample of noncritical spaces returns to their original value.	Y / N
c. Verify that the current operating temperature setpoint in the sample of critical spaces does not change.	Y / N

B. Testing Results.	PASS / FAIL	
Test passes if all answers are yes in Step 1 and Step 2.	☐	☐

C. PASS / FAIL Evaluation (check one):	
☐	PASS: All **Construction Inspection** responses are complete and all **Testing Results** responses are "Pass."
☐	FAIL: Any **Construction Inspection** responses are incomplete *OR* there is one or more "Fail" responses in **Testing Results** section. Provide explanation below. Use and attach additional pages if necessary.

CERTIFICATE OF ACCEPTANCE	**MECH-12A**
Fault Detection and Diagnostics (FDD) for Packaged Direct-Expansion Units	**(Page 1 of 3)**
Project Name/Address:	
System Name or Identification/Tag:	System Location or Area Served:

Enforcement Agency:	Permit Number:
Note: Submit one Certificate of Acceptance for each system that must demonstrate compliance.	Enforcement Agency Use: Checked by/Date

FIELD TECHNICIAN'S DECLARATION STATEMENT

- I certify under penalty of perjury the information provided on this form is true and correct.
- I am the person who performed the acceptance requirements verification reported on this Certificate of Acceptance (Field Technician).
- I certify that the construction/installation identified on this form complies with the acceptance requirements indicated in the plans and specifications approved by the enforcement agency, and conforms to the applicable acceptance requirements and procedures specified in Section E 801.0 through Section E 806.0.
- I have confirmed that the Installation Certificate(s) for the construction/installation identified on this form has been completed and is posted or made available with the building permit(s) issued for the building.

Company Name:		
Field Technician's Name:		Field Technician's Signature:
	Date Signed:	Position with Company (Title):

RESPONSIBLE PERSON'S DECLARATION STATEMENT

- I certify under penalty of perjury that I am the Field Technician, or the Field Technician is acting on my behalf as my employee or my agent and I have reviewed the information provided on this form.
- I am a licensed contractor or registered design professional who is eligible per the requirements of the Authority Having Jurisdiction to take responsibility for the scope of work specified on this document and attest to the declarations in this statement (responsible person).
- I certify that the information provided on this form substantiates that the construction/installation identified on this form complies with the acceptance requirements indicated in the plans and specifications approved by the enforcement agency, and conforms to the applicable acceptance requirements and procedures specified in Section E 801.0 through Section E 806.0.
- I have confirmed that the Installation Certificate(s) for the construction/installation identified on this form has been completed and is posted or made available with the permit(s) issued for the building.
- I will ensure that a completed, signed copy of this Certificate of Acceptance shall be posted, or made available with the building permit(s) issued for the building, and made available to the enforcement agency for all applicable inspections. I understand that a signed copy of this Certificate of Acceptance is required to be included with the documentation the builder provides to the building owner at occupancy.

Company Name:		Phone:
Responsible Person's Name:		Responsible Person's Signature:
License:	Date Signed:	Position With Company (Title):

CERTIFICATE OF ACCEPTANCE	MECH-12A
Fault Detection and Diagnostics (FDD) for Packaged Direct-Expansion Units	**(Page 2 of 3)**
Project Name/Address:	
System Name or Identification/Tag:	System Location or Area Served:

Intent: *The purpose of this test is to verify proper fault detection and reporting for automated fault detection and diagnostics systems for packaged units.*

Construction Inspection
1. Instrumentation to perform test includes, but not limited to: a. List of instrumentation may be needed or included. 2. Installation. ☐ Verify that FDD hardware is installed on equipment by the manufacturer and that equipment make and model include factory-installed FDD hardware that matches the information indicated on copies of the manufacturer's cut sheets and on the plans and specifications.

A. Eligibility Criteria Results.	**Results**
a. A fault detection and diagnostics (FDD) system for direct-expansion packaged units shall contain the following features to be eligible for credit in the performance calculation method:	
b. The unit shall include a factory-installed economizer and shall limit the economizer deadband to no more than 2°F.	Y / N
c. The unit shall include direct-drive actuators on outside air and return air dampers.	Y / N
d. The unit shall include an integrated economizer with either differential dry-bulb or differential enthalpy control.	Y / N
e. The unit shall include a low temperature lockout on the compressor to prevent coil freeze-up or comfort problems.	Y / N
f. Outside air and return air dampers shall have maximum leakage rates conforming to Section E 805.12 through Section E 805.12.2.	Y / N
g. The unit shall have an adjustable expansion control device such as a thermostatic expansion valve (TXV).	Y / N
h. To improve the ability to troubleshoot charge and compressor operation, a high-pressure refrigerant port will be located on the liquid line. A low-pressure refrigerant port will be located on the suction line.	Y / N
i. The following sensors should be permanently installed to monitor system operation and the controller should have the capability of displaying the value of each parameter: ☐ Refrigerant suction pressure ☐ Supply air relative humidity ☐ Return air temp ☐ Supply air relative humidity ☐ Refrigerant suction temp ☐ Outside air relative humidity ☐ Supply air temp ☐ Liquid line pressure ☐ Return air relative humidity ☐ Outside air temp	Y / N
j. The controller will provide system status by indicating the following conditions: ☐ Compressor enabled ☐ Economizer enabled ☐ Free cooling available ☐ Heating enabled ☐ Mixed air low limit cycle active	Y / N
k. The unit controller shall have the capability to manually initiate each operating mode so that the operation of compressors, economizers, fans, and heating system can be independently tested and verified.	Y / N

For SI units: °C = (°F-32)/1.8

CERTIFICATE OF ACCEPTANCE	MECH-12A
Fault Detection and Diagnostics (FDD) for Packaged Direct-Expansion Units	**(Page 3 of 3)**

Project Name/Address:	
System Name or Identification/Tag:	System Location or Area Served:

B. Functional Testing.	**Results**
Step 1: Low airflow test.	
a. Test low airflow condition by replacing the existing filter with a dirty filter or appropriate obstruction.	
b. Verify that the fault detection and diagnostics system reports the fault.	**Y / N**
c. Verify that the system is able to verify the correct refrigerant charge.	**Y / N**
d. Verify that you are able to calibrate the following:	**Y / N**
☐ Outside Air Temperature Sensor. ☐ Return Air Temperature Sensors. ☐ Supply Air Temperature Sensors.	

C. Testing Results	**PASS / FAIL**	
Test passes if all answers are yes under **Eligibility Criteria** and **Functional Testing**.	☐	☐

☐	PASS: All **Construction Inspection** responses are complete and all **Testing Results** responses are "Pass."
☐	FAIL: Any **Construction Inspection** responses are incomplete *OR* there is one or more "Fail" responses in **Testing Results** section. Provide explanation below. Use and attach additional pages if necessary.

CERTIFICATE OF ACCEPTANCE	MECH-13A
Automatic Fault Detection and Diagnostics (FDD) for Packaged Direct-Expansion Units and Zone Terminal Units Acceptance	**(Page 1 of 4)**
Project Name/Address:	
System Name or Identification/Tag:	System Location or Area Served:

Enforcement Agency:	Permit Number:
Note: Submit one Certificate of Acceptance for each system that must demonstrate compliance.	Enforcement Agency Use: Checked by/Date

FIELD TECHNICIAN'S DECLARATION STATEMENT

- I certify under penalty of perjury the information provided on this form is true and correct.
- I am the person who performed the acceptance requirements verification reported on this Certificate of Acceptance (Field Technician).
- I certify that the construction/installation identified on this form complies with the acceptance requirements indicated in the plans and specifications approved by the enforcement agency, and conforms to the applicable acceptance requirements and procedures specified in Section E 801.0 through Section E 806.0.
- I have confirmed that the Installation Certificate(s) for the construction/installation identified on this form has been completed and is posted or made available with the building permit(s) issued for the building.

Company Name:		
Field Technician's Name:		Field Technician's Signature:
	Date Signed:	Position with Company (Title):

RESPONSIBLE PERSON'S DECLARATION STATEMENT

- I certify under penalty of perjury that I am the Field Technician, or the Field Technician is acting on my behalf as my employee or my agent and I have reviewed the information provided on this form.
- I am a licensed contractor or registered design professional who is eligible per the requirements of the Authority Having Jurisdiction to take responsibility for the scope of work specified on this document and attest to the declarations in this statement (responsible person).
- I certify that the information provided on this form substantiates that the construction/installation identified on this form complies with the acceptance requirements indicated in the plans and specifications approved by the enforcement agency, and conforms to the applicable acceptance requirements and procedures specified in Section E 801.0 through Section E 806.0.
- I have confirmed that the Installation Certificate(s) for the construction/installation identified on this form has been completed and is posted or made available with the permit(s) issued for the building.
- I will ensure that a completed, signed copy of this Certificate of Acceptance shall be posted, or made available with the building permit(s) issued for the building, and made available to the enforcement agency for all applicable inspections. I understand that a signed copy of this Certificate of Acceptance is required to be included with the documentation the builder provides to the building owner at occupancy.

Company Name:		Phone:
Responsible Person's Name:		Responsible Person's Signature:
License:	Date Signed:	Position With Company (Title):

CERTIFICATE OF ACCEPTANCE	MECH-13A
Automatic Fault Detection and Diagnostics (FDD) for Packaged Direct-Expansion Units and Zone Terminal Units Acceptance	**(Page 2 of 4)**
Project Name/Address:	
System Name or Identification/Tag:	System Location or Area Served:

Intent: *Verify that the system detects common faults in air handling units and zone terminal units.*

Construction Inspection
1. Instrumentation to perform test includes, but not limited to:
a. No instrumentation is required – changes are implemented at the building automation system control station.
2. Installation.
a. The functional testing verifies proper installation of the controls for FDD for air handling units and zone terminal units. No additional installation checks are required.

A. Eligibility Criteria Results.	Results
Testing of each AHU with FDD controls shall include the following tests:	
Step 1: Sensor Drift/Failure:	
a. Disconnect outside air temperature sensor from unit controller.	Y / N
b. Verify that the FDD system reports a fault.	Y / N
c. Connect OAT sensor to the unit controller.	Y / N
d. Verify that FDD indicates normal system operation.	Y / N
Step 2: Damper/actuator fault.	
a. From the control system workstation, command the mixing box dampers to full open (100% outdoor air).	Y / N
b. Disconnect power to the actuator and verify that a fault is reported at the control workstation.	Y / N
c. Reconnect power to the actuator and command the mixing box dampers to full open.	Y / N
d. Verify that the control system does not report a fault.	Y / N
e. From the control system workstation, command the mixing box dampers to a full-closed position (0% outdoor air).	Y / N
f. Disconnect power to the actuator and verify that a fault is reported at the control workstation.	Y / N
g. Reconnect power to the actuator and command the dampers closed.	Y / N
h. Verify that the control system does not report a fault during normal operation.	Y / N
Step 3: Valve/actuator fault.	
a. From the control system workstation, command the heating and cooling coil valves to full open or closed, then disconnect power to the actuator and verify that a fault is reported at the control workstation.	Y / N
Step 4: Inappropriate simultaneous heating, mechanical cooling, and/or economizing.	
a. From the control system workstation, override the heating coil valve and verify that a fault is reported at the control workstation.	Y / N
b. From the control system workstation, override the cooling coil valve and verify that a fault is reported at the control workstation.	Y / N
c. From the control system workstation, override the mixing box dampers and verify that a fault is reported at the control workstation.	Y / N

CERTIFICATE OF ACCEPTANCE		MECH-13A
Automatic Fault Detection and Diagnostics (FDD) for Packaged Direct-Expansion Units and Zone Terminal Units Acceptance		(Page 3 of 4)
Project Name/Address:		
System Name or Identification/Tag:	System Location or Area Served:	

B. Functional Testing for Zone Terminal Units. Testing shall be performed on one of each type of terminal unit (VAV box) in the project. A minimum of 5% of results the terminal boxes shall be tested.	Results
Step 1: Sensor Drift/Failure:	
a. Disconnect the tubing to the differential pressure sensor of the VAV box.	Y / N
b. Verify that control system detects and reports the fault.	Y / N
c. Reconnect the sensor and verify proper sensor operation.	Y / N
d. Verify that the control system does not report a fault.	Y / N
Step 2: Damper/actuator fault.	
If the damper is stuck open:	
a. Command the damper to be fully open (room temperature above setpoint).	Y / N
b. Disconnect the actuator to the damper.	Y / N
c. Adjust the cooling setpoint so that the room temperature is below the cooling setpoint to command the damper to the minimum position. Verify that the control system reports a fault.	Y / N
d. Reconnect the actuator and restore to normal operation.	Y / N
If the damper is stuck closed:	
a. Set the damper to the minimum position.	Y / N
b. Disconnect the actuator to the damper.	Y / N
c. Set the cooling setpoint below the room temperature to simulate a call for cooling. Verify that the control system reports a fault.	Y / N
d. Reconnect the actuator and restore to normal operation.	Y / N
Step 3: Valve/actuator fault (for systems with hydronic reheat).	
a. Command the reheat coil valve to full open.	Y / N
b. Disconnect power to the actuator. Set the heating setpoint temperature to be lower than the current space temperature, to command the valve closed. Verify that the fault is reported at the control workstation.	Y / N
c. Reconnect the actuator and restore normal operation.	Y / N
Step 4: Feedback loop tuning fault (unstable airflow).	
a. Set the integral coefficient of the box controller to a value 50 times the current value. Lower the space cooling setpoint to simulate a call for cooling.	Y / N
b. The damper cycles continuously and airflow is unstable. Verify that the control system detects and reports the fault.	Y / N
c. Reset the integral coefficient of the controller to the original value to restore normal operation.	Y / N
Step 5: Disconnected inlet duct.	
a. From the control system workstation, command the damper to full closed, then disconnect power to the actuator and verify that a fault is reported at the control workstation.	Y / N

CERTIFICATE OF ACCEPTANCE	MECH-13A
Automatic Fault Detection and Diagnostics (FDD) for Packaged Direct-Expansion Units and Zone Terminal Units Acceptance	**(Page 4 of 4)**

Project Name/Address:	
System Name or Identification/Tag:	System Location or Area Served:

C. Testing Results	**PASS / FAIL**	
Test passes if all answers are yes under **Functional Testing Sections**.	☐	☐

D. PASS / FAIL Evaluation (check one):	
☐	PASS: All **Construction Inspection** responses are complete and all **Testing Results** responses are "Pass."
☐	FAIL: Any **Construction Inspection** responses are incomplete *OR* there is one or more "Fail" responses in **Testing Results** section. Provide explanation below. Use and attach additional pages if necessary.

CERTIFICATE OF ACCEPTANCE			MECH-14A
Distributed Energy Storage DX AC Systems Acceptance			**(Page 1 of 3)**
Project Name/Address:			
System Name or Identification/Tag:		System Location or Area Served:	

Enforcement Agency:	Permit Number:
Note: Submit one Certificate of Acceptance for each system that must demonstrate compliance.	Enforcement Agency Use: Checked by/Date

FIELD TECHNICIAN'S DECLARATION STATEMENT

- I certify under penalty of perjury the information provided on this form is true and correct.
- I am the person who performed the acceptance requirements verification reported on this Certificate of Acceptance (Field Technician).
- I certify that the construction/installation identified on this form complies with the acceptance requirements indicated in the plans and specifications approved by the enforcement agency, and conforms to the applicable acceptance requirements and procedures specified in Section E 801.0 through Section E 806.0.
- I have confirmed that the Installation Certificate(s) for the construction/installation identified on this form has been completed and is posted or made available with the building permit(s) issued for the building.

Company Name:		
Field Technician's Name:		Field Technician's Signature:
	Date Signed:	Position with Company (Title):

RESPONSIBLE PERSON'S DECLARATION STATEMENT

- I certify under penalty of perjury that I am the Field Technician, or the Field Technician is acting on my behalf as my employee or my agent and I have reviewed the information provided on this form.
- I am a licensed contractor or registered design professional who is eligible per the requirements of the Authority Having Jurisdiction to take responsibility for the scope of work specified on this document and attest to the declarations in this statement (responsible person).
- I certify that the information provided on this form substantiates that the construction/installation identified on this form complies with the acceptance requirements indicated in the plans and specifications approved by the enforcement agency, and conforms to the applicable acceptance requirements and procedures specified in Section E 801.0 through Section E 806.0.
- I have confirmed that the Installation Certificate(s) for the construction/installation identified on this form has been completed and is posted or made available with the permit(s) issued for the building.
- I will ensure that a completed, signed copy of this Certificate of Acceptance shall be posted, or made available with the building permit(s) issued for the building, and made available to the enforcement agency for all applicable inspections. I understand that a signed copy of this Certificate of Acceptance is required to be included with the documentation the builder provides to the building owner at occupancy.

Company Name:			Phone:
Responsible Person's Name:		Responsible Person's Signature:	
License:	Date Signed:	Position With Company (Title):	

CERTIFICATE OF ACCEPTANCE	MECH-14A
Distributed Energy Storage DX AC Systems Acceptance	**(Page 2 of 3)**

Project Name/Address:	
System Name or Identification/Tag:	System Location or Area Served:

Intent: *Verify that the system detects common faults in air handling units and zone terminal units.*

Construction Inspection

1. Instrumentation to perform test includes, but not limited to:
   a. No special instrumentation is required to perform these tests.
2. Installation.
   Prior to Performance Testing, verify and document the following:
   - ☐ The water tank is filled to the proper level.
   - ☐ The water tank is sitting on a foundation with adequate structural strength.
   - ☐ The water tank is insulated and the top cover is in place.
   - ☐ The DES/DXAC is installed correctly (refrigerant piping, etc.).
   - ☐ Verify that the correct model number is installed and configured.

A. Functional Testing	Results
Step 1: Simulate no cooling load during a nighttime period by setting system time to between 9:00 p.m. and 6:00 a.m. Raise the space temperature setpoint above the current space temperature. Verify and document the following:	
a. The system charges the tank.	**Y / N**
b. The system does not provide cooling to the building.	**Y / N**
Step 2: Simulate cooling load during daytime period (e.g., by setting time schedule to include actual time and placing thermostat cooling set-point below actual temperature). Verify and document the following:	
a. Supply fan operates continually during occupied hours.	**Y / N**
b. If the DES/DXAC has cooling capacity, DES/DXAC runs to meet the cooling demand (in ice melt mode).	**Y / N / N/A**
c. If the DES/DXAC has no ice and there is a call for cooling, the DES/DXAC runs in direct cooling mode.	**Y / N / N/A**
Step 3: Simulate no cooling load during daytime condition. Verify and document the following:	
a. Supply fan operates as per the facility thermostat or control system.	**Y / N**
b. The DES/DXAC and the condensing unit do not run.	
Step 4: Simulate no cooling load during morning shoulder time period. Verify and document the following:	
a. The DES/DXAC is idle (the condensing unit and the refrigerant pumps remain off).	**Y / N**

B. Calibrating Controls.	Results
a. Verify that you are able to set the proper time and date, as per manufacturer's installation manual for approved installers.	**Y / N**

C. Testing Results.	PASS / FAIL	
Test passes if all answers are yes under **Functional Testing** and **Calibrating Controls**.	☐	☐

For SI units: 1 metric ton = 1000 kg, 1000 British thermal units per hour = 0.293 kW

CERTIFICATE OF ACCEPTANCE	MECH-14A
Distributed Energy Storage DX AC Systems Acceptance	**(Page 3 of 3)**
Project Name/Address:	
System Name or Identification/Tag:	System Location or Area Served:

☐	PASS: All **Construction Inspection** responses are complete and all **Testing Results** responses are "Pass."
☐	FAIL: Any **Construction Inspection** responses are incomplete *OR* there is one or more "Fail" responses in **Testing Results** section. Provide explanation below. Use and attach additional pages if necessary.

CERTIFICATE OF ACCEPTANCE	MECH-15A
Thermal Energy Storage (TES) System Acceptance	**(Page 1 of 3)**
Project Name/Address:	
System Name or Identification/Tag:	System Location or Area Served:

Enforcement Agency:	Permit Number:
Note: Submit one Certificate of Acceptance for each system that must demonstrate compliance.	Enforcement Agency Use: Checked by/Date

FIELD TECHNICIAN'S DECLARATION STATEMENT

- I certify under penalty of perjury the information provided on this form is true and correct.
- I am the person who performed the acceptance requirements verification reported on this Certificate of Acceptance (Field Technician).
- I certify that the construction/installation identified on this form complies with the acceptance requirements indicated in the plans and specifications approved by the enforcement agency, and conforms to the applicable acceptance requirements and procedures specified in Section E 801.0 through Section E 806.0.
- I have confirmed that the Installation Certificate(s) for the construction/installation identified on this form has been completed and is posted or made available with the building permit(s) issued for the building.

Company Name:		
Field Technician's Name:		Field Technician's Signature:
	Date Signed:	Position with Company (Title):

RESPONSIBLE PERSON'S DECLARATION STATEMENT

- I certify under penalty of perjury that I am the Field Technician, or the Field Technician is acting on my behalf as my employee or my agent and I have reviewed the information provided on this form.
- I am a licensed contractor or registered design professional who is eligible per the requirements of the Authority Having Jurisdiction to take responsibility for the scope of work specified on this document and attest to the declarations in this statement (responsible person).
- I certify that the information provided on this form substantiates that the construction/installation identified on this form complies with the acceptance requirements indicated in the plans and specifications approved by the enforcement agency, and conforms to the applicable acceptance requirements and procedures specified in Section E 801.0 through Section E 806.0.
- I have confirmed that the Installation Certificate(s) for the construction/installation identified on this form has been completed and is posted or made available with the permit(s) issued for the building.
- I will ensure that a completed, signed copy of this Certificate of Acceptance shall be posted, or made available with the building permit(s) issued for the building, and made available to the enforcement agency for all applicable inspections. I understand that a signed copy of this Certificate of Acceptance is required to be included with the documentation the builder provides to the building owner at occupancy.

Company Name:		Phone:
Responsible Person's Name:		Responsible Person's Signature:
License:	Date Signed:	Position With Company (Title):

CERTIFICATE OF ACCEPTANCE	MECH-15A
Thermal Energy Storage (TES) System Acceptance	**(Page 2 of 3)**
Project Name/Address:	
System Name or Identification/Tag:	System Location or Area Served:

Intent: *Verify proper operation of distributed energy storage DX systems.*

Construction Inspection

1. Instrumentation to perform test includes, but not limited to:
   a. No special instrumentation is required for the acceptance tests.

A. Certificate of Compliance Information

The following Certificate of Compliance information for both the chiller and the storage tank shall be provided on the plans to document the key TES System parameters and allow plan check comparison to the inputs used in the DOE-2 simulation. DOE-2 keywords are shown in ALL CAPITALS in parentheses.

a. Chiller	Brand and Model:			
	Type (centrifugal, reciprocating, etc):			
	Capacity (tons): (Size)			
	Starting Efficiency (kW/ton): (at beginning of ice production) (COMP-kW/TON-START)			
	Ending Efficiency (kW/ton): (at end of ice production) (COMP-kW/TON-END)			
	Capacity Reduction (% / F): (PER-COMP-REDUCT/F)			
b. Storage Tank	Storage Type (Check): (TES-TYPE)	☐ Chilled Water Storage	☐ Ice-on-Coil	☐ CHS
		☐ Ice Harvester	☐ Brine	
		☐ Ice-Slurry	☐ Eutectic Salt	
	Number of tanks (SIZE)			
	Storage Capacity per Tank (ton-hours)			
	Storage Rate (tons): (COOL-STORE-RATE)			
	Discharge Rate (tons): (COOL-SUPPLY-RATE)			
	Auxiliary Power (watts): (PUMP+AUX-kW)			
	Tank Area (square feet): (CTANK-LOSS-COEFF)			
	Tank Insulation (R-Value): (CTANK-LOSS-COEFF)			

For SI units: 1 metric ton = 1000 kg, 1000 British thermal units per hour = 0.293 kW

CERTIFICATE OF ACCEPTANCE	MECH-15A
Thermal Energy Storage (TES) System Acceptance	**(Page 3 of 3)**
Project Name/Address:	
System Name or Identification/Tag:	System Location or Area Served:

B. Functional Testing	**Results**
Step 1: TES System Design Verification	
a. In the TES System Design Verification part, the installing contractor shall certify the following information, which verifies proper installation of the TES System consistent with system design expectations:	**Y / N**
☐ The TES system is one of the above eligible systems ☐ Initial charge rate of the storage tanks (tons) ☐ Final charge rate of the storage tank (tons) ☐ Tank standby storage losses (UA) ☐ Initial discharge rate of the storage tanks (tons) ☐ Final discharge rate of the storage tank (tons) ☐ Charge test time (hours) ☐ Initial chiller efficiency (kW/ton) during charging ☐ Discharge test time (hours) ☐ Tank storage capacity after charge (ton-hours) ☐ Tank storage capacity after discharge (ton-hours) ☐ Final chiller efficiency (kW/ton) during charging	
Step 2: TES System Controls and Operation Verification	
a. The TES system and the chilled water plant is controlled and monitored by an EMS.	☐ Pass ☐ Fail
b. Force the time between 9:00 p.m. and 9:00 a.m. and simulate a partial or no charge of the tank and simulate no cooling load by setting the indoor temperature setpoint higher than the ambient temperature. Verify that the TES system starts charging (storing energy).	☐ Pass ☐ Fail
c. Force the time to be between 6:00 p.m. and 9:00 p.m. and simulate a partial charge on the tank and simulate a cooling load by setting the indoor temperature set point lower than the ambient temperature. Verify that the TES system starts discharging.	☐ Pass ☐ Fail
d. Force the time to be between noon and 6:00 p.m. and simulate a cooling load by lowering the indoor air temperature set point below the ambient temperature. Verify that the tank starts discharging and the compressor is off. For systems designed to meet partial loads the system should be run until the TES storage is fully depleted. The number of hours of operation must meet or exceed the designed operational hours for the system.	☐ Pass ☐ Fail
e. Force the time to be between 9:00 a.m. to noon, and simulate a cooling load by lowering the indoor air temperature set point below the ambient temperature. Verify that the tank does not discharge and the cooling load is met by the compressor only.	☐ Pass ☐ Fail
f. Force the time to be between 9:00 p.m. and 9:00 a.m. and simulate a full tank charge by changing the output of the sensor to the EMS. Verify that the tank charging is stopped.	☐ Pass ☐ Fail
g. Force the time to be between noon and 6:00 p.m. and simulate no cooling load by setting the indoor temperature set point above the ambient temperature. Verify that the tank does not discharge and the compressor is off.	☐ Pass ☐ Fail

C. PASS / FAIL Evaluation (check one):	
☐	PASS: All **Construction Inspection** responses are complete and all **Testing Results** responses are "Pass."
☐	FAIL: Any **Construction Inspection** responses are incomplete *OR* there is one or more "Fail" responses in **Testing Results** section. Provide explanation below. Use and attach additional pages if necessary.

For SI units: 1 metric ton = 1000 kg, 1000 British thermal units per hour = 0.293 kW

CALIFORNIA MECHANICAL CODE – MATRIX ADOPTION TABLE

APPENDIX F – SIZING OF VENTING SYSTEMS AND OUTDOOR COMBUSTION AND VENTILATION OPENING DESIGN

(Matrix Adoption Tables are non-regulatory, intended only as an aid to the code user. See Chapter 1 for state agency authority and building applications.)

Adopting Agency	BSC	BSC-CG	SFM	HCD			DSA			OSHPD						BSCC	DPH	AGR	DWR	CEC	CA	SL	SLC
				1	2	1-AC	AC	SS	SS/CC	1	1R	2	3	4	5								
Adopt Entire Chapter																							
Adopt Entire Chapter as amended (amended sections listed below)																							
Adopt only those sections that are listed below																							
Chapter/Section																							

This state agency does not adopt sections identified with the following symbol: †

The Office of the State Fire Marshal's adoption of this chapter or individual sections is applicable to structures regulated by other state agencies pursuant to Section 1.11.0.

APPENDIX F
SIZING OF VENTING SYSTEMS AND OUTDOOR COMBUSTION AND VENTILATION OPENING DESIGN

The provisions contained in this appendix are not mandatory unless specifically adopted by a state agency, or referenced in the adopting ordinance.

(The content of this Appendix is based on Annex F and Annex I of NFPA 54)

F 101.0 General.

F 101.1 Applicability. This appendix provides general guidelines for sizing venting systems serving appliances equipped with draft hoods, Category I appliances, and appliances listed for use with Type B vents.

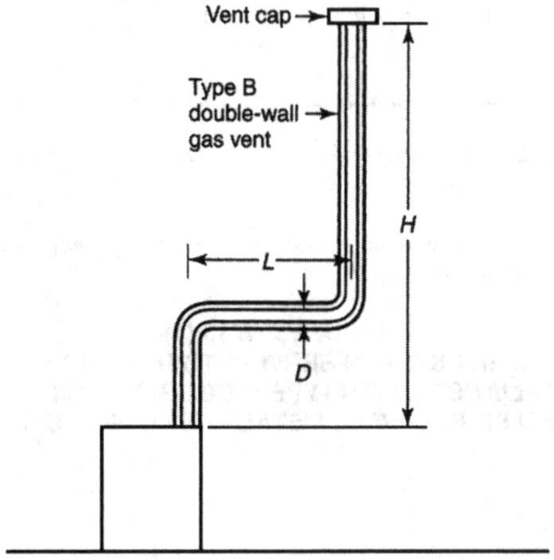

Table 803.1.2(1) is used where sizing a Type B double-wall gas vent connected directly to the appliance.

Note: The appliance is permitted to be either Category I draft hood-equipped or fan-assisted type.

**FIGURE F 101.2(1)
TYPE B DOUBLE-WALL VENT SYSTEM SERVING A SINGLE APPLIANCE WITH A TYPE B DOUBLE-WALL VENT**

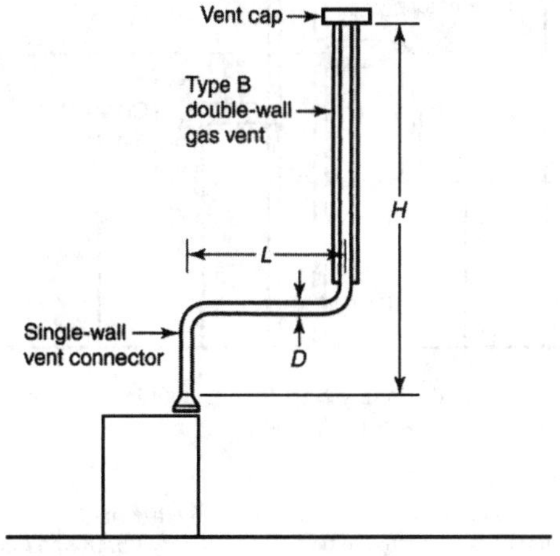

Table 803.1.2(2) is used where sizing a single-wall metal vent connector attached to a Type B double-wall gas vent.

Note: The appliance is permitted to be either Category I draft hood-equipped or fan-assisted type.

**FIGURE F 101.2(2)
TYPE B DOUBLE-WALL VENT SYSTEM SERVING A SINGLE APPLIANCE WITH A SINGLE-WALL METAL VENT CONNECTOR**

F 101.2 Examples Using Single Appliance Venting Tables. See Figure F 101.2(1) through Figure F 101.2(14).

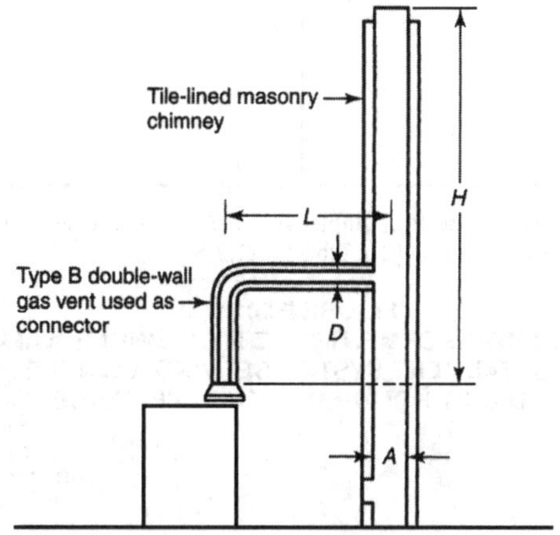

Table 803.1.2(3) is used where sizing a Type B double-wall gas vent connector attached to a tile-lined masonry chimney.

Notes:
1. *A* is the equivalent cross-sectional area of the tile liner.
2. The appliance is permitted to be either Category I draft hood-equipped or fan-assisted type.

**FIGURE F 101.2(3)
VENT SYSTEM SERVING A SINGLE APPLIANCE WITH A MASONRY CHIMNEY AND A TYPE B DOUBLE-WALL VENT CONNECTOR**

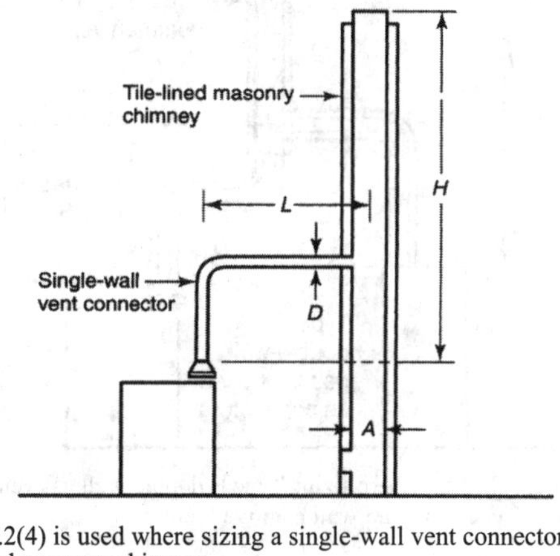

Table 803.1.2(4) is used where sizing a single-wall vent connector attached to a tile-lined masonry chimney.

Notes:
1. *A* is the equivalent cross-sectional area of the tile liner.
2. The appliance is permitted to be either Category I draft hood-equipped or fan-assisted type.

**FIGURE F 101.2(4)
VENT SYSTEM SERVING A SINGLE APPLIANCE USING A MASONRY CHIMNEY AND A SINGLE-WALL METAL VENT CONNECTOR**

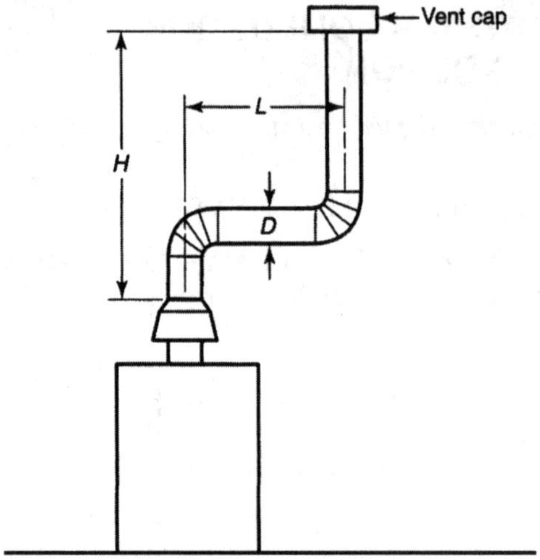

Asbestos cement Type B or single-wall metal vent serving a single draft hood-equipped appliance. [See Table 803.1.2(5)]

FIGURE F 101.2(5)
ASBESTOS CEMENT TYPE B OR SINGLE-WALL METAL VENT SYSTEM SERVING A SINGLE DRAFT HOOD-EQUIPPED APPLIANCE

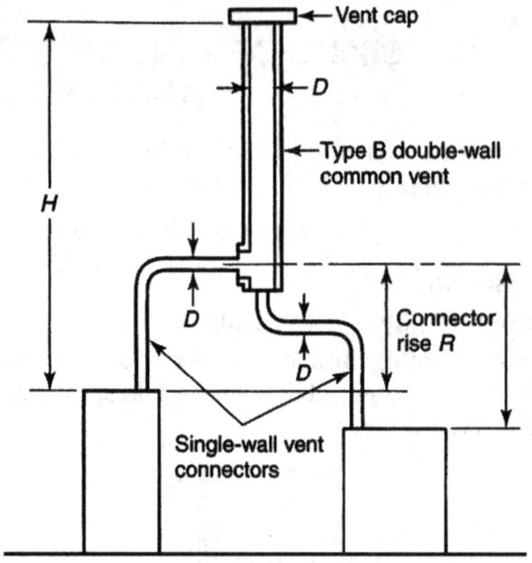

Table 803.2(2) is used where sizing single-wall vent connectors attached to a Type B double-wall common vent.

Note: Each appliance is permitted to be either Category I draft hood-equipped or fan-assisted type.

FIGURE F 101.2(7)
VENT SYSTEM SERVING TWO OR MORE APPLIANCES WITH TYPE B DOUBLE-WALL VENT AND SINGLE-WALL METAL VENT CONNECTORS

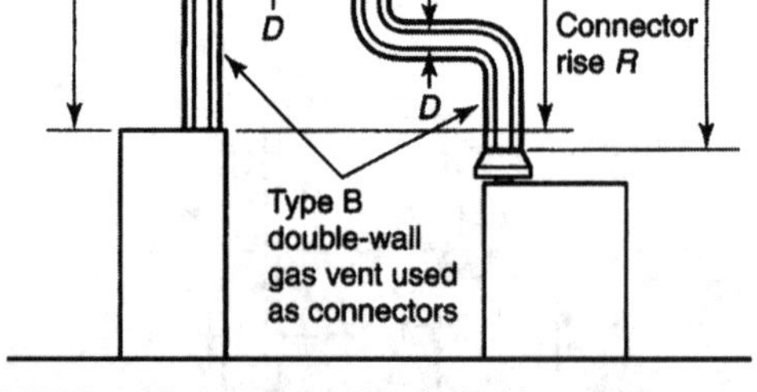

Table 803.2(1) is used where sizing Type B double-wall gas vent connectors attached to a Type B double-wall common vent.

Note: Each appliance is permitted to be either Category I draft hood-equipped or fan-assisted type.

FIGURE F 101.2(6)
VENT SYSTEM SERVING TWO OR MORE APPLIANCES WITH TYPE B DOUBLE-WALL VENT AND TYPE B DOUBLE-WALL VENT CONNECTORS

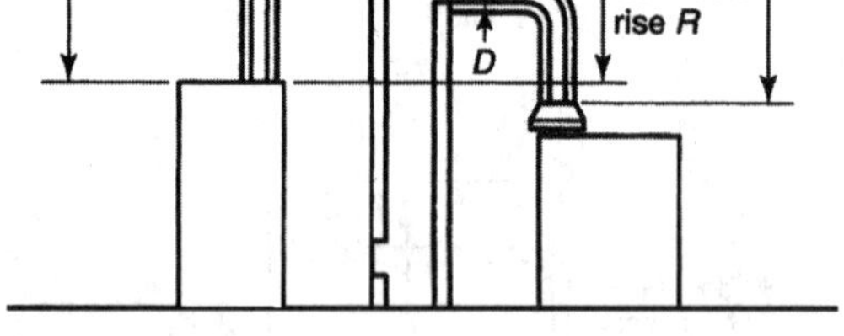

Table 803.2(3) is used where sizing Type B double-wall vent connectors attached to a tile-lined masonry chimney.

Notes:
1. *A* is the equivalent cross-sectional area of the tile liner.
2. The appliance is permitted to be either Category I draft hood-equipped or fan-assisted type.

FIGURE F 101.2(8)
MASONRY CHIMNEY SERVING TWO OR MORE APPLIANCES WITH TYPE B DOUBLE-WALL VENT CONNECTORS

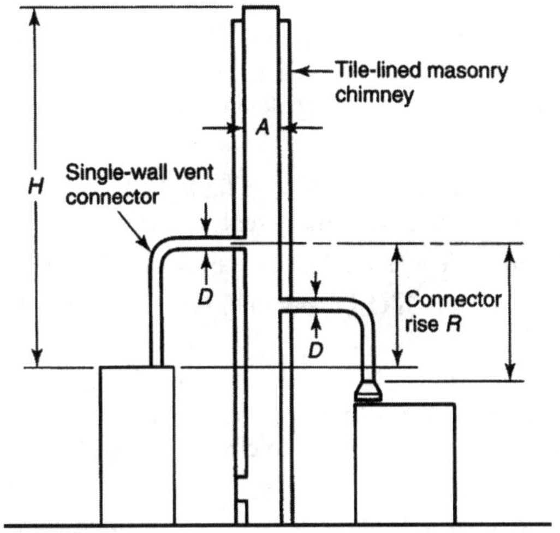

Table 803.2(4) is used where sizing single-wall metal vent connectors attached to a tile-lined masonry chimney.

Notes:

1. A is the equivalent cross-sectional area of the tile liner.
2. Each appliance is permitted to be either Category I draft hood-equipped or fan-assisted type.

FIGURE F 101.2(9)
MASONRY CHIMNEY SERVING TWO OR MORE APPLIANCES WITH SINGLE-WALL METAL VENT CONNECTORS

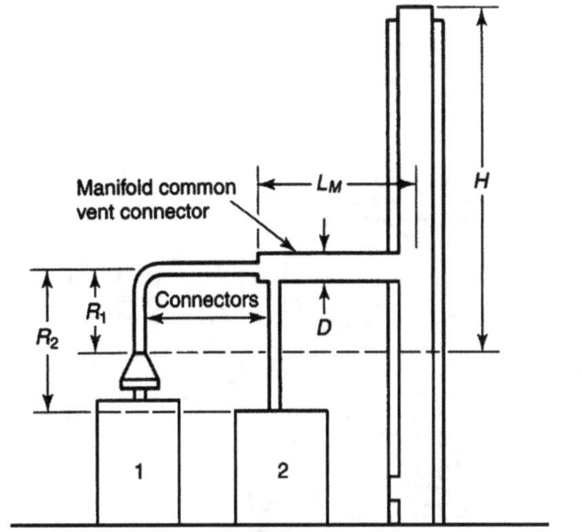

Example: Manifolded common vent connector L_M can be no greater than 18 times the common vent connector manifold inside diameter; that is, a 4 inch (102 mm) inside diameter common vent connector manifold shall not exceed 72 inches (1829 mm) in length. [See Section 803.2.3]

Note: This is an illustration of a typical manifolded vent connector. Different appliance, vent connector, or common vent types are possible. [See Section 803.2]

FIGURE F 101.2(11)
USE OF MANIFOLDED COMMON VENT CONNECTORS

See Section 803.2.4

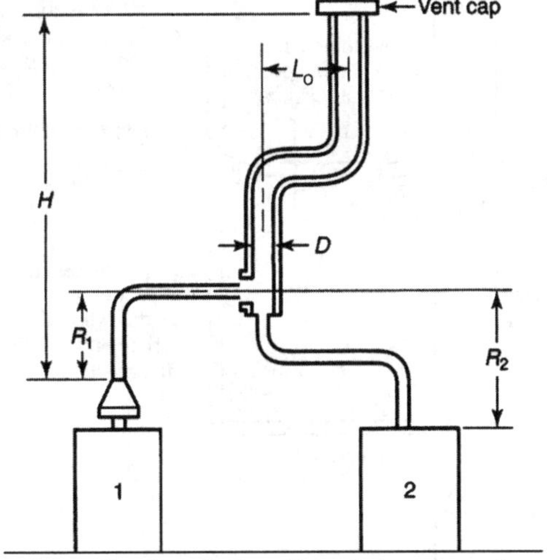

Asbestos cement Type B or single-wall metal pipe vent serving two or more draft hood-equipped appliances. [See Table 803.2(5)]

FIGURE F 101.2(10)
ASBESTOS CEMENT TYPE B OR SINGLE-WALL METAL VENT SYSTEMS SERVING TWO OR MORE DRAFT HOOD-EQUIPPED APPLIANCES

See Section 803.2.4

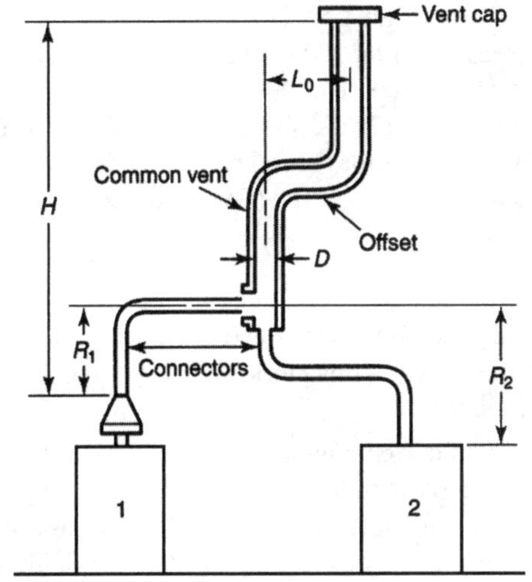

Example: Offset common vent

Note: This is an illustration of a typical offset vent. Different appliance, vent connector, or vent types are possible. [See Section 803.1 and Section 803.2]

FIGURE F 101.2(12)
USE OF OFFSET COMMON VENT

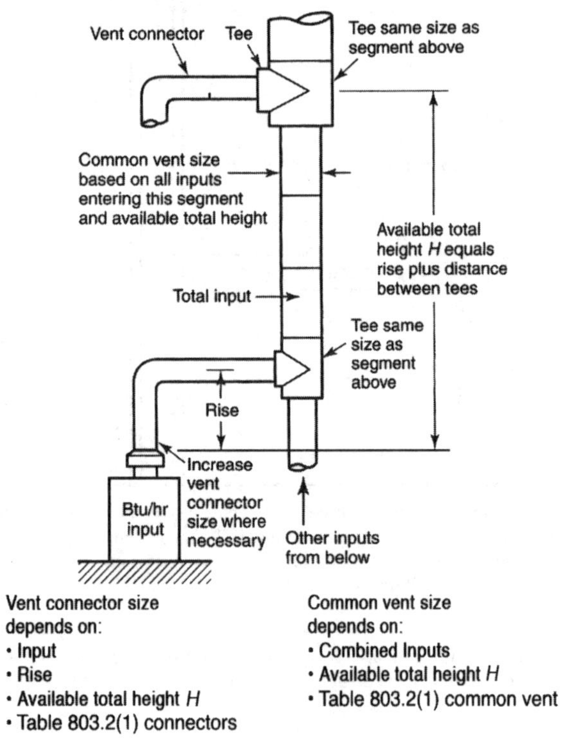

**FIGURE F 101.2(13)
MULTISTORY GAS VENT DESIGN
PROCEDURE FOR EACH SEGMENT OF SYSTEM**

F 101.3 Example 1: Single Draft Hood-Equipped Appliance. An installer has a 120 000 British thermal units per hour (Btu/h) (35 kW) input appliance with a 5 inch (127 mm) diameter draft hood outlet that needs to be vented into a 10 foot (3048 mm) high Type B vent system. What size vent should be used assuming: (1) a 5 foot (1524 mm) lateral single-wall metal vent connector is used with two 90 degree (1.57 rad) elbows or (2) a 5 foot (1524 mm) lateral single-wall metal vent connector is used with three 90 degree (1.57 rad) elbows in the vent system? (See Figure F 101.3)

Solution:

Table 803.1.2(2) shall be used to solve this problem because single-wall metal vent connectors are being used with a Type B vent, as follows:

(1) Read down the first column in Table 803.1.2(2) until the row associated with a 10 foot (3048 mm) height and 5 foot (1524 mm) lateral is found. Read across this row until a vent capacity exceeding 120 000 Btu/h (35 kW) is located in the shaded columns labeled NAT Max for draft hood-equipped appliances. In this case, a 5 inch (127 mm) diameter vent has a capacity of 122 000 Btu/h (35.7 kW) and shall be permitted to be used for this application.

(2) Where three 90 degree (1.57 rad) elbows are used in the vent system, the maximum vent capacity listed in the tables shall be reduced by 10 percent. This implies that the 5 inch (127 mm) diameter vent has an adjusted capacity of only 110 000 Btu/h (32 kW). In this case, the vent system shall be increased to 6 inches (152 mm) in diameter. See the following calculations:

122 000 Btu/h (35.7 kW) x 0.90 = 110 000 Btu/h (32 kW) for 5 inch (127 mm) vent

From Table 803.1.2(2), select 6 inches (152 mm) vent.

186 000 Btu/h (54.5 kW) x 0.90 = 167 000 Btu/h (49 kW)

This figure is exceeding the required 120 000 Btu/h (35 kW). Therefore, use a 6 inch (152 mm) vent and connector where three elbows are used.

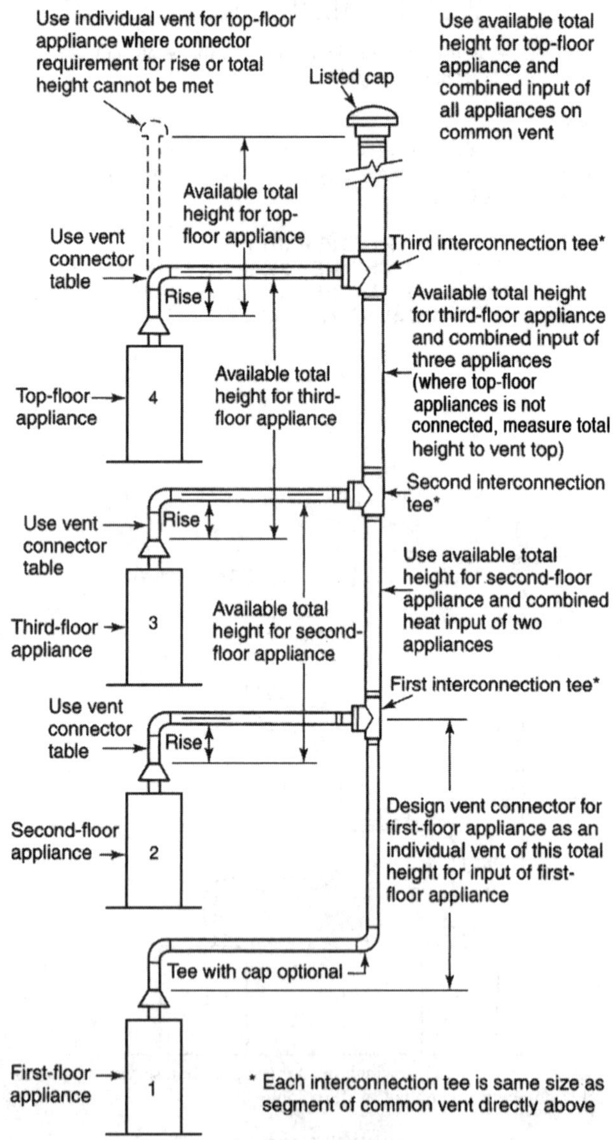

**FIGURE F 101.2(14)
PRINCIPLES OF DESIGN OF MULTISTORY
VENTS USING VENT CONNECTOR AND
COMMON VENT DESIGN TABLES
[See Section 803.2.12 through Section 803.2.15]**

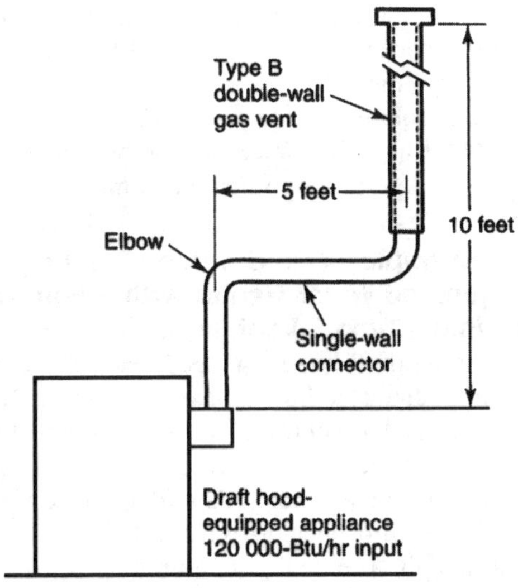

For SI units: 1 foot = 304.8 mm, 1000 British thermal units per hour = 0.293 kW

FIGURE F 101.3
SINGLE DRAFT HOOD-EQUIPPED APPLIANCE
EXAMPLE 1

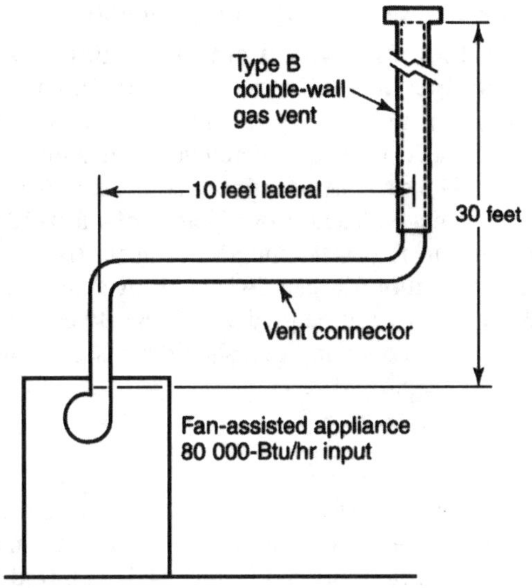

For SI units: 1 foot = 304.8 mm, 1000 British thermal units per hour = 0.293 kW

FIGURE F 101.4
SINGLE FAN-ASSISTED APPLIANCE
EXAMPLE 2

F 101.4 Example 2: Single Fan-Assisted Appliance. An installer has an 80 000 Btu/h (23.4 kW) input fan-assisted appliance that shall be installed using 10 feet (3048 mm) of lateral connector attached to a 30 foot (9144 mm) high Type B vent. Two 90 degree (1.57 rad) elbows are needed for the installation. Is a single-wall metal vent connector permitted to be used for this application? (See Figure F 101.4)

Solution:

Table 803.1.2(2) refers to the use of single-wall metal vent connectors with Type B vent. In the first column find the row associated with a 30 foot (9144 mm) height and a 10 foot (3048 mm) lateral. Read across this row, looking at the FAN Min and FAN Max columns, to find that a 3 inch (76 mm) diameter single-wall metal vent connector is not recommended. Moving to the next larger size single-wall connector [4 inch (102 mm)] we find that a 4 inch (102 mm) diameter single-wall metal connector has a recommended maximum vent capacity of 144 000 Btu/h (42 kW). The 80 000 Btu/h (23.4 kW) fan-assisted appliance is outside this range, so the conclusion is that a single-wall metal connector shall not be used to vent the appliance using a 10 foot (3048 mm) of lateral for the connector. However, if the 80,000 Btu/hr (23.4 kW) input appliance is moved within 5 feet (1524 mm) of the vertical vent, a 4 inch (102 mm) single-wall metal connector shall be used to vent the appliance. Table 803.1.2(2) shows the acceptable range of vent capacities for a 4 inch (102 mm) vent with 5 feet (1524 mm) of lateral to be between 72 000 Btu/h (21.1 kW) and 157 000 Btu/h (46 kW).

Where the appliance cannot be moved closer to the vertical vent, then a Type B vent shall be used as the connector material. In this case, Table 803.1.2(1) shows that, for a 30 foot (9144 mm) high vent with 10 feet (3048 mm) of lateral, the acceptable range of vent capacities for a 4 inch (102 mm) diameter vent attached to a fan-assisted appliance is between 37 000 Btu/h (10.8 kW) and 150 000 Btu/h (44 kW).

F 101.5 Example 3: Interpolating Between Table Values. An installer has an 80 000 Btu/h (23.4 kW) input appliance with a 4 inch (102 mm) diameter draft hood outlet that needs to be vented into a 12 foot (3658 mm) high Type B vent. The vent connector has a 5 foot (1524 mm) lateral length and is also Type B. Is this appliance permitted to be vented using a 4 inch (102 mm) diameter vent?

Solution:

Table 803.1.2(1) is used in the case of an all Type B Vent system. However, since there is no entry in Table 803.1.2(1) for a height of 12 feet (3658 mm), interpolation shall be used. Read down the 4 inch (102 mm) diameter NAT Max column to the row associated with a 10 foot (3048 mm) height and 5 foot (1524 mm) lateral to find the capacity value of 77 000 Btu/h (22.6 kW). Read further down to the 15 foot (4572 mm) height, 5 foot (1524 mm) lateral row to find the capacity value of 87 000 Btu/h (25.5 kW). The difference between the 15 foot (4572 mm) height capacity value and the 10 foot (3048 mm) height capacity value is 10 000 Btu/h (3 kW). The capacity for a vent system with a 12 foot (3658 mm) height is equal to the capacity for a 10 foot (3048 mm) height plus two-fifths of the difference between the 10 foot (3048 mm) and 15 foot (4572 mm) height values, or 77 000 Btu/h (22.6 kW) + $\frac{2}{5}$ x 10 000 Btu/h (3 kW) = 81 000 Btu/h (23.7 kW). Therefore, a 4 inch (102 mm) diameter vent shall be used in the installation.

F 102.0 Examples Using Common Venting Tables.

F 102.1 Example 4: Common Venting Two Draft Hood-Equipped Appliances. A 35 000 Btu/h (10.3 kW) water heater is to be common vented with a 150 000 Btu/h (44 kW) furnace, using a common vent with a total height of 30 feet (9144 mm). The connector rise is 2 feet (610 mm) for the water heater with a horizontal length of 4 feet (1219 mm). The connector rise for the furnace is 3 feet (914 mm) with a horizontal length of 8 feet (2438 mm). Assume single-wall metal connectors will be used with Type B vent. What size connectors and combined vent should be used in this installation? (See Figure F 102.1)

Solution:

Table 803.2(2) shall be used to size single-wall metal vent connectors attached to Type B vertical vents. In the vent connector capacity portion of Table 803.2(2), find the row associated with a 30 foot (9144 mm) vent height. For a 2 foot (610 mm) rise on the vent connector for the water heater, read the shaded columns for draft hood-equipped appliances to find that a 3 inch (76 mm) diameter vent connector has a capacity of 37 000 Btu/h (10.8 kW). Therefore, a 3 inch (76 mm) single-wall metal vent connector shall be used with the water heater. For a draft hood-equipped furnace with a 3 foot (914 mm) rise, read across the row to find that a 5 inch (127 mm) diameter vent connector has a maximum capacity of 120 000 Btu/h (35 kW) (which is too small for the furnace), and a 6 inch (152 mm) diameter vent connector has a maximum vent capacity of 172 000 Btu/h (50 kW). Therefore, a 6 inch (152 mm) diameter vent connector shall be used with the 150 000 Btu/h (44 kW) furnace. Since both vent connector horizontal lengths are less than the maximum lengths listed in Section 803.2.1, the table values shall be used without adjustments.

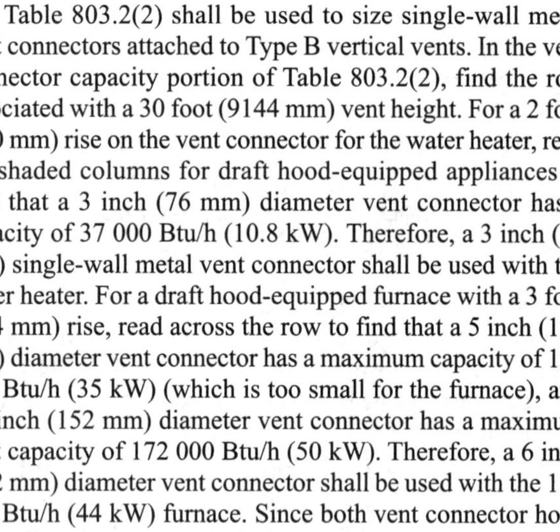

For SI units: 1 foot = 304.8 mm, 1000 British thermal units per hour = 0.293 kW

FIGURE F 102.1
COMMON VENTING TWO DRAFT HOOD-EQUIPPED APPLIANCES
EXAMPLE 4

In the common vent capacity portion of Table 803.2(2), find the row associated with a 30 foot (9144 mm) vent height and read over to the NAT + NAT portion of the 6 inch (152 mm) diameter column to find a maximum combined capacity of 257 000 Btu/h (75 kW). Since the two appliances total 185 000 Btu/h (54 kW), a 6 inch (152 mm) common vent shall be used.

F 102.2 Example 5(a): Common Venting a Draft Hood-Equipped Water Heater with a Fan-Assisted Furnace into a Type B Vent. In this case, a 35 000 Btu/h (10.3 kW) input draft hood-equipped water heater with a 4 inch (102 mm) diameter draft hood outlet, 2 feet (610 mm) of connector rise, and 4 feet (1219 mm) of horizontal length is to be common vented with a 100 000 Btu/h (29 kW) fan-assisted furnace with a 4 inch (102 mm) diameter flue collar, 3 feet (914 mm) of connector rise, and 6 feet (1829 mm) of horizontal length. The common vent consists of a 30 foot (9144 mm) height of Type B vent. What are the recommended vent diameters for each connector and the common vent? The installer would like to use a single-wall metal vent connector. (See Figure F 102.2)

Solution:

Water Heater Vent Connector Diameter. Since the water heater vent connector horizontal length of 4 feet (1219 mm) is less than the maximum value listed in Table 803.2(2), the venting table values shall be used without adjustment. Using the Vent Connector Capacity portion of Table 803.2(2), read down the Total Vent Height (*H*) column to 30 feet (9144 mm) and read across the 2 feet (610 mm) Connector Rise (*R*) row to the first Btu/h rating in the NAT Max column that is equal to or exceeding the water heater input rating. The table shows that a 3 inch (76 mm) vent connector has a maximum input rating of 37 000 Btu/h (10.8 kW). Although this rating is exceeding the water heater input rating, a 3 inch (76 mm) vent connector is prohibited by Section 803.2.18. A 4 inch (102 mm) vent connector has a maximum input rating of 67 000 Btu/h (19.6 kW) and is equal to the draft hood outlet diameter. A 4 inch (102 mm) vent connector is selected. Since the water heater is equipped with a draft hood, there are no minimum input rating restrictions.

Furnace Vent Connector Diameter. Using the Vent Connector Capacity portion of Table 803.2(2), read down the Total Vent Height (*H*) column to 30 feet (9144 mm) and across the 3 feet (914 mm) Connector Rise (*R*) row. Since the furnace has a fan-assisted combustion system, find the first FAN Max column with a Btu/h rating exceeding the furnace input rating. The 4 inch (102 mm) vent connector has a maximum input rating of 119 000 Btu/h (34.9 kW) and a minimum input rating of 85 000 Btu/h (24.9 kW).

The 100 000 Btu/h (29 kW) furnace in this example falls within this range, so a 4 inch (102 mm) connector shall be permitted. Since the furnace vent connector horizontal length of 6 feet (1829 mm) is less than the maximum value listed in Section 803.2.1, the venting table values shall be used without adjustment. Where the furnace had an input rating of 80 000 Btu/h (23.4 kW), then a Type B vent connector shall be needed in order to meet the minimum capacity limit.

Common Vent Diameter. The total input to the common vent is 135 000 Btu/h (40 kW). Using the Common Vent Capacity portion of Table 803.2(2), read down the Vent Height (*H*) column to 30 feet (9144 mm) and across this row to find the smallest vent diameter in the FAN + NAT column that has a Btu/h rating equal to or exceeding 135 000 Btu/h (40 kW). The 4 inch (102 mm) common vent has a capacity of 132 000 Btu/h (39 kW) and the 5 inch (127 mm) common vent has a capacity of 202 000 Btu/h (59 kW). Therefore, the 5 inch (127 mm) common vent shall be used in this example.

Summary: In this example, the installer shall use a 4 inch (102 mm) diameter, single-wall metal vent connector for the water heater and a 4 inch (102 mm) diameter, single-wall metal vent connector for the furnace. The common vent shall be a 5 inch (127 mm) diameter Type B vent.

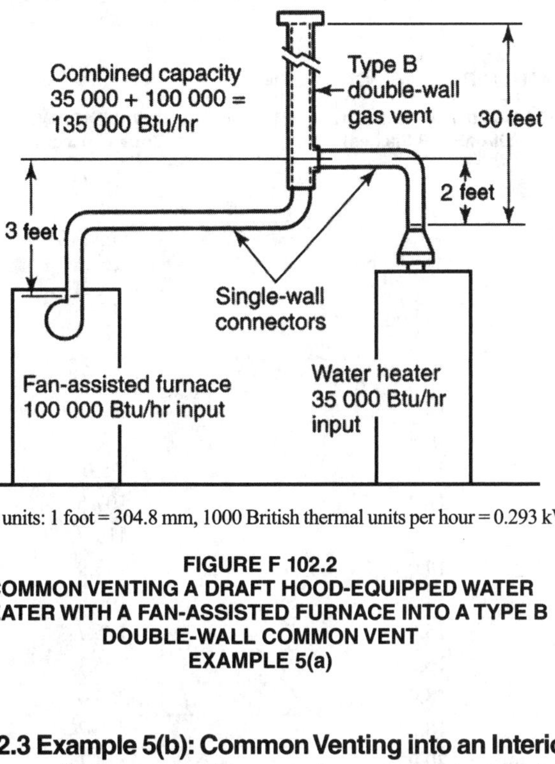

For SI units: 1 foot = 304.8 mm, 1000 British thermal units per hour = 0.293 kW

FIGURE F 102.2
COMMON VENTING A DRAFT HOOD-EQUIPPED WATER HEATER WITH A FAN-ASSISTED FURNACE INTO A TYPE B DOUBLE-WALL COMMON VENT
EXAMPLE 5(a)

F 102.3 Example 5(b): Common Venting into an Interior Masonry Chimney. In this case, the water heater and fan-assisted furnace of Example 5(a) are to be common-vented into a clay-tile-lined masonry chimney with a 30 foot (9144 mm) height. The chimney is not exposed to the outdoors below the roof line. The internal dimensions of the clay tile liner are nominally 8 inches (203 mm) by 12 inches (305 mm). Assuming the same vent connector heights, laterals, and materials found in Example 5(a), what are the recommended vent connector diameters, and is this an acceptable installation?

Solution:

Table 803.2(4) is used to size common venting installations involving single-wall connectors into masonry chimneys.

Water Heater Vent Connector Diameter. Using Table 803.2(4), Vent Connector Capacity, read down the Vent Height (*H*) column to 30 feet (9144 mm), and read across the 2 feet (610 mm) Connector Rise *(R)* row to the first Btu/h rating in the NAT Max column that is equal to or exceeding the water heater input rating. The table shows that a 3 inch (76 mm) vent connector has a maximum input of 31 000 Btu/h (9 kW), while a 4 inch (102 mm) vent connector has a maximum input of 57 000 Btu/h (16.7 kW). A 4 inch (102 mm) vent connector shall be used.

Furnace Vent Connector Diameter. Using the Vent Connector Capacity portion of Table 803.2(4), read down the total Vent Height (*H*) column to 30 feet (9144 mm) and across the 3 feet (914 mm) Connector Rise (*R*) row. Because the furnace has a fan-assisted combustion system, find the first FAN Max column with a Btu/h rating exceeding the furnace input rating. The 4 inch (102 mm) vent connector has a maximum input rating of 127 000 Btu/h (37 kW) and an input rating of not less than 95 000 Btu/h (27.8 kW). The 100 000 Btu/h (29 kW) furnace in this example falls within this range, so a 4 inch (102 mm) connector shall be permitted.

Masonry Chimney. From Table F 102.3, the equivalent area for a nominal liner size of 8 inches (203 mm) by 12 inches (305 mm) is 63.6 of a square inches (0.041 m^2). Using Table 803.2(4), Common Vent Capacity, read down the FAN + NAT column under the Minimum Internal Area of Chimney value of 63 to the row for 30 foot (9144 mm) height to find a capacity value of 739 000 Btu/h (217 kW). The combined input rating of the furnace and water heater, 135 000 Btu/h (40 kW), is less than the table value so this is an acceptable installation.

Section 803.2.17 requires the common vent area to not exceed seven times the smallest listed appliance categorized vent area, flue collar area, or draft hood outlet area. Both appliances in this installation have 4 inch (102 mm) diameter outlets. From Table F 102.3, the equivalent area for an inside diameter of 4 inches (102 mm) is 12.2 of a square inches (0.008 m^2). Seven times 12.2 equals 85.4, which is exceeding 63.6, so this configuration is acceptable.

F 102.4 Example 5(c): Common Venting into an Exterior Masonry Chimney. In this case, the water heater and fan-assisted furnace of Examples 5(a) and 5(b) are to be common-vented into an exterior masonry chimney. The chimney height, clay-tile-liner dimensions, and vent connector heights and laterals are the same as in Example 5(b). This system is being installed in Charlotte, North Carolina. Does this exterior masonry chimney need to be relined? If so, what corrugated metallic liner size is recommended? What vent connector diameters are recommended? [see Table F 102.3 and Figure 803.1.2(6)]

Solution:

According to Section 803.2.20, Type B vent connectors are required to be used with exterior masonry chimneys. Use Table 803.2(8) and Table 803.2(9) to size FAN+NAT common venting installations involving Type-B double-wall connectors into exterior masonry chimneys.

The local 99 percent winter design temperature needed to use Table 803.2(8) and Table 803.2(9) can be found in ASHRAE Handbook – Fundamentals. For Charlotte, North Carolina, this design temperature is 19°F (-7.2°C).

Chimney Liner Requirement. As in Example 5(b), use the 63 square inch (0.04 m^2) internal area columns for this size clay tile liner. Read down the 63 square inches (0.04 m^2) column of Table 803.2(8) to the 30 foot (9144 mm) height row to find that the combined appliance maximum input is 747 000 Btu/h (218.9 kW). The combined input rating of the appliances in this installation, 135 000 Btu/h (40 kW), is less than the maximum value, so this criterion is satisfied. Table 803.2(9), at a 19°F (-7.2°C) design temperature, and at the same vent height and internal area used earlier, shows that the minimum allowable input rating of a space-heating appliance is 470 000 Btu/h (137.7 kW). The furnace input rating of 100 000 Btu/h (29 kW) is less than this minimum value. So this criterion is not satisfied, and an alternative venting design needs to be used, such as a Type B vent shown in Example 5(a) or a listed chimney liner system shown in the remainder of the example.

According to Section 803.2.19, Table 803.2(1) or Table 803.2(2) is used for sizing corrugated metallic liners in masonry chimneys, with the maximum common vent capacities reduced by 20 percent. This example will be continued assuming Type B vent connectors.

Water Heater Vent Connector Diameter. Using Table 803.2(1) Vent Connector Capacity, read down the total Vent Height *(H)* column to 30 feet (9144 mm), and read across the 2 feet (610 mm) Connector Rise *(R)* row to the first Btu/hour rating in the NAT Max column that is equal to or greater than the water heater input rating. The table shows that a 3 inch (76 mm) vent connector has a maximum capacity of 39 000 Btu/h (11.4 kW). Although this rating is greater than the water heater input rating, a 3 inch (76 mm) vent connector is prohibited by Section 803.2.20. A 4 inch (102 mm) vent connector has a maximum input rating of 70 000 Btu/h (20.5 kW)

TABLE F 102.3
MASONRY CHIMNEY LINER DIMENSIONS WITH CIRCULAR EQUIVALENTS*

NOMINAL LINER SIZE (Inches)	INSIDE DIMENSIONS OF LINER (Inches)	INSIDE DIAMETER OR EQUIVALENT DIAMETER (Inches)	EQUIVALENT AREA (Square Inches)
4 x 8	2½ x 6½	4.0	12.2
		5.0	19.6
		6.0	28.3
		7.0	38.3
8 x 8	6¾ x 6¾	7.4	42.7
		8.0	50.3
8 x 12	6½ x 10½	9.0	63.6
		10.0	78.5
12 x 12	9¾ x 9¾	10.4	83.3
		11.0	95.0
12 x 16	9½ x 13½	11.8	107.5
		12.0	113.0
		14.0	153.9
16 x 16	13¼ x 13¼	14.5	162.9
		15.0	176.7
16 x 20	13 x 17	16.2	206.1
		18.0	254.4
20 x 20	16¾ x 16¾	18.2	260.2
		20.0	314.1
20 x 24	16½ x 20½	20.1	314.2
		22.0	380.1
24 x 24	20¼ x 20¼	22.1	380.1
		24.0	452.3
24 x 28	20¼ x 24¼	24.1	456.2
28 x 28	24¼ x 24¼	26.4	543.3
		27.0	572.5
30 x 30	25½ x 25½	27.9	607.0
		30.0	706.8
30 x 36	25½ x 31½	30.9	749.9
		33.0	855.3
36 x 36	31½ x 31½	34.4	929.4
		36.0	1017.9

For SI units, 1 inch. = 25.4 mm, 1 square inch = 0.000645 m^2

* Where liner sizes differ dimensionally from those shown in this table, equivalent diameters shall be permitted to be determined from published tables for square and rectangular ducts of equivalent carrying capacity or by other engineering methods.

and is equal to the draft hood outlet diameter. A 4 inch (102 mm) vent connector is selected.

Furnace Vent Connector Diameter. Using Table 803.2(1), Vent Connector Capacity, read down the total Vent Height *(H)* column to 30 feet (9144 mm), and read across the 3 feet (914 mm) Connector Rise *(R)* row to the first Btu/h rating in the FAN MAX column that is equal to or greater than the furnace input rating. The 100 000 Btu/h (29 kW) furnace in this example falls within this range, so a 4 inch (102 mm) connector is adequate.

Chimney Liner Diameter. The total input to the common vent is 135 000 Btu/h (40 kW). Using the Common Vent Capacity portion of Table 803.2(1), read down the total Vent Height *(H)* column to 30 feet (9144 mm) and across this row to find the smallest vent diameter in the FAN + NAT column that has a Btu/h rating greater than 135 000 Btu/h (40 kW). The 4 inch (102 mm) common vent has a capacity of 138 000 Btu/h (40.4 kW). Reducing the maximum capacity by 20 percent results in a maximum capacity for a 4 inch (102 mm) corrugated liner of 110 000 Btu/h (32 kW), less than the total input of 135 000 Btu/h (40 kW). So a larger liner is needed. The 5 inch (127 mm) common vent capacity listed in Table 803.2(1) is 210 000 Btu/h (62 kW), and after reducing by 20 percent is 168 000 Btu/h (49.2 kW). Therefore, a 5 inch (127 mm) corrugated metal liner should be used in this example.

Single Wall Connectors. Once it has been established that relining the chimney is necessary, Type B double-wall vent connectors are not specifically required. This example could be redone using Table 803.2(2) for single-wall vent connectors. For this case, the vent connector and liner diameters would be the same as found for Type B double-wall connectors.

F 103.0 Example of Combination Indoor and Outdoor Combustion Air Opening Design. Determine the required combination of indoor and outdoor combustion air opening sizes for the following appliance installation example.

Example Installation: A fan-assisted furnace and a draft hood-equipped water heater with the following inputs are located in a 15 foot by 30 foot (4572 mm by 9144 mm) basement with an 8 foot (2438 mm) ceiling. No additional indoor spaces shall be used to help meet the appliance combustion air needs.

Fan-Assisted Furnace Input: 100 000 Btu/h (29 kW)

Draft Hood-Equipped Water Heater Input: 40 000 Btu/h (11.7 kW)

Solution:

(1) Determine the total available room volume.

Appliance room volume.

15 feet by 30 feet (4572 mm by 9144 mm) with an 8 foot (2438 mm) ceiling = 3600 cubic feet (101.94 m^3)

(2) Determine the total required volume.

The standard method to determine combustion air shall be used to calculate the required volume.

The combined input for the appliances located in the basement is calculated as follows:

100 000 Btu/h (29 kW) + 40 000 Btu/h (11.7 kW) = 140 000 Btu/h (41 kW)

The standard method requires that the required volume be determined based on 50 cubic feet per 1000 Btu/h (4.83 m^3/kW).

Using Table F 103.0 the required volume for a 140 000 Btu/h (41 kW) water heater is 7000 cubic feet (198.22 m^3).

Conclusion:

Indoor volume is insufficient to supply combustion air since the total of 3600 cubic feet (101.94 m^3) does not meet the required volume of 7000 cubic feet (198.22 m^3). Therefore, additional combustion air shall be provided from the outdoors.

(3) Determine ratio of the available volume to the required volume.

$$\frac{3600 \text{ cubic feet}}{7000 \text{ cubic feet}} = 0.51$$

(4) Determine the reduction factor to be used to reduce the full outdoor air opening size to the minimum required based on ratio of indoor spaces.

1.00 – 0.51 (from Step 3) = 0.49

(5) Determine the single outdoor combustion air opening size as if combustion air is to come from outdoors. In this example, the combustion air opening directly communicates with the outdoors.

$$\frac{140\,000 \text{ Btu/h}}{3000 \text{ British thermal units per square inch (Btu/in}^2\text{)}} = 47 \text{ square inches } (0.03 \text{ m}^2)$$

(6) Determine the minimum outdoor combustion air opening area.

Outdoor opening area = 0.49 (from Step 4) x 47 square inches (0.03 m^2)
= 23 square inches (0.01 m^2)

Section 701.7.3(3) requires the minimum dimension of the air opening shall be not less than 3 inches (76 mm).

TABLE F 103.0
STANDARD METHOD: REQUIRED VOLUME, ALL APPLIANCES
[NFPA 54: TABLE A.9.3.2.1]

APPLIANCE INPUT (Btu/h)	REQUIRED VOLUME (cubic feet)
5000	250
10 000	500
15 000	750
20 000	1000
25 000	1250
30 000	1500
35 000	1750
40 000	2000
45 000	2250
50 000	2500
55 000	2750
60 000	3000
65 000	3250
70 000	3500
75 000	3750
80 000	4000
85 000	4250
90 000	4500
95 000	4750
100 000	5000
105 000	5250
110 000	5500
115 000	5750
120 000	6000
125 000	6250
130 000	6500
135 000	6750
140 000	7000
145 000	7250
150 000	7500
160 000	8000
170 000	8500
180 000	9000
190 000	9500
200 000	10 000
210 000	10 500
220 000	11 000
230 000	11 500
240 000	12 000
250 000	12 500
260 000	13 000
270 000	13 500
280 000	14 000
290 000	14 500
300 000	15 000

For SI units: 1000 British thermal units per hour = 0.293 kW, 1 cubic foot = 0.0283 m^3

CALIFORNIA MECHANICAL CODE – MATRIX ADOPTION TABLE
APPENDIX G – EXAMPLE CALCULATION OF OUTDOOR AIR RATE

(Matrix Adoption Tables are non-regulatory, intended only as an aid to the code user. See Chapter 1 for state agency authority and building applications.)

Adopting Agency	BSC	BSC-CG	SFM	HCD			DSA			OSHPD						BSCC	DPH	AGR	DWR	CEC	CA	SL	SLC
				1	2	1-AC	AC	SS	SS/CC	1	1R	2	3	4	5								
Adopt Entire Chapter																							
Adopt Entire Chapter as amended (amended sections listed below)																							
Adopt only those sections that are listed below																							
Chapter/Section																							

This state agency does not adopt sections identified with the following symbol: †
The Office of the State Fire Marshal's adoption of this chapter or individual sections is applicable to structures regulated by other state agencies pursuant to Section 1.11.0.

APPENDIX G

EXAMPLE CALCULATION OF OUTDOOR AIR RATE

The provisions contained in this appendix are not mandatory unless specifically adopted by a state agency, or referenced in the adopting ordinance.

G 101.0 Example Calculation of Outdoor Air Rate.

G 101.1 Example Calculation. Determine the outdoor air rate required for a single zone AC unit serving an interior 2000 square feet (185.81 m^2) conference/meeting room with a design occupancy of 100 people. The system supplies and returns air from the ceiling. (See Chapter 4 of this code for guidelines)

Solution:

In accordance with Table 403.2.2, the zone air distribution effectiveness is 1.0 since the system supplies cooling only from the ceiling. Using the rates from Table 402.1 for a conference/meeting room, the minimum system outdoor air rate is calculated to be:

$$V_{ot} = \frac{R_p P_z + R_a A_z}{E_z} \quad \text{(Equation G 101.1)}$$

$$= \frac{5 \text{ x } 100 + 0.06 \text{ x } 2000}{1.0}$$

$$= 620 \text{ cubic feet per minute (ft}^3\text{/min)}$$

Where:

A_z = zone floor area: the net occupiable floor area of the zone in square feet.

P_z = zone population: The largest number of people expected to occupy the zone during typical usage. Where the number of people expected to occupy the zone fluctuates, P_z shall be permitted to be estimated based on averaging approaches described in Section 403.6.1. Where P_z cannot be accurately predicted during design, it shall be estimated based on the zone floor area and the default occupant density in accordance with Table 402.1.

R_p = outdoor airflow rate required per person in accordance with Table 402.1.

R_a = outdoor airflow rate required per unit area in accordance with Table 402.1.

E_z = zone air distribution effectiveness in accordance with Table 403.2.2.

For SI units: 1 square foot = 0.0929 m^2, 1 cubic foot per minute = 0.00047 m^3/s

INDEX

– A –

– D –

– I –

– S –

– W –

HISTORY NOTE APPENDIX

CALIFORNIA MECHANICAL CODE (TITLE 24, PART 4, CALIFORNIA CODE OF REGULATIONS)

For prior code history, see the History Note Appendix to the California Mechanical Code, 2016 Triennial Edition, effective January 1, 2017.

1. *(BSC 03/18, HCD 01/18, DSA-SS 03/18, OSHPD 04/18, SFM 04/18) Adoption by reference of the 2018 Uniform Mechanical Code with necessary amendments to become the 2019 California Mechanical Code, and repeal of the 2015 edition of the Uniform Mechanical Code; effective on January 1, 2020.*